Elliptic Partial Differential Equations from an Elementary Viewpoint

A Fresh Glance at the Classical Theory

Elliptic Partial Differential Equations from an Elementary Viewpoint

A Fresh Glance at the Classical Theory

Serena Dipierro
Enrico Valdinoci
University of Western Australia, Australia

World Scientific

NEW JERSEY • LONDON • SINGAPORE • BEIJING • SHANGHAI • HONG KONG • TAIPEI • CHENNAI • TOKYO

Published by

World Scientific Publishing Co. Pte. Ltd.
5 Toh Tuck Link, Singapore 596224
USA office: 27 Warren Street, Suite 401-402, Hackensack, NJ 07601
UK office: 57 Shelton Street, Covent Garden, London WC2H 9HE

Library of Congress Control Number: 2024007658

British Library Cataloguing-in-Publication Data
A catalogue record for this book is available from the British Library.

ELLIPTIC PARTIAL DIFFERENTIAL EQUATIONS FROM AN ELEMENTARY VIEWPOINT
A Fresh Glance at the Classical Theory

ISBN 978-981-12-9079-4 (hardcover)
ISBN 978-981-12-9162-3 (paperback)
ISBN 978-981-12-9080-0 (ebook for institutions)
ISBN 978-981-12-9081-7 (ebook for individuals)

For any available supplementary material, please visit
https://www.worldscientific.com/worldscibooks/10.1142/13776#t=suppl

Desk Editors: Nandha Kumar/Kwong Lai Fun

Typeset by Stallion Press
Email: enquiries@stallionpress.com

"There's more to life than mathematics," Joan said. "But not much more."

Greg Egan, Glory.

Preface

The notes collected here and in the companion book [DV23] are the outcome of some courses taught to undergraduate and graduate students from the University of Western Australia, the Pontifícia Universidade Católica do Rio de Janeiro, the Indian Institute of Technology Gandhinagar, the Ukrainian Catholic University (Український Католицький Університет) and the Politecnico di Milano in 2021–2023.

Far from aiming at being all-encompassing, the following pages wish to shed some light on a number of selected topics in the theory of elliptic partial differential equations with a style that should be accessible to third-year undergraduate students, possibly under an inspired mentorship, but might also provide some interest to more advanced students and possibly professional researchers. While all the topics presented are of classical flavor, the exposition and the way the material is organized are perhaps rather original, with the intention of addressing several quite difficult points with a style that is as self-contained as possible, rigorous as well as intuitive, and approachable without major prerequisites (indeed, we only assume prior knowledge of the "basic" analysis, giving also references to useful results and theorems whenever the less advanced readers may need to consolidate their backgrounds).

These notes may be used as a first step into the broad realm of elliptic partial differential equations and do not require complicated prerequisites; nevertheless, they are also seen as the natural continuation of the companion book [DV23] in which the fundamentals and motivations are established at an intuitive level. Both this and the companion book [DV23] are essentially self-contained and can be easily read independently from each other. Yet, the two books share a common mathematical theme, despite their apparent stylistic differences: while [DV23] focuses merely on the development of mathematical instinct and intuition, here we rely on rigorous and formal proofs, which are carried out in every single detail, leaving no stone unturned. However, the rigorous approach set forth here heavily relies on intuition as well, and the results and techniques presented are usually explained at an intuitive level, providing examples and counterexamples and comparing, whenever possible, with concrete phenomena and our everyday experiences.

We stress again that the list of topics covered here is far from exhaustive, since we mainly deal with classical problems related to the Laplace operator and their natural counterpart for equations in nondivergence form (the divergence structure case is not really covered here, though the divergence theorem is obviously utilized ubiquitously, the variational structure of the Laplace–Beltrami operator is discussed quite in detail and several pages are devoted to the theory of capacity, which possesses a variational essence). Also, we do not address here nonlinear, singular/degenerate elliptic operators, fully nonlinear equations or fractional/nonlocal elliptic equations. We do not even linger too much on explicit solutions and cheap tricks to find them (e.g., we do not repeat over and over highly specific methods such as the separation of variables) since, all things considered, it is very unlikely that one can solve explicitly[1] a partial differential equation, instead it is often more useful to understand the qualitative and quantitative properties of the solution without solving explicitly the corresponding equation.

The many parts of the theory of elliptic partial differential equations which are missing from these pages are by no means less important than the topics covered: the issue is that we had to make a rather harsh selection of topics just to be able to collect all the material into finitely many pages (arguably, Jorge Luis Borges would remark that a comprehensive treatment of elliptic equations can be found in the Library of Babel [Bor99]).

The interested reader could derive additional fun from the treatment of a few natural problems with geometric flavor, such as the soap bubble theorem and Serrin's overdetermined problem. To treat them, but this is also a general approach used everywhere in the book, we tried to provide different perspectives and different points of view, while also providing physical motivations to trigger creativity and imagination, and favoring a dynamic interplay between geometric and analytic arguments.

These notes also collect a number of classical, albeit not always well-known, topics, which we believe can serve as an excellent training camp to develop some familiarity with elliptic equations.

[1] Moreover, explicit solutions may provide a handy resource to develop an initial theory, but they usually do not exhaust the complexity of the problem. For example, according to [Mur02, p. 452], it is quite common to run into "one of the serious problems with such exact solutions [...]: namely, they often do not determine all possible solutions and indeed, may not even give the most relevant one."

Here are a couple of features of these notes that we hope will be appealing to many readers. First, all is built in an artisanal way, perhaps not by following the most elegant, concise or general approach, but rather by aiming at a possibly slow but conceptually clear, sequence of strategic steps. Second, this set of notes is planned in such a way that one can read them starting at almost any place and freely jumping from one topic to another simply by following a personal stream of thoughts.

Let us now devote some explanatory comments for anyone interested in using these notes for an undergraduate or graduate course. The chapters are grouped in a way which makes it easy to select parts of this project for a specific course, focusing only on some aspects of the theory of elliptic partial differential equations. In particular:

- Chapters 1, 2 and 3 present the Laplace operator as well as the main properties of harmonic and subharmonic functions: they may be efficiently used in an introductory course to elliptic partial differential equations (or to cover the part about elliptic equations in a course presenting partial differential equations from different perspectives). Some parts (such as the ones involving the Laplace–Beltrami operator, the harmonic polynomials and the spherical harmonics) can be skipped in a more essential course (but they provide additional geometric insight for more avid readers and a more comprehensive approach for a more challenging course).
- Chapters 4 and 5 are a thorough introduction to the regularity theory for equations in nondivergence form. Each of these chapters can also be taught independently (one chapter dealing with the regularity theory in Hölder spaces, the other one focusing on Lebesgue spaces).
- A course (or a part of a course) addressing specifically the real analytic setting, which may have somehow a different flavor from the other types of methodologies, could leverage Chapter 9.
- A course (or a part of a course) focused on capacity theory may rely on Chapter 6.
- A course (or a part of a course) dealing with geometric problems, classification theory and rigidity results can benefit from Chapters 7 and 8 (which can also be taught somewhat independently).
- All the above possibilities can be mixed and combined, with great flexibility, according to the preferences of the lecturer.

This book does not contain "exercises": yet, the lecturer teaching the course can easily leave some of the corollaries or side remarks as exercises or assignments. We would not worry too much about the fact that students can find the solutions of these exercises in the book because nowadays the solution of virtually any exercise on a relatively standard topic can be easily found on the internet (instead, reading a book, even if only to browse for solutions after having struggled to find an original one, can be an old-fashioned, but very effective, way to learn, and students may be encouraged to do so).

Though a list of classical and modern references is given at the end, this is not a book focused on the history of mathematics, and hence we will not address topics such as priority, discoveries or progress of subjects over the course of time.

As usual, these notes may contain errors or inconsistencies: if you find any, please let us know. In general, we will be happy to receive comments and criticisms to possibly improve this work. After all, scientific knowledge is based on a dynamic flow of information, and we will certainly cherish and treasure readers' feedback and advice.

All right, enough chitchat! Please fasten your PFD—it's time to start our journey together.

Serena and Enrico

Courtesy of Burbuqe Shaqiri.

About the Authors

Serena Dipierro is a professor of mathematics, an Australian Research Council Future fellow and a fellow of the Australian Mathematical Society. She pursued her academic career in a number of cities around the world, including Santiago de Chile, Edinburgh, Magdeburg, Berlin, Melbourne, Milan and Perth. She has been the recipient of the Australian Mathematical Society Medal, the Mahony–Neumann–Room Prize, the Bartolozzi Prize, the Christopher Heyde Medal, and the Book Prize of the Unione Matematica Italiana.

Enrico Valdinoci is a professor of mathematics and an Australian Laureate fellow. His academic career spanned several cities, including Pisa, Rome, Milan, Berlin, Melbourne and Perth. He is a highly cited researcher and has been awarded the James S. W. Wong Prize, the Mahony–Neumann–Room Prize, the Orazio Arena Prize, the Book Prize of the Unione Matematica Italiana, and the Amerio Gold Medal Prize.

Acknowledgments

The authors are members of AustMS and INdAM. This work was supported by the Australian Research Council DECRA DE180100957 and the Australian Laureate Fellowship FL190100081. It is a pleasure to thank Dario Bambusi, Giancarlo Benettin, Héctor Chang-Lara, Lorenzo D'Ambrosio, Alberto Farina, Giovanni Gallavotti, Massimo Grossi, Damon Haddon, Sergei Kuzenko, Steffen Lauritzen, Jun Lim, Andrew Munyard, Gopalan Nair, David Perrella, David Pfefferlé, Giorgio Poggesi, Fabio Pusateri, Alfonso Sorrentino, Thomas Stemler, Jack Thompson, Clara Torres-Latorre, Josh Troy, Callum Vukovich, and Jingshi Xu for interesting comments.

Contents

Chapter 1

What Is the Laplacian?

The main objective of mathematics is to amuse and delight. However, an interesting, and on occasion useful, byproduct is its capacity to describe, understand and predict natural phenomena and place us at the center of the universe around us.

To achieve these goals, mathematics often relies on partial differential equations as a convenient way to encode quantitatively and unambiguously how quantities of vital importance change in space and time or how they interact with each other.

Yet, it is likely that very few partial differential equations are meaningful in themselves: if we randomly picked derivatives of functions and built equations out of them, we would probably just obtain some garbage. But mathematical intuition and concrete experiments have helped humanity put forth a variety of beautiful partial differential equations which, on several occasions, fit their purposes quite well. In many of these equations, the Laplace operator (or variations of it) plays an important role, in view of its "averaging" features, its "democratic" tendency to reverse the value of a function toward the mean and its elegant geometric interpretations (without the Laplacian, life would be considerably harder).

So, it is not an exaggeration to say that the Laplacian appears ubiquitously in nature; see the companion book [DV23] if you are skeptical about this statement. We thereby enthusiastically dive into some mathematical settings which turn out to be handy when dealing with partial differential equations, in particular those related to the Laplace operator. In this way, we will experience together all the marvelous properties of the Laplacian.

1.1 The Laplacian Under Different Perspectives

Given an open set $\Omega \subseteq \mathbb{R}^n$, a function $u \in C^2(\Omega)$ and a point $x \in \Omega$, we consider the "Laplace operator" Δ defined by

$$\Delta u(x) := \sum_{j=1}^{n} \frac{\partial^2 u}{\partial x_j^2}(x). \tag{1.1.1}$$

We point out that the Laplacian is invariant under translation, namely

$$\Delta\big(u(x+y)\big) = \Delta u(x+y) \tag{1.1.2}$$

for every $x \in \Omega$ and $y \in \mathbb{R}^n$ such that $x + y \in \Omega$, and it possesses a "divergence form" structure since

$$\Delta u = \operatorname{div}(\nabla u). \tag{1.1.3}$$

As a consequence of this and the divergence theorem, if Ω has a boundary of class C^1, the average of the Laplacian can be reconstructed by the normal flow through the boundary of the domain, namely

$$\int_\Omega \Delta u(x)\, dx = \int_{\partial\Omega} \nabla u(x) \cdot \nu(x)\, d\mathcal{H}_x^{n-1}, \tag{1.1.4}$$

where ν denotes the external unit normal of Ω and $\mathcal{H}^{n-1}$ is the $(n-1)$-dimensional Hausdorff measure.

It is also useful to remark that the Laplacian measures the "infinitesimal distance" between the value of a function at a given point and the average nearby, as described by the following theorem.

Theorem 1.1.1. *Let $x_0 \in \mathbb{R}^n$ and $r > 0$. Suppose that $u \in C^2(B_r(x_0))$. Then,*

$$\lim_{\rho \searrow 0} \frac{1}{\rho^2} \left(\fint_{B_\rho(x_0)} u(x)\, dx - u(x_0) \right) = \frac{1}{2(n+2)} \Delta u(x_0). \tag{1.1.5}$$

Proof. We begin by recalling a standard relation between the Lebesgue measure of the unit ball and the $(n-1)$-dimensional Hausdorff measure of the unit sphere. Namely, using polar coordinates, we see that

$$|B_1| = \mathcal{H}^{n-1}(\partial B_1) \int_0^1 t^{n-1}\, dt = \frac{\mathcal{H}^{n-1}(\partial B_1)}{n}. \tag{1.1.6}$$

We now point out a useful cancellation property. If T_j is the reflection across the jth coordinate, i.e.

$$T_j(y_1, \dots, y_n) = (y_1, \dots, y_{j-1}, -y_j, y_{j+1}, \dots, y_n),$$

and $g : B_\rho \to \mathbb{R}$ is such that $g(T_j(y)) = -g(y)$ for all $y \in B_\rho$, using the change of variable $Y := T_j(y)$, it follows that

$$\int_{B_\rho} g(y)\,dy = -\int_{B_\rho} g(T_j(y))\,dy = -\int_{B_\rho} g(Y)\,dY; \tag{1.1.7}$$

therefore,

$$\int_{B_\rho} g(y)\,dy = 0.$$

Applying this observation to the function y_j for every $j \in \{1, \dots, n\}$, we find that

$$\int_{B_\rho} y_j\,dy = 0. \tag{1.1.8}$$

Similarly, applying the previous observation to the function $y_j y_k$ for every $j \neq k \in \{1, \dots, n\}$,

$$\int_{B_\rho} y_j y_k\,dy = 0. \tag{1.1.9}$$

Now, we perform an explicit computation in the ball by taking advantage of the symmetry between coordinate exchanges. Namely, for every $j \in \{1, \dots, n\}$, using polar coordinates, we have that

$$\begin{aligned} n\int_{B_\rho} y_j^2\,dy &= \sum_{k=1}^{n}\int_{B_\rho} y_k^2\,dy = \int_{B_\rho} |y|^2\,dy \\ &= \mathcal{H}^{n-1}(\partial B_1)\int_0^\rho t^{n+1}\,dt = \frac{\mathcal{H}^{n-1}(\partial B_1)\,\rho^{n+2}}{n+2}. \end{aligned} \tag{1.1.10}$$

Now, we focus on the proof of the desired result in (1.1.5). For this, given $\rho > 0$ and $x \in B_\rho(x_0)$, we use the Taylor expansion

$$u(x) = u(x_0) + \nabla u(x_0)\cdot(x - x_0) + \frac{1}{2}D^2u(x_\star)(x - x_0)\cdot(x - x_0), \tag{1.1.11}$$

for a suitable $x_\star \in B_\rho(x_0)$, and we find that

$$\begin{aligned} &⨍_{B_\rho(x_0)} u(x)\,dx - u(x_0) = ⨍_{B_\rho(x_0)} \big(u(x) - u(x_0)\big)\,dx \\ &= ⨍_{B_\rho(x_0)} \left(\nabla u(x_0)\cdot(x - x_0) + \frac{1}{2}D^2u(x_\star)(x - x_0)\cdot(x - x_0)\right) dx \\ &= \sum_{j=1}^{n} ⨍_{B_\rho} \frac{\partial u}{\partial x_j}(x_0)\,y_j\,dy + \frac{1}{2}\sum_{j,k=1}^{n} ⨍_{B_\rho} \frac{\partial^2 u}{\partial x_j \partial x_k}(x_\star)\,y_j y_k\,dy \\ &= \frac{1}{2}\sum_{j,k=1}^{n} ⨍_{B_\rho} \frac{\partial^2 u}{\partial x_j \partial x_k}(x_\star)\,y_j y_k\,dy, \end{aligned}$$

where the latter identity is a consequence of (1.1.8) and (1.1.9).

Consequently, by (1.1.10),

$$\fint_{B_\rho(x_0)} u(x)\,dx - u(x_0) = \frac{1}{2}\sum_{j,k=1}^{n} \fint_{B_\rho} \frac{\partial^2 u}{\partial x_j \partial x_k}(x_\star)\, y_j y_k\, dy$$

$$= \frac{1}{2}\sum_{j=1}^{n} \fint_{B_\rho} \frac{\partial^2 u}{\partial x_j^2}(x_0)\, y_j^2\, dy + \eta(\rho), \tag{1.1.12}$$

where

$$\eta(\rho) := \frac{1}{2}\sum_{j,k=1}^{n} \fint_{B_\rho} \left(\frac{\partial^2 u}{\partial x_j \partial x_k}(x_\star) - \frac{\partial^2 u}{\partial x_j \partial x_k}(x_0)\right) y_j y_k\, dy. \tag{1.1.13}$$

Now, we observe that

$$\lim_{\rho\searrow 0} \frac{\eta(\rho)}{\rho^2} = 0. \tag{1.1.14}$$

Indeed, since $x_\star$ approaches x_0 as $\rho \searrow 0$, given any $\varepsilon > 0$, if ρ is small enough, we have that

$$\left|\frac{\partial^2 u}{\partial x_j \partial x_k}(x_\star) - \frac{\partial^2 u}{\partial x_j \partial x_k}(x_0)\right| \leqslant \varepsilon,$$

and consequently, using again (1.1.10),

$$|\eta(\rho)| \leqslant \frac{\varepsilon}{2}\sum_{j=1}^{n} \fint_{B_\rho} y_j^2\, dy = \frac{\mathcal{H}^{n-1}(\partial B_1)\,\rho^2\,\varepsilon}{2(n+2)\,|B_1|}.$$

From this, the claim in (1.1.14) plainly follows.

The desired result in (1.1.5) is now a direct consequence of (1.1.6), (1.1.10), (1.1.12) and (1.1.14) since

$$\lim_{\rho\searrow 0} \frac{1}{\rho^2}\left(\fint_{B_\rho(x_0)} u(x)\,dx - u(x_0)\right) = \lim_{\rho\searrow 0}\left(\frac{1}{2\rho^2}\sum_{j=1}^{n} \fint_{B_\rho} \frac{\partial^2 u}{\partial x_j^2}(x_0)\, y_j^2\, dy + \frac{\eta(\rho)}{\rho^2}\right)$$

$$= \frac{\mathcal{H}^{n-1}(\partial B_1)}{2n(n+2)\,|B_1|}\sum_{j=1}^{n} \frac{\partial^2 u}{\partial x_j^2}(x_0) = \frac{1}{2(n+2)}\,\Delta u(x_0),$$

as desired. □

For completeness, as a variant of Theorem 1.1.1, we also point out a similar result for the limit of the spherical averages.

Theorem 1.1.2. *Let $x_0 \in \mathbb{R}^n$ and $r > 0$. Suppose that $u \in C^2(B_r(x_0))$. Then,*

$$\lim_{\rho \searrow 0} \frac{1}{\rho^2} \left(\fint_{\partial B_\rho(x_0)} u(x)\, d\mathcal{H}_x^{n-1} - u(x_0) \right) = \frac{1}{2n} \Delta u(x_0).$$

Proof. According to the Taylor expansion in (1.1.11) and two odd cancellation arguments, as $\rho \searrow 0$, we have

$$\begin{aligned}
&\fint_{\partial B_\rho(x_0)} u(x)\, d\mathcal{H}_x^{n-1} - u(x_0) \\
&= \fint_{\partial B_\rho(x_0)} \left(\nabla u(x_0) \cdot (x - x_0) + \frac{1}{2} D^2 u(x_0)(x - x_0) \cdot (x - x_0) \right) d\mathcal{H}_x^{n-1} + o(\rho^2) \\
&= \frac{1}{2} \fint_{\partial B_\rho(x_0)} D^2 u(x_0)(x - x_0) \cdot (x - x_0)\, d\mathcal{H}_x^{n-1} + o(\rho^2) \\
&= \frac{1}{2} \sum_{i,j=1}^{n} \fint_{\partial B_\rho} \partial_{ij} u(x_0) y_i y_j\, d\mathcal{H}_y^{n-1} + o(\rho^2) \\
&= \frac{1}{2} \sum_{i=1}^{n} \fint_{\partial B_\rho} \partial_{ii} u(x_0) y_i^2\, d\mathcal{H}_y^{n-1} + o(\rho^2).
\end{aligned}$$

Now, for every $i \in \{1, \dots, n\}$, we have that

$$\begin{aligned}
n \int_{\partial B_\rho} y_i^2\, d\mathcal{H}_y^{n-1} &= \sum_{k=1}^{n} \int_{\partial B_\rho} y_k^2\, d\mathcal{H}_y^{n-1} = \int_{\partial B_\rho} |y|^2\, d\mathcal{H}_y^{n-1} \\
&= \int_{\partial B_\rho} \rho^2\, d\mathcal{H}_y^{n-1} = \rho^2 \mathcal{H}^{n-1}(\partial B_\rho)
\end{aligned}$$

and consequently

$$\begin{aligned}
\fint_{\partial B_\rho(x_0)} u(x)\, d\mathcal{H}_x^{n-1} - u(x_0) &= \frac{1}{2n} \sum_{i=1}^{n} \partial_{ii} u(x_0) \rho^2 + o(\rho^2) \\
&= \frac{\rho^2}{2n} \Delta u(x_0) + o(\rho^2),
\end{aligned}$$

which plainly leads to the desired result. □

We point out that Theorems 1.1.1 and 1.1.2 are special cases of a more general type of result, known in the literature as Pizzetti's formula, involving the higher-order Laplace operator

$$\Delta^k := \underbrace{\Delta \cdots \Delta}_{k \text{ times}},$$

with $k \in \mathbb{N}$ and the convention that Δ^0 is the identity operator. We recall this setting for the sake of completeness.

Proposition 1.1.3. *Let $x_0 \in \mathbb{R}^n$ and $r > 0$. Suppose that $u \in C^{2N}(B_r(x_0))$. Then, as $\rho \searrow 0$,*

$$\begin{aligned}&\fint_{B_\rho(x_0)} u(x)\,dx\\&\quad = n\Gamma\left(\frac{n}{2}\right)\sum_{k=0}^{N}\frac{\rho^{2k}}{2^{2k+1}\,k!\,\Gamma\left(\frac{n}{2}+k+1\right)}\Delta^k u(x_0) + o(\rho^{2N})\end{aligned} \tag{1.1.15}$$

and

$$\begin{aligned}&\fint_{\partial B_\rho(x_0)} u(x)\,d\mathcal{H}^{n-1}_x\\&\quad = \Gamma\left(\frac{n}{2}\right)\sum_{k=0}^{N}\frac{\rho^{2k}}{2^{2k}\,k!\,\Gamma\left(\frac{n}{2}+k\right)}\Delta^k u(x_0) + o(\rho^{2N}).\end{aligned} \tag{1.1.16}$$

In the above, Γ stands for the Euler gamma function, defined, for every $z \in \mathbb{C}$ with $\Re z > 0$, by

$$\Gamma(z) := \int_0^{+\infty} t^{z-1}e^{-t}\,dt. \tag{1.1.17}$$

Proof. We begin with an observation that can be considered a version of the multinomial theorem for the Laplace operator. We claim that, for every $k \in \mathbb{N}$,

$$\Delta^k u = \sum_{\substack{\alpha\in\mathbb{N}^n\\|\alpha|=k}}\frac{k!}{\alpha!}D^{2\alpha}u. \tag{1.1.18}$$

Here, we are using the multi-index notation for which $\alpha = (\alpha_1, \dots, \alpha_n)$, $|\alpha| = \alpha_1 + \dots + \alpha_n$, $\alpha! = \alpha_1!\cdots\alpha_n!$ and $2\alpha = (2\alpha_1, \dots, 2\alpha_n)$. We prove (1.1.18) by induction over k. Indeed, when $k = 0$, the claim in (1.1.18) reduces to the true identity $u = u$. Then, we suppose that (1.1.18) holds true for the index k and observe that, with e_i being the ith element of the Euclidean basis,

$$\begin{aligned}\Delta^{k+1}u &= \Delta\left(\sum_{\substack{\alpha\in\mathbb{N}^n\\|\alpha|=k}}\frac{k!}{\alpha!}D^{2\alpha}u\right) = \sum_{i=1}^{n}\frac{\partial^2}{\partial x_i^2}\left(\sum_{\substack{\alpha\in\mathbb{N}^n\\|\alpha|=k}}\frac{k!}{\alpha!}D^{2\alpha}u\right) = \sum_{\substack{\alpha\in\mathbb{N}^n\\|\alpha|=k\\1\leqslant i\leqslant n}}\frac{k!}{\alpha!}D^{2(\alpha+e_i)}u\\
&= \sum_{\substack{\alpha\in\mathbb{N}^n\\|\alpha|=k\\1\leqslant i\leqslant n}}\frac{k!\,(\alpha_i+1)}{(\alpha+e_i)!}D^{2(\alpha+e_i)}u = \sum_{\substack{\beta\in\mathbb{N}^n\\|\beta|=k+1\\1\leqslant i\leqslant n}}\frac{k!\,\beta_i}{\beta!}D^{2\beta}u = \sum_{\substack{\beta\in\mathbb{N}^n\\|\beta|=k+1}}\frac{k!\,|\beta|}{\beta!}D^{2\beta}u\\
&= \sum_{\substack{\beta\in\mathbb{N}^n\\|\beta|=k+1}}\frac{k!\,(k+1)}{\beta!}D^{2\beta}u = \sum_{\substack{\beta\in\mathbb{N}^n\\|\beta|=k+1}}\frac{(k+1)!}{\beta!}D^{2\beta}u,\end{aligned}$$

which completes the inductive step and provides the proof of (1.1.18).

Now, we point out that, by odd symmetry of the function $x_i \mapsto x_i^{\alpha_i}$, if α_i is an odd integer, then $\int_{B_\rho} x^\alpha\,dx = 0$ and consequently

$$\int_{B_\rho} x^\alpha\,dx = 0 \quad \text{unless } \alpha = 2\beta \text{ for some } \beta \in \mathbb{N}^n. \tag{1.1.19}$$

In the same way,

$$\int_{\partial B_\rho} x^\alpha\,d\mathcal{H}_x^{n-1} = 0 \quad \text{unless } \alpha = 2\beta \text{ for some } \beta \in \mathbb{N}^n. \tag{1.1.20}$$

It is now convenient to perform some specific calculations related to spherical integrals. So as to achieve this goal, recalling definition (1.1.17) of the Euler gamma function and using the substitution $\tau := t^2$, we point out that, for every $b > -1/2$,

$$\int_0^{+\infty} t^{2b} e^{-t^2}\,dt = \frac{1}{2}\int_0^{+\infty} \tau^{\frac{2b-1}{2}} e^{-\tau}\,d\tau = \frac{1}{2}\Gamma\left(\frac{2b+1}{2}\right). \tag{1.1.21}$$

Consequently, for every $\alpha \in \mathbb{N}^n$,

$$\begin{aligned}\int_{\mathbb{R}^n} x^{2\alpha} e^{-|x|^2}\,dx &= \prod_{i=1}^n \int_{\mathbb{R}} x_i^{2\alpha_i} e^{-x_i^2}\,dx_i \\ &= 2^n \prod_{i=1}^n \int_0^{+\infty} x_i^{2\alpha_i} e^{-x_i^2}\,dx_i = \prod_{i=1}^n \Gamma\left(\frac{2\alpha_i+1}{2}\right).\end{aligned} \tag{1.1.22}$$

On the other hand, using polar coordinates and once again (1.1.21),

$$\begin{aligned}\int_{\mathbb{R}^n} x^{2\alpha} e^{-|x|^2}\,dx &= \int_{\partial B_1}\left(\int_0^{+\infty} r^{2|\alpha|+n-1}\omega^{2\alpha} e^{-r^2}\,dr\right) d\mathcal{H}_\omega^{n-1} \\ &= \frac{1}{2}\Gamma\left(\frac{2|\alpha|+n}{2}\right)\int_{\partial B_1} \omega^{2\alpha}\,d\mathcal{H}_\omega^{n-1}.\end{aligned}$$

Comparing this with (1.1.22), we deduce that

$$\int_{\partial B_1} \omega^{2\alpha}\,d\mathcal{H}_\omega^{n-1} = \frac{2\prod_{i=1}^n \Gamma\left(\frac{2\alpha_i+1}{2}\right)}{\Gamma\left(\frac{2|\alpha|+n}{2}\right)}. \tag{1.1.23}$$

It is also instructive to recall that, integrating by parts,

$$\begin{aligned}\Gamma(z+1) &= \int_0^{+\infty} t^z e^{-t}\,dt = -\int_0^{+\infty} t^z \frac{d}{dt}e^{-t}\,dt \\ &= \int_0^{+\infty} z t^{z-1} e^{-t}\,dt = z\,\Gamma(z),\end{aligned} \tag{1.1.24}$$

from which it follows by induction that

$$\Gamma(j+1) = j! \quad \text{for every } j \in \mathbb{N}.$$

Furthermore, making use of (1.1.21) with $b := 0$,

$$\Gamma\left(\frac{1}{2}\right) = 2\int_0^{+\infty} e^{-t^2}\,dt = \int_{-\infty}^{+\infty} e^{-t^2}\,dt = \sqrt{\pi}. \tag{1.1.25}$$

Now, we claim that, for every $j \in \mathbb{N}$,

$$\Gamma\left(j+1+\frac{1}{2}\right) = \frac{\sqrt{\pi}\,(2j+1)!}{2^{2j+1}\,j!}. \tag{1.1.26}$$

This statement is actually a particular case of the Legendre duplication formula, or the Gauß multiplication formula; however, we provide a direct proof of (1.1.26) for the convenience of the reader. For this, we argue by induction over j. We remark that

$$\Gamma\left(1+\frac{1}{2}\right) = \frac{1}{2}\Gamma\left(\frac{1}{2}\right) = \frac{\sqrt{\pi}}{2}, \tag{1.1.27}$$

thanks to (1.1.24) and (1.1.25), and this gives the claim in (1.1.26) when $j = 0$.

Suppose now that (1.1.26) holds true for j. Then, we use again (1.1.24) to obtain that

$$\begin{aligned}\Gamma\left(j+2+\frac{1}{2}\right) &= \left(j+1+\frac{1}{2}\right)\Gamma\left(j+1+\frac{1}{2}\right) = \left(j+1+\frac{1}{2}\right)\frac{\sqrt{\pi}\,(2j+1)!}{2^{2j+1}\,j!}\\ &= (2j+3)\,\frac{\sqrt{\pi}\,(2j+1)!}{2^{2j+2}\,j!} = (2j+3)\,(2j+2)\,\frac{\sqrt{\pi}\,(2j+1)!}{2^{2j+3}\,(j+1)\,j!}\\ &= \frac{\sqrt{\pi}\,(2j+3)!}{2^{2j+3}\,(j+1)!},\end{aligned}$$

thus completing the inductive step and establishing (1.1.26).

Now, we claim that

$$\Gamma\left(\frac{2\alpha_i+1}{2}\right) = 2^{1-2\alpha_i}\,\varsigma(\alpha_i)\,\sqrt{\pi}\prod_{j=0}^{\alpha_i-1}(\alpha_i+j), \tag{1.1.28}$$

where

$$\varsigma(\alpha_i) := \begin{cases} 1 & \text{if } \alpha_i \neq 0,\\[2ex] \dfrac{1}{2} & \text{if } \alpha_i = 0.\end{cases}$$

Indeed, by (1.1.26), if $\alpha_i \neq 0$, then

$$\Gamma\left(\frac{2\alpha_i+1}{2}\right) = \frac{\sqrt{\pi}\,(2\alpha_i-1)!}{2^{2\alpha_i-1}\,(\alpha_i-1)!} = 2^{1-2\alpha_i}\,\sqrt{\pi}\prod_{j=0}^{\alpha_i-1}(\alpha_i+j).$$

This proves (1.1.28) when $\alpha_i \neq 0$.

If instead $\alpha_i = 0$, we recall that, as usual, the empty product is 1 by convention, and we use (1.1.25) to see that in this case,

$$\Gamma\left(\frac{2\alpha_i + 1}{2}\right) - 2^{1-2\alpha_i}\,\varsigma(\alpha_i)\,\sqrt{\pi}\prod_{j=0}^{\alpha_i - 1}(\alpha_i + j) = \Gamma\left(\frac{1}{2}\right) - 2^0\,\sqrt{\pi} = 0.$$

The proof of (1.1.28) is thereby complete.

We also observe that

$$\prod_{i=1}^{n}\varsigma(\alpha_i) = \frac{1}{2^{\theta(\alpha)}}, \tag{1.1.29}$$

where $\theta(\alpha)$ is the number of vanishing components of the multi-index α.

Thus, by (1.1.28),

$$2\prod_{i=1}^{n}\Gamma\left(\frac{2\alpha_i + 1}{2}\right) = 2^{n+1-2|\alpha|-\theta(\alpha)}\,\pi^{\frac{n}{2}}\prod_{i=1}^{n}\prod_{j=0}^{\alpha_i - 1}(\alpha_i + j);$$

therefore, in light of (1.1.23),

$$\int_{\partial B_1}\omega^{2\alpha}\,d\mathcal{H}^{n-1}_{\omega} = \frac{2^{n+1-2|\alpha|-\theta(\alpha)}\,\pi^{\frac{n}{2}}\displaystyle\prod_{i=1}^{n}\prod_{j=0}^{\alpha_i - 1}(\alpha_i + j)}{\Gamma\left(\dfrac{2|\alpha| + n}{2}\right)}. \tag{1.1.30}$$

Consequently,

$$\begin{aligned}\int_{B_1} x^{2\alpha}\,dx &= \int_{\partial B_1}\left(\int_0^1 r^{2|\alpha|+n-1}\omega^{2\alpha}\,dr\right)d\mathcal{H}^{n-1}_{\omega}\\ &= \frac{1}{2|\alpha| + n}\int_{\partial B_1}\omega^{2\alpha}\,d\mathcal{H}^{n-1}_{\omega}\\ &= \frac{2^{n+1-2|\alpha|-\theta(\alpha)}\,\pi^{\frac{n}{2}}\displaystyle\prod_{i=1}^{n}\prod_{j=0}^{\alpha_i - 1}(\alpha_i + j)}{(2|\alpha| + n)\,\Gamma\left(\dfrac{2|\alpha| + n}{2}\right)}.\end{aligned} \tag{1.1.31}$$

We also note that

$$(2\beta_i)! = 2\varsigma(\beta_i)\,\beta_i!\prod_{j=0}^{\beta_i - 1}(\beta_i + j). \tag{1.1.32}$$

Indeed, if $\beta_i \neq 0$, then

$$(2\beta_i)! = 2\beta_i\,(2\beta_i-1)! = 2\beta_i \prod_{j=1}^{2\beta_i-1} j = 2\beta_i \left(\prod_{k=1}^{\beta_i-1} k\right)\left(\prod_{k=\beta_i}^{2\beta_i-1} k\right)$$
$$= 2\beta_i(\beta_i-1)!\left(\prod_{j=0}^{\beta_i-1}(\beta_i+j)\right) = 2\varsigma(\beta_i)\,\beta_i!\prod_{j=0}^{\beta_i-1}(\beta_i+j),$$

while if $\beta_i = 0$, then

$$(2\beta_i)! - 2\varsigma(\beta_i)\,\beta_i!\prod_{j=0}^{\beta_i-1}(\beta_i+j) = 1 - 2\varsigma(\beta_i) = 0.$$

These observations establish (1.1.32).

From (1.1.29) and (1.1.32), we arrive at

$$\frac{2^{n-\theta(\beta)}\displaystyle\prod_{i=1}^{n}\prod_{j=0}^{\beta_i-1}(\beta_i+j)}{(2\beta)!} = \frac{\displaystyle\prod_{i=1}^{n}\left(2\varsigma(\beta_i)\prod_{j=0}^{\beta_i-1}(\beta_i+j)\right)}{(2\beta)!}$$
$$= \prod_{i=1}^{n}\frac{2\varsigma(\beta_i)\displaystyle\prod_{j=0}^{\beta_i-1}(\beta_i+j)}{(2\beta_i)!} = \prod_{i=1}^{n}\frac{1}{\beta_i!} = \frac{1}{\beta!}. \tag{1.1.33}$$

We now prove (1.1.15). For this, up to a translation, we can reduce to the case of $x_0 = 0$. We exploit the Taylor expansion

$$u(x) = \sum_{\substack{\alpha\in\mathbb{N}^n\\ |\alpha|\leqslant 2N}} \frac{D^\alpha u(0)}{\alpha!}\,x^\alpha + o(x^{2N}), \tag{1.1.34}$$

and we average over B_ρ, thus finding that, for small ρ,

$$\fint_{B_\rho} u(x)\,dx = \sum_{\substack{\alpha\in\mathbb{N}^n\\ |\alpha|\leqslant 2N}} \frac{D^\alpha u(0)}{\alpha!}\fint_{B_\rho} x^\alpha\,dx + o(\rho^{2N})$$
$$= \sum_{\substack{\beta\in\mathbb{N}^n\\ |\beta|\leqslant N}} \frac{D^{2\beta} u(0)}{(2\beta)!}\fint_{B_\rho} x^{2\beta}\,dx + o(\rho^{2N})$$
$$= \sum_{\substack{\beta\in\mathbb{N}^n\\ |\beta|\leqslant N}} \frac{D^{2\beta} u(0)\,\rho^{2|\beta|}}{(2\beta)!}\fint_{B_1} y^{2\beta}\,dy + o(\rho^{2N})$$
$$= \sum_{k=0}^{N}\sum_{\substack{\beta\in\mathbb{N}^n\\ |\beta|=k}} \frac{D^{2\beta} u(0)\,\rho^{2k}}{(2\beta)!}\fint_{B_1} y^{2\beta}\,dy + o(\rho^{2N})$$

$$
= \sum_{k=0}^{N} \sum_{\substack{\beta\in\mathbb{N}^n \\ |\beta|=k}} \frac{D^{2\beta}u(0)\,\rho^{2k}}{(2\beta)!\,|B_1|} \frac{2^{n+1-2k-\theta(\beta)}\,\pi^{\frac{n}{2}} \prod_{i=1}^{n}\prod_{j=0}^{\beta_i-1}(\beta_i+j)}{(2k+n)\,\Gamma\left(\frac{2k+n}{2}\right)} + o(\rho^{2N})
$$

$$
= \sum_{k=0}^{N} \sum_{\substack{\beta\in\mathbb{N}^n \\ |\beta|=k}} \frac{D^{2\beta}u(0)\,\rho^{2k}}{\beta!\,|B_1|} \frac{2^{1-2k}\,\pi^{\frac{n}{2}}}{(2k+n)\,\Gamma\left(\frac{2k+n}{2}\right)} + o(\rho^{2N})
$$

$$
= \sum_{k=0}^{N} \frac{\Delta^k u(0)\,\rho^{2k}}{|B_1|} \frac{2^{1-2k}\,\pi^{\frac{n}{2}}}{k!\,(2k+n)\,\Gamma\left(\frac{2k+n}{2}\right)} + o(\rho^{2N}),
$$

thanks to (1.1.18), (1.1.19), (1.1.31) and (1.1.32).

Hence, since, owing to (1.1.6), (1.1.23) (used here with $\alpha := 0$) and (1.1.25),

$$
\Gamma\left(\frac{n}{2}\right) = \frac{2}{\mathcal{H}^{n-1}(\partial B_1)}\left(\Gamma\left(\frac{1}{2}\right)\right)^n = \frac{2\pi^{\frac{n}{2}}}{\mathcal{H}^{n-1}(\partial B_1)} = \frac{2\pi^{\frac{n}{2}}}{n\,|B_1|}, \tag{1.1.35}
$$

we obtain that

$$
\fint_{B_\rho} u(x)\,dx = n\,\Gamma\left(\frac{n}{2}\right)\sum_{k=0}^{N}\frac{\Delta^k u(0)\,\rho^{2k}}{2^{2k}\,k!\,(2k+n)\,\Gamma\left(\frac{2k+n}{2}\right)} + o(\rho^{2N})
$$

$$
= n\,\Gamma\left(\frac{n}{2}\right)\sum_{k=0}^{N}\frac{\Delta^k u(0)\,\rho^{2k}}{2^{2k+1}\,k!\,\Gamma\left(\frac{n}{2}+k+1\right)} + o(\rho^{2N}),
$$

where (1.1.24) has been used again in the last line. This completes the proof of (1.1.15).

Now, to establish (1.1.16), we exploit (1.1.18), (1.1.20), (1.1.30), (1.1.32) and (1.1.34), and we see that

$$
\fint_{\partial B_\rho} u(x)\,d\mathcal{H}^{n-1}_x = \sum_{\substack{\alpha\in\mathbb{N}^n \\ |\alpha|\leqslant 2N}} \frac{D^\alpha u(0)}{\alpha!} \fint_{\partial B_\rho} x^\alpha\,d\mathcal{H}^{n-1}_x + o(\rho^{2N})
$$

$$
= \sum_{\substack{\beta\in\mathbb{N}^n \\ |\beta|\leqslant N}} \frac{D^{2\beta}u(0)}{(2\beta)!} \fint_{\partial B_\rho} x^{2\beta}\,d\mathcal{H}^{n-1}_x + o(\rho^{2N})
$$

$$
= \sum_{\substack{\beta\in\mathbb{N}^n \\ |\beta|\leqslant N}} \frac{D^{2\beta}u(0)\,\rho^{2|\beta|}}{(2\beta)!} \fint_{\partial B_1} y^{2\beta}\,d\mathcal{H}^{n-1}_y + o(\rho^{2N})
$$

$$
= \sum_{k=0}^{N} \sum_{\substack{\beta\in\mathbb{N}^n \\ |\beta|=k}} \frac{D^{2\beta}u(0)\,\rho^{2k}}{(2\beta)!} \fint_{\partial B_1} y^{2\beta}\,d\mathcal{H}^{n-1}_y + o(\rho^{2N})
$$

$$= \sum_{k=0}^{N} \sum_{\substack{\beta \in \mathbb{N}^n \\ |\beta|=k}} \frac{D^{2\beta} u(0)\, \rho^{2k}}{(2\beta)!\, \mathcal{H}^{n-1}(\partial B_1)} \frac{2^{n+1-2k-\theta(\beta)}\, \pi^{\frac{n}{2}} \prod_{i=1}^{n} \prod_{j=0}^{\beta_i - 1} (\beta_i + j)}{\Gamma\left(\frac{2k+n}{2}\right)} + o(\rho^{2N})$$

$$= \sum_{k=0}^{N} \sum_{\substack{\beta \in \mathbb{N}^n \\ |\beta|=k}} \frac{D^{2\beta} u(0)\, \rho^{2k}}{\beta!\, \mathcal{H}^{n-1}(\partial B_1)} \frac{2^{1-2k}\, \pi^{\frac{n}{2}}}{\Gamma\left(\frac{2k+n}{2}\right)} + o(\rho^{2N})$$

$$= \sum_{k=0}^{N} \frac{\Delta^k u(0)\, \rho^{2k}}{k!\, \mathcal{H}^{n-1}(\partial B_1)} \frac{2^{1-2k}\, \pi^{\frac{n}{2}}}{\Gamma\left(\frac{2k+n}{2}\right)} + o(\rho^{2N}).$$

This, together with (1.1.35), proves (1.1.16), as desired (alternatively, we could have proved any one between (1.1.15) and (1.1.16) with an explicit expression for the remainder and obtained the other by either integration or differentiation in ρ). □

We stress that Theorems 1.1.1 and 1.1.2 are particular cases of (1.1.15) and (1.1.16), respectively, simply corresponding to the case of $N := 1$.

Despite its simple flavor, Theorem 1.1.1 reveals one of the fundamental features of the Laplacian, which will also play an important role in the mean value formula, described in Theorem 2.1.2. Also, it immediately leads to the fact that the Laplace operator is invariant under rotation, as shown by the following corollary.

Corollary 1.1.4. *Let $\mathcal{R} : \mathbb{R}^n \to \mathbb{R}^n$ be a rotation and $u \in C^2(\mathbb{R}^n)$. Let $u_{\mathcal{R}}(x) := u(\mathcal{R}x)$.*

Then,

$$\Delta u_{\mathcal{R}}(x) = \Delta u(\mathcal{R}x).$$

Proof. Let $x_0 \in \mathbb{R}^n$. By Theorem 1.1.1, using the change of variable $y := \mathcal{R}(x - x_0)$ and the notation $v(x) := u(x + \mathcal{R}x_0)$ and exploiting the translation invariance in (1.1.2), we have

$$\begin{aligned} \frac{1}{2(n+2)} \Delta u_{\mathcal{R}}(x_0) &= \lim_{\rho \searrow 0} \frac{1}{\rho^2} \left(\fint_{B_\rho(x_0)} u_{\mathcal{R}}(x)\, dx - u_{\mathcal{R}}(x_0) \right) \\ &= \lim_{\rho \searrow 0} \frac{1}{\rho^2} \left(\fint_{B_\rho(x_0)} u(\mathcal{R}x)\, dx - u(\mathcal{R}x_0) \right) \\ &= \lim_{\rho \searrow 0} \frac{1}{\rho^2} \left(\fint_{B_\rho} u(y + \mathcal{R}x_0)\, dy - u(\mathcal{R}x_0) \right) \end{aligned}$$

$$= \lim_{\rho \searrow 0} \frac{1}{\rho^2} \left(\fint_{B_\rho} v(y)\, dy - v(0) \right)$$
$$= \frac{1}{2(n+2)} \Delta v(0) = \frac{1}{2(n+2)} \Delta u(\mathcal{R} x_0).$$

This proves the desired result (for a proof not relying on Theorem 1.1.1 but rather on a direct computation in matrix form, see e.g. [Kue19, p. 172]). □

Of course, while the integral characterization of the Laplacian presented in Theorem 1.1.1 is conceptually very useful and reveals a deep geometric structure of the operator, the explicit differential structure in (1.1.1) is often simpler to exploit for explicit calculations. As an example, we recall the following Bochner identity.

Lemma 1.1.5. *For a given C^3 function u,*

$$\Delta \left(\frac{|\nabla u|^2}{2} \right) = \nabla(\Delta u) \cdot \nabla u + |D^2 u|^2,$$

where

$$|D^2 u|^2 := \sum_{i,j=1}^{n} (\partial_{ij} u)^2. \tag{1.1.36}$$

Proof. By a direct computation,

$$\Delta \left(\frac{|\nabla u|^2}{2} \right) = \frac{1}{2} \sum_{i,j=1}^{n} \partial_{ii} (\partial_j u)^2 = \sum_{i,j=1}^{n} \partial_i (\partial_j u\, \partial_{ij} u)$$
$$= \sum_{i,j=1}^{n} \Big((\partial_{ij} u)^2 + \partial_j u\, \partial_{iij} u \Big) = |D^2 u|^2 + \sum_{j=1}^{n} \partial_j u\, \partial_j (\Delta u).$$

□

Chapter 2

The Laplace Operator and Harmonic Functions

This chapter is devoted to the analysis of the Laplace operator and of the functions that lie in its kernel.

2.1 The Laplacian and the Mean Value Formula

Among the several properties that a given function may possess, a very relevant one is "harmonicity," corresponding to the vanishing of the trace of the Hessian matrix (in particular, these functions are "saddle-looking" with respect to their tangent planes at every point). The precise setting that we consider is the following.

Definition 2.1.1. Given an open set $\Omega \subseteq \mathbb{R}^n$ and a function $u \in C^2(\Omega)$, we say that u is harmonic in Ω if $\Delta u(x) = 0$ for every $x \in \Omega$.

For example, constant and linear functions are harmonic in all $\mathbb{R}^n$. Also, the functions $u : \mathbb{R}^2 \to \mathbb{R}$ given by

$$\begin{aligned} u(x_1, x_2) &= x_1 x_2, \\ u(x_1, x_2) &= x_1^2 - x_2^2 \\ \text{and} \quad u(x_1, x_2) &= e^{x_1} \sin x_2 \end{aligned}$$

are harmonic.

Other examples of harmonic functions in domains of $\mathbb{R}^2$ can be obtained via complex analysis, identifying $(x, y) \in \mathbb{R}^2$ with $z = x + iy \in \mathbb{C}$, since

$$\text{the real and imaginary parts of holomorphic functions are harmonic,} \tag{2.1.1}$$

see [Rud66, Chapter 11]. In particular, for every $j \in \mathbb{N}$, using the notation $r = |z| = \sqrt{x^2 + y^2}$ and $z = |z|e^{i\vartheta} = re^{i\vartheta}$, the functions

$$\Re z^j = r^j \cos(j\vartheta) \quad \text{and} \quad \Im z^j = r^j \sin(j\vartheta) \tag{2.1.2}$$

are harmonic in all $\mathbb{R}^2$, and so are the functions

$$\Re e^z = e^x \cos y \quad \text{and} \quad \Im e^z = e^x \sin y.$$

In addition, given $\alpha > 0$, the function $z \mapsto z^\alpha := r^\alpha e^{i\alpha\vartheta}$ is well defined and holomorphic in $\vartheta \in (-\pi, \pi)$, and hence

$$\Re z^\alpha = r^\alpha \cos(\alpha\vartheta) \quad \text{and} \quad \Im z^\alpha = r^\alpha \sin(\alpha\vartheta)$$

are harmonic in $\mathbb{R}^2 \setminus \ell$, with $\ell := (-\infty, 0] \times \{0\}$.

One of the most striking properties of harmonic functions is that their value at any point is precisely equal to the average of the values around such point. In this sense, the values attained by harmonic functions happen to be "perfectly balanced," according to the following result.

Theorem 2.1.2. *Given an open set $\Omega \subseteq \mathbb{R}^n$ and a function $u \in L^1_{\mathrm{loc}}(\Omega)$, the following conditions are equivalent:*

(i) *The function u belongs to $C^2(\Omega)$ and is harmonic in Ω.*
(ii) *For almost every $x_0 \in \Omega$ and almost every $r > 0$ such that $B_r(x_0) \Subset \Omega$, we have that*

$$u(x_0) = \fint_{\partial B_r(x_0)} u(x)\, d\mathcal{H}^{n-1}_x.$$

(iii) *For almost every $x_0 \in \Omega$ and almost every $r > 0$ such that $B_r(x_0) \Subset \Omega$, we have that*

$$u(x_0) = \fint_{B_r(x_0)} u(x)\, dx.$$

Additionally, if u satisfies any of the equivalent conditions (i), (ii) or (iii), then[1] *it belongs to $C^\infty(\Omega)$.*

Proof. We begin by showing that

$$\text{if } u \text{ satisfies either (ii) or (iii), then } u \in C^\infty(\Omega), \tag{2.1.3}$$

up to redefining u in a set of null Lebesgue measure. To this end, we use a mollification argument. We[2] take $\tau \in C^\infty_0(B_1, [0, +\infty))$ to be radially symmetric

[1] The equivalence between conditions (i), (ii) and (iii) highlights an interesting regularizing property of the Laplace operator since locally integrable functions satisfying either (ii) or (iii) turn out to belong to $C^2(\Omega)$ and hence their Laplacian can be computed pointwise, and it is equal to zero.

A similar regularizing effect will be highlighted by the forthcoming Lemma 2.2.1.

Also, the last statement in Theorem 2.1.2 concerning the smoothness of u gives that harmonic functions are automatically $C^\infty(\Omega)$: this statement will be strengthened in Theorem 2.14.1, where we show that harmonic functions are, in fact, real analytic.

[2] As customary, here and in the following, the subscript 0 in C^∞_0 means "with compact support in."

and such that $\int_{B_1} \tau(x)\,dx = 1$. Given $\eta > 0$, we let $\tau_\eta(x) := \frac{1}{\eta^n}\tau\left(\frac{x}{\eta}\right)$ and define $u_\eta := u * \tau_\eta$. We pick a point $\overline{x} \in \Omega$ and $R > 0$ such that $B_{2R}(\overline{x}) \Subset \Omega$, and we show that, when $\eta \in (0, R)$,

$$\text{if } u \text{ satisfies either (ii) or (iii), then } u = u_\eta \text{ a.e. in } B_R(\overline{x}). \tag{2.1.4}$$

Indeed, if u satisfies (ii), for each $x \in B_R(\overline{x})$ and $r \in (0, \eta]$, we have that $B_r(x) \subseteq B_\eta(x) \subseteq B_{R+\eta}(\overline{x}) \Subset \Omega$. Hence, using polar coordinates (see e.g. [EG15, Theorem 3.12]), for almost any $x \in B_R(\overline{x})$,

$$\begin{aligned}
u_\eta(x) &= \int_{B_\eta(x)} \tau_\eta(x-y)\,u(y)\,dy \\
&= \int_0^\eta \left[\int_{\partial B_r(x)} \tau_\eta(x-\omega)\,u(\omega)\,d\mathcal{H}^{n-1}_\omega\right] dr \\
&= \int_0^\eta \left[\int_{\partial B_r(x)} \tau_\eta(re_1)\,u(\omega)\,d\mathcal{H}^{n-1}_\omega\right] dr \\
&= \int_0^\eta \left[\tau_\eta(re_1)\,\mathcal{H}^{n-1}(\partial B_r) \fint_{\partial B_r(x)} u(\omega)\,d\mathcal{H}^{n-1}_\omega\right] dr \\
&= u(x) \int_0^\eta \tau_\eta(re_1)\,\mathcal{H}^{n-1}(\partial B_r)\,dr.
\end{aligned}$$

Accordingly, since

$$\begin{aligned}
1 = \int_{B_\eta} \tau_\eta(y)\,dy &= \int_0^\eta \left[\int_{\partial B_r} \tau_\eta(\omega)\,d\mathcal{H}^{n-1}_\omega\right] dr \\
&= \int_0^\eta \left[\int_{\partial B_r} \tau_\eta(re_1)\,d\mathcal{H}^{n-1}_\omega\right] dr \\
&= \int_0^\eta \tau_\eta(re_1)\,\mathcal{H}^{n-1}(\partial B_r)\,dr,
\end{aligned} \tag{2.1.5}$$

we gather that $u_\eta(x) = u(x)$ for almost all $x \in B_R(\overline{x})$, and as a result, (2.1.4) holds true when condition (ii) is satisfied.

Also, if u fulfills condition (iii), then by polar coordinates and (1.1.6), for almost every $x_0 \in \Omega$ and almost every $r > 0$ such that $B_r(x_0) \Subset \Omega$, we have that

$$\begin{aligned}
\fint_{\partial B_r(x_0)} u(y)\,dy &= \frac{1}{r^{n-1}\mathcal{H}^{n-1}(\partial B_1)} \frac{d}{dr} \int_{B_r(x_0)} u(y)\,dy \\
&= \frac{1}{nr^{n-1}} \frac{d}{dr}\left(r^n \fint_{B_r(x_0)} u(y)\,dy\right) = \frac{u(x_0)}{nr^{n-1}} \frac{d}{dr}(r^n) = u(x_0),
\end{aligned}$$

which gives (ii). This reduces to the previous case; therefore, the proof of (2.1.4) is complete.

In turn, we have that (2.1.4) entails (2.1.3), as desired.

Now, we show that (i) implies (ii), which implies (iii), which implies (i).

Let us assume that (i) holds true, and let $B_r(x_0) \Subset \Omega$. Then, u is harmonic in $B_\rho(x_0)$ for every $\rho \in (0, r]$. Accordingly, by the application of the divergence theorem given in equation (1.1.4),

$$\begin{aligned} 0 &= \int_{B_\rho(x_0)} \Delta u(x)\, dx = \int_{\partial B_\rho(x_0)} \nabla u(x) \cdot \nu(x)\, d\mathcal{H}_x^{n-1} \\ &= \frac{1}{\rho} \int_{\partial B_\rho(x_0)} \nabla u(x) \cdot (x - x_0)\, d\mathcal{H}_x^{n-1}. \end{aligned}$$

On the other hand,

$$\begin{aligned} \frac{d}{d\rho}\left(\fint_{\partial B_\rho(x_0)} u(x)\, d\mathcal{H}_x^{n-1} \right) &= \frac{d}{d\rho}\left(\fint_{\partial B_1} u(x_0 + \rho\omega)\, d\mathcal{H}_\omega^{n-1} \right) \\ &= \fint_{\partial B_1} \nabla u(x_0 + \rho\omega) \cdot \omega\, d\mathcal{H}_\omega^{n-1} \\ &= \frac{1}{\rho} \fint_{\partial B_\rho(x_0)} \nabla u(x) \cdot (x - x_0)\, d\mathcal{H}_x^{n-1}. \end{aligned}$$

These observations entail that, for every $\rho \in (0, r]$,

$$\frac{d}{d\rho}\left(\fint_{\partial B_\rho(x_0)} u(x)\, d\mathcal{H}_x^{n-1} \right) = 0,$$

whence

$$\fint_{\partial B_\rho(x_0)} u(x)\, d\mathcal{H}_x^{n-1} \text{ is constant for all } \rho \in (0, r].$$

In particular,

$$\fint_{\partial B_r(x_0)} u(x)\, d\mathcal{H}_x^{n-1} = \lim_{\rho \searrow 0} \fint_{\partial B_\rho(x_0)} u(x)\, d\mathcal{H}_x^{n-1} = u(x_0),$$

and this shows that (ii) holds true.

The fact that (ii) implies (iii) is a consequence of polar coordinates. Indeed, if (ii) is satisfied, then

$$\begin{aligned} \fint_{B_r(x_0)} u(x)\, dx &= \frac{1}{|B_r|} \int_0^r \left(\int_{\partial B_\rho(x_0)} u(x)\, d\mathcal{H}_x^{n-1} \right) d\rho \\ &= \frac{\mathcal{H}^{n-1}(\partial B_1)}{|B_1|\, r^n} \int_0^r \left(\rho^{n-1} \fint_{\partial B_\rho(x_0)} u(x)\, d\mathcal{H}_x^{n-1} \right) d\rho \\ &= \frac{\mathcal{H}^{n-1}(\partial B_1)\, u(x_0)}{|B_1|\, r^n} \int_0^r \rho^{n-1}\, d\rho = \frac{\mathcal{H}^{n-1}(\partial B_1)\, u(x_0)}{|B_1|\, n}. \end{aligned}$$

This and (1.1.6) yield that

$$\fint_{B_r(x_0)} u(x)\,dx = u(x_0),$$

that is, (iii).

Let us now suppose that (iii) holds true. Then, by (2.1.3) and Theorem 1.1.1, for every $x_0 \in \Omega$,

$$0 = \lim_{r \searrow 0} \frac{1}{r^2} \left(\fint_{B_r(x_0)} u(x)\,dx - u(x_0) \right) = \frac{1}{2(n+2)}\,\Delta u(x_0),$$

thus showing the validity of (i). □

For a comprehensive survey on the mean value properties of harmonic and closely related functions, see [NV94].

A simple byproduct of the mean value formula in Theorem 2.1.2 is the following interesting geometric observation.

Corollary 2.1.3. *Let $n \geqslant 2$. A harmonic function does not possess isolated zeroes.*

Proof. Let u be harmonic in some open set $\Omega \subseteq \mathbb{R}^n$, and suppose that $u(x_0) = 0$. Arguing for a contradiction, we suppose that there exists $r > 0$ such that $B_r(x_0) \Subset \Omega$ and $u \neq 0$ in $B_r(x_0) \setminus \{x_0\}$. Thus, by continuity and the fact that $B_r(x_0) \setminus \{x_0\}$ is a connected set when $n \geqslant 2$, by possibly replacing u with $-u$, we can suppose that $u > 0$ in $B_r(x_0) \setminus \{x_0\}$. This and Theorem 2.1.2(iii) yield that

$$0 = u(x_0) = \fint_{B_r(x_0)} u(x)\,dx > 0,$$

which is a contradiction. □

We remark that Corollary 2.1.3 does not hold true when $n = 1$ since the function $u(x) = x$ for all $x \in \mathbb{R}$ is harmonic but possesses an isolated zero.

It can be useful to stress that the "almost every $x_0 \in \Omega$" and "almost every $r > 0$" in Theorem 2.1.2(ii) and (iii) can be replaced with the simpler "every $x_0 \in \Omega$" and "every $r > 0$" thanks to a continuity argument: the details are as follows.

Corollary 2.1.4. *Given an open set $\Omega \subseteq \mathbb{R}^n$ and a function $u \in C(\Omega)$, the following conditions are equivalent:*

(i) *The function u belongs to $C^2(\Omega)$ and is harmonic in Ω.*
(ii) *For every $x_0 \in \Omega$ and every $r > 0$ such that $B_r(x_0) \Subset \Omega$, we have that*

$$u(x_0) = \fint_{\partial B_r(x_0)} u(x)\,d\mathcal{H}^{n-1}_x.$$

(iii) *For every $x_0 \in \Omega$ and every $r > 0$ such that $B_r(x_0) \Subset \Omega$, we have that*

$$u(x_0) = \fint_{B_r(x_0)} u(x)\,dx.$$

Proof. We prove that condition (i) is equivalent to condition (ii) (similarly, one can prove the equivalence between conditions (i) and (iii)). Assume (i) here. Then, condition (i) in Theorem 2.1.2 holds true, which entails condition (ii) in Theorem 2.1.2. This gives that condition (ii) here is satisfied for almost every $x_0 \in \Omega$ and almost every $r > 0$.

Let now $\bar{x} \in \Omega$ and $\bar{r} > 0$ such that $B_{\bar{r}}(\bar{x}) \Subset \Omega$. Let $x^{(j)}$ be a sequence converging to $\bar{x}$ as $j \to +\infty$, with $x^{(j)}$ belonging to the above-mentioned set of full measure, for which (ii) holds true.

Let also $r \in (\bar{r}-3|\bar{x}-x^{(j)}|, \bar{r}-2|\bar{x}-x^{(j)}|)$, and note that $B_r(x^{(j)}) \subseteq B_{\bar{r}}(\bar{x}) \Subset \Omega$. Accordingly, we can find $r_j \in (\bar{r} - 3|\bar{x} - x^{(j)}|, \bar{r} - 2|\bar{x} - x^{(j)}|)$ such that

$$u(x^{(j)}) = \fint_{\partial B_{r_j}(x^{(j)})} u(x)\,d\mathcal{H}^{n-1}_x.$$

By passing to the limit as $j \to +\infty$ and using the continuity of u, we obtain (ii) here, as desired.

Suppose now that (ii) here holds true. Then, condition (ii) in Theorem 2.1.2 holds true, which entails condition (i) in Theorem 2.1.2, which is condition (i) here. □

As a simple variant of Corollary 2.1.4, we have that a function agrees with its average "up to higher orders" if and only if it completely agrees with its average.

Corollary 2.1.5. *Given an open set $\Omega \subseteq \mathbb{R}^n$ and a function $u \in C(\Omega)$, the following conditions are equivalent:*

(a) *For every $x_0 \in \Omega$, we have that*

$$u(x_0) = \lim_{r \searrow 0} \fint_{\partial B_r(x_0)} u(x)\,d\mathcal{H}^{n-1}_x.$$

(b) *For every $x_0 \in \Omega$, we have that*

$$u(x_0) = \lim_{r \searrow 0} \fint_{B_r(x_0)} u(x)\,dx.$$

(c) *Any of the equivalent conditions (i), (ii) or (iii) in Corollary 2.1.4 is satisfied.*

Proof. Assume that (c) holds true. Then, in particular, conditions (ii) and (iii) in Corollary 2.1.4 are satisfied, which clearly imply (a) and (b).

Suppose instead that (a) is satisfied (the case in which (b) is satisfied is similar, obtained by simply changing spherical integrals into volume integrals). Then, we have that

$$u \in C^\infty(\Omega). \tag{2.1.6}$$

The proof of this claim is a variation of that in (2.1.3) and relies on a similar mollification argument (which we apply here with a similar notation): namely, using polar coordinates,

$$\begin{aligned} u_\eta(x) &= \int_{B_\eta(x)} \tau_\eta(x-y)\, u(y)\, dy \\ &= \int_0^\eta \left[\int_{\partial B_r(x)} \tau_\eta(x-\omega)\, u(\omega)\, d\mathcal{H}^{n-1}_\omega \right] dr \\ &= \int_0^\eta \left[\int_{\partial B_r(x)} \tau_\eta(re_1)\, u(\omega)\, d\mathcal{H}^{n-1}_\omega \right] dr \\ &= \int_0^\eta \left[\tau_\eta(re_1)\, \mathcal{H}^{n-1}(\partial B_r) \fint_{\partial B_r(x)} u(\omega)\, d\mathcal{H}^{n-1}_\omega \right] dr. \end{aligned}$$

Thus, using (a) and the dominated convergence theorem, we can take the limit as $r \searrow 0$ in the above equation and conclude that

$$\begin{aligned} u_\eta(x) &= \int_0^\eta \left[\tau_\eta(re_1)\, \mathcal{H}^{n-1}(\partial B_r) \lim_{r\searrow 0} \fint_{\partial B_r(x)} u(\omega)\, d\mathcal{H}^{n-1}_\omega \right] dr \\ &= u(x) \int_0^\eta \tau_\eta(re_1)\, \mathcal{H}^{n-1}(\partial B_r)\, dr. \end{aligned}$$

From this and (2.1.5), we gather that $u_\eta(x) = u(x)$, which establishes (2.1.6). Using Theorem 1.1.2, we thereby infer that $\Delta u = 0$ in Ω, and hence (i) in Corollary 2.1.4 holds true. □

It is instructive to emphasize that the notion of harmonicity is "local": since it relies on the value of the derivative of a function at points, if I is a sets of indices, $\Omega_i \subseteq \mathbb{R}^n$ are open sets for every $i \in I$ and u is harmonic in each of Ω_i, then

$$u \text{ is harmonic in } \bigcup_{i\in I} \Omega_i. \tag{2.1.7}$$

As a result, the mean value formula in Theorem 2.1.2 (or its modification in Corollary 2.1.4) can be also localized, according to this observation.

Corollary 2.1.6. *Given an open set $\Omega \subseteq \mathbb{R}^n$ and a function $u \in C(\Omega)$, the following conditions are equivalent:*

(i) *The function u belongs to $C^2(\Omega)$ and is harmonic in Ω.*

(ii) *For every* $x_0 \in \Omega$, *there exists* $r_0 > 0$ *such that* $B_{r_0}(x_0) \Subset \Omega$, *and for every* $r \in (0, r_0)$, *we have that*

$$u(x_0) = \fint_{\partial B_r(x_0)} u(x)\, d\mathcal{H}_x^{n-1}.$$

(iii) *For every* $x_0 \in \Omega$, *there exists* $r_0 > 0$ *such that* $B_{r_0}(x_0) \Subset \Omega$, *and for every* $r \in (0, r_0)$, *we have that*

$$u(x_0) = \fint_{B_r(x_0)} u(x)\, dx.$$

Proof. We prove that condition (i) is equivalent to condition (ii) (similarly, one can prove the equivalence between conditions (i) and (iii)). On the one hand, condition (i) here coincides with condition (i) in Corollary 2.1.4, which entails condition (ii) in Corollary 2.1.4, which in turn entails condition (ii) here.

On the other hand, if condition (ii) here holds, by Corollary 2.1.4(i), we infer that for every $x_0 \in \Omega$, there exists $r_0(x_0) > 0$ such that $B_{r_0(x_0)}(x_0) \Subset \Omega$ and u is harmonic in $B_{r_0(x_0)}(x_0)$. That is, in view of (2.1.7), u is harmonic in the set

$$\widetilde{\Omega} := \bigcup_{x_0 \in \Omega} B_{r_0(x_0)}(x_0) \supseteq \Omega,$$

which establishes (i). □

A useful consequence of Corollary 2.1.6 is the so-called Schwarz reflection principle, which allows us to extend a harmonic function vanishing on a hyperplane by an odd reflection.

Lemma 2.1.7. *Let* $\Omega \subseteq \mathbb{R}^n$ *be an open set. Let*

$$\begin{aligned} \Omega_+ &:= \Big\{x = (x', x_n) \in \Omega \text{ s.t. } x_n > 0\Big\}, \\ \Omega_0 &:= \Big\{x = (x', x_n) \in \Omega \text{ s.t. } x_n = 0\Big\}, \\ \Omega_- &:= \Big\{x = (x', x_n) \text{ s.t. } x_n < 0 \quad \textit{and} \quad (x', -x_n) \in \Omega_+\Big\} \end{aligned}$$

and $\Omega_\star := \Omega_+ \cup \Omega_0 \cup \Omega_-$.

Let $u \in C^2(\Omega_+) \cap C(\Omega_+ \cup \Omega_0)$ *be harmonic and such that* $u(x) = 0$ *along* Ω_0.

Then, the function

$$\Omega_\star \ni x \mapsto u_\star(x) = u_\star(x', x_n) := \begin{cases} u(x) & \text{if } x \in \Omega_+ \cup \Omega_0, \\ -u(x', -x_n) & \text{if } x \in \Omega_- \end{cases}$$

is harmonic in $\Omega_\star$.

Proof. We observe that

$$u_\star \text{ is continuous and } u_\star(x',0)=0 \text{ for all } x=(x',0)\in\Omega_0. \tag{2.1.8}$$

Furthermore, if $x\in\Omega_-$, then $\Delta u_\star(x) = -\Delta u(x',-x_n) = 0$ and accordingly $u_\star$ is harmonic in $\Omega_+\cup\Omega_-$ due to (2.1.6). This and Corollary 2.1.6(iii) give that

$$\begin{gathered}\text{for every } \bar x\in\Omega_+\cup\Omega_-, \text{ there exists } \bar r>0 \text{ such that for all } r\in(0,\bar r),\\ \text{we have that } \fint_{B_r(\bar x)} u_\star(x)\,dx = u_\star(\bar x).\end{gathered} \tag{2.1.9}$$

Now, we take $x_0\in\Omega_0$ and $r>0$ such that $B_r(x_0)\Subset\Omega_\star$, and we observe that

$$\begin{aligned}\fint_{B_r(x_0)} u_\star(x)\,dx &= \frac{1}{|B_r|}\left(\int_{B_r(x_0)\cap\Omega_+} u_\star(x)\,dx + \int_{B_r(x_0)\cap\Omega_-} u_\star(x)\,dx\right)\\ &= \frac{1}{|B_r|}\left(\int_{B_r(x_0)\cap\Omega_+} u(x)\,dx - \int_{B_r(x_0)\cap\Omega_-} u(x',-x_n)\,dx\right)\\ &= \frac{1}{|B_r|}\left(\int_{B_r(x_0)\cap\Omega_+} u(x)\,dx - \int_{B_r(x_0)\cap\Omega_+} u(x',x_n)\,dx\right)\\ &= 0 = u_\star(x_0),\end{aligned}$$

thanks to (2.1.8).

From this and (2.1.9), it follows that for every $\bar x\in\Omega_\star$, there exists $\bar r>0$ such that for all $r\in(0,\bar r)$, we have that $\fint_{B_r(\bar x)} u_\star(x)\,dx = u_\star(\bar x)$. This and Corollary 2.1.6 give that $u_\star$ is harmonic in $\Omega_\star$, as desired. □

The following classical identities, which are useful variations of the divergence theorem, are often very helpful, and they are named[3] "Green's identities."

[3]One of the cornerstones of the foundation of potential theory was indeed George Green's 1828 essay [Gre28] (see Figure 2.1.1), which founded the mathematical theory of electricity and magnetism. The first edition of the essay was printed for the author and sold on a subscription basis to only 51 people. In 1850–1854, the essay was transcribed in [TG50, Gre52, Gre54], with several typographical corrections and a reference section added. Interestingly, George Green was almost entirely self-taught, having received only about one year of formal schooling, between the ages of 8 and 9.

By the way, the story of George Green is truly amazing. His father, also called George Green, was a baker in Nottingham, a town in the UK linked to the legend of Robin Hood. Actually, this legend does reflect the long-lasting social problems in the area, which also affected the Green family. In particular, at some point, bakers were blamed for the incessant rise in the price of bread, and crowds of people broke into bakers to steal food, and in this circumstance, the Green family's bakery was also attacked.

The bakery was, however, probably doing well from a financial point of view, since, the year after these riots, little George was sent to allegedly the best and most expensive school in Nottingham, where he was taught for four terms, which, as mentioned above, amounted to his whole formal training: at nine, the boy began to work for his father's bakery business.

And this business remained profitable, allowing the Green family to buy land and build a brick wind corn mill (the mill was renovated in 1986 and is now a science center, see Figure 2.1.2). George Green Jr.

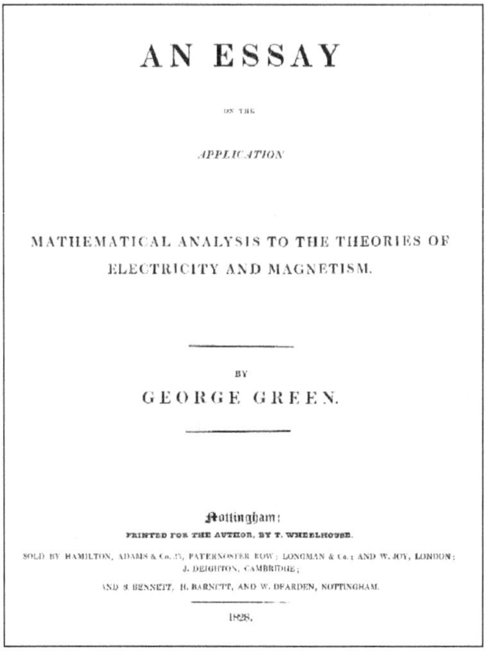

AN ESSAY

ON THE

APPLICATION

MATHEMATICAL ANALYSIS TO THE THEORIES OF ELECTRICITY AND MAGNETISM.

BY

GEORGE GREEN.

Nottingham:

PRINTED FOR THE AUTHOR, BY T. WHEELHOUSE.

SOLD BY HAMILTON, ADAMS & Co. 33, PATERNOSTER ROW; LONGMAN & Co.; AND W. JOY, LONDON;
J. DEIGHTON, CAMBRIDGE;
AND S. BENNETT, H. BARNETT, AND W. DEARDEN, NOTTINGHAM.

1828.

Figure 2.1.1 The title page to Green's original essay.
Source: Public Domain image from Wikipedia.

---

then fell in love with the daughter of the manager of the mill, named Jane Smith. They never married, but they had seven children together.

Green also joined the Nottingham Subscription Library, thus finding access to a few scientific books and articles, as well as some works published in other countries.

Green used to study and do mathematics on the top floor of the mill, and it was probably here that the famous essay by him was conceived and written. The 51 subscribers who bought the first edition of the book (at the price of 7 pounds and 6 pence) were probably, for a vast majority, members of the Nottingham Subscription Library, and their mathematical proficiency was likely insufficient to fully appreciate the content of the essay. However, among these subscribers was also Sir Edward Thomas Bromhead, 2nd Baronet, a wealthy landowner, mathematician and founder of the Analytical Society, a precursor of the Cambridge Philosophical Society. Bromhead realized that the essay was the product of a brilliant scientist and invited Green to send any further papers to the Royal Society of London, the Royal Society of Edinburgh and the Cambridge Philosophical Society. Green took Bromhead's offer as mere politeness and did not respond for two years.

Then, following Bromhead's encouragement, Green wrote three further papers and, having accumulated considerable wealth and land ownership, was able to abandon his miller duties, pursue mathematical studies and, aged nearly 40, enroll as an undergraduate at the University of Cambridge.

There are rumors that, at Cambridge, Green may have succumbed to alcohol, possibly losing the endorsement of his earlier supporters (we forgot to mention that Bromhead was also approached by local ministers to help establish a Temperance Society).

Figure 2.1.2 Green's mill.
Source: Photo by Kev747, image from Wikipedia, licensed under the Creative Commons Attribution-Share Alike 3.0 Unported license.

Lemma 2.1.8. *Let $\Omega \subseteq \mathbb{R}^n$ be a bounded open set of class C^1, with exterior normal ν, and let $\varphi, \psi \in C^2(\Omega) \cap C^1(\overline{\Omega})$.*

Then,

$$\int_\Omega \Big(\varphi(x)\Delta\psi(x) + \nabla\varphi(x) \cdot \nabla\psi(x)\Big)\, dx = \int_{\partial\Omega} \varphi(x)\frac{\partial\psi}{\partial\nu}(x)\, d\mathcal{H}_x^{n-1} \tag{2.1.10}$$

George Green died aged 48, and the Nottingham Review published the following short obituary: "[W]e believe he was the son of a miller, residing near Nottingham, but having a taste for study, he applied his gifted mind to the science of mathematics [...]. Had his life been prolonged, he might have stood eminently high as a mathematician." They obviously did not understand that he had already stood most eminently among his contemporaries and left an essay that would go on to revolutionize science.

Historians of science are unsure how Green managed to acquire his formidable mathematical knowledge with so little formal training. One possibility, however, is that the Leibniz–Newton calculus controversy (the silly dispute over who had first invented calculus) resounded to great disadvantage of the English school, which remained locked into the Newtonian notation of calculus, stubbornly rejecting the notation introduced by Leibnitz but adopted by continental mathematicians, which ultimately proved to be more flexible and effective. It is possible that, being self-taught, Green had the opportunity to get in contact with Leibnitz's notation (or possibly develop his own approach to calculus and analysis) without any rigid bias of the academia or being influenced by superficial politics or sterile nationalism (well, anyway, a good notation is always helpful, but in Green's case, personal talent and inventiveness certainly made the difference).

For further readings on the figure of George Green, see [FP89, CL93, Can01, CS03].

and

$$\int_\Omega \Big(\varphi(x)\Delta\psi(x) - \psi(x)\Delta\varphi(x)\Big)\,dx = \int_{\partial\Omega}\left(\varphi(x)\frac{\partial\psi}{\partial\nu}(x) - \psi(x)\frac{\partial\varphi}{\partial\nu}(x)\right)\,d\mathcal{H}^{n-1}_x. \tag{2.1.11}$$

Proof. According to the divergence theorem,

$$\begin{aligned}\int_\Omega \Big(\varphi(x)\Delta\psi(x) + \nabla\varphi(x)\cdot\nabla\psi(x)\Big)\,dx &= \int_\Omega \operatorname{div}\big(\varphi(x)\nabla\psi(x)\big)\,dx \\ &= \int_{\partial\Omega}\varphi(x)\frac{\partial\psi}{\partial\nu}(x)\,d\mathcal{H}^{n-1}_x,\end{aligned}$$

that is, (2.1.10).

Also, exchanging the roles of φ and ψ in (2.1.10),

$$\int_\Omega \Big(\psi(x)\Delta\varphi(x) + \nabla\psi(x)\cdot\nabla\varphi(x)\Big)\,dx = \int_{\partial\Omega}\psi(x)\frac{\partial\varphi}{\partial\nu}(x)\,d\mathcal{H}^{n-1}_x.$$

Subtracting this from (2.1.10), we obtain (2.1.11). □

We observe that identity (1.1.4) can now be considered a special case of (2.1.10). If creatively exploited, Green's identities are very useful to deduce important integral formulas, which in turn entail structural information on several relevant equations. As a prototype of this idea, we recall the classical Pohožaev identity (see [Poh65]; actually, an early occurrence of the Pohožaev identity was obtained already in [Rel40, equation (2)]).

Theorem 2.1.9. *Let Ω be a bounded open set in $\mathbb{R}^n$ with C^1 boundary and $u \in C^2(\Omega)\cap C^1(\overline{\Omega})$ be a solution of*

$$\begin{cases}\Delta u = f(u) & \text{in } \Omega,\\ u = 0 & \text{on } \partial\Omega,\end{cases} \tag{2.1.12}$$

for some $f \in L^\infty_{\mathrm{loc}}(\mathbb{R})$.

Let also

$$F(r) := \int_0^r f(t)\,dt.$$

Then,

$$\begin{aligned}
&\frac{1}{2}\int_{\partial\Omega}(\partial_\nu u(x))^2\,(x\cdot\nu(x))\,d\mathcal{H}_x^{n-1}\\
&\quad=\frac{n-2}{2}\int_\Omega u(x)\,f(u(x))\,dx-n\int_\Omega F(u(x))\,dx.
\end{aligned}$$

Proof. The idea of the proof is to test the equation against the radial derivative $\nabla u(x)\cdot x$ using suitable integration by parts. Namely, from (2.1.12),

$$\begin{aligned}
\int_\Omega \Delta u(x)\,(\nabla u(x)\cdot x)\,dx &= \int_\Omega f(u(x))(\nabla u(x)\cdot x)\,dx\\
&=\int_\Omega \nabla\big(F(u(x))\big)\cdot x\,dx\\
&=\int_\Omega\Big(\operatorname{div}\big(F(u(x))\,x\big)-nF(u(x))\Big)\,dx.
\end{aligned}$$

This and the divergence theorem, recalling the boundary condition in (2.1.12), give that

$$\begin{aligned}
\int_\Omega \Delta u(x)\,(\nabla u(x)\cdot x)\,dx+n\int_\Omega F(u(x))\,dx &= \int_\Omega \operatorname{div}\big(F(u(x))\,x\big)\,dx\\
&=\int_{\partial\Omega} F(u(x))\,(x\cdot\nu(x))\,d\mathcal{H}_x^{n-1}\\
&=\int_{\partial\Omega} F(0)\,(x\cdot\nu(x))\,d\mathcal{H}_x^{n-1}=0.
\end{aligned}\tag{2.1.13}$$

Furthermore,

$$\begin{aligned}
\Delta u\,(\nabla u\cdot x) &= \operatorname{div}\Big((\nabla u\cdot x)\nabla u\Big)-\nabla(\nabla u\cdot x)\cdot\nabla u\\
&=\operatorname{div}\Big((\nabla u\cdot x)\nabla u\Big)-\sum_{i,j=1}^n \partial_{ij}u\,\partial_i u\,x_j-|\nabla u|^2\\
&=\operatorname{div}\Big((\nabla u\cdot x)\nabla u\Big)-\frac{1}{2}\sum_{j=1}^n \partial_j|\nabla u|^2\,x_j-|\nabla u|^2\\
&=\operatorname{div}\Big((\nabla u\cdot x)\nabla u\Big)-\frac{1}{2}\nabla(|\nabla u|^2)\cdot x-|\nabla u|^2\\
&=\operatorname{div}\Big((\nabla u\cdot x)\nabla u\Big)-\frac{1}{2}\Big[\operatorname{div}\Big(|\nabla u|^2\,x\Big)-n|\nabla u|^2\Big]-|\nabla u|^2\\
&=\operatorname{div}\Big((\nabla u\cdot x)\nabla u\Big)-\frac{1}{2}\operatorname{div}\Big(|\nabla u|^2\,x\Big)+\frac{n-2}{2}|\nabla u|^2.
\end{aligned}$$

Thus, making use of the divergence theorem again and of Green's first identity (2.1.10), we find that

$$\begin{aligned}\int_\Omega \Delta u(x)\,(\nabla u(x)\cdot x)\,dx &= \int_{\partial\Omega} (\nabla u(x)\cdot x)\partial_\nu u(x)\,d\mathcal{H}^{n-1}_x \\ &\quad - \frac{1}{2}\int_{\partial\Omega} |\nabla u(x)|^2\, x\cdot \nu(x)\,d\mathcal{H}^{n-1}_x \\ &\quad + \frac{n-2}{2}\int_\Omega |\nabla u(x)|^2\,dx \\ &= \int_{\partial\Omega} (\nabla u(x)\cdot x)\partial_\nu u(x)\,d\mathcal{H}^{n-1}_x \\ &\quad - \frac{1}{2}\int_{\partial\Omega} |\nabla u(x)|^2\, x\cdot \nu(x)\,d\mathcal{H}^{n-1}_x \\ &\quad - \frac{n-2}{2}\int_\Omega \Delta u(x)\,u(x)\,dx. \end{aligned} \tag{2.1.14}$$

Now, we observe that $\nabla u = \pm|\nabla u|\nu$ on $\partial\Omega$; therefore, for every $x \in \Omega$,

$$(\nabla u(x)\cdot x)\partial_\nu u(x) = |\nabla u(x)|^2(\nu(x)\cdot x)(\nu(x)\cdot\nu(x)) = |\nabla u(x)|^2(\nu(x)\cdot x).$$

Plugging this information into (2.1.14), we find that

$$\begin{aligned}&\int_\Omega \Delta u(x)\,(\nabla u(x)\cdot x)\,dx \\ &= \frac{1}{2}\int_{\partial\Omega} |\nabla u(x)|^2\, x\cdot\nu(x)\,d\mathcal{H}^{n-1}_x - \frac{n-2}{2}\int_\Omega \Delta u(x)\,u(x)\,dx.\end{aligned}$$

Combining this and (2.1.13), we obtain the desired result. □

Equations as in (2.1.12) are often called "semilinear" since they are not linear in u (unless the source term f is linear) but are linear in the second derivative of u. Interestingly, solutions of semilinear equations[4] enjoy the special feature of having constant Laplacian along their level sets, namely if u solves (2.1.12), given any $c \in \mathbb{R}$, we have that $\Delta u = f(c)$ on $\{u = c\}$.

As a consequence of the Pohožaev identity in Theorem 2.1.9, one obtains nonexistence results, as the one in the forthcoming Corollary 2.1.11. For this, we give the following definition.

Definition 2.1.10. Let $\Omega \subseteq \mathbb{R}^n$. Given $x_0 \in \Omega$, we say that Ω is Aped with respect to x_0 if for every $x \in \Omega$, we have that $tx + (1-t)x_0 \in \Omega$ for all $t \in [0, 1]$.

[4]We refer to the footnotes in [DV23, Chapter 20] for motivational comments about semilinear equations.

Furthermore, we say that Ω is star shaped if there exists $x_0 \in \Omega$ such that Ω is star shaped with respect to x_0.

With this, we give the following nonexistence result.

Corollary 2.1.11. *Let $n \geqslant 3$ and $p > \frac{n+2}{n-2}$. Let Ω be a bounded star-shaped open set in $\mathbb{R}^n$ with C^1 boundary. Let $u \in C^2(\Omega) \cap C^1(\overline{\Omega})$ be a solution of*

$$\begin{cases} \Delta u = -|u|^{p-1}u & \text{in } \Omega, \\ u = 0 & \text{on } \partial\Omega. \end{cases}$$

Then, u vanishes identically.

Proof. Up to a translation, we suppose that

$$\Omega \text{ is star shaped with respect to the origin.} \tag{2.1.15}$$

We claim that

$$x \cdot \nu(x) \geqslant 0 \text{ for every } x \in \partial\Omega. \tag{2.1.16}$$

To check this, given $x_0 \in \partial\Omega$, we write Ω in the vicinity of x_0 as the superlevels of some function $\Phi \in C^1(\mathbb{R}^n)$ with $\nabla\Phi(x_0) \neq 0$; that is, we take $\rho > 0$ such that $\Omega \cap B_\rho(x_0) = \{\Phi > 0\} \cap B_\rho(x_0)$. In this way, we have that $\nu = -\frac{\nabla\Phi}{|\nabla\Phi|}$ on $\partial\Omega$.

Also, by (2.1.15), we have that $tx_0 \in \overline{\Omega}$ for every $t \in [0, 1]$. As a result, for $t \in [0, 1]$ sufficiently close to 1, we have $\Phi(tx_0) \geqslant 0$. Therefore,

$$0 \geqslant \lim_{t \nearrow 1} \frac{\Phi(tx_0)}{t-1} = \lim_{t \nearrow 1} \frac{\Phi(tx_0) - \Phi(x_0)}{t-1} = \nabla\Phi(x_0) \cdot x_0 = -|\nabla\Phi(x_0)|\, \nu(x_0) \cdot x_0.$$

This proves (2.1.16).

We now exploit the Pohožaev identity in Theorem 2.1.9 with $f(u) := -|u|^{p-1}u$, and hence $F(r) := -\frac{|r|^{p+1}}{p+1}$. In this way, using (2.1.16), we find that

$$\begin{aligned} 0 &\leqslant \frac{1}{2} \int_{\partial\Omega} (\partial_\nu u(x))^2 \,(x \cdot \nu(x))\, d\mathcal{H}^{n-1}_x \\ &= -\frac{n-2}{2} \int_\Omega |u(x)|^{p+1}\, dx + \frac{n}{p+1} \int_\Omega |u(x)|^{p+1}\, dx \\ &= \frac{p(2-n)+n+2}{2(p+1)} \int_\Omega |u(x)|^{p+1}\, dx \leqslant 0. \end{aligned}$$

In particular,

$$\frac{p(2-n)+n+2}{2(p+1)} \int_\Omega |u(x)|^{p+1}\, dx = 0,$$

from which the desired result follows. □

A natural question is whether or not the average over balls and spheres in the mean value formulas of Theorem 2.1.2 can be substituted with averages on different sets. As we will see in Section 7.4, this is not the case, and in fact the geometry of the balls and spheres play a decisive role in the mean value formula (this classical problem was pioneered in [Eps62, ES65, GO71, Kur72, PS89].

2.2 Weak Solutions

We present here a classical result often referred to as Weyl's lemma:

Lemma 2.2.1. *Let $\Omega \subseteq \mathbb{R}^n$ be an open set, and let $u \in L^1_{\mathrm{loc}}(\Omega)$. Assume that*

$$\int_\Omega u(x)\,\Delta\varphi(x)\,dx = 0 \quad \textit{for every } \varphi \in C_0^\infty(\Omega). \tag{2.2.1}$$

Then, u is harmonic in Ω.

Proof. We stress that the desired claim follows directly from (2.2.1) and the Green's second identity (2.1.11) when $u \in C^2(\Omega)$.

If instead u is merely locally integrable in Ω, we use a mollification argument. To this end, we consider $\tau \in C_0^\infty(B_1, [0, +\infty))$ with $\int_{B_1} \tau(x)\,dx = 1$. Given $\eta > 0$, we let $\tau_\eta(x) := \frac{1}{\eta^n}\tau\left(\frac{x}{\eta}\right)$ and define $u_\eta := u * \tau_\eta$. Then, given $x_0 \in \Omega$ and $\rho > 0$ such that $B_{2\rho}(x_0) \Subset \Omega$, for all $\varphi \in C_0^\infty(B_\rho(x_0))$ and all $\eta \in (0, \rho)$, we have that

$$\begin{aligned}\int_\Omega u_\eta(x)\,\Delta\varphi(x)\,dx &= \iint_{B_{2\rho}(x_0)\times\Omega} u(y)\,\tau_\eta(x-y)\Delta\varphi(x)\,dx\,dy\\ &= \int_\Omega u(x)\,(\tau_\eta * \Delta\varphi)(x)\,dx = \int_\Omega u(x)\,\Delta\varphi_\eta(x)\,dx,\end{aligned}$$

where $\varphi_\eta := \varphi * \tau_\eta \in C_0^\infty(B_{2\rho}(x_0)) \subseteq C_0^\infty(\Omega)$. As a result, from (2.2.1), we deduce that

$$\int_\Omega u_\eta(x)\,\Delta\varphi(x)\,dx = 0 \quad \text{for every } \varphi \in C_0^\infty(B_\rho(x_0)),$$

as long as $\eta \in (0, \rho)$. Since $u_\eta \in C^2(B_\rho(x_0))$, this gives that u_η is harmonic in $B_\rho(x_0)$. Owing to this and Theorem 2.1.2(iii), for every ball $B_r(\overline{x}) \Subset B_\rho(x_0)$, we have that

$$u_\eta(\overline{x}) = \fint_{B_r(\overline{x})} u_\eta(x)\,dx.$$

We now send $\eta \searrow 0$ (see e.g. Theorems 9.6 and 9.13 in [WZ15]), and we conclude that, whenever $B_r(\overline{x}) \Subset B_\rho(x_0)$ and $\overline{x}$ is a Lebesgue density point for u,

$$u(\overline{x}) = \fint_{B_r(\overline{x})} u(x)\,dx. \tag{2.2.2}$$

Furthermore, according to the dominated convergence theorem, for every $\widetilde{x} \in \Omega$ and $r > 0$ such that $B_r(\widetilde{x}) \subset \Omega$,

$$\lim_{p \to \widetilde{x}} \int_{B_r(p)} u(x)\,dx = \int_{B_r(\widetilde{x})} u(x)\,dx.$$

This and (2.2.2) give that, up to continuously extending u in a set of null Lebesgue measure in $B_\rho(x_0)$, we have that

$$u(\overline{x}) = \fint_{B_r(\overline{x})} u(x)\,dx \quad \text{for every } \overline{x} \in B_\rho(x_0),$$

as long as $B_r(\overline{x}) \Subset B_\rho(x_0)$. Using again Theorem 2.1.2, we thereby conclude that u is harmonic in $B_\rho(x_0)$. This gives that $\Delta u(x_0) = 0$ for every $x_0 \in \Omega$, as desired. □

Corollary 2.2.2. *Let $\Omega \subseteq \mathbb{R}^n$ be an open set and u_k be a sequence of harmonic functions in Ω. Suppose that $u_k \to u$ in $L^1_{\mathrm{loc}}(\Omega)$. Then, u is harmonic in Ω.*

Proof. Let $\Omega' \Subset \Omega$ and $\varphi \in C_0^\infty(\Omega')$. By Green's second identity (2.1.11), we know that

$$\begin{aligned}\left|\int_\Omega u(x)\,\Delta\varphi(x)\,dx\right| &\leqslant \int_{\Omega'} |u(x) - u_k(x)|\,|\Delta\varphi(x)|\,dx + \left|\int_\Omega u_k(x)\,\Delta\varphi(x)\,dx\right| \\ &\leqslant \|\varphi\|_{C^2(\Omega')}\,\|u - u_k\|_{L^1(\Omega')} + \left|\int_\Omega \Delta u_k(x)\,\varphi(x)\,dx\right| \\ &= \|\varphi\|_{C^2(\Omega')}\,\|u - u_k\|_{L^1(\Omega')}.\end{aligned}$$

Hence, sending $k \to +\infty$,

$$\int_\Omega u(x)\,\Delta\varphi(x)\,dx = 0.$$

The desired result thus follows from Lemma 2.2.1. □

An alternative proof of Corollary 2.2.2 can also be obtained using directly the mean value formula in Theorem 2.1.2(iii).

See also [Sal15, SV15] and the references therein for a more complete discussion on the role played by "weak" or "distributional" formulations of partial differential equations and a careful discussion of the functional analysis methods involved in such a theory.

A classical application of the weak setting of partial differential equations is provided by Kato's inequality, see [Kat72], as presented in the following result. For this, we use the standard notation, for every $r \in \mathbb{R}$,

$$\operatorname{sign}(r) := \begin{cases} \dfrac{r}{|r|} & \text{if } r \neq 0, \\ 0 & \text{if } r = 0. \end{cases}$$

Theorem 2.2.3. *Let*

$$u \in L^1_{\mathrm{loc}}(\mathbb{R}^n) \tag{2.2.3}$$

be such that there exists $f \in L^1_{\mathrm{loc}}(\mathbb{R}^n)$ satisfying

$$\int_{\mathbb{R}^n} u(x)\, \Delta\psi(x)\, dx = \int_{\mathbb{R}^n} f(x)\, \psi(x)\, dx \quad \text{for all } \psi \in C_0^\infty(\mathbb{R}^n). \tag{2.2.4}$$

Then, for[5] *every $\varphi \in C_0^\infty(\mathbb{R}^n, [0, +\infty))$,*

$$\int_{\mathbb{R}^n} |u(x)|\, \Delta\varphi(x)\, dx \geqslant \int_{\mathbb{R}^n} \operatorname{sign}(u(x))\, \varphi(x)\, f(x)\, dx. \tag{2.2.5}$$

Proof. Given $\eta > 0$, we let $\tau_\eta(x) := \frac{1}{\eta^n}\tau\left(\frac{x}{\eta}\right)$ and define $u_\eta := u * \tau_\eta$. In this setting, by (2.2.3), possibly up to subsequences, we have that u_η converges to u in $L^1_{\mathrm{loc}}(\mathbb{R}^n)$ (see [WZ15, Theorem 9.6]) and almost everywhere, with additionally

---

[5] In jargon, condition (2.2.4) can be rewritten by stating that the Laplacian of u, as defined in the weak sense, is actually a locally integrable function that is denoted by f (and this is of course the case for smooth functions u).

Similarly, equation (2.2.5) can be written as

$$\Delta|u| \geqslant \operatorname{sign}(u)\, \Delta u$$

in the weak sense.

This can be also considered a "limit case" of the following observation: if $\Phi \in C^1(\mathbb{R})$ is a convex function, then

$$\Phi(u(x \pm he_i)) - \Phi(u(x)) \geqslant \Phi'(u(x))(u(x \pm he_i) - u(x))$$

and accordingly

$$\Phi(u(x + he_i)) + \Phi(u(x - he_i)) - 2\Phi(u(x)) \geqslant \Phi'(u(x))(u(x + he_i) + u(x - he_i) - 2u(x)),$$

which leads to

$$\Delta\Big(\Phi(u(x))\Big) \geqslant \Phi'(u(x))\, \Delta u(x).$$

In this respect, Kato's inequality (2.2.5) corresponds, formally, to the limit case in which $\Phi(t) = |t|$, which produces $\Phi'(t) = \operatorname{sign}(t)$ (at least when $t \neq 0$, and the corresponding inequality holding true in the weak sense).

$|u_\eta| \leqslant h$ for a suitable $h \in L^1_{\mathrm{loc}}(\mathbb{R}^n)$ (see [Bre11, Theorem 4.9]). For this reason,

$$\int_{\mathbb{R}^n} |u(x)|\, \Delta\varphi(x)\, dx = \lim_{\eta \searrow 0} \int_{\mathbb{R}^n} |u_\eta(x)|\, \Delta\varphi(x)\, dx. \tag{2.2.6}$$

Furthermore, for every $\psi \in C^\infty_0(\mathbb{R}^n)$,

$$\begin{aligned}
\int_{\mathbb{R}^n} u_\eta(x)\, \Delta\psi(x)\, dx &= \int_{\mathbb{R}^n} \left[\int_{\mathbb{R}^n} u(y)\, \tau_\eta(x-y)\, \Delta\psi(x)\, dx \right] dy \\
&= \int_{\mathbb{R}^n} u(y)\, \Delta\psi_\eta(y)\, dy = \int_{\mathbb{R}^n} f(y)\, \psi_\eta(y)\, dy \\
&= \int_{\mathbb{R}^n} \left[\int_{\mathbb{R}^n} f(y)\, \psi(x)\, \tau_\eta(x-y)\, dy \right] dx \\
&= \int_{\mathbb{R}^n} f_\eta(x)\, \psi(x)\, dx,
\end{aligned}$$

and accordingly $\Delta u_\eta = f_\eta$.

In this way, possibly extracting a subsequence, we deduce that Δu_η converges to f in $L^1_{\mathrm{loc}}(\mathbb{R}^n)$ (see [WZ15, Theorem 9.6]) and almost everywhere, with additionally $|\Delta u_\eta| \leqslant H$ for a suitable $H \in L^1_{\mathrm{loc}}(\mathbb{R}^n)$ (see [Bre11, Theorem 4.9]).

Now, we let $\varepsilon > 0$ and set

$$v_{\varepsilon,\eta}(x) := \sqrt{(u_\eta(x))^2 + \varepsilon^2}.$$

Thus, the function $v_{\varepsilon,\eta}$ belongs to $C^\infty(\mathbb{R}^n)$ and $\operatorname{sign}(v_{\varepsilon,\eta}(x)) = 1$ for all $x \in \mathbb{R}^n$. Furthermore,

$$2v_{\varepsilon,\eta}\nabla v_{\varepsilon,\eta} = \nabla v^2_{\varepsilon,\eta} = \nabla(u^2_\eta + \varepsilon^2) = 2u_\eta \nabla u_\eta.$$

Consequently,

$$|u_\eta|\, |\nabla v_{\varepsilon,\eta}| \leqslant v_{\varepsilon,\eta}\, |\nabla v_{\varepsilon,\eta}| = |u_\eta|\, |\nabla u_\eta|$$

and

$$|\nabla v_{\varepsilon,\eta}|^2 + v_{\varepsilon,\eta}\Delta v_{\varepsilon,\eta} = \operatorname{div}(v_{\varepsilon,\eta}\nabla v_{\varepsilon,\eta}) = \operatorname{div}(u_\eta \nabla u_\eta) = |\nabla u_\eta|^2 + u_\eta \Delta u_\eta.$$

As a result,

$$v_{\varepsilon,\eta}\Delta v_{\varepsilon,\eta} = |\nabla u_\eta|^2 - |\nabla v_{\varepsilon,\eta}|^2 + u_\eta \Delta u_\eta \geqslant u_\eta \Delta u_\eta.$$

We thus define

$$\sigma_{\varepsilon,\eta}(x) := \frac{u_\eta(x)}{v_{\varepsilon,\eta}(x)}$$

and find that $\Delta v_{\varepsilon,\eta} \geqslant \sigma_{\varepsilon,\eta}\Delta u_\eta$, and then

$$\int_{\mathbb{R}^n} v_{\varepsilon,\eta}(x)\, \Delta\varphi(x)\, dx \geqslant \int_{\mathbb{R}^n} \sigma_{\varepsilon,\eta}(x)\, \varphi(x)\, \Delta u_\eta(x)\, dx, \tag{2.2.7}$$

for every $\varphi \in C^\infty_0(\mathbb{R}^n,\ [0, +\infty))$.

It is also helpful to observe that $|v_{\varepsilon,\eta}| = v_{\varepsilon,\eta} \leqslant |u_\eta| + \varepsilon \leqslant h + 1$ and that, for a.e. $x \in \mathbb{R}^n$,

$$\lim_{\eta \searrow 0} v_{\varepsilon,\eta}(x) = \sqrt{(u(x))^2 + \varepsilon}.$$

We can therefore exploit the dominated convergence theorem to find that

$$\lim_{\eta \searrow 0} \int_{\mathbb{R}^n} v_{\varepsilon,\eta}(x)\, \Delta\varphi(x)\, dx = \int_{\mathbb{R}^n} \sqrt{(u(x))^2 + \varepsilon}\, \Delta\varphi(x)\, dx. \tag{2.2.8}$$

Additionally, for a.e. $x \in \mathbb{R}^n$,

$$\lim_{\eta \searrow 0} \sigma_{\varepsilon,\eta}(x) = \frac{u(x)}{\sqrt{(u(x))^2 + \varepsilon}},$$

and $|\sigma_{\varepsilon,\eta}| \leqslant 1$. Hence, according to the dominated convergence theorem,

$$\lim_{\eta \searrow 0} \int_{\mathbb{R}^n} \sigma_{\varepsilon,\eta}(x)\, \varphi(x)\, \Delta u_\eta(x)\, dx = \int_{\mathbb{R}^n} \frac{u(x)}{\sqrt{(u(x))^2 + \varepsilon}}\, \varphi(x)\, f(x)\, dx.$$

Combining this fact with (2.2.8), we can pass (2.2.7) to the limit as $\eta \searrow 0$ and see that, for every $\varphi \in C_0^\infty(\mathbb{R}^n, [0, +\infty))$,

$$\int_{\mathbb{R}^n} \sqrt{(u(x))^2 + \varepsilon}\, \Delta\varphi(x)\, dx \geqslant \int_{\mathbb{R}^n} \frac{u(x)}{\sqrt{(u(x))^2 + \varepsilon}}\, \varphi(x)\, f(x)\, dx.$$

By sending $\varepsilon \searrow 0$, we thereby obtain the desired result in (2.2.5). □

As a consequence of Kato's inequality in Theorem 2.2.3, we present a classification result for global weak solutions of the equation $\Delta u = Vu + cu$, see [RS75] for additional details.

Corollary 2.2.4. *Let $V \in L^2_{\text{loc}}(\mathbb{R}^n, [0, +\infty))$. Let $u \in L^2(\mathbb{R}^n)$, and assume that, for every $\varphi \in C_0^\infty(\mathbb{R}^n)$,*

$$\int_{\mathbb{R}^n} u(x)\, \Delta\varphi(x)\, dx = \int_{\mathbb{R}^n} V(x)\, u(x)\, \varphi(x)\, dx. \tag{2.2.9}$$

Then, u vanishes identically.

Proof. To exploit Theorem 2.2.3, we note that condition (2.2.4) is fulfilled with $f(x) := V(x)\, u(x) \in L^1_{\text{loc}}(\mathbb{R}^n)$, thanks to (2.2.9). Therefore, in light of (2.2.5), for every $\varphi \in C_0^\infty(\mathbb{R}^n, [0, +\infty))$,

$$\int_{\mathbb{R}^n} |u(x)|\, \Delta\varphi(x)\, dx \geqslant \int_{\mathbb{R}^n} \operatorname{sign}(u(x))\, \varphi(x)\, V(x)\, u(x)\, dx. \tag{2.2.10}$$

Now, we take $\tau \in C_0^\infty(B_1, [0,+\infty))$ with $\int_{B_1} \tau(x)\,dx = 1$. Given $\eta > 0$, we let $\tau_\eta(x) := \frac{1}{\eta^n}\tau\left(\frac{x}{\eta}\right)$ and define $w_\eta := |u| * \tau_\eta$. Note that $|u| \in L^2(\mathbb{R}^n)$ and therefore $w_\eta \to |u|$ in $L^2(\mathbb{R}^n)$ (see e.g. [WZ15, Theorem 9.6]). As a result, possibly up to a subsequence, there exists $h \in L^2(\mathbb{R}^n)$ such that

$$|w_\eta(x)| \leqslant h(x) \tag{2.2.11}$$

a.e. $x \in \mathbb{R}^n$ (see e.g. [Bre11, Theorem 4.9]).

Additionally, for every $\varphi \in C_0^\infty(\mathbb{R}^n, [0,+\infty))$,

$$\begin{aligned}\int_{\mathbb{R}^n} w_\eta(x)\,\Delta\varphi(x)\,dx &= \int_{\mathbb{R}^n}\left[\int_{\mathbb{R}^n} |u(y)|\,\tau_\eta(x-y)\,\Delta\varphi(x)\,dx\right]dy\\ &= \int_{\mathbb{R}^n} |u(y)|\,\Delta\varphi_\eta(y)\,dy\\ &\geqslant \int_{\mathbb{R}^n} \operatorname{sign}(u(y))\,\varphi_\eta(y)\,V(y)\,u(y)\,dy\\ &= \int_{\mathbb{R}^n} \varphi_\eta(y)\,V(y)\,|u(y)|\,dy \geqslant 0,\end{aligned}$$

thanks to (2.2.10).

As a result, we find that $\Delta w_\eta \geqslant 0$. Thus, if $R > 1$ and $\xi_R \in C_0^\infty(B_R, [0,1])$ with $\xi_R = 1$ in B_{R-1} and $|\nabla\xi_R| \leqslant 2$, letting $\zeta_R := \xi_R^2$, we find that

$$\begin{aligned}\int_{\mathbb{R}^n} \zeta_R(x)\,|\nabla w_\eta(x)|^2\,dx &= -\int_{\mathbb{R}^n} \operatorname{div}\left(\zeta_R(x)\,\nabla w_\eta(x)\right) w_\eta(x)\,dx\\ &= -\int_{\mathbb{R}^n} \nabla\zeta_R(x)\cdot\nabla w_\eta(x)\,w_\eta(x)\,dx\\ &\quad - \int_{\mathbb{R}^n} \zeta_R(x)\,\Delta w_\eta(x)\,w_\eta(x)\,dx\\ &\leqslant -\int_{\mathbb{R}^n} \nabla\zeta_R(x)\cdot\nabla w_\eta(x)\,w_\eta(x)\,dx\\ &= -2\int_{\mathbb{R}^n} \xi_R(x)\,\nabla\xi_R(x)\cdot\nabla w_\eta(x)\,w_\eta(x)\,dx\\ &\leqslant \frac{1}{2}\int_{\mathbb{R}^n} \zeta_R(x)\,|\nabla w_\eta(x)|^2\,dx\\ &\quad +2\int_{\mathbb{R}^n} |\nabla\xi_R(x)|^2\,(w_\eta(x))^2\,dx.\end{aligned}$$

For this reason, and recalling (2.2.11),

$$\frac{1}{2}\int_{\mathbb{R}^n} \zeta_R(x)\,|\nabla w_\eta(x)|^2\,dx \leqslant 8\int_{\mathbb{R}^n\setminus B_{R-1}} (w_\eta(x))^2\,dx \leqslant 8\int_{\mathbb{R}^n\setminus B_{R-1}} (h(x))^2\,dx$$

and therefore

$$\int_{\mathbb{R}^n} |\nabla w_\eta(x)|^2\,dx = \lim_{R\to+\infty} \int_{B_{R-1}} |\nabla w_\eta(x)|^2\,dx \leqslant \lim_{R\to+\infty} \int_{\mathbb{R}^n} \zeta_R(x)\,|\nabla w_\eta(x)|^2\,dx$$
$$\leqslant 16 \lim_{R\to+\infty} \int_{\mathbb{R}^n\setminus B_{R-1}} (h(x))^2\,dx = 0.$$

This leads to w_η being constant and thus constantly equal to zero due to (2.2.11). From this, taking the limit as $\eta \searrow 0$, we find that u is constantly equal to zero as well. □

2.3 The Laplace–Beltrami Operator

The Laplace operator in $\mathbb{R}^n$ is actually a "special case" of a more general operator acting on functions defined on manifolds embedded in the Euclidean space (or, even more generally, on Riemannian and pseudo-Riemannian manifolds).

For concreteness, though more general settings can be taken into account (see also the comments on p. 62), we consider here the case of a hypersurface $\Sigma = \partial E$ of class C^3, for a bounded and open set $E \subseteq \mathbb{R}^n$, and we denote by ν its unit exterior normal and by d_Σ the signed distance function to Σ (say, with the convention that $d_\Sigma \geqslant 0$ in E and $d_\Sigma \leqslant 0$ in $\mathbb{R}^n \setminus E$).

We point out that d_Σ is also of class C^3 in a suitably small neighborhood $\mathcal{N}$ of Σ, and for every $x \in \mathcal{N}$, there exists a unique point $\pi_\Sigma(x) \in \Sigma$ (often called the "projection of x onto Σ") such that

$$x = \pi_\Sigma(x) - d_\Sigma(x)\,\nu(\pi_\Sigma(x)); \tag{2.3.1}$$

moreover, π_Σ is of class $C^2(\mathcal{N})$ and

$$\nabla d_\Sigma(x) = -\nu(\pi_\Sigma(x)); \tag{2.3.2}$$

see,[6] for example, Lemma 14.16 in [GT01] or Appendix B in [Giu84]. For our purposes, $\mathcal{N}$ will always supposed to be a conveniently small neighborhood of Σ.

[6]The intuition behind (2.3.2) is sketched in Figure 2.3.1. In simple terms, one can consider a point p and measure its distance from Σ by considering the ball centered at p and tangent to Σ. Taking derivatives of the distance function with respect to "tangential directions" corresponds to moving p infinitesimally toward the point p' and considering the ball centered at p' and tangent to Σ: since Σ detaches "quadratically" from its tangent hyperplane at p, this new ball is a small perturbation of the translation of the original ball (thus producing a zero tangential derivative).

Instead, taking normal derivatives of the distance function corresponds to moving p infinitesimally toward the point p'' and considering the ball centered at p'' and tangent to Σ: in this case, the new ball has a radius equal to the one of the old ball, plus the distance between p and p'', up to small perturbations (and this produces a unit normal derivative).

The minus sign in (2.3.2) is due to the fact that the outer normal of E points toward the region in which the sign distance is negative.

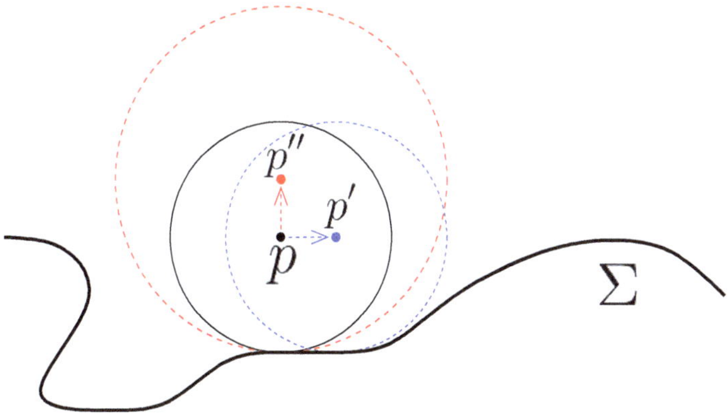

Figure 2.3.1 Taking derivatives of the distance function.

Given a function $u : \Sigma \to \mathbb{R}$, this framework allows us to define the "normal extension of u outside Σ" for all $x \in \mathcal{N}$, as

$$u_{\text{ext}}(x) := u(\pi_\Sigma(x)). \tag{2.3.3}$$

Note that if $p \in \Sigma$ and $|t|$ is sufficiently small such that $p+t\nu(p) \in \mathcal{N}$, then $u_{\text{ext}}(p+t\nu(p)) = u(p)$. As a result,

$$\begin{aligned}0 &= \frac{d}{dt}u(p) = \frac{d}{dt}u_{\text{ext}}(p+t\nu(p)) = \nabla u_{\text{ext}}(p+t\nu(p)) \cdot \nu(p)\\ &= \nabla u_{\text{ext}}(p+t\nu(p)) \cdot \nu_{\text{ext}}(p+t\nu(p)).\end{aligned} \tag{2.3.4}$$

The Laplace–Beltrami operator of a function $u \in C^2(\Sigma)$ is then defined, for each $p \in \Sigma$, by

$$\Delta_\Sigma u(p) := \Delta u_{\text{ext}}(p), \tag{2.3.5}$$

where Δ represents here the standard Laplacian acting on functions in $C^2(\mathcal{N})$.

Interestingly, the Laplace–Beltrami operator is compatible with the gradient structure intrinsic to Σ. To this end, one defines the "tangential gradient" as the projection onto the tangent plane, namely, for every $f \in C^1(\mathcal{N})$ and any $p \in \mathcal{N}$,

$$\nabla_\Sigma f(p) := \nabla f(p) - \big(\nabla f(p) \cdot \nu_{\text{ext}}(p)\big)\, \nu_{\text{ext}}(p). \tag{2.3.6}$$

Also, if $f \in C^1(\Sigma)$, we define its tangential gradient via the tangential gradient of the normal extension, namely

$$\nabla_\Sigma f := \nabla_\Sigma f_{\text{ext}}. \tag{2.3.7}$$

We observe that, in view of (2.3.4), on Σ, we have that

$$\nabla f_{\text{ext}} \cdot \nu = 0 \tag{2.3.8}$$

and thus, by (2.3.6),

$$\nabla_\Sigma f_{\text{ext}} = \nabla f_{\text{ext}}. \tag{2.3.9}$$

In analogy with the tangential gradient defined in (2.3.6), one can introduce the "tangential divergence" of a vector field $F \in C^1(\mathcal{N}, \mathbb{R}^n)$ at the points of $\mathcal{N}$ as

$$\operatorname{div}_\Sigma F := \operatorname{div} F - \nabla(F \cdot \nu_{\text{ext}}) \cdot \nu_{\text{ext}}. \tag{2.3.10}$$

In rough terms, one is "removing" here the normal contribution of the full divergence. Also, if $F \in C^1(\Sigma, \mathbb{R}^n)$, one defines its tangential divergence as that of its normal extension, namely

$$\operatorname{div}_\Sigma F := \operatorname{div}_\Sigma F_{\text{ext}}, \tag{2.3.11}$$

where F_{ext} is the vector field obtained by the normal extension of all the components of F. In this situation, we point out that, for each $p \in \mathcal{N}$,

$$F_{\text{ext}}(p) \cdot \nu_{\text{ext}}(p) = F(\pi_\Sigma(p)) \cdot \nu(\pi_\Sigma(p)) = (F \cdot \nu)_{\text{ext}}(p),$$

whence, in light of (2.3.4), $\nabla\big(F_{\text{ext}} \cdot \nu_{\text{ext}}\big) \cdot \nu = 0$ on Σ. Combining this with (2.3.10) and (2.3.11), it follows that

$$\operatorname{div}_\Sigma F = \operatorname{div} F_{\text{ext}} \quad \text{on } \Sigma. \tag{2.3.12}$$

Concerning the definitions of tangential gradient and divergence, a caveat should be taken into account: namely, given a function $u \in C^1(\mathcal{N})$ (or a vector field $F \in C^1(\mathcal{N}, \mathbb{R}^n)$), one can consider the restriction $u\big|_\Sigma \in C^1(\Sigma)$ (or $F\big|_\Sigma \in C^1(\Sigma, \mathbb{R}^n)$) and then compute the tangential gradient of $u\big|_\Sigma$ (or the tangential divergence of $F\big|_\Sigma$) on Σ, according to definition (2.3.7) (or definition (2.3.11)), that is, using the normal extension defined in (2.3.3). The value obtained in this way coincides with the tangential gradient of u computed via definition (2.3.6) (or the tangential divergence of F computed via definition (2.3.10)) evaluated at Σ. As a matter of fact, the values of a function on Σ suffice to compute its tangential gradient (as well as the values of a vector field on Σ suffice to compute its tangential divergence), according to the following observation.

Lemma 2.3.1. *Let $u, \widetilde{u} \in C^1(\mathcal{N})$ be such that $u = \widetilde{u}$ on Σ, and let $\nabla_\Sigma u$ and $\nabla_\Sigma \widetilde{u}$ be computed as in* (2.3.6)*. Then, on Σ, we have that*

$$\nabla_\Sigma u = \nabla_\Sigma \widetilde{u}. \tag{2.3.13}$$

Furthermore, let F, $\widetilde{F} \in C^1(\mathcal{N}, \mathbb{R}^n)$ *be such that* $F = \widetilde{F}$ *on* Σ, *and let* $\operatorname{div}_\Sigma F$ *and* $\operatorname{div}_\Sigma \widetilde{F}$ *be computed as in* (2.3.10). *Then, on* Σ, *we have that*

$$\operatorname{div}_\Sigma F = \operatorname{div}_\Sigma \widetilde{F}. \tag{2.3.14}$$

Proof. First of all, we observe that, if $F = (F_1, \dots, F_n)$, then

$$\operatorname{div}_\Sigma F = \sum_{j=1}^{n} \nabla_\Sigma F_j \cdot e_j. \tag{2.3.15}$$

Indeed, using (2.3.6) and (2.3.10) and then (2.3.4) as well, we see that

$$\begin{aligned}
&\operatorname{div}_\Sigma F - \sum_{j=1}^{n} \nabla_\Sigma F_j \cdot e_j \\
&= \sum_{j=1}^{n} \partial_j F_j - \nabla\left(\sum_{j=1}^{n} F_j e_j \cdot \nu_{\mathrm{ext}}\right) \cdot \nu_{\mathrm{ext}} - \sum_{j=1}^{n} \left(\nabla F_j - \left(\nabla F_j \cdot \nu_{\mathrm{ext}}\right) \nu_{\mathrm{ext}}\right) \cdot e_j \\
&= -\nabla\left(\sum_{j=1}^{n} F_j e_j \cdot \nu_{\mathrm{ext}}\right) \cdot \nu_{\mathrm{ext}} + \sum_{j=1}^{n} \left(\nabla F_j \cdot \nu_{\mathrm{ext}}\right)\left(\nu_{\mathrm{ext}} \cdot e_j\right) \\
&= -\sum_{j=1}^{n} F_j \nabla\left(e_j \cdot \nu_{\mathrm{ext}}\right) \cdot \nu_{\mathrm{ext}} \\
&= 0,
\end{aligned}$$

thus proving (2.3.15).

Now, we prove (2.3.13). For this, we set $w := u - \widetilde{u}$, and we remark that $w = 0$ on Σ. Given a point p of Σ, we suppose that in a neighborhood of p, the hypersurface Σ is parameterized by the graph of a function $\psi : \mathbb{R}^{n-1} \to \mathbb{R}$ (up to renumbering the variables, we also assume that this graph occurs in the nth coordinate direction, with the set E lying above the graph); namely, there exists $r > 0$ such that

$$B_r(p) \cap E = \{x_n > \psi(x')\} \cap B_r(p). \tag{2.3.16}$$

Note that, on Σ,

$$\nu = \frac{(\nabla'\psi, -1)}{\sqrt{1 + |\nabla'\psi|^2}}, \tag{2.3.17}$$

where the notation

$$\nabla' := (\partial_1, \dots, \partial_{n-1}) \tag{2.3.18}$$

has been used.

In this way, in the vicinity of p, we can write that $w(x',\psi(x'))=0$ and

$$0=\nabla'\Big(w(x',\psi(x'))\Big)=\nabla' w(x',\psi(x'))+\partial_n w(x',\psi(x'))\,\nabla'\psi(x').$$

Consequently, using (2.3.6), we find that, on Σ, in the vicinity of p,

$$\begin{aligned}
&\nabla_\Sigma u-\nabla_\Sigma \widetilde{u}\\
&\quad=\nabla w-\big(\nabla w\cdot\nu\big)\,\nu\\
&\quad=\Big(-\partial_n w\,\nabla'\psi,\partial_n w\Big)-\left(\Big(-\partial_n w\,\nabla'\psi,\partial_n w\Big)\cdot\frac{(\nabla'\psi,-1)}{\sqrt{1+|\nabla'\psi|^2}}\right)\frac{(\nabla'\psi,-1)}{\sqrt{1+|\nabla'\psi|^2}}\\
&\quad=\Big(-\partial_n w\,\nabla'\psi,\partial_n w\Big)+\frac{\partial_n w\,\big(|\nabla'\psi|^2+1\big)}{\sqrt{1+|\nabla'\psi|^2}}\frac{(\nabla'\psi,-1)}{\sqrt{1+|\nabla'\psi|^2}}\\
&\quad=\Big(-\partial_n w\,\nabla'\psi,\partial_n w\Big)+\Big(\partial_n w\,\nabla'\psi,-\partial_n w\Big)\\
&\quad=0,
\end{aligned}$$

which establishes (2.3.13).

To prove (2.3.14), we exploit (2.3.13) (applied to the scalar component functions F_j and $\widetilde{F}_j$) and (2.3.15) to compute that

$$\operatorname{div}_\Sigma F-\operatorname{div}_\Sigma\widetilde{F}=\sum_{j=1}^n\nabla_\Sigma F_j\cdot e_j-\sum_{j=1}^n\nabla_\Sigma\widetilde{F}_j\cdot e_j=0.$$

This completes the proof of (2.3.14). □

In a nutshell, the content of Lemma 2.3.1 is that different extensions of a smooth object defined only on Σ do not alter the tangential "first order" operators, since the tangent hyperplane of Σ "detaches quadratically" from Σ (we will find however that "second order" operators are sensitive to different types of extensions, see (2.5.6)).

We observe that the Laplace–Beltrami operator possesses a "tangential divergence form structure," to be compared with the classical one in (1.1.3), as given in the following lemma.

Lemma 2.3.2. *For every $u\in C^2(\Sigma)$, on Σ, we have that*

$$\Delta_\Sigma u=\operatorname{div}_\Sigma(\nabla_\Sigma u).$$

Proof. Using in order (2.3.5), (2.3.7), (2.3.9), (2.3.10) and (2.3.4), we see that, on Σ,

$$\begin{aligned}
\Delta_\Sigma u-\operatorname{div}_\Sigma(\nabla_\Sigma u)&=\Delta u_{\rm ext}-\operatorname{div}_\Sigma(\nabla_\Sigma u_{\rm ext})=\Delta u_{\rm ext}-\operatorname{div}_\Sigma(\nabla u_{\rm ext})\\
&=\Delta u_{\rm ext}-\operatorname{div}(\nabla u_{\rm ext})-\nabla(\nabla u_{\rm ext}\cdot\nu_{\rm ext})\cdot\nu=0.
\end{aligned}$$ □

For further use, it is now useful to recall an asymptotic result about the sets obtained by "thickening" Σ. For general and precise formulas for computing the tubular neighborhoods of hypersurfaces, see for example [Roc13, Theorem 1].

Lemma 2.3.3. *Let α be a continuous function on Σ and β be a continuous function on $\mathcal{N}$. Let $\varepsilon > 0$ and*

$$\Sigma_\varepsilon(\alpha) := \{p + t\nu(p),\ p \in \Sigma,\ 0 \leqslant t \leqslant \varepsilon\alpha(p)\}. \tag{2.3.19}$$

Then, as $\varepsilon \searrow 0$,

$$\int_{\Sigma_\varepsilon(\alpha)} \beta(y)\, dy = \varepsilon \int_\Sigma \alpha_+(p)\, \beta(p)\, d\mathcal{H}_p^{n-1} + o(\varepsilon), \tag{2.3.20}$$

where $\alpha_+(p) := \max\{\alpha(p), 0\}$.

Proof. In local coordinates, we write a surface element of Σ as a graph of a function

$$\psi : U \subseteq \mathbb{R}^{n-1} \to \mathbb{R}, \tag{2.3.21}$$

say in the nth direction, with normal as in (2.3.17). In this way, points y in this element of $\Sigma_\varepsilon(\alpha)$ are of the form

$$\begin{aligned} &y = (x', \psi(x')) + \frac{t\ (\nabla'\psi(x'), -1)}{\sqrt{1 + |\nabla'\psi(x')|^2}}, \\ &\text{with } x' \in U \text{ and } 0 < t < \varepsilon\alpha(x', \psi(x')). \end{aligned} \tag{2.3.22}$$

That is, one can consider a partition of unity (see e.g. [Boo75, p. 192]) made of functions $\phi_i \in C_0^\infty(\mathcal{N}, [0, 1])$ with $i \in \mathbb{N}$ and finite overlapping supports, each compactly contained in a local chart of Σ, such that $\sum_{i\in\mathbb{N}} \phi_i = 1$ in a given neighborhood $\mathcal{N}'$ of Σ (with $\Sigma \subseteq \mathcal{N}' \Subset \mathcal{N}$, see Figure 2.3.2). Then, letting $\beta_i := \beta\phi_i$, it suffices to prove (2.3.20) with β replaced by β_i, since

$$\int_{\Sigma_\varepsilon(\alpha)} \beta(y)\, dy = \sum_{i\in\mathbb{N}} \int_{\Sigma_\varepsilon(\alpha)} \beta_i(y)\, dy$$

and

$$\int_\Sigma \alpha_+(p)\, \beta(p)\, d\mathcal{H}_p^{n-1} = \sum_{i\in\mathbb{N}} \int_\Sigma \alpha_+(p)\, \beta_i(p)\, d\mathcal{H}_p^{n-1}.$$

Therefore, hereafter, to prove (2.3.20), up to replacing β with β_i, we can suppose that, in the support of β, the hypersurface Σ is a graph of a function ψ as in (2.3.22), say in the nth direction, and for small ε, the tubular neighborhood $\Sigma_\varepsilon(\alpha)$ can be written as the set of points y in (2.3.22).

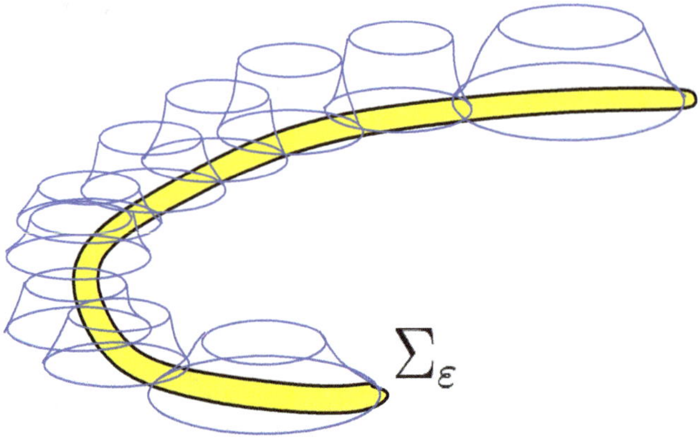

Figure 2.3.2 Local charts for Σ and a partition of unity.

For convenience, one can denote $x_n := t$ and $x := (x', x_n)$ in (2.3.22) and thus describe $\Sigma_\varepsilon(\alpha)$ in the support of β as the collection of points

$$y = (x', \psi(x')) + \frac{x_n \, (\nabla'\psi(x'), -1)}{\sqrt{1 + |\nabla'\psi(x')|^2}},$$

with $x \in U \times [0, \varepsilon\alpha(x', \psi(x'))]$ (if the latter quantity is well defined, i.e. if $\alpha(x', \psi(x')) \geqslant 0$). Note in particular that $x_n = O(\varepsilon)$, and we thus consider, for small ε, the change of variable relating y and x, with

$$\frac{\partial y}{\partial x} = \begin{pmatrix} \partial_{x_1} y_1 & \dots & \partial_{x_{n-1}} y_1 & \partial_{x_n} y_1 \\ & \ddots & & \\ \partial_{x_1} y_{n-1} & \dots & \partial_{x_{n-1}} y_{n-1} & \partial_{x_n} y_{n-1} \\ \partial_{x_1} y_n & \dots & \partial_{x_{n-1}} y_n & \partial_{x_n} y_n \end{pmatrix}$$
$$= \begin{pmatrix} 1 & \dots 0 & \partial_{x_1}\psi/R \\ & \ddots & \\ 0 & \dots 1 & \partial_{x_{n-1}}\psi/R \\ \partial_{x_1}\psi & \dots \partial_{x_{n-1}}\psi & -1/R \end{pmatrix} + O(\varepsilon),$$

with $R := \sqrt{1 + |\nabla'\psi(x')|^2}$.

As a result,

$$\left|\det \frac{\partial y}{\partial x}\right| = \frac{(\partial_{x_1}\psi)^2 + \dots (\partial_{x_{n-1}}\psi)^2 + 1}{R} + O(\varepsilon) = R + O(\varepsilon).$$

We remark that $R\,dx'$ is the surface element on Σ (see e.g. [EG15, p. 125]), and hence we write the volume element of (2.3.20) in the form

$$dy = \left|\det \frac{\partial y}{\partial x}\right| dx = (R(x') + O(\varepsilon))\,dx = d\mathcal{H}^{n-1}_p\,dx_n + O(\varepsilon)\,dx.$$

That is,

$$\begin{aligned}\int_{\Sigma_\varepsilon(\alpha)} \beta(y)\,dy &= \int_{\Sigma_\varepsilon(\alpha)} \big(\beta(\pi_\Sigma(y)) + o(1)\big)\,dy \\ &= \int_\Sigma \left[\int_0^{\varepsilon\alpha_+(p)} \big(\beta(p) + o(1)\big)\,dx_n\right] d\mathcal{H}^{n-1}_p + o(\varepsilon) \\ &= \varepsilon \int_\Sigma \alpha_+(p)\,\beta(p)\,d\mathcal{H}^{n-1}_p + o(\varepsilon),\end{aligned}$$

giving (2.3.20) as desired. □

The tangential differential setting provides another useful form of "integration by parts formula" according to Theorem 2.3.4. Differently from the Euclidean case, this result takes into account an additional term coming from the geometry of Σ. For this, it is useful to introduce the mean curvature at a point $x \in \Sigma$, defined as

$$H(x) := \operatorname{div}_\Sigma \nu(x). \tag{2.3.23}$$

See for example Section 1.2 in [AV14] for a geometric description of the mean curvature. Then, we have the following result, sometimes called the "tangential divergence theorem."

Theorem 2.3.4. *For every $F \in C^1(\Sigma, \mathbb{R}^n)$ and $\varphi \in C^1(\Sigma)$,*

$$\int_\Sigma \operatorname{div}_\Sigma F(x)\,\varphi(x)\,d\mathcal{H}^{n-1}_x = \int_\Sigma F(x)\cdot\Big(H(x)\nu(x)\varphi(x) - \nabla_\Sigma\varphi(x)\Big)\,d\mathcal{H}^{n-1}_x.$$

Proof. Given $\varepsilon > 0$, to be taken conveniently small, we consider a tubular neighborhood of Σ of radius ε; namely, we set Σ_ε as in (2.3.19) with $\alpha := 1$.

Let us now analyze the exterior normal ν_{Σ_ε} along $\partial\Sigma_\varepsilon$. We stress that $\partial\Sigma_\varepsilon = \{d_\Sigma = \varepsilon\} \cup \{d_\Sigma = -\varepsilon\}$, and thus we denote by $\nu^{(\pm)}_\varepsilon$ the exterior normal of Σ_ε along $\{d_\Sigma = \pm\varepsilon\}$ and, for clarity, by ν_Σ the exterior normal of E along Σ, see Figure 2.3.3. In this way, if $x \in \{d_\Sigma = \varepsilon\}$. the exterior normal $\nu^{(+)}_\varepsilon$ at $x \in \{d_\Sigma = \varepsilon\}$ is minus the exterior normal ν_Σ of E at $\pi_\Sigma(x)$, while the exterior normal $\nu^{(-)}_\varepsilon$ at $x \in \{d_\Sigma = -\varepsilon\}$ is plus the exterior normal ν_Σ of E at $\pi_\Sigma(x)$, that is,

$$\begin{aligned}&\nu^{(+)}_\varepsilon(x) = -\nu_\Sigma(\pi_\Sigma(x)) = -\nu_{\text{ext}}(x) \quad \text{if } x \in \{d_\Sigma = \varepsilon\}\\ \text{and}\quad &\nu^{(-)}_\varepsilon(x) = \nu_\Sigma(\pi_\Sigma(x)) = \nu_{\text{ext}}(x) \quad \text{if } x \in \{d_\Sigma = -\varepsilon\}.\end{aligned} \tag{2.3.24}$$

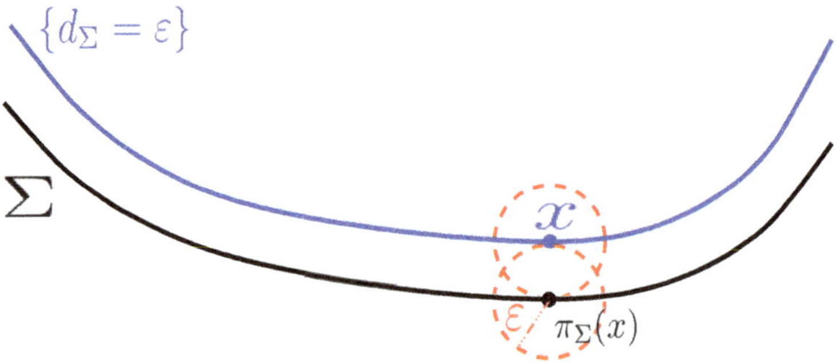

Figure 2.3.3 The hypersurface Σ and the "parallel hypersurface" at distance ε.

Moreover, by (2.3.20), used here with $\alpha := 1$ and $\beta := \operatorname{div} F_{\mathrm{ext}}\, \varphi + F \cdot \nabla\varphi_{\mathrm{ext}}$,

$$\begin{aligned}
&\int_\Sigma \operatorname{div}_\Sigma F(x)\, \varphi(x)\, d\mathcal{H}^{n-1}_x + \int_\Sigma F(x) \cdot \nabla_\Sigma \varphi(x)\, d\mathcal{H}^{n-1}_x \\
&\quad = \int_\Sigma \Big(\operatorname{div} F_{\mathrm{ext}}(x)\, \varphi(x) + F(x) \cdot \nabla\varphi_{\mathrm{ext}}(x) \Big)\, d\mathcal{H}^{n-1}_x \\
&\quad = \lim_{\varepsilon \searrow 0} \frac{1}{\varepsilon} \int_{\Sigma_\varepsilon} \Big(\operatorname{div} F_{\mathrm{ext}}(x)\, \varphi_{\mathrm{ext}}(x) + F_{\mathrm{ext}}(x) \cdot \nabla\varphi_{\mathrm{ext}}(x) \Big)\, dx \\
&\quad = \lim_{\varepsilon \searrow 0} \frac{1}{\varepsilon} \int_{\Sigma_\varepsilon} \operatorname{div} \Big(\varphi_{\mathrm{ext}}(x)\, F_{\mathrm{ext}}(x) \Big)\, dx \\
&\quad = \lim_{\varepsilon \searrow 0} \frac{1}{\varepsilon} \int_{\Sigma_\varepsilon} \operatorname{div} G_{\mathrm{ext}}(x)\, dx,
\end{aligned} \tag{2.3.25}$$

where $G := \varphi F$.

We also point out that, if $\widetilde{G} := G \cdot \nu$,

$$\begin{aligned}
&\operatorname{div} \Big((G_{\mathrm{ext}} \cdot \nu_{\mathrm{ext}}) \nu_{\mathrm{ext}} \Big) - (G_{\mathrm{ext}} \cdot \nu_{\mathrm{ext}}) \operatorname{div} \nu_{\mathrm{ext}} \\
&\quad = \nabla (G_{\mathrm{ext}} \cdot \nu_{\mathrm{ext}}) \cdot \nu_{\mathrm{ext}} = \nabla \widetilde{G}_{\mathrm{ext}} \cdot \nu_{\mathrm{ext}} = 0,
\end{aligned}$$

thanks to (2.3.4). Consequently,

$$\begin{aligned}
\int_\Sigma H(x) F(x) \cdot \nu(x) \varphi(x)\, d\mathcal{H}^{n-1}_x &= \int_\Sigma (G(x) \cdot \nu(x)) \operatorname{div} \nu_{\mathrm{ext}}(x)\, d\mathcal{H}^{n-1}_x \\
&= \int_\Sigma (G_{\mathrm{ext}}(x) \cdot \nu_{\mathrm{ext}}(x)) \operatorname{div} \nu_{\mathrm{ext}}(x)\, d\mathcal{H}^{n-1}_x \\
&= \lim_{\varepsilon \searrow 0} \frac{1}{\varepsilon} \int_{\Sigma_\varepsilon} (G_{\mathrm{ext}}(x) \cdot \nu_{\mathrm{ext}}(x)) \operatorname{div} \nu_{\mathrm{ext}}(x)\, dx \\
&= \lim_{\varepsilon \searrow 0} \frac{1}{\varepsilon} \int_{\Sigma_\varepsilon} \operatorname{div} \Big((G_{\mathrm{ext}}(x) \cdot \nu_{\mathrm{ext}}(x)) \nu_{\mathrm{ext}}(x) \Big)\, dx.
\end{aligned}$$

This and (2.3.25), together with (2.3.24), lead to

$$\begin{aligned}
&\int_\Sigma \operatorname{div}_\Sigma F(x)\,\varphi(x)\,d\mathcal{H}_x^{n-1} + \int_\Sigma F(x)\cdot\nabla_\Sigma\varphi(x)\,d\mathcal{H}_x^{n-1} \\
&\quad - \int_\Sigma H(x)F(x)\cdot\nu(x)\varphi(x)\,d\mathcal{H}_x^{n-1} \\
&= \lim_{\varepsilon\searrow 0}\frac{1}{\varepsilon}\int_{\Sigma_\varepsilon} \operatorname{div}\Big(G_{\text{ext}}(x) - (G_{\text{ext}}(x)\cdot\nu_{\text{ext}}(x))\nu_{\text{ext}}(x)\Big)\,dx \\
&= \lim_{\varepsilon\searrow 0}\frac{1}{\varepsilon}\int_{\partial\Sigma_\varepsilon} \Big(G_{\text{ext}}(x) - (G_{\text{ext}}(x)\cdot\nu_{\text{ext}}(x))\nu_{\text{ext}}(x)\Big)\cdot\nu_{\Sigma_\varepsilon}(x)\,d\mathcal{H}_x^{n-1} \\
&= \lim_{\varepsilon\searrow 0}\frac{1}{\varepsilon}\Bigg(-\int_{\{d_\Sigma=\varepsilon\}} \Big(G_{\text{ext}}(x) - (G_{\text{ext}}(x)\cdot\nu_{\text{ext}}(x))\nu_{\text{ext}}(x)\Big)\cdot\nu_{\text{ext}}(x)\,d\mathcal{H}_x^{n-1} \\
&\quad + \int_{\{d_\Sigma=-\varepsilon\}} \Big(G_{\text{ext}}(x) - (G_{\text{ext}}(x)\cdot\nu_{\text{ext}}(x))\nu_{\text{ext}}(x)\Big)\cdot\nu_{\text{ext}}(x)\,d\mathcal{H}_x^{n-1}\Bigg) \\
&= \lim_{\varepsilon\searrow 0}\frac{1}{\varepsilon}\Bigg(-\int_{\{d_\Sigma=\varepsilon\}} \Big((G_{\text{ext}}(x)\cdot\nu_{\text{ext}}(x)) - (G_{\text{ext}}(x)\cdot\nu_{\text{ext}}(x))\Big)\,d\mathcal{H}_x^{n-1} \\
&\quad + \int_{\{d_\Sigma=-\varepsilon\}} \Big((G_{\text{ext}}(x)\cdot\nu_{\text{ext}}(x)) - (G_{\text{ext}}(x)\cdot\nu_{\text{ext}}(x))\Big)\,d\mathcal{H}_x^{n-1}\Bigg) \\
&= \lim_{\varepsilon\searrow 0}\frac{1}{\varepsilon}\,(0-0) \\
&= 0,
\end{aligned}$$

which gives the desired result. □

The content of Theorem 2.3.4 happens to be a special case of certain "Minkowski integral formulas," see e.g. [MR09, LSZH15]. For instance, a classical identity is the following one.

Corollary 2.3.5. *We have that*

$$\mathcal{H}^{n-1}(\Sigma) = \frac{1}{n-1}\int_\Sigma H(x)\,x\cdot\nu(x)\,d\mathcal{H}_x^{n-1}.$$

Proof. For every $x \in \mathcal{N}$, let $F(x) := x$. Then, by Lemma 2.3.1 and (2.3.10), recalling also (2.3.4), for $x \in \Sigma$. we have that

$$\begin{aligned}
\operatorname{div}_\Sigma F(x) &= \operatorname{div} F(x) - \nabla(F(x)\cdot\nu_{\text{ext}}(x))\cdot\nu(x) = n - \nabla(x\cdot\nu_{\text{ext}}(x))\cdot\nu(x) \\
&= n - 1 - \sum_{j=1}^n x_j\nabla(\nu_{\text{ext}}(x)\cdot e_j)\cdot\nu(x) = n-1.
\end{aligned}$$

The desired result follows by exploiting the tangential divergence theorem (see Theorem 2.3.4) with $\varphi := 1$. □

From the tangential divergence theorem 2.3.4, one also obtains the following.

Corollary 2.3.6. *For every $u \in C^2(\Sigma)$ and $\varphi \in C^1(\Sigma)$,*

$$\int_\Sigma \Delta_\Sigma u(x)\, \varphi(x)\, d\mathcal{H}_x^{n-1} = -\int_\Sigma \nabla_\Sigma u(x) \cdot \nabla_\Sigma \varphi(x)\, d\mathcal{H}_x^{n-1}. \tag{2.3.26}$$

Moreover, for every $u \in C^2(\Sigma)$ and $\varphi \in C^2(\Sigma)$,

$$\int_\Sigma \Delta_\Sigma u(x)\, \varphi(x)\, d\mathcal{H}_x^{n-1} = \int_\Sigma u(x)\, \Delta_\Sigma \varphi(x)\, d\mathcal{H}_x^{n-1}. \tag{2.3.27}$$

Proof. Let $u \in C^2(\Sigma)$ and $F := \nabla_\Sigma u$. Then, by (2.3.7), (2.3.8) and (2.3.9), on Σ, we have that

$$F \cdot \nu = \nabla_\Sigma u_{\text{ext}} \cdot \nu = \nabla u_{\text{ext}} \cdot \nu = 0.$$

Hence, if $\varphi \in C^1(\Sigma)$, by Lemma 2.3.2 and Theorem 2.3.4,

$$\begin{aligned}
&\int_\Sigma \Delta_\Sigma u(x)\, \varphi(x)\, d\mathcal{H}_x^{n-1} + \int_\Sigma \nabla_\Sigma u(x) \cdot \nabla_\Sigma \varphi(x)\, d\mathcal{H}_x^{n-1} \\
&= \int_\Sigma \operatorname{div}_\Sigma F(x)\, \varphi(x)\, d\mathcal{H}_x^{n-1} + \int_\Sigma F(x) \cdot \nabla_\Sigma \varphi(x)\, d\mathcal{H}_x^{n-1} \\
&= \int_\Sigma H(x)\varphi(x)F(x) \cdot \nu(x)\, d\mathcal{H}_x^{n-1} \\
&= 0.
\end{aligned}$$

This establishes (2.3.26).

Now, we also suppose that $\varphi \in C^2(\Sigma)$, and we write (2.3.26), exchanging the roles of u and φ, namely

$$\int_\Sigma \Delta_\Sigma \varphi(x)\, u(x)\, d\mathcal{H}_x^{n-1} = -\int_\Sigma \nabla_\Sigma \varphi(x) \cdot \nabla_\Sigma u(x)\, d\mathcal{H}_x^{n-1}.$$

From this and (2.3.26), the desired result in (2.3.27) plainly follows. □

Now, we give an explicit formula for the mean curvature with respect to the function that describes locally the hypersurface Σ. We use the notation in (2.3.18) and $\operatorname{div}' := \partial_1 + \cdots + \partial_{n-1}$. With this, we have the following theorem.

Theorem 2.3.7. *Let $p = (p', p_n) \in \Sigma$. Assume that there exists $\rho > 0$ such that E in $B_\rho(p)$ can be written as a supergraph of a function $\psi : \mathbb{R}^{n-1} \to \mathbb{R}$, namely*

$$E \cap B_\rho(p) = \{x_n > \psi(x')\} \cap B_\rho(p).$$

Then, at p,

$$H = \operatorname{div}'\left(\frac{\nabla'\psi}{\sqrt{1 + |\nabla'\psi|^2}}\right). \tag{2.3.28}$$

Proof. Given $x = (x', x_n) \in \mathbb{R}^n$, we define

$$\nu_\star(x) := \frac{(\nabla'\psi(x'), -1)}{\sqrt{1 + |\nabla'\psi(x')|^2}}.$$

By (2.3.17), we know that $\nu_\star$ coincides with ν (and thus with $\nu_{\rm ext}$) on Σ in the vicinity of p; therefore, using (2.3.23) and successively (2.3.10), (2.3.11) and (2.3.14), we have that, for all $x \in \Sigma$ in the vicinity of p,

$$H(x) = \mathrm{div}_\Sigma\, \nu(x) = \mathrm{div}_\Sigma\, \nu_{\rm ext}(x) = \mathrm{div}_\Sigma\, \nu_\star = \mathrm{div}\, \nu_\star - \nabla(\nu_\star \cdot \nu_{\rm ext}) \cdot \nu. \quad (2.3.29)$$

We also observe that $\nu_\star(x)$ does not depend on x_n and therefore

$$\mathrm{div}\, \nu_\star = \mathrm{div}'\left(\frac{\nabla'\psi}{\sqrt{1 + |\nabla'\psi|^2}}\right). \quad (2.3.30)$$

Furthermore, for every $j \in \{1, \dots, n\}$,

$$0 = \partial_j \frac{1}{2} = \partial_j \frac{|\nu_\star|^2}{2} = \frac{1}{2}\sum_{m=1}^n \partial_j(\nu_\star \cdot e_m)^2 = \sum_{m=1}^n (\nu_\star \cdot e_m)\, \partial_j(\nu_\star \cdot e_m).$$

That is, on Σ,

$$\sum_{m=1}^n (\nu \cdot e_m)\, \partial_j(\nu_\star \cdot e_m) = 0.$$

Similarly, for every $j \in \{1, \dots, n\}$, on Σ,

$$\sum_{m=1}^n (\nu \cdot e_m)\, \partial_j(\nu_{\rm ext} \cdot e_m) = 0.$$

Hence, on Σ,

$$\begin{aligned}
&\nabla(\nu_\star \cdot \nu_{\rm ext}) \cdot \nu \\
&= \sum_{j=1}^n \partial_j(\nu_\star \cdot \nu_{\rm ext})\,(\nu \cdot e_j) \\
&= \sum_{j,m=1}^n \partial_j\Big((\nu_\star \cdot e_m)(\nu_{\rm ext} \cdot e_m)\Big)\,(\nu \cdot e_j) \\
&= \sum_{j,m=1}^n \Big[(\nu_{\rm ext} \cdot e_m)(\nu \cdot e_j)\,\partial_j(\nu_\star \cdot e_m) + (\nu_\star \cdot e_m)(\nu \cdot e_j)\,\partial_j(\nu_{\rm ext} \cdot e_m)\Big] \\
&= \sum_{j,m=1}^n \Big[(\nu \cdot e_m)(\nu \cdot e_j)\,\partial_j(\nu_\star \cdot e_m) + (\nu \cdot e_m)(\nu \cdot e_j)\,\partial_j(\nu_{\rm ext} \cdot e_m)\Big] \\
&= 0.
\end{aligned}$$

This, (2.3.29) and (2.3.30) yield the desired result. ◻

As a consequence of Theorem 2.3.7, we also point out that the mean curvature is monotone with respect to set inclusions.

Corollary 2.3.8. *Let $E,\ F \subseteq \mathbb{R}^n$ be bounded and open sets with C^3 boundary, with mean curvature $H_{\partial E}$ and $H_{\partial F}$, respectively. Assume that $E \subseteq F$ and that $p \in (\partial E) \cap (\partial F)$. Then, $H_{\partial E}(p) \geqslant H_{\partial F}(p)$.*

Proof. We remark that the external normal $\nu_{\partial E}$ of E coincides with the external normal $\nu_{\partial F}$ of F at p; hence, we can write both E and F as graphs with respect to a common coordinate direction. That is, without loss of generality, we may assume that there exist $\rho > 0$ and two functions $\psi_E,\ \psi_F : \mathbb{R}^{n-1} \to \mathbb{R}$ such that $p = (p', p_n) = (p', \psi_E(p')) = (p', \psi_F(p'))$,

$$E \cap B_\rho(p) = \{x_n > \psi_E(x')\} \cap B_\rho(p)$$

and

$$F \cap B_\rho(p) = \{x_n > \psi_F(x')\} \cap B_\rho(p).$$

Also, since $E \subseteq F$, we have that $\psi_E \geqslant \psi_F$ in the vicinity of p'. Therefore, the function $\phi := \psi_E - \psi_F$ has a local minimum at p', entailing that $\nabla'\phi(p') = 0$ and the Hessian of ϕ at p, which we denote by $D^2_{x'}\phi(p)$, is a nonnegative definite matrix. From this and Theorem 2.3.7, we deduce that at p',

$$\begin{aligned}
H_{\partial E} - H_{\partial F} &= \operatorname{div}'\left(\frac{\nabla'\psi_E}{\sqrt{1+|\nabla'\psi_E|^2}}\right) - \operatorname{div}'\left(\frac{\nabla'\psi_F}{\sqrt{1+|\nabla'\psi_F|^2}}\right)\\
&= \frac{\Delta'\psi_E}{\sqrt{1+|\nabla'\psi_E|^2}} - \frac{\Delta'\psi_F}{\sqrt{1+|\nabla'\psi_F|^2}}\\
&\quad - \sum_{i,j=1}^{n-1}\left(\frac{\partial_{ij}\psi_E\partial_i\psi_E\,\partial_j\psi_E}{(1+|\nabla'\psi_E|^2)^{3/2}} - \frac{\partial_{ij}\psi_F\,\partial_i\psi_F\,\partial_j\psi_F}{(1+|\nabla'\psi_F|^2)^{3/2}}\right)\\
&= \frac{1}{(1+|\nabla'\psi_E|^2)^{3/2}}\left((1+|\nabla'\psi_E|^2)\sum_{i=1}^{n}(D^2_{x'}\phi(p)\,e_i)\cdot e_i\right.\\
&\quad \left. - (D^2_{x'}\phi(p)\,\nabla'\psi_E)\cdot\nabla'\psi_E\right).
\end{aligned} \tag{2.3.31}$$

We now observe that if $M \in \mathrm{Mat}(m \times m)$ is symmetric and nonnegative definite, and $v \in \mathbb{R}^m$, then

$$|v|^2\sum_{i=1}^{m} Me_i \cdot e_i \geqslant Mv \cdot v. \tag{2.3.32}$$

To prove this, we use the spectral theorem to find an orthonormal basis $\{\eta_1, \dots, \eta_m\}$ of $\mathbb{R}^m$ consisting of eigenvectors of M. Namely,

$$M\eta_i = \lambda\eta_i \quad \text{for all } i \in \{1, \dots, m\},$$

with $\lambda_1, \dots, \lambda_m \geqslant 0$. Thus, we write v in terms of this new basis as

$$v = \sum_{i=1}^{m} (v \cdot \eta_i)\eta_i,$$

and we have that

$$\begin{aligned} Mv \cdot v &= \sum_{i,j=1}^{m} (v \cdot \eta_i)(v \cdot \eta_j) M\eta_i \cdot \eta_j \\ &= \sum_{i,j=1}^{m} \lambda_i (v \cdot \eta_i)(v \cdot \eta_j)(\eta_i \cdot \eta_j) = \sum_{i=1}^{m} \lambda_i (v \cdot \eta_i)^2. \end{aligned}$$

Applying this identity with e_k instead of v, we also obtain that, for each $k \in \{1, \dots, m\}$,

$$Me_k \cdot e_k = \sum_{i=1}^{m} \lambda_i (e_k \cdot \eta_i)^2.$$

As a result,

$$\begin{aligned} |v|^2 \sum_{k=1}^{m} Me_k \cdot e_k - Mv \cdot v &= \sum_{i=1}^{m} \lambda_i \left(\sum_{k=1}^{m} |v|^2 (e_k \cdot \eta_i)^2 - (v \cdot \eta_i)^2 \right) \\ &\geqslant |v|^2 \sum_{i=1}^{m} \lambda_i \left(\sum_{k=1}^{m} (e_k \cdot \eta_i)^2 - 1 \right). \end{aligned} \tag{2.3.33}$$

Hence, since

$$1 = |\eta_i|^2 = \left| \sum_{k=1}^{m} (\eta_i \cdot e_k) e_k \right|^2 = \sum_{k=1}^{m} (\eta_i \cdot e_k)^2,$$

we obtain (2.3.32) as a consequence of (2.3.33).

From (2.3.31) and (2.3.32), we conclude that $H_{\partial E} - H_{\partial F} \geqslant 0$, as desired. □

Corollary 2.3.9. *There exists $p \in \Sigma$ such that $H(p) > 0$.*

Proof. Since E is bounded, we have that $E \subseteq B_R$ for some $R > 0$. By decreasing R until the external ball touches ∂E, we can thus suppose that $E \subseteq B_R$ and there exists $p \in (\partial E) \cap (\partial B_R)$. Then, by Corollary 2.3.8, $H(p) \geqslant \frac{n-1}{R} > 0$. □

To further analyze the interplay between the Laplace–Beltrami operator and the mean curvature of Σ, one can also introduce the "tangential Laplacian," defined, for every $p \in \Sigma$ and every $u \in C^2(\mathcal{N})$, as

$$\Delta_T u(p) := \Delta u(p) - \big(D^2 u(p)\, \nu(p)\big) \cdot \nu(p). \tag{2.3.34}$$

That is, the tangential Laplacian is the "$(n-1)$-dimensional Laplacian" on the tangent plane to Σ at p. We stress that the tangential Laplacian does not coincide with the Laplace–Beltrami operator, and in fact their difference is related to the mean curvature of Σ, as pointed out by the following result.

Theorem 2.3.10. *Let $u \in C^2(\mathcal{N})$. Then, on Σ,*

$$\Delta_\Sigma u = \Delta_T u - H \nabla u \cdot \nu.$$

Proof. Using Lemma 2.3.2, (2.3.6) and (2.3.12), we see that, on Σ,

$$\begin{aligned}
\Delta_\Sigma u &= \operatorname{div}_\Sigma (\nabla_\Sigma u) \\
&= \operatorname{div}_\Sigma \Big(\nabla u - (\nabla u \cdot \nu)\, \nu\Big) \\
&= \operatorname{div} \Big(\nabla u - (\nabla u \cdot \nu)\, \nu\Big)_{\rm ext} \\
&= \operatorname{div} \Big((\nabla u)_{\rm ext}\Big) - \nabla\Big((\nabla u \cdot \nu)_{\rm ext}\Big) \cdot \nu - (\nabla u \cdot \nu) \operatorname{div} \nu_{\rm ext}.
\end{aligned}$$

Thus, by (2.3.4), (2.3.12) and (2.3.23),

$$\Delta_\Sigma u = \operatorname{div}_\Sigma (\nabla u) - 0 - (\nabla u \cdot \nu) H.$$

As a consequence, by (2.3.10), and using again (2.3.4), on Σ,

$$\begin{aligned}
\Delta_\Sigma u &= \operatorname{div}(\nabla u) - \nabla(\nabla u \cdot \nu_{\rm ext}) \cdot \nu - (\nabla u \cdot \nu) H \\
&= \Delta u - (D^2 u\, \nu) \cdot \nu - \sum_{j=1}^n \nabla(\nu_{\rm ext} \cdot e_j) \cdot \nu\, \partial_j u - (\nabla u \cdot \nu) H \\
&= \Delta_T u - 0 - (\nabla u \cdot \nu) H,
\end{aligned}$$

as desired. □

An intuitive explanation for the role played by the curvature of Σ in the computation of second-order operator naturally arises from Figure 2.3.4. Namely, near a given point $p \in \Sigma$, the second-order behavior of the hypersurface is well approximated by its osculating sphere. Hence, for simplicity, we present in Figure 2.3.4 the case of $n = 2$, $\Sigma = \partial B_r$ and, up to a rotation, $p = re_1$. We consider a "tangential infinitesimal increment" he_2.

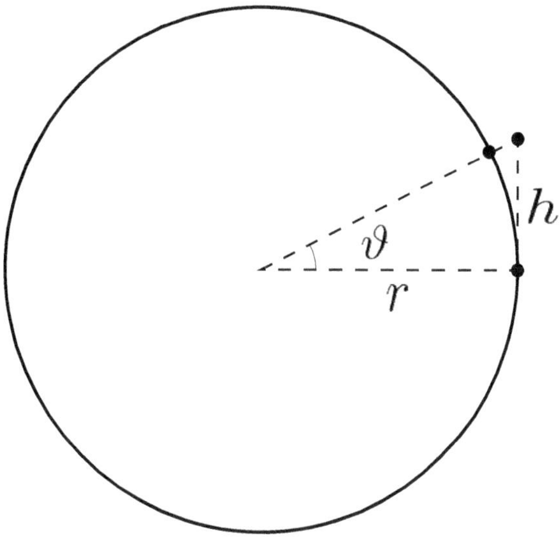

Figure 2.3.4 The role of the curvature in second-order operators.

Note from Figure 2.3.4 that $h = r\tan\vartheta$, and accordingly

$$\cos\vartheta = \frac{r}{\sqrt{r^2+h^2}} = 1 - \frac{h^2}{2r^2} + o(h^2)$$

$$\text{and} \quad \sin\vartheta = \frac{h\cos\vartheta}{r} = \frac{h}{r} + o(h^2).$$

Then, the second-order incremental quotient of u_{ext} in the tangential direction e_2 is therefore

$$\begin{aligned}
&\frac{u_{\text{ext}}(p+he_2)+u_{\text{ext}}(p-he_2)-2u_{\text{ext}}(p)}{h^2}\\
&= \frac{u(r\cos\vartheta, r\sin\vartheta)+u(r\cos\vartheta,-r\sin\vartheta)-2u(r,0)}{h^2}\\
&= \frac{u\left(r-\frac{h^2}{2r}+o(h^2), h+o(h^2)\right)+u\left(r-\frac{h^2}{2r}+o(h^2), -h+o(h^2)\right)-2u(r,0)}{h^2}\\
&= \frac{1}{h^2}\Bigg(\left(-\partial_1 u(r,0)\frac{h^2}{2r}+\partial_2 u(r,0)h+\frac{1}{2}\partial_{22}u(r,0)h^2\right)\\
&\quad+\left(-\partial_1 u(r,0)\frac{h^2}{2r}-\partial_2 u(r,0)h+\frac{1}{2}\partial_{22}u(r,0)h^2\right)+o(h^2)\Bigg)\\
&= -\frac{1}{r}\,\partial_1 u(r,0)+\partial_{22}u(r,0)+o(1),
\end{aligned}$$

which, as $h \searrow 0$, recalling that the curvature of the circle is $\frac{1}{r}$, converges to $-H\partial_\nu u(p) + \partial_{22}u(p)$, and this corresponds to the identity found in Theorem 2.3.10.

A simple consequence of Theorem 2.3.10 is also the expression of the Laplacian along the level sets of the function u, according to the following observation.

Corollary 2.3.11. *Let $u \in C^2(\mathcal{N})$. Suppose that u is constant along Σ. Then, on Σ,*

$$\Delta u = H\,\partial_\nu u + \partial_{\nu\nu} u.$$

Proof. Let $c \in \mathbb{R}$ be such that $u = c$ on Σ. Then, $u_{\text{ext}}(x) = u(\pi_\Sigma(x)) = c$ for all $x \in \mathcal{N}$, whence $\Delta_\Sigma u = 0$. Consequently, one combines Theorem 2.3.10 and the definition of tangential Laplacian in (2.3.34) to see that, on Σ,

$$H\nabla u \cdot \nu = \Delta_\Sigma u + H\nabla u \cdot \nu = \Delta_T u = \Delta u - \left(D^2 u\,\nu\right) \cdot \nu. \qquad \square$$

For completeness, we now give another formula to compute the mean curvature of the boundary of a smooth set. This formula is perhaps not very handy for explicit calculations, but it has the conceptual advantage of expressing the mean curvature as a local average of a set against its complement (thus entailing that a set whose boundary has zero mean curvature has the property that its volume density is well compensated at each boundary point by the volume density of its complement). The precise result is as follows.

Lemma 2.3.12. *For every $p \in \Sigma = \partial E$,*

$$H(p) = \lim_{r\searrow 0} \frac{1}{c r^{n+1}} \int_{B_r(p)} \left(\chi_{\mathbb{R}^n\setminus E}(x) - \chi_E(x)\right) dx, \tag{2.3.35}$$

for a suitable positive constant c depending only on n.

Proof. Up to a translation and a rotation, we assume that the point p is the origin and that E in the vicinity of the origin is the superlevel set of a function ψ, as in (2.3.16), with $\psi(0) = 0$ and $\nabla\psi(0) = 0$. Consequently, given $\varepsilon > 0$, there exists $r_\varepsilon > 0$ such that if $r \in (0, r_\varepsilon)$, then

$$\begin{aligned} U &:= B_r \cap \left\{x_n \geqslant \frac{1}{2} D^2\psi(0)x' \cdot x' + \varepsilon|x'|^2\right\} \subseteq E \cap B_r \\ &\subseteq B_r \cap \left\{x_n \geqslant \frac{1}{2} D^2\psi(0)x' \cdot x' - \varepsilon|x'|^2\right\} =: V, \end{aligned} \tag{2.3.36}$$

see Figure 2.3.5.

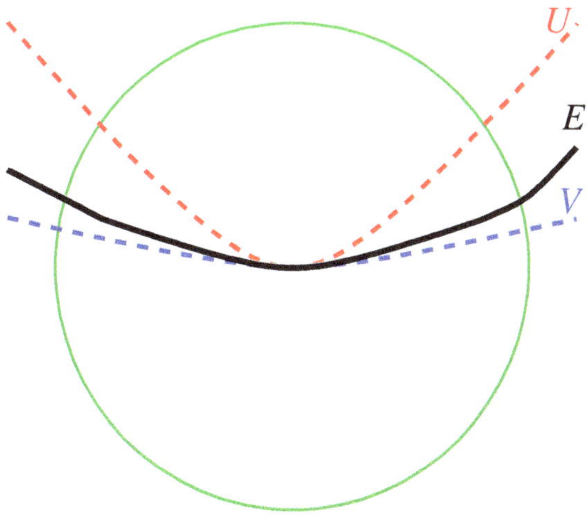

Figure 2.3.5 The sets in (2.3.36).

Accordingly, if $y \in B_r \setminus V$, then $\chi_{\mathbb{R}^n\setminus E}(y) - \chi_E(y) = 1$, and if $y \in U$, then $\chi_{\mathbb{R}^n\setminus E}(y) - \chi_E(y) = -1$. It is therefore convenient to define

$$S := B_r \cap \left\{ x_n - \frac{1}{2} D^2\psi(0)x' \cdot x' \in \big(-\varepsilon|x'|^2, \, \varepsilon|x'|^2 \big) \right\}$$

and note that

$$B_r = U \cup S \cup (B_r \setminus V).$$

These observations give that

$$\int_{B_r} \big(\chi_{\mathbb{R}^n\setminus E}(x) - \chi_E(x)\big)\, dx = |B_r \setminus V| - |U| + \int_S \big(\chi_{\mathbb{R}^n\setminus E}(x) - \chi_E(x)\big)\, dx. \quad (2.3.37)$$

Moreover,

$$|S| \leqslant 2\varepsilon \int_{|x'|<r} |x'|^2 \, dx' = O(\varepsilon r^{n+1}),$$

whence we infer from (2.3.37) that

$$\int_{B_r} \big(\chi_{\mathbb{R}^n\setminus E}(x) - \chi_E(x)\big)\, dx = |B_r \setminus V| - |U| + O(\varepsilon r^{n+1}). \quad (2.3.38)$$

Similarly,

$$|U| = \left| B_r \cap \left\{ x_n \geqslant \frac{1}{2} D^2\psi(0)x' \cdot x' \right\} \right| + O(\varepsilon r^{n+1})$$

and

$$\begin{aligned}|B_r \setminus V| &= \left|B_r \cap \left\{x_n < \frac{1}{2}D^2\psi(0)x' \cdot x'\right\}\right| + O(\varepsilon r^{n+1})\\ &= \left|B_r \cap \left\{x_n > -\frac{1}{2}D^2\psi(0)x' \cdot x'\right\}\right| + O(\varepsilon r^{n+1}),\end{aligned}$$

where the change of variable $x_n \mapsto -x_n$ has been used in the last equality.

From these observations, using the short notation $M(x') := \frac{1}{2}D^2\psi(0)x' \cdot x'$ and (2.3.38), we find that

$$\begin{aligned}&\int_{B_r} \left(\chi_{\mathbb{R}^n\setminus E}(x) - \chi_E(x)\right) dx\\ &= \left|B_r \cap \{x_n > -M(x')\}\right| - \left|B_r \cap \{x_n \geqslant M(x')\}\right| + O(\varepsilon r^{n+1}). \qquad (2.3.39)\end{aligned}$$

Now, letting $C := \frac{1}{2}|D^2\psi(0)|$, we have that

$$\begin{aligned}&\left|B_r \cap \{x_n \geqslant M(x')\}\right|\\ &= \left|B_r \cap \{M(x') \leqslant x_n \leqslant C|x'|^2\}\right| + \left|B_r \cap \{x_n \geqslant C|x'|^2\}\right|\\ &= \left|B_r \cap \{M(x') \leqslant x_n \leqslant C|x'|^2\}\right| + A_r\\ &= E_r - F_r + A_r, \qquad (2.3.40)\end{aligned}$$

where

$$\begin{aligned}A_r &:= \left|B_r \cap \{x_n \geqslant C|x'|^2\}\right|,\\ E_r &:= \left|\{|x'| < r\} \cap \{M(x') \leqslant x_n \leqslant C|x'|^2\}\right|\\ \text{and}\quad F_r &:= \left|\{|x'| < r < |x|\} \cap \{M(x') \leqslant x_n \leqslant C|x'|^2\}\right|.\end{aligned}$$

Additionally, if $x \in F_r$, then

$$\begin{aligned}r^2 &\leqslant |x|^2 = |x'|^2 + x_n^2 \leqslant |x'|^2 + (C|x'|^2)^2 = |x'|^2 + C^2|x'|^4\\ &\leqslant (1 + C^2r^2)|x'|^2 \leqslant (1+\varepsilon)|x'|^2,\end{aligned}$$

as long as r is small enough. This entails that

$$F_r \subseteq \left\{|x'| \in \left[\frac{r}{\sqrt{1+\varepsilon}}, r\right]\right\} \cap \{|x_n| \leqslant C|x'|^2\},$$

whence $|F_r| = O(\varepsilon r^{n+1})$.

From this observation and (2.3.40), we arrive at

$$\begin{aligned}\left|B_r \cap \{x_n \geqslant M(x')\}\right| &= E_r + A_r + O(\varepsilon r^{n+1})\\ &= \left|\{|x'| < r\} \cap \{M(x') \leqslant x_n \leqslant C|x'|^2\}\right| + A_r + O(\varepsilon r^{n+1}).\end{aligned}$$

Thus, replacing M with $-M$ and noting that A_r remains unchanged under this modification,

$$\begin{aligned}&\big|B_r \cap \{x_n > -M(x')\}\big| \\ &\qquad = \big|\{|x'| < r\} \cap \{-M(x') \leqslant x_n \leqslant C|x'|^2\}\big| + A_r + O(\varepsilon r^{n+1}).\end{aligned}$$

Using these remarks and (2.3.39), we thereby conclude that

$$\begin{aligned}&\int_{B_r} \big(\chi_{\mathbb{R}^n\setminus E}(x) - \chi_E(x)\big)\,dx \\ &= \Big(\big|\{|x'| < r\} \cap \{-M(x') \leqslant x_n \leqslant C|x'|^2\}\big| + A_r\Big) \\ &\quad - \Big(\big|\{|x'| < r\} \cap \{M(x') \leqslant x_n \leqslant C|x'|^2\}\big| + A_r\Big) + O(\varepsilon r^{n+1}) \\ &= \big|\{|x'| < r\} \cap \{-M(x') \leqslant x_n \leqslant C|x'|^2\}\big| \\ &\quad - \big|\{|x'| < r\} \cap \{M(x') \leqslant x_n \leqslant C|x'|^2\}\big| + O(\varepsilon r^{n+1}) \\ &= \int_{\{|x'|<r\}} \left(\int_{-M(x')}^{M(x')} dx_n\right) dx' + O(\varepsilon r^{n+1}) \\ &= D^2\psi(0) \int_{\{|x'|<r\}} x' \cdot x'\,dx' + O(\varepsilon r^{n+1}) \\ &= \sum_{i=1}^{n-1} \partial_i^2\psi(0) \int_{\{|x'|<r\}} x_i^2\,dx' + O(\varepsilon r^{n+1}) \\ &= \frac{1}{n-1}\sum_{i=1}^{n-1} \partial_i^2\psi(0) \int_{\{|x'|<r\}} |x'|^2\,dx' + O(\varepsilon r^{n+1}) \\ &= c r^{n+1}\Delta\psi(0) + O(\varepsilon r^{n+1}),\end{aligned}$$

for a suitable $c > 0$, from which (2.3.35) plainly follows in view of (2.3.28). □

2.4 The Laplace–Beltrami Operator in Local Coordinates

Regarding the setting in Section 2.3, it is sometimes useful to express geometric differential operators (such as tangential gradients and divergences as well as the Laplace–Beltrami operator) in "local coordinates" with respect to the manifold Σ. That is, one supposes that an element of Σ is locally parameterized by a diffeomorphism $f : D \to \Sigma$, for some domain $D \subset \mathbb{R}^{n-1}$, see Figure 2.4.1. One considers the metrics $\mathscr{g}$, which can be identified with a symmetric $(n-1)\times(n-1)$ matrix with elements

$$g_{ij} := \frac{\partial f}{\partial \eta_i} \cdot \frac{\partial f}{\partial \eta_j}. \tag{2.4.1}$$

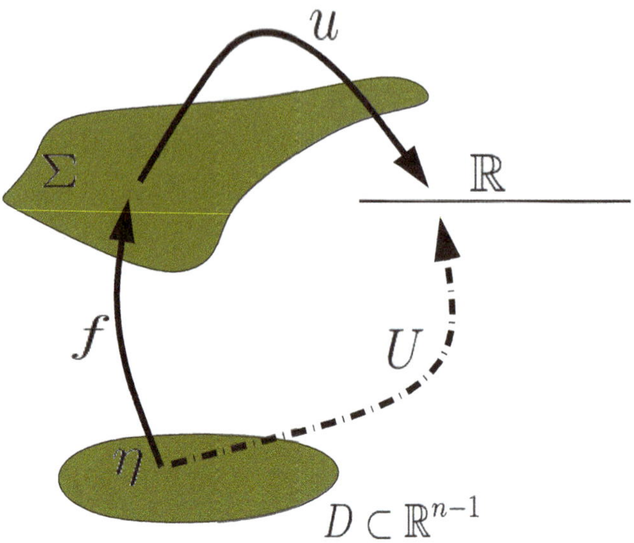

Figure 2.4.1 Local coordinates for Σ.

We observe that the matrix $\mathcal{g}$ is the product of the transpose of $D_\eta f$ and $D_\eta f$ itself, with $(D_\eta f)_{kj} = \frac{\partial f}{\partial \eta_j} \cdot e_k$, for $j \in \{1, \dots, n-1\}$ and $k \in \{1, \dots, n\}$. Therefore, by the Binet–Cauchy formula (see e.g. [EG15, Theorem 3.7]), the square of the determinant of $\mathcal{g}$ is the sum of the squares of the determinants of each $(n-1) \times (n-1)$-submatrix of $D_\eta f$, namely

$$\det \mathcal{g} = \sqrt{\sum_{k=1}^{n} \left[\det \begin{pmatrix} \frac{\partial f}{\partial \eta} \cdot e_1 \\ \vdots \\ \frac{\partial f}{\partial \eta} \cdot e_{k-1} \\ \frac{\partial f}{\partial \eta} \cdot e_{k+1} \\ \vdots \\ \frac{\partial f}{\partial \eta} \cdot e_n \end{pmatrix} \right]^2 }. \tag{2.4.2}$$

For instance, if Σ is locally the graph (say, in the nth direction) of a function ψ, one can take $f(\eta) := (\eta, \psi(\eta))$, note that in this case $\frac{\partial f}{\partial \eta_i} = \left(e_i', \frac{\partial \psi}{\partial \eta_i}\right)$, which are linearly independent, with $\{e_1', \dots, e_{n-1}'\}$ being the Euclidean basis of $\mathbb{R}^{n-1}$, obtain that

$$g_{ij} = \delta_{ij} + \frac{\partial \psi}{\partial \eta_i} \frac{\partial \psi}{\partial \eta_j}$$

and deduce from (2.4.2) that

$$\det \mathscr{g} = \sqrt{1 + |\partial_\eta \psi|^2}.$$

We also remark that, for every $v = (v_1, \ldots, v_{n-1}) \in \mathbb{R}^{n-1}$,

$$(\mathscr{g} v) \cdot v = \sum_{i,j=1}^{n-1} \left(\delta_{ij} + \frac{\partial \psi}{\partial \eta_i} \frac{\partial \psi}{\partial \eta_j} \right) v_i \, v_j = |v|^2 + \left(\frac{\partial \psi}{\partial \eta} \cdot v \right)^2 \geqslant |v|^2,$$

whence $\mathscr{g}$ is positive definite.

In particular, we have that $\mathscr{g}$ is invertible, and we denote the corresponding inverse matrix by g^{ij}. We also set $g := \det \mathscr{g}$.

In this setting, given $u : \Sigma \to \mathbb{R}$, we can consider $U : D \to \mathbb{R}$ defined by $U(\eta) := u(f(\eta))$. Moreover, it is convenient to identify vectors $\tau \in \mathbb{R}^n$ which are tangent to Σ at a given point with suitable vectors in $\mathbb{R}^{n-1}$. This identification should also reconstruct the Euclidean scalar products of tangent vectors in $\mathbb{R}^n$ from a suitable product between the associated vectors in $\mathbb{R}^{n-1}$. To this end, given $p \in \Sigma$, we can assume that $0 \in D$ and $p = f(0)$, and for all $i \in \{1, \ldots, n\}$, we consider the tangent vectors

$$\tau_i := \frac{d}{dt} f(te_i) \Big|_{t=0} = \frac{\partial f}{\partial \eta_i}(0).$$

Given a tangent vector $V \in \mathbb{R}^n$ to Σ at p, we write it in the form

$$V = \sum_{i=1}^{n-1} V^i \, \tau_i. \tag{2.4.3}$$

Thus, we can associate with the tangent vector V its $(n-1)$-dimensional representation

$$\Theta(V) = (\Theta^1(V), \ldots, \Theta^{n-1}(V)) = (V^1, \ldots, V^{n-1}) \in \mathbb{R}^{n-1}. \tag{2.4.4}$$

In this way, considering the scalar product induced by the metrics between $a = (a^1, \ldots, a^{n-1})$, $b = (b^1, \ldots, b^{n-1}) \in \mathbb{R}^{n-1}$, defined as

$$g(a, b) := \sum_{i,j=1}^{n-1} g_{ij} \, a^i \, b^j,$$

it follows that, for every tangent vector $V, W \in \mathbb{R}^n$,

$$V \cdot W = \sum_{i,j=1}^{n} V^i \, W^j \, \tau_i \cdot \tau_j = \sum_{i,j=1}^{n} \Theta^i(V) \, \Theta^j(W) \, g_{ij} = g\big(\Theta(V), \Theta(W)\big). \tag{2.4.5}$$

We also recall a handy way to use the metric to reconstruct the coordinates of a tangent vector with respect to the (not necessarily orthonomal) basis of the tangent space given by $\{\tau_1, \ldots, \tau_{n-1}\}$.

Lemma 2.4.1. *Let V be a tangent vector field on Σ as in* (2.4.3), *and recall the notation in* (2.4.4). *Then, for every $j \in \{1, \dots, n-1\}$,*

$$\Theta^j(V) = \sum_{i=1}^{n-1} g^{ij}\, V \cdot \frac{\partial f}{\partial \eta_i}. \tag{2.4.6}$$

Moreover,

$$V = \sum_{i,j=1}^{n-1} g^{ij}\, V \cdot \frac{\partial f}{\partial \eta_i} \frac{\partial f}{\partial \eta_j}. \tag{2.4.7}$$

Proof. It follows from (2.4.3) that, for every $j \in \{1, \dots, n-1\}$,

$$V \cdot \frac{\partial f}{\partial \eta_j} = \sum_{i=1}^{n-1} V^i \tau_i \cdot \frac{\partial f}{\partial \eta_j} = \sum_{i=1}^{n-1} V^i \frac{\partial f}{\partial \eta_i} \cdot \frac{\partial f}{\partial \eta_j} = \sum_{i=1}^{n-1} g_{ij} V^i.$$

That is, if we consider the vectors $W := \left(V \cdot \frac{\partial f}{\partial \eta_1}, \dots, V \cdot \frac{\partial f}{\partial \eta_{n-1}}\right)$ and $\Theta := \Theta(V) = (V^1, \dots, V^{n-1})$, in matrix notation we have that $W = \mathscr{g}\Theta$ and thus $\Theta = \mathscr{g}^{-1} W$, which gives, for each $j \in \{1, \dots, n-1\}$,

$$\Theta^j = \sum_{i=1}^{n-1} g^{ij} W_i,$$

from which (2.4.6) plainly follows.

In view of (2.4.6), we see that (2.4.3) reduces to

$$V = \sum_{j=1}^{n-1} V^j \frac{\partial f}{\partial \eta_j} = \sum_{i,j=1}^{n-1} g^{ij}\, V \cdot \frac{\partial f}{\partial \eta_i} \frac{\partial f}{\partial \eta_j},$$

which is (2.4.7). □

Formula (2.4.7) can be straightforwardly extended to all (not necessarily tangent) vector fields as follows.

Corollary 2.4.2. *If V is a vector field on Σ,*

$$V = (V \cdot \nu)\nu + \sum_{i,j=1}^{n-1} g^{ij}\, V \cdot \frac{\partial f}{\partial \eta_i} \frac{\partial f}{\partial \eta_j}.$$

Proof. From (2.4.7), considering the tangent vector field $\widetilde{V} := V - (V \cdot \nu)\nu$, we see that

$$\begin{aligned} V - (V \cdot \nu)\nu = \widetilde{V} &= \sum_{i,j=1}^{n-1} g^{ij}\, \widetilde{V} \cdot \frac{\partial f}{\partial \eta_i} \frac{\partial f}{\partial \eta_j} \\ &= \sum_{i,j=1}^{n-1} g^{ij} \Big(V - (V \cdot \nu)\nu\Big) \cdot \frac{\partial f}{\partial \eta_i} \frac{\partial f}{\partial \eta_j} = \sum_{i,j=1}^{n-1} g^{ij}\, V \cdot \frac{\partial f}{\partial \eta_i} \frac{\partial f}{\partial \eta_j}. \end{aligned}$$

□

The above setting allows us to write the tangential gradient in local coordinates, according to the following result.

Lemma 2.4.3. *Let $u \in C^1(\mathcal{N})$. For every $i \in \{1, \dots, n-1\}$, on Σ,*

$$\Theta^i(\nabla_\Sigma u) = \sum_{j=1}^{n-1} g^{ij} \frac{\partial U}{\partial \eta_j}.$$

Proof. We stress that $\nabla_\Sigma u$ is a tangent vector since, by (2.3.6),

$$\nabla_\Sigma u \cdot \nu = \Big(\nabla u - (\nabla u \cdot \nu)\,\nu\Big) \cdot \nu = (\nabla u \cdot \nu) - (\nabla u \cdot \nu)\nu \cdot \nu = 0.$$

Also, by the chain rule, for all $i \in \{1, \dots, n-1\}$,

$$\frac{\partial U}{\partial \eta_i}(\eta) = \frac{\partial}{\partial \eta_i}(u(f(\eta)) = \nabla u(f(\eta)) \cdot \frac{\partial U}{\partial \eta_i}(f(\eta)),$$

and thus, in light of (2.4.6),

$$\begin{aligned}\Theta^j(\nabla_\Sigma u) &= \sum_{i=1}^{n-1} g^{ij}\, \nabla_\Sigma u \cdot \frac{\partial f}{\partial \eta_i} = \sum_{i=1}^{n-1} g^{ij} \Big(\nabla u - (\nabla u \cdot \nu)\,\nu\Big) \cdot \frac{\partial f}{\partial \eta_i} \\ &= \sum_{i=1}^{n-1} g^{ij}\, \nabla u \cdot \frac{\partial f}{\partial \eta_i} = \sum_{i=1}^{n-1} g^{ij} \frac{\partial U}{\partial \eta_i}.\end{aligned}$$

□

Correspondingly to Lemma 2.4.3, one can write the tangential divergence in local coordinates as follows.

Lemma 2.4.4. *Let $F \in C^1(\Sigma, \mathbb{R}^n)$ be a tangent vector field. Then, on Σ,*

$$\operatorname{div}_\Sigma F = \sum_{i=1}^{n-1} \frac{1}{\sqrt{g}} \frac{\partial}{\partial \eta_i} \Big(\sqrt{g}\, \Theta^i(F)\Big).$$

Proof. We consider a smooth function φ supported in a local chart, and let $\Phi(\eta) := \varphi(f(\eta))$. We recall (see [EG15, pp. 125–126]) that the surface element of Σ in local coordinates can be written as $\sqrt{g}\, d\eta$; therefore, by Theorem 2.3.4 and using the fact that F is tangential,

$$\begin{aligned}\operatorname{div}_\Sigma F\, \Phi \sqrt{g}\, d\eta &= \operatorname{div}_\Sigma F\, \varphi\, d\mathcal{H}^{n-1} = HF \cdot \nu\, \varphi\, d\mathcal{H}^{n-1} - F \cdot \nabla_\Sigma \varphi\, d\mathcal{H}^{n-1} \\ &= -F \cdot \nabla_\Sigma \varphi\, d\mathcal{H}^{n-1}.\end{aligned}$$

This and (2.4.5) lead to

$$\operatorname{div}_\Sigma F\, \Phi \sqrt{g}\, d\eta = -\sum_{i,j=1}^{n} g_{ij}\, \Theta^i(F)\, \Theta^j(\nabla_\Sigma \varphi)\, d\mathcal{H}^{n-1}.$$

Since, in light of Lemma 2.4.3,

$$\Theta^j(\nabla_\Sigma \varphi) = \sum_{k=1}^{n-1} g^{jk} \frac{\partial \Phi}{\partial \eta_k},$$

and we thereby obtain that

$$\operatorname{div}_\Sigma F \,\Phi \sqrt{g}\, d\eta = -\sum_{i,j,k=1}^{n} g_{ij}\, g^{jk}\, \Theta^i(F) \frac{\partial \Phi}{\partial \eta_k} \sqrt{g}\, d\eta = -\sum_{i=1}^{n} \Theta^i(F) \frac{\partial \Phi}{\partial \eta_i} \sqrt{g}\, d\eta.$$

The latter term can be integrated by parts, giving

$$\operatorname{div}_\Sigma F \,\Phi \sqrt{g}\, d\eta = \sum_{i=1}^{n} \frac{\partial}{\partial \eta_i}\Big(\Theta^i(F) \sqrt{g}\Big) \Phi \, d\eta.$$

Consequently, since φ (and hence Φ) is an arbitrary test function, we conclude that

$$\operatorname{div}_\Sigma F \sqrt{g} = \sum_{i=1}^{n} \frac{\partial}{\partial \eta_i}\Big(\Theta^i(F) \sqrt{g}\Big). \qquad \square$$

Corollary 2.4.5. *Let $F \in C^1(\Sigma, \mathbb{R}^n)$. Then, on Σ,*

$$\operatorname{div}_\Sigma F = \sum_{i=1}^{n-1} \frac{1}{\sqrt{g}} \frac{\partial}{\partial \eta_i}\Big(\sqrt{g}\, \Theta^i(\widetilde{F})\Big) + H(F \cdot \nu),$$

where $\widetilde{F} := F - (F \cdot \nu)\nu$.

Proof. By (2.3.10), we know that the tangential divergence is linear with respect to the vector fields. Therefore,

$$\operatorname{div}_\Sigma F = \operatorname{div}_\Sigma \widetilde{F} + \operatorname{div}_\Sigma \big((F \cdot \nu)\nu\big). \tag{2.4.8}$$

We also remark that if $\varphi \in C^1(\Sigma)$, on Σ, we have that

$$\begin{aligned}\operatorname{div}_\Sigma(\varphi\nu) &= \operatorname{div}((\varphi\nu)_{\text{ext}}) - \nabla\big((\varphi\nu)_{\text{ext}} \cdot \nu_{\text{ext}}\big) \cdot \nu_{\text{ext}} \\ &= \nabla\varphi_{\text{ext}} \cdot \nu_{\text{ext}} + \varphi \operatorname{div} \nu_{\text{ext}} - \nabla\varphi_{\text{ext}} \cdot \nu_{\text{ext}} = H\varphi,\end{aligned}$$

thanks to (2.3.4) and (2.3.23). This and (2.4.8) entail that

$$\operatorname{div}_\Sigma F = \operatorname{div}_\Sigma \widetilde{F} + H(F \cdot \nu).$$

Since $\widetilde{F}$ is a tangential vector field, we can employ Lemma 2.4.4 and obtain the desired result. $\square$

As a variant of Corollary 2.4.5, we also have the following useful expression in coordinates of the tangential divergence of a vector field.

Corollary 2.4.6. *Let $F \in C^1(\mathcal{N}, \mathbb{R}^n)$. Then, on Σ,*

$$\operatorname{div}_\Sigma F = \sum_{i,j=1}^{n-1} g^{ij} \left(DF \frac{\partial f}{\partial \eta_i} \right) \cdot \frac{\partial f}{\partial \eta_j}.$$

Proof. For every $k \in \{1, \dots, n\}$, by Corollary 2.4.2, applied here to $V := \frac{\partial F}{\partial x_k}$, we see that

$$\frac{\partial F}{\partial x_k} = \left(\frac{\partial F}{\partial x_k} \cdot \nu \right) \nu + \sum_{i,j=1}^{n-1} g^{ij} \frac{\partial F}{\partial x_k} \cdot \frac{\partial f}{\partial \eta_i} \frac{\partial f}{\partial \eta_j}.$$

As a result,

$$\operatorname{div} F = \sum_{k=1}^{n} \frac{\partial F}{\partial x_k} \cdot e_k = \sum_{k=1}^{n} \left(\frac{\partial F}{\partial x_k} \cdot \nu \right) \nu \cdot e_k + \sum_{\substack{1 \leqslant i,j \leqslant n-1 \\ 1 \leqslant k \leqslant n}} g^{ij} \frac{\partial F}{\partial x_k} \cdot \frac{\partial f}{\partial \eta_i} \frac{\partial f}{\partial \eta_j} \cdot e_k.$$

From this, we deduce that

$$\begin{aligned}\sum_{i,j=1}^{n-1} g^{ij} \left(DF \frac{\partial f}{\partial \eta_j} \right) \cdot \frac{\partial f}{\partial \eta_i} &= \sum_{\substack{1 \leqslant i,j \leqslant n-1 \\ 1 \leqslant k \leqslant n}} g^{ij} \frac{\partial F}{\partial x_k} \cdot \frac{\partial f}{\partial \eta_i} \frac{\partial f}{\partial \eta_j} \cdot e_k \\ &= \operatorname{div} F - \sum_{k=1}^{n} \left(\frac{\partial F}{\partial x_k} \cdot \nu \right) \nu \cdot e_k = \operatorname{div} F - (DF\,\nu) \cdot \nu.\end{aligned}$$

That is, recalling the definition of tangential divergence in (2.3.10) and also exploiting (2.3.4),

$$\begin{aligned}&\sum_{i,j=1}^{n-1} g^{ij} \left(DF \frac{\partial f}{\partial \eta_j} \right) \cdot \frac{\partial f}{\partial \eta_i} - \operatorname{div}_\Sigma F \\ &= \sum_{i,j=1}^{n-1} g^{ij} \left(DF \frac{\partial f}{\partial \eta_j} \right) \cdot \frac{\partial f}{\partial \eta_i} - \operatorname{div} F + \nabla(F \cdot \nu_{\text{ext}}) \cdot \nu_{\text{ext}} \\ &= \nabla(F \cdot \nu_{\text{ext}}) \cdot \nu_{\text{ext}} - (DF\,\nu) \cdot \nu = \sum_{k,m=1}^{n} (F \cdot e_m) \left(\frac{\partial \nu_{\text{ext}}}{\partial x_k} \cdot e_m \right) (\nu_{\text{ext}} \cdot e_k) \\ &= \sum_{m=1}^{n} (F \cdot e_m) \nabla(\nu_{\text{ext}} \cdot e_m) \cdot \nu_{\text{ext}} = 0,\end{aligned}$$

as desired. □

As a direct consequence of Lemmata 2.3.2, 2.4.3 and 2.4.4, we obtain that the Laplace–Beltrami operator can be written in terms of these local coordinates, according to the following result.

Corollary 2.4.7. *Let $u \in C^2(\mathcal{N})$. Then,*

$$\Delta_\Sigma u = \sum_{i,j=1}^{n-1} \frac{1}{\sqrt{g}} \frac{\partial}{\partial \eta_i} \left(\sqrt{g}\, g^{ij} \frac{\partial U}{\partial \eta_j} \right). \tag{2.4.9}$$

As a technical remark, we observe that, for the purpose of defining the Laplace–Beltrami operator, the regularity of Σ can be reduced from C^3 to C^2. Indeed, on p. 36, we assumed Σ of class C^3, which was used to deduce that ν was of class C^2 and, accordingly, by (2.3.1), that π_Σ was of class C^2. From this, we deduced that $u_{\rm ext}$ was of class C^2 if so was $u : \Sigma \to \mathbb{R}$, thanks to (2.3.3), and the C^2 regularity of $u_{\rm ext}$ was utilized to give the definition of the Laplace–Beltrami operator in (2.3.5). On the other hand, thanks to Corollary 2.4.7, we can now point out that a C^2 regularity assumption on Σ would suffice for formula (2.4.9). Hence, taking (2.4.9) instead of (2.3.5) as the definition of the Laplace–Beltrami operator would allow us to work with Σ of class C^2 (alternatively, one can define the Laplace–Beltrami operator as in (2.3.5) for Σ of class C^3 and then extend it by approximation when Σ is of class C^2 using (2.4.9) to pass to the limit).

For additional information on the Laplace–Beltrami operator, see e.g. [Heb96, Ros97, Jos17] and the references therein.

2.5 The Laplacian in Spherical Coordinates

Let $u : \mathbb{R}^n \setminus \{0\} \to \mathbb{R}$. For every $r > 0$ and $\vartheta \in \partial B_1$, we define

$$u_0(r, \vartheta) := u(r\vartheta).$$

In this setting, one can compute the Laplace operator in spherical coordinates according to the following formula.

Theorem 2.5.1. *Let $u \in C^2(\mathbb{R}^n \setminus \{0\})$. Then,*

$$\Delta u(x) = \partial_{rr} u_0(r, \vartheta) + \frac{n-1}{r} \partial_r u_0(r, \vartheta) + \Delta_{\partial B_r} u(r\vartheta),$$

for every $x \in \mathbb{R}^n \setminus \{0\}$, where $r := |x|$ and $\vartheta := \frac{x}{|x|}$.

Proof. We take $E := B_r$ and $\Sigma := \partial B_r$. In this way, we have that $\nu(x) = \frac{x}{|x|}$ for every $x \in \partial B_r$. Then, by (2.3.3), $\nu_{\rm ext}(x) = \frac{x}{|x|}$. Hence, we deduce from (2.3.12) and (2.3.23) that, for every $x \in \partial B_r$,

$$H(x) = \operatorname{div}_{\partial B_r} \nu(x) = \operatorname{div} \nu_{\rm ext}(x) = \operatorname{div} \frac{x}{|x|} = \frac{n-1}{|x|} = \frac{n-1}{r}. \tag{2.5.1}$$

In addition, if $x \in \partial B_r$,

$$\begin{aligned}\nabla u(x) \cdot \nu(x) &= \lim_{h\to 0} \frac{u(x + h\nu(x)) - u(x)}{h} \\ &= \lim_{h\to 0} \frac{u_0(r + h, \vartheta) - u_0(r, \vartheta)}{h} = \partial_r u_0(r, \vartheta)\end{aligned}$$

and

$$\begin{aligned}(D^2 u(x)\, \nu(x)) \cdot \nu(x) &= \lim_{h\to 0} \frac{u(x + h\nu(x)) + u(x - h\nu(x)) - 2u(x)}{h^2} \\ &= \lim_{h\to 0} \frac{u_0(r + h, \vartheta) + u_0(r - h, \vartheta) - 2u_0(r, \vartheta)}{h^2} \\ &= \partial_{rr} u_0(r, \vartheta).\end{aligned}$$

Using these identities and (2.5.1) in combination with Theorem 2.3.10, we conclude that, at $x \in \partial B_r$,

$$\begin{aligned}\Delta_{\partial B_r} u = \Delta_T u - H \nabla u \cdot \nu &= \Delta u - (D^2 u\, \nu) \cdot \nu - \frac{n-1}{r}\, \partial_r u_0 \\ &= \Delta u - \partial_{rr} u_0 - \frac{n-1}{r}\, \partial_r u_0,\end{aligned}$$

as desired. □

Given $r > 0$, an equivalent formulation of Theorem 2.5.1 can be given by replacing the Laplace–Beltrami of u along ∂B_r with the Laplace–Beltrami of the map $\partial B_1 \ni \vartheta \mapsto u_0(r, \vartheta)$ along ∂B_1, as follows.

Theorem 2.5.2. *Let $u \in C^2(\mathbb{R}^n \setminus \{0\})$. Then,*

$$\Delta u(x) = \partial_{rr} u_0(r, \vartheta) + \frac{n-1}{r}\, \partial_r u_0(r, \vartheta) + \frac{1}{r^2}\, \Delta_{\partial B_1} u_0(r, \vartheta),$$

for every $x \in \mathbb{R}^n \setminus \{0\}$, where $r := |x|$ and $\vartheta := \frac{x}{|x|}$.

Proof. In light of Theorem 2.5.1, it suffices to show that, given $r > 0$,

$$\Delta_{\partial B_r} u(r\vartheta) = \frac{1}{r^2} \Delta_{\partial B_1} u_0(r, \vartheta) \quad \text{for all } \vartheta \in \partial B_1. \tag{2.5.2}$$

To this end, we perform a careful scaling argument. Given $r > 0$, for every $\vartheta \in \partial B_1$, we let $v^{(r)}(\vartheta) := u_0(r, \vartheta)$. Thus, the normal extension of $v^{(r)}$ outside ∂B_1, as introduced in (2.3.3), is the function

$$w^{(r)}(x) := v^{(r)}\left(\frac{x}{|x|}\right) = u_0\left(r, \frac{x}{|x|}\right) = u\left(\frac{rx}{|x|}\right). \tag{2.5.3}$$

Note that

$$w^{(r)}(\lambda x) = u_0\left(\frac{r\lambda x}{|\lambda x|}\right) = u_0\left(\frac{rx}{|x|}\right) = w^{(r)}(x),$$

whence $w^{(r)}$ is positively homogeneous of degree zero.

Therefore, the function $\phi^{(r)} := \Delta w^{(r)}$ is homogeneous of degree -2. Thus, recalling the Laplace–Beltrami definition in (2.3.5), for every $y \in \partial B_1$,

$$\Delta_{\partial B_1} u_0(r, y) = \Delta_{\partial B_1} v^{(r)}(y) = \Delta w^{(r)}(y) = \phi^{(r)}(y) = r^2\,\phi^{(r)}(ry). \tag{2.5.4}$$

Additionally, we have that the normal extension of u outside ∂B_r is the function

$$\mathbb{R}^n \setminus \{0\} \ni x \mapsto u\left(\frac{rx}{|x|}\right),$$

which coincides with $w^{(r)}(x)$, thanks to (2.5.3).

On this account, by the Laplace–Beltrami definition in (2.3.5), we have that for every $x \in \partial B_r$,

$$\Delta_{\partial B_r} u(x) = \Delta w^{(r)}(x) = \phi^{(r)}(x) = \phi^{(r)}(ry),$$

where $y := \frac{x}{r}$. Hence, comparing with (2.5.4), we conclude that $\Delta_{\partial B_r} u(ry) = r^{-2}\Delta_{\partial B_1} u_0(r, y)$ for every $y \in \partial B_1$, and this completes the proof of (2.5.2). □

As a special case of Theorem 2.5.1, one obtains that if $u \in C^2(\mathbb{R}^n \setminus \{0\})$ is radially symmetric, i.e. $u(x) = u_0(|x|)$ for some $u_0 : (0, +\infty) \to \mathbb{R}$, then

$$\Delta u(x) = u_0''(r) + \frac{n-1}{r}\,u_0'(r), \tag{2.5.5}$$

with $r = |x|$. Of course, this formula can also be obtained through direct computations, see e.g. [Eva98, p. 21].

We also point out that if $\alpha \in \mathbb{R}$ and $u_\alpha(x) := |x|^\alpha$, then, taking $\Sigma := \partial B_1$, we have that $u_\alpha = 1$ on ∂B_1,

$$\begin{aligned} &\Delta u_\alpha = \alpha\,(n + \alpha - 2), \quad \Delta_T u_\alpha = \alpha\,(n-1) \\ &\text{and} \quad \Delta_{\partial B_1} u_\alpha = 0 \quad \text{on } \partial B_1. \end{aligned} \tag{2.5.6}$$

In particular, note that the normal extension of the function identically equal to 1 on ∂B_1 according to definition (2.3.3) is the function identically equal to 1 on $\mathbb{R}^n$, corresponding to u_α with $\alpha := 0$. The other values of α provide different extensions of the function identically equal to 1 on ∂B_1, and in this case, the Laplace–Beltrami operator is not equal to the full Laplacian of these extensions. Namely, geometric second-order operators can be reconstructed using full operators by extension, but they are sensitive to the type of extension chosen (differently from first-order operators, as discussed in Lemma 2.3.1).

2.6 The Kelvin Transform

The Kelvin transform is a useful tool[7] to reduce the analysis of partial differential equations in exterior domains to that of interior ones (and vice versa).

Moreover, this transformation possesses a number of geometric and analytic properties that make it handy in several occasions. To introduce[8] it, we define,

Figure 2.6.1 Meander of the River Kelvin with the Gilmorehill campus of the University of Glasgow. *Source*: Public Domain image from Wikipedia.

---

[7]The Kelvin transform is named after William Thomson, 1st Baron Kelvin. Actually, Thomson himself was named Baron Kelvin after the River Kelvin, which flows past the University of Glasgow, where Thomson used to work. See Figure 2.6.1 for a photochrom print of the river and the university dating back to the end of the nineteenth century. See also Figure 2.6.2 for a caricature of Lord Kelvin (by caricaturist Sir Leslie Matthew Ward, a.k.a. Spy), published in the magazine *Vanity Fair* in 1897.

[8]For the sake of clarity, we present the Kelvin transform with the aim of highlighting its remarkable analytic, algebraic and geometric properties. On the other hand, the Kelvin transform is naturally motivated by some important physical considerations inspired by the method of image charges: the reader who wishes to go straight to this motivation can look at Section 2.8.

Also, for simplicity, we take ∂B_1 as the reference set which remains invariant for the Kelvin transform, but we observe that a similar theory can be derived by leaving invariant ∂B_R (for this, one can either proceed by scaling or changing (2.6.1) into $\mathcal{K}(x) := \frac{R^2 x}{|x|^2}$.

Figure 2.6.2 Caricature of Lord Kelvin.
Source: Public Domain image from Wikipedia.

for all $x \in \mathbb{R}^n \setminus \{0\}$,

$$\mathcal{K}(x) := \frac{x}{|x|^2}. \tag{2.6.1}$$

We list here some interesting properties of the Kelvin transform.

Lemma 2.6.1. *The Kelvin transform is an involution, meaning that* $\mathcal{K}(\mathcal{K}(x)) = x$ *for all* $x \in \mathbb{R}^n \setminus \{0\}$.

Also, for all $x, y \in \mathbb{R}^n \setminus \{0\}$,

$$|\mathcal{K}(x)|\,|x| = 1, \tag{2.6.2}$$

$$\frac{|x|^2}{1 - |x|^2} = \frac{1}{|\mathcal{K}(x)|^2 - 1}, \tag{2.6.3}$$

$$\frac{\mathcal{K}(x)\cdot\mathcal{K}(y)}{|\mathcal{K}(x)|\,|\mathcal{K}(y)|}=\frac{x\cdot y}{|x|\,|y|},\tag{2.6.4}$$

$$|\mathcal{K}(x)-\mathcal{K}(y)|=\frac{|x-y|}{|x|\,|y|}.\tag{2.6.5}$$

Moreover,[9] *for each* $e\in\partial B_1$,

$$|x|\,|\mathcal{K}(x)-e|=|x-e|,\tag{2.6.6}$$

and, for each $i,\ j\in\{1,\dots,n\}$,

$$\partial_{x_j}\mathcal{K}(x)\cdot e_i=\frac{1}{|x|^2}\left(\delta_{ij}-\frac{2x_ix_j}{|x|^2}\right).\tag{2.6.7}$$

Proof. We remark that (2.6.2) is a direct consequence of the definition in (2.6.1). Thus,

$$\mathcal{K}(\mathcal{K}(x))=\frac{\mathcal{K}(x)}{|\mathcal{K}(x)|^2}=|x|^2\mathcal{K}(x)=x,$$

which proves the involution property.

From (2.6.2), we also deduce that

$$\frac{1}{|\mathcal{K}(x)|^2-1}=\frac{1}{|x|^{-2}-1}=\frac{|x|^2}{1-|x|^2},$$

that is, (2.6.3).

The claim in (2.6.4) is also a straightforward byproduct of the definition in (2.6.1) and (2.6.2). In addition, using (2.6.2) in combination with (2.6.4),

$$\begin{aligned}|x|^2|y|^2|\mathcal{K}(x)-\mathcal{K}(y)|^2&=|x|^2|y|^2\big(|\mathcal{K}(x)|^2+|\mathcal{K}(y)|^2-2\mathcal{K}(x)\cdot\mathcal{K}(y)\big)\\&=|x|^2|y|^2\left(|\mathcal{K}(x)|^2+|\mathcal{K}(y)|^2-2|\mathcal{K}(x)|\,|\mathcal{K}(y)|\frac{x\cdot y}{|x|\,|y|}\right)\\&=|x|^2|y|^2\left(\frac{1}{|x|^2}+\frac{1}{|y|^2}-\frac{2x\cdot y}{|x|^2\,|y|^2}\right)\\&=\big(|y|^2+|x|^2-2x\cdot y\big)=|x-y|^2,\end{aligned}$$

which establishes (2.6.5).

---

[9] We point out that (2.6.4) states that the Kelvin transform is angle preserving. Furthermore, (2.6.7) implies that the Kelvin transform is conformal since its Jacobian matrix is a scalar function times an orthogonal matrix: indeed,

$$\sum_{k=1}^{n}\left(\delta_{ik}-\frac{2x_ix_k}{|x|^2}\right)\left(\delta_{kj}-\frac{2x_kx_j}{|x|^2}\right)=\delta_{ij}.$$

Additionally, using (2.6.2) once again, if $e \in \partial B_1$, then

$$\begin{aligned} &|x|^2\,|\mathcal{K}(x) - e|^2 - |x-e|^2 \\ &= |x|^2\Big(|\mathcal{K}(x)|^2 + 1 - 2\mathcal{K}(x)\cdot e\Big) - \Big(|x|^2 + 1 - 2x\cdot e\Big) \\ &= 1 + |x|^2 - 2|x|^2\mathcal{K}(x)\cdot e - \Big(|x|^2 + 1 - 2x\cdot e\Big) = 0, \end{aligned}$$

and thus[10] we have proved (2.6.6).

Finally,

$$\partial_{x_j}\mathcal{K}(x)\cdot e_i = \partial_{x_j}\frac{x\cdot e_i}{|x|^2} = \frac{|x|^2\delta_{ij} - 2x_ix_j}{|x|^4},$$

that is, (2.6.7). □

We remark that the Kelvin transform is an "inversion of the sphere," namely, in view of (2.6.2), we have that $\mathcal{K}(B_1\setminus\{0\}) = \mathbb{R}^n\setminus B_1$, $\mathcal{K}(\mathbb{R}^n\setminus B_1) = B_1\setminus\{0\}$ and $\mathcal{K}(\partial B_1) = \partial B_1$.

Also, the Kelvin transform acts naturally on functions in a nicely compatible way with respect to the Laplace operator. Namely, setting

$$u_{\mathcal{K}}(x) := |x|^{2-n}u(\mathcal{K}(x)), \tag{2.6.8}$$

we have the following theorem.

Theorem 2.6.2. *If $v := u_{\mathcal{K}}$, then $v_{\mathcal{K}} = u$.*

Moreover[11] *if $n \geqslant 2$ and $u \in C^1(\overline{B_1})$ (respectively, if $u \in C^1(\mathbb{R}^n\setminus B_1)$), then, for every $v \in C_0^\infty(B_1)$ (respectively, $v \in C_0^\infty(\mathbb{R}^n\setminus B_1)$),*

$$\int_{\mathbb{R}^n}\nabla u_{\mathcal{K}}(x)\cdot\nabla v_{\mathcal{K}}(x)\,dx = \int_{\mathbb{R}^n}\nabla u(x)\cdot\nabla v(x)\,dx. \tag{2.6.9}$$

[10] Of course, many of the identities in Lemma 2.6.1 can be proved using different strategies. For instance, one could also deduce (2.6.6) directly from (2.6.5) by noting that, when $|e| = 1$,

$$|\mathcal{K}(x) - e| = |\mathcal{K}(x) - \mathcal{K}(e)| = \frac{|x-e|}{|x|\,|e|} = \frac{|x-e|}{|x|}.$$

[11] In the second part of Theorem 2.6.2, it is convenient to exclude the case $n = 1$ to avoid integrability issues (from a technical point of view, this condition is used in (2.6.12) to avoid contributions coming from infinity). As an illustrative example of the loss of integrability in (2.6.9) when $n = 1$, one can consider a function $v \in C_0^\infty((-1,1))$ such that $v(x) = 1$ for each $x \in [-\frac{1}{2}, \frac{1}{2}]$, and note that, in view of (2.6.8), $v_{\mathcal{K}}(x) = |x|v\left(\frac{x}{|x|^2}\right) = |x| = x$ for each $x \geqslant 2$. Accordingly, in this case, $\nabla v_{\mathcal{K}} = 1$ in $(2, +\infty)$, whence

$$\int_{\mathbb{R}^n}|\nabla v_{\mathcal{K}}(x)|^2\,dx \geqslant \int_2^{+\infty}dx = +\infty.$$

However, the claim in (2.6.10) is valid in dimension 1 as well, as it can be checked through differentiation (though it is arguably not very useful in this case).

In addition to that, if $u \in C^2(\overline{B_1})$ (respectively, if $u \in C^2(\mathbb{R}^n \setminus B_1)$), for every $x \in \mathbb{R}^n \setminus B_1$ (respectively, for every $x \in B_1 \setminus \{0\}$), it holds that

$$\Delta u_{\mathcal{K}}(x) = \frac{1}{|x|^{n+2}}\,\Delta u\big(\mathcal{K}(x)\big). \tag{2.6.10}$$

Proof. We observe that, if $v := u_{\mathcal{K}}$, then

$$v(\mathcal{K}(x)) = u_{\mathcal{K}}(\mathcal{K}(x)) = |\mathcal{K}(x)|^{2-n}u\big(\mathcal{K}(\mathcal{K}(x))\big) = |x|^{n-2}u(x),$$

thanks to (2.6.8) and Lemma 2.6.1, and for this reason,

$$v_{\mathcal{K}}(x) = |x|^{2-n}v(\mathcal{K}(x)) = u(x).$$

This proves the desired involution property.

Now, we establish (2.6.9) and (2.6.10). For this, we consider $v \in C_0^\infty(B_1)$ (respectively, $v \in C_0^\infty(\mathbb{R}^n \setminus B_1)$. We let $M(x)$ be the matrix with entries $\delta_{ij} - \frac{2x_ix_j}{|x|^2}$ for all $i, j \in \{1, \dots, n\}$, and we recall that $M(x)$ is orthogonal (see the footnote on p. 67). As a result, we have that $\det M(x) = 1$. Using (2.6.7), we see that the change of variable $y = \mathcal{K}(x)$ leads to

$$dy = |\det D\mathcal{K}(x)|\ dx = \left|\det \frac{M(x)}{|x|^2}\right|\ dx = \frac{dx}{|x|^{2n}}.$$

Moreover, by (2.6.7),

$$\nabla u_{\mathcal{K}}(x) = \nabla\Big(|x|^{2-n}u(\mathcal{K}(x))\Big) = (2-n)|x|^{-n}xu(\mathcal{K}(x)) + |x|^{-n}M(x)\nabla u(\mathcal{K}(x)),$$

and a similar formula holds for v replacing u.

Therefore,

$$\begin{aligned}\int_{\mathbb{R}^n} \nabla u_{\mathcal{K}}(x)\cdot\nabla v_{\mathcal{K}}(x)\,dx = \int_{\mathbb{R}^n} \Big[&(2-n)^2|x|^{2-2n}u(\mathcal{K}(x))v(\mathcal{K}(x))\\ &+ (2-n)|x|^{-2n}u(\mathcal{K}(x))\big(M(x)\nabla v(\mathcal{K}(x))\big)\cdot x\\ &+ (2-n)|x|^{-2n}v(\mathcal{K}(x))\big(M(x)\nabla u(\mathcal{K}(x))\big)\cdot x\\ &+ |x|^{-2n}\big(M(x)\nabla u(\mathcal{K}(x))\big)\cdot\big(M(x)\nabla v(\mathcal{K}(x))\big)\Big]\,dx.\end{aligned} \tag{2.6.11}$$

We also remark that

$$\begin{aligned}&\operatorname{div}\Big(|x|^{2-2n}xu(\mathcal{K}(x))v(\mathcal{K}(x))\Big)\\ &= \operatorname{div}\Big(|x|^{2-2n}x\Big)u(\mathcal{K}(x))v(\mathcal{K}(x))\\ &\quad + |x|^{2-2n}x\cdot\nabla\Big(u(\mathcal{K}(x))\Big)v(\mathcal{K}(x)) + |x|^{2-2n}xu(\mathcal{K}(x))\cdot\nabla\Big(v(\mathcal{K}(x))\Big)\\ &= (2-n)|x|^{2-2n}u(\mathcal{K}(x))v(\mathcal{K}(x))\\ &\quad + |x|^{-2n}x\cdot\big(M(x)\nabla u(\mathcal{K}(x))\big)v(\mathcal{K}(x)) + |x|^{-2n}x\cdot\big(M(x)\nabla v(\mathcal{K}(x))\big)u(\mathcal{K}(x)).\end{aligned}$$

This and the divergence theorem give that

$$\int_{\mathbb{R}^n} \Big[(2-n)^2|x|^{2-2n}u(\mathcal{K}(x))v(\mathcal{K}(x)) \\ + (2-n)|x|^{-2n}u(\mathcal{K}(x))\big(M(x)\nabla v(\mathcal{K}(x))\big)\cdot x \\ + (2-n)|x|^{-2n}v(\mathcal{K}(x))\big(M(x)\nabla u(\mathcal{K}(x))\big)\cdot x\Big]\,dx = 0. \qquad (2.6.12)$$

Plugging this information into (2.6.11), we thus conclude that

$$\int_{\mathbb{R}^n} \nabla u_{\mathcal{K}}(x)\cdot\nabla v_{\mathcal{K}}(x)\,dx = \int_{\mathbb{R}^n} |x|^{-2n}\big(M(x)\nabla u(\mathcal{K}(x))\big)\cdot\big(M(x)\nabla v(\mathcal{K}(x))\big)\,dx.$$

Therefore, using the orthogonality of $M(x)$ and changing the variable,

$$\int_{\mathbb{R}^n} \nabla u_{\mathcal{K}}(x)\cdot\nabla v_{\mathcal{K}}(x)\,dx = \int_{\mathbb{R}^n} |x|^{-2n}\nabla u(\mathcal{K}(x))\cdot\nabla v(\mathcal{K}(x))\,dx \\ = \int_{\mathbb{R}^n} \nabla u(y)\cdot\nabla v(y)\,dy.$$

This proves (2.6.9).

Now, we prove (2.6.10). For this, we utilize (2.6.9) to see that

$$-\int_{\mathbb{R}^n} |\mathcal{K}(y)|^{n+2}\,\Delta u_{\mathcal{K}}(\mathcal{K}(y))\,v(y)\,dy = -\int_{\mathbb{R}^n} |y|^{-n-2}\,\Delta u_{\mathcal{K}}(\mathcal{K}(y))\,v(y)\,dy \\ = -\int_{\mathbb{R}^n} |x|^{2-n}\,\Delta u_{\mathcal{K}}(x)\,v(\mathcal{K}(x))\,dx = -\int_{\mathbb{R}^n} \Delta u_{\mathcal{K}}(x)\,v_{\mathcal{K}}(x)\,dx \\ = \int_{\mathbb{R}^n} \nabla u_{\mathcal{K}}(x)\cdot\nabla v_{\mathcal{K}}(x)\,dx = \int_{\mathbb{R}^n} \nabla u(y)\cdot\nabla v(y)\,dy = -\int_{\mathbb{R}^n} \Delta u(y)\,v(y)\,dy.$$

This yields that $|\mathcal{K}(y)|^{n+2}\,\Delta u_{\mathcal{K}}(\mathcal{K}(y)) = \Delta u(y)$, from which we obtain the desired result in (2.6.10). □

Another interesting geometric property of the Kelvin transform is that it carries spheres and hyperplanes into spheres and hyperplanes.

Proposition 2.6.3. *The Kelvin transform carries*

- *spheres not passing through the origin into spheres not passing through the origin,*
- *spheres passing through the origin into hyperplanes not passing through the origin,*
- *and hyperplanes not passing through the origin into spheres passing through the origin.*

Also, the Kelvin transform leaves invariant all the hyperplanes passing through the origin.

More explicitly, if $p \in \mathbb{R}^n$, $r \in (0, +\infty)$, $\omega \in \partial B_1$, and $c \in \mathbb{R}$, we have that

$$\mathcal{K}\Big(\{x \in \mathbb{R}^n \setminus \{0\} \text{ s.t. } |x-p|^2 = r^2\}\Big) = \begin{cases} \left\{x \in \mathbb{R}^n \setminus \{0\} \text{ s.t. } \left|x - \dfrac{p}{|p|^2 - r^2}\right|^2 = \dfrac{r^2}{(|p|^2 - r^2)^2}\right\} & \text{if } r \neq |p|, \\ \left\{x \in \mathbb{R}^n \setminus \{0\} \text{ s.t. } p \cdot x = \dfrac{1}{2}\right\} & \text{if } r = |p|, \end{cases} \tag{2.6.13}$$

and

$$\mathcal{K}\Big(\{x \in \mathbb{R}^n \setminus \{0\} \text{ s.t. } \omega \cdot x = c\}\Big) = \begin{cases} \left\{x \in \mathbb{R}^n \setminus \{0\} \text{ s.t. } \left|x - \dfrac{\omega}{2c}\right|^2 = \dfrac{1}{4c^2}\right\} & \text{if } c \neq 0, \\ \{x \in \mathbb{R}^n \setminus \{0\} \text{ s.t. } \omega \cdot x = 0\} & \text{if } c = 0. \end{cases} \tag{2.6.14}$$

Proof. Let $x \neq 0$ and $y := \mathcal{K}(x)$. Since, by Lemma 2.6.1, the Kelvin transform is an involution, we have that $x = \mathcal{K}(y) = \frac{y}{|y|^2}$. Therefore, if $|x-p|^2 = r^2$, then

$$\begin{aligned} 0 &= |y|^2 \left(|x-p|^2 - r^2\right) = |y|^2 \left(\left|\frac{y}{|y|^2} - p\right|^2 - r^2\right) \\ &= \left|\frac{y}{|y|} - |y| p\right|^2 - r^2|y|^2 = 1 - 2p \cdot y + (|p|^2 - r^2)|y|^2. \end{aligned}$$

Now, if $r = |p|$, this boils down to $p \cdot y = \frac{1}{2}$. If instead $r \neq |p|$, then

$$\begin{aligned} 0 &= \frac{1}{|p|^2 - r^2} - \frac{2p \cdot y}{|p|^2 - r^2} + |y|^2 \\ &= \frac{1}{|p|^2 - r^2} - \frac{|p|^2}{(|p|^2 - r^2)^2} + \left|y - \frac{p}{|p|^2 - r^2}\right|^2 \\ &= -\frac{r^2}{(|p|^2 - r^2)^2} + \left|y - \frac{p}{|p|^2 - r^2}\right|^2. \end{aligned}$$

Furthermore, we note that, since $y := \mathcal{K}(x) = \frac{x}{|x|^2}$,

$$|y-p|^2 = |y|^2 - 2p \cdot y + |p|^2 = \frac{1}{|x|^2} - \frac{2p \cdot x}{|x|^2} + |p|^2.$$

Thus, if $r = |p|$ and $p \cdot x = \frac{1}{2}$, then

$$|y - p|^2 = \frac{1}{|x|^2} - \frac{1}{|x|^2} + |p|^2 = |p|^2 = r^2.$$

If instead $r \neq |p|$ and $\left|x - \frac{p}{|p|^2 - r^2}\right|^2 = \frac{r^2}{(|p|^2 - r^2)^2}$, then

$$\begin{aligned}|y - p|^2 &= \frac{1}{|x|^2} - \frac{1}{|x|^2}\left(|x|^2(|p|^2 - r^2) + 1\right) + |p|^2 \\ &= \frac{1}{|x|^2} - |p|^2 + r^2 - \frac{1}{|x|^2} + |p|^2 = r^2.\end{aligned}$$

These considerations establish (2.6.13).

Now, if $\omega \cdot x = c$, we find that $\omega \cdot y = c|y|^2$. This reduces to $\omega \cdot y = 0$ if $c = 0$. Instead, if $c \neq 0$,

$$0 = |y|^2 - \frac{\omega \cdot y}{c} = \left|y - \frac{\omega}{2c}\right|^2 - \frac{1}{4c^2}.$$

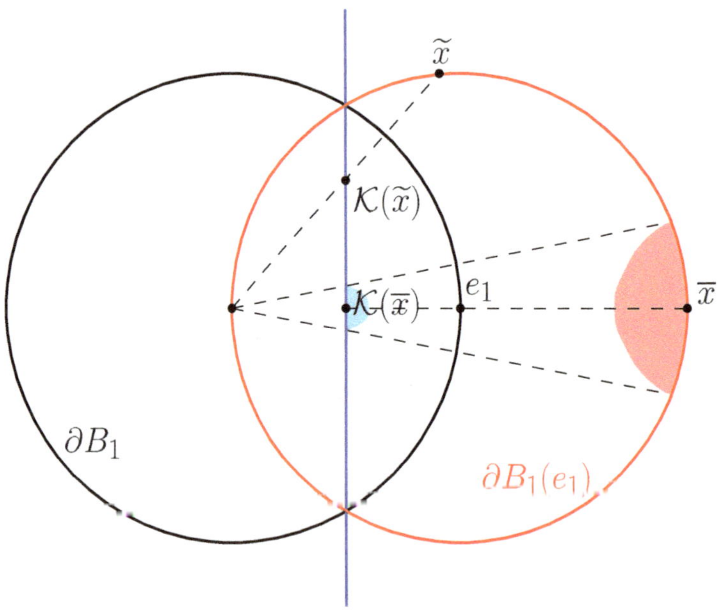

Figure 2.6.3 The Kelvin transform carries the sphere $\partial B_1(e_1)$ into the hyperplane $\{x_1 = \frac{1}{2}\}$. A small neighborhood of $2e_1$ in $B_1(e_1)$ is carried into a small neighborhood of $\frac{e_1}{2}$ in $\{x_1 > \frac{1}{2}\}$.

Vice versa, if $\omega \cdot x = 0$, then $\omega \cdot y = 0 = c$. If instead $\left|x - \frac{\omega}{2c}\right|^2 = \frac{1}{4c^2}$, then

$$\omega \cdot y = \frac{\omega \cdot x}{|x|^2} = \frac{c|x|^2}{|x|^2} = c,$$

thus completing the proof of (2.6.14). □

We refer to Figure 2.6.3 for a graphical representation of the situation discussed in detail in Proposition 2.6.3.

2.7 The Fundamental Solution

Here, we describe the notion of fundamental solution of the Laplace[12] equation, which is physically motivated by the electrostatic (or gravitational) potential produced by a point charge (or a point mass). For this, the isotropy and homogeneity of the ambient space play a crucial role by inducing rotational and translational symmetries.

To begin with, in view of the polar coordinate representation of the Laplacian in (2.5.5), one can explicitly find all the rotationally symmetric harmonic functions in $\mathbb{R}^n \setminus \{0\}$, according to the following observation.

Lemma 2.7.1. *If $v \in C^2(\mathbb{R}^n \setminus \{0\})$ (respectively, if $v \in C^2(B_R \setminus \{0\})$ for some ball $B_R \subset \mathbb{R}^n$) is rotationally symmetric and harmonic in $\mathbb{R}^n \setminus \{0\}$ (respectively, in $B_R \setminus \{0\}$), then there exist $a, b \in \mathbb{R}$ such that, for every $x \in \mathbb{R}^n \setminus \{0\}$ (respectively, for every $x \in B_R \setminus \{0\}$), we have that*

$$v(x) = \begin{cases} \dfrac{a}{|x|^{n-2}} + b & \text{if } n \neq 2, \\ a \ln |x| + b & \text{if } n = 2. \end{cases}$$

Proof. We argue in $\mathbb{R}^n \setminus \{0\}$, with the case in $B_R \setminus \{0\}$ being similar. We use the notation $r := |x|$, and let $v_0 : (0, +\infty) \to \mathbb{R}$ be such that $v(x) = v_0(|x|)$. Then, by (2.5.5), in $\mathbb{R}^n \setminus \{0\}$,

$$0 = \Delta v = v_0'' + \frac{n-1}{r} v_0' = r^{1-n} \frac{d}{dr}\left(r^{n-1} v_0'\right);$$

[12]Usually, though the notation is certainly not uniform across the literature. The equation $\Delta u(x) = 0$ for all x in Ω is often referred to by the name "Laplace equation" and, correspondingly, $\Delta u(x) = f(x)$ by the name "Poisson equation." When the Laplace equation (or, sometimes, the Poisson equation) is complemented with a boundary datum $u = g$ along $\partial\Omega$, it is referred to as the "Dirichlet problem."

therefore, there exists $a \in \mathbb{R}$ such that $r^{n-1}v_0'(r) = a$ for every $r > 0$. Integrating this expression, we obtain that there exists $b \in \mathbb{R}$ such that

$$v_0(r) = \begin{cases} \dfrac{ar^{2-n}}{2-n} + b & \text{if } n \neq 2, \\ a \ln r + b & \text{if } n = 2. \end{cases}$$

Hence, by renaming the constant a,

$$v_0(r) = \begin{cases} \dfrac{a}{r^{n-2}} + b & \text{if } n \neq 2, \\ a \ln r + b & \text{if } n = 2, \end{cases}$$

and the desired result plainly follows. □

The functions introduced in Lemma 2.7.1 play a pivotal role in the development of the theory of elliptic partial differential equations. This is because, as observed in Lemma 2.7.1, they not only satisfy the equation $\Delta v = 0$ pointwise outside the origin, but (as we will soon see) they also provide the "fundamental solutions" of the Laplace operators up to normalization constants; that is, their Laplacian (in a suitable distributional sense, as made precise in Theorem 2.7.2) is (minus) the Dirac delta function at the origin. To clarify this concept, in light of Lemma 2.7.1, it is useful to define

$$\overline{v}(r) = \begin{cases} \dfrac{1}{r^{n-2}} & \text{if } n \neq 2, \\ -\ln r & \text{if } n = 2 \end{cases} \tag{2.7.1}$$

and choose the normalizing constant

$$c_n := \begin{cases} \dfrac{1}{n(n-2)\,|B_1|} & \text{if } n \neq 2, \\ \dfrac{1}{2\pi} & \text{if } n = 2. \end{cases} \tag{2.7.2}$$

The reason for choosing c_n in this way is that, for every $\rho > 0$, the function

$$\overline{\Gamma}_\rho(r) := \begin{cases} \dfrac{\rho^2 - r^2}{2n|B_\rho|} + c_n\,\overline{v}(\rho) & \text{if } r \in (0, \rho), \\ c_n\,\overline{v}(r) & \text{if } r \in [\rho, +\infty) \end{cases} \tag{2.7.3}$$

is such that $\overline{\Gamma}_\rho' \in C^{0,1}((0, +\infty))$. Let also

$$\Gamma_\rho(x) := \overline{\Gamma}_\rho(|x|). \tag{2.7.4}$$

We remark that Γ_ρ is obtained, in simple terms, by "gluing a paraboloid near the origin" to the harmonic function outside the origin that was introduced in

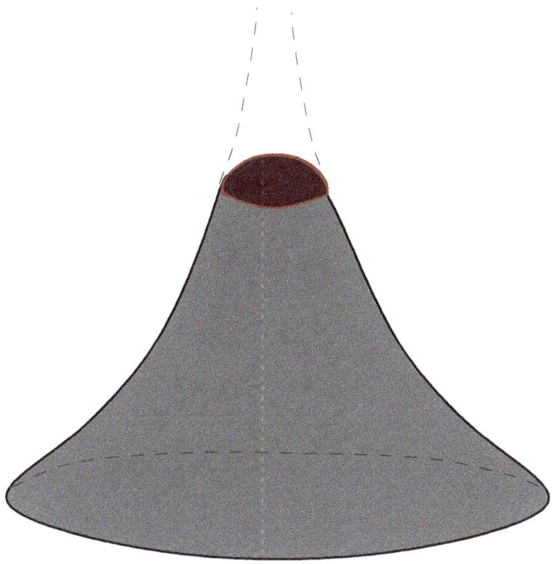

Figure 2.7.1 The function Γ_ρ (if $n \geqslant 3$).

Lemma 2.7.1, see Figure 2.7.1. Moreover, this paraboloid is normalized to have a Laplacian that integrates to -1 in B_ρ; namely, Γ_ρ can be extended to a function in $C^{1,1}(\mathbb{R}^n)$ satisfying

$$\Delta\Gamma_\rho = \begin{cases} -\dfrac{1}{|B_\rho|} & \text{in } B_\rho, \\ 0 & \text{in } \mathbb{R}^n \setminus \overline{B_\rho}. \end{cases} \tag{2.7.5}$$

We then consider the formal limit as $\rho \searrow 0$ of Γ_ρ by defining

$$\Gamma(x) := c_n \overline{v}(|x|) = \begin{cases} \dfrac{c_n}{|x|^{n-2}} & \text{if } n \neq 2, \\ -c_n \ln |x| & \text{if } n = 2. \end{cases} \tag{2.7.6}$$

We stress that when $n = 3$, this function represents, up to a normalizing constant, the electrostatic potential generated by a point charge (as well as the gravitational potential generated by a point mass, or the equilibrium temperature produced by a concentrated heat source), and we can exploit this physical motivation in any dimension as well, at least to facilitate our mathematical intuition. In our setting, the function Γ is the "fundamental solution" of the Laplace operator in the sense made precise by the following result.

Theorem 2.7.2. *For every $\varphi \in C_0^\infty(\mathbb{R}^n)$,*

$$\int_{\mathbb{R}^n} \Gamma(x)\, \Delta\varphi(x)\, dx = -\varphi(0).$$

Proof. Since $\overline{\Gamma}_\rho' \in C^{0,1}((0,+\infty))$, there exists $C > 0$ such that $|\overline{\Gamma}_\rho'| + |\overline{\Gamma}_\rho''| \leqslant C$ a.e. in $\mathbb{R}^n$. As a result, for every $i, j \in \{1,\dots,n\}$ and a.e. $x \in \mathbb{R}^n \setminus B_{\rho/2}$,

$$|\partial_{ij}\Gamma_\rho(x)| = \left| \overline{\Gamma}_\rho''(|x|)\, \frac{x_i x_j}{|x|^2} + \overline{\Gamma}_\rho'(|x|)\, \frac{|x|^2\delta_{ij} - x_i x_j}{|x|^3} \right| \leqslant 4C\left(1 + \frac{1}{\rho}\right).$$

As a result, letting $A_{\rho,\varepsilon} := B_{\rho+\varepsilon} \setminus B_{\rho-\varepsilon}$,

$$\lim_{\varepsilon\searrow 0} \left| \int_{A_{\rho,\varepsilon}} \Delta\Gamma_\rho(x)\, \varphi(x)\, dx \right| \leqslant \lim_{\varepsilon\searrow 0} 4Cn\, \|\varphi\|_{L^\infty(\mathbb{R}^n)} \left(1 + \frac{1}{\rho}\right) |A_{\rho,\varepsilon}| = 0. \quad (2.7.7)$$

Also,

$$\lim_{\varepsilon\searrow 0} \left| \int_{A_{\rho,\varepsilon}} \Gamma_\rho(x)\, \Delta\varphi(x)\, dx \right| \leqslant \lim_{\varepsilon\searrow 0} n\, \|\Gamma_\rho\|_{L^\infty(S)}\, \|D^2\varphi\|_{L^\infty(\mathbb{R}^n)}\, |A_{\rho,\varepsilon}| = 0, \quad (2.7.8)$$

where S is the support of φ.

Now, we point out that if $F : \mathbb{R}^n \to \mathbb{R}$ is a continuous vector field, then

$$\lim_{\varepsilon\searrow 0} \int_{\partial A_{\rho,\varepsilon}} F(x) \cdot \nu(x)\, d\mathcal{H}_x^{n-1} = 0. \quad (2.7.9)$$

For this, given $\eta > 0$, we consider a smooth vector field F_η such that $\|F - F_\eta\|_{L^\infty(B_{2\rho},\mathbb{R}^n)} \leqslant \eta$, and we use the divergence theorem to see that

$$\begin{aligned}
&\lim_{\varepsilon\searrow 0} \left| \int_{\partial A_{\rho,\varepsilon}} F(x) \cdot \nu(x)\, d\mathcal{H}_x^{n-1} \right| \\
&\quad \leqslant \lim_{\varepsilon\searrow 0} \|F - F_\eta\|_{L^\infty(\partial A_{\rho,\varepsilon},\mathbb{R}^n)}\, \mathcal{H}^{n-1}(\partial A_{\rho,\varepsilon}) + \left| \int_{\partial A_{\rho,\varepsilon}} F_\eta(x) \cdot \nu(x)\, d\mathcal{H}_x^{n-1} \right| \\
&\quad \leqslant 2\eta \mathcal{H}^{n-1}(\partial B_\rho) + \lim_{\varepsilon\searrow 0} \left| \int_{A_{\rho,\varepsilon}} \operatorname{div} F_\eta(x)\, dx \right| \\
&\quad \leqslant 2\eta \mathcal{H}^{n-1}(\partial B_\rho) + \lim_{\varepsilon\searrow 0} n\, \|F_\eta\|_{C^1(B_{2\rho},\mathbb{R}^n)}\, |A_{\rho,\varepsilon}| \\
&\quad = 2\eta \mathcal{H}^{n-1}(\partial B_\rho).
\end{aligned}$$

We now send $\eta \searrow 0$, and we conclude the proof of (2.7.9).

As a consequence of (2.7.9), we find that

$$\lim_{\varepsilon\searrow 0} \int_{\partial A_{\rho,\varepsilon}} \Gamma_\rho(x) \frac{\partial\varphi}{\partial\nu}(x)\, d\mathcal{H}_x^{n-1} = 0$$

and

$$\lim_{\varepsilon\searrow 0}\int_{\partial A_{\rho,\varepsilon}}\varphi(x)\frac{\partial\Gamma_\rho}{\partial\nu}(x)\,d\mathcal{H}^{n-1}_x=0.$$

Hence, by (2.7.7), (2.7.8) and the Green's second identity (recall (2.1.11)),

$$\begin{aligned}\int_{\mathbb{R}^n}\Gamma_\rho(x)\,\Delta\varphi(x)\,dx &= \lim_{\varepsilon\searrow 0}\int_{\mathbb{R}^n\setminus A_{\rho,\varepsilon}}\Gamma_\rho(x)\,\Delta\varphi(x)\,dx\\ &= \lim_{\varepsilon\searrow 0}\Bigg[\int_{\mathbb{R}^n\setminus A_{\rho,\varepsilon}}\Delta\Gamma_\rho(x)\,\varphi(x)\,dx\\ &\qquad-\int_{\partial A_{\rho,\varepsilon}}\left(\Gamma_\rho(x)\frac{\partial\varphi}{\partial\nu}(x)-\varphi(x)\frac{\partial\Gamma_\rho}{\partial\nu}(x)\right)d\mathcal{H}^{n-1}_x\Bigg]\\ &= \int_{\mathbb{R}^n}\Delta\Gamma_\rho(x)\,\varphi(x)\,dx. \end{aligned}\tag{2.7.10}$$

Thus, using (2.7.5),

$$\int_{\mathbb{R}^n}\Gamma_\rho(x)\,\Delta\varphi(x)\,dx=-⨍_{B_\rho}\varphi(x)\,dx.$$

The desired result now follows by sending $\rho\searrow 0$ (note that $|\Gamma_\rho|\leqslant\Gamma$, which is locally integrable, and hence the dominated convergence theorem can be utilized here). □

As a side remark concerning the dependence on the dimension in the fundamental solution (2.7.6), we stress that it is usually a challenging task to understand "intuitively" higher dimensions, and the change in the fundamental solution when $n=2$ is a deep feature to keep in mind. To aid the intuition and possibly recover, at least at a heuristic level, lower-dimensional cases from higher-dimensional ones, we propose some reflections about how the electrostatic potentials of a line of uniformly distributed charges in $\mathbb{R}^n$ produce the fundamental solutions in $\mathbb{R}^{n-1}$ (up to some renormalization that plays a role in low dimension). For this, we consider a uniform distribution of charges on the line $L:=\{x=(x',x_n)\in\mathbb{R}^{n-1}\times\mathbb{R}$ s.t. $x'=0\}$, which (up to physical constant) produces an electrostatic potential at the point $p=(p',0)\in\mathbb{R}^{n-1}\times\{0\}$ of the type

$$\begin{aligned}U(r)&:=\frac{1}{c_n}\int_L\Gamma_n(p-x)\,d\mathcal{H}^{n-1}_x=\int_L\overline{v}(|p-x|)\,d\mathcal{H}^{n-1}_x\\ &=\int_{-\infty}^{+\infty}\overline{v}_n\left(\sqrt{|p'|^2+x_n^2}\right)dx_n=\int_{-\infty}^{+\infty}\overline{v}_n\left(\sqrt{r^2+x_n^2}\right)dx_n,\end{aligned}$$

where $r:=|p'|$, and $\overline{v}_n$ and Γ_n are as in (2.7.1) and (2.7.6), with the subscript n used to underline the dependence of these functions on the dimension.

Strangely enough, the easiest case to understand is the high-dimensional one: namely, when $n \geqslant 4$, we obtain

$$U(r) = \int_{-\infty}^{+\infty} \left(r^2 + x_n^2\right)^{\frac{2-n}{2}} dx_n = r^{1-n} \int_{-\infty}^{+\infty} \left(1 + t^2\right)^{\frac{2-n}{2}} dt = cr^{3-n} = c\overline{v}_{n-1}(r),$$

where $c > 0$, showing that the potential of the charged line produces the fundamental solution in one dimension less, up to a normalizing constant.

The lower-dimensional cases $n \in \{2, 3\}$ are instead more tricky, since the corresponding constant c would diverge. To make the argument work (at least at a heuristic level), one needs to perform a "renormalization" procedure, formally "subtracting infinity" to the potential (as a matter of fact, potentials are always defined "up to additive constants" since the physical forces come from their derivatives). Concretely, when $n \in \{2, 3\}$, one has to replace the previous definition of U with the following renormalized one:

$$U(r) := \lim_{R \to +\infty} \left(\int_{-R}^{+R} \overline{v}_n \left(\sqrt{r^2 + x_n^2} \right) dx_n - \phi_n(R) \right),$$

for a suitable renormalization $\phi_n(R)$. The previous computations give that one can choose ϕ_n to be identically zero when $n \geqslant 3$; however, as we will see now, the cases $n \in \{2, 3\}$ do require a more specific choice.

If $n = 3$, we choose $\phi(R) := 2\ln(2R)$, and we thereby obtain

$$\begin{aligned}
U(r) &= \lim_{R \to +\infty} \left(\int_{-R}^{+R} \frac{1}{\sqrt{r^2 + x_n^2}} \, dx_n - 2\ln(2R) \right) \\
&= \lim_{R \to +\infty} \left(\ln\left(1 + \frac{2R(\sqrt{r^2 + R^2} + R)}{r^2} \right) - \ln(4R^2) \right) \\
&= \lim_{R \to +\infty} \ln \frac{r^2 + 2R(\sqrt{r^2 + R^2} + R)}{4r^2 R^2} \\
&= \ln \frac{1}{r^2} \\
&= 2\overline{v}_2(r).
\end{aligned}$$

If instead $n = 2$, we choose $\phi(R) := -2R(\ln R + 1)$, and we have

$$\begin{aligned}
U(r) &= \lim_{R \to +\infty} \left(\frac{1}{2} \int_{-R}^{+R} \ln\left(r^2 + x_n^2\right) dx_n + 2R(\ln R + 1) \right) \\
&= \lim_{R \to +\infty} \left(-2r \arctan \frac{R}{r} - R\ln(r^2 + R^2) + 2R\ln R \right) \\
&= \lim_{R \to +\infty} \left(-2r \arctan \frac{R}{r} - R \ln \frac{r^2 + R^2}{R^2} \right)
\end{aligned}$$

$$
\begin{aligned}
&= \lim_{R\to+\infty}\left(-2r\arctan\frac{R}{r} - R\ln\left(1+\frac{r^2}{R^2}\right)\right)\\
&= -\pi r\\
&= -\pi\overline{v}_1(r).
\end{aligned}
$$

The negative sign here above is consistent with the fact that the constant in (2.7.2) is negative when $n = 1$.

Related approaches for constructing lower-dimensional effective potentials deal with the interaction of a charged line with a "test line" (rather than a "test point").

Another approach to recover dimension 1 directly from dimension 3 is to consider the renormalized electrostatic potential of a charged plate (say, $\mathbb{R}^2 \times \{0\}$) at the point $(0, 0, r)$: in this case, the computation, up to dimensional constants, would involve a corrector of the type $2\pi R$ and proceeds as follows:

$$
\begin{aligned}
U(r) &= \lim_{R\to+\infty}\left[\int_0^R\left(\int_{\{|x'|=\rho,\ x_3=0\}}\frac{d\mathcal{H}^2_x}{|x-(0,0,r)|}\right)d\rho - 2\pi R\right]\\
&= \lim_{R\to+\infty}\left[\int_0^R\frac{2\pi\rho\, d\rho}{\sqrt{\rho^2+r^2}} - 2\pi R\right]\\
&= \lim_{R\to+\infty} 2\pi\sqrt{R^2+r^2} - 2\pi r - 2\pi R\\
&= -2\pi r\\
&= -2\pi\overline{v}_1(r).
\end{aligned}
$$

A variant[13] of Theorem 2.7.2 is given as follows.

Theorem 2.7.3. *Let Ω be a bounded open set in $\mathbb{R}^n$ with C^1 boundary. Let $x_0 \in \Omega$.*

---

[13]For completeness, we observe that Theorem 2.7.3 provides an alternative approach toward the mean value formula in Theorem 2.1.2. For instance, by using (2.7.11) with $\varphi := -1$ and $\Omega := B_r(x_0)$, we find that

$$
1 = -\int_{\partial B_r(x_0)}\frac{\partial\Gamma}{\partial\nu}(x-x_0)\,d\mathcal{H}^{n-1}_x = -\left.\frac{\partial\Gamma}{\partial\nu}\right|_{\partial B_r}\mathcal{H}^{n-1}(\partial B_r).
$$

Therefore, if φ is harmonic in $B_R(x_0)$ and $r \in (0, R)$, it follows from (2.7.11) that

$$
\begin{aligned}
\varphi(x_0) &= \int_{\partial B_r(x_0)}\left(\Gamma(x-x_0)\frac{\partial\varphi}{\partial\nu}(x) - \varphi(x)\frac{\partial\Gamma}{\partial\nu}(x-x_0)\right)d\mathcal{H}^{n-1}_x\\
&= \Gamma\big|_{\partial B_r}\int_{\partial B_r(x_0)}\frac{\partial\varphi}{\partial\nu}(x)\,d\mathcal{H}^{n-1}_x - \left.\frac{\partial\Gamma}{\partial\nu}\right|_{\partial B_r}\int_{\partial B_r(x_0)}\varphi(x)\,d\mathcal{H}^{n-1}_x\\
&= \Gamma\big|_{\partial B_r}\int_{B_r(x_0)}\Delta\varphi(x)\,dx + \frac{1}{\mathcal{H}^{n-1}(\partial B_r)}\int_{\partial B_r(x_0)}\varphi(x)\,d\mathcal{H}^{n-1}_x\\
&= \fint_{\partial B_r(x_0)}\varphi(x)\,d\mathcal{H}^{n-1}_x,
\end{aligned}
$$

which corresponds to Theorem 2.1.2(ii).

Then, for every $\varphi \in C^2(\Omega) \cap C^1(\overline{\Omega})$,

$$\int_\Omega \Gamma(x - x_0)\, \Delta\varphi(x)\, dx - \int_{\partial\Omega} \left(\Gamma(x - x_0) \frac{\partial \varphi}{\partial \nu}(x) - \varphi(x) \frac{\partial \Gamma}{\partial \nu}(x - x_0) \right) d\mathcal{H}^{n-1}_x = -\varphi(x_0). \tag{2.7.11}$$

Proof. Up to replacing $\varphi(x)$ with $\widetilde{\varphi}(x) := \varphi(x + x_0)$, we can assume that $x_0 = 0$. We take $\rho > 0$ so small that $B_\rho \Subset \Omega$, and thus $\Gamma_\rho = \Gamma$ in a neighborhood of $\partial\Omega$. Then, we replace (2.7.10) in this framework by

$$\begin{aligned}
&\int_\Omega \Gamma_\rho(x)\, \Delta\varphi(x)\, dx \\
&\quad = \lim_{\varepsilon \searrow 0} \int_{\Omega \setminus A_{\rho,\varepsilon}} \Gamma_\rho(x)\, \Delta\varphi(x)\, dx \\
&\quad = \lim_{\varepsilon \searrow 0} \left[\int_{\Omega \setminus A_{\rho,\varepsilon}} \Delta\Gamma_\rho(x)\, \varphi(x)\, dx + \int_{\partial\Omega} \left(\Gamma_\rho(x) \frac{\partial \varphi}{\partial \nu}(x) - \varphi(x) \frac{\partial \Gamma_\rho}{\partial \nu}(x) \right) d\mathcal{H}^{n-1}_x \right. \\
&\qquad \left. - \int_{\partial A_{\rho,\varepsilon}} \left(\Gamma_\rho(x) \frac{\partial \varphi}{\partial \nu}(x) - \varphi(x) \frac{\partial \Gamma_\rho}{\partial \nu}(x) \right) d\mathcal{H}^{n-1}_x \right] \\
&\quad = \int_\Omega \Delta\Gamma_\rho(x)\, \varphi(x)\, dx + \int_{\partial\Omega} \left(\Gamma(x) \frac{\partial \varphi}{\partial \nu}(x) - \varphi(x) \frac{\partial \Gamma}{\partial \nu}(x) \right) d\mathcal{H}^{n-1}_x,
\end{aligned}$$

and we conclude as in the proof of Theorem 2.7.2 by sending $\rho \searrow 0$. □

The identity (2.7.11) is sometimes called "Green's representation formula." Interestingly, it allows one to reconstruct the pointwise value of a function from its Laplacian in a given domain and its value and the values of its normal derivative at the boundary of the domain.

The fundamental solution can be used to construct regular solutions of the Poisson equation with a sufficiently regular right-hand side. As an example, we provide the following result (a more precise version will follow from the Schauder estimates in Proposition 4.2.3).

Proposition 2.7.4. *Let* $f \in C^1_0(\mathbb{R}^n)$. *Then, the function* $v := -\Gamma * f$ *belongs to* $C^2(\mathbb{R}^n)$ *and satisfies* $\Delta v = f$ *in* $\mathbb{R}^n$.

Proof. We note that v is well defined since Γ is locally integrable and f is bounded and with bounded support. In fact, $v \in L^\infty(\mathbb{R}^n)$. Also, by Theorem 2.7.2, for every $\varphi \in C_0^\infty(\mathbb{R}^n)$ and every $x \in \mathbb{R}^n$, we know that

$$\int_{\mathbb{R}^n} \Gamma(Y-x)\,\Delta\varphi(Y)\,dY = \int_{\mathbb{R}^n} \Gamma(y)\,\Delta\varphi(x+y)\,dy = -\varphi(x)$$

and therefore

$$\begin{aligned}\int_{\mathbb{R}^n} v(Y)\,\Delta\varphi(Y)\,dY &= -\iint_{\mathbb{R}^{2n}} \Gamma(Y-x)\,f(x)\,\Delta\varphi(Y)\,dx\,dY \\ &= \int_{\mathbb{R}^n} \varphi(x)\,f(x)\,dx. \end{aligned} \tag{2.7.12}$$

We observe that if $\phi \in L^\infty(\mathbb{R}^n) \cap L^1(\mathbb{R}^n)$ then, for all $i \in \{1, \dots, n\}$,

$$\partial_i(\Gamma * \phi) = (\partial_i\Gamma) * \phi. \tag{2.7.13}$$

To check this, given $\rho > 0$, we use the regularization Γ_ρ of the fundamental solution introduced in (2.7.3) and (2.7.4). We define $\psi := \Gamma * \phi$ and $\psi_\rho := \Gamma_\rho * \phi$, and we observe that, for every $x \in \mathbb{R}^n$,

$$\begin{aligned}|\psi_\rho(x) - \psi(x)| &\leqslant \frac{c_n}{2n|B_\rho|}\int_{B_\rho} (\rho^2 - |y|^2)|\phi(y)|\,dy \\ &\leqslant \frac{C\,\|\phi\|_{L^\infty(\mathbb{R}^n)}}{\rho^n}\int_0^\rho (\rho^2 - r^2)\,r^{n-1}\,dr \leqslant C\,\|\phi\|_{L^\infty(\mathbb{R}^n)}\,\rho^2\end{aligned} \tag{2.7.14}$$

for some constant $C > 0$ depending only on n and possibly varying from line to line.

Additionally,

$$\begin{aligned}\left|\partial_i\psi_\rho(x) - (\partial_i\Gamma) * \phi(x)\right| &= \left|\partial_i\Gamma_\rho * \phi(x) - (\partial_i\Gamma) * \phi(x)\right| \\ &\leqslant \frac{C}{\rho^n}\int_{B_\rho} |y|\,|\phi(x-y)|\,dy + C\int_{B_\rho} |y|^{1-n}\,|\phi(x-y)|\,dy \\ &\leqslant C\,\|\phi\|_{L^\infty(\mathbb{R}^n)}\,\rho.\end{aligned}$$

From this and (2.7.14), we infer that ψ_ρ converges uniformly to ψ and its derivative to $(\partial_i\Gamma) * \phi$, whence it follows that $\partial_i\psi = (\partial_i\Gamma) * \phi$, thus establishing (2.7.13).

We also remark that $\partial_j(\Gamma * f) = \Gamma * (\partial_j f)$ (see e.g. [WZ15, Theorem 9.3]); therefore, it follows from (2.7.13) that $\partial^2_{ij}v = -(\partial_i\Gamma) * (\partial_j f)$. Since $\partial_j f$ is continuous and compactly supported and $\partial_i\Gamma$ is locally integrable, we have that $\partial^2_{ij}v \in L^\infty(\mathbb{R}^n)$. Moreover, we suppose that the support of f is contained in a bounded set Ω, and for every $x \in \mathbb{R}^n$, we denote by Ω_x the bounded set containing all the points y such that $x - y$ belong to $\bigcup_{p\in\Omega} B_1(p)$. Then, if $x \in \mathbb{R}^n$ and $x_k \to x$ as $k \to +\infty$, we have that

$$|\partial_i\Gamma(y)\partial_j f(x_k - y)| \leqslant \|f\|_{C^1(\mathbb{R}^n)}\,|\nabla\Gamma(y)|\,\chi_{\Omega_x}(y),$$

and the latter is an integrable function of $y \in \mathbb{R}^n$. As a result, according to the dominated convergence theorem,

$$\begin{aligned}\lim_{k\to+\infty} \partial^2_{ij} v(x_k) &= -\lim_{k\to+\infty} \int_{\mathbb{R}^n} \partial_i \Gamma(y) \partial_j f(x_k - y)\, dy \\ &= \int_{\mathbb{R}^n} \partial_i \Gamma(y) \partial_j f(x - y)\, dy = \partial^2_{ij} v(x),\end{aligned}$$

whence $v \in C^2(\mathbb{R}^n)$. This and (2.7.12) yield that $\Delta v = f$. □

The function $-\Gamma * f$ in Proposition 2.7.4 (or sometimes, up to a sign convention, the function $\Gamma * f$) is often referred to by the name "Newtonian potential."

A neat intuition for the result in Proposition 2.7.4 comes from physical motivations: namely, since the fundamental solution $-\Gamma(x - y)$ (disregarding physical constants and possible sign conventions) corresponds to the gravitational field at the point x generated by a pointwise mass located at the point y, it follows that the quantity $-\Gamma * f(x)$ (which agrees with the the Newtonian potential) corresponds to the gravitational field at the point x generated by a distribution f of masses, since it can be seen as the superposition of $\Gamma(x - y) f(y)\, dy$. By Gauß' law, the flux of the field through a given surface $\partial\Omega$ (seen as the boundary of a bounded set Ω) corresponds to the total mass comprised inside $\partial\Omega$, namely

$$\int_\Omega f(x)\, dx = -\int_{\partial\Omega} \nabla(\Gamma * f)(x) \cdot \nu(x)\, d\mathcal{H}^{n-1}_x.$$

Using the divergence theorem on the latest integral, we thereby find that

$$\int_\Omega f(x)\, dx = -\int_\Omega \operatorname{div}\Big(\nabla(\Gamma * f)(x)\Big)\, dx = -\int_\Omega \Delta(\Gamma * f(x))\, dx.$$

Hence, since Ω is an arbitrary domain, we obtain that $f = -\Delta(\Gamma * f)$, which is precisely the content of Proposition 2.7.4. In this sense, Proposition 2.7.4 is a mathematically structured statement of Gauß' law (with some care devoted to the regularity of the distribution of mass f required for the result to hold true).

As a byproduct of our analysis, one can also deal with the gravitational potential of a homogeneous ball and show that at all external points, this potential is equal to the potential of the material point of the same mass placed at its center (see (B.1.3) for an application to physical geodesy).

Lemma 2.7.5. *Let $R > 0$ and $x \in \mathbb{R}^n \setminus \overline{B_R}$. Then,*

$$\int_{B_R} \Gamma(x - y)\, dy = |B_R|\, \Gamma(x).$$

Proof. Given $y \in B_R$, we have that $\Delta\Gamma(x-y) = 0$ for every $x \in \mathbb{R}^n \setminus B_R$. In particular, by Green's identity (2.1.11), if $r \in (0, R)$ and $\psi(y) := \frac{|y|^2 - r^2}{2n}$,

$$\begin{aligned}\int_{B_r} \Gamma(x-y)\,dy &= \int_{B_r} \Gamma(x-y)\Delta\psi(y)\,dy \\ &= \int_{B_r} \psi(y)\Delta\Gamma(x-y)\,dy \\ &\quad + \int_{\partial B_r} \left(\Gamma(x-y)\frac{\partial\psi}{\partial\nu}(y) - \psi(y)\frac{\partial\Gamma}{\partial\nu}(x-y)\right) d\mathcal{H}^{n-1}_y \\ &= 0 + \int_{\partial B_r} \left(\frac{r}{n}\Gamma(x-y) - 0\right) d\mathcal{H}^{n-1}_y = \frac{r}{n}\int_{\partial B_r} \Gamma(x-y)\,d\mathcal{H}^{n-1}_y.\end{aligned}$$

As a result, defining

$$\Psi(r) := r^{-n}\int_{B_r} \Gamma(x-y)\,dy,$$

and using polar coordinates (see e.g. [EG15, Theorem 3.12]), for $r \in (0, R)$, we have

$$\Psi'(r) = -nr^{-n-1}\int_{B_r} \Gamma(x-y)\,dy + r^{-n}\int_{\partial B_r} \Gamma(x-y)\,d\mathcal{H}^{n-1}_y = 0.$$

This gives that Ψ is constant in $(0, R]$, and thus

$$\Gamma(x) = \lim_{r\searrow 0} \fint_{B_r} \Gamma(x-y)\,dy = \lim_{r\searrow 0} \frac{\Psi(r)}{|B_1|} = \lim_{r\nearrow R} \frac{\Psi(r)}{|B_1|} = \fint_{B_R} \Gamma(x-y)\,dy,$$

as desired. □

2.8 Back to the Kelvin Transform: The Method of Image Charges

We discuss here a simple, but very influential, technique from electrostatics that naturally leads to the Kelvin transform introduced in Section 2.6. This method uses the fundamental solution showcased in Section 2.7, interpreting it as the electrostatic potential generated by a point charge (the method can also be considered an inspiration for the construction of the Green function of the ball in the forthcoming Theorem 2.10.4).

The details of this motivation are as follows. Suppose that we have a positive unit point charge located at some point in the ball B_1. Is it possible to place a negative point charge (not necessarily a unit charge) somewhere outside of the ball in order to make ∂B_1 a surface with constant, say zero, potential? If so, can we determine the position and intensity of this auxiliary charge?

For a mathematical setting for this situation, we suppose[14] that $n \neq 2$. In this way, up to a normalizing constant, we can suppose that the electrostatic potential at some point $x \in \mathbb{R}^n$ generated by a positive unit charge located at $x_0 \in B_1$ is equal to $\frac{1}{|x-x_0|^{n-2}}$, recall (2.7.6). So, suppose that we place a charge of some intensity $-\alpha(x_0) \in (-\infty, 0)$ at some point $T(x_0) \in \mathbb{R}^n \setminus B_1$. The potential generated by this auxiliary charge is equal to $-\frac{\alpha(x_0)}{|x-T(x_0)|^{n-2}}$.

Therefore, the condition that ∂B_1 is a surface with zero potential boils down to the relation

$$\frac{1}{|x-x_0|^{n-2}} - \frac{\alpha(x_0)}{|x-T(x_0)|^{n-2}} = 0 \qquad \text{for every } x \in \partial B_1. \tag{2.8.1}$$

Setting $\mu(x_0) := (\alpha(x_0))^{\frac{2}{n-2}}$, we therefore obtain that, for every $x \in \partial B_1$,

$$\begin{aligned} 1 + |T(x_0)|^2 - 2x \cdot T(x_0) &= |x|^2 + |T(x_0)|^2 - 2x \cdot T(x_0) = |x - T(x_0)|^2 \\ &= (\alpha(x_0))^{\frac{2}{n-2}} |x - x_0|^2 = \mu(x_0)\,|x - x_0|^2 \\ &= \mu(x)\left(|x| + |x_0|^2 - 2x \cdot x_0\right) \\ &= \mu(x_0)\left(1 + |x_0|^2 - 2x \cdot x_0\right), \end{aligned} \tag{2.8.2}$$

that is, for every $x \in \partial B_1$,

$$1 + |T(x_0)|^2 - \mu(x_0)\left(1 + |x_0|^2\right) = 2x \cdot \left(T(x_0) - \mu(x_0)\,x_0\right). \tag{2.8.3}$$

Now, given $x \in \partial B_1$, we exploit (2.8.3) both for x and $-x \in \partial B_1$, finding that

$$2x \cdot \left(T(x_0) - \mu(x_0)\,x_0\right) = 1 + |T(x_0)|^2 - \mu(x_0)\left(1 + |x_0|^2\right) = -2x \cdot \left(T(x_0) - \mu(x_0)\,x_0\right).$$

Consequently,

$$x \cdot \left(T(x_0) - \mu(x_0)\,x_0\right) = 0 \quad \text{for every } x \in \partial B_1, \tag{2.8.4}$$

and thus (2.8.3) reduces to

$$1 + |T(x_0)|^2 - \mu(x_0)\left(1 + |x_0|^2\right) = 0. \tag{2.8.5}$$

[14]The case $n = 2$ requires some conceptual modification: when $n = 2$, rather than imposing that ∂B_1 is a surface with zero potential, one requires the weaker condition that it is an equipotential surface, and the value of the potential may depend on x_0. To compensate for this weaker potential condition, however, in dimension 2, one can additionally impose that the image charge also has unit intensity (but it is negatively charged); that is, up to a sign, both the original charge and the mirror charge have the same intensity. In this way, when $n = 2$, equation (2.8.1) is replaced by

$$\ln|x - x_0| - \ln|x - T(x_0)| = \beta(x_0) \qquad \text{for every } x \in \partial B_1,$$

for a suitable $\beta(x_0)$. That is, setting $\gamma(x_0) := e^{\beta(x_0)}$,

$$1 + |x_0|^2 - 2x \cdot x_0 = |x - x_0|^2 = (\gamma(x_0))^2 |x - T(x_0)|^2 = (\gamma(x_0))^2 (1 + |T(x_0)|^2 - 2x \cdot T(x_0)).$$

This identity is the counterpart of (2.8.2) when $n = 2$ (with the notation $\mu(x_0) = \frac{1}{(\gamma(x_0))^2}$), and hence the computation provided in these pages (and leading to (2.8.7)) would give $\gamma(x_0) = |x_0|$ and thus $\beta(x_0) = \ln|x_0|$.

As a matter of fact, by choosing x as an element of the Euclidean basis in (2.8.4), we obtain that

$$T(x_0) \cdot e_i = \mu(x_0)\, x_0 \cdot e_i \quad \text{for every } i \in \{1, \dots, n\} \tag{2.8.6}$$

and therefore

$$|T(x_0)|^2 = \sum_{i=1}^{n} (T(x_0) \cdot e_i)^2 = (\mu(x_0))^2 \sum_{i=1}^{n} (x_0 \cdot e_i)^2 = (\mu(x_0))^2\, |x_0|^2.$$

Substituting this information into (2.8.5), we find that

$$0 = 1 + (\mu(x_0))^2\, |x_0|^2 - \mu(x_0)\left(1 + |x_0|^2\right),$$

and accordingly either $\mu(x_0) = 1$ or $\mu(x_0) = \frac{1}{|x_0|^2}$. However, the possibility $\mu(x_0) = 1$ must be ruled out; otherwise, by (2.8.6), we would have $\mathbb{R}^n \setminus B_1 \ni T(x_0) = x_0 \in B_1$, which is a contradiction.

In this way, we have established that

$$\mu(x_0) = \frac{1}{|x_0|^2}, \tag{2.8.7}$$

and thus, necessarily, $x_0 \neq 0$ if we want our problem to have a solution, whence

$$\alpha(x_0) = \frac{1}{|x_0|^{n-2}}. \tag{2.8.8}$$

Furthermore, we obtain from (2.8.6) that

$$T(x_0) = \frac{x_0}{|x_0|^2}. \tag{2.8.9}$$

Thus, conditions (2.8.8) and (2.8.9) provide the solution to our image electrostatic charge problem, specifying, respectively, the intensity of the image charge and its spatial location.

Remarkably, the spatial location of the image charge in (2.8.9) coincides with the Kelvin transform in (2.6.1), and the charge intensity in (2.8.8) coincides with the multiplicative factor in the functional action of the Kelvin transform as defined in (2.6.8).

2.9 Maximum Principles

Maximum principles are one of the cornerstones of the theory of elliptic partial differential equations, and they can also be seen as the real analysis counterpart of the maximum modulus principle, which is in turn one of the backbones of complex analysis (see e.g. [Rud66]). Broadly, the main idea is that if $\Delta u = 0$, then each point of the graph of u is necessarily a "saddle point" and therefore u cannot have local maxima or minima. In this section, we obtain the simplest possible versions of the maximum principle for the Laplace operator. First of all, in the following, we present a statement, often called the "strong maximum principle."

Theorem 2.9.1. *Let $\Omega \subseteq \mathbb{R}^n$ be open and connected, and let $u \in C^2(\Omega)$.*

(i) *If $\Delta u \geqslant 0$ in Ω and there exists $\overline{x} \in \Omega$ such that $u(\overline{x}) = \sup_\Omega u$, then u is constant.*
(ii) *If $\Delta u \leqslant 0$ in Ω and there exists $\underline{x} \in \Omega$ such that $u(\underline{x}) = \inf_\Omega u$, then u is constant.*
(iii) *If u is harmonic in Ω, then it cannot attain an interior maximum or minimum value unless it is constant.*

Proof. We observe that the claim in (ii) follows from (i) by changing u into $-u$. Also, the claim in (iii) follows from (i) and (ii). Therefore, it suffices to prove (i). To this end, we define

$$\mathcal{U} := \left\{ x \in \Omega \text{ s.t. } u(x) = \sup_\Omega u \right\}. \tag{2.9.1}$$

We remark that $\overline{x} \in \mathcal{U}$ and thus $\mathcal{U} \neq \varnothing$. The continuity of u also gives that $\mathcal{U}$ is closed in Ω.

Moreover, if $r > 0$ is such that $B_r(\overline{x}) \Subset \Omega$, we let $H(x) := \Gamma(x - \overline{x}) - \Gamma(re_1)$, and we deduce from Green's representation formula (2.7.11) and the divergence theorem (recall also the setting of the fundamental solution in (2.7.2) and (2.7.6), as well as the measure theoretic identity in (1.1.6)) that

$$\begin{aligned} 0 &\leqslant \int_{B_r(\overline{x})} H(x)\,\Delta u(x)\,dx \\ &= \int_{B_r(\overline{x})} \Gamma(x-\overline{x})\,\Delta u(x)\,dx - \Gamma(re_1) \int_{B_r(\overline{x})} \Delta u(x)\,dx \\ &= \int_{\partial B_r(\overline{x})} \left(\Gamma(x-\overline{x}) \frac{\partial u}{\partial \nu}(x) - u(x) \frac{\partial \Gamma}{\partial \nu}(x-\overline{x}) \right) d\mathcal{H}^{n-1}_x \\ &\quad - u(\overline{x}) - \Gamma(re_1) \int_{\partial B_r(\overline{x})} \frac{\partial u}{\partial \nu}(x)\,d\mathcal{H}^{n-1}_x \\ &= \frac{r}{n\,|B_1|} \int_{\partial B_r(\overline{x})} \frac{u(x)}{|x-\overline{x}|^n}\,d\mathcal{H}^{n-1}_x - \sup_\Omega u \\ &= \fint_{\partial B_r(\overline{x})} u(x)\,d\mathcal{H}^{n-1}_x - \sup_\Omega u. \end{aligned} \tag{2.9.2}$$

Now, we take $\rho > 0$ such that $B_\rho(\overline{x}) \Subset \Omega$, and we claim that

$$\text{for every } p \in B_\rho(\overline{x}), \text{ it holds that } u(p) = \sup_\Omega u. \tag{2.9.3}$$

Indeed, suppose not. Then, we have that

$$\fint_{B_\rho(\overline{x})} u(x)\,dx < \sup_\Omega u.$$

Thus, using polar coordinates and (2.9.2),

$$\begin{aligned}
\sup_\Omega u &> \frac{1}{|B_\rho|}\int_0^\rho \left(\int_{\partial B_r(\overline{x})} u(x)\,d\mathcal{H}_x^{n-1}\right) dr \\
&\geqslant \frac{\mathcal{H}^{n-1}(\partial B_1)}{|B_1|\rho^n}\int_0^\rho \left(r^{n-1}\sup_\Omega u\right) dr \\
&= \frac{\mathcal{H}^{n-1}(\partial B_1)}{n\,|B_1|}\sup_\Omega u \\
&= \sup_\Omega u.
\end{aligned}$$

This contradiction proves (2.9.3).

As a consequence of (2.9.3), we have that $\mathcal{U}$ is open. Then, by the connectedness of Ω, it follows that $\mathcal{U} = \Omega$ and consequently $u(x) = \sup_\Omega u$ for all $x \in \Omega$. □

A more general version of this result, without assuming the function u to be smooth, will be given in Lemma 3.2.5.

An interesting consequence of the strong maximum principle in Theorem 2.9.1 is the observation, originally due to Samuel Earnshaw, according to which

$$\text{a collection of point charges cannot be maintained in a stable stationary electrostatic equilibrium.} \tag{2.9.4}$$

Of course, if we place a positive charge at e_1 and another at $-e_1$, a positive charge at the origin would be in equilibrium (by symmetry), but this would be an unstable equilibrium, since a small perturbation of the position of the charge at the origin in the e_2 direction would make it drift away.

Given a collection of point charges in $\mathbb{R}^n$, with $n \geqslant 2$, the corresponding harmonic electrostatic potential u would generate a force field $F = -\nabla u$ (recall the discussion on electromagnetic fields in [DV23, Chapter 18]). The equilibria correspond to the zeros of F, and the notion of stability adopted here means that if we move the particle slightly away from the equilibrium, the electrostatic force will tend to bring it back to the equilibrium.

Hence, if, say, the origin were a stable equilibrium in the above setting, then $F(0) = 0$, and if x is close enough to the origin, then $F(x)\cdot x \leqslant 0$; that is, $\nabla u(0) = 0$, and if x is close enough to the origin, $\nabla u(x)\cdot x \geqslant 0$. In particular,

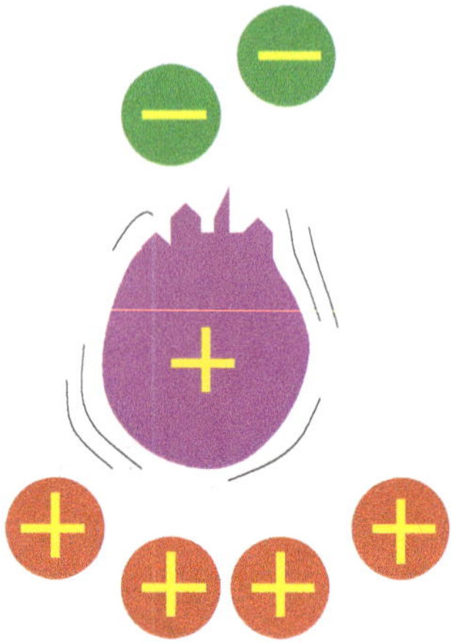

Figure 2.9.1 Electromagnetic levitation, after René Magritte's the Castle of the Pyrenees.

there exists $\rho > 0$ such that if $x \in B_\rho \setminus \{0\}$ and $\omega := \frac{x}{|x|}$, then

$$u(x) - u(0) = u(|x|\omega) - u(0) = \int_0^{|x|} \frac{d}{dr} u(r\omega)\, dr = \int_0^{|x|} \nabla u(r\omega) \cdot \omega\, dr$$
$$= \frac{1}{r} \int_0^{|x|} \nabla u(r\omega) \cdot (r\omega)\, dr \geqslant 0.$$

In particular, a stable equilibrium at the origin would produce a local minimum for u, which would violate Theorem 2.9.1(iii). This proves Earnshaw's theorem in (2.9.4).

Actually, Theorem 2.9.1(iii) states, more generally, that the electrostatic potential cannot have a maximum or a minimum value at any point in space not occupied by an electric charge (saddle points are instead possible and would correspond to unstable equilibria, as discussed above, see Figure 2.9.2).

We stress that the notion of stability used here is the one formulated by Maxwell in [Max54], according to which an equilibrium is stable if the field at points close to the equilibrium pushes back toward the equilibrium itself. This notion of stability is not the same as Lyapunov stability (but a version of Earnshaw's theorem also holds in terms of Lyapunov stability, see [Rau14, Theorem 1.2]).

An interesting consequence of Earnshaw's theorem is that it is not possible to achieve a stable static levitation by only using a combination of fixed magnets and electric charges, see Figure 2.9.1 (and note that including a constant gravitation field would just modify the potential by an additional harmonic term gx_n,

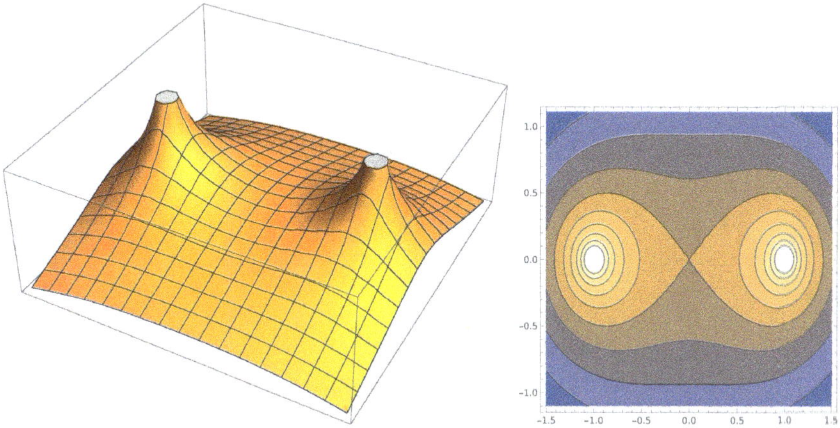

Figure 2.9.2 Electrostatic potential generated by two point charges in the plane located at e_1 and $-e_1$, corresponding to $-\frac{1}{2\pi}(\ln|x-e_1|+\ln|x+e_1|)$, and the plot of the corresponding level sets. Note that the origin is a saddle point.

with g being the constant gravitational acceleration). Also, if one wants to trap charged particles by only using an electromagnetic field, the particles should necessarily be dynamically kept in motion (with no stable rest equilibrium being allowed), which is one of the inspiring principles for the construction of a tokamak, see Figure 2.9.3 (and recall [DV23, Chapter 11] for some basic notions of plasma models).

See [Wei76] and the references therein for more information on Earnshaw's theorem.

As a byproduct of the strong maximum principle in Theorem 2.9.1, we have a weak maximum principle, as follows.

Corollary 2.9.2. *Let $\Omega \subseteq \mathbb{R}^n$ be open, bounded and connected, and let $u \in C^2(\Omega)\cap C(\overline{\Omega})$.*

(i) *If $\Delta u \geqslant 0$ in Ω, then $\sup_\Omega u = \sup_{\partial\Omega} u$.*
(ii) *If $\Delta u \leqslant 0$ in Ω, then $\inf_\Omega u = \inf_{\partial\Omega} u$.*
(iii) *If u is harmonic in Ω, then, for every $x \in \Omega$,*

$$\inf_{\partial\Omega} u \leqslant u(x) \leqslant \sup_{\partial\Omega} u.$$

Proof. Since claim (ii) follows from (i) up to changing a sign in u and claim (iii) is a consequence of (i) and (ii), we focus on the proof of (i). To this end, we consider a point $p \in \overline{\Omega}$ such that $u(p) = \max_{\overline{\Omega}} u$. If $p \in \partial\Omega$, then the claim in (i) plainly follows. Hence, we can assume that $p \in \Omega$. In this way, we have

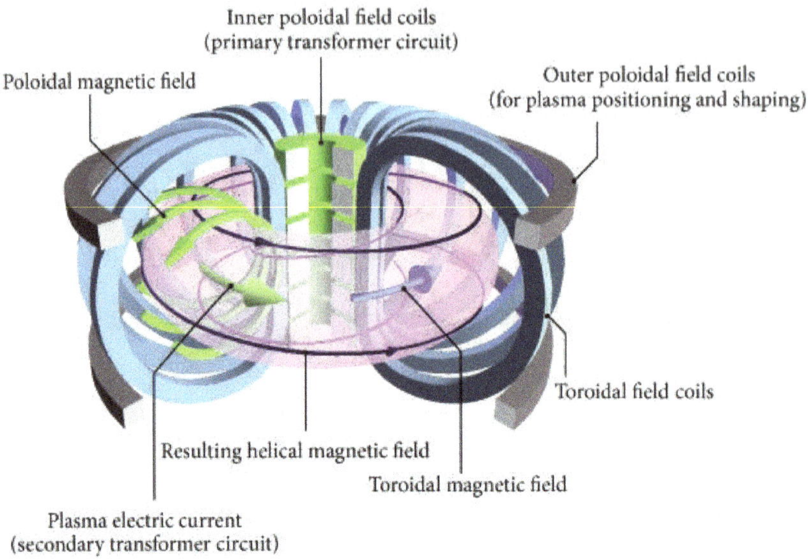

Figure 2.9.3 Schematic diagram of a tokamak chamber.
Source: Image from Wikipedia, taken from [LJRX14], licensed under the Creative Commons Attribution-Share Alike 4.0 International license.

that $\sup_\Omega u = u(p)$, and accordingly, in light of Theorem 2.9.1, we have that u is constant in Ω, and this gives claim (i) as desired. □

A sharper version of Corollary 2.9.2 not relying on smoothness assumptions on u will be presented later on in Corollary 3.2.6.

A useful uniqueness result follows straightforwardly from the weak maximum principle in Corollary 2.9.2:

Corollary 2.9.3. *Let $\Omega \subseteq \mathbb{R}^n$ be open, bounded and connected, and let $u, v \in C^2(\Omega) \cap C(\overline{\Omega})$.*

Assume that $\Delta u = \Delta v$ in Ω and that $u = v$ on $\partial\Omega$. Then, $u(x) = v(x)$ for every $x \in \Omega$.

Proof. The function $w := u - v$ is harmonic in Ω and vanishes along $\partial\Omega$. The desired result then plainly follows from Corollary 2.9.2(iii). □

In general, the maximum principle fails in the case of unbounded domains. For instance, for all $c_1, c_2, c_3 \in \mathbb{R}$, the function $u(x) := c_1 x_n + c_2 x_{n-1} x_n + c_3(x_n^3 - 3x_{n-1}^2 x_n)$ is harmonic in the halfspace $\{x_n > 0\}$ and vanishes along $\{x_n = 0\}$, without being identically zero (unless $c_1 = c_2 = c_3 = 0$). Similarly, for all $k \in \mathbb{N}$, $k \geqslant 1$, the function $u(x) := e^{kx_{n-1}} \sin(kx_n)$ is harmonic in the slab $\{x_n \in (0, \pi)\}$ and vanishes on the boundary of the slab, without being identically zero. In addition, in the external domain $\mathbb{R}^n \setminus B_1$, the function $u(x) := \Gamma(e_1) - \Gamma(x)$ is harmonic with $u = 0$ on $\partial(\mathbb{R}^n \setminus B_1)$, but u is not identically zero (here, Γ is the fundamental solution presented in Section 2.7).

However, there are suitable assumptions on the solution and on the domain that guarantee the validity of a suitable maximum principle for unbounded regions as well, see e.g. [BNV94, Cab95, BCN96, CV02] and the references therein (also, some of the maximum principles for unbounded domains can be related to the so-called Phragmén–Lindelöf principle, see e.g. [Ahl37]). Here, we only recall a prototypical situation, as follows.

Proposition 2.9.4. *Let Ω be a halfspace. Let $u \in C^2(\Omega)$ be uniformly continuous, bounded from above and harmonic in Ω, with $u \leqslant 0$ on $\partial\Omega$.*

Then, $u \leqslant 0$ in Ω.

Proof. Up to a rigid motion, we can suppose that

$$\Omega = \{x_n > 0\}. \tag{2.9.5}$$

Given $\varepsilon > 0$, we define $v(x) := u(x) - \varepsilon x_n$. We observe that $v \leqslant u$ in $\overline{\Omega}$, thanks to (2.9.5), and therefore

$$v \leqslant 0 \text{ on } \partial\Omega = \{x_n = 0\}. \tag{2.9.6}$$

Now, we take σ to be the supremum of v. Let also $p_k \in \Omega$ be a sequence such that $v(p_k) \to \sigma$ as $k \to +\infty$. We write $p_k = (p'_k, p_{k,n}) \in \mathbb{R}^{n-1} \times [0, +\infty)$, and up to disregarding a finite number of indices, we can assume that $v(p_k) \geqslant \sigma - 1$.

Then, we have that $\sigma - 1 \leqslant u(p_k) - \varepsilon p_{k,n} \leqslant \sup_\Omega u - \varepsilon p_{k,n}$, from which it follows that $p_{k,n}$ is a bounded sequence. As a result, up to a subsequence, we assume that $p_{k,n} \to q$ for some $q \in [0, +\infty)$ as $k \to +\infty$.

Thus, we define $v_k(x) := v(x' + p'_k, x_n)$. Additionally, v_k is uniformly equicontinuous, thanks to our uniform continuity assumption on u. Moreover,

$$\begin{aligned} |v_k(0, q)| &\leqslant |v_k(0, p_{k,n}) - v_k(0, q)| + |v_k(0, p_{k,n})| \\ &\leqslant |u(p'_k, p_{k,n}) - u(p'_k, q)| + \varepsilon|p_{k,n} - q| + |v(p_k)| \leqslant 1 + \varepsilon + \sigma + 1, \end{aligned}$$

for large k, which, together with the equicontinuity, gives that v_k is locally equibounded. Consequently, according to the Arzelà–Ascoli theorem, up to a subsequence, we can assume that v_k locally uniformly converges to some v_∞ as $k \to +\infty$.

As a consequence,

$$\sigma \geqslant v_\infty(0,q) = \lim_{k\to+\infty} v_k(0,q) \geqslant \lim_{k\to+\infty} \big(v_k(0,p_{k,n}) - |v_k(0,q) - v_k(0,p_{k,n})|\big)$$
$$= \lim_{k\to+\infty} v(p'_k, p_{k,n}) - 0 = \sigma,$$

that is,

$$v_\infty(0,q) = \max_{\overline{\Omega}} v_\infty = \sigma. \tag{2.9.7}$$

Furthermore, by (2.9.6), for all $x' \in \mathbb{R}^{n-1}$,

$$v_\infty(x',0) = \lim_{k\to+\infty} v(x'+p'_k, 0) \leqslant 0. \tag{2.9.8}$$

Moreover, for every $\varphi \in C_0^\infty(\Omega)$,

$$\int_\Omega v_\infty(x)\,\Delta\varphi(x)\,dx = \lim_{k\to+\infty} \int_\Omega v_k(x)\,\Delta\varphi(x)\,dx = 0,$$

and thus v_∞ is harmonic in Ω due to Weyl's lemma in (2.2.1).

From this, (2.9.7) and the maximum principle in Theorem 2.9.1(iii), it follows that either

$$v_\infty(x) = \sigma \text{ for all } x \in \Omega \tag{2.9.9}$$

or $(0,q) \in \partial\Omega = \{x_n = 0\}$, and therefore

$$q = 0. \tag{2.9.10}$$

In any case, from either (2.9.9) or (2.9.10), it follows that $v_\infty(0,q) = v_\infty(0,0)$. Consequently, by (2.9.7) and (2.9.8), we deduce that $\sigma = v_\infty(0,q) = v_\infty(0,0) \leqslant 0$.

Therefore, we find that $u(x) \leqslant \varepsilon x_n$ for every $x \in \Omega$, and thus the desired result follows by sending $\varepsilon \searrow 0$. □

For further reference, we also point out a maximum principle in the Sobolev space setting (for the sake of simplicity, we focus on a concrete result, though the method of proof is more general, see e.g. [GT01, Theorem 8.1]). As customary, here and in the following, we denote by $\mathcal{D}^{1,2}(\Omega)$ the closure of $C_0^\infty(\Omega)$ with respect to the seminorm $[u]_{H^1(\Omega)} := \sqrt{\int_\Omega |\nabla u(x)|^2\,dx}$.

Lemma 2.9.5. *Let $n \geqslant 3$ and $\mathcal{U} \subset \mathbb{R}^n$ be open and with uniformly Lipschitz boundary. Let $u \in C^2(\mathcal{U}) \cap \mathcal{D}^{1,2}(\mathbb{R}^n)$. Suppose that u is harmonic in $\mathcal{U}$ and that $u \geqslant 0$ on $\partial\mathcal{U}$ (in the trace sense). Suppose also that $|\mathcal{U}| = +\infty$.*

Then, $u \geqslant 0$ a.e. in $\mathcal{U}$.

Proof. Let $v := \max\{-u, 0\}$. Since v belongs to $\mathcal{D}^{1,2}(\mathbb{R}^n)$, we take a sequence of functions $v_j \in C_0^\infty(\mathbb{R}^n)$ such that $\nabla v_j \to \nabla v$ in $L^2(\mathbb{R}^n)$. Thus, using the fact that u is harmonic and the Green's first identity in (2.1.10), we see that

$$\int_{\mathcal{U}} \nabla u(x) \cdot \nabla v_j(x)\, dx = 0$$

and therefore, sending $j \to +\infty$,

$$\int_{\mathcal{U}} \nabla u(x) \cdot \nabla v(x)\, dx = 0.$$

As a consequence, since $\nabla v = -\nabla u \chi_{\{u<0\}}$ (see e.g. [LL01, Corollary 6.18]),

$$0 = -\int_{\mathcal{U}\cap\{u<0\}} |\nabla u(x)|^2\, dx = -\int_{\mathcal{U}} |\nabla v(x)|^2\, dx.$$

This gives that v is necessarily constant in $\mathcal{U}$ (see e.g. [LL01, Theorem 6.11]), say $v(x) = c$ a.e. $x \in \mathcal{U}$, for some $c \in \mathbb{R}$.

Hence, using the fact that $u \in L^{\frac{2n}{n-2}}(\mathbb{R}^n)$ (thanks to the Gagliardo–Nirenberg–Sobolev inequality, see e.g. [Leo09, Theorem 12.4]),

$$+\infty > \int_{\mathcal{U}} |u(x)|^{\frac{2n}{n-2}}\, dx \geqslant \int_{\mathcal{U}} |v(x)|^{\frac{2n}{n-2}}\, dx = \int_{\mathcal{U}} |c|^{\frac{2n}{n-2}}\, dx.$$

Since $\mathcal{U}$ has infinite measure, we thus infer that $c = 0$.

Recapitulating, we have proven that $v = \max\{-u, 0\} = 0$ a.e. in $\mathcal{U}$, from which we obtain the desired result. □

2.10 The Green Function

Let $n \geqslant 2$. Given an open set $\Omega \subset \mathbb{R}^n$ with C^1 boundary and a point $x_0 \in \Omega$, we consider the fundamental solution Γ presented in Theorem 2.7.3 and set $\Gamma^{(x_0)}(x) := \Gamma(x - x_0)$. We also consider a function $\Psi^{(x_0)} \in C^2(\Omega) \cap C(\overline{\Omega})$ that satisfies

$$\begin{cases} \Delta \Psi^{(x_0)} = 0 & \text{in } \Omega, \\ \Psi^{(x_0)} = \Gamma^{(x_0)} & \text{on } \partial\Omega. \end{cases} \tag{2.10.1}$$

In the literature, the function $\Psi^{(x_0)}$ is sometimes called the Robin function.

We stress that $\Gamma^{(x_0)}$ is always finite on $\partial\Omega$, and hence the right-hand side of the boundary condition in (2.10.1) is well defined. Also, if Ω is bounded, then the solution to (2.10.1) is unique due to the maximum principle in Corollary 2.9.3.

When Ω is unbounded, the solution to (2.10.1) is not unique: for instance, if $\Omega = \{x_n > 0\}$ and $\Psi^{(x_0)}$ is a solution of (2.10.1), then, for example, so is $\Psi^{(x_0)} + c_1 x_n + c_2 x_{n-1} x_n + c_3(x_n^3 - 3x_{n-1}^2 x_n)$, for all $c_1, c_2, c_3 \in \mathbb{R}$.

In our setting, the function $\Psi^{(x_0)}$ acts as a "corrector for the fundamental solution": namely, for all $x \in \mathbb{R}^n \setminus \{x_0\}$, we define

$$G(x, x_0) := \Gamma(x - x_0) - \Psi^{(x_0)}(x). \tag{2.10.2}$$

The function G is called the Green function of Ω.

In the following lemma, we provide a symmetry result for the Green function.

Lemma 2.10.1. *For every $x, y \in \mathbb{R}^n$ with $x \neq y$,*

$$G(x, y) = G(y, x).$$

Proof. Let $a, b \in \Omega$ with $a \neq b$. Let $\rho > 0$ sufficiently small such that $B_\rho(a) \cup B_\rho(b) \Subset \Omega$ and $B_\rho(a) \cap B_\rho(b) = \varnothing$.

Let

$$\begin{aligned} \Omega_\rho &:= \Omega \setminus (B_\rho(a) \cup B_\rho(b)), \\ \alpha(x) &:= G(x, a) \\ \text{and} \quad \beta(x) &:= G(x, b). \end{aligned}$$

Note that, if $x \in \partial\Omega$, then $\alpha(x) = \Gamma(x - a) - \Psi^{(a)}(x) = \Gamma(x - a) - \Gamma^{(a)}(x) = 0$, and similarly $\beta(x) = 0$.

Thus, making use of Green's second identity (2.1.11), we have

$$\begin{aligned} &\int_{\Omega_\rho} \Big(\alpha(x)\Delta\beta(x) - \beta(x)\Delta\alpha(x)\Big)\,dx \\ &= \int_{\partial\Omega_\rho} \left(\alpha(x)\frac{\partial\beta}{\partial\nu}(x) - \beta(x)\frac{\partial\alpha}{\partial\nu}(x)\right) d\mathcal{H}^{n-1}_x \\ &= \int_{\partial B_\rho(a)} \left(\beta(x)\frac{\partial\alpha}{\partial\nu}(x) - \alpha(x)\frac{\partial\beta}{\partial\nu}(x)\right) d\mathcal{H}^{n-1}_x \\ &\quad + \int_{\partial B_\rho(b)} \left(\beta(x)\frac{\partial\alpha}{\partial\nu}(x) - \alpha(x)\frac{\partial\beta}{\partial\nu}(x)\right) d\mathcal{H}^{n-1}_x. \end{aligned}$$

Since in Ω_ρ it holds that $\Delta\alpha = \Delta\Gamma^{(a)} - \Delta\Psi^{(a)}(x) = 0$ and similarly $\Delta\beta = 0$, we conclude that

$$\begin{aligned} 0 &= \int_{\partial B_\rho(a)} \left(\beta(x)\frac{\partial\alpha}{\partial\nu}(x) - \alpha(x)\frac{\partial\beta}{\partial\nu}(x)\right) d\mathcal{H}^{n-1}_x \\ &\quad + \int_{\partial B_\rho(b)} \left(\beta(x)\frac{\partial\alpha}{\partial\nu}(x) - \alpha(x)\frac{\partial\beta}{\partial\nu}(x)\right) d\mathcal{H}^{n-1}_x. \end{aligned} \tag{2.10.3}$$

In addition,

$$\begin{aligned}\lim_{\rho\searrow 0} \fint_{\partial B_\rho(a)} \beta(x)\frac{\partial}{\partial\nu}\Psi^{(a)}(x)\, d\mathcal{H}^{n-1}_x &= \beta(a)\frac{\partial}{\partial\nu}\Psi^{(a)}(a)\\ &= \big(\Gamma(a-b) - \Psi^{(b)}(a)\big)\Psi^{(a)}(a)\end{aligned}$$

and

$$\begin{aligned}\lim_{\rho\searrow 0} \fint_{\partial B_\rho(a)} \Psi^{(a)}(x)\frac{\partial\beta}{\partial\nu}(x)\, d\mathcal{H}^{n-1}_x &= \Psi^{(a)}(a)\frac{\partial\beta}{\partial\nu}(a)\\ &= \Psi^{(a)}(a)\frac{\partial\Gamma}{\partial\nu}(a-b) - \Psi^{(a)}(a)\frac{\partial\Psi^{(b)}}{\partial\nu}(a),\end{aligned}$$

whence

$$\lim_{\rho\searrow 0} \int_{\partial B_\rho(a)} \beta(x)\frac{\partial}{\partial\nu}\Psi^{(a)}(x)\, d\mathcal{H}^{n-1}_x = 0$$

and

$$\lim_{\rho\searrow 0} \int_{\partial B_\rho(a)} \Psi^{(a)}(x)\frac{\partial\beta}{\partial\nu}(x)\, d\mathcal{H}^{n-1}_x = 0.$$

For this reason,

$$\begin{aligned}&\lim_{\rho\searrow 0} \int_{\partial B_\rho(a)} \left(\beta(x)\frac{\partial\alpha}{\partial\nu}(x) - \alpha(x)\frac{\partial\beta}{\partial\nu}(x)\right) d\mathcal{H}^{n-1}_x\\ &\quad= \lim_{\rho\searrow 0} \int_{\partial B_\rho(a)} \Bigg(\beta(x)\frac{\partial}{\partial\nu}\big(\Gamma(x-a) - \Psi^{(a)}(x)\big)\\ &\quad\quad - \big(\Gamma(x-a) - \Psi^{(a)}(x)\big)\frac{\partial\beta}{\partial\nu}(x)\Bigg) d\mathcal{H}^{n-1}_x\\ &\quad= \lim_{\rho\searrow 0} \int_{\partial B_\rho(a)} \left(\beta(x)\frac{\partial\Gamma}{\partial\nu}(x-a) - \Gamma(x-a)\frac{\partial\beta}{\partial\nu}(x)\right) d\mathcal{H}^{n-1}_x. \qquad (2.10.4)\end{aligned}$$

Now, since, by (2.7.6),

$$\int_{\partial B_\rho(a)} \Gamma(x-a)\frac{\partial\beta}{\partial\nu}(x)\, d\mathcal{H}^{n-1}_x = \begin{cases} \dfrac{c_n}{\rho^{n-2}} \displaystyle\int_{\partial B_\rho(a)} \frac{\partial\beta}{\partial\nu}(x)\, d\mathcal{H}^{n-1}_x & \text{if } n \neq 2,\\ -c_n \ln\rho \displaystyle\int_{\partial B_\rho(a)} \frac{\partial\beta}{\partial\nu}(x)\, d\mathcal{H}^{n-1}_x & \text{if } n = 2,\end{cases}$$

we find that

$$\lim_{\rho\searrow 0} \int_{\partial B_\rho(a)} \Gamma(x-a)\frac{\partial\beta}{\partial\nu}(x)\, d\mathcal{H}^{n-1}_x = 0. \qquad (2.10.5)$$

Besides,

$$\begin{aligned}&\lim_{\rho\searrow 0}\int_{\partial B_\rho(a)}\beta(x)\frac{\partial\Gamma}{\partial\nu}(x-a)\,d\mathcal{H}^{n-1}_x\\&=\lim_{\rho\searrow 0}\frac{\widetilde{c}_n}{\rho^{n-1}}\int_{\partial B_\rho(a)}\beta(x)\,d\mathcal{H}^{n-1}_x=C_n\,\beta(a)\end{aligned}$$

for some $\widetilde{c}_n$, $C_n\in\mathbb{R}$.

Plugging this information and (2.10.5) into (2.10.4), we find that

$$\lim_{\rho\searrow 0}\int_{\partial B_\rho(a)}\left(\beta(x)\frac{\partial\alpha}{\partial\nu}(x)-\alpha(x)\frac{\partial\beta}{\partial\nu}(x)\right)\,d\mathcal{H}^{n-1}_x=C_n\,\beta(a). \tag{2.10.6}$$

Exchanging the roles of a and b (and of α and β), we also have that

$$\lim_{\rho\searrow 0}\int_{\partial B_\rho(b)}\left(\alpha(x)\frac{\partial\beta}{\partial\nu}(x)-\beta(x)\frac{\partial\alpha}{\partial\nu}(x)\right)\,d\mathcal{H}^{n-1}_x=C_n\,\alpha(b).$$

From this and (2.10.6), we deduce from (2.10.3) that $\alpha(b)=\beta(a)$. That is,

$$G(b,a)=\alpha(b)=\beta(a)=G(a,b).$$ □

The importance of the Green function in the theory of partial differential equations lies in the possibility of reconstructing a solution from its boundary value. Specifically, we already know from Green's representation formula in (2.7.11) how to reconstruct a function from its Laplacian and the boundary values of the function *and of its normal derivative*: the Green function allows us to "minimize the necessary information" reconstructing the function merely from its Laplacian and the boundary values (without the necessity of knowing additionally the values of the normal derivative at the boundary). Indeed, we have the following.

Theorem 2.10.2. *Let Ω be a bounded open set in $\mathbb{R}^n$ with C^1 boundary. Then, for every $u\in C^2(\Omega)\cap C^1(\overline{\Omega})$,*

$$u(x)=-\int_\Omega G(x,y)\,\Delta u(y)\,dy-\int_{\partial\Omega}u(y)\,\frac{\partial G}{\partial\nu}(x,y)\,d\mathcal{H}^{n-1}_y,$$

where, for every $x\in\Omega$ and $y\in\partial\Omega$,

$$\frac{\partial G}{\partial\nu}(x,y):=\nabla_y G(x,y)\cdot\nu(y)=\sum_{i=1}^n\frac{\partial G}{\partial y_i}(x,y)\,\nu(y)\cdot e_i. \tag{2.10.7}$$

Proof. We note that $\Psi^{(x)}(y) = \Psi^{(y)}(x)$ due to the symmetry property of the Green function in Lemma 2.10.1, one of the fundamental solution in (2.7.6) and the setting in (2.10.2). As a consequence, in the notation of (2.10.7), for all $x \in \Omega$ and $y \in \partial\Omega$,

$$\begin{aligned}
\frac{\partial \Psi^{(x)}}{\partial \nu}(y) &= \sum_{i=1}^{n} \frac{\partial}{\partial y_i}\big(\Psi^{(x)}(y)\big)\, \nu(y)\cdot e_i = \sum_{i=1}^{n} \frac{\partial}{\partial y_i}\big(\Psi^{(y)}(x)\big)\, \nu(y)\cdot e_i \\
&= \sum_{i=1}^{n} \frac{\partial}{\partial y_i}\big(\Gamma(x-y) - G(x,y)\big)\, \nu(y)\cdot e_i \\
&= -\nabla\Gamma(x-y)\cdot \nu(y) - \sum_{i=1}^{n} \frac{\partial}{\partial y_i} G(x,y)\, \nu(y)\cdot e_i \\
&= -\frac{\partial \Gamma}{\partial \nu}(x-y) - \frac{\partial G}{\partial \nu}(x,y).
\end{aligned}$$

Thus, we use Green's representation formula (2.7.11) combined with (2.10.1), recalling also Green's second identity (2.1.11), to see that

$$\begin{aligned}
&\int_{\Omega} G(x,y)\, \Delta u(y)\, dy + \int_{\partial\Omega} u(y)\, \frac{\partial G}{\partial \nu}(x,y)\, d\mathcal{H}^{n-1}_y \\
&= \int_{\Omega} \Gamma(x-y)\, \Delta u(y)\, dy - \int_{\Omega} \Psi^{(y)}(x)\, \Delta u(y)\, dy \\
&\quad - \int_{\partial\Omega} u(y)\, \frac{\partial \Psi^{(x)}}{\partial \nu}(y)\, d\mathcal{H}^{n-1}_y - \int_{\partial\Omega} u(y)\, \frac{\partial \Gamma}{\partial \nu}(x-y)\, d\mathcal{H}^{n-1}_y \\
&= \int_{\Omega} \Gamma(x-y)\, \Delta u(y)\, dy - \int_{\partial\Omega} \Psi^{(x)}(y)\, \frac{\partial u}{\partial \nu}(y)\, d\mathcal{H}^{n-1}_y \\
&\quad - \int_{\partial\Omega} u(y)\, \frac{\partial \Gamma}{\partial \nu}(x-y)\, d\mathcal{H}^{n-1}_y \\
&= \int_{\Omega} \Gamma(y-x)\, \Delta u(y)\, dy - \int_{\partial\Omega} \Gamma(y-x)\, \frac{\partial u}{\partial \nu}(y)\, d\mathcal{H}^{n-1}_y \\
&\quad + \int_{\partial\Omega} u(y)\, \frac{\partial \Gamma}{\partial \nu}(y-x)\, d\mathcal{H}^{n-1}_y \\
&= -u(x),
\end{aligned}$$

as desired. □

A straightforward consequence of Theorem 2.10.2 is a representation result for the solutions of the boundary value problem

$$\begin{cases} -\Delta u = f & \text{in } \Omega, \\ u = g & \text{on } \partial\Omega, \end{cases} \tag{2.10.8}$$

as detailed[15] in the following result.

Corollary 2.10.3. *Let Ω be a bounded open set in $\mathbb{R}^n$ with C^1 boundary. If $u \in C^2(\Omega) \cap C^1(\overline{\Omega})$ solves* (2.10.8) *for some $f : \Omega \to \mathbb{R}$ and $g : \partial\Omega \to \mathbb{R}$, then, for every $x \in \Omega$,*

$$u(x) = \int_\Omega f(y)\, G(x,y)\, dy - \int_{\partial\Omega} g(y)\, \frac{\partial G}{\partial \nu}(x,y)\, d\mathcal{H}^{n-1}_y,$$

where the notation in (2.10.7) *has been used.*

Given the result in Corollary 2.10.3, to obtain an explicit expression of the solution of the boundary value problem (2.10.8), it would be desirable to know an explicit expression of the Green function G. Unfortunately, this is rarely possible in concrete cases; nonetheless, there are at least two interesting situations in which the Green function can be written in terms of elementary functions. These two examples are given by the cases in which the domain is either a ball or a halfspace, as discussed in the forthcoming Theorems 2.10.4 and 2.10.5.

[15] In particular, by Corollary 2.10.3, the solutions of

$$\begin{cases} -\Delta u = f & \text{in } \Omega, \\ u = 0 & \text{on } \partial\Omega, \end{cases}$$

take the form

$$u(x) = \int_\Omega f(y)\, G(x,y)\, dy.$$

This can be seen as the continuous counterpart of [DV23, equation (31.22)], which was obtained in the discrete setting.

Theorem 2.10.4. *The Green function*[16] *of the ball B_R is*

$$G(x,x_0) = \begin{cases} \Gamma(x-x_0) - \Gamma\left(\frac{|x_0|}{R}\left(x - \frac{R^2 x_0}{|x_0|^2}\right)\right) & \text{if } x_0 \in B_R \setminus \{0\}, \\ \Gamma(x) - \Gamma(Re_1) & \text{if } x_0 = 0. \end{cases} \tag{2.10.9}$$

Proof. We remark that the function defined in (2.10.9) when $x_0 \in B_R \setminus \{0\}$ is continuously extended to $x_0 = 0$ since

$$\begin{aligned} \frac{|x_0|}{R}\left|x - \frac{R^2 x_0}{|x_0|^2}\right| &= \frac{|x_0|}{R}\frac{\left||x_0|^2 x - R^2 x_0\right|}{|x_0|^2} = R\left|\frac{|x_0|x}{R^2} - \frac{x_0}{|x_0|}\right| \\ &= R\sqrt{1 - \frac{2x\cdot x_0}{R^2} + \frac{|x_0|^2|x|^2}{R^4}} = R(1 + O(|x_0|)). \end{aligned}$$

Thus, to prove the claim in Theorem 2.10.4, we consider

$$\Psi^{(x_0)}(x) := \begin{cases} \Gamma\left(\frac{|x_0|}{R}\left(x - \frac{R^2 x_0}{|x_0|^2}\right)\right) & \text{if } x_0 \in B_R \setminus \{0\}, \\ \Gamma(Re_1) & \text{if } x_0 = 0, \end{cases}$$

and we need to show that

$$\Psi^{(x_0)} \text{ satisfies (2.10.1) with } \Omega := B_R. \tag{2.10.10}$$

---

[16]The physical intuition underlying Theorem 2.10.4 can be grasped by taking, up to scaling, $R := 1$ and by recalling the method of electrostatic image charges, as discussed in Section 2.8. Namely, by construction, we have that, in the setting of Theorem 2.10.4,

$$-\Delta G(x,x_0) = \Delta\Psi^{(x_0)}(x) - \Delta\Gamma(x-x_0) = \delta_{x_0}(x)$$

for all $x \in B_1$. In this sense, $G(x,x_0)$ corresponds, up to normalizing constants, to the potential of a unit electrostatic charge located at x_0. On the other hand, we know that, if $x \in \partial B_1$, then $G(x,x_0) = \Gamma(x-x_0) - \Psi^{(x_0)}(x) = 0$, whence ∂B_1 is an equipotential surface.

The determination of the Green function of B_1 is therefore reduced to that of an electrostatic potential generated by a unit charge in x_0 and an auxiliary charge outside B_1 that makes ∂B_1 an equipotential surface. We know from Section 2.8 (see in particular (2.8.8) and (2.8.9), comparing with the Kelvin transform in (2.6.1)), that to produce such an electrostatic potential (say, when $n \neq 2$), it suffices to place a charge of intensity $-\frac{1}{|x_0|^{n-2}}$ at the position $\mathcal{K}(x_0)$. Thus, the corresponding electrostatic potential takes the form

$$\frac{1}{|x-x_0|^{n-2}} - \frac{1}{|x_0|^{n-2}\,|x - \mathcal{K}(x_0)|^{n-2}},$$

which agrees with (2.10.9) up to normalizing constants, and this gives a concrete justification for the result in Theorem 2.10.4 (say, when $n \neq 2$, and the case of $n = 2$ is discussed in the footnote on p. 84). See also Figure 2.10.1 for graphical images of the Green function of a ball (to be compared also with Figure 2.10.2 that represents a Green function for a halfspace).

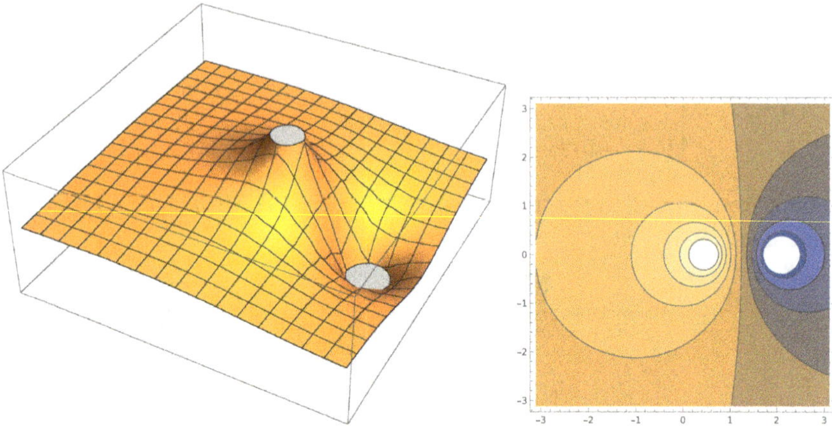

Figure 2.10.1 Plot of the Green function of the unit ball in dimension 2 when $x_0 = \left(\frac{1}{2}, 0\right)$, and its level sets. Note that ∂B_1 is a zero-level set.

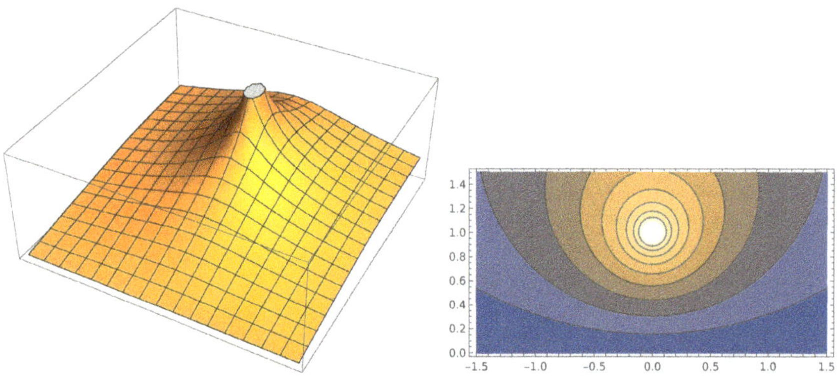

Figure 2.10.2 Plot of the Green function of the halfplane when $x_0 = (0, 1)$, and its level sets. Note that $\mathbb{R} \times \{0\}$ is a zero-level set.

As a matter of fact, since $\left|\frac{R^2 x_0}{|x_0|^2}\right| = \frac{R^2}{|x_0|} > R$, we have that $\Delta\Psi^{(x_0)}(x) = 0$ for every $x \in B_R$. Moreover, if $x \in \partial B_R$, recalling (2.6.1) and (2.6.6), we see that

$$\begin{aligned}\left|\frac{|x_0|}{R}\left(x - \frac{R^2 x_0}{|x_0|^2}\right)\right| &= |x_0|\left|\frac{x}{R} - \frac{R x_0}{|x_0|^2}\right| \\ &= |x_0|\left|\frac{x}{R} - \mathcal{K}\left(\frac{x_0}{R}\right)\right| = R\left|\frac{x}{R} - \frac{x_0}{R}\right| = |x - x_0|. \qquad (2.10.11)\end{aligned}$$

Therefore, for every $x \in \partial B_R$, we have that $\Psi^{(x_0)}(x) = \Gamma(x - x_0) = \Gamma^{(x_0)}(x)$, and these observations establish (2.10.10). $\square$

Theorem 2.10.5. *A Green function*[17] *of the halfspace* $\{x_n > 0\}$ *is*

$$G(x, x_0) = \Gamma(x - x_0) - \Gamma(\mathcal{R}(x) - x_0),$$

where $\mathcal{R}(x_1, \dots, x_{n-1}, x_n) = (x_1, \dots, x_{n-1}, -x_n)$.

Proof. We let $\Psi^{(x_0)}(x) := \Gamma(\mathcal{R}(x) - x_0)$, and we need to show that

$$\Psi^{(x_0)} \text{ satisfies (2.10.1) with } \Omega := \{x_n > 0\}. \tag{2.10.12}$$

For this, we observe that $|\mathcal{R}(x) - x_0| = |x - \mathcal{R}(x_0)|$, hence $\Psi^{(x_0)}(x) = \Gamma(x - \mathcal{R}(x_0))$ and consequently $\Delta\Psi^{(x_0)} = 0$ in $\{x_n > 0\}$. Furthermore, if $x_n = 0$, then $\mathcal{R}(x) = x$ and $\Psi^{(x_0)}(x) = \Gamma(x - x_0)$. These remarks prove the validity of (2.10.12). □

For an extensive treatment of Green functions, see [Duf01] and the references therein. We now recall an interesting relation about the surface area of the level sets of the Green function (we focus on the case of the Laplace operator, but the approach to the problem is rather general, see [Wei62] for full details).

Lemma 2.10.6. *Let* Ω *be a bounded and open subset of* $\mathbb{R}^n$*, and let* $x_0 \in \Omega$*. Let* G *be the Green function of* Ω*, and for every* $t \geqslant 0$*, set*

$$v(t) := \Big|\big\{x \in \Omega \text{ s.t. } G(x, x_0) > t\big\}\Big|. \tag{2.10.13}$$

Then,

$$\big(v(t)\big)^{\frac{2-n}{n}} - |\Omega|^{\frac{2-n}{n}} \geqslant n(n-2)|B_1|^{\frac{2}{n}} t.$$

Proof. Let $S(t) := \mathcal{H}^{n-1}\big(\big\{x \in \Omega \text{ s.t. } G(x, x_0) = t\big\}\big)$. By the Cauchy–Schwarz inequality,

$$\begin{aligned} S(t) &= \int_{\{G(\cdot, x_0) = t\}} d\mathcal{H}^{n-1}_x \\ &= \int_{\{G(\cdot, x_0) = t\}} \frac{\sqrt{|\nabla G(x, x_0)|}\, d\mathcal{H}^{n-1}_x}{\sqrt{|\nabla G(x, x_0)|}} \\ &\leqslant \sqrt{\int_{\{G(\cdot, x_0) = t\}} |\nabla G(x, x_0)|\, d\mathcal{H}^{n-1}_x \int_{\{G(\cdot, x_0) = t\}} \frac{d\mathcal{H}^{n-1}_x}{|\nabla G(x, x_0)|}}. \end{aligned} \tag{2.10.14}$$

[17] The physical intuition underlying Theorem 2.10.5 is that the method of electrostatic image charges, as described in Section 2.8 when the domain is a ball, dramatically simplifies when the domain is a halfspace, since in this case, it suffices to place an image charge with the same intensity in a symmetrical position with respect to the boundary of the halfspace.

Additionally, by the Coarea formula (see e.g. [EG15]),

$$
\begin{aligned}
v(t) &= \int_{\Omega} \frac{\chi_{(t,+\infty)}(G(x,x_0))\,|\nabla G(x,x_0)|}{|\nabla G(x,x_0)|}\,dx \\
&= \int_{\mathbb{R}} \left[\int_{\{G(\cdot,x_0)=s\}} \frac{\chi_{(t,+\infty)}(G(x,x_0))}{|\nabla G(x,x_0)|}\,d\mathcal{H}_x^{n-1}\right] ds \\
&= \int_t^{+\infty} \left[\int_{\{G(\cdot,x_0)=s\}} \frac{d\mathcal{H}_x^{n-1}}{|\nabla G(x,x_0)|}\right] ds
\end{aligned}
$$

and therefore, for a.e. $t \geqslant 0$,

$$
-v'(t) = \int_{\{G(\cdot,x_0)=t\}} \frac{d\mathcal{H}_x^{n-1}}{|\nabla G(x,x_0)|}.
$$

By combining this and (2.10.14), we obtain

$$
-v'(t) \geqslant \left(\int_{\{G(\cdot,x_0)=t\}} |\nabla G(x,x_0)|\,d\mathcal{H}_x^{n-1}\right)^{-1} \big(S(t)\big)^2. \tag{2.10.15}
$$

By the isoperimetric inequality (see e.g. [Giu84]),

$$
\frac{S(t)}{\big(v(t)\big)^{\frac{n-1}{n}}} \geqslant \frac{\mathcal{H}^{n-1}(\partial B_1)}{|B_1|^{\frac{n-1}{n}}},
$$

which, together with (2.10.15), gives that

$$
-v'(t) \geqslant \left(\int_{\{G(\cdot,x_0)=t\}} |\nabla G(x,x_0)|\,d\mathcal{H}_x^{n-1}\right)^{-1} \frac{\big(\mathcal{H}^{n-1}(\partial B_1)\big)^2}{|B_1|^{\frac{2(n-1)}{n}}} \big(v(t)\big)^{\frac{2(n-1)}{n}}. \tag{2.10.16}
$$

We stress that, a.e. $t \geqslant 0$, the set $\{G(\cdot,x_0) = t\}$ is a manifold of class C^1 which corresponds to the boundary of $\{G(\cdot,x_0) > t\}$ because $\nabla G(\cdot,x_0) \neq 0$ along such a set (as a consequence of the Morse–Sard theorem, see e.g. Theorem 4.3 in [Mor39]). As a consequence, a.e. $t \geqslant 0$, along $\{G(\cdot,x_0) = t\}$, we have that

$$
|\nabla G(\cdot,x_0)| = -\nabla G(\cdot,x_0)\cdot\nu(\cdot).
$$

Note that in this setting, $\nu(\cdot)$ denotes the exterior unit normal to the set $\{G(\cdot,x_0) > t\}$.

Also, since $G(x_0, x_0) = +\infty$, we have that $x_0 \in \{G(\cdot, x_0) > t\}$. Consequently, using the fact that G is harmonic in $\Omega \setminus \{x_0\} \supseteq \{G(\cdot, x_0) < t\}$,

$$\begin{aligned}
\int_{\{G(\cdot,x_0)=t\}} |\nabla G(x,x_0)|\, d\mathcal{H}_x^{n-1} &= -\int_{\{G(\cdot,x_0)=t\}} \nabla G(x,x_0)\cdot \nu(x)\, d\mathcal{H}_x^{n-1} \\
&= \int_{\partial\Omega} \nabla G(x,x_0)\cdot \nu(x)\, d\mathcal{H}_x^{n-1} - \int_{\{G(\cdot,x_0)=t\}} \nabla G(x,x_0)\cdot \nu(x)\, d\mathcal{H}_x^{n-1} \\
&\quad - \int_{\partial\Omega} \nabla G(x,x_0)\cdot \nu(x)\, d\mathcal{H}_x^{n-1} \\
&= \int_{\{G(\cdot,x_0)<t\}} \Delta G(x,x_0)\, d\mathcal{H}_x^{n-1} - \int_{\partial\Omega} \nabla G(x,x_0)\cdot \nu(x)\, d\mathcal{H}_x^{n-1} \\
&= -\int_{\partial\Omega} \nabla G(x,x_0)\cdot \nu(x)\, d\mathcal{H}_x^{n-1}.
\end{aligned}$$

Combining this with Theorem 2.10.2 (used here with $u := 1$), we conclude that

$$\int_{\{G(\cdot,x_0)=t\}} |\nabla G(x,x_0)|\, d\mathcal{H}_x^{n-1} = 1.$$

This and (2.10.16) lead to

$$-v'(t) \geqslant \frac{\big(\mathcal{H}^{n-1}(\partial B_1)\big)^2}{|B_1|^{\frac{2(n-1)}{n}}} \big(v(t)\big)^{\frac{2(n-1)}{n}}. \tag{2.10.17}$$

Since G is nonnegative by the maximum principle, we also know that

$$v(0) = \Big|\big\{x \in \Omega \text{ s.t. } G(x,x_0) > 0\big\}\Big| = |\Omega|.$$

From this and (2.10.17), it follows that, for every $T \geqslant 0$,

$$\begin{aligned}
\big(v(T)\big)^{\frac{2-n}{n}} - |\Omega|^{\frac{2-n}{n}} &= \big(v(T)\big)^{\frac{2-n}{n}} - \big(v(0)\big)^{\frac{2-n}{n}} = \int_0^T \frac{d}{dt}\big(v(t)\big)^{\frac{2-n}{n}}\, dt \\
&= -\frac{n-2}{n}\int_0^T \big(v(t)\big)^{\frac{2(1-n)}{n}} v'(t)\, dt \geqslant \frac{(n-2)\big(\mathcal{H}^{n-1}(\partial B_1)\big)^2 T}{n\,|B_1|^{\frac{2(n-1)}{n}}},
\end{aligned}$$

yielding the desired result in view of (1.1.6). □

In relation to Lemma 2.10.6, we also recall an interesting bound on the Robin function evaluated at the central point x_0 (as pointed out in equation (2.14) of [PS89]).

Lemma 2.10.7. *Let $n \geqslant 3$. Let Ω be a bounded and open subset of $\mathbb{R}^n$. and let $x_0 \in \Omega$. Then, the Robin function of Ω satisfies*[18]

$$(n-2)\Psi^{(x_0)}(x_0) \geqslant \frac{1}{n|B_1|^{\frac{2}{n}}|\Omega|^{\frac{n-2}{n}}}.$$

Proof. Without loss of generality, up to a translation, we suppose that $x_0 = 0$. By (2.7.2) and (2.7.6), we have that $\Gamma(x) = \frac{1}{n(n-2)\,|B_1|\,|x|^{n-2}}$; therefore, recalling (2.10.2),

$$\begin{aligned} |x| &= \frac{1}{\big(n(n-2)\,|B_1|\big)^{\frac{1}{n-2}}\,\big(\Gamma(x)\big)^{\frac{1}{n-2}}} \\ &= \frac{1}{\big(n(n-2)\,|B_1|\big)^{\frac{1}{n-2}}\,\big(G(x,0)+\Psi^{(0)}(x)\big)^{\frac{1}{n-2}}}. \end{aligned} \tag{2.10.18}$$

Let v be as in (2.10.13). We have, for a.e. $t > 0$,

$$\begin{aligned} v(t) &= \frac{1}{n}\int_{\{G(\cdot,0)>t\}} \operatorname{div}(x)\,dx \\ &= \frac{1}{n}\int_{\{G(\cdot,0)=t\}} x\cdot\nu(x)\,d\mathcal{H}^{n-1}_x \\ &\leqslant \frac{1}{n}\int_{\{G(\cdot,0)=t\}} |x|\,d\mathcal{H}^{n-1}_x \\ &= \frac{1}{n}\int_{\{G(\cdot,0)=t\}} \frac{d\mathcal{H}^{n-1}_x}{\big(n(n-2)\,|B_1|\big)^{\frac{1}{n-2}}\,\big(G(x,0)+\Psi^{(0)}(x)\big)^{\frac{1}{n-2}}} \\ &= \frac{1}{n}\int_{\{G(\cdot,0)=t\}} \frac{d\mathcal{H}^{n-1}_x}{\big(n(n-2)\,|B_1|\big)^{\frac{1}{n-2}}\,\big(t+\Psi^{(0)}(x)\big)^{\frac{1}{n-2}}}. \end{aligned}$$

We point out that, as $t \to +\infty$, the set $\{G(\cdot,0) \geqslant t\}$ shrinks toward the origin, and thus, if $x \in \{G(\cdot,0) = t\}$,

$$\big(t+\Psi^{(0)}(x)\big)^{\frac{1}{2-n}} = t^{\frac{1}{2-n}}\left(1+\frac{\Psi^{(0)}(x)}{t}\right)^{\frac{1}{2-n}} = t^{\frac{1}{2-n}}\left(1+\frac{1}{2-n}\,\frac{\Psi^{(0)}(x)}{t}+O\left(\frac{1}{t^2}\right)\right).$$

[18] We observe that the inequality in Lemma 2.10.7 is sharp since equality is attained when $\Omega = B_1$ and $x_0 = 0$; indeed, in this case, by (2.7.2), (2.7.6) and (2.10.9),

$$\begin{aligned} (n-2)\Psi^{(x_0)}(x_0) - \frac{1}{n|B_1|^{\frac{2}{n}}|\Omega|^{\frac{n-2}{n}}} &= (n-2)\Psi^{(0)}(0) - \frac{1}{n|B_1|} \\ &= (n-2)\Gamma(e_1) - \frac{1}{n|B_1|} = c_n(n-2) - \frac{1}{n|B_1|} = 0. \end{aligned}$$

Therefore,

$$v(t) \leqslant \frac{t^{\frac{1}{2-n}}}{n\big(n(n-2)\,|B_1|\big)^{\frac{1}{n-2}}} \int_{\{G(\cdot,0)=t\}} \left(1+\frac{\Psi^{(0)}(x)}{(2-n)t}+O\left(\frac{1}{t^2}\right)\right)\,d\mathcal{H}_x^{n-1}. \tag{2.10.19}$$

Now, we let

$$\rho_0 := \frac{1}{\big(n(n-2)\,|B_1|\big)^{\frac{1}{n-2}}}$$

and use the notation $\varepsilon := 1/t$. For every $(\rho,\omega,\varepsilon) \in [0,+\infty) \times (\partial B_1) \times [0,1]$, we consider the function

$$F(\rho,\omega,\varepsilon) := \rho - \frac{1}{\big(n(n-2)\,|B_1|\big)^{\frac{1}{n-2}} \left(1+\varepsilon\Psi^{(0)}\big(\varepsilon^{\frac{1}{n-2}}\rho\omega\big)\right)^{\frac{1}{n-2}}}, \tag{2.10.20}$$

and we note that $F(\rho_0,\omega,0)=0$. Since $\partial_\rho F(\rho,\omega,0)=1$, according to the implicit function theorem, we find a smooth function $\rho=\rho(\omega,\varepsilon)$ such that $\rho(\omega,\varepsilon) = \rho_0+\rho_1(\omega)\varepsilon+o(\varepsilon)$, for some $\rho_1 : \partial B_1 \to \mathbb{R}$, and $F\big(\rho(\omega,\varepsilon),\omega,\varepsilon\big)=0$, that is,

$$\rho(\omega,\varepsilon) = \frac{1}{\big(n(n-2)\,|B_1|\big)^{\frac{1}{n-2}} \left(1+\varepsilon\Psi^{(0)}\big(\varepsilon^{\frac{1}{n-2}}\rho(\omega,\varepsilon)\omega\big)\right)^{\frac{1}{n-2}}}. \tag{2.10.21}$$

We also observe that

$$\rho_1(\omega) = \rho_1 := -\frac{\rho_0\,\Psi^{(0)}(0)}{n-2}. \tag{2.10.22}$$

Indeed, by (2.10.21),

$$\begin{aligned}
&\rho_0+\rho_1(\omega)\varepsilon+o(\varepsilon)\\
&= \frac{1}{\big(n(n-2)\,|B_1|\big)^{\frac{1}{n-2}} \left(1+\varepsilon\Psi^{(0)}\big(\varepsilon^{\frac{1}{n-2}}(\rho_0+\rho_1(\omega)\varepsilon+o(\varepsilon))\omega\big)\right)^{\frac{1}{n-2}}}\\
&= \frac{\rho_0}{\left(1+\varepsilon\Psi^{(0)}\big(\varepsilon^{\frac{1}{n-2}}(\rho_0+\rho_1(\omega)\varepsilon+o(\varepsilon))\omega\big)\right)^{\frac{1}{n-2}}}\\
&= \frac{\rho_0}{\left(1+\varepsilon\Psi^{(0)}(0)+o(\varepsilon)\right)^{\frac{1}{n-2}}}\\
&= \rho_0\left(1-\frac{\varepsilon\,\Psi^{(0)}(0)}{n-2}+o(\varepsilon)\right),
\end{aligned}$$

from which the desired result in (2.10.22) follows.

Thus, we consider the map

$$\partial B_1 \ni \omega \mapsto \zeta(\omega,\varepsilon) := \varepsilon^{\frac{1}{n-2}}\rho(\omega,\varepsilon)\omega = \varepsilon^{\frac{1}{n-2}}(\rho_0+\rho_1\varepsilon+o(\varepsilon))\omega,$$

and we observe that, by (2.10.18),

$$\begin{aligned}\left(G(\zeta(\omega,\varepsilon),0)+\Psi^{(0)}(\zeta(\omega,\varepsilon))\right)^{\frac{1}{n-2}} &= \frac{1}{(n(n-2)\,|B_1|)^{\frac{1}{n-2}}\,|\zeta(\omega,\varepsilon)|}\\ &= \frac{1}{(n(n-2)\,|B_1|)^{\frac{1}{n-2}}\,\varepsilon^{\frac{1}{n-2}}\rho(\omega,\varepsilon)}\\ &= \frac{\left(1+\varepsilon\Psi^{(0)}\left(\varepsilon^{\frac{1}{n-2}}\rho(\omega,\varepsilon)\omega\right)\right)^{\frac{1}{n-2}}}{\varepsilon^{\frac{1}{n-2}}}\\ &= \left(\frac{1}{\varepsilon}+\Psi^{(0)}\left(\varepsilon^{\frac{1}{n-2}}\rho(\omega,\varepsilon)\omega\right)\right)^{\frac{1}{n-2}}\\ &= \left(\frac{1}{\varepsilon}+\Psi^{(0)}(\zeta(\omega,\varepsilon))\right)^{\frac{1}{n-2}}\end{aligned}$$

and accordingly $G(\zeta(\omega,\varepsilon),0) = 1/\varepsilon$. This states that

$$\{G(\cdot,0)=1/\varepsilon\} \supseteq \zeta(\partial B_1,\varepsilon). \tag{2.10.23}$$

We also claim that

$$\begin{aligned}&\text{if } x\in\{G(\cdot,0)=1/\varepsilon\}, \text{ then there exists } \omega\in\partial B_1\\ &\text{such that } x=\varepsilon^{\frac{1}{n-2}}\rho(\omega,\varepsilon)\omega.\end{aligned} \tag{2.10.24}$$

To prove this, we let $x \in \{G(\cdot,0) = 1/\varepsilon\}$, and we define $\omega := x/|x|$ and $\rho := |x|/\varepsilon^{\frac{1}{n-2}}$. In this way, $x = \varepsilon^{\frac{1}{n-2}}\rho\omega$. Hence, recalling (2.10.18) and (2.10.20), we have that

$$\begin{aligned}F(\rho,\omega,\varepsilon) &= \rho - \frac{1}{(n(n-2)\,|B_1|)^{\frac{1}{n-2}}\left(1+\varepsilon\Psi^{(0)}\left(\varepsilon^{\frac{1}{n-2}}\rho\omega\right)\right)^{\frac{1}{n-2}}}\\ &= \frac{|x|}{\varepsilon^{\frac{1}{n-2}}} - \frac{1}{(n(n-2)\,|B_1|)^{\frac{1}{n-2}}\left(1+\varepsilon\Psi^{(0)}\left(\varepsilon^{\frac{1}{n-2}}\rho\omega\right)\right)^{\frac{1}{n-2}}}\\ &= \frac{1}{\varepsilon^{\frac{1}{n-2}}(n(n-2)\,|B_1|)^{\frac{1}{n-2}}\left(G(x,0)+\Psi^{(0)}(x)\right)^{\frac{1}{n-2}}}\\ &\quad - \frac{1}{(n(n-2)\,|B_1|)^{\frac{1}{n-2}}\left(1+\varepsilon\Psi^{(0)}\left(\varepsilon^{\frac{1}{n-2}}\rho\omega\right)\right)^{\frac{1}{n-2}}}\end{aligned}$$

$$
\begin{aligned}
&= \frac{1}{\varepsilon^{\frac{1}{n-2}}\big(n(n-2)\,|B_1|\big)^{\frac{1}{n-2}}\big(1/\varepsilon+\Psi^{(0)}(x)\big)^{\frac{1}{n-2}}} \\
&\quad - \frac{1}{\big(n(n-2)\,|B_1|\big)^{\frac{1}{n-2}}\left(1+\varepsilon\Psi^{(0)}(x)\right)^{\frac{1}{n-2}}} \\
&= 0.
\end{aligned}
$$

As a consequence of this and the uniqueness statement in the implicit function theorem, we obtain that $\rho = \rho(\omega,\varepsilon)$, and therefore $x = \varepsilon^{\frac{1}{n-2}}\rho(\omega,\varepsilon)\omega$. This establishes (2.10.24).

From (2.10.23) and (2.10.24), we deduce that

$$
\{G(\cdot,0)=t\} = \{G(\cdot,0)=1/\varepsilon\} = \zeta(\partial B_1,\varepsilon),
$$

which is a dilation of ∂B_1 by a factor

$$
\varepsilon^{\frac{1}{n-2}}\big(\rho_0+\rho_1\varepsilon+o(\varepsilon)\big) = t^{\frac{1}{2-n}}\left(\rho_0+\frac{\rho_1}{t}+o\left(\frac{1}{t}\right)\right) =: \rho(t).
$$

Note that $\rho(t)\to 0$ as $t\to+\infty$.

We insert this information into (2.10.19), and we see that

$$
\begin{aligned}
v(t) &\leqslant \frac{t^{\frac{1}{2-n}}}{n\big(n(n-2)\,|B_1|\big)^{\frac{1}{n-2}}}\int_{\partial B_{\rho(t)}}\left(1+\frac{\Psi^{(0)}(x)}{(2-n)t}+o\left(\frac{1}{t}\right)\right)d\mathcal{H}^{n-1}_x \\
&= \frac{t^{\frac{1}{2-n}}\left(1+o\left(\frac{1}{t}\right)\right)}{n\big(n(n-2)\,|B_1|\big)^{\frac{1}{n-2}}}\left[(\rho(t))^{n-1}\mathcal{H}^{n-1}(\partial B_1)+\frac{1}{(2-n)t}\int_{\partial B_{\rho(t)}}\Psi^{(0)}(x)\,d\mathcal{H}^{n-1}_x\right] \\
&= \frac{t^{\frac{1}{2-n}}(\rho(t))^{n-1}\mathcal{H}^{n-1}(\partial B_1)\left(1+o\left(\frac{1}{t}\right)\right)}{n\big(n(n-2)\,|B_1|\big)^{\frac{1}{n-2}}}\left[1+\frac{1}{(2-n)t}\fint_{\partial B_{\rho(t)}}\Psi^{(0)}(x)\,d\mathcal{H}^{n-1}_x\right] \\
&= \frac{t^{\frac{1}{2-n}}(\rho(t))^{n-1}\mathcal{H}^{n-1}(\partial B_1)\left(1+o\left(\frac{1}{t}\right)\right)}{n\big(n(n-2)\,|B_1|\big)^{\frac{1}{n-2}}}\left[1+\frac{1}{(2-n)t}\big(\Psi^{(0)}(0)+o(1)\big)\right].
\end{aligned}
\tag{2.10.25}
$$

Now, we observe that

$$
\begin{aligned}
t^{\frac{1}{2-n}}(\rho(t))^{n-1} &= t^{\frac{1}{2-n}}\left(t^{\frac{1}{2-n}}\left(\rho_0+\frac{\rho_1}{t}+o\left(\frac{1}{t}\right)\right)\right)^{n-1} \\
&= t^{\frac{n}{2-n}}\rho_0^{n-1}\left(1+\frac{\rho_1}{\rho_0 t}+o\left(\frac{1}{t}\right)\right)^{n-1} \\
&= t^{\frac{n}{2-n}}\rho_0^{n-1}\left(1+\frac{(n-1)\rho_1}{\rho_0 t}+o\left(\frac{1}{t}\right)\right).
\end{aligned}
$$

From this, (1.1.6) and (2.10.25), we deduce that

$$\begin{aligned} v(t) &\leqslant \frac{t^{\frac{n}{2-n}}\rho_0^{n-1}|B_1|\left(1+\frac{(n-1)\rho_1}{\rho_0 t}+o\left(\frac{1}{t}\right)\right)\left(1+o\left(\frac{1}{t}\right)\right)}{\left(n(n-2)\,|B_1|\right)^{\frac{1}{n-2}}} \\ &\quad\times\left[1+\frac{1}{(2-n)t}\left(\Psi^{(0)}(0)+o(1)\right)\right] \\ &= t^{\frac{n}{2-n}}\rho_0^{n}|B_1|\left(1+\frac{(n-1)\rho_1}{\rho_0 t}+o\left(\frac{1}{t}\right)\right)\left[1+\frac{1}{(2-n)t}\left(\Psi^{(0)}(0)+o(1)\right)\right] \\ &= t^{\frac{n}{2-n}}\rho_0^{n}|B_1|\left[1+\frac{1}{t}\left(\frac{(n-1)\rho_1}{\rho_0}+\frac{\Psi^{(0)}(0)}{2-n}+o(1)\right)\right]. \end{aligned}$$

Recalling (2.10.22), we observe that

$$\frac{(n-1)\rho_1}{\rho_0}+\frac{\Psi^{(0)}(0)}{2-n}=\frac{(n-1)\Psi^{(0)}(0)}{2-n}+\frac{\Psi^{(0)}(0)}{2-n}=\frac{n\Psi^{(0)}(0)}{2-n},$$

and therefore

$$v(t)\leqslant t^{\frac{n}{2-n}}\rho_0^{n}|B_1|\left[1+\frac{1}{t}\left(\frac{n\Psi^{(0)}(0)}{2-n}+o(1)\right)\right].$$

Accordingly,

$$\begin{aligned} \left(v(t)\right)^{\frac{2-n}{n}} &\leqslant t\rho_0^{2-n}|B_1|^{\frac{2-n}{n}}\left[1+\frac{1}{t}\left(\frac{n\Psi^{(0)}(0)}{2-n}+o(1)\right)\right]^{\frac{2-n}{n}} \\ &= t\rho_0^{2-n}|B_1|^{\frac{2-n}{n}}\left[1+\frac{1}{t}\left(\Psi^{(0)}(0)+o(1)\right)\right] \\ &= t\,n(n-2)|B_1|^{\frac{2}{n}}\left[1+\frac{1}{t}\left(\Psi^{(0)}(0)+o(1)\right)\right]. \end{aligned}$$

From this and Lemma 2.10.6, we deduce that

$$|\Omega|^{\frac{2-n}{n}}+n(n-2)|B_1|^{\frac{2}{n}}t\leqslant t\,n(n-2)|B_1|^{\frac{2}{n}}\left[1+\frac{1}{t}\left(\Psi^{(0)}(0)+o(1)\right)\right].$$

Simplifying one term, we conclude that

$$|\Omega|^{\frac{2-n}{n}}\leqslant n(n-2)|B_1|^{\frac{2}{n}}\left(\Psi^{(0)}(0)+o(1)\right).$$

Sending $t\longrightarrow+\infty$, we obtain the desired result. □

2.11 The Poisson Kernel

The Poisson kernel is defined as (minus) the normal derivative of the Green function. More specifically, given an open set $\Omega \subset \mathbb{R}^n$ with C^1 boundary and a point $x_0 \in \partial\Omega$, we let $\Omega \ni x \mapsto G(x, x_0)$ be the Green function of Ω, as presented in (2.10.2). Then, for every $x \in \Omega$, we define the Poisson kernel of Ω by

$$P(x, x_0) := -\frac{\partial G}{\partial \nu}(x, x_0), \tag{2.11.1}$$

where the notation of (2.10.7) has been used.

We observe that if $x \neq x_0$,

$$\begin{aligned} -\Delta_x P(x, x_0) &= -\Delta_x \left(\nabla_y G(x, y) \cdot \nu(y)\right) \\ &= -\Delta_x \left(\nabla_y \Gamma(x - y) \cdot \nu(y) - \nabla_y \Psi^{(y)}(x) \cdot \nu(y)\right) \\ &= -\nabla_y - \Delta_x \Gamma(x - y) \cdot \nu(y) + \nabla_y - \Delta_x \Psi^{(y)}(x) \cdot \nu(y) = 0, \end{aligned} \tag{2.11.2}$$

thanks to (2.10.1) and (2.10.2).

The importance of this kernel lies in the following representation result for harmonic functions with prescribed boundary values.

Theorem 2.11.1. *If $\Omega \subseteq \mathbb{R}^n$ is an open and bounded set with C^1 boundary and $u \in C^2(\Omega) \cap C(\overline{\Omega})$ solves*

$$\begin{cases} \Delta u = 0 & \text{in } \Omega, \\ u = g & \text{on } \partial\Omega, \end{cases} \tag{2.11.3}$$

for some $g : \partial\Omega \to \mathbb{R}$, then, for every $x \in \Omega$,

$$u(x) = \int_{\partial\Omega} g(y)\, P(x, y)\, d\mathcal{H}^{n-1}_y. \tag{2.11.4}$$

Moreover,

$$\int_{\partial\Omega} P(x, y)\, d\mathcal{H}^{n-1}_y = 1. \tag{2.11.5}$$

Furthermore, if u is defined as in (2.11.4) *and g is continuous, then $u \in C^2(\Omega) \cap C(\overline{\Omega})$ and u satisfies* (2.11.3).

Proof. When $u \in C^1(\overline{\Omega})$, this is in fact a particular case of Corollary 2.10.3, used here with $f := 0$ (and, *en passant*, this observation also proves (2.11.5) since the function u, identically equal to 1, solves (2.11.3) with $g := 1$, and hence (2.11.5) follows from (2.11.4) applied to a function as regular as we wish).

If instead one wants to prove (2.11.4) by only assuming that $u \in C(\overline{\Omega})$, one can reduce to the case of $u \in C^1(\overline{\Omega})$ via a mollification argument. For this, we take $\tau \in C_0^\infty(B_1, [0,+\infty))$ with $\int_{B_1} \tau(x)\,dx = 1$. Given $\eta > 0$, we let $\tau_\eta(x) := \frac{1}{\eta^n}\tau\left(\frac{x}{\eta}\right)$ and define $u_\eta := u * \tau_\eta$ (recall e.g. [WZ15, Theorem 9.8] for the uniform approximation properties of u_η to u when $\eta \searrow 0$). We also consider Ω_η to be the points p in Ω for which $B_{2\eta}(p) \Subset \Omega$. Then, $u_\eta \in C^2(\overline{\Omega_\eta})$; therefore, setting $g_\eta(x) := u_\eta(x)$ for all $x \in \partial\Omega_\eta$, we can write that, for all $x \in \Omega_\eta$,

$$u_\eta(x) = \int_{\partial\Omega_\eta} g_\eta(y)\,P(x,y)\,d\mathcal{H}^{n-1}_y.$$

Thus, for every $x \in \Omega$,

$$\begin{aligned}
u(x) - \int_{\partial\Omega} g(y)\,P(x,y)\,d\mathcal{H}^{n-1}_y &= \lim_{\eta\searrow 0} u_\eta(x) - \int_{\partial\Omega} u(y)\,P(x,y)\,d\mathcal{H}^{n-1}_y \\
&= \lim_{\eta\searrow 0} \int_{\partial\Omega_\eta} g_\eta(y)\,P(x,y)\,d\mathcal{H}^{n-1}_y \\
&\quad - \int_{\partial\Omega_\eta} u(y)\,P(x,y)\,d\mathcal{H}^{n-1}_y \\
&= \lim_{\eta\searrow 0} \int_{\partial\Omega_\eta} \big(u_\eta(y) - u(y)\big)\,P(x,y)\,d\mathcal{H}^{n-1}_y = 0,
\end{aligned}$$

thanks to (2.11.5), which establishes (2.11.4) in the claimed generality.

We now prove that if u is defined as in (2.11.4) and g is continuous, then $u \in C^2(\Omega) \cap C(\overline{\Omega})$ and u satisfies (2.11.3). We begin by noting that u is harmonic since so is the Poisson kernel (since so are the Green function and the fundamental solution outside its singularity, recall (2.10.2) and (2.11.1)). Thus, we focus on proving that u attains the boundary datum g continuously. For this, the idea is to exploit the fact that the Poisson kernel has a single singularity and integrates to 1. The details of the proof are as follows. Let $x_0 \in \partial\Omega$ and $\varepsilon > 0$. We use the continuity of g to pick a $\delta > 0$ such that $|g(x) - g(x_0)| \leqslant \varepsilon$ for each $x \in (\partial\Omega) \cap B_\delta(x_0)$. We also take a sequence $x_k \in \Omega$ converging to x_0 as $k \to +\infty$. Without loss of generality, we can suppose that $x_k \in B_\delta(x_0)$. Then, by (2.11.4) and (2.11.5),

$$\begin{aligned}
|u(x_k) - g(x_0)| &= \left| \int_{\partial\Omega} g(y)\,P(x_k,y)\,d\mathcal{H}^{n-1}_y - g(x_0) \right| \\
&= \left| \int_{\partial\Omega} \big(g(y) - g(x_0)\big)\,P(x_k,y)\,d\mathcal{H}^{n-1}_y \right|
\end{aligned}$$

$$\leqslant \varepsilon \int_{(\partial\Omega)\cap B_\delta(x_0)} P(x_k,y)\,d\mathcal{H}^{n-1}_y$$
$$+ \int_{(\partial\Omega)\setminus B_\delta(x_0)} \big|g(y)-g(x_0)\big|\,P(x_k,y)\,d\mathcal{H}^{n-1}_y$$
$$\leqslant \varepsilon + \int_{(\partial\Omega)\setminus B_\delta(x_0)} \big|g(y)-g(x_0)\big|\,P(x_k,y)\,d\mathcal{H}^{n-1}_y.$$

Sending $k \to +\infty$, we thereby deduce that

$$\limsup_{k\to+\infty} |u(x_k)-g(x_0)| \leqslant \varepsilon.$$

Hence, taking now ε as small as we wish,

$$\lim_{k\to+\infty} |u(x_k)-g(x_0)| = 0,$$

as desired. □

Since, in light of (2.7.2) and (2.7.6), we know that $\nabla\Gamma(x) = -\frac{x}{n\,|B_1|\,|x|^n}$, we straightforwardly deduce from Theorems 2.10.4 and 2.10.5 the explicit expression of the Poisson kernels of the ball and the halfspace (see Figure 2.11.1 for a plot of the case of the unit ball and Figure 2.11.2 for the case of the halfspace).

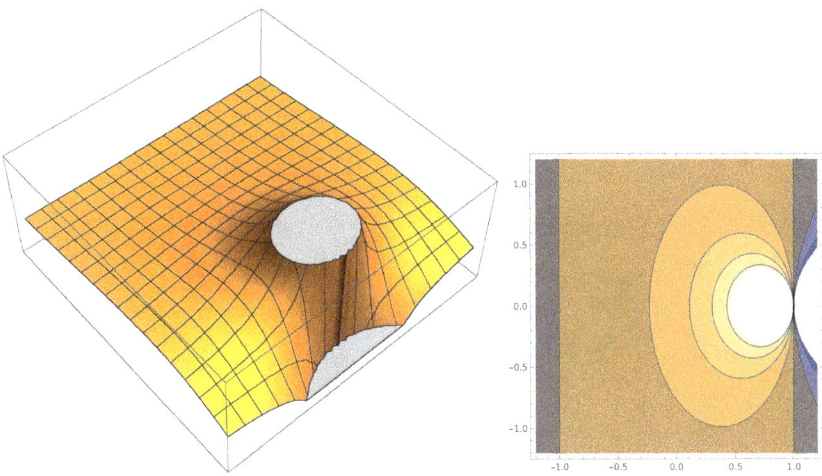

Figure 2.11.1 The Poisson kernel for the unit ball and its level sets.

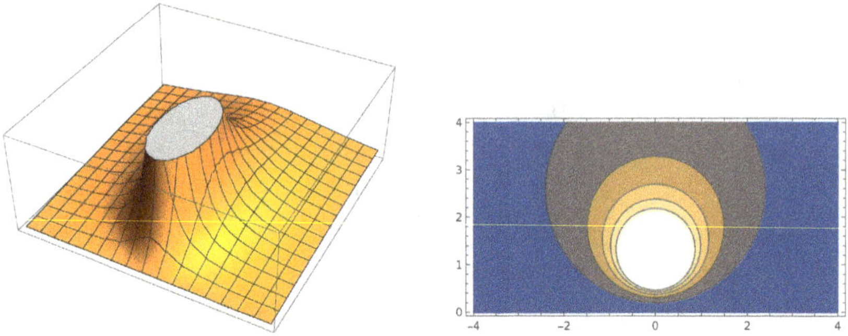

Figure 2.11.2 The Poisson kernel for the halfplane and its level sets.

Theorem 2.11.2. *The Poisson kernel of the ball B_R is*

$$P(x,x_0)=\frac{R^2-|x|^2}{n\,|B_1|\,R\,|x-x_0|^n},$$

for every $x\in B_R$ and $x_0\in\partial B_R$.

Proof. We recall the notation in (2.11.1) and (2.10.7). Thus, from Theorem 2.10.4, (2.11.1) and (2.10.11), we have that, if $x\in B_R\setminus\{0\}$,

$$\begin{aligned}
-P(x,x_0)&=\nabla_{x_0}G(x,x_0)\cdot\frac{x_0}{R}=\nabla_{x_0}\big(G(x_0,x)\big)\cdot\frac{x_0}{R}\\
&=\nabla_{x_0}\left[\Gamma(x_0-x)-\Gamma\left(\frac{|x|}{R}\left(x_0-\frac{R^2x}{|x|^2}\right)\right)\right]\cdot\frac{x_0}{R}\\
&=-\frac{x_0}{n\,|B_1|\,R}\cdot\left(\frac{x_0-x}{|x_0-x|^n}-\frac{\frac{|x|^2}{R^2}\left(x_0-\frac{R^2x}{|x|^2}\right)}{\left|\frac{|x|}{R}\left(x_0-\frac{R^2x}{|x|^2}\right)\right|^n}\right)\\
&=-\frac{x_0}{n\,|B_1|\,R\,|x-x_0|^n}\cdot\left(x_0-x-\frac{|x|^2}{R^2}\left(x_0-\frac{R^2x}{|x|^2}\right)\right)\\
&=-\frac{x_0}{n\,|B_1|\,R\,|x-x_0|^n}\cdot\left(x_0-\frac{|x|^2x_0}{R^2}\right)\\
&=-\frac{|x_0|^2}{n\,|B_1|\,R\,|x-x_0|^n}\left(1-\frac{|x|^2}{R^2}\right)\\
&=-\frac{R^2-|x|^2}{n\,|B_1|\,R\,|x-x_0|^n},
\end{aligned}$$

which is the desired result when $x\in B_R\setminus\{0\}$.

If instead $x = 0$, then Theorem 2.10.4 and (2.11.1) lead to

$$-P(0,x_0) = \nabla_{x_0} G(0,x_0) \cdot \frac{x_0}{R} = \nabla_{x_0}\big(G(x_0,0)\big) \cdot \frac{x_0}{R} = \nabla_{x_0}\big(\Gamma(x_0) - \Gamma(R)\big) \cdot \frac{x_0}{R}$$
$$= -\frac{x_0}{n\,|B_1|\,|x_0|^n} \cdot \frac{x_0}{R} = -\frac{R^2}{n\,|B_1|\,R\,|x_0|^n},$$

which completes the proof of Theorem 2.11.2. □

Theorem 2.11.3. *A Poisson kernel of the halfspace $\{x_n > 0\}$ is*

$$P(x,x_0) = \frac{2}{n|B_1|} \frac{x \cdot e_n}{|x - x_0|^n}, \tag{2.11.6}$$

for every $x \in \{x_n > 0\}$ and $x_0 \in \{x_n = 0\}$.

Also, if $u \in C^2(\{x_n > 0\})$ is a bounded and uniformly continuous solution of

$$\begin{cases} \Delta u = 0 & \text{in } \mathbb{R}^{n-1} \times (0,+\infty), \\ u = g & \text{on } \mathbb{R}^{n-1} \times \{0\}, \end{cases} \tag{2.11.7}$$

for some $g : \partial\Omega \to \mathbb{R}$, then, for every $x = (x', x_n) \in \mathbb{R}^{n-1} \times (0,+\infty)$,

$$u(x) = \frac{2}{n|B_1|} \int_{\mathbb{R}^{n-1}} \frac{x_n\, g(y)}{\big(|x' - y|^2 + x_n^2\big)^{\frac{n}{2}}}\, dy. \tag{2.11.8}$$

Proof. From Theorem 2.10.5, for every x_0 with $x_0 \cdot e_n = 0$,

$$\begin{aligned} P(x,x_0) &= \nabla_{x_0}\big(\Gamma(x - x_0) - \Gamma(\mathcal{R}(x) - x_0)\big) \cdot e_n \\ &= \big(-\nabla\Gamma(x - x_0) + \nabla\Gamma(\mathcal{R}(x) - x_0)\big) \cdot e_n \\ &= \frac{1}{n\,|B_1|\,|x - x_0|^n} \big(x - x_0 - (\mathcal{R}(x) - x_0)\big) \cdot e_n \\ &= \frac{2x \cdot e_n}{n\,|B_1|\,|x - x_0|^n}, \end{aligned}$$

which proves (2.11.6).

Now, we prove (2.11.8). To this end, we first check that the function introduced in (2.11.8) is indeed a solution of (2.11.7). For this, we take $x = (x', x_n) \in \mathbb{R}^{n-1} \times \mathbb{R}$, and we define

$$\Theta(x) := \frac{x_n}{\big(|x'|^2 + x_n^2\big)^{\frac{n}{2}}} = \frac{x_n}{|x|^n}.$$

We observe that, by the expression of the Laplace operator in spherical coordinates (2.5.5),

$$\begin{aligned} \Delta\Theta(x) &= \Delta(x_n\,|x|^{-n}) = x_n\,\Delta|x|^{-n} + 2\nabla x_n \cdot \nabla|x|^{-n} \\ &= x_n\,\big(n(n+1)|x|^{-n-2} - n(n-1)|x|^{-n-2}\big) - 2n|x|^{-n-2}x_n = 0. \end{aligned} \tag{2.11.9}$$

Furthermore, for each $i \in \{1, \dots, n\}$ and $h \in \left(0, \frac{x_n}{2}\right)$,

$$
\begin{aligned}
&|\Theta(x + he_i) + \Theta(x - he_i) - 2\Theta(x)| \leqslant \sup_{\xi \in B_h(x)} |D^2\Theta(\xi)| \\
&\leqslant C \sup_{\xi = (\xi', \xi_n) \in B_h(x)} \frac{\xi_n}{|\xi|^{n+2}} \\
&\leqslant \frac{C x_n}{\left((|x'| - h)^2 + x_n^2\right)^{\frac{n+2}{2}}} \\
&\leqslant \begin{cases} \dfrac{C}{x_n^{n+1}} & \text{if } |x'| \leqslant 8x_n, \\ \dfrac{C x_n}{\left(|x'|^2 + x_n^2\right)^{\frac{n+2}{2}}} & \text{if } |x'| \geqslant 8x_n, \end{cases} \\
&\leqslant \frac{C x_n}{\left(|x'|^2 + x_n^2\right)^{\frac{n+2}{2}}}
\end{aligned}
$$

for some constant $C > 0$, possibly varying from line to line.

Consequently, since, for a given $x_n > 0$, the function

$$
\mathbb{R}^{n-1} \ni x' \longmapsto \frac{x_n}{\left(|x'|^2 + x_n^2\right)^{\frac{n+2}{2}}}
$$

belongs to $L^1(\mathbb{R}^{n-1})$, recalling that g is bounded, we deduce from the dominated convergence theorem and (2.11.9) that

$$
\begin{aligned}
&\Delta\left(\int_{\mathbb{R}^{n-1}} g(y)\Theta(x' - y, x_n)\, dy\right) \\
&= \lim_{h \searrow 0} \frac{1}{h^2} \sum_{i=1}^{n} \int_{\mathbb{R}^{n-1}} g(y)\Big(\Theta\big((x' - y, x_n) + he_i\big) \\
&\quad + \Theta\big((x' - y, x_n) - he_i\big) - 2\Theta(x' - y, x_n)\Big)\, dy \\
&= \lim_{h \searrow 0} \frac{1}{h^2} \sum_{i=1}^{n} \int_{\mathbb{R}^{n-1}} g(x' - \zeta)\Big(\Theta\big((\zeta, x_n) + he_i\big) \\
&\quad + \Theta\big((\zeta, x_n) - he_i\big) - 2\Theta(\zeta, x_n)\Big)\, d\zeta \\
&= \sum_{i=1}^{n} \int_{\mathbb{R}^{n-1}} g(x' - \zeta) \frac{\partial^2\Theta}{\partial x_i^2}(\zeta, x_n)\, d\zeta \\
&= \int_{\mathbb{R}^{n-1}} g(x' - \zeta)\Delta\Theta(\zeta, x_n)\, d\zeta \\
&= 0.
\end{aligned}
$$

This shows that u satisfies the differential equation in (2.11.7).

Besides, using the notation of the Euler gamma function, recalling (1.1.35),

$$\Gamma\left(\frac{n}{2}\right) = \frac{2\pi^{\frac{n}{2}}}{\mathcal{H}^{n-1}(\partial B_1)} = \frac{2\pi^{\frac{n}{2}}}{n\,|B_1|};$$

hence, if B_1' is the unit ball in dimension $n-1$,

$$\frac{\mathcal{H}^{n-2}(\partial B_1')}{n\,|B_1|} = \frac{\mathcal{H}^{n-2}(\partial B_1')\,\Gamma\left(\frac{n}{2}\right)}{2\pi^{\frac{n}{2}}} = \frac{2\pi^{\frac{n-1}{2}}\,\Gamma\left(\frac{n}{2}\right)}{2\pi^{\frac{n}{2}}\,\Gamma\left(\frac{n-1}{2}\right)} = \frac{\Gamma\left(\frac{n}{2}\right)}{\sqrt{\pi}\,\Gamma\left(\frac{n-1}{2}\right)}.$$

Therefore, using the substitution $t := \frac{x'-y}{\eta}$ and the dominated convergence theorem, for each $x' \in \mathbb{R}^{n-1}$,

$$\begin{aligned}
\lim_{\eta\searrow 0} u(x',\eta) &= \lim_{\eta\searrow 0} \frac{2}{n|B_1|}\int_{\mathbb{R}^{n-1}} \frac{\eta\, g(y)}{\left(|x'-y|^2+\eta^2\right)^{\frac{n}{2}}}\,dy \\
&= \lim_{\eta\searrow 0} \frac{2}{n|B_1|}\int_{\mathbb{R}^{n-1}} \frac{g(x'-\eta t)}{\left(|t|^2+1\right)^{\frac{n}{2}}}\,dt \\
&= \frac{2}{n|B_1|}\, g(x') \int_{\mathbb{R}^{n-1}} \frac{dt}{\left(|t|^2+1\right)^{\frac{n}{2}}} \\
&= \frac{2\mathcal{H}^{n-2}(\partial B_1')}{n|B_1|}\, g(x') \int_0^{+\infty} \frac{\rho^{n-2}\,d\rho}{\left(\rho^2+1\right)^{\frac{n}{2}}} \\
&= \frac{2\Gamma\left(\frac{n}{2}\right)}{\sqrt{\pi}\,\Gamma\left(\frac{n-1}{2}\right)}\, g(x') \int_0^{+\infty} \frac{\rho^{n-2}\,d\rho}{\left(\rho^2+1\right)^{\frac{n}{2}}}. \qquad (2.11.10)
\end{aligned}$$

Now, we claim that, for all $n \in \mathbb{N}$, $n \geqslant 2$, it holds that

$$\int_0^{+\infty} \frac{\rho^{n-2}\,d\rho}{\left(\rho^2+1\right)^{\frac{n}{2}}} = \frac{\sqrt{\pi}\,\Gamma\left(\frac{n-1}{2}\right)}{2\Gamma\left(\frac{n}{2}\right)}. \qquad (2.11.11)$$

To prove this, we denote by $\mathcal{J}_n$ the left-hand side of (2.11.11), and we observe that

$$\frac{d}{d\rho}\frac{1}{\left(\rho^2+1\right)^{\frac{n}{2}}} = -\frac{n\rho}{\left(\rho^2+1\right)^{\frac{n+2}{2}}}$$

and consequently, integrating by parts,

$$\begin{aligned}
\mathcal{J}_{n+2} &= \int_0^{+\infty} \frac{\rho^n\,d\rho}{\left(\rho^2+1\right)^{\frac{n+2}{2}}} = -\frac{1}{n}\int_0^{+\infty} \rho^{n-1}\left(\frac{d}{d\rho}\frac{1}{\left(\rho^2+1\right)^{\frac{n}{2}}}\right)d\rho \\
&= \frac{n-1}{n}\int_0^{+\infty} \frac{\rho^{n-2}}{\left(\rho^2+1\right)^{\frac{n}{2}}}\,d\rho = \frac{n-1}{n}\,\mathcal{J}_n. \qquad (2.11.12)
\end{aligned}$$

In addition,

$$\mathcal{J}_2 = \int_0^{+\infty} \frac{d\rho}{\rho^2+1} = \frac{\pi}{2}, \tag{2.11.13}$$

and using the substitution $r := \rho^2 + 1$,

$$\mathcal{J}_3 = \int_0^{+\infty} \frac{\rho\, d\rho}{\left(\rho^2+1\right)^{\frac{3}{2}}} = \frac{1}{2}\int_1^{+\infty} \frac{dr}{r^{\frac{3}{2}}} = 1. \tag{2.11.14}$$

We stress that (2.11.13) is precisely (2.11.11) when $n = 2$, in light of (1.1.25). Similarly we have that (2.11.14) is precisely (2.11.11) when $n = 3$, owing to (1.1.27).

Therefore, to complete the proof of (2.11.11), we argue by induction. Having established the basis of the induction for $n \in \{2, 3\}$ in (2.11.13) and (2.11.14), we can suppose that (2.11.11) holds true for all $n \leqslant N$ with $N \in \mathbb{N} \cap [3, +\infty)$, and we prove it for $N + 1$. To this end, we make use of (2.11.12), we recall the basic property of the Euler gamma function in (1.1.24) and we gather that

$$\mathcal{J}_{N+1} = \frac{N-2}{N-1}\mathcal{J}_{N-1} = \frac{N-2}{N-1}\,\frac{\sqrt{\pi}\,\Gamma\left(\frac{N-2}{2}\right)}{2\Gamma\left(\frac{N-1}{2}\right)} = \frac{\frac{N-2}{2}}{\frac{N-1}{2}}\,\frac{\sqrt{\pi}\,\Gamma\left(\frac{N-2}{2}\right)}{2\Gamma\left(\frac{N-1}{2}\right)} = \frac{\sqrt{\pi}\,\Gamma\left(\frac{N}{2}\right)}{2\Gamma\left(\frac{N+1}{2}\right)},$$

which completes the proof of (2.11.11).

Now, from (2.11.10) and (2.11.11), we conclude that

$$\lim_{\eta\searrow 0} u(x',\eta) = g(x'),$$

which verified that u satisfies the boundary condition in (2.11.7).

Additionally, we remark that the solution to (2.11.7) in the class of bounded and uniformly continuous functions is unique, thanks to the maximum principle in Proposition 2.9.4, and this establishes (2.11.8), as desired. □

We refer to [Nee94] for several beautiful geometric observations on the Poisson kernels of the ball and the halfplane, also in connection with conformal and hyperbolic geometries. See also [Vig92, Nee97] for interesting geometric connections with complex analysis.

An additional interesting physical motivation of the Poisson kernel can be provided in terms of the "double layer" potential, which is the electrostatic potential associated with a dipole distribution around the smooth surface $\Sigma := \partial\Omega$. In simple terms, if we place some positive (respectively, negative) charges in the vicinity of the external (respectively, internal) side of Σ, these charges will almost "cancel" each other, but the first nonnegligible order can be related directly to the Poisson kernel, up to a harmonic correction (a precise statement will be given in equation (2.11.16)). To make this idea more explicit, let g be a smooth function defined on Σ and ν be the external normal along Σ. Given $\varepsilon > 0$, we endow the surface $\Sigma_+ := \{y + \varepsilon\nu(y),\ y \in \Sigma\}$ with a charge density modeled on g and the surface $\Sigma_- := \{y - \varepsilon\nu(y),\ y \in \Sigma\}$ with a charge density modeled on $-g$. In view of the comment immediately after (2.7.6), for each $x \in \Omega$, the electrostatic potential obtained in this way can be written as

$$\begin{aligned} u_\varepsilon(x) := &\int_\Sigma g(y)\,\Gamma\big(x - (y + \varepsilon\nu(y))\big)\,d\mathcal{H}^{n-1}_y \\ &- \int_\Sigma g(y)\,\Gamma\big(x - (y - \varepsilon\nu(y))\big)\,d\mathcal{H}^{n-1}_y. \end{aligned} \tag{2.11.15}$$

We observe that

$$\begin{aligned} &\int_\Sigma g(y)\,\Gamma\big(x - (y \pm \varepsilon\nu(y))\big)\,d\mathcal{H}^{n-1}_y \\ &\quad = \int_\Sigma g(y)\,\Big[\Gamma(x - y) \mp \varepsilon\nabla\Gamma(x - y)\cdot\nu(y)\Big]\,d\mathcal{H}^{n-1}_y + o(\varepsilon). \end{aligned}$$

This and (2.11.15) lead to

$$u_\varepsilon(x) = -2\varepsilon\int_\Sigma g(y)\nabla\Gamma(x - y)\cdot\nu(y)\,d\mathcal{H}^{n-1}_y + o(\varepsilon).$$

Therefore, recalling the definitions of the Robin function in (2.10.1), the Green function in (2.10.2), and the Poisson kernel in (2.11.1), and making use of the notation in (2.10.7) and the symmetry of the fundamental solution, we see that (up to the factor 2 that we omit), the kernel related to the double layer potential has the form

$$\begin{aligned} -\nabla\Gamma(x - y)\cdot\nu(y) &= \nabla_y\Gamma(x - y)\cdot\nu(y) \\ &= \nabla_y G(x, y)\cdot\nu(y) - \nabla_y\Psi^{(y)}(x)\cdot\nu(y) \\ &= \frac{\partial G}{\partial\nu}(x, y) - \nabla_y\Psi^{(y)}(x)\cdot\nu(y) \\ &= -P(x, y) - \nabla_y\Psi^{(y)}(x)\cdot\nu(y). \end{aligned} \tag{2.11.16}$$

Since the Robin function is harmonic, this equation expresses the Poisson kernel as a suitable harmonic correction of the double layer potential. It is indeed a classical approach to use the double layer potential to solving boundary value problems for elliptic equations, see e.g. [DiB95, Chapter III].

2.12 Point Singularities for Harmonic Functions

A natural question is what types of blowups can harmonic functions exhibit at a given point. For instance, the fundamental solution in (2.7.6) provides an example of a harmonic function in $B_1 \setminus \{0\}$ which, when $n \geqslant 2$, diverges at the origin. It turns out that milder blowups than the one offered by the fundamental solution are not allowed, as pointed out in the following result.

Theorem 2.12.1. *Let $n \geqslant 2$, $r > 0$, and suppose that u is harmonic in $B_r \setminus \{0\}$. Assume also that*

$$\lim_{x \to 0} \frac{u(x)}{\Gamma(x)} = 0, \tag{2.12.1}$$

where Γ is the fundamental solution in (2.7.6). *Then, u can be extended to a harmonic function in the whole of B_r.*

Proof. Let $\rho \in (0, r)$. Using the Poisson kernel in Theorems 2.11.1 and 2.11.2, we can construct a harmonic function v in B_ρ such that $v = u$ along ∂B_ρ. Let $\varepsilon > 0$. Define $\alpha := u - v + \varepsilon\Gamma + \varepsilon$ and $\beta := u - v - \varepsilon\Gamma - \varepsilon$. By (2.12.1), we can find $\delta \in (0, \rho) \cap (0, \varepsilon)$ such that if $|x| \in (0, \delta]$, then $|u(x)| \leqslant \frac{\varepsilon}{2}\Gamma(x)$. Since Γ diverges at the origin, possibly by taking smaller δ, we can also suppose that if $|x| \in (0, \delta]$, then $\|v\|_{L^\infty(B_\rho)} \leqslant \frac{\varepsilon}{2}\Gamma(x)$.

As a result, if $x \in \partial B_\delta$,

$$\alpha(x) \geqslant \varepsilon \quad \text{and} \quad \beta(x) \leqslant -\varepsilon.$$

Moreover, if $x \in \partial B_\rho$,

$$\alpha(x) = \varepsilon\Gamma(x) + \varepsilon \geqslant \varepsilon \quad \text{and} \quad \beta(x) = -\varepsilon\Gamma(x) - \varepsilon \leqslant -\varepsilon.$$

Since α and β are harmonic in $B_\rho \setminus B_\delta$, we thereby deduce from the maximum principle in Corollary 2.9.2 that $\alpha \geqslant \varepsilon$ and $\beta \leqslant -\varepsilon$ in $B_\rho \setminus B_\delta$. Accordingly, in $B_\rho \setminus B_\delta$,

$$u - v + \varepsilon\Gamma + \varepsilon \geqslant \varepsilon \quad \text{and} \quad u - v - \varepsilon\Gamma - \varepsilon \leqslant -\varepsilon.$$

By taking ε as small as we wish, we thus conclude that in B_ρ

$$u - v \geqslant 0 \quad \text{and} \quad u - v \leqslant 0.$$

That is, u coincides with the harmonic function v in B_ρ, and the desired result is established. □

We stress that assumption (2.12.1) cannot be removed from the statement of Theorem 2.12.1: for instance, if we consider the function

$$\begin{aligned} \mathbb{R}^2 \setminus \{0\} \ni (x, y) \longmapsto u(x, y) &:= \Re\left(\exp\left(\frac{1}{x + iy}\right)\right) \\ &= e^{\frac{x}{x^2+y^2}} \cos\left(\frac{y}{x^2 + y^2}\right), \end{aligned} \tag{2.12.2}$$

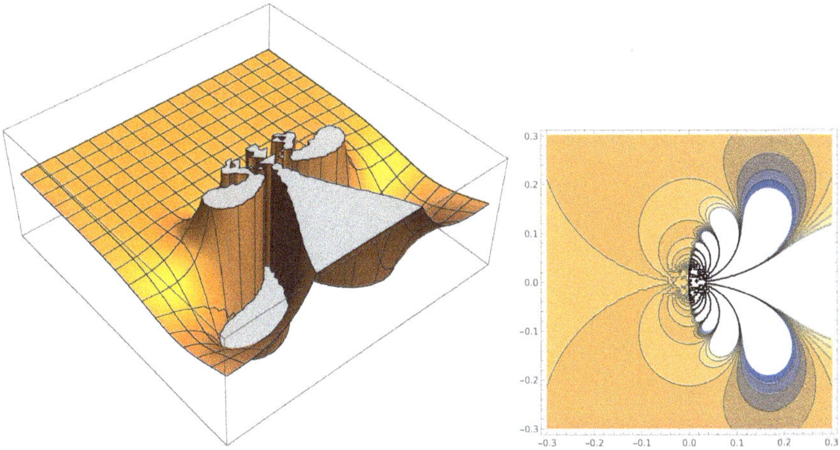

Figure 2.12.1 Plot of the function in (2.12.2) and of its level sets.

we see that u is harmonic in $\mathbb{R}^2 \setminus \{0\}$ (being the real part of a holomorphic function outside the origin), but it presents an essential singularity at the origin (see Figure 2.12.1 for a diagram of this function).

The study of removable singularities is an expansive and classical topic of investigation in partial differential equations, see e.g. [KV86] for further information.

2.13 Further Geometric Insights on Harmonic Functions in Balls

Here, we present a number of fascinating results focused on some geometric properties of harmonic functions. The core of the discussion begins only on p. 128 (and the expert reader is welcome to jump directly there): before that, we begin by discussing a change of variable formula for a spherical integral in $\mathbb{R}^n$.

Lemma 2.13.1. *Let $n \geqslant 2$, $R > 1 \geqslant r > 0$, and Q be a diffeomorphism of $B_R \setminus B_r$ such that $Q(\partial B_\rho) = \partial B_\rho$ for each $\rho \in [r, R]$. Let $g : \mathbb{R}^n \to \mathbb{R}$ be a continuous function.*

Then,

$$\int_{\partial B_1} g(\omega)\, d\mathcal{H}^{n-1}_\omega = \int_{\partial B_1} g\left(Q(\omega)\right) |\det DQ(\omega)|\, d\mathcal{H}^{n-1}_\omega. \tag{2.13.1}$$

Proof. The change of variable $x := Q(y)$ leads to

$$\begin{aligned}
&\int_{\partial B_1} g(\omega)\, d\mathcal{H}^{n-1}_\omega \\
&= \frac{n}{R^n - r^n} \int_r^R \left[\int_{\partial B_1} \rho^{n-1} g(\omega)\, d\mathcal{H}^{n-1}_\omega \right] d\rho \\
&= \frac{n}{R^n - r^n} \int_{B_R \setminus B_r} g\left(\frac{x}{|x|}\right) dx \\
&= \frac{n}{R^n - r^n} \int_{B_R \setminus B_r} g\left(\frac{Q(y)}{|Q(y)|}\right) |\det DQ(y)|\, dy \\
&= \frac{n}{R^n - r^n} \int_r^R \left[\int_{\partial B_1} \rho^{n-1} g\left(\frac{Q(\rho\omega)}{|Q(\rho\omega)|}\right) |\det DQ(\rho\omega)|\, d\mathcal{H}^{n-1}_\omega \right] d\rho.
\end{aligned}$$

In particular, picking $R := 1 + \varepsilon$ and $r := 1$,

$$\begin{aligned}
&\int_{\partial B_1} g(\omega)\, d\mathcal{H}^{n-1}_\omega \\
&= \frac{n}{(1+\varepsilon)^n - 1} \int_1^{1+\varepsilon} \left[\int_{\partial B_1} \rho^{n-1} g\left(\frac{Q(\rho\omega)}{|Q(\rho\omega)|}\right) |\det DQ(\rho\omega)|\, d\mathcal{H}^{n-1}_\omega \right] d\rho.
\end{aligned}$$

Taking the limit as $\varepsilon \searrow 0$, this equation and the L'Hôpital's rule lead to

$$\begin{aligned}
&\int_{\partial B_1} g(\omega)\, d\mathcal{H}^{n-1}_\omega \\
&= \lim_{\varepsilon \searrow 0} \frac{1}{(1+\varepsilon)^{n-1}} \int_{\partial B_1} (1+\varepsilon)^{n-1} g\left(\frac{Q((1+\varepsilon)\omega)}{|Q((1+\varepsilon)\omega)|}\right) |\det DQ((1+\varepsilon)\omega)|\, d\mathcal{H}^{n-1}_\omega \\
&= \int_{\partial B_1} g\left(\frac{Q(\omega)}{|Q(\omega)|}\right) |\det DQ(\omega)|\, d\mathcal{H}^{n-1}_\omega \\
&= \int_{\partial B_1} g\left(Q(\omega)\right) |\det DQ(\omega)|\, d\mathcal{H}^{n-1}_\omega,
\end{aligned}$$

as desired. □

We now specify Lemma 2.13.1 for a particular choice of the map Q (for neat geometric interpretations in the plane, see [Duf57, Nee94]). Namely, given $P \in B_1$, for all $y \in B_2 \setminus B_{(1+|P|)/2}$, we define

$$Q(y) := y - \frac{2(P-y)\cdot y}{|P-y|^2}\,(P-y). \tag{2.13.2}$$

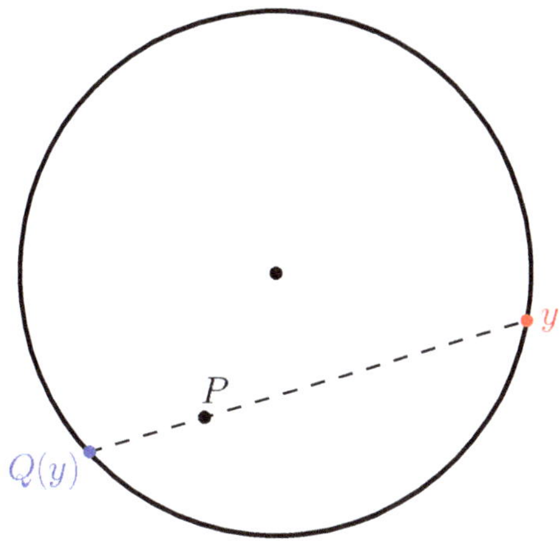

Figure 2.13.1 The map $Q : \partial B_1 \to \partial B_1$ in (2.13.2).

We observe that $Q(y)$ belongs to the straight line $\{y + t(P-y),\ t \in \mathbb{R}\}$ passing through y and P. Furthermore,

$$|Q(y)|^2 = |y|^2 + \frac{4((P-y)\cdot y)^2}{|P-y|^2} - \frac{4(P-y)\cdot y}{|P-y|^2}\,(P-y)\cdot y = |y|^2$$

and therefore Q maps ∂B_ρ into itself, for each $\rho \in \left[\frac{1+|P|}{2}, 2\right]$; see Figure 2.13.1 for a representation of the action of Q on the unit circumference in the plane. In this setting, we have the following lemma.

Lemma 2.13.2. *Let $n \geqslant 2$, $P \in B_1$ and Q be as in* (2.13.2).

Then, for every $\omega \in \partial B_1$, we have that

$$|\det DQ(\omega)| = \left(\frac{1-|P|^2}{|P-\omega|^2}\right)^{n-1}. \tag{2.13.3}$$

Moreover, for every continuous function $g : \mathbb{R}^n \to \mathbb{R}$,

$$\int_{\partial B_1} g(\omega)\, d\mathcal{H}^{n-1}_\omega = \int_{\partial B_1} g\left(Q(\omega)\right) \left(\frac{1-|P|^2}{|P-\omega|^2}\right)^{n-1} d\mathcal{H}^{n-1}_\omega. \tag{2.13.4}$$

Proof. In light of (2.13.2), we point out that

$$Q(y) - P = -\left(1 + \frac{2(P-y)\cdot y}{|P-y|^2}\right)(P-y) = -\frac{|P|^2-|y|^2}{|P-y|^2}\,(P-y); \tag{2.13.5}$$

therefore, for every $\omega \in \partial B_1$,

$$\frac{|P-Q(\omega)|}{|P-\omega|} = \frac{1-|P|^2}{|P-\omega|^2}.$$

We now pick a point $\omega \in \partial B_1$ and consider the plane containing the origin, P and ω. Up to a rotation, we suppose that this plane is the one given by the first coordinates, namely $\Pi := \{x_3 = \cdots = x_n = 0\}$ (of course, if $n = 2$, this part simplifies since Π would coincide with the whole plane of interest $\mathbb{R}^2$). Consequently, by (2.13.2), we know that if $y \in \Pi$, then $Q(y) \in \Pi$. Since $\omega, e_1, e_2 \in \Pi$, this gives that $Q(\omega+\varepsilon e_1)$ and $Q(\omega + \varepsilon e_2)$ belong to Π for all $\varepsilon \in \mathbb{R}$. Accordingly, $\partial_1 Q(\omega)$ and $\partial_2 Q(\omega)$ belong to Π, that is,

$$\partial_1 Q(\omega) = \big(\partial_1 Q_1(\omega), \partial_1 Q_2(\omega), 0, \dots, 0\big)$$

and

$$\partial_2 Q(\omega) = \big(\partial_2 Q_1(\omega), \partial_2 Q_2(\omega), 0, \dots, 0\big).$$

For this reason,

$$DQ(\omega) = \begin{pmatrix} \partial_1 Q_1(\omega) & \partial_1 Q_2(\omega) & 0 & \dots & 0 \\ \partial_2 Q_1(\omega) & \partial_2 Q_2(\omega) & 0 & \dots & 0 \\ \partial_3 Q_1(\omega) & \partial_3 Q_2(\omega) & \partial_3 Q_3(\omega) & \dots & \partial_3 Q_n(\omega) \\ & & & \ddots & \\ \partial_n Q_1(\omega) & \partial_n Q_2(\omega) & \partial_n Q_3(\omega) & \dots & \partial_n Q_n(\omega) \end{pmatrix},$$

which entails that

$$\begin{aligned} &|\det DQ(\omega)| \\ &= \left|\det \begin{pmatrix} \partial_1 Q_1(\omega) & \partial_1 Q_2(\omega) \\ \partial_2 Q_1(\omega) & \partial_2 Q_2(\omega) \end{pmatrix}\right| \left|\det \begin{pmatrix} \partial_3 Q_3(\omega) & \dots & \partial_3 Q_n(\omega) \\ & \ddots & \\ \partial_n Q_3(\omega) & \dots & \partial_n Q_n(\omega) \end{pmatrix}\right|. \end{aligned} \tag{2.13.6}$$

Also, by (2.13.5), for every $j \in \{1, \dots, n\}$,

$$\begin{aligned} &\partial_j Q(\omega) \\ &= \frac{2\omega_j}{|P-\omega|^2}(P-\omega) + \frac{2(1-|P|^2)(P_j-\omega_j)}{|P-\omega|^4}(P-\omega) - \frac{1-|P|^2}{|P-\omega|^2} e_j. \end{aligned} \tag{2.13.7}$$

Therefore, using again that $\omega_3 = \cdots = \omega_n = P_3 = \cdots = P_n = 0$, we deduce that, for every $j \in \{3, \dots, n\}$,

$$\partial_j Q(\omega) = -\frac{1-|P|^2}{|P-\omega|^2} e_j.$$

As a result, the matrix

$$\begin{pmatrix} \partial_3 Q_3(\omega) & \dots & \partial_3 Q_n(\omega) \\ & \ddots & \\ \partial_n Q_3(\omega) & \dots & \partial_n Q_n(\omega) \end{pmatrix}$$

equals $-\frac{1-|P|^2}{|P-\omega|^2}$ times the identity matrix (intended here as a square $(n-2)\times(n-2)$ matrix). This observation and (2.13.6) lead to

$$|\det DQ(\omega)| = \left(\frac{1-|P|^2}{|P-\omega|^2}\right)^{n-2} \left|\det\begin{pmatrix}\partial_1 Q_1(\omega) & \partial_1 Q_2(\omega)\\ \partial_2 Q_1(\omega) & \partial_2 Q_2(\omega)\end{pmatrix}\right|. \tag{2.13.8}$$

Now, we rewrite (2.13.7) in the form

$$\partial_j Q(\omega) = -\frac{1-|P|^2}{|P-\omega|^2}\, e_j + \alpha_j(\omega)\,(P-\omega),$$

where

$$\alpha_j(\omega) := \frac{2\omega_j}{|P-\omega|^2} + \frac{2(1-|P|^2)(P_j-\omega_j)}{|P-\omega|^4}.$$

Consequently, letting $v(\omega) := (P_1-\omega_1, P_2-\omega_2)$

$$\begin{aligned}
&\det\begin{pmatrix}\partial_1 Q_1(\omega) & \partial_1 Q_2(\omega)\\ \partial_2 Q_1(\omega) & \partial_2 Q_2(\omega)\end{pmatrix}\\
&= \det\begin{pmatrix}-\frac{1-|P|^2}{|P-\omega|^2}\,(1,0) + \alpha_1(\omega)\,v(\omega)\\ -\frac{1-|P|^2}{|P-\omega|^2}\,(0,1) + \alpha_2(\omega)\,v(\omega)\end{pmatrix}\\
&= \left(\frac{1-|P|^2}{|P-\omega|^2}\right)^2 \det\begin{pmatrix}1 & 0\\ 0 & 1\end{pmatrix} - \alpha_2(\omega)\frac{1-|P|^2}{|P-\omega|^2}\det\begin{pmatrix}(1,0)\\ v(\omega)\end{pmatrix}\\
&\quad - \alpha_1(\omega)\frac{1-|P|^2}{|P-\omega|^2}\det\begin{pmatrix}v(\omega)\\ (0,1)\end{pmatrix}\\
&= \left(\frac{1-|P|^2}{|P-\omega|^2}\right)^2 - \alpha_2(\omega)\,(P_2-\omega_2)\frac{1-|P|^2}{|P-\omega|^2} - \alpha_1(\omega)\,(P_1-\omega_1)\frac{1-|P|^2}{|P-\omega|^2}\\
&= \left(\frac{1-|P|^2}{|P-\omega|^2}\right)\left[\left(\frac{1-|P|^2}{|P-\omega|^2}\right) - \alpha_2(\omega)\,(P_2-\omega_2) - \alpha_1(\omega)\,(P_1-\omega_1)\right].
\end{aligned} \tag{2.13.9}$$

We also point out that

$$\begin{aligned}
&\alpha_1(\omega)\,(P_1-\omega_1) + \alpha_2(\omega)\,(P_2-\omega_2)\\
&\quad = \sum_{j=1}^{2}\left(\frac{2\omega_j(P_j-\omega_j)}{|P-\omega|^2} + \frac{2(1-|P|^2)(P_j-\omega_j)^2}{|P-\omega|^4}\right)\\
&\quad = \frac{2(P\cdot\omega-1)}{|P-\omega|^2} + \frac{2(1-|P|^2)}{|P-\omega|^2}\\
&\quad = \frac{2(P\cdot\omega-|P|^2)}{|P-\omega|^2}
\end{aligned}$$

and therefore

$$\frac{1-|P|^2}{|P-\omega|^2} - \alpha_1(\omega)\,(P_1-\omega_1) - \alpha_2(\omega)\,(P_2-\omega_2) = \frac{1-2P\cdot\omega+|P|^2}{|P-\omega|^2} = 1.$$

Plugging this information into (2.13.9), we find that

$$\det\begin{pmatrix}\partial_1 Q_1(\omega) & \partial_1 Q_2(\omega)\\ \partial_2 Q_1(\omega) & \partial_2 Q_2(\omega)\end{pmatrix} = \frac{1-|P|^2}{|P-\omega|^2}.$$

Hence, recalling (2.13.8), we obtain (2.13.3). This and (2.13.1) give (2.13.4), as desired. □

Another interesting change of variable, which can be seen as a slight modification of the one discussed in Lemma 2.13.2, is the following. For every $P\in B_1$ and every $e\in\partial B_1$, we consider the points $Q_+(e)$ and $Q_-(e)$ lying in the intersection between the straight line of direction e passing through P and ∂B_1; namely, let $r_+(e)>0>r_-(e)$ be such that

$$Q_+(e) = P + r_+(e)\,e \in \partial B_1 \quad\text{and}\quad Q_-(e) = P + r_-(e)\,e \in \partial B_1. \tag{2.13.10}$$

See Figure 2.13.2 for a sketch of this configuration. Also, by (2.13.10),

$$1 = |Q_\pm(e)|^2 = |P + r_\pm(e)\,e|^2 = |P|^2 + (r_\pm(e))^2 + 2P\cdot e\, r_\pm(e). \tag{2.13.11}$$

Therefore,

$$r_\pm(e) = -P\cdot e \pm \sqrt{D(e)}, \quad\text{where}\quad D(e) := (P\cdot e)^2 - |P|^2 + 1. \tag{2.13.12}$$

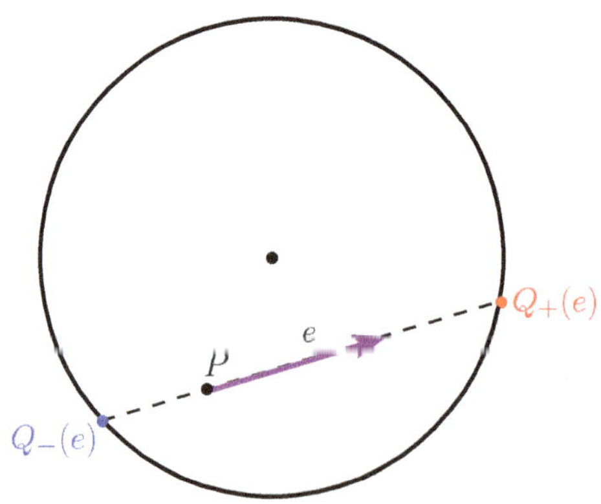

Figure 2.13.2 The maps $Q_\pm : \partial B_1 \to \partial B_1$ in (2.13.10).

The maps $Q_\pm$ can also be extended in $\mathbb{R}^n \setminus \{0\}$ by considering the homogeneous extension of degree 1: in this way, for every $x \in \mathbb{R}^n \setminus \{0\}$, we have that

$$Q_\pm(x) = |x|\, Q_\pm\left(\frac{x}{|x|}\right). \tag{2.13.13}$$

This and (2.13.11) give that, for each $\rho > 0$, if $x \in \partial B_\rho$ then $|Q_\pm(x)| = \rho \left|Q_\pm\left(\frac{x}{|x|}\right)\right| = \rho$, and hence $Q_\pm$ maps ∂B_ρ into itself. Thus, in analogy with Lemma 2.13.2, we have the following result.

Lemma 2.13.3. *Let $n \geqslant 2$, $P \in B_1$ and $Q_\pm$ be as in* (2.13.10).

Then, for every $\omega \in \partial B_1$, we have that

$$|\det DQ_\pm(\omega)| = \frac{(\pm r_\pm(\omega))^n}{1 - |P|^2 - r_\pm(\omega) P \cdot \omega}. \tag{2.13.14}$$

Moreover, for every continuous function $g : \mathbb{R}^n \to \mathbb{R}$,

$$\int_{\partial B_1} g(\omega)\, d\mathcal{H}^{n-1}_\omega = \int_{\partial B_1} g\left(Q_\pm(\omega)\right) \frac{(\pm r_\pm(\omega))^n}{1 - |P|^2 - r_\pm(\omega) P \cdot \omega}\, d\mathcal{H}^{n-1}_\omega. \tag{2.13.15}$$

Proof. Let $\omega \in \partial B_1$, and let O denote the origin. Up to a rotation, we can suppose that the points O, P and $P + \omega$ lie on the plane $\{x_3 = \cdots = x_n = 0\}$. Also, up to a further rotation in this plane, we can suppose that $\omega = e_1$. Thus, by (2.13.13),

$$\begin{aligned} Q_\pm(\omega + \varepsilon e_1) &= Q_\pm((1+\varepsilon)e_1) = (1+\varepsilon)\, Q_\pm(e_1) \\ &= (1+\varepsilon)\big(Q_{\pm,1}(e_1), Q_{\pm,2}(e_1), 0, \ldots, 0\big). \end{aligned}$$

Consequently,

$$\partial_1 Q_\pm(\omega) = \big(Q_{\pm,1}(e_1), Q_{\pm,2}(e_1), 0, \ldots, 0\big).$$

Let now $j \in \{2, \ldots, n\}$. We have that $|\omega + \varepsilon e_j| = |e_1 + \varepsilon e_j| = 1 + o(\varepsilon)$; therefore, recalling (2.13.10) and (2.13.12),

$$\begin{aligned} & Q_\pm(\omega + \varepsilon e_j) \\ &= (1 + o(\varepsilon))\, Q_\pm\left(\frac{\omega + \varepsilon e_j}{|\omega + \varepsilon e_j|}\right) \\ &= P + r_\pm\left(\frac{\omega + \varepsilon e_j}{|\omega + \varepsilon e_j|}\right) \frac{\omega + \varepsilon e_j}{|\omega + \varepsilon e_j|} + o(\varepsilon) \\ &= P + \left(-P \cdot \frac{\omega + \varepsilon e_j}{|\omega + \varepsilon e_j|} \pm \sqrt{\left(P \cdot \frac{\omega + \varepsilon e_j}{|\omega + \varepsilon e_j|}\right)^2 - |P|^2 + 1}\right)(\omega + \varepsilon e_j) + o(\varepsilon) \\ &= P + \left(-P \cdot (\omega + \varepsilon e_j) \pm \sqrt{\big(P \cdot (\omega + \varepsilon e_j)\big)^2 - |P|^2 + 1}\right)(\omega + \varepsilon e_j) + o(\varepsilon) \end{aligned}$$

$$
\begin{aligned}
&= P + \left(-P\cdot\omega - \varepsilon P\cdot e_j \pm \sqrt{(P\cdot\omega)^2 + 2\varepsilon(P\cdot\omega)(P\cdot e_j) - |P|^2 + 1}\right)\\
&\quad \times(\omega+\varepsilon e_j) + o(\varepsilon)\\
&= P + \left(-P\cdot e_1 - \varepsilon P\cdot e_j \pm \sqrt{(P\cdot e_1)^2 - |P|^2 + 1} \pm \frac{\varepsilon(P\cdot e_1)(P\cdot e_j)}{\sqrt{(P\cdot e_1)^2 - |P|^2 + 1}}\right)\\
&\quad \times(e_1+\varepsilon e_j) + o(\varepsilon).
\end{aligned}
$$

Taking the first order in ε, we find that

$$
\begin{aligned}
&\partial_j Q_\pm(\omega)\\
&= \left(-P\cdot e_j \pm \frac{(P\cdot e_1)(P\cdot e_j)}{\sqrt{(P\cdot e_1)^2 - |P|^2+1}}\right) e_1 + \left(-P\cdot e_1 \pm \sqrt{(P\cdot e_1)^2 - |P|^2 + 1}\right) e_j\\
&= \alpha_j e_1 + r_\pm(e_1) e_j,
\end{aligned}
$$

where

$$
\begin{aligned}
\alpha_j &:= -P\cdot e_j \pm \frac{(P\cdot e_1)(P\cdot e_j)}{\sqrt{(P\cdot e_1)^2 - |P|^2 + 1}}\\
&= -\frac{P\cdot e_j\left(\sqrt{D(e_1)} \mp P\cdot e_1\right)}{\sqrt{D(e_1)}} = \mp\frac{P\cdot e_j\, r_\pm(e_1)}{\sqrt{D(e_1)}}.
\end{aligned}
$$

These observations lead to

$$
DQ_\pm(\omega) = \begin{pmatrix}
Q_{\pm,1}(e_1) & Q_{\pm,2}(e_1) & 0 & 0 & 0 & \cdots & 0\\
\alpha_2 & r_\pm(e_1) & 0 & 0 & 0 & \cdots & 0\\
\alpha_3 & 0 & r_\pm(e_1) & 0 & 0 & \cdots & 0\\
\alpha_4 & 0 & 0 & r_\pm(e_1) & 0 & \cdots & 0\\
\alpha_5 & 0 & 0 & 0 & r_\pm(e_1) & \cdots & 0\\
 & & & \ddots & & & \\
\alpha_n & 0 & 0 & 0 & 0 & \cdots & r_\pm(e_1)
\end{pmatrix}
$$

and therefore

$$
|\det DQ_\pm(\omega)| = \left| r_\pm(e_1)^{n-2} \det\begin{pmatrix} Q_{\pm,1}(e_1) & Q_{\pm,2}(e_1)\\ \alpha_2 & r_\pm(e_1)\end{pmatrix}\right|. \tag{2.13.16}
$$

We also note that

$$
\begin{aligned}
Q_\pm(e_1) &= (P\cdot e_1, P\cdot e_2, 0,\dots,0) + \left(-P\cdot e_1 \pm \sqrt{(P\cdot e_1)^2 - |P|^2 + 1}\right) e_1\\
&= \left(\pm\sqrt{(P\cdot e_1)^2 - |P|^2 + 1}, P\cdot e_2, 0,\dots,0\right)\\
&= \left(\pm\sqrt{D(e_1)}, P\cdot e_2, 0,\dots,0\right)
\end{aligned}
$$

and consequently

$$\begin{aligned}
&\det\begin{pmatrix} Q_{\pm,1}(e_1) & Q_{\pm,2}(e_1) \\ \alpha_2 & r_\pm(e_1) \end{pmatrix} = Q_{\pm,1}(e_1)r_\pm(e_1) - Q_{\pm,2}(e_1)\alpha_2 \\
&= \pm\sqrt{D(e_1)}r_\pm(e_1) \pm \frac{(P\cdot e_2)^2\, r_\pm(e_1)}{\sqrt{D(e_1)}} \\
&= \pm\frac{r_\pm(e_1)}{\sqrt{D(e_1)}}\Big(D(e_1) + (P\cdot e_2)^2\Big) = \pm\frac{r_\pm(e_1)}{\sqrt{D(e_1)}}\Big((P\cdot e_1)^2 - |P|^2 + 1 + (P\cdot e_2)^2\Big) \\
&= \pm\frac{r_\pm(e_1)}{\sqrt{D(e_1)}} = \frac{r_\pm(e_1)}{r_\pm(e_1) + P\cdot e_1} = \frac{(r_\pm(e_1))^2}{(r_\pm(e_1))^2 + P\cdot e_1\, r_\pm(e_1)}.
\end{aligned}$$

Since

$$\begin{aligned}
&(r_\pm(e_1))^2 + P\cdot e_1\, r_\pm(e_1) \\
&\quad = (P\cdot e_1)^2 + D(e_1) \mp 2P\cdot e_1\sqrt{D(e_1)} + P\cdot e_1\, r_\pm(e_1) \\
&\quad = (P\cdot e_1)^2 + D(e_1) \mp 2P\cdot e_1\sqrt{D(e_1)} - (P\cdot e_1)^2 \pm (P\cdot e_1)\sqrt{D(e_1)} \\
&\quad = D(e_1) \mp P\cdot e_1\sqrt{D(e_1)} \\
&\quad = 1 - |P|^2 + (P\cdot e_1)^2 \mp P\cdot e_1\sqrt{D(e_1)} \\
&\quad = 1 - |P|^2 - (P\cdot e_1)\Big(-P\cdot e_1 \pm \sqrt{D(e_1)}\Big) \\
&\quad = 1 - |P|^2 - (P\cdot e_1)r_\pm(e_1),
\end{aligned} \tag{2.13.17}$$

we arrive at

$$\det\begin{pmatrix} Q_{\pm,1}(e_1) & Q_{\pm,2}(e_1) \\ \alpha_2 & r_\pm(e_1) \end{pmatrix} = \frac{(r_\pm(e_1))^2}{1 - |P|^2 - (P\cdot e_1)r_\pm(e_1)}.$$

Thus, retaking (2.13.16),

$$|\det DQ_\pm(\omega)| = \left|\frac{(r_\pm(e_1))^n}{1 - |P|^2 - (P\cdot e_1)r_\pm(e_1)}\right|. \tag{2.13.18}$$

It is also useful to note that

$$\begin{aligned}
&1 - |P|^2 - (P\cdot e_1)r_\pm(e_1) \\
&\quad = 1 - (P\cdot e_1)^2 - (P\cdot e_2)^2 - (P\cdot e_1)\Big(-(P\cdot e_1) \pm \sqrt{D(e_1)}\Big) \\
&\quad = 1 - (P\cdot e_1)^2 - (P\cdot e_2)^2 - (P\cdot e_1)\big(-(P\cdot e_1) \pm \sqrt{1 - (P\cdot e_2)^2}\big) \\
&\quad = 1 - (P\cdot e_2)^2 \mp (P\cdot e_1)\sqrt{1 - (P\cdot e_2)^2} \\
&\quad = \sqrt{1 - (P\cdot e_2)^2}\,\Big(\sqrt{1 - (P\cdot e_2)^2} \mp (P\cdot e_1)\Big) \\
&\quad \geqslant \sqrt{1 - (P\cdot e_2)^2}\,\Big(\sqrt{1 - (P\cdot e_2)^2} - |P\cdot e_1|\Big)
\end{aligned}$$

$$= \frac{\sqrt{1-(P\cdot e_2)^2}}{\sqrt{1-(P\cdot e_2)^2}+|P\cdot e_1|}\left(1-(P\cdot e_2)^2-(P\cdot e_1)^2\right)$$
$$= \frac{\sqrt{1-(P\cdot e_2)^2}}{\sqrt{1-(P\cdot e_2)^2}+|P\cdot e_1|}\left(1-|P|^2\right) \geqslant 0,$$

which, together with (2.13.18), establishes (2.13.14). This and (2.13.1) lead to (2.13.15). □

Now, we are ready to dive into the analysis of some geometric features of harmonic functions. In this framework, a classical result, sometimes referred to by the name Malmheden's theorem, provides an elegant geometric algorithm for solving the Dirichlet problem in a ball, see [Neu84, Mal34, Duf57, Nee94, AKS10] (see also [Bôc06] for related results which emerged in the analysis of Fourier series, especially in relation to the Gibbs overshooting phenomenon at jump discontinuities). In a nutshell, the main idea of this algorithm is:

- to consider a point P in a given ball, whose boundary is endowed by some continuous datum f;
- then, to take an arbitrary chord passing through the point P and calculate the value at P of the linear function that interpolates the values of f at the endpoints of the chord L; and
- finally, to compute the average of these values over all possible chords through P.

Quite remarkably, this procedure produces the harmonic function in the ball[19] with datum f on the boundary.

The details of this construction are as follows. Let $f : \partial B_1 \to \mathbb{R}$ be a continuous function. Let L_e be the segment joining $Q_-(e)$ to $Q_+(e)$, as defined in (2.13.10), and

$$\begin{aligned}&\text{let } \ell_e \text{ be the linear (or, better to say, affine) function on } L_e\\ &\text{such that } \ell_e(Q_-(e)) = f(Q_-(e)) \text{ and } \ell_e(Q_+(e)) = f(Q_+(e)).\end{aligned} \tag{2.13.19}$$

Let also

$$u(P) := \fint_{\partial B_1} \ell_e(P)\, d\mathcal{H}^{n-1}_e. \tag{2.13.20}$$

Note that the function u is constructed by averaging the linear interpolations of the boundary datum in each direction.

[19]We stress that this procedure essentially works only with balls and cannot be generalized to other domains, see [Wei64] and Section 4 in [AKS10]. Instead, a generalization of this procedure to cross-sections of higher dimensions is provided in Section 3 of [AKS10].

Theorem 2.13.4. *The function u in* (2.13.20) *is harmonic in B_1, continuous in $\overline{B_1}$, and $u = f$ on ∂B_1.*

We observe that when P is the origin, Theorem 2.13.4 reduces to the mean value formula in Theorem 2.1.2(ii). Also, when $n = 1$, Theorem 2.13.4 boils down to the fact that harmonic functions are linear (hence, Theorem 2.13.4 is interesting only when $n \geqslant 2$).

Following [AKS10], we give two proofs of Theorem 2.13.4: one now, exploiting Lemma 2.13.3, and the other on p. 213, relying on harmonic polynomials.

Proof of Theorem 2.13.4. First, we show that, for every $\zeta \in \partial B_1$,

$$\lim_{P \to \zeta} u(P) = f(\zeta). \tag{2.13.21}$$

To this end, we point out that the linear function ℓ_e on $L_e := \{P + se,\ s \in [r_-(e), r_+(e)]\}$ such that $\ell_e(Q_-(e)) = f(Q_-(e))$ has the form

$$\ell_e(P + se) = \frac{\big(f(Q_+(e)) - f(Q_-(e))\big)s + r_+(e) f(Q_-(e)) - r_-(e) f(Q_+(e))}{r_+(e) - r_-(e)}. \tag{2.13.22}$$

Furthermore, in view of (2.13.12), the quantities $Q_\pm(e)$ and $r_\pm(e)$ depend continuously on P, with either $Q_-(e)$ or $Q_+(e)$ approaching ζ and correspondingly either to $r_-(e)$ or $r_+(e)$ approaching 0 as $P \to \zeta$. As a result, by (2.13.22),

$$\lim_{P \to \zeta} \ell_e(P) = \lim_{P \to \zeta} \frac{r_+(e) f(Q_-(e)) - r_-(e) f(Q_+(e))}{r_+(e) - r_-(e)} = f(\zeta).$$

From this and (2.13.20), we obtain (2.13.21), as desired.

Furthermore, in light of (2.13.20) and (2.13.22),

$$\begin{aligned} u(P) &= \fint_{\partial B_1} \frac{r_+(e) f(Q_-(e)) - r_-(e) f(Q_+(e))}{r_+(e) - r_-(e)}\, d\mathcal{H}_e^{n-1} \\ &= \fint_{\partial B_1} \frac{r_+(e) f(Q_-(e))}{r_+(e) - r_-(e)}\, d\mathcal{H}_e^{n-1} - \fint_{\partial B_1} \frac{r_-(e) f(Q_+(e))}{r_+(e) - r_-(e)}\, d\mathcal{H}_e^{n-1} \\ &= 2 \fint_{\partial B_1} \frac{r_+(e) f(Q_-(e))}{r_+(e) - r_-(e)}\, d\mathcal{H}_e^{n-1}. \end{aligned} \tag{2.13.23}$$

Moreover, recalling (2.13.12),

$$r_+(e) r_-(e) = (P \cdot e)^2 - D(e) = |P|^2 - 1. \tag{2.13.24}$$

Consequently, owing also to (2.13.12) and (2.13.17),

$$\begin{aligned}\frac{2r_+(e)}{r_+(e)-r_-(e)} &= \frac{r_+(e)}{\sqrt{D(e)}} = -\frac{r_+(e)}{r_-(e)+P\cdot e}\\ &= -\frac{r_+(e)r_-(e)}{(r_-(e))^2+P\cdot e\, r_-(e)} = -\frac{r_+(e)r_-(e)}{1-|P|^2-(P\cdot e)r_-(e)}\\ &= \frac{1-|P|^2}{1-|P|^2-(P\cdot e)r_-(e)}.\end{aligned} \tag{2.13.25}$$

Thus, since $|P-Q_-(e)| = |r_-(e)| = -r_-(e)$, we deduce from (2.13.15), applied here with $g(e) := \frac{f(e)\,(1-|P|^2)}{|P-e|^n}$, and (2.13.23) that

$$\begin{aligned}u(P) &= \fint_{\partial B_1} \frac{f(Q_-(e))\,(1-|P|^2)}{1-|P|^2-(P\cdot e)r_-(e)}\, d\mathcal{H}_e^{n-1}\\ &= \fint_{\partial B_1} \frac{f(Q_-(e))\,(1-|P|^2)}{|P-Q_-(e)|^n}\,\frac{(-r_-(e))^n}{1-|P|^2-(P\cdot e)r_-(e)}\, d\mathcal{H}_e^{n-1}\\ &= \fint_{\partial B_1} g(Q_-(e))\,\frac{(-r_-(e))^n}{1-|P|^2-(P\cdot e)r_-(e)}\, d\mathcal{H}_e^{n-1}\\ &= \fint_{\partial B_1} g(e)\, d\mathcal{H}_e^{n-1}\\ &= \fint_{\partial B_1} \frac{f(e)\,(1-|P|^2)}{|P-e|^n}\, d\mathcal{H}_e^{n-1}.\end{aligned}$$

That is, recalling (1.1.6),

$$u(P) = \frac{1}{n\,|B_1|}\int_{\partial B_1} \frac{f(e)\,(1-|P|^2)}{|P-e|^n}\, d\mathcal{H}_e^{n-1}.$$

Recognizing the Poisson kernel of the ball, in view of Theorems 2.11.1 and 2.11.2, we obtain that u is harmonic in B_1, as desired. □

Harmonic functions in the plane enjoy special geometric features (see e.g. [Nee94]). In particular, the following result holds true.

Theorem 2.13.5. *Let $n=2$, $f:\partial B_1\to\mathbb{R}$ be a continuous function and $P\in B_1$. For every $\omega\in\partial B_1$, let*

$$Q(\omega) := \omega - \frac{2(P-\omega)\cdot\omega}{|P-\omega|^2}\,(P-\omega).$$

Let

$$u(P) := \fint_{\partial B_1} f(Q(\omega))\, d\mathcal{H}^1_\omega. \tag{2.13.26}$$

Then, u is harmonic in B_1, continuous in $\overline{B_1}$, and $u = f$ on ∂B_1.

We observe that when P is the origin, Theorem 2.13.5 reduces to the mean value formula in Theorem 2.1.2(ii). Also, Theorem 2.13.5 has a neat geometric interpretation, which was probably a source of inspiration for Hermann Schwarz's classic work [Sch72]. Namely, suppose that the temperature on the boundary of a planar unit disk is equal to 1 along a given arc Σ and 0 elsewhere. Then, given a point P on the disk, the temperature at P can be calculated precisely via a simple, and purely geometric, protocol: namely, one projects the arc Σ through the focal point P, obtaining a "conjugated arc" Σ', and the temperature at P is exactly equal to the length of Σ', divided by 2π (note indeed that this is the content of (2.13.26) in the limiting case in which $f = \chi_\Sigma$), see Figure 2.13.3.

Proof of Theorem 2.13.5. Setting $\varpi := Q(\omega)$, we exploit the spherical change of variable in (2.13.4) (recall that here $n = 2$). In this way, we deduce from (2.13.26) that

$$u(P) = \fint_{\partial B_1} \frac{f(\varpi)(1 - |P|^2)}{|P - \varpi|^2}\, d\mathcal{H}^1_\varpi.$$

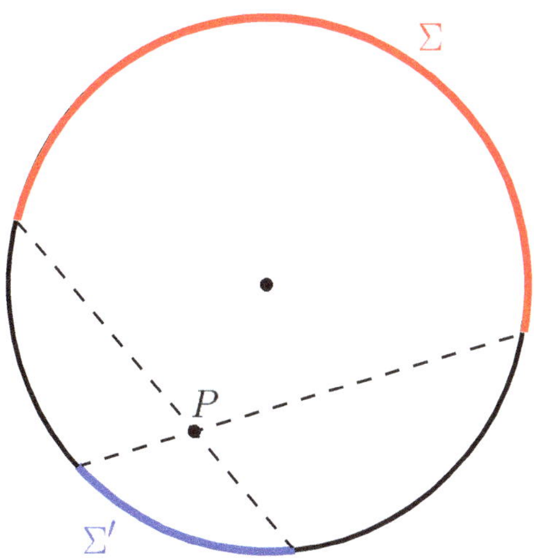

Figure 2.13.3 Finding the temperature in a disk simply by using a ruler.

This expression coincides with the one obtained using the Poisson kernel representation in Theorems 2.11.1 and 2.11.2; therefore, u is harmonic in B_1 and attains the datum f along ∂B_1. □

Another Proof of Theorem 2.13.5. The idea of the proof is to reduce it to Theorem 2.13.4. We use the notation in (2.13.10) and (2.13.19): hence, recalling (2.13.22),

$$\ell_e(P) = \frac{r_+(e)f(Q_-(e)) - r_-(e)f(Q_+(e))}{r_+(e) - r_-(e)}.$$

Note also that $Q(Q_\pm(e)) = Q_\mp(e)$. By (2.13.20) and Theorem 2.13.4, we know that the harmonic function in B_1 with boundary datum equal to f can be written as

$$\begin{aligned} v(P) &:= \fint_{\partial B_1} \ell_e(P)\, d\mathcal{H}^1_e \\ &= \fint_{\partial B_1} \frac{r_+(e)f(Q_-(e))}{r_+(e) - r_-(e)}\, d\mathcal{H}^1_e - \fint_{\partial B_1} \frac{r_-(e)f(Q_+(e))}{r_+(e) - r_-(e)}\, d\mathcal{H}^1_e. \end{aligned}$$

Our goal is thus to show that $v(P) = u(P)$, with which the desired result would be proved. To this end, we exploit (2.13.15), applied with $g(\omega) := f(Q(\omega))$, and (2.13.25) to see that

$$\begin{aligned} u(P) &= \fint_{\partial B_1} f(Q(\omega))\, d\mathcal{H}^1_\omega = \fint_{\partial B_1} g(\omega)\, d\mathcal{H}^1_\omega \\ &= \fint_{\partial B_1} g\left(Q_-(\omega)\right) \frac{(r_-(\omega))^2}{1 - |P|^2 - r_-(\omega)P \cdot \omega}\, d\mathcal{H}^1_\omega \\ &= \fint_{\partial B_1} f\left(Q(Q_-(\omega))\right) \frac{(r_-(\omega))^2}{1 - |P|^2 - r_-(\omega)P \cdot \omega}\, d\mathcal{H}^1_\omega \\ &= \fint_{\partial B_1} f(Q_+(\omega)) \frac{(r_-(\omega))^2}{1 - |P|^2 - r_-(\omega)P \cdot \omega}\, d\mathcal{H}^1_\omega \\ &= \fint_{\partial B_1} f(Q_+(\omega)) \frac{2r_+(\omega)(r_-(\omega))^2}{(r_+(\omega) - r_-(\omega))(1 - |P|^2)}\, d\mathcal{H}^1_\omega. \end{aligned}$$

This and (2.13.24) lead to

$$u(P) = -\fint_{\partial B_1} f(Q_+(\omega)) \frac{2r_-(\omega)}{r_+(\omega) - r_-(\omega)}\, d\mathcal{H}^1_\omega = v(P),$$

as desired. □

We stress that Theorem 2.13.5 only holds on the plane. To construct a counterexample in dimension $n \geqslant 3$, one can consider the function $f(x_1, \dots, x_n) := \chi_{(-\infty,0)}(x_n)$ and the corresponding function u, as in (2.13.26) (technically, since f is discontinuous, one should first mollify f and then pass to the limit). We observe that u is not harmonic when $n \geqslant 3$. Indeed, let $P := (0, \dots, 0, 1-\varepsilon) = (1-\varepsilon)e_n$, for some small $\varepsilon > 0$. Then, up to a multiplicative constant,

$$u(P) \text{ coincides with the surface area of the spherical cap obtained by projecting the lower halfsphere } \{x_n < 0\} \cap \partial B_1 \text{ through the point } P. \tag{2.13.27}$$

One can estimate this surface area using elementary geometry, see Figure 2.13.4. Indeed, with respect to this picture, if α denotes the angle $\widehat{EPD}$, we see that this angle is also equal to the angle $\widehat{OPe_1}$; therefore, looking at the triangle with vertices O, P and e_1,

$$\tan\alpha = \frac{1}{\overline{OP}} = \frac{1}{1-\varepsilon} = 1 + \varepsilon + o(\varepsilon).$$

This leads to

$$\alpha = \arctan(1 + \varepsilon + o(\varepsilon)) = \frac{\pi}{4} + \frac{\varepsilon}{2} + o(\varepsilon).$$

As a consequence,

$$\cos\alpha = \frac{\sqrt{2}}{2} - \frac{\sqrt{2}\,\varepsilon}{4} + o(\varepsilon)$$
$$\text{and} \quad \sin\alpha = \frac{\sqrt{2}}{2} + \frac{\sqrt{2}\,\varepsilon}{4} + o(\varepsilon).$$

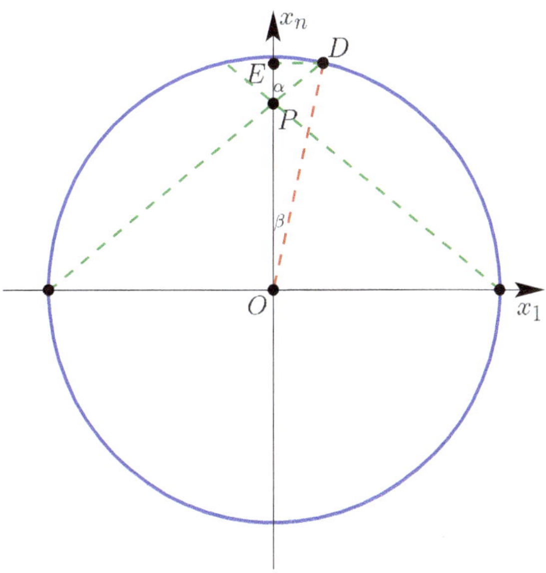

Figure 2.13.4 A counterexample to Theorem 2.13.5 when $n \geqslant 3$.

Also, applying the law of cosines to the triangle with vertices O, P and D,

$$\begin{aligned} 1 = \overline{OD}^2 &= \overline{OP}^2 + \overline{PD}^2 - 2\,\overline{OP}\,\overline{PD}\cos(\pi - \alpha) \\ &= (1-\varepsilon)^2 + \overline{PD}^2 + 2(1-\varepsilon)\overline{PD}\cos\alpha \\ &= 1 - 2\varepsilon + \overline{PD}^2 + \left(\sqrt{2} - \frac{3\sqrt{2}\,\varepsilon}{2}\right)\overline{PD} + o(\varepsilon). \end{aligned}$$

As a result,

$$\overline{PD}^2 + \left(\sqrt{2} - \frac{3\sqrt{2}\,\varepsilon}{2}\right)\overline{PD} - 2\varepsilon + o(\varepsilon) = 0.$$

By solving this quadratic equation, we find that

$$\overline{PD} = \frac{\sqrt{4 + 4\varepsilon} + 3\varepsilon - 2}{2\sqrt{2}} + o(\varepsilon) = \sqrt{2}\,\varepsilon + o(\varepsilon).$$

Consequently, if β denotes the angle $\widehat{POD}$, by applying the law of sines to the triangle with vertices O, P and D, we deduce that

$$\sin\beta = \frac{\overline{PD}\,\sin(\pi - \alpha)}{\overline{OD}} = \overline{PD}\,\sin\alpha = \sqrt{2}\,\varepsilon\left(\frac{\sqrt{2}}{2} + \frac{\sqrt{2}\,\varepsilon}{4}\right) + o(\varepsilon) = \varepsilon + o(\varepsilon).$$

Accordingly, we have that $\beta = \varepsilon + o(\varepsilon)$ and thus, by (2.13.27),

$$u(P) \leqslant C\varepsilon^{n-1}, \tag{2.13.28}$$

for some $C > 0$ depending only on n.

On the other hand, a lower bound for $u(P)$ can be obtained directly from the Poisson kernel of the ball. Namely, in light of Theorems 2.11.1 and 2.11.2,

$$\begin{aligned} u(P) &= \int_{\{y_n<0\}\cap\partial B_1} \frac{1 - |P|^2}{n\,|B_1|\,|P - y|^n}\, d\mathcal{H}_y^{n-1} \\ &= \int_{\{y_n<0\}\cap\partial B_1} \frac{1 - (1-\varepsilon)^2}{n\,|B_1|\left(|y'|^2 + (1 - \varepsilon - y_n)^2\right)^{\frac{n}{2}}}\, d\mathcal{H}_y^{n-1} \\ &\geqslant \int_{\{y_n<0\}\cap\partial B_1} \frac{\varepsilon}{n\,|B_1|\,4^{\frac{n}{2}}}\, d\mathcal{H}_y^{n-1} = c\varepsilon, \end{aligned}$$

for a suitable $c > 0$ depending only on n. Comparing this with (2.13.28), we find that $c \leqslant C\varepsilon^{n-2}$, which, for small ε, cannot hold true when $n \geqslant 3$.

2.14 Analyticity of Harmonic Functions

A direct consequence of the Poisson kernel representation obtained in Theorem 2.11.2 is the interior regularity of harmonic functions, as pointed out in the following result.

Theorem 2.14.1. *Harmonic functions are real analytic.*

Proof. Let $\Omega \subseteq \mathbb{R}^n$ be an open set, and let u be harmonic in Ω. Let $p \in \Omega$, and pick $R > 0$ such that $B_R(p) \Subset \Omega$. Up to a translation, we suppose that $p = 0$, and thus we exploit Theorems 2.11.1 and 2.11.2 to write, for every $x \in B_R$,

$$u(x) = \int_{\partial B_R} \frac{(R^2 - |x|^2)\, u(y)}{n\,|B_1|\, R\, |x - y|^n}\, d\mathcal{H}^{n-1}_y. \tag{2.14.1}$$

Now, if $y \in \partial B_R$, we have that

$$\text{the function } B_{R/2} \ni x \mapsto |x - y|^2 = |x|^2 + |y|^2 + 2x \cdot y \text{ is real analytic} \tag{2.14.2}$$

since it is a polynomial.

Also, for every $x \in B_{R/2}$ and $y \in \partial B_R$,

$$\begin{aligned} &|x - y|^2 \geqslant (|y| - |x|)^2 \geqslant \frac{R^2}{4} \\ \text{and} \quad &|x - y|^2 \leqslant (|x| + |y|)^2 \leqslant \frac{9R^2}{4}. \end{aligned} \tag{2.14.3}$$

Additionally,

$$\text{the function } \left(\tfrac{R^2}{8}, \tfrac{9R^2}{2}\right) \ni t \mapsto t^{-\frac{n}{2}} \text{ is real analytic.} \tag{2.14.4}$$

Indeed, we can write $\sigma := \frac{16}{R^2}\left(t - \frac{R^2}{16}\right) \in (1, 72)$ and, by applying the binomial series, find that the function $(1 + \sigma)^{-\frac{n}{2}}$ is real analytic, namely

$$(1 + \sigma)^{-\frac{n}{2}} = \sum_{k=0}^{+\infty} \frac{1}{k!} \prod_{j=0}^{k-1} \left(-\frac{n}{2} - j\right) \sigma^k.$$

Since σ^k can be written as a binomial expansion in powers of t, we have in fact established (2.14.4).

Thus, by (2.14.2), (2.14.3) and (2.14.4), we deduce that the function

$$B_{R/2} \ni x \mapsto (|x - y|^2)^{-\frac{n}{2}} = |x - y|^{-n}$$

is real analytic (see e.g. the composition result in [KP02, Proposition 1.6.7]).

In this way, we write

$$|x - Re_1|^{-n} = \sum_{\alpha \in \mathbb{N}^n} c_\alpha x^\alpha,$$

and this series is uniformly convergent for $x \in B_\rho$, for some $\rho \in \left(0, \frac{R}{4}\right)$.

As a matter of fact, considering a rotation $\mathcal{R}_y$ such that $\frac{y}{|y|} = \mathcal{R}_y e_1$ and setting $X := \mathcal{R}_y^{-1} x$, for all $x \in B_\rho$, we have that $X \in B_\rho$, and so we obtain the uniformly convergent series expression

$$\begin{aligned} |x - y|^{-n} &= |\mathcal{R}_y X - R \mathcal{R}_y e_1|^{-n} = |X - Re_1|^{-n} = \sum_{\alpha \in \mathbb{N}^n} c_\alpha X^\alpha \\ &= \sum_{\alpha \in \mathbb{N}^n} c_\alpha (\mathcal{R}_y^{-1} x)^\alpha = \sum_{\alpha \in \mathbb{N}^n} C_\alpha(y) x^\alpha \end{aligned}$$

for suitable coefficients $C_\alpha(y)$.

This and (2.14.1) give that, if $x \in B_\rho$,

$$u(x) = \int_{\partial B_R} \sum_{\alpha \in \mathbb{N}^n} \frac{C_\alpha(y)\,(R^2 - |x|^2)\,u(y)\,x^\alpha}{n\,|B_1|\,R}\,d\mathcal{H}_y^{n-1} = \sum_{\alpha \in \mathbb{N}^n} C_\alpha\,(R^2 - |x|^2)\,x^\alpha,$$

where

$$C_\alpha := \int_{\partial B_R} \frac{C_\alpha(y)\,u(y)}{n\,|B_1|\,R}\,d\mathcal{H}_y^{n-1}.$$

This shows that u can be written as a Taylor series in a neighborhood the origin, which thus establishes the desired result. □

For a different proof of Theorem 2.14.1, see e.g. [Eva98, pp. 31–32].

2.15 The Harnack Inequality

As a consequence of the Poisson kernel obtained in Theorem 2.11.2, we have the following result that provides a sharp and explicit control on the oscillation of harmonic functions:

Theorem 2.15.1. *Let $x_0 \in \mathbb{R}^n$ and $R > r > 0$. Assume that u is nonnegative and harmonic in $B_R(x_0)$. Then, for every $x \in B_r(x_0)$,*

$$\left(\frac{R}{R+r}\right)^{n-2} \frac{R-r}{R+r}\,u(x_0) \leqslant u(x) \leqslant \left(\frac{R}{R-r}\right)^{n-2} \frac{R+r}{R-r}\,u(x_0). \tag{2.15.1}$$

Proof. Up to a translation, we suppose $x_0 = 0$. We also take $\rho \in (r, R)$. Thus, by Theorems 2.11.1 and 2.11.2 and recalling (1.1.6), for every $x \in B_\rho$

we have that

$$\begin{aligned} u(x) &= \int_{\partial B_\rho} \frac{(\rho^2 - |x|^2)\, u(y)}{n\,|B_1|\,\rho\,|x-y|^n}\, d\mathcal{H}^{n-1}_y \\ &= \int_{\partial B_\rho} \frac{(\rho^2 - |x|^2)\, u(y)}{\mathcal{H}^{n-1}(\partial B_1)\,\rho\,|x-y|^n}\, d\mathcal{H}^{n-1}_y \\ &= \rho^{n-2} ⨍_{\partial B_\rho} \frac{(\rho^2 - |x|^2)\, u(y)}{|x-y|^n}\, d\mathcal{H}^{n-1}_y. \end{aligned} \tag{2.15.2}$$

Additionally, if $x \in B_r$ and $y \in \partial B_\rho$,

$$\frac{\rho^2 - r^2}{(\rho + r)^n} \leqslant \frac{\rho^2 - |x|^2}{(|y| + |x|)^n} \leqslant \frac{\rho^2 - |x|^2}{|x-y|^n}$$

$$\text{and} \quad \frac{\rho^2 - |x|^2}{|x-y|^n} \leqslant \frac{\rho^2 - |x|^2}{(|y| - |x|)^n} = \frac{\rho^2 - |x|^2}{(\rho - |x|)^n} = \frac{\rho + |x|}{(\rho - |x|)^{n-1}} \leqslant \frac{\rho + r}{(\rho - r)^{n-1}}.$$

These inequalities, (2.15.2) and the mean value formula in Theorem 2.1.2(ii) give that, if $x \in B_r$,

$$\begin{aligned} \left(\frac{\rho}{\rho + r}\right)^{n-2} \frac{\rho - r}{\rho + r}\, u(0) &= \frac{\rho^{n-2}(\rho^2 - r^2)}{(\rho + r)^n}\, u(0) \\ &= \frac{\rho^{n-2}(\rho^2 - r^2)}{(\rho + r)^n} ⨍_{\partial B_\rho} u(y)\, d\mathcal{H}^{n-1}_y \\ &\leqslant \rho^{n-2} ⨍_{\partial B_\rho} \frac{(\rho^2 - |x|^2)\, u(y)}{|x-y|^n}\, d\mathcal{H}^{n-1}_y = u(x) \end{aligned}$$

and also

$$\begin{aligned} u(x) &\leqslant \frac{\rho^{n-2}(\rho + r)}{(\rho - r)^{n-1}} ⨍_{\partial B_\rho} u(y)\, d\mathcal{H}^{n-1}_y = \frac{\rho^{n-2}(\rho + r)}{(\rho - r)^{n-1}}\, u(0) \\ &= \left(\frac{\rho}{\rho - r}\right)^{n-2} \frac{\rho + r}{\rho - r}\, u(0). \end{aligned}$$

Sending $\rho \nearrow R$, we obtain the desired result. □

We point out that the estimate in Theorem 2.15.1 is optimal. Indeed, when $n = 1$, the function $u(x) := x - x_0 + R$ satisfies

$$\frac{R - r}{R}\, u(x_0) = R - r = u(x_0 - r)$$

and

$$u(x_0 + r) = R + r = \frac{R + r}{R}\, u(x_0),$$

attaining the bounds presented in (2.15.1) when $n = 1$. Moreover, when $n \geqslant 2$, in light of Theorem 2.11.2, given $\varepsilon > 0$, we can consider the Poisson kernel $P_\varepsilon(\cdot, \cdot)$

for $B_{R+\varepsilon}$ and define

$$u(x) := P_\varepsilon(x, (R+\varepsilon)e_1) = \frac{(R+\varepsilon)^2 - |x|^2}{n\,|B_1|\,(R+\varepsilon)\,|x-(R+\varepsilon)e_1|^n}. \tag{2.15.3}$$

We observe that u is harmonic in B_R, thanks to (2.11.2). Moreover, if $x := re_1$,

$$\frac{u(x)}{u(0)} = \frac{\big((R+\varepsilon)^2 - r^2\big)(R+\varepsilon)^n}{(R+\varepsilon)^2|r-(R+\varepsilon)|^n},$$

which, as $\varepsilon \searrow 0$, attains the right-hand side in (2.15.1).

Similarly, if $x := -re_1$,

$$\frac{u(x)}{u(0)} = \frac{\big((R+\varepsilon)^2 - r^2\big)(R+\varepsilon)^n}{(R+\varepsilon)^2|r+(R+\varepsilon)|^n},$$

which, as $\varepsilon \searrow 0$, attains the left-hand side in (2.15.1). These observations show that the bounds in (2.15.1) are sharp and cannot be improved.

As a straightforward consequence[20] of Theorem 2.15.1, one obtains the following result, which is often referred to by the name[21] Harnack inequality:

---

[20]We mention that an alternative proof of Corollary 2.15.2 that does not rely on Theorem 2.15.1 can be obtained directly from the mean value formula in Theorem 2.1.2(iii): namely, given two points p, $q \in B_{R/8}(x_0)$, we have that $B_{R/8}(q) \subseteq B_{R/2}(p) \subseteq B_{3R/4}(x_0)$ and therefore, for a nonnegative and harmonic u in $B_R(x_0)$,

$$\begin{aligned} u(p) &= \fint_{B_{R/2}(p)} u(x)\,dx \geqslant \frac{1}{|B_{R/2}|}\int_{B_{R/8}(q)} u(x)\,dx \\ &= \frac{|B_{R/8}|}{|B_{R/2}|}\fint_{B_{R/8}(q)} u(x)\,dx = \frac{|B_{1/8}|}{|B_{1/2}|}\,u(q) \end{aligned}$$

and consequently

$$\sup_{B_{R/8}(x_0)} u \leqslant \frac{|B_{1/2}|}{|B_{1/8}|}\inf_{B_{R/8}(x_0)} u,$$

from which Corollary 2.15.2 would follow from a covering argument.

In our presentation, the relevance of Theorem 2.15.1 is however to produce sharp constants in (2.15.1).

[21]Results of this type are named after the author of the pioneering work [Har87]. The gruesome fate of Carl Gustav Axel Harnack (the mathematician) seems to be that of being too often mistaken for his twin brother Carl Gustav Adolf von Harnack (a theologian, historian, author of many religious publications, signatory of a public statement in support of German military actions in World War I and ennobled with the addition of von to his name; see also a footnote in [DV23, Chapter 24]). For a while, Wikipedia erroneously displayed a picture of Adolf in place of that of Axel: the error was spotted since Adolf's picture was representing a rather senior person, while Axel prematurely died at the age of 37 after suffering from health problems (Adolf long outlived him and died at the age of 79, confirming that personal interests and physical conditions are not always akin between twin brothers).

Quite surprisingly for us, the prestigious Harnack Medal of the Max Planck Society is named after Adolf, not after Axel. This is possibly due to the fact that Adolf was one of the founders, as well as the first president, of the Kaiser–Wilhelm–Gesellschaft zur Förderung der Wissenschaften (Society for the Advancement of Science, named in honor of the German Emperor Wilhelm II), whose functions were later taken over by the Max Planck Society.

To partially compensate for his sorry fate, a picture of Axel is given here in Figure 2.15.1.

Figure 2.15.1 Carl Gustav Axel Harnack (Public Domain image from Wikipedia).

Corollary 2.15.2. *There exists a constant $C > 1$, depending only on n, such that for every $x_0 \in \mathbb{R}^n$, every $R > 0$ and every nonnegative, harmonic function u in $B_R(x_0)$, we have that*

$$\sup_{B_{R/2}(x_0)} u \leqslant C \inf_{B_{R/2}(x_0)} u.$$

Proof. Take $r := R/2$ in Theorem 2.15.1. □

A more general version of the Harnack inequality in Corollary 2.15.2 can be obtained by a covering argument for arbitrary connected domains (with a constant dependent on the domain).

Corollary 2.15.3. *Let $\Omega \subseteq \mathbb{R}^n$ be open. Assume that u is nonnegative and harmonic in Ω, and let $\Omega' \Subset \Omega$ be open, connected and bounded.*

Then, there exists a constant $C > 1$ depending only on n, Ω' and Ω such that

$$\sup_{\Omega'} u \leqslant C \inf_{\Omega'} u.$$

Proof. We let D be the distance between Ω' and $\partial\Omega$ and set $R := D/4$ (if D is finite; if $D = +\infty$, simply take $R := 1$). Let $p, q \in \overline{\Omega'}$ be such that $\max_{\overline{\Omega'}} u = u(p)$ and $\min_{\overline{\Omega'}} u = u(q)$. By the connectedness of Ω', we can take a closed arc γ joining p and q that lies in Ω'. Since

$$\overline{\Omega'} \subseteq \bigcup_{x\in\Omega'} B_{R/2}(x),$$

by compactness we can find $x^{(1)}, \ldots, x^{(N)} \in \Omega'$ such that

$$\Omega' \subseteq \bigcup_{i=1}^{N} B_{R/2}(x^{(i)}),$$

with N depending only on Ω and Ω'. We let $i_1, \ldots, i_{N'} \in \{1, \ldots, N\}$ be such that $\gamma \subseteq B_{R/2}(x^{(i_1)}) \cup \cdots \cup B_{R/2}(x^{(i_{N'})})$. Without loss of generality, we can consider that all $x^{(i_j)}$ are different. Additionally, we can conveniently reorder these balls (and possibly disregard balls that do not contribute to the covering of γ). Namely, we define $B^{(1)} := B_{R/2}(x^{(i_1)})$. If γ is all contained in $B^{(1)}$, we stop; otherwise, by the continuity of the arc γ, there must be a point P_1 of γ on $\partial B^{(1)}$. Hence, there must exist a ball of the collection $\{B_{R/2}(x^{(i_2)}), \ldots, B_{R/2}(x^{(i_M)})$ that covers P_1, and we call this ball $B^{(2)}$. Again, if γ is contained in $B^{(1)} \cup B^{(2)}$, we stop; otherwise, we consider a point $P_2 \in \partial(B^{(1)} \cup B^{(2)})$ and a ball of the remaining collection, which we denote by $B^{(3)}$ and which contains P_3. Recursively, we have found balls $B^{(1)}, \ldots, B^{(M)}$ such that $M \leqslant N' \leqslant N$, with

$$\gamma \subseteq B^{(1)} \cup \cdots \cup B^{(M)}$$

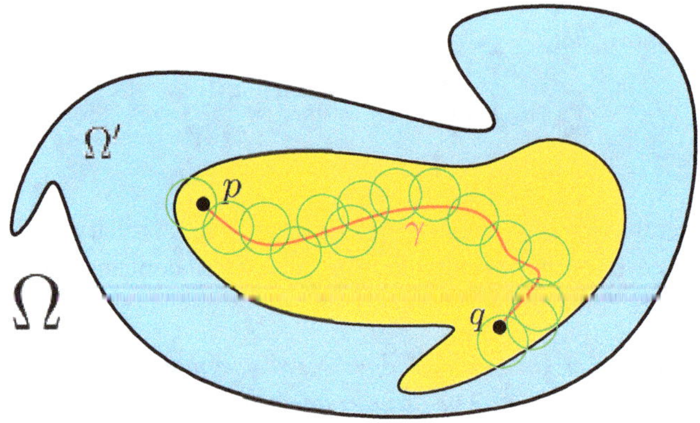

Figure 2.15.2 Proof of Corollary 2.15.3.

and with the additional property that, for all $j \in \{1, \dots, M-1\}$,

$$\left(B^{(1)} \cup \cdots \cup B^{(j)}\right) \cap B^{(j+1)} \neq \varnothing,$$

see Figure 2.15.2.

In particular, taking ζ_j in the intersection above and applying Corollary 2.15.2, we see that

$$\sup_{B^{(j+1)}} u \leqslant C \inf_{B^{(j+1)}} u \leqslant Cu(\zeta_j) \leqslant C \sup_{B^{(1)}\cup\cdots\cup B^{(j)}} u \tag{2.15.4}$$

and

$$\inf_{B^{(1)}\cup\cdots\cup B^{(j)}} u \leqslant u(\zeta_j) \leqslant \sup_{B^{(j+1)}} u \leqslant C \inf_{B^{(j+1)}} u, \tag{2.15.5}$$

for some $C > 1$ depending only on n.

Now, we claim that, for every $j \in \{1, \dots, M\}$,

$$\sup_{B^{(1)}\cup\cdots\cup B^{(j)}} u \leqslant C^{j-1} \sup_{B^{(1)}} u. \tag{2.15.6}$$

To prove this, we argue by induction. When $j = 1$, the claim is obvious, and hence we suppose the claim to be true for some index $j \in \{1, \dots, M-1\}$, and we aim at establishing it for the index $j+1$. To this end, we make use of (2.15.4) and the inductive assumption to find that

$$\begin{aligned}
\sup_{B^{(1)}\cup\cdots\cup B^{(j+1)}} u &= \max\left\{ \sup_{B^{(1)}\cup\cdots\cup B^{(j)}} u,\ \sup_{B^{(j+1)}} u \right\} \\
&\leqslant C \sup_{B^{(1)}\cup\cdots\cup B^{(j)}} u \\
&\leqslant C^j \sup_{B^{(1)}} u.
\end{aligned}$$

This completes the inductive step and establishes (2.15.6).

In analogy with (2.15.6), we claim that

$$\inf_{B^{(1)}\cup\cdots\cup B^{(j)}} u \geqslant C^{1-j} \inf_{B^{(1)}} u. \tag{2.15.7}$$

Once again, for this, we argue by induction. Since the desired result is obvious when $j = 1$, we suppose that the claim in (2.15.7) holds true for some index $j \in \{1, \dots, M-1\}$, and we aim at establishing it for the index $j+1$. For this, using (2.15.5) and the inductive assumption, we see that

$$\begin{aligned}\inf_{B^{(1)}\cup\cdots\cup B^{(j+1)}} u &= \min\left\{\inf_{B^{(1)}\cup\cdots\cup B^{(j)}} u,\ \inf_{B^{(j+1)}} u\right\}\\ &\geqslant C^{-1} \inf_{B^{(1)}\cup\cdots\cup B^{(j)}} u\\ &\geqslant C^{-j} \inf_{B^{(1)}} u,\end{aligned}$$

which finishes the proof of (2.15.7).

Now, we take $j_q \in \{1, \ldots, M\}$ such that $q \in B^{(j_q)}$ and $j_p \in \{1, \ldots, M\}$ such that $p \in B^{(j_p)}$. Combining (2.15.6) and (2.15.7) and applying Corollary 2.15.2 once more, we thus deduce that

$$\begin{aligned}\max_{\overline{\Omega'}} u = u(p) &\leqslant \sup_{B^{(j_p)}} u \leqslant \sup_{B^{(1)}\cup\cdots\cup B^{(j_p)}} u \leqslant C^{j_p-1} \sup_{B^{(1)}} u \leqslant C^{j_p} \inf_{B^{(1)}} u\\ &\leqslant C^{j_p+j_q-1} \inf_{B^{(1)}\cup\cdots\cup B^{(j_q)}} u \leqslant C^{j_p+j_q-1} \inf_{B^{(j_q)}} u \leqslant C^{2N-1} u(q) = C^{2N-1} \min_{\overline{\Omega'}} u,\end{aligned}$$

as desired. □

The Harnack inequality, in its various forms, is certainly of paramount importance since, in broad terms, it states that in the interior, all the values of harmonic functions are essentially comparable (and this, in particular, provides a quantitative version of the maximum principle in Theorem 2.9.1(iii)).

Here, we present a variant of Corollary 2.2.2 that exploits the Harnack inequality, which is sometimes called the Harnack convergence theorem.

Corollary 2.15.4. *Let $\Omega \subseteq \mathbb{R}^n$ be open and connected. Let u_k be a monotone increasing sequence of harmonic functions in Ω, and suppose that there exists $y \in \Omega$ such that the sequence $u_k(y)$ is bounded from above.*

Then, the sequence u_k converges locally uniformly in Ω to a harmonic function.

Proof. Let $\Omega' \Subset \Omega$ be open and bounded. Let $\Omega'' \Subset \Omega$ be connected, open, bounded and such that $\Omega' \cup \{y\} \subseteq \Omega''$.

By construction, the sequence $u_k(y)$ is convergent. In particular, given $\varepsilon > 0$, there exists $k_\varepsilon \in \mathbb{N}$ such that for all $k,\ k' \geqslant k_\varepsilon$, say with $k \geqslant k'$, we have that $|u_k(y) - u_{k'}(y)| \leqslant \varepsilon$.

We observe, in addition, that the function $u_k - u_{k'}$ is harmonic and nonnegative. Thus, for every $x \in \Omega''$,

$$\begin{aligned}|u_k(x) - u_{k'}(x)| = u_k(x) - u_{k'}(x) &\leqslant \sup_{\Omega''}(u_k - u_{k'}) \leqslant C \inf_{\Omega''}(u_k - u_{k'})\\ &\leqslant C\,(u_k(y) - u_{k'}(y)) \leqslant C\varepsilon,\end{aligned}$$

as long as $k \geqslant k' \geqslant k_\varepsilon$, thanks to Corollary 2.15.3.

For this reason, we have that u_k converges uniformly in Ω'' (and therefore in Ω') to some function u. Hence, recalling Corollary 2.2.2, we obtain that u is harmonic. □

A boundary counterpart of the Harnack inequality is briefly discussed in Section 2.19.

2.16 The Hopf Lemma

Concerning the boundary behavior of harmonic functions, we recall here a (simple version of a) classical result, often referred to by the name Hopf lemma.

Lemma 2.16.1. *Let Ω be a bounded, connected, open set with C^1 boundary. Let $u \in C^2(\Omega) \cap C^1(\overline{\Omega})$ be harmonic in Ω. Assume that $x_0 \in \partial\Omega$ is such that*

$$\max_{\partial\Omega} u = u(x_0).$$

Suppose also that there exists a ball $B \subseteq \Omega$ such that $x_0 \in \partial B$.

Then, either $\partial_\nu u(x_0) > 0$ or u is constant in Ω.

Proof. By the weak maximum principle in Corollary 2.9.2(iii), we already know that $u(x_0) = \max_{\partial\Omega} u = \max_{\overline{\Omega}} u$. Hence, $u(x_0) = \max_{\partial B} u$.

Consequently, if $\max_{\partial B} u = \min_{\partial B} u$, it follows that u is constantly equal to $u(x_0)$ along ∂B. In particular, u attains an interior maximum in Ω, and accordingly, using the strong maximum principle in Theorem 2.9.1(iii), we obtain that u is constant, and we are done.

For this reason, we can suppose that $\max_{\partial B} u > \min_{\partial B} u$. As a result, by the continuity of u, we can take $x_\star \in \partial B$ such that $u(x_\star) = \min_{\partial B} u$ and also $\rho \in \left(0, \frac{|x_0 - x_\star|}{2}\right)$ and $a > 0$ such that for all $y \in (\partial B) \cap B_\rho(x_\star)$, we have

$$u(y) + a \leqslant \max_{\partial B} u = \max_{\partial\Omega} u. \tag{2.16.1}$$

Now, up to a translation, we can assume that the ball B is centered at the origin, and hence we have that $B = B_R$ for some $R > 0$. Furthermore, up to a rotation, since $x_0 \in \partial B = \partial B_R$, we can suppose that $x_0 = Re_1$. Then, the exterior normal of B_R at x_0 coincides with the exterior normal of Ω at x_0, and thus, without any ambiguity, we can write $\nu(x_0) = e_1$. Let also $t > 0$, which is to be taken as small as we wish in the following, and $x_t := x_0 - t\nu(x_0) = (R-t)e_1$. Thus, we exploit the Poisson kernel representation in Theorems 2.11.1 and 2.11.2, as well as (2.16.1), to obtain that

$$\begin{aligned} u(x_0) - u(x_t) &= \max_{\partial\Omega} u - \int_{\partial B_R} u(y)\, P(x_t, y)\, d\mathcal{H}_y^{n-1} \\ &= \int_{\partial B_R} \left(\max_{\partial\Omega} u - u(y) \right) P(x_t, y)\, d\mathcal{H}_y^{n-1} \\ &= \frac{R^2 - |x_t|^2}{n\,|B_1|\,R} \int_{\partial B_R} \left(\max_{\partial\Omega} u - u(y) \right) \frac{d\mathcal{H}_y^{n-1}}{|x_t - y|^n} \\ &\geqslant \frac{a\,(R^2 - |x_t|^2)}{n\,|B_1|\,R} \int_{(\partial B_R)\cap B_\rho(x_\star)} \frac{d\mathcal{H}_y^{n-1}}{|x_t - y|^n}. \end{aligned}$$

Hence, since, if $y \in B_\rho(x_\star)$,

$$\begin{aligned} |x_t - y| &= |x_0 - t\nu(x_0) - y| \geqslant |x_0 - t\nu(x_0) - x_\star| - |x_\star - y| \\ &\geqslant |x_0 - x_\star| - t - \rho \geqslant \frac{|x_0 - x_\star|}{2} - t \geqslant \frac{|x_0 - x_\star|}{4} \end{aligned}$$

and

$$R^2 - |x_t|^2 = R^2 - (R-t)^2 = 2Rt - t^2 \geqslant Rt$$

as long as t is small enough, we conclude that

$$\begin{aligned} u(x_0) - u(x_t) &\geqslant \frac{4^n a t}{n\,|B_1|} \int_{(\partial B_R)\cap B_\rho(x_\star)} \frac{d\mathcal{H}_y^{n-1}}{|x_0 - x_\star|^n} \\ &= \frac{4^n a\, \mathcal{H}^{n-1}((\partial B_R) \cap B_\rho(x_\star))\, t}{n\,|B_1|\,|x_0 - x_\star|^n}. \end{aligned}$$

Dividing by t and sending $t \searrow 0$, we obtain the desired result. □

Different proofs of Lemma 2.16.1 can be obtained by constructing suitable barriers and utilizing the maximum principle, see e.g. [GT01] (as a matter of fact, we use a barrier method here to prove a useful generalization of Lemma 2.16.1, as given in Lemma 2.16.3).

We stress that the interior ball condition at x_0 in Lemma 2.16.1 (that is, the existence of the ball B tangent from inside to $\partial\Omega$ and containing x_0 on its boundary, see Figure 2.16.1) can be relaxed but[22] not completely dropped, see e.g. [AN16]

[22] Also, the assumption that the boundary of Ω is of class C^1 cannot be removed from Lemma 2.16.1. Not only is this assumption needed to properly define the exterior normal ν, but it is also necessary to avoid cases in which u could be smooth in the vicinity of the point x_0, but $\nabla u(x_0) = 0$. As an example, one can consider $\Omega := (0,1)^2 \subset \mathbb{R}^2$, $x_0 := (0,0)$ and $u(x,y) := -xy$, since in this situation one has that $x_0 \in \partial\Omega$, $u(x_0) = 0 = \max_{\overline{\Omega}} u$, but $\nabla u(x_0) = 0$.

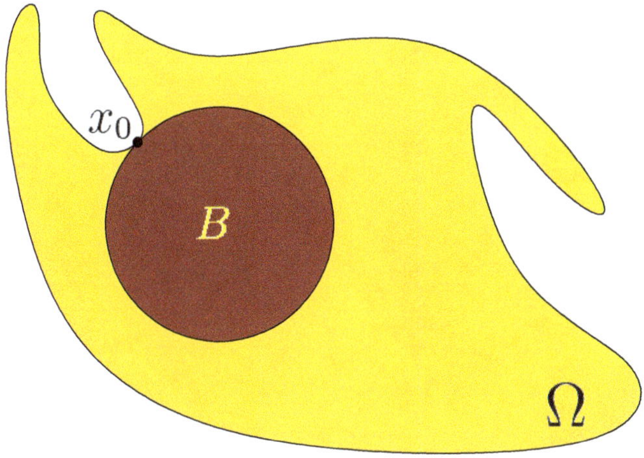

Figure 2.16.1 The interior ball condition in Lemma 2.16.1.

and the references therein (as well as Appendix A and Lemma 8.3.1 in this set of notes) for further details. We also remark that this interior ball condition is satisfied if the boundary of Ω is of class $C^{1,1}$, see e.g. [ROV16, Lemma A.1]. See also [Jos13, PS07b] and the references therein for additional details on the Hopf lemma and related topics.

In terms of generality of the methods used here, it is worth pointing out that the strong maximum principle in (2.9.1) fails if the Laplacian is replaced with more general operators with zero-order terms; that is, the solutions of equations of the form $\Delta u(x) + c(x)u(x) = 0$ may develop interior minima and maxima, even for "nice" coefficients c (as an example, we observe that the function $\mathbb{R} \ni x \mapsto u(x) := \cos x$ satisfies $\Delta u + u = 0$ in $\mathbb{R}$, and it possesses interior minima and maxima; another example is provided by the function $\mathbb{R}^2 \ni x = (x_1, x_2) \mapsto u(x) = \sin x_1 \sin x_2$, which satisfies $\Delta u + 2u = 0$ in $\mathbb{R}^2$, see Figure 2.16.2). However, a variation of Theorem 2.9.1 holds true under suitable sign assumptions (e.g. on the coefficient c or on the solution). In this spirit, we point out here a result that will be employed in Section 8.4.

Theorem 2.16.2. *Let $\Omega \subseteq \mathbb{R}^n$ be open and connected.*

Let a_{ij}, b_i, $c \in C(\overline{\Omega})$. Assume that for every $\xi = (\xi_1, \dots, \xi_n) \in \partial B_1$

$$\sum_{i,j=1}^{n} a_{ij}(x)\xi_i\xi_j \in [\lambda, \Lambda], \tag{2.16.2}$$

for some $\Lambda \geqslant \lambda > 0$.

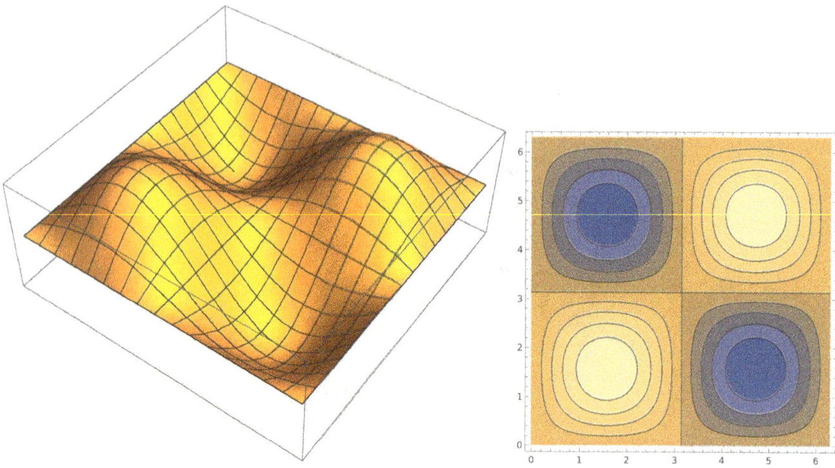

Figure 2.16.2 Graph and level sets of the function $u(x_1, x_2) = \sin x_1 \sin x_2$.

Let $u \in C^2(\Omega)$ be such that

$$\begin{cases} \displaystyle\sum_{i,j=1}^{n} a_{ij}(x)\partial_{ij}u(x) + \sum_{i=1}^{n} b_i(x)\partial_i u(x) + c(x)u(x) \geqslant 0 & \text{for all } x \in \Omega, \\ u(x) \leqslant 0 & \text{for all } x \in \Omega. \end{cases} \tag{2.16.3}$$

Then, $u < 0$ in Ω, unless it is constantly equal to zero.

The reason for which we consider quite general operators in (2.16.3) (and not just, say, the Laplacian plus possibly a zero-order term) is that these operators pop up naturally when we deal with soap bubbles in Section 8.2: compare, in particular, with equation (8.2.14); fortunately, this additional generality does not really cause major complications in the proofs, whose structures would remain pretty much the same even in the case of a Laplacian plus a zero-order term.

As we will see more specifically in Section 4.4, the assumption in (2.16.2) can be seen as an "ellipticity requirement" on the leading coefficients of the equation (recall also the discussion about elliptic, parabolic and hyperbolic equations in [DV23, Chapter 1]): in rough terms, this type of condition is usually taken to ensure some kinship to the Laplace operator and, for instance, guarantee[23] the validity of a suitable maximum principle.

The proof of Theorem 2.16.2 makes use of a generalization of the Hopf lemma presented in Lemma 2.16.1, as follows.

Lemma 2.16.3. *Let B be an open ball in $\mathbb{R}^n$ with outward normal ν.*

Let a_{ij}, b_i, $c \in C(\overline{B})$. Assume the ellipticity condition in (2.16.2), *and suppose also that $c \leqslant 0$ in B.*

Let $u \in C^2(B) \cap C^1(\overline{B})$ be a solution of

$$\sum_{i,j=1}^{n} a_{ij}(x)\partial_{ij}u(x) + \sum_{i=1}^{n} b_i(x)\partial_i u(x) + c(x)u(x) \geqslant 0 \quad \text{for all } x \in B.$$

Assume that $x_0 \in \partial B$ is such that $u(x_0) \geqslant 0$ and $u(x) < u(x_0)$ for every $x \in B$.

Then, $\partial_\nu u(x_0) > 0$.

Proof. Let $\widetilde{B}$ be a ball contained in B, with a radius half of that of B and such that $x_0 \in \partial\widetilde{B}$. Up to a translation, we suppose that $\widetilde{B}$ is centered at the origin, namely $\widetilde{B} = B_r$ for some $r > 0$. Let $S := B_r \cap B_{r/2}(x_0)$ and note that $B_r \cap (\partial B_{r/2}(x_0)) \Subset B$, see Figure 2.16.3; therefore,

$$\sup_{B_r \cap (\partial B_{r/2}(x_0))} u < u(x_0).$$

[23]To see why condition (2.16.2) is advantageous when dealing with a maximum principle, we point out that such a requirement "behaves nicely" with respect to the maxima of a function. For instance, if u has a local maximum at some point $\overline{x}$, then

$$\sum_{i,j=1}^{n} a_{ij}(\overline{x})\partial_{ij}u(\overline{x}) \leqslant 0.$$

Since we repeatedly exploit this observation, let us give a quick proof for it. By construction, the Hessian matrix of u at $\overline{x}$ is nonpositive; therefore, we can denote by $M = \{M_{ij}\}_{i,j\in\{1,\dots,n\}}$ the square root (in the matrix sense, see e.g. [Gen07, p. 125]) of the negative of the Hessian matrix of u at $\overline{x}$. In this way, we have that

$$-\partial_{ij}u(\overline{x}) = \sum_{k=1}^{n} M_{ik}M_{jk}.$$

We also let $\zeta_k := (M_{1k}, \dots, M_{nk})$ so that

$$|\zeta_k|^2 = \sum_{\ell=1}^{n} M_{\ell k}^2.$$

As a result, from the ellipticity assumption in (2.16.2), we arrive at

$$-\sum_{i,j=1}^{n} a_{ij}(\overline{x})\partial_{ij}u(\overline{x}) = \sum_{i,j,k=1}^{n} a_{ij}(\overline{x})M_{ik}M_{jk} \geqslant \lambda\sum_{k=1}^{n}|\zeta_k|^2 = \lambda\sum_{k,\ell=1}^{n}|M_{\ell k}|^2 \geqslant 0.$$

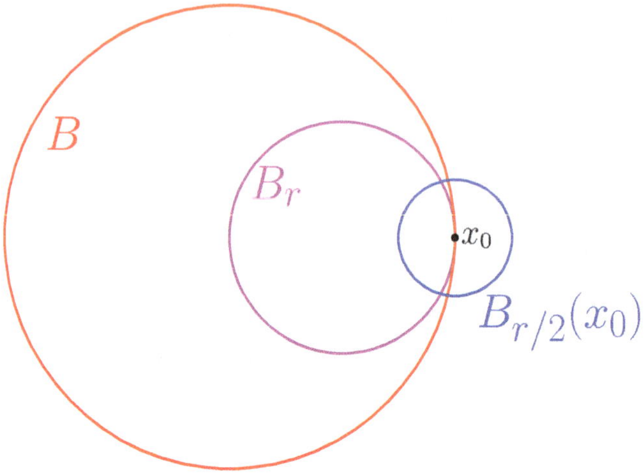

Figure 2.16.3 The geometry involved in the proof of Lemma 2.16.3.

Hence, there exists $\delta_0 > 0$ such that

$$\sup_{B_r \cap (\partial B_{r/2}(x_0))} u + \delta_0 \leqslant u(x_0). \tag{2.16.4}$$

Let also

$$h(x) := e^{-\alpha|x|^2} - e^{-\alpha r^2}, \tag{2.16.5}$$

with $\alpha > 1$ to be conveniently chosen. We point out that, for all $x \in B_r$, $h(x) \geqslant 0$, and using the sign of c and the ellipticity condition in (2.16.2), we find that, for all $x \in B_r \setminus B_{r/4}$,

$$\begin{aligned}
&\sum_{i,j=1}^{n} a_{ij}(x)\partial_{ij}h(x) + \sum_{i=1}^{n} b_i(x)\partial_i h(x) + c(x)h(x) \\
&\quad = 2\alpha e^{-\alpha|x|^2} \sum_{i,j=1}^{n} a_{ij}(x)\big(2\alpha x_i x_j - \delta_{ij}\big) \\
&\qquad - 2\alpha e^{-\alpha|x|^2} \sum_{i=1}^{n} b_i(x)x_i + c(x)\big(e^{-\alpha|x|^2} - e^{-\alpha r^2}\big) \\
&\quad \geqslant 2\alpha e^{-\alpha|x|^2} \sum_{i,j=1}^{n} a_{ij}(x)\big(2\alpha x_i x_j - \delta_{ij}\big) - 2\alpha e^{-\alpha|x|^2} \sum_{i=1}^{n} b_i(x)x_i + c(x)e^{-\alpha|x|^2} \\
&\quad = 2\alpha e^{-\alpha|x|^2} \left[\sum_{i,j=1}^{n} a_{ij}(x)\big(2\alpha x_i x_j - \delta_{ij}\big) - \sum_{i=1}^{n} b_i(x)x_i + \frac{c(x)}{2\alpha}\right]
\end{aligned}$$

$$\begin{aligned}
&\geqslant 2\alpha e^{-\alpha|x|^2}\left[2\alpha\lambda|x|^2 - n^2 \sup_{i,j\in\{1,\dots,n\}} \|a_{ij}\|_{L^\infty(B)}\right.\\
&\qquad \left. - nr \sup_{i\in\{1,\dots,n\}} \|b_i\|_{L^\infty(B)} - \frac{\|c\|_{L^\infty(B)}}{2}\right]\\
&\geqslant 2\alpha e^{-\alpha|x|^2}\left[2\alpha\lambda|x|^2 - C\,(1+r)\right]\\
&\geqslant 2\alpha e^{-\alpha|x|^2}\left[\frac{\alpha\lambda r^2}{8} - C\,(1+r)\right],
\end{aligned}$$

where $C > 0$ depends only on n and on the structural bounds on the coefficients of the equation.

As a result, for all $x \in B_r \setminus B_{r/4}$,

$$\sum_{i,j=1}^{n} a_{ij}(x)\partial_{ij}h(x) + \sum_{i=1}^{n} b_i(x)\partial_i h(x) + c(x)h(x) \geqslant 2\alpha e^{-\alpha|x|^2}, \tag{2.16.6}$$

as long as α is sufficiently large (and this determines α once and for all).

Now, let $\varepsilon \in (0, \delta_0)$, to be chosen conveniently small in what follows, and $v_\varepsilon := u + \varepsilon h$. We point out that

$$v_\varepsilon(x_0) = u(x_0) \geqslant 0. \tag{2.16.7}$$

We claim that

$$\sup_S v_\varepsilon = \sup_{\partial S} v_\varepsilon. \tag{2.16.8}$$

Indeed, suppose not. Then, v_ε possesses an interior maximum x_ε in S. For this reason, $\nabla v_\varepsilon(x_\varepsilon) = 0$ and $D^2 v_\varepsilon(x_\varepsilon) \leqslant 0$, which, using again the ellipticity condition in (2.16.2), yields that

$$\sum_{i,j=1}^{n} a_{ij}(x_\varepsilon)\partial_{ij}v_\varepsilon(x_\varepsilon) + \sum_{i=1}^{n} b_i(x_\varepsilon)\partial_i v_\varepsilon(x_\varepsilon) + c(x_\varepsilon)v_\varepsilon(x_\varepsilon) \leqslant c(x_\varepsilon)v_\varepsilon(x_\varepsilon).$$

Moreover, $v_\varepsilon(x_\varepsilon) \geqslant v_\varepsilon(x_0) \geqslant 0$, thanks to (2.16.7). Using these observations, the sign of c and (2.16.6), we infer that

$$\begin{aligned}
0 &\geqslant c(x_\varepsilon)v_\varepsilon(x_\varepsilon)\\
&\geqslant \sum_{i,j=1}^{n} a_{ij}(x_\varepsilon)\partial_{ij}v_\varepsilon(x_\varepsilon) + \sum_{i=1}^{n} b_i(x_\varepsilon)\partial_i v_\varepsilon(x_\varepsilon) + c(x_\varepsilon)v_\varepsilon(x_\varepsilon)\\
&\geqslant \varepsilon\left[\sum_{i,j=1}^{n} a_{ij}(x_\varepsilon)\partial_{ij}h(x_\varepsilon) + \sum_{i=1}^{n} b_i(x_\varepsilon)\partial_i h(x_\varepsilon) + c(x_\varepsilon)h(x_\varepsilon)\right]\\
&\geqslant 2\alpha\varepsilon e^{-\alpha|x_\varepsilon|^2}.
\end{aligned}$$

But the latter term is strictly positive, and this contradiction establishes (2.16.8).

We also observe that if $x \in \partial B_r$, then $h(x) = 0$ and therefore $v_\varepsilon(x) = u(x) \leqslant u(x_0) \leqslant v_\varepsilon(x_0)$. This gives that

$$\sup_{(\partial B_r)\cap B_{r/2}(x_0)} v_\varepsilon = v_\varepsilon(x_0). \tag{2.16.9}$$

Furthermore, by (2.16.4), if $x \in B_r \cap (\partial B_{r/2}(x_0))$, then

$$v_\varepsilon(x) \leqslant u(x) + \varepsilon \|h\|_{L^\infty(B_r)} \leqslant u(x) + \delta_0 \leqslant u(x_0) \leqslant v_\varepsilon(x_0),$$

provided that ε is chosen suitably small (and this determines ε once and for all). Accordingly,

$$\sup_{B_r \cap (\partial B_{r/2}(x_0))} v_\varepsilon \leqslant v_\varepsilon(x_0).$$

With this, (2.16.8) and (2.16.9), we see that

$$\sup_S v_\varepsilon = \max\left\{ \sup_{(\partial B_r)\cap B_{r/2}(x_0)} v_\varepsilon, \sup_{B_r \cap (\partial B_{r/2}(x_0))} v_\varepsilon \right\} = v_\varepsilon(x_0).$$

As a consequence,

$$0 \leqslant \partial_\nu v_\varepsilon(x_0) = \partial_\nu u(x_0) + \varepsilon \partial_\nu h(x_0) = \partial_\nu u(x_0) - 2\alpha\varepsilon |x_0| e^{-\alpha |x_0|^2} < \partial_\nu u(x_0),$$

as desired. □

A consequence of Lemma 2.16.3 is that, if the solution u has a sign, then the sign assumption on c can be dropped, according to the following statement.

Corollary 2.16.4. *Let B be an open ball in $\mathbb{R}^n$ with outward normal ν.*

Let $a_{ij}, b_i, c \in C(\overline{B})$. Assume the ellipticity condition in (2.16.2).

Let $u \in C^2(B) \cap C^1(\overline{B})$ be a solution of

$$\sum_{i,j=1}^n a_{ij}(x)\partial_{ij}u(x) + \sum_{i=1}^n b_i(x)\partial_i u(x) + c(x)u(x) \geqslant 0 \quad \text{for all } x \in B.$$

Assume that $x_0 \in \partial B$ is such that $u(x_0) = 0$ and $u(x) < 0$ for every $x \in B$.

Then, $\partial_\nu u(x_0) > 0$.

Proof. We note that, for every $x \in B$,

$$\begin{aligned} 0 &\leqslant \sum_{i,j=1}^n a_{ij}(x)\partial_{ij}u(x) + \sum_{i=1}^n b_i(x)\partial_i u(x) + c(x)u(x) \\ &= \sum_{i,j=1}^n a_{ij}(x)\partial_{ij}u(x) + \sum_{i=1}^n b_i(x)\partial_i u(x) + c^+(x)u(x) - c^-(x)u(x) \\ &\leqslant \sum_{i,j=1}^n a_{ij}(x)\partial_{ij}u(x) + \sum_{i=1}^n b_i(x)\partial_i u(x) - c^-(x)u(x). \end{aligned}$$

Accordingly, the assumptions in Lemma 2.16.3 are satisfied, taking $c := -c^- \leqslant 0$, and therefore $\partial_\nu u(x_0) > 0$, as desired. □

Proof of Theorem 2.16.2. Let $\mathcal{U} := \{x \in \Omega \text{ s.t. } u(x) = 0\}$, and note that $\mathcal{U}$ is a closed set in Ω. If $\mathcal{U} = \varnothing$, then $u < 0$ in Ω, and we are done. Hence, we can suppose that $\mathcal{U} \neq \varnothing$. Our goal is to show that $\mathcal{U} = \Omega$, since in this way we would get that u vanishes identically, as claimed in Theorem 2.16.2. For this, we argue by contradiction and assume that $\mathcal{U}$ is strictly contained in Ω. In particular, there exists $p \in \mathcal{U}$ such that for all $j \in \mathbb{N}$, we have that $B_{1/j}(p) \setminus \mathcal{U} \neq \varnothing$ (otherwise, $\mathcal{U}$ would be also open and then coincide with Ω by connectedness).

As a consequence, we can find a point $q \in \Omega \setminus \mathcal{U}$ with $\text{dist}(q, \partial\Omega) > |q - p|$. That is,

$$\text{dist}(q, \partial\Omega) > |q - p| \geqslant \text{dist}(q, \mathcal{U}) =: r,$$

yielding that

$$B_r(q) \text{ is contained in } \Omega \setminus \mathcal{U} \text{ and there exists } x_0 \in (\partial B_r(q)) \cap \mathcal{U}.$$

See Figure 2.16.4 for a sketch of this configuration.

Now, we let

$$c^+ := \max\{c, 0\} \quad \text{and} \quad c^- := \max\{-c, 0\}. \tag{2.16.10}$$

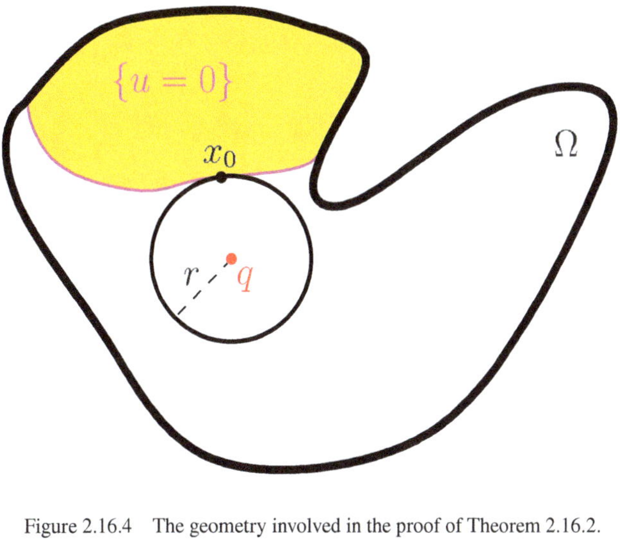

Figure 2.16.4 The geometry involved in the proof of Theorem 2.16.2.

In this way, $c^+ \geqslant 0$, $c^- \geqslant 0$ and $c = c^+ - c^-$. Therefore, using the sign of u, for every $x \in \Omega$,

$$\begin{aligned}
&\sum_{i,j=1}^{n} a_{ij}(x)\partial_{ij}u(x) + \sum_{i=1}^{n} b_i(x)\partial_i u(x) - c^-(x)u(x) \\
&\quad = \sum_{i,j=1}^{n} a_{ij}(x)\partial_{ij}u(x) + \sum_{i=1}^{n} b_i(x)\partial_i u(x) + c(x)u(x) - c^+(x)u(x) \\
&\quad \geqslant -c^+(x)u(x) \\
&\quad \geqslant 0.
\end{aligned}$$

Given the sign of c^-, we are therefore in a position to use Lemma 2.16.3 in this context. With the aid of Lemma 2.16.3 (used here with c replaced by $-c^-$ and B replaced by $B_r(q)$), we thus deduce that either u is constant (and thus constantly equal to zero, since $\mathcal{U}$ is nonvoid) or $\nabla u(x_0) \neq 0$. But the latter condition cannot hold (since x_0 is an interior maximum, and hence $\nabla u(x_0) = 0$), whence the proof of the desired result is complete. □

Putting together Theorem 2.16.2 and Lemma 2.16.3, we have the following.

Corollary 2.16.5. *Let B be an open ball in $\mathbb{R}^n$ with outward normal ν. Let a_{ij}, $b_i \in C(\overline{B})$, and assume the ellipticity condition in* (2.16.2).

Let $u \in C^2(B) \cap C^1(\overline{B})$. Suppose that

$$\begin{cases} \displaystyle\sum_{i,j=1}^{n} a_{ij}(x)\partial_{ij}u(x) + \sum_{i=1}^{n} b_i(x)\partial_i u(x) \geqslant 0 & \textit{for all } x \in B, \\ u(x) \leqslant 0 & \textit{for all } x \in B. \end{cases}$$

Let $x_0 \in \partial B$ be such that $u(x_0) = 0$. Then, either $\partial_\nu u(x_0) > 0$ or u vanishes identically.

Proof. Suppose that u does not vanish identically in B. Then, by Theorem 2.16.2, we have that $u < 0$ in B. Consequently, we can use Lemma 2.16.3 and infer that $\partial_\nu u(x_0) > 0$. □

A related result is the so-called maximum principle for narrow domains: this result focuses on domains which are constrained in a sufficiently thin strip. It is presented as follows.

Theorem 2.16.6. *Let Ω be an open and bounded subset of $\mathbb{R}^n$ with boundary of class C^1. Let $c \in C(\overline{\Omega})$ and assume that*

$$\Omega \Subset \{x = (x_1, \dots, x_n) \in \mathbb{R}^n \text{ s.t. } x_1 \in (0, d)\}$$
$$\text{for some } d > 0 \text{ with } \|c^+\|^{1/2}_{L^\infty(\Omega)}\, d \in [0, \pi]. \tag{2.16.11}$$

Let $u \in C^2(\Omega) \cap C^1(\overline{\Omega})$ be a solution of

$$\begin{cases} \Delta u(x) + c(x)u(x) \geqslant 0 & \text{for all } x \in \Omega, \\ u(x) \leqslant 0 & \text{for all } x \in \partial\Omega. \end{cases}$$

Then, $u \leqslant 0$ in Ω.

Also, if Ω is connected, then $u < 0$ in Ω, unless it is constantly equal to zero.

From a geometric point of view, condition (2.16.11) requires the domain Ω to lie within a slab of sufficiently small width, which justifies the name of maximum principle for narrow domains (the position of the slab is irrelevant, up to a translation and a rotation, recall (1.1.2) and Corollary 1.1.4).

Proof of Theorem 2.16.6. First of all, we show that

$$u(x) \leqslant 0 \quad \text{for every } x \in \Omega. \tag{2.16.12}$$

For this, up to reducing our analysis to each connected component of Ω, we can assume that

$$\Omega \text{ is connected.} \tag{2.16.13}$$

Let

$$w(x) := \sin \frac{\pi x_1}{d},$$

and observe that $w > 0$ in $\overline{\Omega}$. Thus, if

$$\Lambda := \frac{\|u\|_{L^\infty(\Omega)}}{\min\limits_{\overline{\Omega}} w} + 1$$

we see that, for all $x \in \overline{\Omega}$,

$$\Lambda w(x) - u(x) \geqslant \Lambda \min_{\overline{\Omega}} w - \|u\|_{L^\infty(\Omega)} > 0.$$

We thereby define

$$\Lambda_0 := \inf \{\Lambda \text{ s.t. } \Lambda w(x) - u(x) > 0 \text{ for every } x \in \Omega\}. \tag{2.16.14}$$

Now, we claim that

$$\Lambda_0 \leqslant 0. \tag{2.16.15}$$

Indeed, suppose not, namely $\Lambda_0 > 0$. Let $v := u - \Lambda_0 w$, and observe that, for all $x \in \Omega$,

$$\Delta v(x) + c(x)v(x) \geqslant -\Lambda_0 \Big(\Delta w(x) + c(x)w(x)\Big) = -\Lambda_0 \left(-\frac{\pi^2}{d^2} + c(x)\right) w(x).$$

From this and the fact (recall (2.16.11)) that

$$c(x) \leqslant c^+(x) \leqslant \|c^+\|_{L^\infty(\Omega)} \leqslant \frac{\pi^2}{d^2},$$

we conclude that $\Delta v(x) + c(x)v(x) \geqslant 0$ for all $x \in \Omega$. Also, $v \leqslant 0$ in Ω due to (2.16.14); therefore, in the setting of (2.16.13), we are in a position to use Theorem 2.16.2 and deduce that

$$\text{either } v < 0 \text{ in } \Omega \quad \text{or} \quad v \text{ vanishes identically.} \tag{2.16.16}$$

We also remark that, on $\partial\Omega$, it holds that $v < u \leqslant 0$. Combining this information with (2.16.16), we infer that $v < 0$ in $\overline{\Omega}$. As a result,

$$\varepsilon_0 := -\max_{\overline{\Omega}} v > 0;$$

therefore, for all $x \in \Omega$ and $\Lambda \in (\Lambda_0 - \varepsilon_0, \Lambda_0]$,

$$\Lambda w(x) - u(x) = (\Lambda - \Lambda_0)w(x) - v(x) \geqslant \varepsilon_0 - |\Lambda - \Lambda_0| = \varepsilon_0 - \Lambda_0 + \Lambda > 0.$$

This is in contradiction with the infimum property in (2.16.14), and so it completes the proof of (2.16.15).

From (2.16.15), we deduce (2.16.12).

This and Theorem 2.16.2 give the desired result. □

A variant of the maximum principle for narrow domains in Theorem 2.16.6 is the so-called maximum principle for small volume domains, which is given as follows.

Theorem 2.16.7. *There exists $c_0 \in (0, 1)$, depending only on n, such that the following statement holds true.*

Let Ω be an open and bounded subset of $\mathbb{R}^n$ with boundary of class C^1. Let $c \in C(\overline{\Omega})$, and assume that

$$\|c^+\|_{L^\infty(\Omega)}\, |\Omega|^{\frac{2}{n}} < c_0. \tag{2.16.17}$$

Let $u \in C^2(\Omega) \cap C^1(\overline{\Omega})$ be a solution to

$$\begin{cases} \Delta u(x) + c(x)u(x) \geqslant 0 & \textit{for all } x \in \Omega, \\ u(x) \leqslant 0 & \textit{for all } x \in \partial\Omega. \end{cases}$$

Then, $u \leqslant 0$ in Ω.

Also, if Ω is connected, then $u < 0$ in Ω, unless it is constantly equal to zero.

This result is actually a particular case of a more general statement. The core of the idea, dating back to Guido Stampacchia [Sta66], is that suitable maximum principles continue to remain valid for "small" linear perturbations of the Laplacian. Note that the main structural condition of Theorem 2.16.7 (namely hypothesis (2.16.17)) is satisfied provided that the Lebesgue measure of the domain Ω is sufficiently small (possibly depending on the size of c^+), and this justifies the name of maximum principle for small volume domains. The reader can appreciate the similarities and differences between the narrow domain condition (2.16.11) and the small volume condition (2.16.17).

Proof of Theorem 2.16.7. We employ[24] a classical inequality introduced by John Nash [Nas58], stating that for all $v \in L^1(\mathbb{R}^n) \cap W^{1,2}(\mathbb{R}^n)$, we have that

$$\|v\|_{L^2(\mathbb{R}^n)}^{\frac{n+2}{n}} \leqslant C\,\|v\|_{L^1(\mathbb{R}^n)}^{\frac{2}{n}}\,\|\nabla v\|_{L^2(\mathbb{R}^n)}, \tag{2.16.18}$$

for some $C > 0$ depending only on n. For completeness, we recall the elegant proof of this inequality (for another elegant proof based on rearrangements, see [LL01, Theorem 8.13]). We can assume that v does not vanish identically (otherwise (2.16.18) is obvious), and we take the Fourier transform $\widehat{v}(\xi)$ of $v(x)$, and we note that $\partial_j v(x)$ coincides with the inverse Fourier transform of $2\pi i\xi_j\widehat{v}(\xi)$, for all $j \in \{1, \dots, n\}$. Consequently, the Plancherel theorem gives that, for all $\rho > 0$,

$$\begin{aligned}\int_{\mathbb{R}^n\setminus B_\rho} |\widehat{v}(\xi)|^2\,d\xi &\leqslant \frac{1}{\rho^2}\int_{\mathbb{R}^n\setminus B_\rho} |\xi|^2\,|\widehat{v}(\xi)|^2\,d\xi \leqslant \frac{1}{\rho^2}\int_{\mathbb{R}^n} |\xi|^2\,|\widehat{v}(\xi)|^2\,d\xi\\ &= \frac{1}{4\pi^2\rho^2}\int_{\mathbb{R}^n} |\nabla v(x)|^2\,dx \leqslant \frac{\|\nabla v\|^2_{L^2(\mathbb{R}^n)}}{4\pi^2\rho^2}.\end{aligned}$$

Additionally,

$$|\widehat{v}(\xi)| = \left|\int_{\mathbb{R}^n} v(x)\,e^{2\pi i x\cdot\xi}\,dx\right| \leqslant \int_{\mathbb{R}^n} |v(x)|\,dx = \|v\|_{L^1(\mathbb{R}^n)};$$

therefore,

$$\int_{B_\rho} |\widehat{v}(\xi)|^2\,d\xi \leqslant |B_1|\,\rho^n\,\|v\|^2_{L^1(\mathbb{R}^n)}.$$

[24] When $n \geqslant 3$, one can instead use the Sobolev–Gagliardo–Nirenberg inequality (see e.g. [Bre11, Theorem 9.9]), as done in [Bre99, Lemma 1]. Using the Nash inequality here, we can treat all the dimensions n at the same time.

From these observations and using the Plancherel theorem again, it follows that

$$\|v\|^2_{L^2(\mathbb{R}^n)} = \int_{\mathbb{R}^n} |\widehat{v}(\xi)|^2\, d\xi \leqslant \frac{\|\nabla v\|^2_{L^2(\mathbb{R}^n)}}{4\pi^2\rho^2} + |B_1|\,\rho^n\,\|v\|^2_{L^1(\mathbb{R}^n)}.$$

With this, the claim in (2.16.18) follows by picking $\rho := \left(\frac{\|\nabla v\|_{L^2(\mathbb{R}^n)}}{\|v\|_{L^1(\mathbb{R}^n)}}\right)^{\frac{2}{n+2}}$.

Now, if $v \in W^{1,2}_0(\Omega)$, we deduce from (2.16.18) and the Cauchy–Schwarz inequality that

$$\|v\|^{\frac{n+2}{n}}_{L^2(\mathbb{R}^n)} \leqslant C \left(|\Omega| \int_\Omega |v(x)|^2\, dx\right)^{\frac{1}{n}} \|\nabla v\|_{L^2(\mathbb{R}^n)} = C\,|\Omega|^{\frac{1}{n}}\,\|v\|^{\frac{2}{n}}_{L^2(\mathbb{R}^n)}\,\|\nabla v\|_{L^2(\mathbb{R}^n)}$$

and therefore[25]

$$\|v\|_{L^2(\mathbb{R}^n)} \leqslant C\,|\Omega|^{\frac{1}{n}}\,\|\nabla v\|_{L^2(\mathbb{R}^n)}. \tag{2.16.19}$$

Now, we show that

$$u(x) \leqslant 0 \quad \text{for every } x \in \Omega. \tag{2.16.20}$$

To this end, we note that

$$u^+ = 0 \text{ along } \partial\Omega, \tag{2.16.21}$$

and thus $u^+ \in W^{1,2}(\mathbb{R}^n)$, once we extend it by zero outside Ω, see e.g. [Bre11, Proposition 9.18], and

$$\begin{aligned}
0 &\leqslant \int_\Omega \big(\Delta u(x) + c(x)u(x)\big)\, u^+(x)\, dx \\
&= \int_\Omega \Delta u(x)\, u^+(x)\, dx + \int_\Omega c(x)\,(u^+(x))^2\, dx \\
&\leqslant \int_\Omega \Big(\operatorname{div}\big(u^+(x)\,\nabla u(x)\big) - \nabla u^+(x)\cdot\nabla u(x)\Big)\, dx + \int_\Omega c^+(x)\,(u^+(x))^2\, dx \\
&= \int_{\partial\Omega} u^+(x)\,\partial_\nu u(x)\, d\mathcal{H}^{n-1}_x - \int_\Omega |\nabla u^+(x)|^2\, dx + \int_\Omega c^+(x)\,(u^+(x))^2\, dx \\
&= -\int_\Omega |\nabla u^+(x)|^2\, dx + \int_\Omega c^+(x)\,(u^+(x))^2\, dx.
\end{aligned} \tag{2.16.22}$$

Besides, by (2.16.19),

$$\int_\Omega c^+(x)\,(u^+(x))^2\, dx \leqslant \|c^+\|_{L^\infty(\Omega)}\,\|u^+\|^2_{L^2(\Omega)} \leqslant C\,\|c^+\|_{L^\infty(\Omega)}\,|\Omega|^{\frac{2}{n}}\,\|\nabla u^+\|^2_{L^2(\mathbb{R}^n)}.$$

[25] Equation (2.16.19) can also be seen as a version of the Poincaré inequality with an explicit dependence of the constant on the volume of the domain. Compare, for example, with [Leo09, Theorem 12.17].

From this and (2.16.22), we arrive at

$$\begin{aligned}\|\nabla u^+\|^2_{L^2(\mathbb{R}^n)} &= \int_\Omega |\nabla u^+(x)|^2\,dx \\ &\leqslant \int_\Omega c^+(x)\,(u^+(x))^2\,dx \leqslant C\,\|c^+\|_{L^\infty(\Omega)}\,|\Omega|^{\frac{2}{n}}\,\|\nabla u^+\|^2_{L^2(\mathbb{R}^n)}.\end{aligned}$$

This and (2.16.17) give that $\|\nabla u^+\|_{L^2(\Omega)} = 0$; therefore, u^+ is constant in each connected component of Ω. From this and (2.16.21), we obtain (2.16.20).

The desired result then follows from (2.16.20) and Theorem 2.16.2. □

Several of the results presented here actually hold true in further generality, see e.g. [Ser71, GNN79, Bre99, PS07b] and the references therein. See also [HL11, Theorem 2.32] for a formulation of the maximum principle for small volume domains due to Sathamangalam Ranga Iyengar Srinivasa Varadhan.

2.17 Cauchy's Estimates

Now, we deduce from the regularity result in Theorem 2.14.1 and the mean value formula in Theorem 2.1.2(iii) a useful set of explicit bounds on the derivative of harmonic functions, often referred to with the name[26] of Cauchy's estimates:

Theorem 2.17.1. *Let $\Omega \subseteq \mathbb{R}^n$ be open, and let u be harmonic in Ω. Assume that $B_r(x_0) \Subset \Omega$. Then, for every $k \in \mathbb{N}$ and every $\alpha \in \mathbb{N}^n$ with $|\alpha| = k$, we have that*

$$\left|\frac{\partial^\alpha u}{\partial x^\alpha}(x_0)\right| \leqslant \frac{C_k\,\|u\|_{L^1(B_r(x_0))}}{r^{n+k}}.$$

The constant C_k above can be taken of the form

$$C_0 := \frac{1}{|B_1|} \quad \text{and} \quad C_k := \frac{(2^{n+1}nk)^k}{|B_1|} \quad \text{for all } k \geqslant 1. \tag{2.17.1}$$

Proof. We stress that, by the regularity result in Theorem 2.14.1, we know that u is differentiable as many times as we want and all its derivatives are harmonic. We suppose, up to a translation, that $x_0 = 0$, and we argue by induction over k.

Thus, we use the mean value formula in Theorem 2.1.2(iii) to see that

$$|u(0)| = \left|\fint_{B_r} u(x)\,dx\right| \leqslant \fint_{B_r} |u(x)|\,dx = \frac{\|u\|_{L^1(B_r)}}{|B_1|\,r^n},$$

and this is the desired estimate for $k = 0$.

[26]For related Cauchy's estimates in the complex variable setting, see e.g. [ST18, Lemma 10.7].

When $k = 1$, we define $v_i := \partial_{x_i} u$ for every $i \in \{1, \dots, n\}$, and we observe that v_i is harmonic in Ω. Hence, we can exploit the mean value formula in Theorem 2.1.2(iii) (applied here to v_i) and the divergence Theorem to see that

$$\begin{aligned}\partial_{x_i} u(0) = v_i(0) &= \fint_{B_{r/2}} v_i(x)\,dx = \fint_{B_{r/2}} \partial_{x_i} u(x)\,dx \\ &= \frac{2^n}{|B_1|\,r^n} \int_{B_{r/2}} \operatorname{div}(u(x)e_i)\,dx = \frac{2^{n+1}}{|B_1|\,r^{n+1}} \int_{\partial B_{r/2}} u(x) x_i \,d\mathcal{H}^{n-1}_x.\end{aligned} \tag{2.17.2}$$

Now, we observe that for every $q \in \partial B_{r/2}$, we have that $B_{r/2}(q) \subseteq B_r$. Consequently, using Cauchy's estimate for $k = 0$,

$$|u(q)| \leqslant \frac{C_0\, 2^n \,\|u\|_{L^1(B_{r/2}(q))}}{r^n} \leqslant \frac{2^n \,\|u\|_{L^1(B_r)}}{|B_1|\,r^n}.$$

Plugging this information into (2.17.2), we thereby obtain that

$$|\partial_{x_i} u(0)| \leqslant \frac{2^{n+1} n \,\|u\|_{L^1(B_r)}}{|B_1|\,r^{n+1}},$$

which proves the desired estimate when $k = 1$.

Now, we argue recursively, and we suppose that the desired estimate holds true for the index $k \in \mathbb{N}$. We pick $\alpha = (\alpha_1, \dots, \alpha_n) \in \mathbb{N}^n$ with $|\alpha| = k + 1$, and up to exchanging the order of the variables, we assume that $\alpha_1 \neq 0$. Hence, we can write $\alpha = \beta + e_1$, with $\beta := (\alpha_1 - 1, \alpha_2, \dots, \alpha_n)$. Note that $|\beta| = k$. Let also $w := \partial^\beta_x u$ and $v := \partial^\alpha_x u = \partial_{x_1} w$. Since v is harmonic in Ω, applying on v the mean value formula in Theorem 2.1.2(iii) and the divergence theorem, for all $\rho \in (0, r]$,

$$\begin{aligned}\partial^\alpha_x u(0) = v(0) &= \fint_{B_\rho} v(x)\,dx = \fint_{B_\rho} \partial_{x_1} w(x)\,dx \\ &= \frac{1}{|B_1|\,\rho^n} \int_{B_\rho} \operatorname{div}(w(x)e_1)\,dx = \frac{1}{|B_1|\,\rho^{n+1}} \int_{\partial B_\rho} w(x) x_1 \,d\mathcal{H}^{n-1}_x.\end{aligned} \tag{2.17.3}$$

Now, we choose $\rho := \frac{r}{k+1}$ and $R := k\rho$: in this way, for every $q \in \partial B_\rho$, we have that $B_R(q) \subseteq B_r$. Thus, the inductive assumption gives that

$$|w(q)| = \left|\frac{\partial^\beta u}{\partial x^\beta}(q)\right| \leqslant \frac{C_k \,\|u\|_{L^1(B_R(q))}}{R^{n+k}} \leqslant \frac{C_k \,\|u\|_{L^1(B_r)}}{R^{n+k}}.$$

As a result,

$$\int_{\partial B_\rho} |w(x)|\,|x_1| \,d\mathcal{H}^{n-1}_x \leqslant \rho \int_{\partial B_\rho} |w(x)| \,d\mathcal{H}^{n-1}_x \leqslant \frac{C_k\, \rho^n\, \mathcal{H}^{n-1}(\partial B_1)\, \|u\|_{L^1(B_r)}}{R^{n+k}}.$$

Combined with (2.17.3) and recalling (1.1.6), this information leads to

$$
\begin{aligned}
|\partial_x^\alpha u(0)| &\leqslant \frac{C_k\,\mathcal{H}^{n-1}(\partial B_1)\,\|u\|_{L^1(B_r)}}{\rho\,|B_1|\,R^{n+k}} \\
&= \frac{C_k\,n\,\|u\|_{L^1(B_r)}}{\rho\,R^{n+k}} \\
&= \frac{C_k\,(k+1)^{n+k+1}\,n\,\|u\|_{L^1(B_r)}}{k^{n+k}r^{n+k+1}} \\
&= \frac{(2^{n+1}nk)^k\,(k+1)^{n+k+1}\,n\,\|u\|_{L^1(B_r)}}{|B_1|\,k^{n+k}r^{n+k+1}} \\
&\leqslant \frac{(2^{n+1}n(k+1))^{k+1}\,\|u\|_{L^1(B_r)}}{|B_1|\,r^{n+k+1}} \\
&= \frac{C_{k+1}\,\|u\|_{L^1(B_r)}}{r^{n+k+1}},
\end{aligned}
$$

which completes the inductive step. □

For completeness, we observe that Cauchy's estimates (and in particular the explicit value of C_k determined in (2.17.1)) can be exploited to obtain a proof of the real analyticity of harmonic functions as an alternative to the one presented for Theorem 2.14.1 here (see e.g. [Eva98]).

For further reference, we also use Cauchy's estimates in Theorem 2.17.1 to deduce a maximum principle in Lebesgue spaces for unbounded domains, which can be seen as a counterpart of Corollary 2.9.2.

Lemma 2.17.2. *Let $p \geqslant 1$ and $M \geqslant 0$. Let $\Omega \subseteq \mathbb{R}^n$ be open and connected and $u \in C^2(\Omega) \cap L^p(\Omega)$ be harmonic in Ω.*

Suppose that

$$\mathbb{R}^n \setminus \Omega \text{ is a bounded set} \tag{2.17.4}$$

and

$$\limsup_{\Omega \ni y \to z} u(y) \leqslant M \quad \text{for all } z \in \partial\Omega. \tag{2.17.5}$$

Then,

$$\sup_\Omega u \leqslant M.$$

Proof. We argue for a contradiction and suppose that

$$\sup_\Omega u \geqslant M + a, \tag{2.17.6}$$

for some $a > 0$. We also take a sequence $x_k \in \Omega$ such that

$$\lim_{k\to+\infty} u(x_k) = \sup_{\Omega} u.$$

We distinguish two cases: either x_k is bounded or not.

If x_k is bounded, up to a subsequence, we can assume that $x_k \to \overline{x}$, for some $\overline{x} \in \overline{\Omega}$. By (2.17.5), we infer that $\overline{x} \in \Omega$. Thus, if $\rho > 0$ is so small that $B_\rho(\overline{x}) \subset \Omega$, we have that $u \in C(B_\rho(\overline{x}))$ and thus

$$u(\overline{x}) = \lim_{k\to+\infty} u(x_k) = \sup_{\Omega} u.$$

This and the strong maximum principle in Theorem 2.9.1(iii) yield that u is constant in Ω, say $u = c$ in Ω for some $c \in \mathbb{R}$.

On the one hand, it follows from (2.17.5) that $c \leqslant M$. On the other hand, by (2.17.6), we have that $c \geqslant M + a > M$, which is a contradiction.

We can therefore focus on the case in which x_k is unbounded. By (2.17.4), we can suppose that $\mathbb{R}^n \setminus \Omega \Subset B_R$, for some $R > 0$, hence $B_{r_k}(x_k) \Subset \Omega$ for $r_k := |x_k| - (R+1)$ (note that $r_k > 0$ if k is large enough). As a result, using Cauchy's Estimates in Theorem 2.17.1 and the Hölder Inequality, we see that, for every $x \in B_{r_k/2}(x_k)$,

$$|\nabla u(x)| \leqslant \frac{C\,\|u\|_{L^1(B_{r_k}(x_k))}}{r_k^{n+1}} \leqslant \frac{C\,\|u\|_{L^p(\Omega)}\, r_k^{\frac{n(p-1)}{p}}}{r_k^{n+1}} = \frac{C\,\|u\|_{L^p(\Omega)}}{r_k^{1+\frac{n}{p}}}$$

for some constant $C > 0$, depending only on n and possibly varying from step to step.

Consequently, taking k sufficiently large such that

$$r_k \geqslant 2 \quad \text{and} \quad u(x_k) \geqslant \sup_{\Omega} u - \frac{a}{2},$$

for every $x \in B_1(x_k)$, we have that

$$u(x) \geqslant u(x_k) - \|\nabla u\|_{L^\infty(B_1(x_k))} \geqslant \sup_{\Omega} u - \frac{a}{2} - \frac{C\,\|u\|_{L^p(\Omega)}}{r_k^{1+\frac{n}{p}}} \geqslant M + \frac{a}{2} - \frac{C\,\|u\|_{L^p(\Omega)}}{r_k^{1+\frac{n}{p}}}.$$

That is, if k is sufficiently large, for every $x \in B_1(x_k)$,

$$u(x) \geqslant M + \frac{a}{4} \geqslant \frac{a}{4},$$

from which we obtain a contradiction since $u \in L^p(\Omega)$. The desired result is thereby proved. □

2.18 The Weak Harnack Inequality

In this section, we present a series of results that are often called in the literature, possibly under different forms, the "weak Harnack inequality." In this jargon, the word "weak" should not be considered reductive. Instead, it highlights the fact that the estimates obtained do not rely only on "strong" notions, such as the pointwise values of a solution (as it happens, for instance, in the Harnack inequality presented in Section 2.15), but rather on "weak" notions, such as the measures of level sets and integrals of the solution. We do not address here the most general statement of this theory: for a comprehensive overview, see [DG57, Nas58, Mos60, Mos61, KS79, CC95, GT01, Kas07, HL11, CGP+15] and the references therein.

To begin this topic, we show, in broad terms, that a harmonic function bounded from above and with a positive measure set of values below its maximum is necessarily pointwise well separated from its maximum inside the domain. The quantitative formulation of this statement is given as follows.

Theorem 2.18.1. *Let $R > r > 0$ and $\eta > 0$. Let Ω be an open subset of $\mathbb{R}^n$ with $B_R \Subset \Omega$. Let $u \in C^2(\Omega)$ be harmonic in Ω and such that*

$$\sup_{B_R} u \leqslant 1. \tag{2.18.1}$$

Assume that

$$\left|\{x \in B_R \text{ s.t. } u(x) \leqslant 0\}\right| \geqslant \eta. \tag{2.18.2}$$

Then, there exists $c > 0$, depending only on n, η, r and R, such that

$$u \leqslant 1 - c \quad \text{in } B_r. \tag{2.18.3}$$

Proof. By the mean value formula in Theorem 2.1.2(iii), (2.18.1) and (2.18.2),

$$\begin{aligned} u(0) &= \fint_{B_R} u(x)\,dx = \frac{1}{|B_R|}\left(\int_{B_R\cap\{u\leqslant 0\}} u(x)\,dx + \int_{B_R\cap\{u>0\}} u(x)\,dx\right) \\ &\leqslant \frac{1}{|B_R|}\int_{B_R\cap\{u>0\}} u(x)\,dx \leqslant \frac{|B_R|-\eta}{|B_R|} = 1 - c_\star, \end{aligned}$$

where $c_\star := \frac{\eta}{|B_R|} > 0$.

Now, we use Cauchy's estimate in Theorem 2.17.1 with $k = 1$. In particular, we take C_1 to be the constant introduced in (2.17.1) with $k := 1$, we set

$r_0 := \min\left\{\frac{R}{2}, \frac{c_\star R}{2^{n+2}C_1|B_1|}\right\}$ and we deduce from Theorem 2.17.1 that, for every $x \in B_{r_0}$,

$$u(x) \leqslant u(0) + r_0 \sup_{B_{r_0}} |\nabla u| \leqslant 1 - c_\star + \frac{C_1 r_0 \|u\|_{L^1(B_R)}}{(R/2)^{n+1}}$$
$$\leqslant 1 - c_\star + \frac{2^{n+1}C_1 r_0|B_1|}{R} \leqslant 1 - \frac{c_\star}{2}. \tag{2.18.4}$$

If $r_0 \geqslant r$, this proves the desired result. If instead $r_0 < r$, we can iterate the previous inequality. To this end, we define recursively

$$r_{k+1} := \frac{(R - r_k)r_0 + Rr_k}{R}. \tag{2.18.5}$$

Note that

$$r_{k+1} \leqslant R. \tag{2.18.6}$$

Indeed, the claim is true for r_0 and then, proceeding inductively,

$$r_{k+1} = \frac{Rr_0 + (R - r_0)r_k}{R} \leqslant \frac{Rr_0 + (R - r_0)R}{R} = R,$$

which establishes (2.18.6).

Now, we claim that, for every $k \in \mathbb{N}$, there exists $c_k > 0$ such that

$$\sup_{B_{r_k}} u \leqslant 1 - c_k. \tag{2.18.7}$$

We prove this by induction. Indeed, when $k = 0$, the claim in (2.18.7) follows from (2.18.4) by taking $c_0 := \frac{c_\star}{2}$. Suppose now that (2.18.7) holds true for some index k, and let $p \in B_{r_k}$. We define

$$v(x) := \frac{1}{c_k}\left[u\left(p + \frac{(R - r_k)x}{R}\right) - 1 + c_k\right],$$

and we observe that if $x \in B_R$, then

$$\left|p + \frac{(R - r_k)x}{R}\right| \leqslant r_k + \frac{(R - r_k)|x|}{R} < R,$$

and thus v is harmonic in B_R. Additionally, we have that v is bounded from above by 1. Furthermore, we observe that

$$\lim_{\varepsilon \searrow 0} \frac{r_k + (R - r_k)\varepsilon}{R} r_k + \frac{(R - r_k)\varepsilon}{R} = \frac{r_k}{R} r_k < r_k,$$

whence we can take $\varepsilon_k \in (0, 1)$ so small that

$$\frac{r_k + (R - r_k)\varepsilon_k}{R} r_k + \frac{(R - r_k)\varepsilon_k}{R} < r_k.$$

With this choice, it follows that, if $q_k := (\varepsilon_k - 1)p$ and $x \in B_{\varepsilon_k}(q_k)$, we have that

$$\begin{aligned}\left|p + \frac{(R-r_k)x}{R}\right| &= \left|p + \frac{(R-r_k)(x-q_k)}{R} + \frac{(R-r_k)q_k}{R}\right| \\ &\leqslant \left|p + \frac{(R-r_k)(\varepsilon_k - 1)p}{R}\right| + \left|\frac{(R-r_k)(x-q_k)}{R}\right| \\ &< \frac{r_k + (R-r_k)\varepsilon_k}{R}\, r_k + \frac{(R-r_k)\varepsilon_k}{R} \\ &< r_k\end{aligned}$$

and therefore $p + \frac{(R-r_k)x}{R} \in B_{r_k}$.

Consequently, by the inductive hypothesis,

$$\begin{aligned}\{x \in B_R \text{ s.t. } v(x) \leqslant 0\} &= \left\{x \in B_R \text{ s.t. } u\left(p + \frac{(R-r_k)x}{R}\right) \leqslant 1 - c_k\right\} \\ &\supseteq \left\{x \in B_R \text{ s.t. } p + \frac{(R-r_k)x}{R} \in B_{r_k}\right\} \supseteq B_{\varepsilon_k}(q_k)\end{aligned}$$

and, as a result,

$$\eta_k := \left|\{x \in B_R \text{ s.t. } v(x) \leqslant 0\}\right| > 0.$$

We can therefore apply inequality (2.18.4) to v and find that

$$\sup_{B_{r_0}} v \leqslant 1 - \widetilde{c}_k,$$

for some $\widetilde{c}_k > 0$. This gives that, for every $x \in B_{r_0}$,

$$u\left(p + \frac{(R-r_k)x}{R}\right) \leqslant c_k(1 - \widetilde{c}_k) + 1 - c_k = 1 - c_k\,\widetilde{c}_k$$

and therefore

$$\sup_{B_{(R-r_k)r_0/R}(p)} u \leqslant 1 - c_k\,\widetilde{c}_k.$$

Since p is an arbitrary point in B_{r_k}, we thus infer that

$$\sup_{B_{((R-r_k)r_0+Rr_k)/R}} u \leqslant 1 - c_k\,\widetilde{c}_k,$$

which, in light of the definition of r_{k+1} in (2.18.5), concludes the inductive step and establishes (2.18.7).

We also stress that, by (2.18.5) and (2.18.6),

$$r_{k+1} - r_k = \frac{(R - r_k)r_0}{R} \geqslant 0,$$

and hence r_{k+1} is a bounded and monotone increasing sequence. We can therefore define $R_\star$ to be its limit, and we deduce from (2.18.5) that

$$R_\star = \frac{(R - R_\star)r_0 + RR_\star}{R},$$

leading to $(R - R_\star)r_0 = 0$ and thus $R = R_\star$.

This observation gives that

$$\lim_{k \to +\infty} r_k = R > r,$$

and hence we can choose $k_\star \in \mathbb{N}$ such that $r_{k_\star} > r$. In this way, the desired claim in (2.18.3) follows from (2.18.7) with $k := k_\star$. □

As a variant of Theorem 2.18.1, one can control the supremum of a harmonic function by its L^p-norm (interestingly, the case $p \in (0, 1]$ is also allowed). This result will be stated precisely in Theorem 2.18.3, and its proof will also rely on the following very useful observation, which allows us to "reabsorb a term" on the left-hand side of a scaled inequality.

Lemma 2.18.2. *Let $a > b$ and $\varphi : [a, b] \to \mathbb{R}$ be a bounded function. Assume that there exist $\vartheta \in [0, 1)$, $A \geqslant 0$, $B \geqslant 0$ and $\gamma > 0$ such that, for every r, $R \in [a, b]$ with $r < R$, we have that*

$$\varphi(r) \leqslant \vartheta\varphi(R) + \frac{A}{(R - r)^\gamma} + B. \tag{2.18.8}$$

Then, there exists $C > 0$, depending only on ϑ and γ such that for every r, $R \in [a, b]$ with $r < R$, we have that

$$\varphi(r) \leqslant C\left(\frac{A}{(R - r)^\gamma} + B\right).$$

Proof. Fix r, $R \in [a, b]$ with $r < R$, and let $\tau := \frac{1+\vartheta^{\frac{1}{\gamma}}}{2}$. Note that

$$\tau \in \left(\vartheta^{\frac{1}{\gamma}}, 1\right) \tag{2.18.9}$$

since $\vartheta < 1$. We also define $r_0 := r$ and, recursively, for each $k \in \mathbb{N}$,

$$r_{k+1} := r_k + (1 - \tau)\,\tau^k\,(R - r).$$

We point out that the sequence r_k is monotone increasing, thanks to (2.18.9), and can be written in closed form as

$$\begin{aligned} r_{k+1} &= r_{k-1} + (1-\tau)\,\tau^{k-1}\,(R-r) + (1-\tau)\,\tau^{k}\,(R-r) \\ &= r_{k-2} + (1-\tau)\,\tau^{k-2}\,(R-r)r_{k-1} + (1-\tau)\,\tau^{k-1}\,(R-r) + (1-\tau)\,\tau^{k}\,(R-r) \\ &= \cdots \\ &= r_0 + \sum_{j=0}^{k} (1-\tau)\,\tau^{j}\,(R-r) \\ &= r + \frac{(1-\tau)(R-r)(1-\tau^{k+1})}{1-\tau}. \end{aligned}$$

As a result,

$$\lim_{k\to+\infty} r_k = r + \frac{(1-\tau)(R-r)}{1-\tau} = R.$$

Now, we claim that, for every $k \in \mathbb{N}$,

$$\varphi(r) \leqslant \vartheta^k \varphi(r_k) + \left(\frac{A}{(1-\tau)^\gamma (R-r)^\gamma} + B\right) \sum_{i=0}^{k-1} \left(\frac{\vartheta}{\tau^\gamma}\right)^i. \tag{2.18.10}$$

Indeed, when $k = 0$, the above formula is obvious. When $k = 1$, we apply (2.18.8) (used here with r_1 in place of R) and see that

$$\varphi(r) \leqslant \vartheta\varphi(r_1) + \frac{A}{(r_1 - r)^\gamma} + B = \vartheta\varphi(r_1) + \frac{A}{(1-\tau)^\gamma (R-r)^\gamma} + B,$$

which is precisely (2.18.10) with $k = 1$.

Thus, to complete the proof of (2.18.10), we argue recursively, assuming that this formula holds true for the index k, and we prove it for the index $k+1$. To this end, we use the inductive assumption and (2.18.8) (utilized here with r_k in place of r and r_{k+1} in place of R) and see that

$$\begin{aligned} \varphi(r) &\leqslant \vartheta^k \varphi(r_k) + \left(\frac{A}{(1-\tau)^\gamma (R-r)^\gamma} + B\right) \sum_{i=0}^{k-1} \left(\frac{\vartheta}{\tau^\gamma}\right)^i \\ &\leqslant \vartheta^k \left(\vartheta\varphi(r_{k+1}) + \frac{A}{(r_{k+1} - r_k)^\gamma} + B\right) \\ &\quad + \left(\frac{A}{(1-\tau)^\gamma (R-r)^\gamma} + B\right) \sum_{i=0}^{k-1} \left(\frac{\vartheta}{\tau^\gamma}\right)^i \\ &= \vartheta^{k+1}\varphi(r_{k+1}) + \vartheta^k \left(\frac{A}{(1-\tau)^\gamma\, \tau^{\gamma k}\, (R-r)^\gamma} + B\right) \\ &\quad + \left(\frac{A}{(1-\tau)^\gamma (R-r)^\gamma} + B\right) \sum_{i=0}^{k-1} \left(\frac{\vartheta}{\tau^\gamma}\right)^i \end{aligned}$$

$$\begin{aligned}
&\leqslant \vartheta^{k+1}\varphi(r_{k+1}) + \vartheta^k\left(\frac{A}{(1-\tau)^\gamma\,\tau^{\gamma k}\,(R-r)^\gamma} + \frac{B}{\tau^{\gamma k}}\right)\\
&\quad + \left(\frac{A}{(1-\tau)^\gamma(R-r)^\gamma} + B\right)\sum_{i=0}^{k-1}\left(\frac{\vartheta}{\tau^\gamma}\right)^i\\
&= \vartheta^{k+1}\varphi(r_{k+1}) + \left(\frac{A}{(1-\tau)^\gamma(R-r)^\gamma} + B\right)\sum_{i=0}^{k}\left(\frac{\vartheta}{\tau^\gamma}\right)^i,
\end{aligned}$$

which proves (2.18.10).

We also observe that $\frac{\vartheta}{\tau^\gamma} < 1$, thanks to (2.18.9), and we can therefore send $k \to +\infty$ in (2.18.10), concluding that

$$\varphi(r) \leqslant \left(\frac{A}{(1-\tau)^\gamma(R-r)^\gamma} + B\right)\sum_{i=0}^{+\infty}\left(\frac{\vartheta}{\tau^\gamma}\right)^i = \left(\frac{A}{(1-\tau)^\gamma(R-r)^\gamma} + B\right)\frac{1}{1-\frac{\vartheta}{\tau^\gamma}},$$

which proves the desired result. □

Theorem 2.18.3. *Let $R > r > 0$ and $p > 0$. Let Ω be an open subset of $\mathbb{R}^n$ with $B_R \Subset \Omega$. Let $u \in C^2(\Omega)$ be harmonic in Ω.*

Then, there exists $C(n, p) > 0$, depending only on n and p, such that

$$\sup_{B_r} u_+ \leqslant \frac{C(n,p)}{(R-r)^{\frac{n}{p}}}\left(\int_{B_R}(u_+(x))^p\,dx\right)^{\frac{1}{p}}. \tag{2.18.11}$$

Proof. We first establish (2.18.11) when $p \geqslant 1$.

For this, let $\zeta \in B_R$. We use the mean value formula in Theorem 2.1.2(iii) to see that, if $\rho \in (0, R - |\zeta|]$, then

$$u(\zeta) = \fint_{B_\rho(\zeta)} u(x)\,dx \leqslant \fint_{B_\rho(\zeta)} u_+(x)\,dx.$$

Thus, if $p \geqslant 1$, one can use the Hölder inequality and find that, for all $\zeta \in B_r$,

$$\begin{aligned}
u(\zeta) &\leqslant \fint_{B_{R-r}(\zeta)} u_+(x)\,dx \leqslant \left(\fint_{B_{R-r}(\zeta)} (u_+(x))^p\,dx\right)^{\frac{1}{p}}\\
&\leqslant \left(\frac{1}{|B_{R-r}|}\int_{B_R}(u_+(x))^p\,dx\right)^{\frac{1}{p}}.
\end{aligned}$$

Since the above estimate is also obviously true when $u(\zeta) \leqslant 0$, we thus deduce that

$$u_+(\zeta) \leqslant \left(\frac{1}{|B_{R-r}|}\int_{B_R}(u_+(x))^p\,dx\right)^{\frac{1}{p}},$$

from which we obtain (2.18.11) when $p \geqslant 1$.

Suppose now that $p \in (0,1)$. In this case, we have that

$$\int_{B_R} u_+(x)\,dx = \int_{B_R} (u_+(x))^{p+(1-p)}\,dx \leqslant \left(\sup_{B_R} u_+\right)^{1-p} \int_{B_R} (u_+(x))^p\,dx.$$

Therefore, using (2.18.11) with $p = 1$ and setting

$$\varphi(t) := \sup_{B_t} u_+,$$

we find that

$$\begin{aligned}\varphi(r) = \sup_{B_r} u_+ &\leqslant \frac{C(n,1)}{(R-r)^n} \int_{B_R} u_+(x)\,dx \\ &\leqslant \frac{C(n,1)}{(R-r)^n} (\varphi(R))^{1-p} \int_{B_R} (u_+(x))^p\,dx.\end{aligned}$$

We thus exploit Young's inequality with exponents $\frac{1}{1-p}$ and $\frac{1}{p}$, obtaining that

$$\varphi(r) \leqslant \frac{\varphi(R)}{2} + \frac{C}{(R-r)^{\frac{n}{p}}} \left(\int_{B_R} (u_+(x))^p\,dx\right)^{\frac{1}{p}}.$$

We can therefore utilize Lemma 2.18.2 and conclude, up to renaming C, that

$$\varphi(r) \leqslant \frac{C}{(R-r)^{\frac{n}{p}}} \left(\int_{B_R} (u_+(x))^p\,dx\right)^{\frac{1}{p}}.$$

□

A counterpart of Theorem 2.18.3 (actually relying merely on the Harnack inequality presented in Section 2.15) is given by the following result, in which the L^p-norm of a nonnegative harmonic function is bounded from above by the infimum in a smaller ball.

Theorem 2.18.4. *Let $R_0 > R > r > 0$ and $p > 0$. Let Ω be an open subset of $\mathbb{R}^n$ with $B_{R_0} \Subset \Omega$. Let $u \in C^2(\Omega)$ be nonnegative and harmonic in Ω.*

Then, there exists $c > 0$, depending only on n, p, r, R and R_0 such that

$$\inf_{B_r} u \geqslant c \left(\int_{B_R} (u(x))^p\,dx\right)^{\frac{1}{p}}. \tag{2.18.12}$$

Proof. By the Harnack inequality in Theorem 2.15.1, for every $x \in B_R$,

$$u(x) \leqslant \left(\frac{R_0}{R_0 - R}\right)^{n-2} \frac{R_0 + R}{R_0 - R} u(0).$$

As a result,

$$\int_{B_R} (u(x))^p\,dx \leqslant \left(\frac{R_0}{R_0 - R}\right)^{(n-2)p} \frac{(R_0+R)^p}{(R_0-R)^p} (u(0))^p\,|B_R|. \tag{2.18.13}$$

Also, using again the Harnack inequality in Theorem 2.15.1,

$$\left(\frac{R_0}{R_0+r}\right)^{n-2}\frac{R_0-r}{R_0+r}\,u(0)\leqslant \inf_{B_r} u.$$

Combining this and (2.18.13), we conclude that

$$\int_{B_R}(u(x))^p\,dx\leqslant \frac{(R_0+R)^p\,(R_0+r)^{(n-1)p}}{(R_0-R)^{(n-1)p}\,(R_0-r)^p}\,|B_R|\left(\inf_{B_r} u\right)^p. \qquad \square$$

We observe that, when $p\in(0,1]$, Theorem 2.18.4 also holds true[27] for $R_0=R$. Instead, for large p, it is necessary to assume $R_0>R$ since, if u is as in (2.15.3), then, for small ε,

$$\begin{aligned}\int_{B_R}(u(x))^p\,dx &\geqslant \int_{B_\varepsilon((R-\varepsilon)e_1)}\frac{((R+\varepsilon)^2-|x|^2)^p}{n^p\,|B_1|^p\,(R+\varepsilon)^p\,|x-(R+\varepsilon)e_1|^{np}}\,dx\\ &\geqslant \int_{B_\varepsilon((R-\varepsilon)e_1)}\frac{((R+\varepsilon)^2-R^2)^p}{n^p\,|B_1|^p\,(2R)^p\,(3\varepsilon)^{np}}\,dx\\ &\geqslant \frac{c}{\varepsilon^{np-p-n}},\end{aligned}$$

for some $c>0$ depending only on n, p and R, and the latter term in the above chain of inequalities diverges as $\varepsilon\searrow 0$ when $n\geqslant 2$ and p is sufficiently large (for further discussions on the range of p in more general contexts, see e.g. [GT01, Theorem 8.18 and Problem 9.12] and [HL11, Theorem 4.15]).

---

[27] Indeed, if $p\in(0,1]$, one considers $P:=\frac{1}{p}\geqslant 1$ and uses the Hölder inequality and the mean value formula in Theorem 2.1.2(iii) to see that

$$⨍_{B_R}(u(x))^p\,dx\leqslant\left(⨍_{B_R}u(x)\,dx\right)^{\frac{1}{P}}=(u(0))^p.$$

Thus, since, by the Harnack inequality in Theorem 2.15.1,

$$\left(\frac{R}{R+r}\right)^{n-2}\frac{R-r}{R+r}\,u(0)\leqslant\inf_{B_r}u,$$

it follows that, say when $n\geqslant 2$,

$$\begin{aligned}\left(⨍_{B_R}(u(x))^p\,dx\right)^{\frac{1}{p}}&\leqslant u(0)\leqslant\left(\frac{R+r}{R}\right)^{n-2}\frac{R+r}{R-r}\inf_{B_r}u\\ &\leqslant\left(\frac{2R}{R}\right)^{n-2}\frac{2R}{R-r}\inf_{B_r}u=\frac{2^{n-1}R}{R-r}\inf_{B_r}u,\end{aligned}$$

and this gives (2.18.12) when $p\in(0,1]$ with $R_0=R$.

2.19 The Boundary Harnack Inequality

For completeness, and without aiming for exhaustiveness, we recall here a boundary version of the Harnack inequality. For this, we use the notation

$$B_r^+ := \{x = (x_1, \dots, x_n) \text{ s.t. } x_n > 0\}.$$

Theorem 2.19.1. *Let u, $v \in C^2(B_2^+) \cap C^1(\overline{B_2^+})$ be harmonic functions in B_2^+. Assume that u, $v > 0$ in B_2^+ and $u = v = 0$ along $\{x_n = 0\}$. Then, there exists a constant $C > 1$ only depending on n such that, for every $x \in B_{1/8}^+$,*

$$\frac{1}{C}\,\frac{u(e_n/8)}{v(e_n/8)} \leqslant \frac{u(x)}{v(x)} \leqslant C\,\frac{u(e_n/8)}{v(e_n/8)}. \tag{2.19.1}$$

Proof. We observe that, by possibly replacing u with $\frac{u}{u(e_n/8)}$ and v with $\frac{v}{v(e_n/8)}$, we can assume that

$$u\left(\frac{e_n}{8}\right) = 1 = v\left(\frac{e_n}{8}\right). \tag{2.19.2}$$

Let

$$B_r^- := \{x = (x_1, \dots, x_n) \text{ s.t. } x_n < 0\}.$$

Given $x = (x_1, \dots, x_n)$, we let $x_\star := (x_1, \dots, -x_n)$. For every $x \in B_2$, let also

$$u_\star(x) := \begin{cases} u(x) & \text{if } x \in \overline{B_2^+}, \\ -u(x_\star) & \text{if } x \in \overline{B_2^-}. \end{cases}$$

Note that this is a good definition since $u = 0$ and $x_\star = x$ along $\{x_n = 0\}$. We observe that

$$u_\star \text{ is harmonic in } B_2. \tag{2.19.3}$$

Indeed, if $\varphi \in C_0^\infty(B_2)$, we let $\psi(x) := \varphi(x_\star)$. Thus, the use of Green's second identity (2.1.11) leads to

$$\begin{aligned}
&\int_{B_2} u_\star(x)\,\Delta\varphi(x)\,dx \\
&= \lim_{\varepsilon \searrow 0} \int_{B_2 \cap \{x_n > \varepsilon\}} u_\star(x)\,\Delta\varphi(x)\,dx + \int_{B_2 \cap \{x_n < -\varepsilon\}} u_\star(x)\,\Delta\varphi(x)\,dx \\
&= \lim_{\varepsilon \searrow 0} \int_{B_2 \cap \{x_n > \varepsilon\}} u(x)\,\Delta\varphi(x)\,dx - \int_{B_2 \cap \{x_n > \varepsilon\}} u(x)\,\Delta\psi(x)\,dx \\
&= \lim_{\varepsilon \searrow 0} \int_{B_2 \cap \{x_n = \varepsilon\}} \big(u(x)\,\partial_\nu\varphi(x) - \varphi(x)\,\partial_\nu u(x)\big)\,dx \\
&\quad - \lim_{\varepsilon \searrow 0} \int_{B_2 \cap \{x_n = \varepsilon\}} \big(u(x)\,\partial_\nu\psi(x) - \psi(x)\,\partial_\nu u(x)\big)\,dx
\end{aligned}$$

$$= \lim_{\varepsilon \searrow 0} \int_{B_2 \cap \{x_n = \varepsilon\}} (\psi(x) - \varphi(x))\, \partial_\nu u(x)\, dx$$
$$= 0.$$

In view of Weyl's lemma (i.e. recall Lemma 2.2.1), this proves (2.19.3).

As a result, we can exploit the Poisson kernel representation (recall Theorems 2.11.1 and 2.11.2) and obtain that, for all $x \in B_1^+$,

$$\begin{aligned} n\,|B_1|\,u(x) &= \int_{\partial B_1} \frac{(1-|x|^2)\,u_\star(y)}{|x-y|^n}\,d\mathcal{H}_y^{n-1} \\ &= \int_{\partial B_1^+} \frac{(1-|x|^2)\,u(y)}{|x-y|^n}\,d\mathcal{H}_y^{n-1} - \int_{\partial B_1^-} \frac{(1-|x|^2)\,u(y_\star)}{|x-y|^n}\,d\mathcal{H}_y^{n-1} \\ &= \int_{\partial B_1^+} \frac{(1-|x|^2)\,u(y)}{|x-y|^n}\,d\mathcal{H}_y^{n-1} - \int_{\partial B_1^+} \frac{(1-|x|^2)\,u(y)}{|x-y_\star|^n}\,d\mathcal{H}_y^{n-1} \\ &= \int_{\partial B_1^+} \left(\frac{1}{|x-y|^n} - \frac{1}{|x-y_\star|^n}\right)(1-|x|^2)\,u(y)\,d\mathcal{H}_y^{n-1}. \end{aligned} \tag{2.19.4}$$

Now, we point out that, if $a, b \in B_2 \setminus B_{1/2}$, then

$$\begin{aligned} |a|^{-n} - |b|^{-n} &= |b + (a-b)|^{-n} - |b|^{-n} \\ &= -n \int_0^1 |b + t(a-b)|^{-n-2}(b + t(a-b)) \cdot (a-b)\,dt. \end{aligned} \tag{2.19.5}$$

Also, if $|a - b| \leqslant 1/4$, for each $t \in [0, 1]$, we have that

$$|b + t(a-b)| \geqslant |b| - |a-b| \geqslant \frac{1}{2} - \frac{1}{4} = \frac{1}{4}.$$

Consequently,

$$|a|^{-n} - |b|^{-n} \leqslant n \int_0^1 |b + t(a-b)|^{-n-1}|a-b|\,dt \leqslant 4^{n+1} n\,|a-b|. \tag{2.19.6}$$

Also, if $y \in \partial B_1$ with $y_n \geqslant 3/4$ and $x \in B_{1/4}$, for every $t \in [0, 1]$, we have that

$$x_n + y_n - 2tx_n \geqslant \frac{3}{4} + (1-2t)x_n \geqslant \frac{1}{2},$$

whence we can use (2.19.5) with $a := x - y$ and $b := x_\star - y$ and find that, for every $x \in B^+_{1/4}$,

$$\begin{aligned}|x-y|^{-n} - |x-y_\star|^{-n} &= |x-y|^{-n} - |x_\star - y|^{-n}\\ &= -2nx_n \int_0^1 |x_\star - y + 2tx_n e_n|^{-n-2}(x_\star - y + 2tx_n e_n)\cdot e_n\,dt\\ &= 2nx_n \int_0^1 |x_\star - y + 2tx_n e_n|^{-n-2}(x_n + y_n - 2tx_n)\,dt\\ &\geqslant \frac{nx_n}{4^{n+2}}.\end{aligned} \tag{2.19.7}$$

Thus, for every $x \in B^+_{1/8}$, we can use (2.19.6) with $a := x - y$ and $b := x_\star - y$ to obtain that

$$\int_{\partial B_1^+}\left(\frac{1}{|x-y|^n} - \frac{1}{|x-y_\star|^n}\right)(1-|x|^2)\,u(y)\,d\mathcal{H}^{n-1}_y \leqslant Cx_n \int_{\partial B_1^+} u(y)\,d\mathcal{H}^{n-1}_y. \tag{2.19.8}$$

Besides, for every $x \in B^+_{1/8}$, we note that, for every $y \in B^+_1$, it holds that $|x_n - y_n| \leqslant |x_n| + |y_n| = x_n + y_n$, whence

$$|x-y_\star|^2 = \sum_{i=1}^{n-1}(x_i - y_i)^2 + (x_n + y_n)^2 \geqslant \sum_{i=1}^{n-1}(x_i-y_i)^2 + (x_n - y_n)^2 = |x-y|^2.$$

Accordingly, we can utilize (2.19.7) to obtain that

$$\begin{aligned}&\int_{\partial B_1^+}\left(\frac{1}{|x-y|^n} - \frac{1}{|x-y_\star|^n}\right)(1-|x|^2)\,u(y)\,d\mathcal{H}^{n-1}_y\\ &\quad\geqslant \int_{(\partial B_1^+)\cap\{y_n\geqslant 3/4\}}\left(\frac{1}{|x-y|^n} - \frac{1}{|x-y_\star|^n}\right)(1-|x|^2)\,u(y)\,d\mathcal{H}^{n-1}_y\\ &\quad\geqslant \frac{x_n}{C}\int_{(\partial B_1^+)\cap\{y_n\geqslant 3/4\}} u(y)\,d\mathcal{H}^{n-1}_y,\end{aligned} \tag{2.19.9}$$

for a suitable $C > 1$.

Now, we claim that

$$\sup_{[-1,1]^{n-1}\times\left[0,\frac{1}{32}\right]} u \leqslant C, \tag{2.19.10}$$

for some $C > 1$ (possibly larger than the previous C). To check this, given $p = (p', p_n) \in \left[-\frac{33}{32}, \frac{33}{32}\right]^{n-1} \times \left[0, \frac{1}{16}\right]$, we apply the Harnack inequality of Corollary 2.15.2 in $B_{p_n/2}(p)$ to see that

$$u(p', p_n) \leqslant \sup_{B_{p_n/2}(p)} u \leqslant C \inf_{B_{p_n/2}(p)} u \leqslant Cu\left(p', \frac{3}{2}p_n\right)$$

for some $C > 1$. We can therefore iterate this inequality, letting $\eta(p_n) := \frac{\ln(1/(8p_n))}{\ln(3/2)}$ and choosing $k \in \mathbb{N} \cap [\eta(p_n), \eta(p_n) + 1)$, thus finding that

$$u(p) \leqslant C^k u\left(p', \left(\frac{3}{2}\right)^k p_n\right).$$

Since

$$\begin{aligned} \frac{1}{8p_n} = e^{\ln(1/(8p_n))} = \left(\frac{3}{2}\right)^{\eta(p_n)} &\leqslant \left(\frac{3}{2}\right)^k \\ \text{and} \quad \left(\frac{3}{2}\right)^k \leqslant \left(\frac{3}{2}\right)^{\eta(p_n)+1} &= \frac{3}{16p_n}, \end{aligned}$$

we thereby obtain that

$$u(p) \leqslant C^{\eta(p_n)+1} \sup_{\left[-\frac{33}{32}, \frac{33}{32}\right]^{n-1} \times \left[\frac{1}{8}, \frac{3}{16}\right]} u \leqslant \frac{C}{p_n^{C_\star}} \sup_{\left[-\frac{33}{32}, \frac{33}{32}\right]^{n-1} \times \left[\frac{1}{8}, \frac{3}{16}\right]} u, \tag{2.19.11}$$

for some $C_\star > 0$, up to renaming $C > 1$ at each step.

Now, let $q \in [-1, 1]^{n-1} \times \left[0, \frac{1}{32}\right]$. We exploit the local boundedness result in Theorem 2.18.3 (specifically, the estimate in (2.18.11), used here with $p := 1/(2C_\star)$), and making use of (2.19.11), we find that, possibly renaming C,

$$\begin{aligned} u(q) &= u_\star(q) \\ &\leqslant C\left(\int_{\left[-\frac{33}{32}, \frac{33}{32}\right]^{n-1} \times \left[-\frac{1}{16}, \frac{1}{16}\right]} |u_\star(p)|^{\frac{1}{2C_\star}} \, dp\right)^{2C_\star} \\ &= C\left(2\int_{\left[-\frac{33}{32}, \frac{33}{32}\right]^{n-1} \times \left[0, \frac{1}{16}\right]} |u(p)|^{\frac{1}{2C_\star}} \, dp\right)^{2C_\star} \\ &\leqslant C\left(\int_{\left[-\frac{33}{32}, \frac{33}{32}\right]^{n-1} \times \left[0, \frac{1}{16}\right]} \frac{1}{\sqrt{p_n}} \, dp\right)^{2C_\star} \sup_{\left[-\frac{33}{32}, \frac{33}{32}\right]^{n-1} \times \left[\frac{1}{8}, \frac{3}{16}\right]} u \\ &\leqslant C \sup_{\left[-\frac{33}{32}, \frac{33}{32}\right]^{n-1} \times \left[\frac{1}{8}, \frac{3}{16}\right]} u. \end{aligned} \tag{2.19.12}$$

Additionally, recalling the normalization in (2.19.2), using the Harnack inequality in Corollary 2.15.3 with $\Omega := \left(-\frac{35}{32}, \frac{35}{32}\right)^{n-1} \times \left(\frac{1}{16}, \frac{1}{4}\right)$ and $\Omega' := \left(-\frac{33}{32}, \frac{33}{32}\right)^{n-1} \times \left(\frac{1}{8}, \frac{3}{16}\right)$, and possibly renaming C,

$$\sup_{[-\frac{33}{32}, \frac{33}{32}]^{n-1} \times [\frac{1}{8}, \frac{3}{16}]} u \leqslant C \inf_{[-\frac{33}{32}, \frac{33}{32}]^{n-1} \times [\frac{1}{8}, \frac{3}{16}]} u \leqslant C u\left(\frac{e_n}{8}\right) = C.$$

This and (2.19.12) yield (2.19.10), as desired.

In addition, using the Harnack inequality in Corollary 2.15.3 with $\Omega := B^+_{3/2} \cap \left\{x_n > \frac{1}{64}\right\}$ and $\Omega' := B^+_1 \cap \left\{x_n > \frac{1}{32}\right\}$,

$$\sup_{B^+_1 \cap \{x_n > \frac{1}{32}\}} u \leqslant C \inf_{B^+_1 \cap \{x_n > \frac{1}{32}\}} u.$$

This information and the normalization in (2.19.2) lead to

$$1 = u\left(\frac{e_n}{8}\right) \leqslant \sup_{B^+_1 \cap \{x_n > \frac{1}{32}\}} u \leqslant C \inf_{B^+_1 \cap \{x_n > \frac{1}{32}\}} u \tag{2.19.13}$$

and

$$\sup_{B^+_1 \cap \{x_n > \frac{1}{32}\}} u \leqslant C \inf_{B^+_1 \cap \{x_n > \frac{1}{32}\}} u \leqslant C u\left(\frac{e_n}{8}\right) = C. \tag{2.19.14}$$

As a byproduct of (2.19.10) and (2.19.14), we have that

$$\int_{\partial B^+_1} u(y)\, d\mathcal{H}^{n-1}_y \leqslant C,$$

and thus, in light of (2.19.8), if $x \in B^+_{1/8}$,

$$\int_{\partial B^+_1} \left(\frac{1}{|x-y|^n} - \frac{1}{|x-y_\star|^n} \right) (1-|x|^2)\, u(y)\, d\mathcal{H}^{n-1}_y \leqslant C x_n, \tag{2.19.15}$$

up to renaming C.

Similarly, from (2.19.9) and (2.19.13), and possibly renaming C once again, if $x \in B^+_{1/8}$,

$$\int_{\partial B^+_1} \left(\frac{1}{|x-y|^n} - \frac{1}{|x-y_\star|^n} \right) (1-|x|^2)\, u(y)\, d\mathcal{H}^{n-1}_y \geqslant \frac{x_n}{C}. \tag{2.19.16}$$

Therefore, gathering (2.19.4), (2.19.15) and (2.19.16), and possibly renaming C, we deduce that, for all $x \in B^+_{1/8}$,

$$\frac{x_n}{C} \leqslant u(x) \leqslant C x_n,$$

and a similar estimate holds for v as well.

Consequently, for all $x \in B^+_{1/8}$,

$$\frac{u(x)}{v(x)} \leqslant \frac{u(x)}{x_n/C} \leqslant \frac{C x_n}{x_n/C} = C^2$$

and

$$\frac{u(x)}{v(x)} \geqslant \frac{u(x)}{Cx_n} \geqslant \frac{x_n/C}{Cx_n} = \frac{1}{C^2}.$$

These observations complete the proof of (2.19.1). □

For a thorough presentation of boundary Harnack inequalities (also known as the Carleson estimates) see [Kem72, Dah77, Wu78, Anc78, CFMS81, JK82, AC85, CS05, Aik08, DSS15, DSS16, DSS20] and the references therein.

2.20 Liouville's Theorem

A classical result known by the name[28] of Liouville's theorem states that bounded harmonic functions in the whole of $\mathbb{R}^n$ are necessarily constant. As a matter

[28]Liouville's contributions to science are variegate and cover a large number of topics such as electrodynamics, potential theory, spectral theory, the theory of heat, fractional calculus, number theory, rational mechanics and differential geometry. Nowadays, several different classical results are presented under the name of "Liouville's theorem." The one related to harmonic functions (and complex analysis) was probably found around 1844 by Liouville while working on doubly periodic meromorphic functions (nowadays called "elliptic functions" in the complex analysis jargon), and its initial proof relied on trigonometric series (later on, he also proposed several different approaches).

The claim about Liouville's theorem is however controversial. In particular, Cauchy, eager to secure his position in the field of complex analysis, called attention to some of his results from 1843–1844 stating that "if a function $f(z)$ of the real or imaginary variable z always remain continuous, and consequently always finite, it is reduced to a simple constant." In hindsight, Cauchy's statement, though perhaps foregoing or more general than Liouville's, would be considered today as poorly stated, both because of the ambiguity between "finite" and "bounded" and, more importantly, because at this stage Cauchy seems to erroneously believe that continuity is equivalent to holomorphy (*aliquando bonus dormitat Homerus*: this mistake seems to have been repeated by Cauchy on other occasions as well).

In any case, the debate on the claim to Liouville's theorem remains unsettled (but, after all, who cares?). See [Pei83, Lüt90] for more information about Liouville's biography and mathematics, as well as about the history of Liouville's theorem. See also Figure 2.20.1 for a portrait of Liouville and Figure 9.1.6 for a portrait of Cauchy (next to a rather obscure figure, Libri; see also footnote 8 on p. 585 for some crazy stories about this Libri; here, it suffices to say that Liouville discovered some plagiarism, and errors, by Libri, who at some point became Liouville's arch-enemy; see [Lüt90] for more details about these events).

Interestingly, the first application that Liouville made of his/Cauchy's result was an elegant proof of the fundamental theorem of algebra, which is still standard in today's textbooks; yet, Liouville had left this proof unpublished (this proof can be read from Liouville's manuscripts available at the Bibliothèque de l'Institut de France, Ms 3617(5), p. 85).

We also recall that in 1836, Liouville founded a prominent scientific journal of mathematics named "*Journal de Mathématiques Pures et Appliquées*." The journal is still published on a monthly basis, and it is often nicknamed "Liouville's Journal" after its creator. See Figure 2.20.2 for the title page of the journal's first volume.

Figure 2.20.1 Portrait of Joseph Liouville.
Source: Public Domain image from Wikipedia.

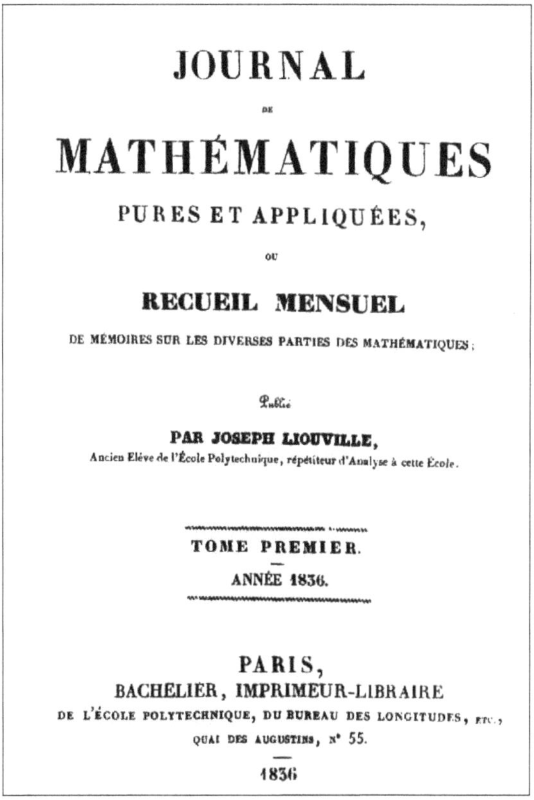

JOURNAL

DE

MATHÉMATIQUES

PURES ET APPLIQUÉES,

OU

RECUEIL MENSUEL

DE MÉMOIRES SUR LES DIVERSES PARTIES DES MATHÉMATIQUES;

Publié

PAR JOSEPH LIOUVILLE,

Ancien Elève de l'École Polytechnique, répétiteur d'Analyse à cette École.

TOME PREMIER.

ANNÉE 1836.

PARIS,

BACHELIER, IMPRIMEUR-LIBRAIRE

DE L'ÉCOLE POLYTECHNIQUE, DU BUREAU DES LONGITUDES, ETC.,

QUAI DES AUGUSTINS, N° 55.

1836

Figure 2.20.2 Title page of the first volume of Liouville's Journal.
Source: Public Domain image from Wikipedia.

of fact, to obtain such a result, a one-sided bound (i.e. from either above or below) is sufficient, as we discuss in the following result.

Theorem 2.20.1. *If u is harmonic in $\mathbb{R}^n$ and either bounded from above or from below, then u is constant.*

Proof. Up to replacing u with $-u$, we can assume that u is bounded from below. Then, up to replacing u with $u - \inf_{\mathbb{R}^n} u$, we can suppose that

$$u \geqslant 0 \text{ in } \mathbb{R}^n. \tag{2.20.1}$$

We now exploit the first-derivative Cauchy's estimate in Theorem 2.17.1. Namely, given $x_0 \in \mathbb{R}^n$ and $r > 0$, we have that

$$|\nabla u(x_0)| \leqslant \frac{C\,\|u\|_{L^1(B_r(x_0))}}{r^{n+1}},$$

for some $C > 0$ depending only on n. This, (2.20.1) and the mean value formula in Theorem 2.1.2(iii) lead to

$$|\nabla u(x_0)| \leqslant \frac{C}{r^{n+1}} \int_{B_r(x_0)} u(x)\,dx = \frac{C\,|B_1|}{r} ⨍_{B_r(x_0)} u(x)\,dx = \frac{C\,|B_1|\,u(x_0)}{r}.$$

Sending $r \to +\infty$, we thereby find that $\nabla u(x_0) = 0$. Since this is valid for all $x_0 \in \mathbb{R}^n$, we conclude that u is constant, as desired. □

An alternative proof of Liouville's theorem is given as follows.

Another Proof of Theorem 2.20.1. As noted in (2.20.1), we can suppose that $u \geqslant 0$ in $\mathbb{R}^n$. Let $a, b \in \mathbb{R}^n$. Let also $R > 0$, and note that $B_R(a) \subseteq B_{R+|a-b|}(b)$. Then, by the mean value formula for harmonic functions (recall Theorem 2.1.2), we have that

$$\begin{aligned} u(a) &= ⨍_{B_R(a)} u(x)\,dx = \frac{1}{|B_1|\,R^n} \int_{B_R(a)} u(x)\,dx \leqslant \frac{1}{|B_1|\,R^n} \int_{B_{R+|a-b|}(b)} u(x)\,dx \\ &= \frac{(R+|a-b|)^n}{R^n} ⨍_{B_{R+|a-b|}(b)} u(x)\,dx = \frac{(R+|a-b|)^n}{R^n}\,u(b). \end{aligned}$$

Consequently, sending $R \to +\infty$, we find that $u(a) \leqslant u(b)$. By exchanging a and b, we also obtain that $u(b) \leqslant u(a)$, whence u is constant. □

A variant of Theorem 2.20.1 goes as follows.

Theorem 2.20.2. *Let u be harmonic in $\mathbb{R}^n$, and suppose that*

$$|u(x)| \leqslant M\,(|x|+1)^k$$

for some $M \geqslant 0$ and $k \in \mathbb{N}$.

Then, u is a polynomial of degree at most k.

Proof. Let $x_0 \in \mathbb{R}^n$, and pick $r > |x_0| + 1$. We note that

$$\|u\|_{L^1(B_r(x_0))} \leqslant M \int_{B_r(x_0)} (|x|+1)^k\,dx \leqslant M \int_{B_{2r}} (|x|+1)^k\,dx \leqslant Cr^{k+n},$$

for some $C > 0$ depending only on M, k and n.

Therefore, by Cauchy's estimate in Theorem 2.17.1, if $\alpha \in \mathbb{N}^n$ with $|\alpha| = k+1$,

$$\left|\frac{\partial^\alpha u}{\partial x^\alpha}(x_0)\right| \leqslant \frac{C_{k+1}\,\|u\|_{L^1(B_r(x_0))}}{r^{n+k+1}} \leqslant \frac{C_{k+1}\,C}{r}.$$

Hence, we can send $r \to +\infty$ and deduce that $\frac{\partial^\alpha u}{\partial x^\alpha}(x_0) = 0$.

Since this is valid for all $x_0 \in \mathbb{R}^n$, we conclude that the derivatives of order $k+1$ of u are identically zero. As a result, we use a Taylor expansion near the origin to see that

$$u(x) = \sum_{|\alpha| \leqslant k} \frac{\partial^\alpha u(0)}{k!}\,x^\alpha.$$

In particular, u coincides with a polynomial of degree at most k near the origin. Since u is real analytic (recall Theorem 2.14.1), making use of the unique continuation principle for analytic functions (see e.g. [KP02, p. 67]), we conclude that u coincides with this polynomial in the whole of $\mathbb{R}^n$. □

2.21 Harmonic Polynomials and Spherical Harmonics

Harmonic polynomials are polynomials $p : \mathbb{R}^n \to \mathbb{R}$ which are harmonic, i.e. $\Delta p = 0$ in $\mathbb{R}^n$. The restrictions of homogeneous[29] harmonic polynomials

[29] As customary, a polynomial is called homogeneous of degree k if all its nonzero terms have the same degree k, namely if it is of the form

$$P(x_1, \dots, x_n) = \sum_{\substack{a_1,\dots,a_n \in \mathbb{N} \\ a_1+\dots+a_n=k}} c_{a_1,\dots,a_n}\, x_1^{a_1} \dots x_n^{a_n},$$

for some coefficients $c_{a_1,\dots,a_n} \in \mathbb{R}$. We stress that, with this notation, zero is a homogeneous polynomial of any degree (it suffices to take all the coefficients to be zero). This setting is convenient since it makes the family of homogeneous polynomials of degree k a vector space.

to the sphere ∂B_1 are called spherical harmonics (hereinafter, in this section, we suppose $n \geqslant 2$ to avoid the trivial case of $n = 1$ in which ∂B_1 consists of two points). More specifically, the restrictions of homogeneous harmonic polynomials of degree k to the sphere ∂B_1 are called spherical harmonics of degree k.

We begin with an approximation result for polynomials on the sphere.

Lemma 2.21.1. *The set of polynomials is dense in $L^2(\partial B_1)$.*

Proof. Let $f \in L^2(\partial B_1)$ and $\varepsilon > 0$. We can find a continuous function g such that $\|f - g\|_{L^2(\partial B_1)} < \varepsilon$ (indeed, up to a partition of unity, we can work in a local chart and reduce the approximation problem to that in a domain of $\mathbb{R}^{n-1}$, where standard convolution methods, such as in [WZ15, Theorem 9.6], can be applied). Then, let $\psi \in C_0^\infty(B_4 \setminus B_{1/4})$ with $\psi = 1$ in $B_2 \setminus B_{1/2}$ and $h(x) := \psi(x)\, g\left(\frac{x}{|x|}\right)$. We observe that h is a continuous function on $\mathbb{R}^n$. Hence, according to the Stone–Weierstraß theorem (see e.g. [DSV17, Lemma 2.1]), there exists a polynomial p such that $\|h - p\|_{L^\infty(B_4)} < \varepsilon$. As a result,

$$\begin{aligned}\|f - p\|_{L^2(\partial B_1)} &\leqslant \|f - g\|_{L^2(\partial B_1)} + \|g - p\|_{L^2(\partial B_1)}\\ &= \|f - g\|_{L^2(\partial B_1)} + \|h - p\|_{L^2(\partial B_1)}\\ &< \varepsilon + \sqrt{\int_{\partial B_1} |h(\omega) - p(\omega)|^2\, d\mathcal{H}^{n-1}_\omega}\\ &< \left(1 + \sqrt{\mathcal{H}^{n-1}(\partial B_1)}\right)\varepsilon. \end{aligned}$$ □

Now, we show that polynomials on the sphere can be harmonically and polynomially extended to the whole of $\mathbb{R}^n$.

Lemma 2.21.2. *Let p be a polynomial of degree at most k. Then, there exists a unique harmonic function h in $\mathbb{R}^n$ such that $h = p$ on ∂B_1. Moreover, h is a polynomial of degree at most k.*

More specifically, $h = p - (1 - |x|^2)\widetilde{p}$, for a suitable polynomial $\widetilde{p}$ of degree at most $k - 2$.

Proof. For every polynomial q of degree at most k, we define $Tq(x) = \Delta\big((1 - |x|^2)q(x)\big)$, and we remark that Tq is also a polynomial of degree at most k. Note that T is linear.

Furthermore, T is injective: to check this, suppose that $Tq(x) = 0$ for every $x \in \mathbb{R}^n$, and set $\widetilde{q}(x) := (1 - |x|^2)q(x)$. Note that $\widetilde{q}$ is harmonic in B_1 and vanishes along ∂B_1, whence, by the maximum principle in Corollary 2.9.2(iii), necessarily, $\widetilde{q}$ vanishes identically in B_1. This entails that q vanishes identically in B_1 and also in $\mathbb{R}^n$ by the unique continuation principle for analytic functions (see e.g. [KP02, p. 67]).

Since the space of polynomials of degree at most k is finite dimensional, we thus deduce that T is also surjective. Hence, taking $h := p - (1 - |x|^2)T^{-1}(\Delta p)$, we have that h is a polynomial of degree at most k, that $h = p$ along ∂B_1, and that

$$\Delta h = \Delta p - \Delta\big((1 - |x|^2)T^{-1}(\Delta p)\big) = \Delta p - T\big(T^{-1}(\Delta p)\big) = 0.$$

Accordingly, h satisfies the desired claim.

Suppose now that there exists another harmonic function f such that $f = p$ on ∂B_1, and let $u := f - h$. We have that $\Delta u = 0$ in B_1 and $u = 0$ along ∂B_1. Accordingly, by the maximum principle in Corollary 2.9.2(iii), we find that u vanishes identically and thus $f = h$, which shows the uniqueness of h. □

In the following Corollaries 2.21.3 and 2.21.4, we deduce some useful consequence of Lemma 2.21.2.

Corollary 2.21.3. *Let p be a homogeneous polynomial of degree $k \geqslant 2$. Then, there exist a harmonic homogeneous polynomial h of degree k and a homogeneous polynomial q of degree $k - 2$ such that*

$$p = h + |x|^2 q. \tag{2.21.1}$$

Also, this representation is unique.

Proof. By Lemma 2.21.2, we can write $p = H - (1 - |x|^2)\widetilde{p}$, where H is a harmonic polynomial of degree at most k, $H = p$ on ∂B_1 and $\widetilde{p}$ is a polynomial of degree at most $k - 2$. We let h to be the homogeneous part of H of degree k, namely we write $H = h + \widetilde{h}$, where h is a harmonic homogeneous polynomial of degree k and $\widetilde{h}$ is a polynomial of degree at most $k - 1$.

We also take q to be the homogeneous part of $\widetilde{p}$ of degree $k - 2$; namely, we write $\widetilde{p} = q + \widetilde{q}$, where q is a homogeneous polynomial of degree $k - 2$ and $\widetilde{q}$ is a polynomial of degree at most $k - 3$. As a side remark, we stress that p, h and q are also allowed to be zero in our setting (recall the footnote about homogeneous polynomials on p. 177).

As a result,

$$p = h + \widetilde{h} - (1 - |x|^2)(q + \widetilde{q}) = h + |x|^2 q + P, \tag{2.21.2}$$

with $P := \widetilde{h} - q - \widetilde{q} + |x|^2\widetilde{q}$. Note that P is a polynomial of degree at most $k - 1$. On the other hand, by construction, $p - h - |x|^2 q$ is a homogeneous polynomial of degree k. These observations and (2.21.2) yield that P is necessarily zero. From this and (2.21.2), we obtain the desired claim in (2.21.1).

To complete the desired result, it remains to check that the representation in (2.21.1) is unique. For this, by taking the difference between the two representations, it suffices to prove that

$$\begin{aligned}&\text{if a harmonic function } \overline{h}\\ &\text{and a homogeneous polynomial } \overline{q} \text{ of degree } k-2 \text{ are such that}\\ &\overline{h}(x)+|x|^2\overline{q}(x)=0 \text{ for all } x\in\mathbb{R}^n,\\ &\text{then both } \overline{h} \text{ and } \overline{q} \text{ are identically zero.}\end{aligned} \tag{2.21.3}$$

As a matter of fact, to prove this, it is enough to show that

$$\begin{aligned}&\text{if a harmonic function } \overline{h}\\ &\text{and a homogeneous polynomial } \overline{q} \text{ of degree } k-2 \text{ are such that}\\ &\overline{h}(x)+|x|^2\overline{q}(x)=0 \text{ for all } x\in\mathbb{R}^n,\\ &\text{then for every } j\in\mathbb{N}\cap\left[1,\tfrac{k}{2}\right]\\ &\text{there exists a homogeneous polynomial } q_j \text{ of degree } k-2j \text{ such that}\\ &\overline{h}(x)+|x|^{2j}q_j(x)=0 \text{ for all } x\in\mathbb{R}^n.\end{aligned} \tag{2.21.4}$$

Indeed, if (2.21.4) holds true, then we can choose

$$j_\star := \begin{cases} \dfrac{k}{2} & \text{if } k \text{ is even,}\\[2mm] \dfrac{k-1}{2} & \text{if } k \text{ is odd,}\end{cases}$$

and note that $k-2j_\star\in\{0,1\}$. Accordingly, the degree of the polynomial $q_{j_\star}$ is either zero or one. Then, we have that $q_{j_\star}(x)=a\cdot x+b$ for some $a\in\mathbb{R}^n$ and $b\in\mathbb{R}$ and, as a result,

$$\overline{h}(x)=-|x|^{2j_\star}q_{j_\star}(x)=-|x|^{2j_\star}(a\cdot x+b). \tag{2.21.5}$$

From this, we find that

$$\begin{aligned}0&=\Delta\overline{h}\\ &=-\Delta\Big(-|x|^{2j_\star}(a\cdot x+b)\Big)\\ &=2j_\star(2j_\star+n-2)|x|^{2j_\star-2}(a\cdot x+b)+2j_\star|x|^{2j_\star-2}x\cdot a\\ &=2j_\star(2j_\star+n-1)|x|^{2j_\star-2}a\cdot x+2j_\star(2j_\star+n-2)b|x|^{2j_\star-2}.\end{aligned}$$

This gives that $a=0$ and $b=0$, which, in turn, combined with (2.21.5), yields that $\overline{h}$ vanishes identically. From this and the fact that $|x|^2\overline{q}=-\overline{h}$, we conclude that $\overline{q}$ vanishes identically as well, thus showing the validity of (2.21.3).

In view of these observations, to show the validity of (2.21.3) and thus complete the proof of Corollary 2.21.3, it only remains to check (2.21.4).

The proof of (2.21.4) is by induction over j. When $j = 1$, one takes $q_j := \overline{q}$. Then, suppose that there exists a homogeneous polynomial q_j of degree $k - 2j$ with $j \leqslant \frac{k-2}{2}$ and such that $\overline{h}(x) + |x|^{2j} q_j(x) = 0$ for all $x \in \mathbb{R}^n$. We observe that

$$\begin{aligned} 0 &= \Delta\big(\overline{h} + |x|^{2j} q_j\big) = \Delta\big(|x|^{2j} q_j\big) \\ &= 2j(2j + n - 2)|x|^{2j-2} q_j + |x|^{2j} \Delta q_j + 4j|x|^{2j-2} x \cdot \nabla q_j. \end{aligned} \tag{2.21.6}$$

Also, by homogeneity,

$$x \cdot \nabla q_j(x) = \frac{d}{dt} q_j(tx)\bigg|_{t=1} = \frac{d}{dt}\Big(t^{k-2j} q_j(x)\Big)\bigg|_{t=1} = (k - 2j) q_j(x),$$

and thus, from (2.21.6),

$$\begin{aligned} 0 &= 2j(2j + n - 2)|x|^{2j-2} q_j + |x|^{2j} \Delta q_j + 4j|x|^{2j-2}(k - 2j) q_j \\ &= 2j(n - 2 + 2k - 2j)|x|^{2j-2} q_j + |x|^{2j} \Delta q_j. \end{aligned}$$

Therefore, since $n - 2 + 2k - 2j \geqslant n - 2 + 2k - (k - 2) > 0$, we can define

$$q_{j+1} := -\frac{\Delta q_j}{2j(n - 2 + 2k - 2j)}$$

and find that

$$0 = q_j + \frac{|x|^2 \Delta q_j}{2j(n - 2 + 2k - 2j)} = q_j - |x|^2 q_{j+1}.$$

In this way,

$$\overline{h} + |x|^{2(j+1)} q_{j+1} = \overline{h} + |x|^{2j} q_j + |x|^{2j}\left(|x|^2 q_{j+1} - q_j\right) = \overline{h} + |x|^{2j} q_j = 0,$$

which shows the validity of (2.21.4), as desired. □

Corollary 2.21.4. *Harmonic polynomials are dense in $L^2(\partial B_1)$.*

Proof. Let $f \in L^2(\partial B_1)$ and $\varepsilon > 0$. By Lemma 2.21.1, we can find a polynomial p such that $\|f - p\|_{L^2(\partial B_1)} < \varepsilon$. Also, by Lemma 2.21.2, there exists a harmonic polynomial h such that $\|p - h\|_{L^2(\partial B_1)} < \varepsilon$. As a byproduct, $\|f - h\|_{L^2(\partial B_1)} < 2\varepsilon$. □

It is also useful to remark that spherical harmonics of different degrees are orthogonal. For this, we first observe that there is a strong relation between harmonic polynomials and the eigenfunctions of the Laplace–Beltrami operator.

Lemma 2.21.5. *The spherical harmonics of degree k are precisely the eigenfunctions of the Laplace–Beltrami operator on ∂B_1 with*[30] *eigenvalue $-k(n+k-2)$.*

That is, let $h \in C^\infty(\partial B_1)$. Then, the following conditions are equivalent:

(i) *h is a spherical harmonic of degree k.*
(ii) *h is an eigenfunction of the Laplace–Beltrami operator on the sphere with eigenvalue $-k(n+k-2)$.*

Proof. Let $h : \partial B_1 \to \mathbb{R}$ be a spherical harmonic of degree k. Then, there exists a homogeneous harmonic polynomial $p : \mathbb{R}^n \to \mathbb{R}$ of degree k such that $h = p$ on ∂B_1. In particular, by (2.3.5), we have that

$$\Delta_{\partial B_1} h = \Delta_{\partial B_1} p. \tag{2.21.7}$$

This and Theorem 2.5.1 yield that, on ∂B_1,

$$0 = \Delta p = \partial_{rr} p_0 + (n-1)\,\partial_r p_0 + \Delta_{\partial B_1} h,$$

where

$$p_0(r,\vartheta) := p(r\vartheta) \text{ for each } r > 0 \text{ and } \vartheta \in \partial B_1. \tag{2.21.8}$$

By homogeneity, we know that $p_0(r,\vartheta) = r^k p(\vartheta)$; therefore,

$$\begin{aligned} \partial_r p_0(r,\vartheta) &= kr^{k-1} p(\vartheta) = kr^{k-1} h(\vartheta) \\ \text{and} \quad \partial_{rr} p_0(r,\vartheta) &= k(k-1)r^{k-2} p(\vartheta) = k(k-1)r^{k-2} h(\vartheta). \end{aligned} \tag{2.21.9}$$

Consequently, on ∂B_1,

$$0 = k(k-1)h + (n-1)kh + \Delta_{\partial B_1} h = k(n+k-2)h + \Delta_{\partial B_1} h.$$

Accordingly, we have that h is an eigenfunction of the Laplace–Beltrami operator with eigenvalue $-k(n+k-2)$.

Vice versa, suppose now that h is an eigenfunction of the Laplace–Beltrami operator on ∂B_1 with eigenvalue $-k(n+k-2)$. Let $p(x) := |x|^k h\left(\frac{x}{|x|}\right)$. Let p_0 be as in (2.21.8), and note that p is homogeneous of degree k. Thus, exploiting Theorem 2.5.1, (2.21.7) and (2.21.9), we see that, on ∂B_1,

$$\begin{aligned} \Delta p &= \partial_{rr} p_0 + (n-1)\,\partial_r p_0 + \Delta_{\partial B_1} h \\ &= k(k-1)r^{k-2}h + (n-2)kr^{k-2}h - k(n+k-2)h = 0. \end{aligned}$$

Since, by homogeneity, for every $\lambda > 0$ and $x \in \mathbb{R}^n$,

$$\frac{\partial^2 p}{\partial x_i \partial x_j}(\lambda x) = \frac{1}{\lambda^2}\frac{\partial^2}{\partial x_i \partial x_j}(p(\lambda x)) = \frac{1}{\lambda^2}\frac{\partial^2}{\partial x_i \partial x_j}(\lambda^k p(x)) = \lambda^{k-2}\frac{\partial^2 p}{\partial x_i \partial x_j}(x),$$

[30] We will see in Theorem 2.21.8 that the eigenvalues listed in Lemma 2.21.5 are indeed all the eigenvalues of the Laplace–Beltrami operator on the sphere.

we thus conclude that, for each $x \in \mathbb{R}^n \setminus \{0\}$,

$$\Delta p(x) = \Delta p\left(|x| \frac{x}{|x|}\right) = |x|^{k-2} \Delta p\left(\frac{x}{|x|}\right) = 0,$$

and so, by continuity, $\Delta p = 0$ in $\mathbb{R}^n$.

In addition, for every $x \in \mathbb{R}^n$,

$$|p(x)| \leqslant |x|^k \max_{\partial B_1} |h|,$$

and hence, according to Liouville's theorem (recall Theorem 2.20.2), it follows that p is a polynomial of degree k.

Since p coincides with h on ∂B_1, this gives that h is a spherical harmonics of degree k. □

With this, we are in a position to prove that spherical harmonics are orthogonal.

Lemma 2.21.6. *Let h and g be spherical harmonics of different degrees. Then,*

$$\int_{\partial B_1} h(x)\, g(x)\, d\mathcal{H}_x^{n-1} = 0.$$

Proof. Let k and m be the degrees of h and g, respectively. Let also $\phi(t) := t(t+n-2)$. Since $\phi'(t) = 2t + n - 2 \geqslant n > 0$ for all $t \geqslant 1$, it follows that ϕ is strictly increasing in $[1, +\infty)$, and in particular, if $k \neq m$, then

$$k(n+k-2) \neq m(m+n-2). \tag{2.21.10}$$

Moreover, by Lemma 2.21.5 and the self-adjointness property pointed out in (2.3.27),

$$\begin{aligned} -k(n+k-2) \int_{\partial B_1} h(x)\, g(x)\, d\mathcal{H}_x^{n-1} &= \int_{\partial B_1} \Delta_{\partial B_1} h(x)\, g(x)\, d\mathcal{H}_x^{n-1} \\ = \int_{\partial B_1} h(x)\, \Delta_{\partial B_1} g(x)\, d\mathcal{H}_x^{n-1} &= -m(m+n-2) \int_{\partial B_1} h(x)\, g(x)\, d\mathcal{H}_x^{n-1}. \end{aligned}$$

This and (2.21.10) entail the desired result. □

With this, we can actually strengthen Corollary 2.21.4 by approximating any function in $L^2(\partial B_1)$ using the harmonic polynomials of increasing degree and modifying each step of the approximation by a polynomial of higher degree.

Corollary 2.21.7. *For every $f \in L^2(\partial B_1)$, there exists a sequence h_k of spherical harmonics of degree equal to k such that*

$$\lim_{N \to +\infty} \left\| f - \sum_{k=0}^{N} h_k \right\|_{L^2(\partial B_1)} = 0.$$

Proof. The proof exploits some classical methods from functional analysis and Fourier series. We give a self-contained argument for the facility of the reader. For every $k \in \mathbb{N} \setminus \{0\}$, we consider the vector space of spherical harmonics of degree precisely equal to k. This space is finite dimensional (since so is the space of polynomials); therefore, we can consider an orthonormal basis (with respect to the scalar product in $L^2(\partial B_1)$) that we denote by $\{\eta_{1,k}, \dots, \eta_{n_k,k}\}$. By Lemma 2.21.6, the set $\{\eta_{j,k}\}_{\substack{k\in\mathbb{N}\\ 1\leqslant j\leqslant n_k}}$ is an orthonormal family, that is,

$$\langle \eta_{j,k}, \eta_{i,m} \rangle_{L^2(\partial\Omega)} = \delta_{km}\delta_{ij}. \tag{2.21.11}$$

Let

$$h_k := \sum_{1\leqslant j\leqslant n_k} \langle f, \eta_{j,k} \rangle_{L^2(\partial\Omega)}\, \eta_{j,k},$$

and note that h_k is a spherical harmonic of degree equal to k.

Now, we take $\varepsilon > 0$ and employ Corollary 2.21.4 to find a linear combination of spherical harmonics $h^{(\varepsilon)}$ such that $\|f - h^{(\varepsilon)}\|^2_{L^2(\partial\Omega)} < \varepsilon$. We write

$$h^{(\varepsilon)} = \sum_{k=0}^{N^{(\varepsilon)}} h_k^{(\varepsilon)},$$

for suitable spherical harmonics $h_k^{(\varepsilon)}$ of degree precisely equal to k and some $N^{(\varepsilon)} \in \mathbb{N}$. We also express each $h_k^{(\varepsilon)}$ in terms of the above orthonormal basis by writing

$$h_k^{(\varepsilon)} = \sum_{1\leqslant j\leqslant n_k} a_{jk}^{(\varepsilon)}\, \eta_{j,k},$$

for suitable $a_{jk}^{(\varepsilon)} \in \mathbb{N}$. All in all, we find that

$$h^{(\varepsilon)} = \sum_{\substack{0\leqslant k\leqslant N^{(\varepsilon)}\\ 1\leqslant j\leqslant n_k}} a_{jk}^{(\varepsilon)}\, \eta_{j,k},$$

and accordingly

$$h^{(\varepsilon)} \pm \sum_{k=0}^{N^{(\varepsilon)}} h_k = \sum_{\substack{0\leqslant k\leqslant N^{(\varepsilon)}\\ 1\leqslant j\leqslant n_k}} \left(a_{jk}^{(\varepsilon)} \pm \langle f, \eta_{j,k} \rangle_{L^2(\partial\Omega)} \right) \eta_{j,k}.$$

In light of (2.21.11), we thereby have that

$$
\begin{aligned}
&\left\| f - \sum_{k=0}^{N^{(\varepsilon)}} h_k \right\|_{L^2(\partial\Omega)}^2 - \varepsilon \\
&< \left\| f - \sum_{k=0}^{N^{(\varepsilon)}} h_k \right\|_{L^2(\partial\Omega)}^2 - \| f - h^{(\varepsilon)} \|_{L^2(\partial\Omega)}^2 \\
&= \| f \|_{L^2(\partial\Omega)}^2 + \left\langle \sum_{m=0}^{N^{(\varepsilon)}} h_m, \sum_{k=0}^{N^{(\varepsilon)}} h_k \right\rangle_{L^2(\partial\Omega)} - 2 \left\langle f, \sum_{k=0}^{N^{(\varepsilon)}} h_k \right\rangle_{L^2(\partial\Omega)} \\
&\quad - \| f \|_{L^2(\partial\Omega)}^2 - \| h^{(\varepsilon)} \|_{L^2(\partial\Omega)}^2 + 2 \left\langle f, h^{(\varepsilon)} \right\rangle_{L^2(\partial\Omega)} \\
&= \left\langle \sum_{\substack{0 \leqslant m \leqslant N^{(\varepsilon)} \\ 1 \leqslant i \leqslant n_m}} \langle f, \eta_{i,m} \rangle_{L^2(\partial\Omega)} \, \eta_{i,m}, \sum_{\substack{0 \leqslant k \leqslant N^{(\varepsilon)} \\ 1 \leqslant j \leqslant n_k}} \langle f, \eta_{j,k} \rangle_{L^2(\partial\Omega)} \, \eta_{j,k} \right\rangle_{L^2(\partial\Omega)} \\
&\quad - 2 \left\langle f, \sum_{\substack{0 \leqslant k \leqslant N^{(\varepsilon)} \\ 1 \leqslant j \leqslant n_k}} \langle f, \eta_{j,k} \rangle_{L^2(\partial\Omega)} \, \eta_{j,k} \right\rangle_{L^2(\partial\Omega)} \\
&\quad - \left\langle \sum_{\substack{0 \leqslant m \leqslant N^{(\varepsilon)} \\ 1 \leqslant i \leqslant n_k}} a_{im}^{(\varepsilon)} \, \eta_{i,m}, \sum_{\substack{0 \leqslant k \leqslant N^{(\varepsilon)} \\ 1 \leqslant j \leqslant n_k}} a_{jk}^{(\varepsilon)} \, \eta_{j,k} \right\rangle_{L^2(\partial\Omega)} \\
&\quad + 2 \left\langle f, \sum_{\substack{0 \leqslant k \leqslant N^{(\varepsilon)} \\ 1 \leqslant j \leqslant n_k}} a_{jk}^{(\varepsilon)} \, \eta_{j,k} \right\rangle_{L^2(\partial\Omega)} \\
&= - \sum_{\substack{0 \leqslant k \leqslant N^{(\varepsilon)} \\ 1 \leqslant j \leqslant n_k}} \langle f, \eta_{j,k} \rangle_{L^2(\partial\Omega)}^2 - \sum_{\substack{0 \leqslant k \leqslant N^{(\varepsilon)} \\ 1 \leqslant j \leqslant n_k}} (a_{jk}^{(\varepsilon)})^2 \\
&\quad + 2 \sum_{\substack{0 \leqslant k \leqslant N^{(\varepsilon)} \\ 1 \leqslant j \leqslant n_k}} a_{jk}^{(\varepsilon)} \, \langle f, \eta_{j,k} \rangle_{L^2(\partial\Omega)} \\
&= - \sum_{\substack{0 \leqslant k \leqslant N^{(\varepsilon)} \\ 1 \leqslant j \leqslant n_k}} \left(\langle f, \eta_{j,k} \rangle_{L^2(\partial\Omega)} - a_{jk}^{(\varepsilon)} \right)^2 \\
&\leqslant 0,
\end{aligned}
$$

establishing the desired result. □

We now establish that the eigenvalues found in Lemma 2.21.5 are indeed all the possible eigenvalues of the Laplace–Beltrami operator on the sphere.

Theorem 2.21.8. *The eigenvalues of the Laplace–Beltrami operator on* ∂B_1 *are* $-k(n+k-2)$ *with* $k \in \mathbb{N} \setminus \{0\}$.

Proof. We recall that we are assuming $n \geqslant 2$, and we use the notation $x = (x_1, \dots, x_n)$ and $z = x_1 + ix_2 \in \mathbb{C}$. By (2.1.2), we know that, for every $k \in \mathbb{N}$, the following function is harmonic in $\mathbb{R}^2$ (and hence in $\mathbb{R}^n$ by trivial extension in the variables $(x_3, \dots, x_n)$):

$$\begin{aligned} p_k(x_1, x_2) &:= \Re z^k = \Re(x_1 + ix_2)^k = \Re\left(\sum_{m=0}^{k} \binom{k}{m} i^m x_1^{k-m} x_2^m\right) \\ &= \sum_{\substack{0 \leqslant m \leqslant k \\ m \in 2\mathbb{Z}}} \binom{k}{m} (-1)^{\frac{m}{2}} x_1^{k-m} x_2^m. \end{aligned}$$

Note that p_k is a homogeneous polynomial of degree k and thus, in light of Lemma 2.21.5, the corresponding spherical harmonic defined by $h_k(x) := p_k(x)$ for all $x \in \partial B_1$ is an eigenfunction of the Laplace–Beltrami operator on ∂B_1 with eigenvalue $-k(n+k-2)$.

This shows that the set of eigenvalues of the Laplace–Beltrami operator on ∂B_1 contains $\{-k(n+k-2),\ k \in \mathbb{N} \setminus \{0\}\}$.

Suppose now that λ is an eigenvalue of the Laplace–Beltrami operator on ∂B_1. We aim to show that $\lambda \in \{-k(n+k-2),\ k \in \mathbb{N} \setminus \{0\}\}$. To this end, we argue by contradiction and suppose that

$$\lambda \notin \{-k(n+k-2),\ k \in \mathbb{N} \setminus \{0\}\}. \tag{2.21.12}$$

For all $k \in \mathbb{N} \setminus \{0\}$, we observe that the set of spherical harmonics of degree k is a vector space, and it is finite dimensional (since so is the space of polynomials). Thus, we consider an orthonormal basis for such a space (with respect to the scalar product in $L^2(\partial B_1)$), denoted by $\{E_{1,k}, \dots, E_{n_k,k}\}$, and the projection

$$\begin{aligned} \Pi_k \varphi(x) &:= \sum_{j=1}^{n_k} \int_{\partial B_1} \varphi(y)\, E_{j,k}(y)\, d\mathcal{H}_y^{n-1}\, E_{j,k}(x) \\ &= \sum_{j=1}^{n_k} \langle \varphi, E_{j,k} \rangle_{L^2(\partial B_1)}\, E_{j,k}(x). \end{aligned} \tag{2.21.13}$$

Furthermore, by Lemma 2.21.5 and the self-adjointness property in (2.3.27), if φ is an eigenfunction corresponding to λ, say normalized such that

$$\|\varphi\|_{L^2(\partial B_1)} = 1, \tag{2.21.14}$$

then

$$\begin{aligned}-k(n+k-2)\langle\varphi, E_{j,k}\rangle_{L^2(\partial B_1)} &= -\int_{\partial B_1}\varphi(y)\,k(n+k-2)E_{j,k}(y)\,d\mathcal{H}^{n-1}_y\\ &= \int_{\partial B_1}\varphi(y)\,\Delta_{\partial B_1}E_{j,k}(y)\,d\mathcal{H}^{n-1}_y\\ &= \int_{\partial B_1}\Delta_{\partial B_1}\varphi(y)\,E_{j,k}(y)\,d\mathcal{H}^{n-1}_y\\ &= \lambda\int_{\partial B_1}\varphi(y)\,E_{j,k}(y)\,d\mathcal{H}^{n-1}_y = \lambda\langle\varphi, E_{j,k}\rangle_{L^2(\partial B_1)}\end{aligned}$$

and accordingly

$$\left(\frac{\lambda}{k(n+k-2)}+1\right)\langle\varphi, E_{j,k}\rangle_{L^2(\partial B_1)} = 0.$$

With this information and (2.21.12), we deduce that $\langle\varphi, E_{j,k}\rangle_{L^2(\partial B_1)} = 0$ for each $j \in \{1,\dots,n_k\}$. This and (2.21.13) yield that $\Pi_k\varphi = 0$ for every $k \in \mathbb{N}\setminus\{0\}$.

Now, for each $\varepsilon > 0$, we use Corollary 2.21.4 to pick a spherical harmonic h_ε such that $\|\varphi - h_\varepsilon\|_{L^2(\partial B_1)} < \varepsilon$. We let k_ε be the degree of h_ε, and we observe that the projection of h_ε on the space of spherical harmonics with degree k_ε is h_ε itself. As a result, we have that

$$\begin{aligned}\|h_\varepsilon\|_{L^2(\partial B_1)} &= \|h_\varepsilon - \Pi_{k_\varepsilon}\varphi\|_{L^2(\partial B_1)} = \|\Pi_{k_\varepsilon}h_\varepsilon - \Pi_{k_\varepsilon}\varphi\|_{L^2(\partial B_1)}\\ &= \|\Pi_{k_\varepsilon}(h_\varepsilon - \varphi)\|_{L^2(\partial B_1)} \leqslant \|h_\varepsilon - \varphi\|_{L^2(\partial B_1)} < \varepsilon.\end{aligned}$$

Consequently,

$$\|\varphi\|_{L^2(\partial B_1)} \leqslant \|\varphi - h_\varepsilon\|_{L^2(\partial B_1)} + \|h_\varepsilon\|_{L^2(\partial B_1)} < 2\varepsilon.$$

Since ε is arbitrary, this entails that $\|\varphi\|_{L^2(\partial B_1)} = 0$, in contradiction with the normalization in (2.21.14). □

We can now compute the dimension of the vector spaces of spherical harmonics.

Theorem 2.21.9. *The space of harmonic homogeneous polynomials of degree k has dimension*[31]

$$\begin{cases}1 & \text{if } k=0,\\ n & \text{if } k=1,\\ \binom{k+n-1}{n-1}-\binom{k+n-3}{n-1} & \text{if } k\geqslant 2.\end{cases} \tag{2.21.15}$$

[31]When $k \geqslant 1$, after a simple algebraic computation, one can alternatively write the quantity in (2.21.15) as

$$\frac{n+2k-2}{k}\binom{k+n-3}{k-1}.$$

Proof. We consider separately the cases $k = 0$, $k = 1$ and $k \geqslant 2$.

When $k = 0$, the only homogeneous polynomials are the constants, thus producing a one-dimensional vector space.

When $k = 1$, we only have linear functions (which are the homogeneous polynomials of degree one and are also harmonic), thus producing an n-dimensional vector space.

Thus, we can focus now on the proof of (2.21.15) for the case of $k \geqslant 2$. We denote by D_k the dimension of the space of the homogeneous polynomials in $\mathbb{R}^n$ of degree k. Let also $D_k^\star$ be the dimension of the space of the harmonic homogeneous polynomials in $\mathbb{R}^n$ of degree k.

In view of (2.21.1), we know that $D_k = D_k^\star + D_{k-2}$; therefore,

$$\begin{gathered}\text{the dimension } D_k^\star \text{ of the space}\\ \text{of harmonic homogeneous polynomials of degree } k\\ \text{is equal to } D_k - D_{k-2}.\end{gathered} \tag{2.21.16}$$

Now, we prove that

$$D_k = \binom{k+n-1}{n-1}. \tag{2.21.17}$$

This is an algebraically (or, actually, combinatorially) separate statement on polynomials (and nothing to do with harmonic functions). To prove it, we observe that for every $a_1, \dots, a_n \in \mathbb{N}$ with $a_1 + \dots + a_n = k$, the homogeneous polynomials, or actually monomials, of the form $x_1^{a_1} \cdots x_n^{a_n}$ are all linearly independent due to the identity principle for polynomials. Since every homogeneous polynomial is written as the sum of such monomials, we have that

$$D_k = \#\Big\{a_1, \dots, a_n \in \mathbb{N} \text{ with } a_1 + \dots + a_n = k\Big\}. \tag{2.21.18}$$

Now, we claim that

$$\#\Big\{a_1, \dots, a_n \in \mathbb{N} \text{ with } a_1 + \dots + a_n = k\Big\} = \binom{k+n-1}{n-1}. \tag{2.21.19}$$

To prove this, we define

$$\mathcal{A}_{n,k} := \Big\{(a_1, \dots, a_n) \in \mathbb{N}^n \text{ with } a_1 + \dots + a_n = k\Big\}. \tag{2.21.20}$$

We let $\mathcal{T} : \mathcal{A}_{n,k} \to \mathcal{A}_{n-1,k} \cup \mathcal{A}_{n,k-1}$ defined as

$$\mathcal{T}(a_1, \dots, a_n) := \begin{cases} (a_1, \dots, a_{n-1}) & \text{if } a_n = 0, \\ (a_1, \dots, a_{n-1}, a_n - 1) & \text{if } a_n \neq 0. \end{cases}$$

Note indeed that, for each $(a_1, \ldots, a_n) \in \mathcal{A}_{n,k}$, if $a_n = 0$, then $\mathcal{T}(a_1, \ldots, a_n) = (a_1, \ldots, a_{n-1}) \in \mathcal{A}_{n-1,k}$, while if $a_n \neq 0$, then $\mathcal{T}(a_1, \ldots, a_n) = (a_1, \ldots, a_{n-1}, a_n - 1) \in \mathcal{A}_{n,k-1}$.

Furthermore,

$$\mathcal{T} \text{ is injective.} \tag{2.21.21}$$

Indeed, let $(a_1, \ldots, a_n), (b_1, \ldots, b_n) \in \mathcal{A}_{n,k}$ and suppose that $\mathcal{T}(a_1, \ldots, a_n) = \mathcal{T}(b_1, \ldots, b_n) =: \tau$. If $\tau \in \mathcal{A}_{n-1,k}$, then necessarily $a_n = b_n = 0$ and $\tau = (a_1, \ldots, a_{n-1}) = (b_1, \ldots, b_{n-1})$, leading to $(a_1, \ldots, a_n) = (b_1, \ldots, b_n)$.

If instead $\tau \in \mathcal{A}_{n,k-1}$, then necessarily $a_n \neq 0$ and $b_n \neq 0$; moreover, $\tau = (a_1, \ldots, a_{n-1}, a_n - 1) = (b_1, \ldots, b_{n-1}, b_n - 1)$, showing again that $(a_1, \ldots, a_n) = (b_1, \ldots, b_n)$. These observations complete the proof of (2.21.21).

Additionally, we have that

$$\mathcal{T} \text{ is surjective.} \tag{2.21.22}$$

Indeed, if $(\vartheta_1, \ldots, \vartheta_{n-1}) \in \mathcal{A}_{n-1,k}$, we let $a_i := \vartheta_i$ for each $i \in \{1, \ldots, n-1\}$ and $a_n := 0$; in this way, we see that $(a_1, \ldots, a_n) \in \mathcal{A}_{n,k}$ and $\mathcal{T}(a_1, \ldots, a_n) = (\vartheta_1, \ldots, \vartheta_{n-1})$. If instead $(\vartheta_1, \ldots, \vartheta_n) \in \mathcal{A}_{n,k-1}$, then we set $a_i := \vartheta_i$ for each $i \in \{1, \ldots, n-1\}$ and $a_n := \vartheta_n + 1$; in this way, we see that $(a_1, \ldots, a_n) \in \mathcal{A}_{n,k}$ and $\mathcal{T}(a_1, \ldots, a_n) = (\vartheta_1, \ldots, \vartheta_n)$. These considerations prove (2.21.22).

As a result, we deduce from (2.21.21) and (2.21.22) that

$$\#\mathcal{A}_{n,k} = \#\mathcal{A}_{n-1,k} + \#\mathcal{A}_{n,k-1}. \tag{2.21.23}$$

Now, we claim that

$$\#\mathcal{A}_{n,k} = \binom{k+n-1}{n-1}. \tag{2.21.24}$$

We prove this by induction over $N := k + n$. When $N = 1$, we have that necessarily $n = 1$ and $k = 0$, and thus $\mathcal{A}_{n,k} = \mathcal{A}_{1,0}$ reduces to the zero element, whence

$$\#\mathcal{A}_{1,0} = 1 = \binom{0}{0}.$$

As for the inductive step, we consider $N = k + n \geqslant 2$ and assume (2.21.24) to be true for the index $N - 1 \geqslant 1$. Thus, exploiting (2.21.23),

$$\begin{aligned}
\#\mathcal{A}_{n,k} &= \#\mathcal{A}_{n-1,k} + \#\mathcal{A}_{n,k-1} = \binom{k+n-2}{n-2} + \binom{k+n-2}{n-1} \\
&= \frac{(k+n-2)!}{k!\,(n-2)!} + \frac{(k+n-2)!}{(k-1)!\,(n-1)!} = \frac{(k+n-2)!}{k!\,(n-1)!}\big((n-1)+k\big) \\
&= \frac{(k+n-1)!}{k!\,(n-1)!} = \binom{k+n-1}{n-1}.
\end{aligned}$$

This completes the inductive step and establishes (2.21.24), as desired.

The claim in (2.21.19) now plainly follows from (2.21.20) and (2.21.24).

Accordingly, the desired claim in (2.21.15) is a straightforward consequence of (2.21.16) and (2.21.19). □

In view of Theorem 2.21.9, we now aim at constructing explicit bases of the space of harmonic homogeneous polynomials of degree k (this is of clear practical importance, since one can reduce a number of general computations to those related to a specific basis). This is indeed the content of the forthcoming Section 2.22.

2.22 Spherical Harmonics and Legendre Polynomials

We now turn our attention to some special classes of spherical harmonics with explicit and convenient algebraic properties. To this end, we consider a basis for the space of harmonic homogeneous polynomials of degree k and the corresponding spherical harmonics of degree k. After a Gram–Schmidt process, we find a maximal linearly independent set of spherical harmonics $\{Y_{k,1}, \dots, Y_{k,N_k}\}$ that are orthonormal with respect to the scalar product in $L^2(\partial B_1)$ (here N_k is as in (2.21.15), thanks to Theorem 2.21.9). We stress that, in light of Lemma 2.21.6,

$$\int_{\partial B_1} Y_{k,j}(x)\, Y_{m,i}(x)\, d\mathcal{H}_x^{n-1} = \delta_{km}\, \delta_{ji}. \tag{2.22.1}$$

For each $x, y \in \partial B_1$, we also define

$$F_k(x, y) := \sum_{j=1}^{N_k} Y_{k,j}(x)\, Y_{k,j}(y), \tag{2.22.2}$$

and we point out a symmetry property of this function as follows.

Lemma 2.22.1. *The function F_k is invariant under rotation. That is, if $\mathcal{R}$ is a rotation matrix, then*

$$F_k(\mathcal{R}x, \mathcal{R}y) = F_k(x, y) \quad \textit{for every } x, y \in \partial B_1.$$

Proof. If p is a harmonic homogeneous polynomial of degree k, then so is $\widetilde{p}(x) := p(\mathcal{R}x)$, thanks to the invariance under rotation of the Laplace equation (recall Corollary 1.1.4). Accordingly, rotations preserve spherical harmonics, and $\widetilde{Y}_{k,j}(x) := Y_{k,j}(\mathcal{R}x)$ is also a spherical harmonic of degree k. Accordingly, we can write $\widetilde{Y}_{k,j}$ in terms of the orthonormal basis $\{Y_{k,1}, \dots, Y_{k,N_k}\}$ as

$$\widetilde{Y}_{k,j} = \sum_{i=1}^{N_k} M_{ij}\, Y_{k,i},$$

for some $M_{ij} \in \mathbb{R}$. We let M be the $(N_k \times N_k)$ matrix produced by M_{ij}, and we note that

$$\text{the matrix } M \text{ is orthogonal,} \tag{2.22.3}$$

since, by (2.22.1),

$$\begin{aligned}\sum_{i=1}^{N_k} M_{ij} M_{im} &= \sum_{i,\ell=1}^{N_k} M_{ij} M_{\ell m} \delta_{\ell i} = \sum_{i,\ell=1}^{N_k} M_{ij} M_{\ell m} \int_{\partial B_1} Y_{k,\ell}(x)\, Y_{k,i}(x)\, d\mathcal{H}^{n-1}_x \\ &= \int_{\partial B_1} \widetilde{Y}_{k,j}(x)\, \widetilde{Y}_{k,m}(x)\, d\mathcal{H}^{n-1}_x = \int_{\partial B_1} Y_{k,j}(\mathcal{R}x)\, Y_{k,m}(\mathcal{R}x)\, d\mathcal{H}^{n-1}_x \\ &= \int_{\partial B_1} Y_{k,j}(y)\, Y_{k,m}(y)\, d\mathcal{H}^{n-1}_y = \delta_{jm}.\end{aligned}$$

Hence, using (2.22.3), we have that the transpose of M is orthogonal too and therefore

$$\begin{aligned}F_k(\mathcal{R}x, \mathcal{R}y) &= \sum_{j=1}^{N_k} Y_{k,j}(\mathcal{R}x)\, Y_{k,j}(\mathcal{R}y) = \sum_{j=1}^{N_k} \widetilde{Y}_{k,j}(x)\, \widetilde{Y}_{k,j}(y) \\ &= \sum_{j,i,\ell=1}^{N_k} M_{ij}\, M_{\ell j}\, Y_{k,i}(x)\, Y_{k,\ell}(y) = \sum_{i,\ell=1}^{N_k} \delta_{i\ell}\, Y_{k,i}(x)\, Y_{k,\ell}(y) \\ &= \sum_{j=1}^{N_k} Y_{k,j}(x)\, Y_{k,j}(y) = F_k(x, y).\end{aligned}$$

□

Corollary 2.22.2. *There exists*[32] *a polynomial P_k such that, for every $x, y \in \partial B_1$,*

$$F_k(x, y) = P_k(x \cdot y).$$

Proof. First of all, we note that, with $Y_{k,j}$ being a spherical harmonic, we can consider it as the restriction to ∂B_1 of a homogeneous polynomial of degree k, say

$$Y_{k,j}(x) = \sum_{\substack{\alpha^{(j)} \in \mathbb{N}^n \\ |\alpha^{(j)}|=k}} c_{[\alpha^{(j)},j,k]}\, x^{\alpha^{(j)}}$$

for suitable $c_{[\alpha^{(j)},j,k]} \in \mathbb{R}$.

[32]Up to multiplicative constants, this polynomial is precisely the Legendre polynomial that will be introduced in Theorem 2.22.4. This point is addressed specifically in the forthcoming Theorem 2.22.5.

This and (2.22.2) give that F_k can also be considered the restriction to $(\partial B_1)\times(\partial B_1)$ of a suitable polynomial in $\mathbb{R}^n \times \mathbb{R}^n$ of the form

$$F_k(x,y) = \sum_{\substack{1\leqslant j\leqslant N_k\\ \alpha^{(j)}\in\mathbb{N}^n\\ |\alpha^{(j)}|=k\\ \beta^{(j)}\in\mathbb{N}^n\\ |\beta^{(j)}|=k}} c_{[\alpha^{(j)},j,k]}\, c_{[\beta^{(j)},j,k]}\, x^{\alpha^{(j)}}\, x^{\beta^{(j)}} \tag{2.22.4}$$

Now, let $t := x\cdot y$. We consider a rotation $\mathcal{R}_{(x,y)}$ such that $\mathcal{R}_{(x,y)}y = e_1$ and $\mathcal{R}_{(x,y)}x = (t,\sqrt{1-t^2},0,\dots,0) = te_1 + \sqrt{1-t^2}\,e_2$, see Figure 2.22.1. By Lemma 2.22.1, we know that

$$F_k(x,y) = F_k(\mathcal{R}_{(x,y)}x,\, \mathcal{R}_{(x,y)}y) = F_k\Big(te_1 + \sqrt{1-t^2}\,e_2, e_1\Big). \tag{2.22.5}$$

In particular, by (2.22.4), one can write $F_k(x,y)$ as a polynomial in t and $\tau := \sqrt{1-t^2}$, say

$$\begin{aligned} F_k(x,y) &= \sum_{\substack{1\leqslant j\leqslant N_k\\ \alpha^{(j)}\in\mathbb{N}^n\\ |\alpha^{(j)}|=k\\ \beta^{(j)}\in\mathbb{N}^n\\ |\beta^{(j)}|=k}} c_{[\alpha^{(j)},j,k]}\, c_{[\beta^{(j)},j,k]}\, (te_1+\tau e_2)^{\alpha^{(j)}}\, e_1^{\beta^{(j)}} \\ &= \sum_{\substack{1\leqslant j\leqslant N_k\\ \alpha^{(j)}\in\mathbb{N}^2\\ |\alpha^{(j)}|=k}} c_{[\alpha^{(j)},j,k]}\, c_{[ke_1,j,k]}\, t^{\alpha_1^{(j)}}\, \tau^{\alpha_2^{(j)}} \\ &= \sum_{\substack{0\leqslant i,j\leqslant k\\ i+j=k}} c_{ij}\, t^i\, \tau^j, \end{aligned} \tag{2.22.6}$$

for suitable $c_{ij}\in\mathbb{R}$.

Now, we compose $\mathcal{R}_{(x,y)}$ with a further rotation $\mathcal{R}$ of π radians around the first axis, and let $\mathcal{R}'_{(x,y)} := \mathcal{R}\mathcal{R}_{(x,y)}$. In this way, $\mathcal{R}'_{(x,y)}y = \mathcal{R}e_1 = e_1$ and $\mathcal{R}'_{(x,y)}x = \mathcal{R}(t,\sqrt{1-t^2},0,\dots,0) = (t,-\sqrt{1-t^2},0,\dots,0) = te_1 - \sqrt{1-t^2}\,e_2$, see again Figure 2.22.1. Arguing as in (2.22.5), we thus obtain that

$$F_k(x,y) = F_k\Big(te_1 - \sqrt{1-t^2}\,e_2, e_1\Big).$$

That is, in the notation of (2.22.6),

$$F_k(x,y) = \sum_{\substack{0\leqslant i,j\leqslant k\\ i+j=k}} c_{ij}\, t^i\, (-\tau)^j.$$

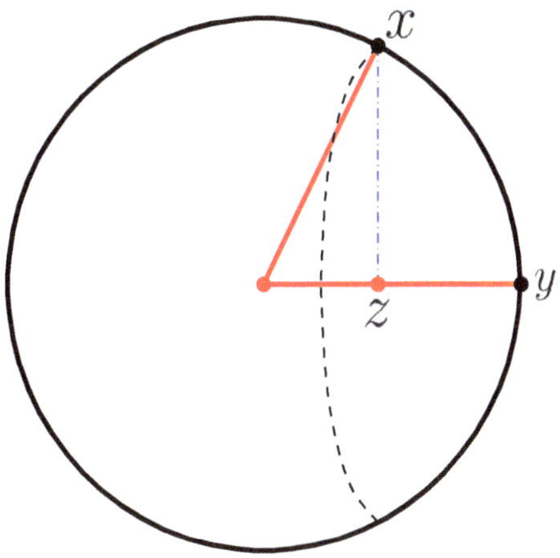

Figure 2.22.1 A rotation used in the proof of Corollary 2.22.2: note that $|z| = x \cdot y = t$ and $|y - z| = \sqrt{1 - t^2}$.

Comparing this and (2.22.6), we obtain that

$$\sum_{\substack{0 \leqslant i, j \leqslant k \\ i+j=k}} c_{ij}\, t^i\, \tau^j = \sum_{\substack{0 \leqslant i, j \leqslant k \\ i+j=k}} (-1)^j c_{ij}\, t^i\, \tau^j,$$

and thus

$$\sum_{\substack{0 \leqslant i, j \leqslant k \\ i+j=k}} d_{ij}\, t^i\, \tau^j = 0, \tag{2.22.7}$$

where

$$d_{ij} := (1 - (-1)^j) c_{ij} = \begin{cases} 0 & \text{if } j \text{ is even,} \\ 2c_{ij} & \text{if } j \text{ is odd.} \end{cases} \tag{2.22.8}$$

We claim that

$$d_{ij} = 0 \text{ for each } i, j \in \{1, \dots, k\} \text{ with } i + j = k. \tag{2.22.9}$$

Indeed, suppose not, and take $d_{IJ} \neq 0$ with J as small as possible and $I \in \{1, \dots, k\}$ with $I + J = k$. That is, from (2.22.7),

$$0 = \sum_{\substack{1 \leqslant j \leqslant k \\ i=k-j}} d_{ij}\, t^i\, \tau^j = \sum_{\substack{J \leqslant j \leqslant k \\ i=k-j}} d_{ij}\, t^i\, \tau^j = d_{IJ}\, t^{k-J}\, \tau^J + \sum_{\substack{J+1 \leqslant j \leqslant k \\ i=k-j}} d_{ij}\, t^i\, \tau^j.$$

Dividing by τ^J, we conclude that

$$0 = d_{IJ}\, t^{k-J} + \sum_{\substack{J+1\leqslant j\leqslant k\\ i=k-j}} d_{ij}\, t^i\, \tau^{j-J}.$$

Hence, sending $\tau \searrow 0$ (which corresponds to $t := 1$), we find that $0 = d_{IJ}$, against our original assumption.

This contradiction proves (2.22.9), and thus, recalling (2.22.8), we obtain that $c_{ij} = 0$ whenever j is odd.

As a result, from (2.22.6),

$$F_k(x, y) = \sum_{\substack{0\leqslant i,j\leqslant k\\ i+j=k\\ j\in 2\mathbb{N}}} c_{ij}\, t^i\, \tau^j = \sum_{\substack{0\leqslant m\leqslant k/2\\ i=k-2m}} c_{ij}\, t^i\, (1-t^2)^m.$$

Since the latter is a polynomial in t, the desired result is established. □

Corollary 2.22.3. *For each $\omega \in \partial B_1$,*

$$F_k(\omega, \omega) = \frac{N_k}{\mathcal{H}^{n-1}(\partial B_1)}.$$

Proof. Using (2.22.1) and (2.22.2),

$$\int_{\partial B_1} F_k(\omega, \omega)\, d\mathcal{H}^{n-1}_\omega = \sum_{j=1}^{N_k} \int_{\partial B_1} \left(Y_{k,j}(\omega)\right)^2 d\mathcal{H}^{n-1}_\omega = N_k.$$

In addition, owing to Corollary 2.22.2, we see that $F_k(\omega, \omega) = P_k(\omega \cdot \omega) = P_k(1)$, which is a constant. On this account, we conclude that

$$\mathcal{H}^{n-1}(\partial B_1)\, F_k(\omega, \omega) = \int_{\partial B_1} F_k(\omega, \omega)\, d\mathcal{H}^{n-1}_\omega = N_k.$$ □

In view of Lemma 2.22.1 and, more generally, in view of the obvious symmetries of the sphere, it is natural to investigate the spherical harmonics that remain invariant under rotations[33] about a given (say, the first) coordinate axis. These are known in jargon as Legendre polynomials (or sometimes, for $n \geqslant 4$, as Gegenbauer or ultraspherical polynomials) and are characterized by the following result.

[33]To investigate this type of symmetry, it is convenient to consider $n \geqslant 3$ since in the plane, the only rotation which fixes an axis is obviously the identity.

Theorem 2.22.4. *There exists a unique spherical harmonic L_k of degree k such that*

$$\begin{aligned} & L_k(e_1) = 1 \\ \text{and} \quad & L_k(\mathcal{R}x) = L_k(x) \\ & \text{for all } x \in \partial B_1 \text{ and all rotations } \mathcal{R} \text{ such that } \mathcal{R}e_1 = e_1. \end{aligned} \tag{2.22.10}$$

Moreover,

$$L_k \text{ is a polynomial in } x_1; \tag{2.22.11}$$

namely, there exists a polynomial $P_k : \mathbb{R} \to \mathbb{R}$ such that $L_k(x) = P_k(x_1)$ for all $x \in \partial B_1$.

If $j \neq k$, we also have that

$$\int_{-1}^{1} P_j(t)\, P_k(t)\, (1-t^2)^{\frac{n-3}{2}}\, dt = 0 \tag{2.22.12}$$

and

$$\begin{aligned} & \{P_0, \dots, P_k\} \text{ is a basis for the space of polynomials} \\ & \text{in one variable with degree less than or equal to } k. \end{aligned} \tag{2.22.13}$$

Additionally, the following recurrence relation holds true:

$$(n+k-2)P_{k+1}(t) = (n+2k-2)tP_k(t) - kP_{k-1}(t). \tag{2.22.14}$$

Furthermore,

$$P_k \text{ has degree equal to } k, \tag{2.22.15}$$

and it satisfies the differential equation

$$\frac{d}{dt}\left[(1-t^2)^{\frac{n-1}{2}} \frac{dP_k}{dt}(t)\right] + k(n+k-2)(1-t^2)^{\frac{n-3}{2}} P_k(t) = 0. \tag{2.22.16}$$

In addition[34]

$$P_k(t) = \frac{(-1)^k}{(1-t^2)^{\frac{n-3}{2}} \displaystyle\prod_{j=1}^{k}(n+2j-3)} \frac{d^k}{dt^k}(1-t^2)^{\frac{n+2k-3}{2}}. \tag{2.22.17}$$

Moreover, $P_k(t)$ corresponds to the coefficients in a formal expansion in powers of τ of the generating function

$$\left(1 - 2t\tau + \tau^2\right)^{\frac{2-n}{2}} = \sum_{k=0}^{+\infty} \mu_k\, P_k(t)\, \tau^k, \tag{2.22.18}$$

[34] The identity in (2.22.17) is often referred to by the name "Rodrigues' formula." Moreover, (2.22.16) is sometimes called "Legendre's differential equation" and (2.22.14) "Bonnet's recursion formula."

where[35]

$$\mu_k := \prod_{j=1}^{k} \frac{j+n-3}{j}. \tag{2.22.19}$$

Besides,

$$P_k(-t) = (-1)^k P_k(t). \tag{2.22.20}$$

Proof. First, we construct an example of spherical harmonic of degree k satisfying (2.22.10) (just to be sure[36] that we are not speaking about the empty set!). This will also highlight the relation with the fundamental solution, as presented in (2.22.18). We take $x, y \in \mathbb{R}^n$ with $0 < r := |x| < |y| =: \varrho$. We also define ϑ to be the angle between x and y, that is, $\vartheta := \arccos \frac{x \cdot y}{r\varrho}$. Let also $\tau := \frac{r}{\varrho}$ and $t := \cos\vartheta$. Then,

$$|x-y|^{2-n} = \left(|x|^2 - 2x \cdot y + |y|^2\right)^{\frac{2-n}{2}} = \frac{1}{\varrho^{n-2}|\tau^2 - 2t\tau + 1|^{\frac{n-2}{2}}}. \tag{2.22.21}$$

Now, if $z \in \mathbb{C}$, then

$$z^2 - 2tz + 1 = z^2 - 2\cos\vartheta\, z + 1 = (z - e^{i\vartheta})(z - e^{-i\vartheta})$$

and thus, if $|z| < 1$,

$$|z^2 - 2tz + 1| = |z - e^{i\vartheta}|\,|z - e^{-i\vartheta}| \geqslant (|e^{i\vartheta}| - |z|)(|e^{-i\vartheta}| - |z|) = (1 - |z|)^2 > 0.$$

As a consequence, we can expand (2.22.21) in power series of τ (which converges as $\tau \in [0, 1)$, or even as $\tau \in \mathbb{C}$ with $|\tau| < 1$), thus obtaining that

$$|x-y|^{2-n} = \frac{1}{\varrho^{n-2}} \sum_{k=0}^{+\infty} C_k(t)\, \tau^k \tag{2.22.22}$$

for suitable coefficients $C_k(t)$.

More precisely,

$$k!\, C_k(t) = \frac{d^k}{d\tau^k} \frac{1}{|\tau^2 - 2t\tau + 1|^{\frac{n-2}{2}}}\Bigg|_{\tau=0}, \tag{2.22.23}$$

---

[35] We note that, when $n = 3$, the quantity μ_k in (2.22.19) is equal to 1.

[36] In strict terms, this step is arguably not even necessary since the proof of the uniqueness statement will also produce a natural candidate in the forthcoming formulas (2.22.30), (2.22.31) and (2.22.32). However, we believe that it is extremely instructive to relate the Legendre polynomial directly to the fundamental solution to develop a direct intuition of the role they play in the Laplace equation. Furthermore, the uniqueness result and the generating formula identity in (2.22.18) will be used to deduce other identities in a rather short and handy fashion.

which shows that $C_k(t)$ depends polynomially on t, and actually $C_k(t)$ has degree less than or equal to k. Thus, in the setting of (2.22.19), we can define $P_k := \frac{C_k}{\mu_k}$ and obtain that[37]

$$P_k(t) \text{ is polynomial of } t \text{ of degree less than or equal to } k. \tag{2.22.24}$$

Additionally, by (2.22.22), if $y := \varrho e_1$, letting $Q_k(x) := |x|^k P_k\left(\frac{x_1}{|x|}\right)$,

$$|x-y|^{2-n} = \frac{1}{\varrho^{n-2}} \sum_{k=0}^{+\infty} \mu_k P_k(t) \left(\frac{r}{\varrho}\right)^k = \frac{1}{\varrho^{n-2}} \sum_{k=0}^{+\infty} \frac{\mu_k}{\varrho^k} Q_k(x). \tag{2.22.25}$$

We now claim that

$$Q_k \text{ is a homogeneous polynomial of degree } k. \tag{2.22.26}$$

For this, since the homogeneity is obvious from the definition of Q_k, it remains to show that Q_k is a polynomial. To this end, we observe that, by (2.22.23), one can write $C_k(t) = \frac{d^k}{d\tau^k}\phi(t,\tau)\Big|_{\tau=0}$, for a suitable function ϕ such that $\phi(-t,\tau) = \phi(t,-\tau)$. For this reason,

$$C_k(-t) = \frac{d^k}{d\tau^k}\phi(-t,\tau)\bigg|_{\tau=0} = \frac{d^k}{d\tau^k}\phi(t,-\tau)\bigg|_{\tau=0} = (-1)^k C_k(t),$$

and correspondingly

$$P_k(-t) = (-1)^k P_k(t). \tag{2.22.27}$$

Hence, each P_k is the linear combination of monomials of the type t^j, with $j \leqslant k$ and j even if k is even and j odd if k is odd. Consequently, $Q_k(x)$ is the linear combination of terms of the type $|x|^k \left(\frac{x_1}{|x|}\right)^j = |x|^{k-j} x_1^j = |x|^m x_1^j$, where $m := k-j$ is even, and this completes the proof of (2.22.26).

Now, exploiting (2.22.25), for every $\varphi \in C_0^\infty(B_\varrho)$, we have that

$$\begin{aligned}
0 &= \int_{\mathbb{R}^n} \varphi(x)\,\Delta|x-y|^{2-n}\,dx = \int_{\mathbb{R}^n} \Delta\varphi(x)\,|x-y|^{2-n}\,dx \\
&= \int_{\mathbb{R}^n} \Delta\varphi(x)\,\frac{1}{\varrho^{n-2}} \sum_{k=0}^{+\infty} \frac{\mu_k}{\varrho^k} Q_k(x)\,dx = \frac{1}{\varrho^{n-2}} \sum_{k=0}^{+\infty} \frac{\mu_k}{\varrho^k} \int_{\mathbb{R}^n} \Delta\varphi(x)\,Q_k(x)\,dx \\
&= \frac{1}{\varrho^{n-2}} \sum_{k=0}^{+\infty} \frac{\mu_k}{\varrho^k} \int_{\mathbb{R}^n} \varphi(x)\,\Delta Q_k(x)\,dx.
\end{aligned}$$

[37]As a matter of fact, the statement in (2.22.24) will be sharpened once we prove (2.22.15).

As a consequence, by the identity principle for power series, we deduce that

$$\mu_k \int_{\mathbb{R}^n} \varphi(x)\, \Delta Q_k(x)\, dx = 0.$$

Thus, recalling (2.22.19),

$$\int_{\mathbb{R}^n} \varphi(x)\, \Delta Q_k(x)\, dx = 0,$$

which gives that $\Delta Q_k(x) = 0$ for every $x \in B_\varrho$. Since this identity is now valid for all $\varrho > 0$, we thus conclude that $\Delta Q_k(x) = 0$ for every $x \in \mathbb{R}^n$. This and (2.22.26) give that the function

$$\partial B_1 \ni x \mapsto L_k(x) := Q_k(x) = P_k(x_1) \tag{2.22.28}$$

is a spherical harmonic. We are left to prove (2.22.10). For this, we exploit (2.22.25) with $x := re_1$ and $y := \varrho e_1$, with $0 < r < \varrho = 1$, which produces $\vartheta = 0$ and thus $t = 1$, to see that

$$(1-r)^{2-n} = \sum_{k=0}^{+\infty} \mu_k\, P_k(1)\, r^k.$$

Comparing every order in r with (2.22.19), we thus conclude that $1 = P_k(1) = Q_k(e_1) = L_k(e_1)$. Moreover, if $\mathcal{R}$ is a rotation that fixes e_1, we have that $L_k(\mathcal{R}x) = P_k(\mathcal{R}x \cdot e_1) = P_k(x_1) = L_k(x)$, completing the proof of (2.22.10).

We also observe that (2.22.18) holds true in this setting, thanks to (2.22.21) and (2.22.25).

Having established one example of spherical harmonics with the desired properties (as well as the validity of (2.22.18) for this example), we now turn our attention to the uniqueness claim in the statement of Theorem 2.22.4 and to the proof of (2.22.11), (2.22.12), (2.22.13), (2.22.14), (2.22.15), (2.22.16), (2.22.17) and (2.22.20).

More specifically, for the uniqueness claim, we argue as follows. We take a spherical harmonic L_k of degree k satisfying (2.22.10), and we consider the corresponding homogeneous and harmonic polynomial Q_k which coincides with L_k along ∂B_1. We extrapolate from every monomial of Q_k the factor depending on x_1; namely, we write

$$Q_k(x) = \sum_{j=0}^{k} x_1^j\, h_{k-j,k}(x_2, \dots, x_n), \tag{2.22.29}$$

with $h_{i,k}$ being homogeneous polynomials in $x_2, \ldots, x_n$ of degree i. Thus, if $\mathcal{R}$ is a rotation that fixes the first axis and we denote by $\widetilde{\mathcal{R}}$ its action[38] on the variables $X := (x_2, \ldots, x_n) \in \mathbb{R}^{n-1}$,

$$\begin{aligned} 0 &= |x|^k \big(L_k(\mathcal{R}x) - L_k(x)\big) = |\mathcal{R}x|^k L_k(\mathcal{R}x) - |x|^k L_k(x) = Q_k(\mathcal{R}x) - Q_k(x) \\ &= \sum_{j=0}^{k} x_1^j \left(h_{k-j,k}(\widetilde{\mathcal{R}}X) - h_{k-j,k}(X)\right) \end{aligned}$$

and consequently $h_{k-j,k}(\widetilde{\mathcal{R}}X) = h_{k-j,k}(X)$ for all $X \in \mathbb{R}^{n-1}$. That is, each $h_{i,k}$ only depends on $|X|$, and hence we write $h_{i,k}(X) = \widetilde{h}_{i,k}(|X|)$ for some $\widetilde{h}_{i,k} : \mathbb{R} \to \mathbb{R}$. As a matter of fact, by the homogeneity of $h_{i,k}$, for every $r \geqslant 0$,

$$\widetilde{h}_{i,k}(r) = h_{i,k}(re_n) = r^i h_{i,k}(e_n) = c_{i,k} r^i,$$

where $c_{i,k} := h_{i,k}(e_n) \in \mathbb{R}$.

Thus, since $h_{i,k}$ is a polynomial and $h_{i,k}(X) = \widetilde{h}_{i,k}(|X|) = c_{i,k}\,|X|^i$, necessarily $c_{i,k} = 0$ whenever i is odd. As a result,

$$\begin{aligned} Q_k(x) &= \sum_{j=0}^{k} c_{k-j,k}\, x_1^j\, |X|^{k-j} = \sum_{i=1}^{k} c_{i,k}\, x_1^{k-i}\, |X|^i \\ &= \sum_{\substack{\ell\in\mathbb{N} \\ 0\leqslant\ell\leqslant k/2}} c_{2\ell,k}\, x_1^{k-2\ell}\, |X|^{2\ell}. \end{aligned} \tag{2.22.30}$$

To show the desired uniqueness claim, it thus suffices to show that these $c_{2\ell,k}$ are uniquely determined. First of all, by (2.22.10) and (2.22.30),

$$1 = L_k(e_1) = Q_k(e_1) = c_{0,k}. \tag{2.22.31}$$

[38] To be precise, one can define $\overline{\mathcal{R}} : \mathbb{R}^{n-1} \to \mathbb{R}^n$ as $\overline{\mathcal{R}}X := \mathcal{R}(0, X)$ and note that $\overline{\mathcal{R}}X \cdot e_1 = (0, X) \cdot e_1 = 0$. Accordingly $\overline{\mathcal{R}} : \mathbb{R}^{n-1} \to \{0\} \times \mathbb{R}^{n-1}$; therefore, defining

$$\widetilde{\mathcal{R}}X := \sum_{i=2}^{n} \big((\overline{\mathcal{R}}X) \cdot e_2\big)e_2,$$

we have that $\widetilde{\mathcal{R}} : \mathbb{R}^{n-1} \to \mathbb{R}^{n-1}$ and $\overline{\mathcal{R}}X = (0, \widetilde{\mathcal{R}}X)$. Furthermore, if $x = (x_1, X) \in \mathbb{R} \times \mathbb{R}^{n-1}$,

$$\mathcal{R}x = \mathcal{R}(x_1 e_1 + (0, X)) = x_1 \mathcal{R}e_1 + \overline{\mathcal{R}}(0, X) = x_1 e_1 + (0, \widetilde{\mathcal{R}}X) = (x_1, \widetilde{\mathcal{R}}X).$$

We also remark that, conversely, every rotation $\widetilde{\mathcal{R}}$ on $\mathbb{R}^{n-1}$ produces a rotation $\mathcal{R}$ on $\mathbb{R}^n$ that fixes the first coordinate axis, simply by defining $\mathcal{R}x := (x_1, \widetilde{\mathcal{R}}X)$.

Additionally, from the harmonicity of Q_k,

$$\begin{aligned}
0 &= \Delta Q_k(x) \\
&= \sum_{\substack{\ell\in\mathbb{N}\\ 0\leqslant\ell\leqslant k/2}} c_{2\ell,k}\,\Delta(x_1^{k-2\ell}\,|X|^{2\ell}) \\
&= \sum_{\substack{\ell\in\mathbb{N}\\ 0\leqslant\ell\leqslant k/2}} c_{2\ell,k}\,\partial^2_{x_1}(x_1^{k-2\ell}\,|X|^{2\ell}) + \sum_{\substack{\ell\in\mathbb{N}\\ 0\leqslant\ell\leqslant k/2\\ 2\leqslant i\leqslant n}} c_{2\ell,k}\,\partial^2_{x_i}(x_1^{k-2\ell}\,|X|^{2\ell}) \\
&= \sum_{\substack{\ell\in\mathbb{N}\\ 0\leqslant\ell\leqslant(k-2)/2}} (k-2\ell)(k-2\ell-1)\,c_{2\ell,k}\,x_1^{k-2\ell-2}\,|X|^{2\ell} \\
&\quad + \sum_{\substack{\ell\in\mathbb{N}\\ 1\leqslant\ell\leqslant k/2\\ 2\leqslant i\leqslant n}} 2\ell(2\ell+n-3)\,c_{2\ell,k}\,x_1^{k-2\ell}\,|X|^{2\ell-2}.
\end{aligned}$$

That is, changing the indices' names,

$$0 = \sum_{\substack{m\in\mathbb{N}\\ 1\leqslant m\leqslant k/2}} \Big((k-2m+2)(k-2m+1)\,c_{2m-2,k} + 2m(2m+n-3)\,c_{2m,k}\Big)\,x_1^{k-2m}\,|X|^{2m-2}.$$

This produces the recurrence relation

$$c_{2m,k} = \frac{(k-2m+2)(k-2m+1)\,c_{2m-2,k}}{2m(2m+n-3)}, \tag{2.22.32}$$

which, when joined with (2.22.31), determines uniquely all the coefficients. This completes the proof of the uniqueness claim in the statement of Theorem 2.22.4.

We observe that in light of the example constructed and the uniqueness statement, the unique spherical harmonic L_k of degree k satisfying (2.22.10) is given by (2.22.28). Hence, (2.22.11) is also satisfied.

Moreover, we remark that (2.22.18) and (2.22.20) have been already proved in this special case (recall also (2.22.27)), and hence they follow from the uniqueness statement as well.

Furthermore, if $f \in L^\infty([-1,1])$, writing

$$\partial B_1 \cap \{x_1 > 0\} = \Big\{x_1 = \sqrt{1-|X|^2} \text{ with } X \in \mathbb{R}^{n-1} \text{ and } |X| < 1\Big\},$$

and letting ω_{n-2} be the $(n-2)$-dimensional Hausdorff measure of the unit sphere in $\mathbb{R}^{n-1}$, we see that

$$\begin{aligned}
\int_{\partial B_1\cap\{x_1>0\}} f(x_1)\,d\mathcal{H}^{n-1}_x &= \int_{\substack{X\in\mathbb{R}^{n-1}\\ |X|<1}} \frac{f(\sqrt{1-|X|^2})}{\sqrt{1-|X|^2}}\,dX \\
&= \omega_{n-2}\int_0^1 \frac{f(\sqrt{1-\rho^2})}{\sqrt{1-\rho^2}}\,\rho^{n-2}\,d\rho.
\end{aligned}$$

Thus, the substitution $t := \sqrt{1-\rho^2}$ leads to

$$\int_{\partial B_1 \cap \{x_1>0\}} f(x_1)\, d\mathcal{H}_x^{n-1} = \omega_{n-2} \int_0^1 f(t)\,(1-t^2)^{\frac{n-3}{2}}\, dt.$$

Applying this also to $\widetilde{f}(t) := f(-t)$, we obtain the following useful integral formula on the sphere for functions of one variable:

$$\begin{aligned}
\int_{\partial B_1} f(x_1)\, d\mathcal{H}_x^{n-1} &= \int_{\partial B_1 \cap \{x_1>0\}} f(x_1)\, d\mathcal{H}_x^{n-1} + \int_{\partial B_1 \cap \{x_1<0\}} f(x_1)\, d\mathcal{H}_x^{n-1} \\
&= \omega_{n-2} \int_0^1 f(t)\,(1-t^2)^{\frac{n-3}{2}}\, dt + \int_{\partial B_1 \cap \{x_1>0\}} \widetilde{f}(x_1)\, d\mathcal{H}_x^{n-1} \\
&= \omega_{n-2} \left[\int_0^1 f(t)\,(1-t^2)^{\frac{n-3}{2}}\, dt + \int_0^1 \widetilde{f}(t)\,(1-t^2)^{\frac{n-3}{2}}\, dt \right] \\
&= \omega_{n-2} \int_{-1}^1 f(t)\,(1-t^2)^{\frac{n-3}{2}}\, dt.
\end{aligned}$$

As a result,

$$\begin{aligned}
\omega_{n-2} \int_{-1}^1 P_j(t)\, P_k(t)\,(1-t^2)^{\frac{n-3}{2}}\, dt &= \int_{\partial B_1} P_j(x_1)\, P_k(x_1)\, d\mathcal{H}_x^{n-1} \\
&= \int_{\partial B_1} L_j(x_1)\, L_k(x_1)\, d\mathcal{H}_x^{n-1}.
\end{aligned}$$

The latter quantity is zero, thanks to the orthogonality condition for spherical harmonics of different degrees in Lemma 2.21.6, and this proves (2.22.12).

Also, the set $\{P_0, \dots, P_k\}$ contains $k+1$ elements, and they are necessarily linearly independent, thanks to (2.22.12), whence (2.22.13) plainly follows.

Now, we focus on the proof of (2.22.14). For this, we differentiate (2.22.18) with respect to τ, finding that

$$\frac{(2-n)(\tau-t)}{(1-2t\tau+\tau^2)^{\frac{n}{2}}} = \sum_{k=1}^{+\infty} k\mu_k\, P_k(t)\, \tau^{k-1}.$$

Comparing with (2.22.18), we thereby gather that

$$\frac{(2-n)(\tau-t)}{1-2t\tau+\tau^2} \sum_{k=0}^{+\infty} \mu_k\, P_k(t)\, \tau^k = \sum_{k=1}^{+\infty} k\mu_k\, P_k(t)\, \tau^{k-1}.$$

As a result,

$$\begin{aligned}
&(2-n) \sum_{k=1}^{+\infty} \mu_{k-1}\, P_{k-1}(t)\, \tau^k - (2-n) \sum_{k=0}^{+\infty} \mu_k\, t\, P_k(t)\, \tau^k \\
&= (2-n) \sum_{k=0}^{+\infty} \mu_k\, P_k(t)\, \tau^{k+1} - (2-n) \sum_{k=0}^{+\infty} \mu_k\, t\, P_k(t)\, \tau^k
\end{aligned}$$

$$
\begin{aligned}
&= (2-n)(\tau - t)\sum_{k=0}^{+\infty} \mu_k P_k(t)\,\tau^k = (1 - 2t\tau + \tau^2)\sum_{k=1}^{+\infty} k\mu_k P_k(t)\,\tau^{k-1} \\
&= \sum_{k=1}^{+\infty} k\mu_k P_k(t)\,\tau^{k-1} - 2\sum_{k=1}^{+\infty} k\mu_k t P_k(t)\,\tau^k + \sum_{k=1}^{+\infty} k\mu_k P_k(t)\,\tau^{k+1} \\
&= \sum_{k=0}^{+\infty}(k+1)\mu_{k+1} P_{k+1}(t)\,\tau^k - 2\sum_{k=1}^{+\infty} k\mu_k t P_k(t)\,\tau^k \\
&\quad + \sum_{k=2}^{+\infty}(k-1)\mu_{k-1} P_{k-1}(t)\,\tau^k.
\end{aligned}
$$

Taking the kth order of this identity, it follows that

$$
\begin{aligned}
&(2-n)\mu_{k-1} P_{k-1} - (2-n)\mu_k t P_k \\
&\quad = (k+1)\mu_{k+1} P_{k+1} - 2k\mu_k t P_k + (k-1)\mu_{k-1} P_{k-1}.
\end{aligned} \tag{2.22.33}
$$

Since from (2.22.19) we have that

$$
\mu_{k+1} = \frac{n+k-2}{k+1}\,\mu_k \quad \text{and} \quad \mu_{k-1} = \frac{n}{n+k-3}\,\mu_k,
$$

the identity in (2.22.33) leads to (2.22.14), as desired.

To prove (2.22.15), we argue by induction. First of all, using (2.22.23), we see that the desired claim holds true for $k \in \{0, 1\}$. Indeed,

$$
\begin{aligned}
P_0(t) &= \frac{C_0(t)}{\mu_0} = 1 \\
\text{and} \quad P_1(t) &= \frac{C_1(t)}{\mu_1} = \frac{(n-2)t}{n-2} = t.
\end{aligned} \tag{2.22.34}
$$

To perform the inductive step, we suppose that P_j has degree j for all $j \in \mathbb{N} \cap [0, k]$. Then, we exploit (2.22.14) to deduce that P_{k+1} has degree $k+1$, which establishes (2.22.15).

Now, we prove (2.22.16). To this end, we note that

$$
\begin{aligned}
&\frac{d}{dt}\left[(1-t^2)^{\frac{n-1}{2}} \frac{dP_k}{dt}(t)\right] \\
&= -(n-1)t(1-t^2)^{\frac{n-3}{2}} \frac{dP_k}{dt}(t) + (1-t^2)^{\frac{n-1}{2}} \frac{d^2P_k}{dt^2}(t) \\
&= (1-t^2)^{\frac{n-3}{2}}\, T_k(t),
\end{aligned} \tag{2.22.35}
$$

where

$$T_k(t) := (1-n)t\,\frac{dP_k}{dt}(t) + (1-t^2)\,\frac{d^2P_k}{dt^2}(t).$$

We also observe that T_k is a polynomial of degree less than or equal to k, owing to (2.22.24). Consequently, in light of (2.22.13), we can write T_k as a linear combination of $\{P_0, \dots, P_k\}$; namely, there exist $s_{j,k} \in \mathbb{R}$ such that

$$T_k = \sum_{j=0}^{k} s_{j,k}\,P_j. \tag{2.22.36}$$

As a result, for each $i \in \{0, \dots, k\}$, recalling (2.22.12), we get that

$$\begin{aligned}
&\int_{-1}^{1} P_i(t)\,T_k(t)\,(1-t^2)^{\frac{n-3}{2}}\,dt \\
&= \sum_{j=0}^{k} s_{j,k} \int_{-1}^{1} P_i(t)\,P_j(t)\,(1-t^2)^{\frac{n-3}{2}}\,dt \\
&= s_{i,k} \int_{-1}^{1} (P_i(t))^2\,(1-t^2)^{\frac{n-3}{2}}\,dt.
\end{aligned} \tag{2.22.37}$$

In addition,

$$\lim_{t\to\pm1} (1-t^2)^{\frac{n-1}{2}} = 0.$$

Therefore, we can exploit (2.22.35) and (2.22.37) and integrate by parts twice to find that

$$\begin{aligned}
&s_{i,k} \int_{-1}^{1} (P_i(t))^2\,(1-t^2)^{\frac{n-3}{2}}\,dt \\
&= \int_{-1}^{1} (1-t^2)^{\frac{n-3}{2}}\,T_k(t)\,P_i(t)\,dt \\
&= \int_{-1}^{1} \frac{d}{dt}\left[(1-t^2)^{\frac{n-1}{2}}\,\frac{dP_k}{dt}(t)\right] P_i(t)\,dt \\
&= -\int_{-1}^{1} (1-t^2)^{\frac{n-1}{2}}\,\frac{dP_k}{dt}(t)\,\frac{dP_i}{dt}(t)\,dt \\
&= \int_{-1}^{1} \frac{d}{dt}\left[(1-t^2)^{\frac{n-1}{2}}\,\frac{dP_i}{dt}(t)\right] P_k(t)\,dt \\
&= \int_{-1}^{1} (1-t^2)^{\frac{n-3}{2}}\,T_i(t)\,P_k(t)\,dt.
\end{aligned} \tag{2.22.38}$$

We stress that (2.22.35) was exploited again in the last identity but with the index i instead of k. Thus, on account of (2.22.12), (2.22.36) and (2.22.38), for every $i \in \{0, \dots, k-1\}$,

$$
\begin{aligned}
s_{i,k} & \int_{-1}^{1} (P_i(t))^2 (1-t^2)^{\frac{n-3}{2}} \, dt \\
& = s_{j,i} \sum_{j=1}^{i} \int_{-1}^{1} (1-t^2)^{\frac{n-3}{2}} P_j(t) \, P_k(t) \, dt = 0. \qquad (2.22.39)
\end{aligned}
$$

Since, by (2.22.10), we know that $P_i(1) = L_i(e_1) = 1$, we deduce from (2.22.39) that $s_{i,k} = 0$ whenever $i \in \{0, \dots, k-1\}$. As a byproduct, we infer from (2.22.36) that $T_k = s_{k,k} P_k$; consequently, (2.22.35) boils down to

$$
\frac{d}{dt}\left[(1-t^2)^{\frac{n-1}{2}} \frac{dP_k}{dt}(t)\right] = s_{k,k} (1-t^2)^{\frac{n-3}{2}} P_k(t). \qquad (2.22.40)
$$

Now, we denote by d_k the highest power of t in the monomial expansion of P_k; namely, owing to (2.22.15), we write $P_k(t) = d_k t^k + \widetilde{P}_k(t)$, where $d_k \neq 0$, and $\widetilde{P}_k$ is a polynomial of degree strictly lower that k (and possibly zero). Therefore, as $t \longrightarrow +\infty$,

$$
\begin{aligned}
& \frac{d}{dt}\left[(1-t^2)^{\frac{n-1}{2}} \frac{dP_k}{dt}(t)\right] \\
& \quad = \frac{d}{dt}\left[(1-t^2)^{\frac{n-1}{2}} \left(k d_k t^{k-1} + \frac{d\widetilde{P}_k}{dt}(t)\right)\right] \\
& \quad = -(n-1)t(1-t^2)^{\frac{n-3}{2}} \left(k d_k t^{k-1} + \frac{d\widetilde{P}_k}{dt}(t)\right) \\
& \qquad + (1-t^2)^{\frac{n-1}{2}} \left(k(k-1) d_k t^{k-2} + \frac{d^2\widetilde{P}_k}{dt^2}(t)\right) \\
& \quad = \left((-1)^{\frac{n-1}{2}} (n-1) t^{n-2} + O(t^{n-3})\right) \left(k d_k t^{k-1} + O(t^{k-2})\right) \\
& \qquad + \left((-1)^{\frac{n-1}{2}} t^{n-1} + O(t^{n-2})\right)\left(k(k-1) d_k t^{k-2} + O(t^{k-3})\right) \\
& \quad = (-1)^{\frac{n-1}{2}} k d_k (n+k-2)\, t^{n+k-3} + O(t^{n+k-4})
\end{aligned}
$$

and

$$
\begin{aligned}
(1-t^2)^{\frac{n-3}{2}} P_k(t) &= \left((-1)^{\frac{n-3}{2}} t^{n-3} + O(t^{n-4})\right)\left(d_k t^k + O(t^{k-1})\right) \\
&= (-1)^{\frac{n-3}{2}} d_k t^{n+k-3} + O(t^{n+k-4}).
\end{aligned}
$$

From these observations and (2.22.40), we deduce that, for large t,

$$(-1)^{\frac{n-1}{2}} k d_k (n+k-2)\, t^{n+k-3} = (-1)^{\frac{n-3}{2}} s_{k,k} d_k t^{n+k-3} + O(t^{n+k-4})$$

and therefore $s_{k,k} = -k(n+k-2)$. Combining this and (2.22.40), we obtain the desired result in (2.22.16).

Now, we prove (2.22.17). For this, we note that, for all $\alpha \in \mathbb{R}$ and $\ell \in \mathbb{N}$,

$$\frac{d^\ell}{dt^\ell}(1-t^2)^{\frac{\alpha}{2}} = (1-t^2)^{\frac{\alpha-2\ell}{2}} p_{\alpha,\ell}(t), \tag{2.22.41}$$

for a suitable polynomial $p_{\alpha,\ell}$ of degree less than or equal to ℓ. To check this, we argue by induction. When $\ell = 0$, the claim in (2.22.41) is obviously true with $p_{\alpha,0} := 1$. Suppose now that (2.22.41) is true for the index ℓ. Then, taking one more derivative, we have that

$$\begin{aligned} \frac{d^{\ell+1}}{dt^{\ell+1}}(1-t^2)^{\frac{\alpha}{2}} &= \frac{d}{dt}\Big((1-t^2)^{\frac{\alpha-2\ell}{2}} p_{\alpha,\ell}(t)\Big) \\ &= (2\ell-\alpha)t(1-t^2)^{\frac{\alpha-2(\ell+1)}{2}} p_{\alpha,\ell}(t) + (1-t^2)^{\frac{\alpha-2\ell}{2}} \frac{dp_{\alpha,\ell}}{dt}(t) \\ &= (1-t^2)^{\frac{\alpha-2(\ell+1)}{2}} p_{\alpha,\ell+1}(t), \end{aligned}$$

with $p_{\alpha,\ell+1} := (2\ell-\alpha)tp_{\alpha,\ell} + (1-t^2)\frac{dp_{\alpha,\ell}}{dt}$, that is, a polynomial of degree up to $\ell+1$. This completes the inductive step and establishes (2.22.41), as desired.

Now, we define

$$\zeta_k := \frac{(-1)^k}{\prod_{j=1}^{k}(n+2j-3)} \quad \text{and} \quad P_k^\star(t) := \zeta_k\,(1-t^2)^{\frac{3-n}{2}} \frac{d^k}{dt^k}(1-t^2)^{\frac{n+2k-3}{2}}.$$

Observe that $P_k^\star$ is precisely the right-hand side of (2.22.17).

Now, we claim that, if $j < k$ and $i \in \{0,\dots,j\}$, then

$$\begin{aligned} &\int_{-1}^{1} \frac{d^j}{dt^j}(1-t^2)^{\frac{n+2j-3}{2}} \frac{d^k}{dt^k}(1-t^2)^{\frac{n+2k-3}{2}} (1-t^2)^{\frac{3-n}{2}}\,dt \\ &= \int_{-1}^{1} q_{i,j}(t) \frac{d^{k-i}}{dt^{k-i}}(1-t^2)^{\frac{n+2k-3}{2}}\,dt. \end{aligned} \tag{2.22.42}$$

Here, $q_{i,j}$ is a suitable polynomial of degree less than or equal to $j-i$. To prove this, we argue by induction over i. When $i = 0$, we define

$$q_{0,j}(t) := \frac{d^j}{dt^j}(1-t^2)^{\frac{n+2j-3}{2}} (1-t^2)^{\frac{3-n}{2}},$$

and we have that the identity in (2.22.42) holds true. Moreover, $q_{0,j}$ is a polynomial of degree less than or equal to j, thanks to (2.22.41).

To perform the inductive step, we suppose that the desired claim is true for all indices up to some i, with $i \leqslant j-1$, and we aim at proving it for the index $i+1$. To this end, we let

$$\psi_{i,k}(t) := \frac{d^{k-i-1}}{dt^{k-i-1}}(1-t^2)^{\frac{n+2k-3}{2}}.$$

Using (2.22.41), we have that $\psi_{i,k}(t) = (1-t^2)^{\frac{n+2i-1}{2}}\widetilde{q}_{i,k}(t)$, for a suitable polynomial $\widetilde{q}_{i,k}$. In particular,

$$\psi_{i,k}(\pm 1) = (1-(\pm 1)^2)^{\frac{n+2i-1}{2}}\widetilde{q}_{i,k}(\pm 1) = 0.$$

Thus, by the inductive assumption and an integration by parts,

$$\begin{aligned}
&\int_{-1}^{1} \frac{d^j}{dt^j}(1-t^2)^{\frac{n+2j-3}{2}} \frac{d^k}{dt^k}(1-t^2)^{\frac{n+2k-3}{2}} (1-t^2)^{\frac{3-n}{2}}\,dt = \int_{-1}^{1} q_{i,j}(t)\,\frac{d\psi_{i,k}}{dt}(t)\,dt\\
&= q_{i,j}(1)\psi_{i,k}(1) - q_{i,j}(-1)\psi_{i,k}(-1) - \int_{-1}^{1} \frac{dq_{i,j}}{dt}(t)\,\psi_{i,k}(t)\,dt\\
&= 0 + \int_{-1}^{1} q_{i+1,j}(t)\,\frac{d^{k-i-1}}{dt^{k-i-1}}(1-t^2)^{\frac{n+2k-3}{2}}\,dt
\end{aligned}$$

with $q_{i+1,j} := -\frac{dq_{i,j}}{dt}$, and the latter is a polynomial with degree at most $j-i-1$. The proof of (2.22.42) is thereby complete.

We also observe that, if $j < k$,

$$\left.\frac{d^{k-j-1}}{dt^{k-j-1}}(1-t^2)^{\frac{n+2k-3}{2}}\right|_{t=\pm 1} = 0. \tag{2.22.43}$$

Indeed, by (2.22.41), we know that $\frac{d^{k-j-1}}{dt^{k-j-1}}(1-t^2)^{\frac{n+2k-3}{2}}$ is equal to some polynomial times $(1-t^2)^{\frac{n+2j-1}{2}}$, and the latter function vanishes at $t = \pm 1$, thus proving (2.22.43).

Now, we exploit (2.22.42) with $i := j$. Note that the polynomial $q_{j,j}(t)$ has degree zero; hence, it is a constant (still denoted by $q_{j,j}$ for simplicity). Therefore, if $j < k$, using also (2.22.43), we have that

$$\begin{aligned}
&\int_{-1}^{1} P_j^\star(t)\,P_k^\star(t)\,(1-t^2)^{\frac{n-3}{2}}\,dt\\
&= \zeta_j\,\zeta_k \int_{-1}^{1} \frac{d^j}{dt^j}(1-t^2)^{\frac{n+2j-3}{2}} \frac{d^k}{dt^k}(1-t^2)^{\frac{n+2k-3}{2}} (1-t^2)^{\frac{3-n}{2}}\,dt\\
&= \zeta_j\,\zeta_k\,q_{j,j} \int_{-1}^{1} \frac{d^{k-j}}{dt^{k-j}}(1-t^2)^{\frac{n+2k-3}{2}}\,dt
\end{aligned}$$

$$= \zeta_j\, \zeta_k\, q_{j,j} \left(\frac{d^{k-j-1}}{dt^{k-j-1}} (1-t^2)^{\frac{n+2k-3}{2}} \Big|_{t=1} - \frac{d^{k-j-1}}{dt^{k-j-1}} (1-t^2)^{\frac{n+2k-3}{2}} \Big|_{t=-1} \right)$$
$$= 0,$$

and therefore, up to exchanging j and k,

$$\int_{-1}^{1} P_j^\star(t)\, P_k^\star(t)\, (1-t^2)^{\frac{n-3}{2}}\, dt = 0 \quad \text{whenever } j \neq k. \tag{2.22.44}$$

Now, we claim that

$$P_k^\star(t) = P_k(t) \text{ for all } k \in \mathbb{N}. \tag{2.22.45}$$

To this end, we argue by induction. Recalling (2.22.34), we remark that $P_0^\star(t) = 1 = P_0(t)$, and hence (2.22.45) is valid for $k = 0$. Let us now suppose that (2.22.45) holds true for all indices $\{0, \dots, k-1\}$ with $k \geqslant 1$, and we prove it for the index k. For this, we remark that $P_k^\star$ is a polynomial of degree at most k, owing to (2.22.41). Hence, by (2.22.13), we can write $P_k^\star$ as a linear combination of $\{P_0, \dots, P_k\}$; namely, there exist $\sigma_{j,k} \in \mathbb{R}$ such that

$$P_k^\star = \sum_{j=0}^{k} \sigma_{j,k}\, P_j.$$

This and the inductive assumption lead to

$$P_k^\star = \sigma_{k,k}\, P_k + \sum_{j=0}^{k-1} \sigma_{j,k}\, P_j^\star.$$

Therefore, for each $i \in \{1, \dots, k-1\}$, exploiting (2.22.12) and (2.22.44), we have that

$$0 = \int_{-1}^{1} P_i^\star(t)\, P_k^\star(t)\, (1-t^2)^{\frac{n-3}{2}}\, dt$$
$$= \sigma_{k,k} \int_{-1}^{1} P_i^\star(t)\, P_k(t)\, (1-t^2)^{\frac{n-3}{2}}\, dt + \sum_{j=0}^{k-1} \sigma_{j,k} \int_{-1}^{1} P_i^\star(t) P_j^\star(t)\, (1-t^2)^{\frac{n-3}{2}}\, dt$$
$$= \sigma_{k,k} \int_{-1}^{1} P_i(t)\, P_k(t)\, (1-t^2)^{\frac{n-3}{2}}\, dt + \sigma_{i,k} \int_{-1}^{1} (P_i(t))^2\, (1-t^2)^{\frac{n-3}{2}}\, dt$$
$$= \sigma_{i,k} \int_{-1}^{1} (P_i(t))^2\, (1-t^2)^{\frac{n-3}{2}}\, dt.$$

Thus, since $P_i(1) = 1$ due to (2.22.10), we gather that $\sigma_{i,k} = 0$ for all $i \in \{1, \dots, k-1\}$, and accordingly

$$P_k^\star = \sigma_{k,k}\, P_k. \tag{2.22.46}$$

Consequently, to complete the proof of (2.22.45), it remains to show that

$$\sigma_{k,k} = 1. \tag{2.22.47}$$

For this, we note that if $j \in \{1, \ldots, k\}$,

$$\left|\lim_{t\nearrow 1}(1-t)^{\frac{3-n}{2}} \frac{d^{k-j}}{dt^{k-j}}(1-t)^{\frac{n+2k-3}{2}}\right| = \left|\lim_{y\searrow 0} y^{\frac{3-n}{2}} \frac{d^{k-j}}{dy^{k-j}} y^{\frac{n+2k-3}{2}}\right| = 0.$$

Thus, by the general Leibniz rule for higher-order derivatives,

$$\begin{aligned}
&\lim_{t\nearrow 1}(1-t^2)^{\frac{3-n}{2}} \frac{d^k}{dt^k}(1-t^2)^{\frac{n+2k-3}{2}} \\
&\quad = \lim_{t\nearrow 1}(1+t)^{\frac{3-n}{2}}(1-t)^{\frac{3-n}{2}} \frac{d^k}{dt^k}\left((1+t)^{\frac{n+2k-3}{2}}(1-t)^{\frac{n+2k-3}{2}}\right) \\
&\quad = 2^{\frac{3-n}{2}} \lim_{t\nearrow 1}(1-t)^{\frac{3-n}{2}} \sum_{j=0}^{k} \binom{k}{j} \frac{d^j}{dt^j}(1+t)^{\frac{n+2k-3}{2}} \frac{d^{k-j}}{dt^{k-j}}(1-t)^{\frac{n+2k-3}{2}} \\
&\quad = 2^{\frac{3-n}{2}} \lim_{t\nearrow 1}(1-t)^{\frac{3-n}{2}}(1+t)^{\frac{n+2k-3}{2}} \frac{d^k}{dt^k}(1-t)^{\frac{n+2k-3}{2}} \\
&\quad = 2^k \lim_{t\nearrow 1}(1-t)^{\frac{3-n}{2}} \frac{d^k}{dt^k}(1-t)^{\frac{n+2k-3}{2}} = (-1)^k\, 2^k \prod_{i=0}^{k-1} \frac{n+2(k-i)-3}{2} \\
&\quad = (-1)^k \prod_{i=0}^{k-1}(n+2(k-i)-3) = (-1)^k \prod_{j=1}^{k}(n+2j-3) = \frac{1}{\zeta_k}.
\end{aligned}$$

This shows that $P_k^\star(1) = 1$, and thus, comparing with (2.22.46), we find that $\sigma_{k,k} = 1$. This proves (2.22.47), and thus (2.22.45), from which we obtain (2.22.17), as desired. □

For a list of the first 10 Legendre polynomials in dimension 3, see for example the footnote on p. 620 (at this level, we will not make much use of the explicit values of the Legendre polynomials but rather their fundamental algebraic structure).

Now, we point out an additional identity for Legendre polynomials in terms of the function introduced in (2.22.2) (as a byproduct, this characterizes the Legendre polynomials precisely as the polynomials for which Corollary 2.22.2 holds true, up to normalization constants).

Theorem 2.22.5. *Let P_k be the Legendre polynomials in Theorem 2.22.4 and F_k be as in* (2.22.2), *with N_k as in* (2.21.15).

Then, for every $x,\ y \in \partial B_1$,

$$P_k(x \cdot y) = \frac{\mathcal{H}^{n-1}(\partial B_1)}{N_k} F_k(x, y).$$

Proof. For every $\omega \in \partial B_1$, we define

$$L_k^\star(\omega) := \frac{\mathcal{H}^{n-1}(\partial B_1)}{N_k} F_k(\omega, e_1).$$

By (2.22.2), we know that $L_\omega^\star$ is a spherical harmonic. Moreover, by Corollary 2.22.3,

$$L_k^\star(e_1) = \frac{\mathcal{H}^{n-1}(\partial B_1)}{N_k} F_k(e_1, e_1) = 1.$$

Additionally, if $\mathcal{R}$ is a rotation on $\mathbb{R}^n$ such that $\mathcal{R}e_1 = e_1$, we infer from Lemma 2.22.1 that

$$\begin{aligned} L_k^\star(\mathcal{R}\omega) &= \frac{\mathcal{H}^{n-1}(\partial B_1)}{N_k} F_k(\mathcal{R}\omega, e_1) \\ &= \frac{\mathcal{H}^{n-1}(\partial B_1)}{N_k} F_k(\mathcal{R}\omega, \mathcal{R}e_1) = \frac{\mathcal{H}^{n-1}(\partial B_1)}{N_k} F_k(\omega, e_1) = L_k^\star(\omega). \end{aligned}$$

These observations and the uniqueness claim in Theorem 2.22.4 entail that $L_k^\star = L_k$.

Therefore, if $x,\ y \in \partial B_1$ and $\mathcal{R}_y$ is a rotation on $\mathbb{R}^n$ such that $\mathcal{R}_y y = e_1$, we let $\omega := \mathcal{R}_y x$. In this way, noting that

$$\omega \cdot e_1 = \mathcal{R}_y x \cdot e_1 = x \cdot \mathcal{R}_y^{-1} e_1 = x \cdot y$$

and using Lemma 2.22.1 once again, we conclude that

$$\begin{aligned} \frac{\mathcal{H}^{n-1}(\partial B_1)}{N_k} F_k(x, y) &= \frac{\mathcal{H}^{n-1}(\partial B_1)}{N_k} F_k(\mathcal{R}_y x, \mathcal{R}_y y) = \frac{\mathcal{H}^{n-1}(\partial B_1)}{N_k} F_k(\omega, e_1) \\ &= L_k^\star(\omega) = L_k(\omega) = P_k(\omega_1) = P_k(x \cdot y), \end{aligned}$$

which establishes the desired result. □

Corollary 2.22.6. *We have that*

$$\max_{[-1,1]} |P_k| \leqslant 1 \tag{2.22.48}$$

and, for every $x \in \partial B_1$,

$$\int_{\partial B_1} \big(P_k(x \cdot y)\big)^2 \, d\mathcal{H}_y^{n-1} = \frac{\mathcal{H}^{n-1}(\partial B_1)}{N_k}. \tag{2.22.49}$$

Theorem 2.22.8. *Let Y_k be a spherical harmonic of degree k. Then, there exist $a_1, \dots, a_{N_k} \in \mathbb{R}$ and $x^{(1)}, \dots, x^{(N_k)} \in \partial B_1$ such that, for all $x \in \partial B_1$,*

$$Y_k(x) = \sum_{j=1}^{N_k} a_j \, P_k(x \cdot x^{(j)}). \tag{2.22.52}$$

Furthermore, for all $x \in \partial B_1$,

$$Y_k(x) = \frac{N_k}{\mathcal{H}^{n-1}(\partial B_1)} \int_{\partial B_1} Y_k(\omega) \, P_k(x \cdot \omega) \, d\mathcal{H}^{n-1}_\omega. \tag{2.22.53}$$

Proof. Using Theorem 2.22.5 and Lemma 2.22.7, for every $x \in \partial B_1$, for all $j \in \{1, \dots, N_k\}$,

$$P_k(x \cdot x^{(j)}) = \frac{\mathcal{H}^{n-1}(\partial B_1)}{N_k} F_k(x, x^{(j)}) = \frac{\mathcal{H}^{n-1}(\partial B_1)}{N_k} \sum_{m=1}^{N_k} Y_{k,m}(x) \, Y_{k,m}(x^{(j)}),$$

that is,

$$\begin{pmatrix} P_k(x \cdot x^{(1)}) \\ \vdots \\ P_k(x \cdot x^{(N_k)}) \end{pmatrix} = \frac{\mathcal{H}^{n-1}(\partial B_1)}{N_k} \begin{pmatrix} Y_{k,1}(x^{(1)}) & \dots & Y_{k,N_k}(x^{(1)}) \\ \vdots & & \\ Y_{k,1}(x^{(N_k)}) & \dots & Y_{k,N_k}(x^{(N_k)}) \end{pmatrix} \begin{pmatrix} Y_{k,1}(x) \\ \vdots \\ Y_{k,N_k}(x) \end{pmatrix}, \tag{2.22.54}$$

and the matrix on the right-hand side of (2.22.54) is invertible. Hence, inverting this matrix in (2.22.54), we obtain that, for each $m \in \{1, \dots, N_k\}$,

$$Y_{k,m}(x) = \sum_{j=1}^{N_k} c_{j,k,m} \, P_k(x \cdot x^{(j)}),$$

for suitable $c_{j,k,m} \in \mathbb{R}$. Also, since $\{Y_{k,1}, \dots, Y_{k,N_k}\}$ is a basis,

$$Y_k(x) = \sum_{m=1}^{N_k} a_{k,m} \, Y_{k,m}(x), \tag{2.22.55}$$

for suitable $a_{k,m} \in \mathbb{R}$, and we thereby conclude that

$$Y_k(x) = \sum_{j=1}^{N_k} \left(\sum_{m=1}^{N_k} a_{k,m} \, c_{j,k,m} \right) P_k(x \cdot x^{(j)}),$$

which gives the desired result in (2.22.52) with $a_j := \sum_{m=1}^{N_k} a_{k,m} \, c_{j,k,m}$.

Now, we prove (2.22.53). To this end, we use again (2.22.55), combined with the additional identity in Theorem 2.22.5 and the orthogonality condition in (2.22.1), to find that

$$\begin{aligned}
&\frac{N_k}{\mathcal{H}^{n-1}(\partial B_1)} \int_{\partial B_1} Y_k(\omega)\, P_k(x\cdot\omega)\, d\mathcal{H}^{n-1}_\omega \\
&= \int_{\partial B_1} Y_k(\omega)\, F_k(x,\omega)\, d\mathcal{H}^{n-1}_\omega \\
&= \sum_{j=1}^{N_k} \int_{\partial B_1} Y_k(\omega)\, Y_{k,j}(x)\, Y_{k,j}(\omega)\, d\mathcal{H}^{n-1}_\omega \\
&= \sum_{j,m=1}^{N_k} \int_{\partial B_1} a_{k,m}\, Y_{k,m}(\omega)\, Y_{k,j}(x)\, Y_{k,j}(\omega)\, d\mathcal{H}^{n-1}_\omega \\
&= \sum_{m=1}^{N_k} a_{k,m}\, Y_{k,m}(x) \\
&= Y_k(x),
\end{aligned}$$

and this completes the proof of (2.22.53). □

For a concrete application of spherical harmonics and Legendre polynomials to the evaluation of the gravitational field, see Appendix B. See also [Kal95, ABR01, Sau06, EF14] for an extensive treatment of spherical harmonics.

Following [AKS10], we now give another proof of Theorem 2.13.4. This proof is interesting since it uses harmonic polynomials (and moreover, as pointed out in [AKS10], it is flexible enough to address the case of higher-order operators as well).

Another Proof of Theorem 2.13.4. We know that the function f is attained by u at the boundary, owing to (2.13.21).

We show that

$$u \text{ is harmonic in } B_1. \tag{2.22.56}$$

For this, we consider a homogeneous harmonic polynomial F, say of degree $m \geqslant 1$, and we define $f(y) := F(y-P)$ for every $y \in \partial B_1$. Furthermore, we let

$$u_F(P) := \fint_{\partial B_1} \ell_e(P)\, d\mathcal{H}^{n-1}_e. \tag{2.22.57}$$

Note that this definition coincides with that in (2.13.20), but we emphasize here the dependence of the function u on F. We also observe that

$$F(0) = 0 \tag{2.22.58}$$

and, recalling the notation in (2.13.10),

$$f(Q_\pm(e)) = F(Q_\pm(e) - P) = F\big(r_\pm(e)\, e\big) = (r_\pm(e))^m F(e).$$

As a result, recalling (2.13.22),

$$\begin{aligned} \ell_e(P) &= \frac{r_+(e) f(Q_-(e)) - r_-(e) f(Q_+(e))}{r_+(e) - r_-(e)} \\ &= \frac{r_+(e)(r_-(e))^m - r_-(e)(r_+(e))^m}{r_+(e) - r_-(e)}\, F(e) \\ &= r_+(e) r_-(e)\, \frac{(r_-(e))^{m-1} - (r_+(e))^{m-1}}{r_+(e) - r_-(e)}\, F(e). \end{aligned} \tag{2.22.59}$$

Now, we use that

$$(r_-(e) - r_+(e)) \sum_{j=0}^{m-2} (r_-(e))^j (r_+(e))^{m-2-j} = (r_-(e))^{m-1} - (r_+(e))^{m-1}.$$

With this and (2.13.24), we write (2.22.59) in the form

$$\ell_e(P) = (1 - |P|^2)\, F(e) \sum_{j=0}^{m-2} (r_-(e))^j (r_+(e))^{m-2-j}. \tag{2.22.60}$$

Now, we remark that, if $j \in \mathbb{N} \cap \left[0, \frac{m-2}{2}\right]$,

$$\begin{aligned} &(r_-(e))^j (r_+(e))^{m-2-j} + (r_-(e))^{m-2-j} (r_+(e))^j \\ &\quad = (r_-(e)\, r_+(e))^j \Big((r_+(e))^{m-2-2j} + (r_-(e))^{m-2-2j} \Big) \\ &\quad = (|P|^2 - 1)^j \Big((r_+(e))^{m-2-2j} + (r_-(e))^{m-2-2j} \Big). \end{aligned}$$

As a consequence, using the change of index $J := m - 2 - j$,

$$\begin{aligned} &\sum_{j=0}^{m-2} (r_-(e))^j (r_+(e))^{m-2-j} \\ &\quad = \sum_{0 \leqslant j < (m-2)/2} (r_-(e))^j (r_+(e))^{m-2-j} + \sum_{(m-2)/2 < j \leqslant m-2} (r_-(e))^j (r_+(e))^{m-2-j} \\ &\qquad + \chi_{2\mathbb{N}}(m) (r_-(e))^{\frac{m-2}{2}} (r_+(e))^{\frac{m-2}{2}} \end{aligned}$$

$$
\begin{aligned}
&= \sum_{0\leqslant j<(m-2)/2} (r_-(e))^j (r_+(e))^{m-2-j} + \sum_{0\leqslant J<(m-2)/2} (r_-(e))^{m-2-J} (r_+(e))^j \\
&\quad + \chi_{2\mathbb{N}}(m)(r_-(e))^{\frac{m-2}{2}} (r_+(e))^{\frac{m-2}{2}} \\
&= \sum_{0\leqslant j<(m-2)/2} \Big((r_-(e))^j (r_+(e))^{m-2-j} + (r_-(e))^{m-2-J} (r_+(e))^j \Big) \\
&\quad + \chi_{2\mathbb{N}}(m)(r_-(e))^{\frac{m-2}{2}} (r_+(e))^{\frac{m-2}{2}} \\
&= \sum_{0\leqslant j<(m-2)/2} (|P|^2-1)^j \Big((r_+(e))^{m-2-2j} + (r_-(e))^{m-2-2j} \Big) \\
&\quad + \chi_{2\mathbb{N}}(m)(|P|^2-1)^{\frac{m-2}{2}} \Big((r_+(e))^{\frac{m-2}{2}} + (r_-(e))^{\frac{m-2}{2}} \Big) \\
&= \sum_{j\in\mathbb{N}\cap\left[0,\frac{m-2}{2}\right]} C_{j,P} \Big((r_+(e))^{m-2-2j} + (r_-(e))^{m-2-2j} \Big),
\end{aligned}
$$

for suitable coefficients $C_{j,P}$ not depending on e. Plugging this information into (2.22.60), we find that

$$
\ell_e(P) = F(e) \sum_{j\in\mathbb{N}\cap\left[0,\frac{m-2}{2}\right]} C_{j,P} \Big((r_+(e))^{m-2-2j} + (r_-(e))^{m-2-2j} \Big), \qquad (2.22.61)
$$

up to renaming the coefficients $C_{j,P}$.

We also point out that, exploiting (2.13.12), for each $j \in \mathbb{N} \cap \left[0, \frac{m-2}{2}\right]$, we have that

$$
\begin{aligned}
&(r_+(e))^{m-2-2j} + (r_-(e))^{m-2-2j} \\
&= \Big(-P\cdot e + \sqrt{D(e)} \Big)^{m-2-2j} + \Big(-P\cdot e - \sqrt{D(e)} \Big)^{m-2-2j} \\
&= \sum_{k=0}^{m-2-2j} \binom{m-2-2j}{k} \Big((-P\cdot e)^{m-2-2j-k} \big(\sqrt{D(e)}\big)^k \\
&\quad + (-P\cdot e)^{m-2-2j-k} \big(-\sqrt{D(e)}\big)^k \Big) \\
&= \sum_{\substack{0\leqslant k\leqslant m-2-2j \\ k\in 2\mathbb{N}}} \binom{m-2-2j}{k} \Big((-P\cdot e)^{m-2-2j-k} \big(\sqrt{D(e)}\big)^k \\
&\quad + (-P\cdot e)^{m-2-2j-k} \big(-\sqrt{D(e)}\big)^k \Big) \\
&= 2 \sum_{0\leqslant i\leqslant (m-2-2j)/2} \binom{m-2-2j}{2i} (-P\cdot e)^{m-2-2j-2i} (D(e))^i \\
&= 2 \sum_{0\leqslant i\leqslant (m-2-2j)/2} \binom{m-2-2j}{2i} (-P\cdot e)^{m-2-2j-2i} \big((P\cdot e)^2 - |P|^2 + 1 \big)^i,
\end{aligned}
$$

which is a polynomial of degree at most $m-2$ in the variable $P \cdot e$ (and therefore in the variable e), which we denote by $\Pi_{j,P}$. We infer from this and (2.22.61) that

$$\ell_e(P) = \Pi_P(e)\, F(e), \tag{2.22.62}$$

for a suitable polynomial Π_P of degree at most $m-2$.

Now, we exploit Lemma 2.21.2 to find a harmonic polynomial H_P of degree at most $m-2$ such that $H_P = \Pi_P$ along ∂B_1. Hence, since F was supposed to be a homogeneous harmonic polynomial of degree m, we deduce from Lemma 2.21.6 that

$$\int_{\partial B_1} F(e)\, \Pi_P(e)\, d\mathcal{H}_e^{n-1} = \int_{\partial B_1} F(e)\, H_P(e)\, d\mathcal{H}_e^{n-1} = 0.$$

This, (2.22.57), (2.22.58) and (2.22.62) lead to

$$u_F(P) = \fint_{\partial B_1} \ell_e(P)\, d\mathcal{H}_e^{n-1} = \fint_{\partial B_1} \Pi_P(e)\, F(e)\, d\mathcal{H}_e^{n-1} = 0 = F(0). \tag{2.22.63}$$

We note that F is the unique (by Corollary 2.9.3) harmonic function in $B_1(-P)$ with its datum on $\partial B_1(-P)$. We can thus reconstruct its values from the Poisson kernel in Theorems 2.11.1 and 2.11.2; namely, we write, for all $x \in B_1(-P)$,

$$F(x) = \int_{\partial B_1(-P)} F(y)\, P_{B_1(-P)}(x, y)\, d\mathcal{H}_y^{n-1},$$

where $P_{B_1(-P)}$ denotes the Poisson kernel of $B_1(-P)$.

This and (2.22.63) give that

$$u_F(P) = \int_{\partial B_1(-P)} F(y)\, P_{B_1(-P)}(0, y)\, d\mathcal{H}_y^{n-1}. \tag{2.22.64}$$

We stress that this result was obtained by assuming that F is a homogeneous harmonic polynomial, but we now extend (2.22.64) to any continuous function F on $\partial B_1(-P)$. To this end, we first remark that, by linearity, we see that (2.22.64) holds true when F is a harmonic polynomial (not necessarily homogeneous). Accordingly, we make use of Corollary 2.21.4, and we find a sequence of harmonic polynomials F_j that converge to F in $L^2(\partial B_1(-P))$: we thus deduce from (2.22.64) that

$$\begin{aligned}\lim_{j\to+\infty} u_{F_j}(P) &= \lim_{j\to+\infty} \int_{\partial B_1(-P)} F_j(y)\, P_{B_1(-P)}(0,y)\, d\mathcal{H}^{n-1}_y \\ &= \int_{\partial B_1(-P)} F(y)\, P_{B_1(-P)}(0,y)\, d\mathcal{H}^{n-1}_y \\ &= \int_{\partial B_1} F(y+P)\, P_{B_1(-P)}(0,y+P)\, d\mathcal{H}^{n-1}_y \\ &= \int_{\partial B_1} f(y)\, P_{B_1}(P,y)\, d\mathcal{H}^{n-1}_y, \end{aligned} \tag{2.22.65}$$

where P_{B_1} denotes the Poisson kernel of B_1.

On the other hand, we recall (2.13.22) and write

$$\begin{aligned}\ell_e(P) &= \frac{r_+(e)f(Q_-(e)) - r_-(e)f(Q_+(e))}{r_+(e)-r_-(e)} \\ &= \frac{r_+(e)F(Q_-(e)-P) - r_-(e)F(Q_+(e)-P)}{r_+(e)-r_-(e)}.\end{aligned}$$

This and (2.22.57) give that

$$\begin{aligned}&\big|u_F(P) - u_{F_j}(P)\big|\,\mathcal{H}^{n-1}(\partial B_1) \\ &\leqslant \int_{\partial B_1}\Bigg(\frac{r_+(e)}{r_+(e)-r_-(e)}\big|F(Q_-(e)-P) - F_j(Q_-(e)-P)\big| \\ &\quad + \frac{-r_-(e)}{r_+(e)-r_-(e)}\big|F(Q_+(e)-P) - F_j(Q_+(e)-P)\big|\Bigg)\, d\mathcal{H}^{n-1}_e \\ &\leqslant \sqrt{\int_{\partial B_1}\left(\frac{r_+(e)}{r_+(e)-r_-(e)}\right)^2 d\mathcal{H}^{n-1}_e} \\ &\quad \cdot\sqrt{\int_{\partial B_1}\big|F(Q_-(e)-P) - F_j(Q_-(e)-P)\big|^2\, d\mathcal{H}^{n-1}_e} \\ &\quad + \sqrt{\int_{\partial B_1}\left(\frac{-r_-(e)}{r_+(e)-r_-(e)}\right)^2 d\mathcal{H}^{n-1}_e} \\ &\quad \cdot\sqrt{\int_{\partial B_1}\big|F(Q_+(e)-P) - F_j(Q_+(e)-P)\big|^2\, d\mathcal{H}^{n-1}_e}.\end{aligned}$$

We note that $r_+(e), -r_-(e) \leqslant 2$ and that

$$r_+(e) - r_-(e) = 2\sqrt{D(e)} = 2\sqrt{(P\cdot e)^2 - |P|^2 + 1} \geqslant 2\sqrt{1-|P|^2}.$$

As a consequence,

$$\begin{aligned}&\left|u_F(P)-u_{F_j}(P)\right|\\&\leqslant C_P\left(\sqrt{\int_{\partial B_1}\left|F(Q_-(e)-P)-F_j(Q_-(e)-P)\right|^2\,d\mathcal{H}^{n-1}_e}\right.\\&\qquad\left.+\sqrt{\int_{\partial B_1}\left|F(Q_+(e)-P)-F_j(Q_+(e)-P)\right|^2\,d\mathcal{H}^{n-1}_e}\right),\end{aligned}\tag{2.22.66}$$

for some $C_P>0$ depending on P and n.

Also, exploiting (2.13.15) with

$$g(\omega):=\left|F(\omega-P)-F_j(\omega-P)\right|^2\frac{1-|P|^2-r_-\left(\frac{P-Q_-(\omega)}{|P-Q_-(\omega)|}\right)P\cdot\frac{P-Q_-(\omega)}{|P-Q_-(\omega)|}}{\left|r_-\left(\frac{P-Q_-(\omega)}{|P-Q_-(\omega)|}\right)\right|^n},$$

we obtain that

$$\begin{aligned}&\int_{\partial B_1}\left|F(Q_-(e)-P)-F_j(Q_-(e)-P)\right|^2\,d\mathcal{H}^{n-1}_e\\&=\int_{\partial B_1}g(Q_-(e))\,\frac{|r_-(e)|^n}{1-|P|^2-r_-(e)P\cdot e}\,d\mathcal{H}^{n-1}_e=\int_{\partial B_1}g(e)\,d\mathcal{H}^{n-1}_e.\end{aligned}$$

Now, we observe that

$$\frac{1-|P|^2-r_-\left(\frac{P-Q_-(\omega)}{|P-Q_-(\omega)|}\right)P\cdot\frac{P-Q_-(\omega)}{|P-Q_-(\omega)|}}{\left|r_-\left(\frac{P-Q_-(\omega)}{|P-Q_-(\omega)|}\right)\right|^n}\leqslant\frac{4}{(1-|P|)^n},$$

which gives that

$$\begin{aligned}&\int_{\partial B_1}\left|F(Q_-(e)-P)-F_j(Q_-(e)-P)\right|^2\,d\mathcal{H}^{n-1}_e\\&\leqslant C_P\int_{\partial B_1}\left|F(e-P)-F_j(e-P)\right|^2\,d\mathcal{H}^{n-1}_e\\&=C_P\int_{\partial B_1(-P)}\left|F(\zeta)-F_j(\zeta)\right|^2\,d\mathcal{H}^{n-1}_\zeta,\end{aligned}$$

up to renaming $C_P>0$. Similarly,

$$\begin{aligned}&\int_{\partial B_1}\left|F(Q_+(e)-P)-F_j(Q_+(e)-P)\right|^2\,d\mathcal{H}^{n-1}_e\\&\leqslant C_P\int_{\partial B_1(-P)}\left|F(\zeta)-F_j(\zeta)\right|^2\,d\mathcal{H}^{n-1}_\zeta.\end{aligned}$$

Therefore, recalling (2.22.66), we obtain that

$$\left|u_F(P) - u_{F_j}(P)\right| \leqslant C_P \int_{\partial B_1(-P)} \left|F(\zeta) - F_j(\zeta)\right|^2 d\mathcal{H}_\zeta^{n-1},$$

up to renaming C_P once again. Consequently,

$$\lim_{j\to+\infty} u_{F_j}(P) = u_F(P).$$

From this and (2.22.65), we deduce that

$$u_F(P) = \int_{\partial B_1} f(y)\, P_{B_1}(P, y)\, d\mathcal{H}_y^{n-1}.$$

As a result, by the Poisson kernel representation in Theorems 2.11.1 and 2.11.2, we have that u_F is harmonic, and this completes the proof of (2.22.56).

The proof of Theorem 2.13.4 is thus completed in view of (2.13.21) and (2.22.56). □

2.23 Almansi's Formula

As a short digression, we now point out that the representation method introduced in Lemma 2.21.2 has a natural counterpart in the classical Almansi's formula, according to which any function satisfying $\Delta^k u = 0$ can be represented by functions that are harmonic (that is, polyharmonic functions can be algebraically reconstructed from harmonic ones). The details are as follows.

Lemma 2.23.1. *Let $k \in \mathbb{N}$, $k \geqslant 2$. Let Ω be a star-shaped open subset of $\mathbb{R}^n$. Then, for every $u \in C^{2k}(\overline{\Omega})$ such that*

$$\Delta^k u = 0 \quad \text{in } \Omega, \tag{2.23.1}$$

there exist $v, w \in C^{2(k-1)}(\overline{\Omega})$ such that

$$\Delta^{k-1} v = \Delta^{k-1} w = 0 \quad \text{in } \Omega \tag{2.23.2}$$

and

$$u(x) = |x|^2\, v(x) + w(x) \quad \text{for every } x \in \Omega. \tag{2.23.3}$$

Proof. First of all, we claim that, for every $i \in \mathbb{N}$, $i \geqslant 1$, and any smooth function f,

$$\Delta^i(|x|^2 f) = c_i \Delta^{i-1} f + d_i x \cdot \nabla\Delta^{i-1} f + |x|^2 \Delta^i f, \tag{2.23.4}$$

with

$$c_i := 2ni + 4i(i-1) \quad \text{and} \quad d_i := 4i.$$

To check this, we argue by induction. When $i = 1$, the claim in (2.23.4) reduces to $\Delta(|x|^2 f) = 2nf + 4x \cdot \nabla f + |x|^2 \Delta f$, which is indeed true. Let us now suppose that (2.23.4) holds true for some index i. Then, we find that

$$\begin{aligned}\Delta^{i+1}(|x|^2 f) &= \Delta\Big(c_i \Delta^{i-1} f + d_i x \cdot \nabla\Delta^{i-1} f + |x|^2 \Delta^i f\Big)\\ &= c_i \Delta^i f + d_i x \cdot \nabla \Delta^i f + 2d_i \Delta^i f + 2n\Delta^i f + 4x \cdot \nabla\Delta^i f + |x|^2 \Delta^{i+1} f\\ &= (c_i + 2d_i + 2n)\Delta^i f + (d_i + 4)x \cdot \nabla\Delta^i f + |x|^2 \Delta^{i+1} f\\ &= (2ni + 4i(i-1) + 8i + 2n)\Delta^i f + (4i+4)x \cdot \nabla\Delta^i f + |x|^2\Delta^{i+1} f\\ &= c_{i+1}\Delta^i f + d_{i+1} x \cdot \nabla\Delta^i f + |x|^2 \Delta^{i+1} f,\end{aligned}$$

which completes the inductive step and establishes the validity of (2.23.4).

Now, we define

$$\alpha := \frac{c_{k-1}}{d_{k-1}} \in (0, +\infty). \tag{2.23.5}$$

Up to a translation, we can assume that Ω is star-shaped with respect to the origin, and then, for every $x \in \Omega$, we can define

$$T(x) := \frac{1}{d_{k-1}} \int_0^1 t^{\alpha-1} \Delta^{k-1} u(tx)\, dt.$$

We remark that, for all $x \in \Omega$,

$$\Delta T(x) = \frac{1}{d_{k-1}} \int_0^1 t^{\alpha+1} \Delta^k u(tx)\, dt = 0, \tag{2.23.6}$$

thanks to (2.23.1); furthermore,

$$\begin{aligned}x \cdot \nabla T(x) &= \frac{1}{d_{k-1}} \int_0^1 t^\alpha x \cdot \nabla\Delta^{k-1} u(tx)\, dt\\ &= \frac{1}{d_{k-1}} \int_0^1 \left(\frac{d}{dt}\Big(t^\alpha \Delta^{k-1} u(tx)\Big) - \alpha t^{\alpha-1}\Delta^{k-1} u(tx)\right) dt\\ &= \frac{1}{d_{k-1}} \Delta^{k-1} u(x) - \alpha T(x).\end{aligned} \tag{2.23.7}$$

Now, we take a function ζ_1 such that $\Delta\zeta_1 = T$ in Ω: for this, for instance, we can define

$$\widetilde{T} := \begin{cases} T & \text{in } \Omega,\\ 0 & \text{in } \mathbb{R}^n \setminus \Omega,\end{cases}$$

and exploit the fundamental solution in Theorem 2.7.3 and set $\zeta_1 := -\Gamma * \widetilde{T}$. Similarly, we take a function ζ_2 such that $\Delta\zeta_2 = \zeta_1$ in Ω and, recursively for all $j \in \{2, \dots, k-2\}$, find functions ζ_j such that $\Delta\zeta_j = \zeta_{j-1}$ in Ω. In particular, choosing $v := \zeta_{k-2}$, we find that, in Ω,

$$\Delta^{k-2} v = \Delta^{k-2} \zeta_{k-2} = \Delta^{k-3}(\Delta \zeta_{k-2}) = \Delta^{k-3} \zeta_{k-3} = \cdots = \Delta \zeta_1 = T. \tag{2.23.8}$$

Let also $w(x) := u(x) - |x|^2 v(x)$. In this way, the claim in (2.23.3) is satisfied. Moreover, by (2.23.6) and (2.23.8),

$$\Delta^{k-1} v = \Delta T = 0, \tag{2.23.9}$$

and also, recalling in addition (2.23.4), (2.23.5) and (2.23.7),

$$\begin{aligned} \Delta^{k-1} w &= \Delta^{k-1} u - \Delta^{k-1}(|x|^2 v) \\ &= \Delta^{k-1} u - c_{k-1} \Delta^{k-2} v - d_{k-1} x \cdot \nabla \Delta^{k-2} v - |x|^2 \Delta^{k-1} v \\ &= \Delta^{k-1} u - c_{k-1} T - d_{k-1} x \cdot \nabla T \\ &= d_{k-1} \left(\frac{1}{d_{k-1}} \Delta^{k-1} u - \alpha T - x \cdot \nabla T \right) \\ &= 0. \end{aligned} \tag{2.23.10}$$

Thus, the claim in (2.23.2) follows from (2.23.9) and (2.23.10). □

Corollary 2.23.2. *Let $k \in \mathbb{N}$, $k \geqslant 1$. Let Ω be a star-shaped open subset of $\mathbb{R}^n$. Then, every solution $u \in C^{2k}(\overline{\Omega})$ of*

$$\Delta^k u = 0 \quad \text{in } \Omega$$

can be written in the form

$$u(x) = \sum_{j=0}^{k-1} |x|^{2j} v^{(j)}(x), \tag{2.23.11}$$

with $v^{(0)}, \ldots, v^{(k-1)} \in C^2(\overline{\Omega})$ and harmonic in Ω.

Proof. We argue by induction over k. When $k = 1$, it suffices to define $v^{(0)} := u$. Suppose now that the desired claim holds true for some index $k - 1$, with $k \geqslant 2$, and take $u \in C^{2k}(\overline{\Omega})$ such that $\Delta^k u = 0$ in Ω. We invoke Lemma 2.23.1, and we find $v, w \in C^{2(k-1)}(\overline{\Omega})$ such that $\Delta^{k-1} v = \Delta^{k-1} w = 0$ in Ω and $u(x) = |x|^2 v(x) + w(x)$. We can apply the inductive assumption on v and w and obtain functions $\alpha^{(0)}, \ldots, \alpha^{(k-2)}, \beta^{(0)}, \ldots, \beta^{(k-2)} \in C^2(\overline{\Omega})$ and harmonic in Ω such that

$$v(x) = \sum_{j=0}^{k-2} |x|^{2j} \alpha^{(j)}(x) \quad \text{and} \quad w(x) = \sum_{j=0}^{k-2} |x|^{2j} \beta^{(j)}(x).$$

In this way, we have that

$$\begin{aligned} u(x) &= |x|^2 \sum_{j=0}^{k-2} |x|^{2j} \alpha^{(j)}(x) + \sum_{j=0}^{k-2} |x|^{2j} \beta^{(j)}(x) \\ &= \sum_{j=1}^{k-1} |x|^{2j} \alpha^{(j-1)}(x) + \sum_{j=0}^{k-2} |x|^{2j} \beta^{(j)}(x). \end{aligned}$$

Hence, the claim in (2.23.11) holds true with

$$v^{(j)} := \begin{cases} \beta^{(0)} & \text{if } j = 0, \\ \alpha^{(j-1)} + \beta^{(j)} & \text{if } j \in \{1, \dots, k-2\}, \\ \alpha^{(k-2)} & \text{if } j = k-1, \end{cases}$$

which completes the proof of the desired result. □

We recall here one of the the main results, motivated by elasticity theory, that inspired the above discussions (see [Alm99, pp. 15–16] and also [Sch91] and the references therein).

Corollary 2.23.3. *Let $g, h \in C(\partial B_1)$. Then, the solution $u \in C^4(\overline{B_1})$ of*

$$\begin{cases} \Delta^2 u = 0 & \text{in } B_1, \\ u = g & \text{on } \partial B_1, \\ \partial_\nu u = h & \text{on } \partial B_1 \end{cases}$$

can be explicitly written as

$$u(x) = -\frac{(1-|x|^2)^2}{4n\,|B_1|} \int_{\partial B_1} \left[\frac{g(y)}{|x-y|^{n+2}} \Big((n-4)|x-y|^2 - n(1-|x|^2) \Big) + \frac{2h(y)}{|x-y|^n} \right] d\mathcal{H}_y^{n-1}.$$

Proof. We exploit Lemma 2.23.1 with $k := 2$ and set $W := v + w$. Note that W is harmonic and, in light of (2.23.3),

$$u(x) = (|x|^2 - 1)v(x) + W(x). \tag{2.23.12}$$

In particular, we have that $g = u = W$ along ∂B_1, whence we can exploit the Poisson kernel representation in Theorems 2.11.1 and 2.11.2 and conclude that, for every $x \in B_1$,

$$W(x) = \int_{\partial B_1} \frac{(1-|x|^2)\, g(y)}{n\,|B_1|\,|x-y|^n}\, d\mathcal{H}_y^{n-1}. \tag{2.23.13}$$

Additionally, setting $V(x) := 2v(x) + x \cdot \nabla W(x)$, we have that, on ∂B_1,

$$h = \partial_\nu u = \partial_\nu\big((|x|^2 - 1)v + W\big) = 2v + \partial_\nu W = 2v + x \cdot \nabla W = V.$$

Thus, since $\Delta V = 2\Delta v + x \cdot \nabla \Delta W + 2\Delta W = 0$, we can again use the Poisson kernel representation in Theorems 2.11.1 and 2.11.2 and find that, for every $x \in B_1$,

$$V(x) = \int_{\partial B_1} \frac{(1-|x|^2)\, h(y)}{n\,|B_1|\,|x-y|^n}\, d\mathcal{H}_y^{n-1}$$

As a result, recalling (2.23.13),

$$\begin{aligned} v(x) &= \frac{V(x) - x \cdot \nabla W(x)}{2} \\ &= \frac{1}{2}\left(\int_{\partial B_1} \frac{(1-|x|^2)\, h(y)}{n\,|B_1|\,|x-y|^n}\, d\mathcal{H}_y^{n-1} - x \cdot \nabla \left(\int_{\partial B_1} \frac{(1-|x|^2)\, g(y)}{n\,|B_1|\,|x-y|^n}\, d\mathcal{H}_y^{n-1}\right)\right) \\ &= \frac{1}{2n\,|B_1|}\left(\int_{\partial B_1} \frac{(1-|x|^2)\, h(y)}{|x-y|^n}\, d\mathcal{H}_y^{n-1} + \int_{\partial B_1} \frac{2|x|^2\, g(y)}{|x-y|^n}\, d\mathcal{H}_y^{n-1} \right. \\ &\quad \left. + n \int_{\partial B_1} \frac{x \cdot (x-y)\,(1-|x|^2)\, g(y)}{|x-y|^{n+2}}\, d\mathcal{H}_y^{n-1}\right). \end{aligned}$$

This and (2.23.12), exploiting (2.23.13) once more, give that

$$\begin{aligned} u(x) &= \frac{|x|^2-1}{2n\,|B_1|}\left(\int_{\partial B_1} \frac{(1-|x|^2)\, h(y)}{|x-y|^n}\, d\mathcal{H}_y^{n-1} + \int_{\partial B_1} \frac{2|x|^2\, g(y)}{|x-y|^n}\, d\mathcal{H}_y^{n-1} \right. \\ &\quad \left. + n \int_{\partial B_1} \frac{x \cdot (x-y)\,(1-|x|^2)\, g(y)}{|x-y|^{n+2}}\, d\mathcal{H}_y^{n-1}\right) + \int_{\partial B_1} \frac{(1-|x|^2)\, g(y)}{n\,|B_1|\,|x-y|^n}\, d\mathcal{H}_y^{n-1} \\ &= \frac{|x|^2-1}{2n\,|B_1|}\left(\int_{\partial B_1} \frac{(1-|x|^2)\, h(y)}{|x-y|^n}\, d\mathcal{H}_y^{n-1} + \int_{\partial B_1} \frac{2|x|^2\, g(y)}{|x-y|^n}\, d\mathcal{H}_y^{n-1} \right. \\ &\quad \left. + n \int_{\partial B_1} \frac{x \cdot (x-y)\,(1-|x|^2)\, g(y)}{|x-y|^{n+2}}\, d\mathcal{H}_y^{n-1} - \int_{\partial B_1} \frac{2g(y)}{|x-y|^n}\, d\mathcal{H}_y^{n-1}\right) \\ &= \frac{|x|^2-1}{2n\,|B_1|}\left(\int_{\partial B_1} \frac{(1-|x|^2)\, h(y)}{|x-y|^n}\, d\mathcal{H}_y^{n-1} \right. \\ &\quad \left. + \int_{\partial B_1} \frac{(|x|^2-1)(2|x-y|^2 - nx \cdot (x-y))\, g(y)}{|x-y|^{n+2}}\, d\mathcal{H}_y^{n-1}\right) \\ &= -\frac{(1-|x|^2)^2}{2n\,|B_1|}\left(\int_{\partial B_1} \frac{h(y)}{|x-y|^n}\, d\mathcal{H}_y^{n-1} \right. \\ &\quad \left. - \int_{\partial B_1} \frac{(2|x-y|^2 - nx \cdot (x-y))\, g(y)}{|x-y|^{n+2}}\, d\mathcal{H}_y^{n-1}\right). \end{aligned}$$

Thus, since, for all $y \in \partial B_1$,

$$\begin{aligned} -2(2|x-y|^2 - nx \cdot (x-y)) &= -4|x-y|^2 + 2nx \cdot (x-y) \\ &= -4|x-y|^2 + n(2|x|^2 - 2x \cdot y) \end{aligned}$$

$$
\begin{aligned}
&= -4|x-y|^2 + n(|x|^2 + |x|^2 + |y|^2 - 2x \cdot y - 1) \\
&= -4|x-y|^2 + n(|x|^2 - 1 + |x-y|^2) \\
&= (n-4)|x-y|^2 - n(1-|x|^2),
\end{aligned}
$$

we obtain the desired result. □

We stress that the treatment of higher-order operators that we present in these notes is far from exhaustive, and the results presented here should not leave the false impression that higher-order operators can be generally reduced to the case of the Laplacian. As a matter of fact, despite some similarities (such as the existence of a convenient Green function for the ball, see [Bog05]), even for the biharmonic operator, important structural differences arise. Just to mention one, we stress that, differently from harmonic functions, biharmonic functions do not satisfy in general the maximum principle: for instance (see [ST94]), as an instructive example, one can consider, for a suitably small $\varepsilon > 0$, the function

$$
u(x,y) := \Big((4-3x)(1-x)^2 - \varepsilon\Big)(x^2 + 25y^2 - 1)^2 \tag{2.23.14}
$$

on the ellipse $\Omega := \{(x,y) \in \mathbb{R}^2 \text{ s.t. } x^2 + 25y^2 < 1\}$. One can compute explicitly that $\Delta^2 u > 0$ in Ω. Since also $u = 0$ on $\partial\Omega$, the validity of a maximum principle (compare, for example, with Corollary 2.9.2) would entail that u should not change the sign in Ω, while, by inspection $u(0,0) > 0 > u(1-\varepsilon, 0)$, thus showing that the maximum principle is violated[39] in this case.

[39]It is interesting to remark that the example in (2.23.14) also satisfies $\nabla u = 0$ along $\partial\Omega$, and hence, even with adding such a "natural" boundary prescription, the maximum principle is still violated.

Instead, under the boundary condition $\Delta u = 0$ on $\partial\Omega$, the maximum principle for the biharmonic operator holds true since it boils down to the case of the Laplacian: namely, if

$$
\begin{cases} \Delta^2 u \geqslant 0 & \text{in } \Omega, \\ u = \Delta u = 0 & \text{on } \partial\Omega, \end{cases}
$$

it suffices to define $v := \Delta u$, observe that

$$
\begin{cases} \Delta v \geqslant 0 & \text{in } \Omega, \\ v = 0 & \text{on } \partial\Omega, \end{cases}
$$

deduce from Theorem 2.9.1 and Corollary 2.9.2 that $v \leqslant 0$ in Ω (and in fact either $v(x) = 0$ for all $x \in \Omega$, or $v(x) < 0$ for all $x \in \Omega$), infer that

$$
\begin{cases} \Delta u \leqslant 0 & \text{in } \Omega, \\ u = 0 & \text{on } \partial\Omega, \end{cases}
$$

apply again Theorem 2.9.1 and Corollary 2.9.2 and conclude that $u \geqslant 0$ in Ω (and in fact either $u(x) = 0$ for all $x \in \Omega$, or $u(x) > 0$ for all $x \in \Omega$).

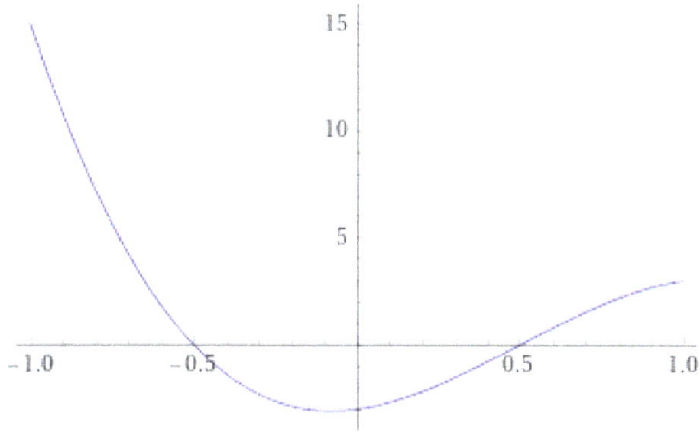

Figure 2.23.1 Plot of the function in (2.23.15).

As a one-dimensional example, one can also take into account the cubic function

$$\begin{aligned}(-1,1) \ni x \mapsto u(x) &= (1-2x)(1+2x)(2x-3) \\ &= -8x^3 + 12x^2 + 2x - 3 \qquad (2.23.15)\end{aligned}$$

which satisfies $u(-1) = 15 > 0$, $u(1) = 3 > 0$ and $\Delta^2 u = u'''' = 0$; since $u(0) = -3 < 0$, the maximum principle is violated in this case as well (see Figure 2.23.1 for a plot of this function).

See [Gar51, HJS02, RG12] for other counterexamples to maximum principles for the biLaplace operator. See [Swe16a, Swe19] for counterexamples to maximum principles for the triLaplacian and quadriLaplacian. See also [Swe16b, AJS21] for more insight on the lack of maximum principles for higher-order operators, as well as [Kry96, GGS10] and the references therein for a thorough discussion of polyharmonic problems.

Chapter 3

Upper Semicontinuity and Subharmonicity

In this chapter, we introduce and discuss the notion of subharmonic functions, that is, in broad terms, functions that "stay below" the corresponding harmonic functions with the same boundary data. The essence of the theory of subharmonic functions is extremely beautiful, and their structure is pivotal to understanding the behavior of the solutions of elliptic equations. Nevertheless, if one aims for a theory of subharmonic functions that remains "closed" under useful and natural operations, then a number of subtleties arise, and a convenient functional framework becomes necessary. For this, in the forthcoming section, we recall the notion of upper-semicontinuous functions, which will provide a convenient scenario for dealing with subharmonic functions in Section 3.2.

3.1 Semicontinuous Functions

An often useful ancillary tool of mathematical analysis is provided by semi-continuous functions. This is a weaker notion than continuity since it allows "discontinuities in one direction." In simple terms, a function is upper (respectively, lower) semi-continuous at a point if the values of the function nearby the point are not much higher (respectively, lower) than the value attained precisely at the point. Differently from continuous functions, upper (respectively, lower) semi-continuous functions may possess values near the point that are much lower (respectively, higher) than the value at the point. The notion of semicontinuity is exploited in Section 3.2 to deal with subharmonic and superharmonic functions. Before that, we introduce the formal definition of semicontinuity, discuss the main properties of semicontinuous functions and develop some intuition about this setting (for this, we mainly focus on upper semicontinuity since the case of lower semicontinuity is then easily obtained by a sign change).

Definition 3.1.1. Given an open set $\Omega \subseteq \mathbb{R}^n$ and a function $u : \Omega \to \mathbb{R} \cup \{-\infty\}$, we say that u is upper semicontinuous in Ω if

$$\limsup_{x \to x_0} u(x) \leqslant u(x_0), \tag{3.1.1}$$

for each $x_0 \in \Omega$. Also, we say that $u : \Omega \to \mathbb{R} \cup \{+\infty\}$ is lower semicontinuous in Ω if the function $-u$ is upper semicontinuous.

As an example, we have that the function

$$u(x) := \begin{cases} 1 & \text{if } x \geqslant 0, \\ 0 & \text{if } x < 0 \end{cases}$$

is lower semicontinuous. As is easily seen, a function is continuous in Ω if and only if it is both upper and lower semicontinuous. Also, upper- and lower-semicontinuous functions are measurable (see e.g. [WZ15, Corollary 4.16]). Moreover, semicontinuity is preserved under suitably monotone limits, as remarked in the following result.

Lemma 3.1.2. *The pointwise limit of a decreasing sequence of upper-semicontinuous functions is upper semicontinuous.*

Proof. Let u_k be a decreasing sequence of upper-semicontinuous functions in some domain Ω, and let u be its pointwise limit. Then, for every $j \in \mathbb{N}$ and $x \in \Omega$, we have that $u_{k+j}(x) \leqslant u_k(x)$; therefore,

$$u(x) = \lim_{j \to +\infty} u_{k+j}(x) \leqslant u_k(x).$$

As a result, using the upper semicontinuity of u_k, for each $x_0 \in \Omega$,

$$\limsup_{x \to x_0} u(x) \leqslant \limsup_{x \to x_0} u_k(x) \leqslant u_k(x_0).$$

Thus, taking the limit as $k \to +\infty$,

$$\limsup_{x \to x_0} u(x) \leqslant \lim_{k \to +\infty} u_k(x_0) = u(x_0).$$ □

We stress that the decreasing monotonicity assumption on the sequence of functions in Lemma 3.1.2 cannot be removed: for instance, one can consider the sequence of function $u_k(x) := -e^{-k|x|^2}$ and observe that it converges pointwise to $u(x) := -\chi_{\{0\}}(x)$, which is not upper semicontinuous.

Decreasing sequences, such as the ones in Lemma 3.1.2, can indeed be a useful tool to approximate upper-semicontinuous functions, as explicitly observed in the following result.

Lemma 3.1.3. *If u is upper semicontinuous in Ω and bounded above, then there exists a decreasing sequence of Lipschitz continuous functions that converges to u at every point of Ω.*

Proof. We can assume that u is not identically $-\infty$ (otherwise, we consider the decreasing sequence of constant functions $u_j := -j$, and we are done). We also consider the extension of u to the whole of $\mathbb{R}^n$ given, for every $x \in \mathbb{R}^n$, by

$$\overline{u}(x) := \begin{cases} u(x) & \text{if } x \in \Omega, \\ \limsup\limits_{\Omega \ni y \to x} u(y) & \text{if } x \in \partial\Omega, \\ -\infty & \text{if } x \in \mathbb{R}^n \setminus \overline{\Omega}. \end{cases}$$

We remark that

$$\overline{u} \text{ is upper semicontinuous in } \mathbb{R}^n. \tag{3.1.2}$$

Indeed, $\overline{u}$ is obviously upper semicontinuous in Ω and in $\mathbb{R}^n \setminus \overline{\Omega}$. Furthermore, we consider a point $x_0 \in \partial\Omega$, and we take a sequence $x_k \to x_0$ as $k \to +\infty$. Up to a subsequence, we can suppose that the limit of $\overline{u}(x_k)$ exists and realizes the lim sup at the point x_0, that is,

$$\lim_{k \to +\infty} \overline{u}(x_k) = \limsup_{x \to x_0} \overline{u}(x).$$

Hence, to prove (3.1.2), we aim at showing that

$$\lim_{k \to +\infty} \overline{u}(x_k) \leqslant \overline{u}(x_0). \tag{3.1.3}$$

To accomplish this goal, we distinguish two cases: either $x_k \in \partial\Omega$ for finitely many indices k or $x_k \in \partial\Omega$ for infinitely many indices k.

Let us first deal with the case in which $x_k \in \partial\Omega$ for finitely many indices k. In this situation, if x_k belonged to Ω only for finitely many indices k, we would have that, for all k large enough, $x_k \in \mathbb{R}^n \setminus \overline{\Omega}$, and thus $u(x_k) = -\infty$, making the claim in (3.1.3) obvious. This allows us to suppose that x_k belongs to Ω for infinitely many indices k, which we denote by k_j. In this setting, we conclude that

$$\limsup_{x \to x_0} \overline{u}(x) = \lim_{k \to +\infty} \overline{u}(x_k) = \lim_{j \to +\infty} \overline{u}(x_{k_j}) = \lim_{j \to +\infty} u(x_{k_j}) \leqslant \limsup_{\Omega \ni y \to x_0} u(y) = \overline{u}(x_0),$$

which is (3.1.3).

Hence it remains to consider the case in which $x_k \in \partial\Omega$ for infinitely many indices k. In this case, by the definition of $\overline{u}$, we have that there exists $y_k \in B_{1/k}(x_k) \cap \Omega$ such that $\overline{u}(x_k) \leqslant u(y_k) + \frac{1}{k}$. Also,

$$|y_k - x_0| \leqslant |y_k - x_k| + |x_k - x_0| \leqslant \frac{1}{k} + |x_k - x_0|,$$

which gives that $y_k \longrightarrow x_0$ as $k \longrightarrow +\infty$. As a consequence,

$$\overline{u}(x_0) = \limsup_{\Omega \ni y \to x} u(y) \geqslant \lim_{k \to +\infty} u(y_k) \geqslant \lim_{k \to +\infty} \left(\overline{u}(x_k) - \frac{1}{k} \right) = \lim_{k \to +\infty} \overline{u}(x_k).$$

These considerations complete the proof of (3.1.3) and thus of (3.1.2).

Moreover, $\overline{u}$ is bounded above. In particular,

$$\overline{u} \leqslant M, \tag{3.1.4}$$

for some $M \geqslant 0$.

For every $x \in \mathbb{R}^n$, we utilize a "sup-convolution method" by setting

$$u_j(x) := \sup_{y \in \mathbb{R}^n} \overline{u}(y) - j\,|x - y|. \tag{3.1.5}$$

We observe that, for every $x, y \in \mathbb{R}^n$,

$$\overline{u}(y) - (j+1)\,|x-y| = -|x-y| + \big(\overline{u}(y) - j\,|x-y|\big) \leqslant -|x-y| + u_j(x) \leqslant u_j(x),$$

hence $u_{j+1}(x) \leqslant u_j(x)$, and thus u_j is a decreasing sequence.

Moreover, picking $y := x$ in (3.1.5),

$$u_j(x) \geqslant \overline{u}(x) \qquad \text{for every } x \in \mathbb{R}^n. \tag{3.1.6}$$

In addition, given two points $x, x_0 \in \mathbb{R}^n$, for every $y \in \mathbb{R}^n$,

$$u_j(x_0) \geqslant \overline{u}(y) - j\,|y - x_0| \geqslant \overline{u}(y) - j\,|y - x| - j\,|x - x_0|.$$

Taking the supremum in y, we thus see that

$$u_j(x_0) \geqslant u_j(x) - j\,|x - x_0|,$$

whence, exchanging the roles of x and x_0,

$$|u_j(x) - u_j(x_0)| \leqslant j\,|x - x_0|.$$

This proves that u_j is Lipschitz continuous.

Now, we show that

$$\lim_{j \to +\infty} u_j(x) = \overline{u}(x) \qquad \text{for every } x \in \mathbb{R}^n. \tag{3.1.7}$$

By (3.1.6), it suffices to show that

$$\limsup_{j \to +\infty} u_j(x) \leqslant \overline{u}(x) \qquad \text{for every } x \in \mathbb{R}^n.$$

Suppose not. Then, there exist $x \in \mathbb{R}^n$ and $a > 0$ such that

$$\limsup_{j \to +\infty} u_j(x) > -\infty \tag{3.1.8}$$

and

$$\limsup_{j\to+\infty} u_j(x) \geqslant a + \overline{u}(x). \tag{3.1.9}$$

For each $j \in \mathbb{N}$, we use the definition in (3.1.5) and take $y_j \in \mathbb{R}^n$ such that

$$u_j(x) \leqslant \overline{u}(y_j) - j\,|x - y_j| + \frac{a}{2}. \tag{3.1.10}$$

This and (3.1.9) lead to

$$\limsup_{j\to+\infty} \Big(\overline{u}(y_j) - j\,|x - y_j|\Big) \geqslant \frac{a}{2} + \overline{u}(x). \tag{3.1.11}$$

Now, we exploit the upper semicontinuity of $\overline{u}$, and we take $\rho > 0$ sufficiently small such that if $y \in B_\rho(x)$, then $\overline{u}(y) \leqslant \overline{u}(x) + \frac{a}{4}$. We stress that $y_j \notin B_\rho(x)$, otherwise

$$\overline{u}(y_j) - j\,|x - y_j| \leqslant \overline{u}(y_j) \leqslant \overline{u}(x) + \frac{a}{4},$$

in contradiction with (3.1.11).

But then, recalling (3.1.4),

$$\overline{u}(y_j) - j\,|x - y_j| \leqslant M - j\rho \to -\infty \qquad \text{as } j \to +\infty.$$

This and (3.1.10) give that

$$\lim_{j\to+\infty} u_j(x) = -\infty,$$

which is in contradiction with (3.1.8). This establishes (3.1.7), which in turn yields the desired result. □

Also, upper-semicontinuous functions attain their maximum in compact sets.

Lemma 3.1.4. *If u is upper semicontinuous in Ω and $K \Subset \Omega$ is compact, then the maximum of u in K is attained and it is finite.*

Proof. One takes a maximizing sequence and proceeds as in the classical proof of Weierstraß' extreme value theorem, exploiting (3.1.1). □

By combining Lemmata 3.1.3 and 3.1.4, we obtain the following.

Corollary 3.1.5. *Let u be upper semicontinuous in Ω. Let $\Omega' \Subset \Omega$ be an open set. Then, there exists a decreasing sequence of Lipschitz continuous functions that converges to u at every point of Ω'.*

See Figure 3.1.1 for a visualization of a regular and decreasing approximation of an upper-semicontinuous function (note that one cannot obtain a similar approximation from below).

We can thus characterize upper-semicontinuous functions as follows.

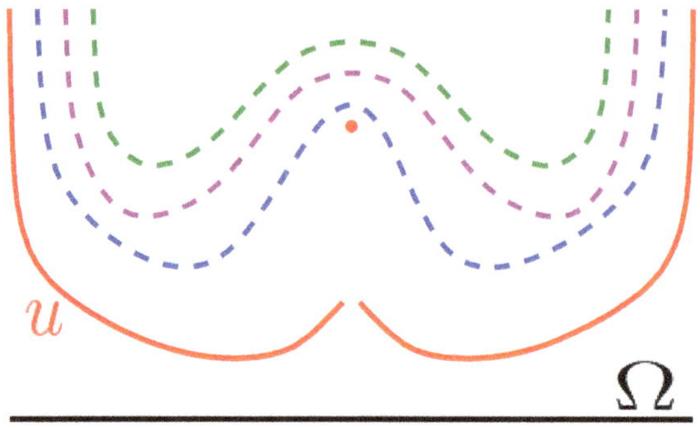

Figure 3.1.1 Regular approximation from above of an upper-semicontinuous function.

Corollary 3.1.6. *The following conditions are equivalent:*

(i) *u is upper semicontinuous in Ω.*

(ii) *For each $\alpha \in [-\infty, +\infty)$, the supergraph $\{u \geqslant \alpha\}$ is closed in Ω.*

(iii) *For every open set $\Omega' \Subset \Omega$, there exists a decreasing sequence of functions $u_j \in C^\infty(\Omega')$ that converges to u at every point of Ω'.*

Proof. We check that (i) and (iii) are equivalent. Indeed, if (i) holds true, given $\Omega' \Subset \Omega$, we consider Ω'' such that $\Omega' \Subset \Omega'' \Subset \Omega$, and we utilize Corollary 3.1.5 to find a decreasing sequence of Lipschitz continuous functions v_j that converges to u at every point of Ω''. We define $w_j := v_j + \frac{1}{2^j}$, and we consider $\tau \in C_0^\infty(B_1, [0, +\infty))$ with $\int_{B_1} \tau(x)\,dx = 1$. Given $\varepsilon > 0$, we let $\tau_\varepsilon(x) := \frac{1}{\varepsilon^n}\tau\left(\frac{x}{\varepsilon}\right)$ and $w_{j,\varepsilon} := w_j * \tau_\varepsilon$. We stress that $w_{j,\varepsilon} \in C^\infty(\Omega')$ and $w_{j,\varepsilon}$ converges to w_j uniformly in Ω' as $\varepsilon \searrow 0$ (see e.g. [WZ15, Theorem 9.8]). Consequently, we can find ε_j such that $\|w_{j,\varepsilon_j} - w_j\|_{L^\infty(\Omega')} \leqslant \frac{1}{2^{j+2}}$. In this way, setting $u_j := w_{j,\varepsilon_j}$, we have that, for every $x \in \Omega'$,

$$\begin{aligned}\lim_{j\to+\infty} u_j(x) &= \lim_{j\to+\infty} \Big(\big(w_{j,\varepsilon_j} - w_j(x)\big) + w_j(x) \Big) = \lim_{j\to+\infty} w_j(x) \\ &= \lim_{j\to+\infty} \left(v_j(x) + \frac{1}{2^j} \right) = \lim_{j\to+\infty} v_j(x) = u(x);\end{aligned}$$

that is, u_j converges to u at each point of Ω'. It only remains to prove that the convergence occurs in a monotonically decreasing way. To this end, we point out that

$$u_j = w_{j,\varepsilon_j} \leqslant w_j + \frac{1}{2^{j+2}} = v_j + \frac{1}{2^j} + \frac{1}{2^{j+2}} = v_j + \frac{5}{2^{j+2}} \tag{3.1.12}$$

and

$$u_j = w_{j,\varepsilon_j} \geqslant w_j - \frac{1}{2^{j+2}} = v_j + \frac{1}{2^j} - \frac{1}{2^{j+2}} = v_j + \frac{3}{2^{j+2}} \geqslant v_{j+1} + \frac{3}{2^{j+2}}.$$

We rephrase the latter inequality as

$$u_{j-1} \geqslant v_j + \frac{3}{2^{j+1}},$$

that is,

$$v_j \leqslant u_{j-1} - \frac{3}{2^{j+1}}.$$

As a byproduct of this and (3.1.12),

$$u_j \leqslant v_j + \frac{5}{2^{j+2}} \leqslant u_{j-1} - \frac{3}{2^{j+1}} + \frac{5}{2^{j+2}} = u_{j-1} - \frac{1}{2^{j+2}} < u_{j-1},$$

confirming thereby that u_j is a decreasing sequence. This shows that (i) implies (iii).

Moreover, if (iii) holds true, then we can apply Lemma 3.1.2 and conclude that u is upper semicontinuous. This argument shows that (iii) implies (i).

Let us now check that (i) implies (ii). To this end, given $\alpha \in [-\infty, +\infty)$, we distinguish two cases. If $\alpha = -\infty$, then $\{u \geqslant \alpha\} = \Omega$, which is closed in Ω, and then (ii) is fulfilled in this case.

If instead $\alpha \in \mathbb{R}$, we take a sequence $x_k \in \Omega$ such that $u(x_k) \geqslant \alpha$ and suppose that $x_k \to x_\star \in \Omega$ as $k \to +\infty$. Then, by (3.1.1),

$$\alpha \leqslant \lim_{k\to+\infty} u(x_k) \leqslant \limsup_{x\to x_\star} u(x) \leqslant u(x_\star),$$

and hence $x_\star \in \{u \geqslant \alpha\}$, thereby establishing the validity of (ii).

Thus, to complete the proof of Corollary 3.1.6, we now check that (ii) implies (i). For this, let $x_\star \in \Omega$ and $\alpha := u(x_\star) \in [-\infty, +\infty)$. Pick also a sequence $x_k \in \Omega$ such that $x_k \to x_\star$ as $k \to +\infty$ and

$$\lim_{k\to+\infty} u(x_k) = \limsup_{x\to x_\star} u(x).$$

Hence, to check that (i) holds true, we need to show that

$$\lim_{k\to+\infty} u(x_k) \leqslant u(x_\star). \tag{3.1.13}$$

Suppose not, and distinguish whether $\alpha \in \mathbb{R}$ or $\alpha = -\infty$. Suppose first that $\alpha \in \mathbb{R}$: then, if (3.1.13) is violated, we have that $u(x_k) \geqslant u(x_\star) + b = \alpha + b$, for some $b > 0$ and infinitely many indices k. In particular, $x_k \in \{u \geqslant \alpha + b\}$ and thus, by (ii), $x_\star \in \{u \geqslant \alpha + b\}$ as well. This gives that $\alpha = u(x_\star) \geqslant \alpha + b$, which is a contradiction.

Suppose now that $\alpha = -\infty$. If (3.1.13) did not hold, then $u(x_k) \geqslant d$, for some $d \in \mathbb{R}$ and infinitely many indices k. In particular, $x_k \in \{u \geqslant d\}$ and thus, by (ii), $x_\star \in \{u \geqslant d\}$ as well. This gives that $-\infty = \alpha = u(x_\star) \geqslant d$, which is a contradiction. The proof of (3.1.13), and thus of Corollary 3.1.6, is thereby complete. □

We emphasize that the convergence in Corollary 3.1.6(iii) is in the everywhere sense, not only almost everywhere: as an example, one can consider the decreasing sequence of functions $u_j(x) := \frac{1}{j}$, which converge almost everywhere to $u(x) := -\chi_{\{0\}}(x)$, but u is not upper semicontinuous.

We also point out that upper semicontinuity is preserved by taking infima (as well as finitely many maxima), as stated in the following two results.

Lemma 3.1.7. *The maximum of finitely many upper-semicontinuous functions is upper semicontinuous.*

Proof. Let $u_1, \dots, u_N$ be upper semicontinuous in a given $\Omega \subseteq \mathbb{R}^n$. We aim at showing that $u := \max\{u_1, \dots, u_N\}$ is also upper semicontinuous. Suppose not. Then, there exist $x_\star \in \Omega$, $a > 0$ and a sequence of points x_k converging to $x_\star$ as $k \to +\infty$ such that $u(x_k) \geqslant u(x_\star) + a$. Note, in particular, that $x_k \neq x_\star$, and hence $\{x_k\}_{k\in\mathbb{N}}$ is an infinite set. As a result, with the set $\{1, \dots, N\}$ being finite, there must exist $i \in \{1, \dots, N\}$ such that $u_i(x_k) = u(x_k) \geqslant u(x_\star) + a$ for infinitely many indices k. Consequently $u_i(x_k) \geqslant u_i(x_\star) + a$ for infinitely many indices k and, for this reason,

$$\limsup_{x\to x_\star} u_i(x) \geqslant \limsup_{k\to+\infty} u_i(x_k) \geqslant u_i(x_\star) + a,$$

and this is in contradiction with the upper semicontinuity of u_i. □

Lemma 3.1.8. *The infimum of a family of upper-semicontinuous functions is upper semicontinuous.*

Proof. Let Ω be an open set. We consider a set of indices $\mathcal{J}$ and, for each $j \in \mathcal{J}$, an upper-semicontinuous function $u_j : \Omega \to [-\infty, +\infty)$. We let $u(x) := \inf_{j\in\mathcal{J}} u_j(x)$, and we check that u is upper semicontinuous. For this, we observe that, for every $\alpha \in [-\infty, +\infty)$,

$$\{u \geqslant \alpha\} = \bigcap_{j\in\mathcal{J}} \{u_j \geqslant \alpha\}.$$

Since, by Corollary 3.1.6(ii), the set $\{u_j \geqslant \alpha\}$ is closed, we thus infer that $\{u \geqslant \alpha\}$ is also closed. Thus, using again Corollary 3.1.6(ii), we obtain that u is upper semicontinuous. □

Regarding Lemmata 3.1.7 and 3.1.8, we note that it is not true in general that the supremum of an infinite family of upper-semicontinuous functions is upper semicontinuous: as a counterexample, one can take $\Omega := (-1,1)$ and $u_j(x) := -e^{-\frac{|x|^2}{j}}$: indeed, in this case,

$$\sup_j u_j(x) = -\chi_{\{0\}}(x),$$

which is not upper semicontinuous.

We also mention that Lemma 3.1.8 has a natural counterpart, leading to another characterization of upper-semicontinuous functions (in addition to the ones described in Corollary 3.1.6).

Corollary 3.1.9. *The following conditions are equivalent:*

(i) *u is upper semicontinuous in Ω.*
(ii) *u is locally the infimum of a family of upper-semicontinuous functions.*
(iii) *u is locally the infimum of a family of C^∞ functions.*

Proof. We show that (iii) implies (ii), which implies (i), which implies (iii). Indeed: since C^∞ functions are upper semicontinuous, obviously (iii) implies (ii). Moreover, as a consequence of Lemma 3.1.8, we have that (ii) implies (i). Finally, in view of Corollary 3.1.6(iii), we know that (i) implies (iii). □

It is also useful to define the upper-semicontinuous envelope of a function $u : \Omega \to [-\infty, +\infty)$ that is locally bounded above[1] by

$$\begin{aligned} U_u(x) := \inf \big\{ & v(x), \text{ among all } v \text{ which are} \\ & \text{upper semicontinuous on } \Omega \text{ and } v \geqslant u \big\}. \end{aligned} \tag{3.1.14}$$

We point out that, by construction,

$$U_u \geqslant u. \tag{3.1.15}$$

Then, we have the following lemma.

Lemma 3.1.10. *The upper-semicontinuous envelope of a locally bounded-above function u is an upper-semicontinuous function.*

Moreover, u coincides with its upper-semicontinuous envelope if and only if u is upper semicontinuous.

[1]The assumption that u is locally bounded above is convenient to ensure that the set on the right-hand side of (3.1.14) is nonempty, namely that there exists at least one upper-semicontinuous function above u (for instance, the constant function equal to the supremum of u).

Proof. Let $x_\star \in \Omega$. Let $x_k \in \Omega$ be such that $x_k \longrightarrow x_\star$ as $k \longrightarrow +\infty$. Given $\varepsilon > 0$, we take v_ε, which is upper semicontinuous on Ω, $v_\varepsilon \geqslant u$ and $v_\varepsilon(x_\star) \leqslant U_u(x_\star)+\varepsilon$. In this way, since $U_u \leqslant v_\varepsilon$, we have that

$$\lim_{k\to+\infty} U_u(x_k) \leqslant \lim_{k\to+\infty} v_\varepsilon(x_k) \leqslant v_\varepsilon(x_\star) \leqslant U_u(x_\star) + \varepsilon.$$

Sending $\varepsilon \searrow 0$, we get that

$$\lim_{k\to+\infty} U_u(x_k) \leqslant U_u(x_\star),$$

and this shows that U_u is upper semicontinuous.

As a byproduct, one also has that if u coincides with U_u, then u is upper semicontinuous.

Now, suppose that u is upper semicontinuous. Then, one can choose $v := u$ in the definition (3.1.14) of the upper-semicontinuous envelope, thus concluding that

$$\begin{aligned} U_u(x) &\leqslant \inf \big\{ v(x), \text{ among all } v \text{ which are upper semicontinuous on } \Omega \text{ and } v \geqslant u \big\} \\ &\leqslant u(x). \end{aligned}$$

This and (3.1.15) entail that $U_u = u$, as desired. □

Alternative definitions of the upper-semicontinuous envelope are as follows.

Lemma 3.1.11. *Let $u : \Omega \to [-\infty, +\infty)$ be locally bounded, and let U_u be its upper-semicontinuous envelope, as in* (3.1.14)*. Then, for each $x \in \Omega$,*

$$U_u(x) = \inf_{r>0} \sup_{\Omega\cap B_r(x)} u = \max\left\{ \limsup_{y\to x} u(y),\ u(x) \right\}.$$

Proof. First, we claim that

$$U_u(x) \leqslant \inf_{r>0} \sup_{\Omega\cap B_r(x)} u. \tag{3.1.16}$$

For this, let $\varepsilon > 0$ and take $r_\varepsilon > 0$ such that

$$\varepsilon + \inf_{r>0} \sup_{\Omega\cap B_r(x)} u \geqslant \sup_{\Omega\cap B_{r_\varepsilon}(x)} u. \tag{3.1.17}$$

For all $y \in \Omega$, let also

$$v(y) := \begin{cases} \displaystyle\sup_{\Omega\cap B_{r_\varepsilon}(x)} u & \text{if } y \in \Omega \cap B_{r_\varepsilon}(x), \\ \displaystyle\sup_{\Omega} u & \text{if } y \in \Omega \setminus B_{r_\varepsilon}(x). \end{cases}$$

We observe that v is upper semicontinuous, and it is larger than or equal to u; therefore, by (3.1.14), we have that

$$U_u(x) \leqslant v(x) = \sup_{\Omega\cap B_{r_\varepsilon}(x)} u.$$

From this and (3.1.17), we conclude that

$$U_u(x) \leqslant \varepsilon + \inf_{r>0} \sup_{\Omega \cap B_r(x)} u,$$

whence (3.1.16) follows by sending $\varepsilon \searrow 0$.

Now, we show that

$$\inf_{r>0} \sup_{\Omega \cap B_r(x)} u \leqslant \max\left\{\limsup_{y \to x} u(y),\ u(x).\right\} \tag{3.1.18}$$

For this, we argue by contradiction and suppose that there exists $a > 0$ such that, for every $k \in \mathbb{N}$,

$$\sup_{\Omega \cap B_{1/k}(x)} u \geqslant a + \max\left\{\limsup_{y \to x} u(y),\ u(x)\right\}.$$

Let also $x_k \in \Omega \cap B_{1/k}(x)$ be such that $u(x_k) \geqslant \sup_{\Omega \cap B_{1/k}(x)} u - \frac{a}{2}$. In this way,

$$u(x_k) \geqslant \frac{a}{2} + \max\left\{\limsup_{y \to x} u(y),\ u(x)\right\}. \tag{3.1.19}$$

In particular, we have that $u(x_k) \geqslant \frac{a}{2} + u(x)$, which shows that $x_k \neq x$, and consequently, recalling [Wal04, Definition 2.6], since $|x_k - x| \leqslant \frac{1}{k}$,

$$\limsup_{k \to +\infty} u(x_k) \leqslant \limsup_{y \to x} u(y).$$

This provides a contradiction with (3.1.19), thus proving (3.1.18).

Now, let v be upper semicontinuous and $v \geqslant u$. Then,

$$v(x) \geqslant \limsup_{y \to x} v(y) \geqslant \limsup_{y \to x} u(y).$$

Taking the infimum over v, we thereby deduce that $U_u(x) \geqslant \limsup_{y \to x} u(y)$. This and (3.1.15) yield that

$$\max\left\{\limsup_{y \to x} u(y),\ u(x)\right\} \leqslant U_u(x).$$

Combining this inequality with (3.1.16) and (3.1.18), the desired result plainly follows. □

As a side remark, we point out that if two functions are locally bounded and coincide almost everywhere, it is not necessarily For instance, let u be the function identically zero on $\mathbb{R}$, and $\widetilde{u} := \chi_{\mathbb{Q}}$. Then, by Lemma 3.1.11 and the density of the rationals, for every $x \in \mathbb{R}$,

$$U_{\widetilde{u}}(x) \geqslant \limsup_{y \to x} \chi_{\mathbb{Q}}(y) = 1 > 0 = U_u(x).$$

3.2 Subharmonic and Superharmonic Functions

We now consider a class of functions that extends one of the harmonic functions introduced in Definition 2.1.1. In rough terms, harmonic functions are exceptionally special since they lie precisely in the kernel of the Laplacian and satisfy equality with their means. Of course, an equality is made by the common validity of two inequalities with different signs; hence, we consider here functions that satisfy one of the opposite inequalities characterizing the harmonic functions (but maybe not both of them). On the one hand, these functions carry "half of the information" with respect to harmonic functions. On the other hand, we will see that these functions are still quite special and enjoy a number of important properties (and moreover, they play a crucial role in the development of the theory of elliptic partial differential equations, especially in view of a one-sided maximum principle that these functions satisfy).

We begin with an equivalence result that characterizes the main properties enjoyed by this pivotal class of functions.

Theorem 3.2.1. *Let $\Omega \subseteq \mathbb{R}^n$ be an open set and $u : \Omega \to \mathbb{R} \cup \{-\infty\}$. The following conditions are equivalent:*

(i) *The function u is upper semicontinuous, and for every $x \in \Omega$ and $r > 0$ with $B_r(x) \Subset \Omega$, we have that*

$$u(x) \leqslant \fint_{B_r(x)} u(y)\, dy.$$

(ii) *The function u is upper semicontinuous, and for every $x \in \Omega$, there exists $r_0 > 0$ with $B_{r_0}(x) \Subset \Omega$ such that, for each $r \in (0, r_0]$,*

$$u(x) \leqslant \fint_{B_r(x)} u(y)\, dy.$$

(iii) *The function u is upper semicontinuous, and for every $x \in \Omega$ and $r > 0$ with $B_r(x) \Subset \Omega$, we have that*

$$u(x) \leqslant \fint_{\partial B_r(x)} u(y)\, d\mathcal{H}^{n-1}_y.$$

(iv) *The function u is upper semicontinuous, and for every $x \in \Omega$ there exists $r_0 > 0$ with $B_{r_0}(x) \Subset \Omega$ such that, for each $r \in (0, r_0]$,*

$$u(x) \leqslant \fint_{\partial B_r(x)} u(y)\, d\mathcal{H}^{n-1}_y.$$

(v) *Either u is constantly*[2] *equal to $-\infty$ in a connected component of Ω or $u \in L^1_{loc}(\Omega)$, and for every $\varphi \in C^\infty_0(\Omega, [0,+\infty))$,*

$$\int_\Omega u(x)\,\Delta\varphi(x)\,dx \geqslant 0. \tag{3.2.1}$$

(vi) *Either u is constantly equal to $-\infty$ in a connected component of Ω or $u \in L^1_{loc}(\Omega)$, and for every $x \in \Omega$, there exists $r_0 > 0$ such that for every $\varphi \in C^\infty_0(B_{r_0}(x), [0,+\infty))$,*

$$\int_{B_{r_0}(x)} u(y)\,\Delta\varphi(y)\,dy \geqslant 0.$$

(vii) *The function u is upper semicontinuous, and for every bounded open sets $\Omega' \Subset \Omega$ and every harmonic function $h \in C^2(\Omega') \cap C(\overline{\Omega'})$ such that $u \leqslant h$ on $\partial\Omega'$, we have that $u \leqslant h$ in Ω'.*

(viii) *The function u is upper semicontinuous, and for every $x \in \Omega$, there exists $r_0 > 0$ such that $\overline{B_{r_0}(x)} \Subset \Omega$, and for every harmonic function $h \in C^2(B_{r_0}(x)) \cap C(\overline{B_{r_0}(x)})$ such that $u \leqslant h$ on $\partial B_{r_0}(x)$, we have that $u \leqslant h$ in $B_{r_0}(x)$.*

Proof. To begin with, we prove that

$$\text{the statements (i), (iii), (v) and (vii) are all equivalent.} \tag{3.2.2}$$

The strategy that we adopt for this purpose is to show that (i) implies (v), (v) implies (vii), (vii) implies (iii) and (iii) implies (i).

First, we show that statement (i) implies statement (v). To this end, we can suppose that

$$u \text{ is not constantly equal to } -\infty \text{ in any connected component,} \tag{3.2.3}$$

otherwise we are done. We claim that

$$u \in L^1_{loc}(\Omega). \tag{3.2.4}$$

For this, we consider a connected component $\widetilde{\Omega}$ of Ω, and we define Ω_0 to be a collection of points $y \in \widetilde{\Omega}$ such that there exists $\rho > 0$ such that $B_\rho(y) \Subset \widetilde{\Omega}$ and $u \in L^1(B_\rho(y))$. Our goal is to show that

$$\Omega_0 = \widetilde{\Omega}. \tag{3.2.5}$$

[2]Note that we are allowing u to be possibly constantly $-\infty$ in Ω, though of course this provides a trivial case in Theorem 3.2.1. Other settings are possible as well, see e.g. [AG01, Definition 3.1.2]. A slight abuse of notation is used in (v) and (vi) since, in the usual sense of Lebesgue spaces, the function u in (v) and (vi) is defined only up to sets of null measure. More precisely, when needed, we will be able to define u as the limit of its volume average. See in particular (3.2.10) and (3.2.11). The setting in (v) is sometimes denoted by "weakly subharmonic functions," see in particular [Lit59, Lit63, Ser11]. See also [Fri61a] for a parabolic analogue.

The proof is by contradiction: suppose not, and let $y_0 \in \widetilde{\Omega} \setminus \Omega_0$. Choose $\rho_0 > 0$ sufficiently small such that $B_{3\rho_0}(y_0) \Subset \widetilde{\Omega}$. Let now $\xi \in B_{\rho_0}(y_0)$. We observe that

$$u \notin L^1(B_{2\rho_0}(\xi)); \tag{3.2.6}$$

otherwise, since $B_{\rho_0}(y_0) \subseteq B_{2\rho_0}(\xi)$, we would have that $u \in L^1(B_{\rho_0}(y_0))$, contradicting the assumption that $y_0 \notin \Omega_0$.

Using (3.2.6) and the fact that u is bounded from above (recall Lemma 3.1.4), we find that

$$\int_{B_{2\rho_0}(\xi)} u(y)\,dy = -\infty.$$

This and the setting in (i) give that $u(\xi) = -\infty$. Since ξ was an arbitrary point in $B_{\rho_0}(y_0)$, we have thereby established that $u = -\infty$ in $B_{\rho_0}(y_0)$. In particular, we have that $B_{\rho_0}(y_0) \subseteq \widetilde{\Omega} \setminus \Omega_0$. As a result, since y_0 is taken to be any point of $\widetilde{\Omega} \setminus \Omega_0$, we have that $\widetilde{\Omega} \setminus \Omega_0$ is open. Since $\widetilde{\Omega} \setminus \Omega_0$ is also closed in the topology of $\widetilde{\Omega}$, we have thus shown that $\widetilde{\Omega} \setminus \Omega_0 = \widetilde{\Omega}$, that is,

$$\Omega_0 = \varnothing. \tag{3.2.7}$$

We now exploit (3.2.3), and we take $\zeta_0 \in \widetilde{\Omega}$ such that $u(\zeta_0) \neq -\infty$. This and (i) give that, if $r_0 > 0$ is sufficiently small such that $B_{r_0}(\zeta_0) \Subset \widetilde{\Omega}$, then

$$\int_{B_{r_0}(\zeta_0)} u(y)\,dy > -\infty.$$

This and the boundedness from above of u (recall again Lemma 3.1.4) yield that $u \in L^1(B_{r_0}(\zeta_0))$, which is in contradiction with (3.2.7). This contradiction establishes (3.2.5), from which (3.2.4) plainly follows.

Hence, by (3.2.4), to complete the proof of (v), we take $\varphi \in C_0^\infty(\Omega, [0, +\infty))$. We let

$$f_\rho(x) := \frac{1}{\rho^2}\left(\fint_{B_\rho(x)} \varphi(y)\,dy - \varphi(x)\right),$$

and we employ Theorem 1.1.1 to see that, for every $x \in \mathbb{R}^n$,

$$\lim_{\rho \searrow 0} f_\rho(x) = \frac{1}{2(n+2)}\,\Delta\varphi(x).$$

In addition, using a Taylor expansion and an odd symmetry cancellation, for small $\rho > 0$, we have that

$$\left|\int_{B_\rho(x)} (\varphi(y) - \varphi(x))\, dy\right| = \left|\int_{B_\rho(x)} \Big(\nabla\varphi(x)\cdot(y-x) + O(|y-x|^2)\Big)\, dy\right| \\ = \left|\int_{B_\rho(x)} O(|y-x|^2)\, dy\right| = O(\rho^{n+2}),$$

with a bound depending only on n and $\|\varphi\|_{C^2(\mathbb{R}^n)}$. This gives that $f_\rho \in L^\infty(\mathbb{R}^n)$. As a consequence, recalling (3.2.4), we can utilize the dominated convergence theorem and conclude that, for every $\Omega' \Subset \Omega$,

$$\lim_{\rho\searrow 0} \int_{\Omega'} u(x)\, f_\rho(x)\, dx = \int_{\Omega'} \left(\lim_{\rho\searrow 0} u(x)\, f_\rho(x)\right) dx \\ = \frac{1}{2(n+2)} \int_{\Omega'} u(x)\, \Delta\varphi(x)\, dx. \tag{3.2.8}$$

Furthermore, for all $\rho > 0$ sufficiently small and choosing Ω' such that the support of φ is contained in Ω',

$$\int_{\Omega'} u(x)\, f_\rho(x)\, dx = \frac{1}{\rho^2}\int_{\Omega'} u(x) \left(⨍_{B_\rho(x)} \varphi(y)\, dy - \varphi(x)\right) dx \\ = \frac{1}{\rho^2\,|B_\rho|}\int_{\mathbb{R}^n} \left(\int_{B_\rho(y)} u(x)\,\varphi(y)\, dx\right) dy \\ - \frac{1}{\rho^2}\int_{\Omega'} u(x)\,\varphi(x)\, dx \\ \geqslant \frac{1}{\rho^2}\int_{\mathbb{R}^n} u(y)\,\varphi(y)\, dy - \frac{1}{\rho^2}\int_{\Omega'} u(x)\,\varphi(x)\, dx = 0,$$

thanks to (i). Combining this with (3.2.8), we find that

$$\int_{\Omega} u(x)\, \Delta\varphi(x)\, dx = \int_{\Omega'} u(x)\, \Delta\varphi(x)\, dx \geqslant 0,$$

which is the desired claim in (v).

Now, we show that (v) implies (vii). To this end, given $\Omega' \Subset \Omega$, we can restrict our analysis to the connected component of Ω containing Ω' and assume that u is not identically equal to $-\infty$ there, otherwise (vii) is obvious. Hence, we can suppose that $u \in L^1_{\text{loc}}(\Omega)$. We also remark that, if $B_r(x) \Subset \Omega$, then

$$\lim_{p\to x} ⨍_{B_r(p)} u(y)\, dy = ⨍_{B_r(x)} u(y)\, dy. \tag{3.2.9}$$

To check this, let p_j be a sequence of points converging to x as $j \to +\infty$, and let $f_j := u\chi_{B_r(p_j)}$. Note that $|f_j| \leqslant u \in L^1_{\mathrm{loc}}(\Omega)$. Consequently, according to the dominated convergence theorem,

$$\lim_{j\to+\infty} \int_{B_r(p_j)} u(y)\,dy = \int_{B_r(x)} u(y)\,dy,$$

which proves (3.2.9).

The focus now is to consider $x \in \Omega$ and, for every $r > 0$ such that $B_r(x) \Subset \Omega$, the volume average

$$\mathcal{V}(r) := \fint_{B_r(x)} u(y)\,dy.$$

We claim that

$$\text{if (3.2.1) holds true, the function } \mathcal{V} \text{ is monotone increasing in } r. \tag{3.2.10}$$

To check this, we let $\varrho \geqslant \rho > 0$ such that $B_\varrho(x) \Subset \Omega$. We consider the regularizations Γ_ϱ and Γ_ρ of the fundamental solution that was introduced in (2.7.4), and we observe that $\Gamma_\varrho \leqslant \Gamma_\rho$. Thus, setting, for each $y \in \mathbb{R}^n$,

$$\varphi_{\varrho,\rho}(y) := \Gamma_\rho(y-x) - \Gamma_\varrho(y-x),$$

we have that $\varphi_{\varrho,\rho} \in C^{1,1}_0(B_\varrho(x))$.

Now, we utilize the statement in (v). For this, we remark that, by density, the inequality (3.2.1) holds true for all $\varphi \in C^{1,1}_0(\Omega)$; therefore, recalling the basic property of the regularized fundamental solution (2.7.5),

$$\begin{aligned} 0 &\leqslant \int_\Omega u(y)\,\Delta\varphi_{\varrho,\rho}(y)\,dy \\ &= \int_{B_\rho(x)} u(y)\,\Delta\Gamma_\rho(y-x)\,dy - \int_{B_\varrho(x)} u(y)\,\Delta\Gamma_\varrho(y-x)\,dy \\ &= -\fint_{B_\rho(x)} u(y)\,dy + \fint_{B_\varrho(x)} u(y)\,dy, \end{aligned}$$

which establishes (3.2.10).

As a consequence of (3.2.10), the limit as $r \searrow 0$ of $\mathcal{V}(r)$ exists (and this, for every given $x \in \Omega$). Hence (recall Lebesgue's differentiation theorem, see e.g. [WZ15, Theorem 7.15]), we can redefine u in a set of null measure and suppose that, for every $x \in \Omega$,

$$u(x) = \lim_{r\searrow 0} \fint_{B_r(x)} u(y)\,dy = \lim_{r\searrow 0} \mathcal{V}(r). \tag{3.2.11}$$

With this setting, we have that u is the pointwise limit of the decreasing sequence of continuous functions $u_j(x) := \fint_{B_{1/j}(x)} u(y)\,dy$ (the continuity being a consequence of (3.2.9)), and hence

$$u \text{ is upper semicontinuous,} \tag{3.2.12}$$

thanks to Lemma 3.1.2.

An interesting byproduct of the monotonicity of the volume average $\mathcal{V}$ with respect to r and the integral in polar coordinates (see e.g. [EG15, Theorem 3.12]) is that, possibly excluding a set of zero Lebesgue measure of the values of r,

$$\begin{aligned} 0 &\leqslant \frac{d}{dr}\left(\fint_{B_r(x)} u(y)\,dy\right) = \frac{1}{|B_r|}\frac{d}{dr}\left(\int_{B_r(x)} u(y)\,dy\right) - \frac{n}{r}\fint_{B_r(x)} u(y)\,dy \\ &= \frac{1}{|B_r|}\int_{\partial B_r(x)} u(y)\,d\mathcal{H}^{n-1}_y - \frac{n}{r}\fint_{B_r(x)} u(y)\,dy. \end{aligned}$$

As a result,

$$\fint_{B_r(x)} u(y)\,dy \leqslant \frac{r}{n\,|B_r|}\int_{\partial B_r(x)} u(y)\,d\mathcal{H}^{n-1}_y = \fint_{\partial B_r(x)} u(y)\,d\mathcal{H}^{n-1}_y.$$

Consequently, using again (3.2.11) and the increasing monotonicity of $\mathcal{V}$, for almost every r,

$$u(x) \leqslant \fint_{B_r(x)} u(y)\,dy \leqslant \fint_{\partial B_r(x)} u(y)\,d\mathcal{H}^{n-1}_y. \tag{3.2.13}$$

Now, let h be as required in (vii). Given $\varepsilon > 0$, we let $h_\varepsilon := h + \varepsilon$. Note that $h_\varepsilon \geqslant u + \varepsilon$ on $\partial\Omega'$. We claim that there exists $\delta_\varepsilon > 0$ such that

$$h_\varepsilon \geqslant u + \frac{\varepsilon}{2} \text{ in } \Omega'_\varepsilon := \bigcup_{p\in\partial\Omega'} \left(B_{\delta_\varepsilon}(p) \cap \overline{\Omega'}\right). \tag{3.2.14}$$

Indeed, suppose not. Then, for every j, there exists $q_j \in \overline{\Omega'}$ at distance less than $\frac{1}{j}$ from $\partial\Omega'$ and such that $h_\varepsilon(q_j) < u(q_j) + \frac{\varepsilon}{2}$. Since $\overline{\Omega'}$ is compact, up to a subsequence, we can suppose that $q_j \to q \in \overline{\Omega'}$ as $j \to +\infty$. By construction $q \in \partial\Omega'$. Hence, by the upper semicontinuity of u,

$$h(q) + \varepsilon = h_\varepsilon(q) = \lim_{j\to+\infty} h_\varepsilon(q_j) \leqslant \limsup_{j\to+\infty} u(q_j) + \frac{\varepsilon}{2} \leqslant u(q) + \frac{\varepsilon}{2},$$

hence $h(q) < u(q)$, in contradiction with the assumptions in (vii). This completes the proof of (3.2.14).

Now, we consider a radial function $\tau \in C^\infty_0(B_1, [0, +\infty))$ with $\int_{B_1} \tau(x)dx = 1$. Given $\eta > 0$, we let $\tau_\eta(x) := \frac{1}{\eta^n}\tau\left(\frac{x}{\eta}\right)$ and define

$$u_\eta := u * \tau_\eta. \tag{3.2.15}$$

We claim that, if $\eta > 0$ is sufficiently small,

$$u_\eta \geqslant u \text{ in } \overline{\Omega'}. \tag{3.2.16}$$

Indeed, integrating in polar coordinates (see e.g. [EG15, Theorem 3.12]),

$$\begin{aligned} u_\eta(x) &= \int_{B_\eta} u(x-y)\,\tau_\eta(y)\,dy \\ &= \int_0^\eta \left(\int_{\partial B_\rho} u(x-\zeta)\,\tau_\eta(\zeta)\,d\mathcal{H}^{n-1}_\zeta \right) d\rho \\ &= \int_0^\eta \tau_\eta(\rho e_1) \left(\int_{\partial B_\rho} u(x-\zeta)\,d\mathcal{H}^{n-1}_\zeta \right) d\rho. \end{aligned} \tag{3.2.17}$$

Consequently, recalling (3.2.13),

$$u_\eta(x) \geqslant u(x) \int_0^\eta \tau_\eta(\rho e_1)\mathcal{H}^{n-1}(\partial B_\rho)\,d\rho = u(x) \int_{B_\eta} \tau_\eta(y)\,dy = u(x),$$

which proves (3.2.16).

Now, we observe that, for small $\eta > 0$,

$$\Delta u_\eta \geqslant 0 \text{ in } \overline{\Omega'}. \tag{3.2.18}$$

Indeed, for every $\varphi \in C_0^\infty(\Omega, [0,+\infty))$, we let $\varphi_\eta := \varphi * \tau_\eta$, and we remark that $\varphi_\eta \in C_0^\infty(\Omega, [0,+\infty))$ as long as η is sufficiently small. Then, using (v),

$$\begin{aligned} 0 \leqslant \int_\Omega u(y)\,\Delta\varphi_\eta(y)\,dy &= \int_{\mathbb{R}^n} \left(\int_{\mathbb{R}^n} u(y)\,\varphi(z)\,\Delta\tau_\eta(y-z)\,dz \right) dy \\ &= \int_{\mathbb{R}^n} \Delta u_\eta(z)\varphi(z),\, dz \end{aligned}$$

and thus (3.2.18) follows since Δu_η is continuous in $\overline{\Omega'}$.

We also set $h_{\varepsilon,\eta} := h_\varepsilon * \tau_\eta$, and we observe that, if $\eta > 0$ is conveniently small, then $\Delta h_{\varepsilon,\eta} = (\Delta h_\varepsilon) * \tau_\eta = (\Delta h) * \tau_\eta = 0$ in $\overline{\Omega'}$. From this and (3.2.18), setting $v_{\varepsilon,\eta} := u_\eta - h_{\varepsilon,\eta}$, we deduce that

$$\Delta v_{\varepsilon,\eta} \geqslant 0 \text{ in } \overline{\Omega'}. \tag{3.2.19}$$

Also, if η is sufficiently small, possibly depending on ε, we know from (3.2.14) that, for every $x \in \partial\Omega'$,

$$v_{\varepsilon,\eta}(x) = (u - h_\varepsilon) * \tau_\eta(x) = \int_{B_\eta(x)} \big(u(y) - h_\varepsilon(y)\big)\,\tau_\eta(x-y)\,dy < 0.$$

Consequently, by (3.2.19) and the maximum principle in Corollary 2.9.2(i), we conclude that $v_{\varepsilon,\eta}(x) \leqslant 0$ for every $x \in \Omega'$. For this reason and recalling (3.2.16), by taking limits, for every $x \in \Omega'$,

$$\begin{aligned} 0 &\geqslant \lim_{\varepsilon\searrow 0}\left(\lim_{\eta\searrow 0} v_{\varepsilon,\eta}(x)\right) = \lim_{\varepsilon\searrow 0}\left(\lim_{\eta\searrow 0}(u_\eta(x) - h_{\varepsilon,\eta}(x))\right) \\ &\geqslant \lim_{\varepsilon\searrow 0}\left(\lim_{\eta\searrow 0}(u(x) - h_{\varepsilon,\eta}(x))\right) \\ &= \lim_{\varepsilon\searrow 0}(u(x) - h_\varepsilon(x)) = u(x) - h(x). \end{aligned}$$

This proves that (v) implies (vii).

Now, we prove that (vii) implies (iii). For this, pick a ball $B_r(x) \Subset \Omega$. Let also $R > r$ be such that $B_R(x) \Subset \Omega$. Given $j \in \mathbb{N}$, we exploit Corollary 3.1.6(iii) to find a decreasing sequence of functions $u_j \in C^\infty(B_R(x))$ that converges to u at every point of $B_R(x)$. Let h_j be the harmonic function in $B_r(x)$ that coincides with u_j on $\partial B_r(x)$. The existence of h_j is warranted, for instance, by the Poisson kernel representation in Theorems 2.11.1 and 2.11.2. Also, we know that $h_j \in C^2(B_r(x)) \cap C(\overline{B_r(x)})$. Since $h_j = u_j \geqslant u$ on $\partial B_r(x)$, we deduce from (vii) and the mean value formula for harmonic functions (recall Theorem 2.1.2(ii)) that

$$\begin{aligned} u(x) \leqslant h_j(x) &= \lim_{\rho\searrow r} ⨍_{\partial B_\rho(x)} h_j(y)\, d\mathcal{H}^{n-1}_y \\ &= ⨍_{\partial B_r(x)} h_j(y)\, d\mathcal{H}^{n-1}_y = ⨍_{\partial B_r(x)} u_j(y)\, d\mathcal{H}^{n-1}_y. \end{aligned} \tag{3.2.20}$$

Now, we pass $j \to +\infty$. For this, we stress that $u_j \leqslant u_1$ and that u_j converges to u everywhere in $B_r(x)$ in a monotone-decreasing way; therefore, we can employ the monotone convergence theorem (see e.g. [WZ15, Theorem 10.27(ii)], and note that a convergence of u_j in the almost everywhere sense with respect to the Lebesgue measure would not suffice to exploit this result). In this way, we infer that

$$\lim_{j\to+\infty} ⨍_{\partial B_r(x)} u_j(y)\, d\mathcal{H}^{n-1}_y = ⨍_{\partial B_r(x)} u(y)\, d\mathcal{H}^{n-1}_y.$$

By combining this with (3.2.20), we have established that (iii) holds true, thus proving that (vii) implies (iii).

Now, we prove that (iii) implies (i). For this, we use polar coordinates (see e.g. [EG15, Theorem 3.12]) to find that

$$\begin{aligned} ⨍_{B_r(x)} (u(y) - u(x))\, dy &= \frac{1}{|B_r|}\int_0^r \left(\int_{\partial B_\rho(x)} (u(\zeta) - u(x))\, d\mathcal{H}^{n-1}_\zeta\right) d\rho \\ &= \frac{\mathcal{H}^{n-1}(\partial B_1)}{|B_r|}\int_0^r \left(\rho^{n-1} ⨍_{\partial B_\rho(x)} (u(\zeta) - u(x))\, d\mathcal{H}^{n-1}_\zeta\right) d\rho. \end{aligned}$$

Thus, if (iii) holds true, then the latter integrand is positive, and one obtains (i), as desired. This completes the proof of (3.2.2).

As a matter of fact, given a ball $B_{r_0}(x) \Subset \Omega$, we can now apply (3.2.2) to this ball instead of Ω: in this way, we obtain that

$$\text{the statements (ii), (iv), (vi) and (viii) are all equivalent.} \tag{3.2.21}$$

In light of (3.2.2) and (3.2.21), to complete the proof of Theorem 3.2.1, it is enough to show that the statements (v) and (vi) are equivalent. As a matter of fact, since (v) obviously implies (vi), it suffices to prove that

$$\text{statement (vi) implies statement (v).} \tag{3.2.22}$$

To check this, we can suppose that $u \in L^1_{\text{loc}}(\Omega)$, otherwise we are done, and we let $\varphi \in C^\infty_0(\Omega, [0, +\infty))$, as required in (v). We can therefore denote by K a compact subset of Ω such that $\varphi = 0$ outside K. For each $x \in \Omega$, we let $r_0 = r_0(x) > 0$ as in (vi), and we note that

$$K \subseteq \bigcup_{x \in K} B_{r_0(x)}(x).$$

In view of the compactness of K, we can take a finite subcover, thus finding $x_1, \ldots, x_N \in \Omega$ such that

$$K \subseteq \bigcup_{j=1}^{N} B_{r_0(x_j)}(x_j).$$

Then, we consider a partition of unity made of functions $\phi_1 \in C^\infty(B_{r_0(x_1)}(x_1), [0, 1]), \ldots, \phi_N \in C^\infty(B_{r_0(x_N)}(x_N), [0, 1])$, with finite overlapping supports, such that

$$\sum_{j=1}^{N} \phi_j = 1 \quad \text{in} \quad \bigcup_{j=1}^{N} B_{r_0(x_j)/2}(x_j).$$

Note in particular that

$$\sum_{j=1}^{N} \nabla\phi_j = 0 \quad \text{and} \quad \sum_{j=1}^{N} \Delta\phi_j = 0 \quad \text{in} \quad \bigcup_{j=1}^{N} B_{r_0(x_j)/2}(x_j).$$

Thus, if we define $\varphi_j := \varphi\phi_j$, we see that

$$\Delta\varphi = \Delta\varphi\chi_K = \sum_{j=1}^{N} \Delta\varphi\phi_j = \sum_{j=1}^{N} \Big(\Delta(\varphi\phi_j) - \varphi\Delta\phi_j - 2\nabla\varphi \cdot \nabla\phi_j\Big) = \sum_{j=1}^{N} \Delta\varphi_j. \tag{3.2.23}$$

Also, since $\varphi_j \in C^\infty(B_{r_0(x_j)}(x_j),\ [0,1])$, by (vi), we know that

$$\int_\Omega u(y)\,\Delta\varphi_j(y)\,dy = \int_{B_{r_0(x_j)}(x_j)} u(y)\,\Delta\varphi_j(y)\,dy \geqslant 0.$$

This and (3.2.23) yield that

$$\int_\Omega u(y)\,\Delta\varphi(y)\,dy = \sum_{j=1}^{N} \int_\Omega u(y)\,\Delta\varphi_j(y)\,dy \geqslant 0,$$

that is, statement (v). This establishes (3.2.22) and thus completes the proof of Theorem 3.2.1. □

Definition 3.2.2. Given an open set $\Omega \subseteq \mathbb{R}^n$, an upper-semicontinuous function $u : \Omega \to \mathbb{R} \cup \{-\infty\}$ is said to be subharmonic if any of the equivalent properties listed in Theorem 3.2.1 holds true.

A function $u : \Omega \to \mathbb{R} \cup \{+\infty\}$ is said to be superharmonic if $-u$ is subharmonic.

The name "subharmonic" is clearly a legacy of Theorem 3.2.1(vii). It is also interesting to note that the statements (ii), (iv), (vi) and (viii) in Theorem 3.2.1 are simply the localized versions of the statements (i), (iii), (v) and (vii): in this sense, a function is subharmonic in Ω if and only if it is subharmonic in any subdomain of Ω.

As a consequence of Theorems 2.7.2 and 3.2.1(v), we also have that

$$-\Gamma \text{ is subharmonic,} \tag{3.2.24}$$

with Γ being the fundamental solution in (2.7.6) (extended as $\Gamma(0) := +\infty$ when $n \geqslant 2$, which also clarifies a good reason for allowing upper-semicontinuous functions with values in $[-\infty, +\infty)$).

Moreover, we observe that, in virtue of the local integrability property in Theorem 3.2.1(v), we have that if u is subharmonic in Ω, then $\{|u| = +\infty\}$ has null measure in Ω, unless $u = -\infty$ in the whole of the connected component of Ω.

When the function is smooth, the notion of subharmonicity boils down to an inequality in the corresponding Laplace equation, as clarified by the following observation.

Lemma 3.2.3. *Given an open set $\Omega \subseteq \mathbb{R}^n$, a function $u \in C^2(\Omega, \mathbb{R} \cup \{-\infty\})$ is subharmonic if and only if $\Delta u \geqslant 0$ in Ω.*

Proof. Using Theorem 3.2.1(v) and Green's second identity (2.1.11), we have that a function $u :\in C^2(\Omega, \mathbb{R} \cup \{-\infty\})$ is subharmonic if and only if

$$\int_\Omega \Delta u(x)\, \varphi(x)\, dx \geqslant 0$$

for every $\varphi \in C_0^\infty(\Omega,\ [0, +\infty))$, from which the desired result follows. □

As a byproduct of the above observations, we also have that the averages of a subharmonic function increase with radius, as given in the following.

Corollary 3.2.4. *Let u be subharmonic in Ω. Let $x \in \Omega$ and $R > r > 0$ be such that $B_R(x) \Subset \Omega$. Then,*

$$\fint_{B_r(x)} u(y)\, dy \leqslant \fint_{B_R(x)} u(y)\, dy$$
$$\textit{and} \quad \fint_{\partial B_r(x)} u(y)\, d\mathcal{H}_y^{n-1} \leqslant \fint_{\partial B_R(x)} u(y)\, d\mathcal{H}_y^{n-1}. \tag{3.2.25}$$

Proof. Without loss of generality, we can reduce to the case of a connected domain Ω. Also, we can suppose that u is not identically $-\infty$ in Ω, otherwise the desired claims are obviously satisfied. Hence, we exploit Theorem 3.2.1(v), and we conclude that $u \in L^1_{\text{loc}}(\Omega)$ and (3.2.1) holds true. This and (3.2.10) give the first inequality in (3.2.25).

Now, let $R' > R$ be such that $B_{R'}(x) \Subset \Omega$. We consider the convolution u_η introduced in (3.2.15). Given $\varrho \in (r, R')$, we take $h_{\eta,\varrho}$ to be the harmonic function in $B_\varrho(x)$ that coincides with u_η along $\partial B_\varrho(x)$. Note that $h_{\eta,\varrho} \in C^2(B_\varrho(x)) \cap C(\overline{B_\varrho(x)})$. Since u_η is subharmonic in $B_\varrho(x)$, thanks to (3.2.18) and Lemma 3.2.3, we thereby deduce from Theorem 3.2.1(vii) that

$$u_\eta \leqslant h_{\eta,\varrho} \text{ in } B_\varrho(x). \tag{3.2.26}$$

Now, we take $h_{\eta,r}$ to be the harmonic function in $B_r(x)$ that coincides with u_η along $\partial B_r(x)$. Note that $h_{\eta,r} = u_\eta \leqslant h_{\eta,\varrho}$ along $\partial B_r(x)$, due to (3.2.26). From this and Theorem 3.2.1(vii), we conclude that $h_{\eta,r} \leqslant h_{\eta,\varrho}$ in $B_r(x)$. As a consequence, using the mean value formula in Theorem 2.1.2(ii),

$$\begin{aligned}\fint_{\partial B_r(x)} u_\eta(y)\, d\mathcal{H}_y^{n-1} &= \fint_{\partial B_r(x)} h_{\eta,r}(y)\, d\mathcal{H}_y^{n-1} = h_{\eta,r}(x) \leqslant h_{\eta,\varrho}(x) \\ &= \fint_{\partial B_\varrho(x)} h_{\eta,\varrho}(y)\, d\mathcal{H}_y^{n-1} = \fint_{\partial B_\varrho(x)} u_\eta(y)\, d\mathcal{H}_y^{n-1}.\end{aligned}$$

This and (3.2.16) yield that

$$\fint_{\partial B_r(x)} u(y)\, d\mathcal{H}_y^{n-1} \leqslant \fint_{\partial B_\varrho(x)} u_\eta(y)\, d\mathcal{H}_y^{n-1},$$

that is,

$$\varrho^{n-1}\mathcal{H}^{n-1}(\partial B_1)\fint_{\partial B_r(x)} u(y)\,d\mathcal{H}^{n-1}_y \leqslant \int_{\partial B_\varrho(x)} u_\eta(y)\,d\mathcal{H}^{n-1}_y.$$

We now integrate this inequality in $\varrho \in (R-\delta, R+\delta)$, with $\delta \in (0, R'-R)$, using polar coordinates on the right-hand side (see e.g. [EG15, Theorem 3.12]), thus finding that

$$\begin{aligned}&\frac{(R+\delta)^n-(R-\delta)^n}{n}\,\mathcal{H}^{n-1}(\partial B_1)\fint_{\partial B_r(x)} u(y)\,d\mathcal{H}^{n-1}_y\\ &\qquad\leqslant \int_{B_{R+\delta}(x)\setminus B_{R-\delta}(x)} u_\eta(y)\,dy.\end{aligned}$$

Now, we exploit that $u_\eta \to u$ in $L^1(B_{R'}(x)$ (see e.g. [WZ15, Theorem 9.6]) and thus conclude that

$$\begin{aligned}&\frac{(R+\delta)^n-(R-\delta)^n}{n}\,\mathcal{H}^{n-1}(\partial B_1)\fint_{\partial B_r(x)} u(y)\,d\mathcal{H}^{n-1}_y\\ &\qquad\leqslant \int_{B_{R+\delta}(x)\setminus B_{R-\delta}(x)} u(y)\,dy.\end{aligned}$$

Hence, we combine polar coordinates (see e.g. [EG15, Theorem 3.12]) and L'Hôpital's rule to find that

$$\begin{aligned}2\int_{\partial B_R(x)} u(y)\,d\mathcal{H}^{n-1}_y &= \lim_{\delta\searrow 0}\frac{d}{d\delta}\left(\int_{B_{R+\delta}(x)} u(y)\,dy - \int_{B_{R-\delta}(x)} u(y)\,dy\right)\\
&= \lim_{\delta\searrow 0}\frac{\frac{d}{d\delta}\int_{B_{R+\delta}(x)\setminus B_{R-\delta}(x)} u(y)\,dy}{\frac{d}{d\delta}\delta}\\
&= \lim_{\delta\searrow 0}\frac{\int_{B_{R+\delta}(x)\setminus B_{R-\delta}(x)} u(y)\,dy}{\delta}\\
&\geqslant \lim_{\delta\searrow 0}\frac{(R+\delta)^n-(R-\delta)^n}{n\delta}\,\mathcal{H}^{n-1}(\partial B_1)\\
&\qquad\times\fint_{\partial B_r(x)} u(y)\,d\mathcal{H}^{n-1}_y\\
&= 2R^{n-1}\mathcal{H}^{n-1}(\partial B_1)\fint_{\partial B_r(x)} u(y)\,d\mathcal{H}^{n-1}_y\\
&= 2\mathcal{H}^{n-1}(\partial B_R)\fint_{\partial B_r(x)} u(y)\,d\mathcal{H}^{n-1}_y,\end{aligned}$$

which is the second inequality in (3.2.25). □

Using Theorem 3.2.1, we can also strengthen the strong maximum principle given in Theorem 2.9.1 by removing the smoothness assumption on u.

Lemma 3.2.5. *Let $\Omega \subseteq \mathbb{R}^n$ be open and connected, and let u be subharmonic in Ω. If there exists $\overline{x} \in \Omega$ such that $u(\overline{x}) = \sup_\Omega u$, then u is constant.*

Proof. Suppose that $u(\overline{x}) = \sup_\Omega u$ for some $\overline{x} \in \Omega$. We argue as in the proof of Theorem 3.2.1, considering the set $\mathcal{U}$ in (2.9.1), which is nonvoid since $\overline{x} \in \mathcal{U}$.

Also,

$$\text{the set } \mathcal{U} \text{ is closed in } \Omega, \tag{3.2.27}$$

since if $x_k \in \mathcal{U}$ and $x_k \to \widetilde{x}$ as $k \to +\infty$, we deduce from the upper-semicontinuity condition (3.1.1) that

$$\sup_\Omega u = \limsup_{k\to+\infty} u(x_k) \leqslant u(\widetilde{x}) \leqslant \sup_\Omega u,$$

showing that $\widetilde{x} \in \mathcal{U}$ and establishing (3.2.27).

Hence, to prove the claim in Lemma 3.2.5, it suffices to show that

$$\text{the set } \mathcal{U} \text{ is also open.} \tag{3.2.28}$$

To this end, we consider $r > 0$ such that $B_r(\overline{x}) \Subset \Omega$, and we utilize Theorem 3.2.1 to see that

$$\sup_\Omega u = u(\overline{x}) \leqslant \fint_{B_r(\overline{x})} u(y)\,dy \leqslant \sup_\Omega u,$$

and accordingly $u = \sup_\Omega u$ a.e. in $B_r(\overline{x})$.

For this reason, for every $\widetilde{p} \in B_r(\overline{x})$, we can take a sequence $p_k \in B_r(\overline{x})$ such that $u(p_k) = \sup_\Omega u$ and $p_k \to \widetilde{p}$ as $k \to +\infty$. We thereby deduce from the upper semicontinuity of u (recall (3.1.1)) that

$$\sup_\Omega u = \limsup_{k\to+\infty} u(p_k) \leqslant u(\widetilde{p}) \leqslant \sup_\Omega u,$$

and accordingly $u(\widetilde{p}) = \sup_\Omega u$. This shows that $u = \sup_\Omega u$ everywhere in $B_r(\overline{x})$ which completes the proof of (3.2.28), as desired. □

Corollary 3.2.6. *Let $\widetilde{\Omega} \subseteq \mathbb{R}^n$ be open and connected, and let u be subharmonic in $\widetilde{\Omega}$. Let also $\Omega \Subset \widetilde{\Omega}$ be bounded. Then,*

$$\sup_\Omega u = \sup_{\partial\Omega} u.$$

Proof. The proof of Corollary 3.2.6 proceeds by applying Lemma 3.2.5 in lieu of Theorem 2.9.1. Indeed, one uses Lemma 3.1.4 to ensure that u attains a maximum in $\overline{\Omega}$; that is, there exists $p \in \overline{\Omega}$ such that $u(p) = \max_{\overline{\Omega}} u$. If $p \in \partial\Omega$, then the desired claim follows.

If instead $p \in \Omega$, then we use Lemma 3.2.5 to say that u is constant in Ω. Now, we take $\widetilde{x} \in \partial\Omega$ and a sequence of points $x_k \in \Omega$ such that $x_k \to \widetilde{x}$ as $k \to +\infty$. Accordingly, we exploit the upper semicontinuity of u to obtain that

$$\max_{\overline{\Omega}} u = u(p) = \lim_{k \to +\infty} u(x_k) \leqslant \limsup_{x \to \widetilde{x}} u(x) \leqslant u(\widetilde{x}) \leqslant \max_{\overline{\Omega}} u,$$

which also gives the desired result in this case. □

In addition, it follows from Theorem 3.2.1 that the set of subharmonic functions is endowed with a conical structure; namely, if u and v are subharmonic in Ω and a, $b \in [0, +\infty)$, then the function $au + bv$ is subharmonic in Ω as well.

An additional interesting feature is that the spherical mean of subharmonic functions always converges to the value at a point (and this holds everywhere, not only almost everywhere as warranted by Lebesgue's differentiation theorem, see e.g. [WZ15, Theorem 7.15]). Indeed, we have the following observation.

Lemma 3.2.7. *If u is subharmonic in Ω and $x \in \Omega$, then*

$$\lim_{r \searrow 0} \fint_{B_r(x)} u(y)\, dy = \lim_{r \searrow 0} \fint_{\partial B_r(x)} u(y)\, d\mathcal{H}_y^{n-1} = \limsup_{y \to x} u(y) = u(x).$$

Proof. Given $\varepsilon > 0$, in view of (3.1.1), we know that there exists $\delta > 0$ such that, for every $y \in B_\delta(x)$, $u(y) \leqslant u(x) + \varepsilon$. Thus, if $r \in (0, \delta)$,

$$\fint_{B_r(x)} u(y)\, dy \leqslant u(x) + \varepsilon \quad \text{and} \quad \fint_{\partial B_r(x)} u(y)\, d\mathcal{H}_y^{n-1} \leqslant u(x) + \varepsilon.$$

As a result,

$$\limsup_{r \searrow 0} \fint_{B_r(x)} u(y)\, dy \leqslant u(x) + \varepsilon$$

and

$$\limsup_{r \searrow 0} \fint_{\partial B_r(x)} u(y)\, d\mathcal{H}_y^{n-1} \leqslant u(x) + \varepsilon.$$

Therefore, since ε is arbitrary,

$$\begin{aligned} &\limsup_{r \searrow 0} \fint_{B_r(x)} u(y)\, dy \leqslant u(x) \\ \text{and} \quad &\limsup_{r \searrow 0} \fint_{\partial B_r(x)} u(y)\, d\mathcal{H}_y^{n-1} \leqslant u(x). \end{aligned} \tag{3.2.29}$$

On the other hand, by Theorem 3.2.1(i) and (iii),

$$\liminf_{r \searrow 0} \fint_{B_r(x)} u(y)\, dy \geqslant u(x) \quad \text{and} \quad \liminf_{r \searrow 0} \fint_{\partial B_r(x)} u(y)\, d\mathcal{H}_y^{n-1} \geqslant u(x).$$

From this and (3.2.29), we conclude that

$$\lim_{r\searrow 0} \fint_{B_r(x)} u(y)\,dy = \lim_{r\searrow 0} \fint_{\partial B_r(x)} u(y)\,d\mathcal{H}^{n-1}_y = u(x). \tag{3.2.30}$$

In addition, by Theorem 3.2.1(i), for every $k \in \mathbb{N}$ (say, sufficiently large such that $B_{1/k}(x) \Subset \Omega$), there exists $x_k \in B_{1/k}(x)$ such that $u(x_k) \geqslant u(x)$. As a consequence,

$$\limsup_{y\to x} u(y) \geqslant \lim_{k\to+\infty} u(x_k) \geqslant u(x),$$

which, combined with (3.1.1), leads to

$$\limsup_{y\to x} u(y) = u(x).$$

From this and (3.2.30), the desired result plainly follows. □

Corollary 3.2.8. *Two subharmonic functions that coincide almost everywhere necessarily coincide everywhere.*

Proof. Let u and v be subharmonic and equal almost everywhere. Then, for every x and every $r > 0$ such that $B_r(x)$ lies in the subharmonicity domain of u and v,

$$\int_{B_r(x)} u(y)\,dy = \int_{B_r(x)} v(y)\,dy.$$

Thus, recalling Lemma 3.2.7, for every point x, we have that

$$u(x) = \lim_{r\searrow 0} \fint_{B_r(x)} u(y)\,dy = \lim_{r\searrow 0} \fint_{B_r(x)} v(y)\,dy = v(x).$$ □

The one-dimensional subharmonic functions reduce to the convex ones, as remarked in the following observation.

Lemma 3.2.9. *If $n = 1$ and u is subharmonic in an open interval I, then u is convex in I.*

Proof. Let $a, b \in I$ with $a < b$, and for every $x \in [a, b]$, let

$$h(x) := \frac{(u(b) - u(a))(x - a)}{b - a} + u(a).$$

Since $h'' = 0$, we have that h is harmonic, and thus we can use Theorem 3.2.1(vii), deducing that, for every $t \in [0, 1]$, if $x := tb + (1 - t)a$,

$$u(tb + (1 - t)a) = u(x) \leqslant h(x) = t\,(u(b) - u(a)) + u(a) = t\,u(b) + (1 - t)\,u(a),$$

which establishes the desired convexity property. □

The converse of Lemma 3.2.9 holds true in every dimension, as given by the following.

Lemma 3.2.10. *If u is convex in Ω, then it is subharmonic in Ω.*

Proof. We recall that a convex function in $\mathbb{R}^n$ is continuous, see e.g. [EG15, Theorem 6.7(i)]. Moreover, let $x \in \Omega$, and let $r_0 > 0$ be such that $B_{r_0}(x) \Subset \Omega$. Take the supporting plane for u at x; namely, consider $\omega \in \mathbb{R}^n$ such that $u(y) \geqslant \omega \cdot (y - x) + u(x)$ for all y in Ω. Then, for all $r \in (0, r_0)$,

$$\fint_{B_r(x)} u(y)\,dy \geqslant \fint_{B_r(x)} \big(\omega \cdot (y-x) + u(x)\big)\,dy = \omega \cdot \fint_{B_r} \eta\,d\eta + u(x) = 0 + u(x),$$

thanks to an odd symmetry cancellation. This and Theorem 3.2.1(ii) yield the desired result. □

We now turn our attention to some operations that preserve subharmonicity.

Lemma 3.2.11. *The pointwise maximum of two subharmonic functions is subharmonic.*

Proof. If u and v are subharmonic in $\Omega \Supset B_r(x)$ and $w := \max\{u, v\}$, then we know that w is upper semicontinuous due to Lemma 3.1.7 and

$$\begin{aligned}\fint_{B_r(x)} w(y)\,dy &\geqslant \max\left\{\fint_{B_r(x)} u(y)\,dy,\ \fint_{B_r(x)} v(y)\,dy\right\}\\ &\geqslant \max\{u(x), v(x)\} = w(x),\end{aligned}$$

yielding the desired result. □

Of course, the previous result can be extended inductively to the pointwise maximum of finitely many subharmonic functions. The case of infinitely many subharmonic functions requires the extra assumption of upper semicontinuity.

Lemma 3.2.12. *Let I be a set of indices, and let u_i be subharmonic in Ω for all $i \in I$. Let $u := \sup_{i \in I} u_i$, and suppose that u is bounded above. Then, the upper-semicontinuous envelope of u is subharmonic in Ω.*

Also, if u is upper semicontinuous, then u is also subharmonic in Ω.

Proof. We recall that the upper-semicontinuous envelope U_u of u was introduced in (3.1.14). Up to reducing to a connected component, we can suppose that Ω is connected. We can assume that at least one u_i is not identically equal to $-\infty$ in Ω; otherwise, u would also be $-\infty$ in Ω, thus trivially leading to the desired result.

Hence, by Theorem 3.2.1(vi), we can assume that

$$u \in L^1_{\mathrm{loc}}(\Omega). \tag{3.2.31}$$

We take a connected open set $\Omega' \Subset \Omega$ and $r \in (0, \mathrm{dist}(\Omega', \partial\Omega))$. Let $x \in \Omega'$, then, for all $i \in \mathcal{I}$,

$$u_i(x) \leqslant \fint_{B_r(x)} u_i(y)\, dy \leqslant \fint_{B_r(x)} u(y)\, dy;$$

therefore, taking the supremum,

$$u(x) \leqslant \fint_{B_r(x)} u(y)\, dy. \tag{3.2.32}$$

This is not sufficient to state that u is subharmonic (unless u is upper semicontinuous); nevertheless, for all $x \in \Omega'$, we deduce from (3.2.32) that $u(x) \leqslant v_r(x)$, where

$$v_r(x) := \fint_{B_r(x)} u(y)\, dy.$$

Also, by (3.2.31) and the dominated convergence theorem, we infer that v_r is continuous. Consequently, by the definition of the upper-semicontinuous envelope U_u in (3.1.14), we deduce that $U_u \leqslant v_r$ in Ω'. From this and (3.1.15), we obtain that, for all $x \in \Omega'$,

$$U_u(x) \leqslant v_r(x) = \fint_{B_r(x)} u(y)\, dy \leqslant \fint_{B_r(x)} U_u(y)\, dy. \tag{3.2.33}$$

This and the fact that U_u is upper semicontinuous (recall Lemma 3.1.10) show that U_u is subharmonic.

Moreover, if u is upper semicontinuous, then u coincides with U_u, thanks to Lemma 3.1.10, and therefore u is subharmonic as well. □

The assumption that u is bounded above in Lemma 3.2.12 is taken to avoid problematic examples such as the following one: consider the fundamental solution Γ in (2.7.6), extended as $\Gamma(0) := +\infty$, and let $u_j(x) := -\Gamma(x) + j$. Then, by (3.2.24), we know that u_j is subharmonic, but

$$\sup_j u_j(x) = \begin{cases} -\infty & \text{if } x = 0, \\ +\infty & \text{otherwise,} \end{cases}$$

and for such a function, one cannot define an upper-semicontinuous envelope (see the footnote on p. 235).

Further, the assumption that u is upper semicontinuous cannot be removed from Lemma 3.2.12. Indeed, when $n \geqslant 2$, we can[3] define $u_j(x) := -\frac{\Gamma(x)}{j}$. Recalling (3.2.24), we have that u_j is subharmonic in B_1 and

$$\sup_j u_j(x) = \begin{cases} -\infty & \text{if } x = 0, \\ 0 & \text{otherwise,} \end{cases}$$

which is not upper semicontinuous (however, its upper-semicontinuous envelope is the function identically zero, which is subharmonic, in agreement with Lemma 3.2.12). Despite such examples, it is worth observing that the "pathological" points in which the supremum of subharmonic functions does not agree with its upper-semicontinuous envelope are negligible in the measure-theoretic sense.

Corollary 3.2.13. *Let $\mathcal{I}$ be a set of indices, and let u_i be subharmonic in Ω for all $i \in \mathcal{I}$. Let $u := \sup_{i\in\mathcal{I}} u_i$, and suppose that u is bounded above. Then, u coincides almost everywhere in Ω with its upper-semicontinuous envelope.*

In particular, u coincides almost everywhere in Ω with a subharmonic function.

Proof. As discussed in (3.2.31), we can focus on the case in which $u \in L^1_{\text{loc}}(\Omega)$. We let U_u be the upper-semicontinuous envelope of u. Thus, using (3.2.33),

$$U_u(x) \leqslant \fint_{B_r(x)} u(y)\,dy;$$

hence, if x is a Lebesgue density point for u,

$$U_u(x) \leqslant \lim_{r\searrow 0} \fint_{B_r(x)} u(y)\,dy = u(x).$$

From this and (3.1.15), it follows that $U_u = u$ at all Lebesgue density points for u and hence almost everywhere in Ω (see e.g. [WZ15, Theorem 7.13]). Since, in light of Lemma 3.2.12, we already know that U_u is subharmonic, we obtain as a byproduct that u coincides almost everywhere in Ω with a subharmonic function, as desired. □

For further details about the set in which the supremum of superharmonic functions may differ from its upper-semicontinuous envelope, see e.g. [AG01, Theorems 3.7.5 and 5.7.1].

Now, we analyze the pointwise limits of subharmonic functions.

[3]The case $n = 1$ does not lead to similar counterexamples, see Lemma 3.2.9, and recall that the supremum of convex functions is convex and hence continuous. In a sense, the semicontinuity setting is superfluous in dimension 1 since subharmonicity in that case automatically entails continuity.

Theorem 3.2.14. *The pointwise limit of a decreasing sequence of subharmonic functions is subharmonic.*

Proof. Let u_k be a decreasing sequence of subharmonic functions in some domain Ω, and let u be its pointwise limit. By Lemma 3.1.2, we already know that u is upper semicontinuous.

Now, suppose that $B_r(x) \Subset \Omega$. Then, by Lemma 3.1.4, we have that u_1 is bounded from above in $B_r(x)$, say $u_1(y) \leqslant M$, for every $y \in B_r(x)$. As a result, $u_k(y) \leqslant M$ as well for every $y \in B_r(x)$. We can therefore employ the monotone convergence theorem (see e.g. [WZ15, Theorem 5.32(ii)]) and conclude that

$$\lim_{k\to+\infty} \fint_{B_r(x)} u_k(y)\,dy = \fint_{B_r(x)} u(y)\,dy. \tag{3.2.34}$$

Also, $u_k \geqslant u_{k+j}$ for all $j \in \mathbb{N}$, whence $u_k \geqslant u$. This, Theorem 3.2.1(i) and (3.2.34) yield that

$$u(x) \leqslant \lim_{k\to+\infty} u_k(x) \leqslant \lim_{k\to+\infty} \fint_{B_r(x)} u_k(y)\,dy = \fint_{B_r(x)} u(y)\,dy. \qquad \square$$

As a confirming example for Theorem 3.2.14, one can consider the sequence $u_k : \mathbb{R}^n \to \mathbb{R}$ given by

$$u_k(x) := -\sum_{j=1}^{k} \frac{1}{2^j}\Gamma\left(x - \frac{e_n}{j}\right), \tag{3.2.35}$$

where Γ is the fundamental solution in (2.7.6) (extended as $\Gamma(0) := +\infty$ when $n \geqslant 2$), see Figure 3.2.1. Note that u_k is subharmonic, thanks to (3.2.24), the sequence u_k is decreasing, and

$$\lim_{k\to+\infty} u_k(x) = u(x) := -\sum_{j=1}^{+\infty} \frac{1}{2^j}\Gamma\left(x - \frac{e_n}{j}\right). \tag{3.2.36}$$

We observe that

$$|u(0)| \leqslant \sum_{j=1}^{+\infty} \frac{j^{n-2} + \ln j}{2^j} < +\infty,$$

while, when $n \geqslant 2$,

$$u(x) = -\infty \text{ if } x \in \bigcup_{j=1}^{+\infty} \frac{e_n}{j}.$$

We observe that u is upper semicontinuous, due to Lemma 3.1.2, but not continuous when $n \geqslant 2$, since

$$\liminf_{x\to 0} u(x) \leqslant \lim_{j\to+\infty} u\left(\frac{e_n}{j}\right) = -\infty < u(0). \tag{3.2.37}$$

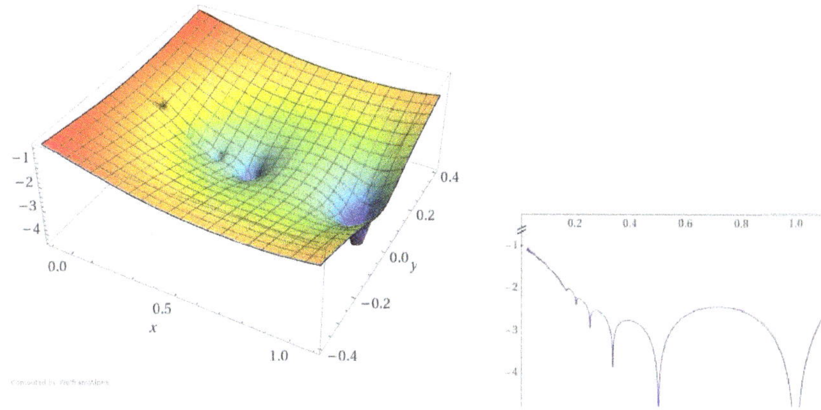

Figure 3.2.1 A computer plot of the function u_k when $n = 2$ for some "large" k, and a section of the graph in the e_2-direction.

This example[4] also clarifies a structural reason to include the upper-semicontinuity assumption in the definition of subharmonic functions (however, an approach to subharmonicity without upper semicontinuity is also possible, see [CR97]; see also [AG01, Chapter 7] for further details on possible discontinuity points of subharmonic functions and [Kel67] for a theory of subharmonic functions under continuity assumptions).

Interestingly, the example in (3.2.36) can be generalized to produce rather surprising and highly discontinuous subharmonic functions. For instance, if we enumerate the rationals in B_1 as $\mathbb{Q} \cap B_1 = \{q_j\}_{j\in\mathbb{N}}$, we can define

$$u_k(x) := -\sum_{j=1}^{+\infty} \frac{1}{2^j}\,\Gamma(x - q_j)$$

---

[4]The example constructed in (3.2.35) is that of a discontinuous, but unbounded, subharmonic function. If instead one wants to construct an example of a locally bounded and discontinuous subharmonic function, it suffices to consider u as in (3.2.36) and define $\widetilde{u}(x) := \max\{u(x),\, u(0) - 1\}$. In this way, we have that $\widetilde{u}$ is subharmonic by Lemma 3.2.11, and when $n \geqslant 2$, formula (3.2.37) gets replaced by

$$\liminf_{x\to 0} \widetilde{u}(x) \leqslant \lim_{j\to+\infty} \widetilde{u}\left(\frac{e_n}{j}\right) = \max\{-\infty,\, u(0) - 1\} = u(0) - 1 < u(0) = \widetilde{u}(0).$$

We also observe that when $n \geqslant 3$, we have that $u(0) - 1 \leqslant \widetilde{u} \leqslant 0$, and hence $\widetilde{u}$ is globally bounded. On the other hand, in dimension 2, one cannot construct global, discontinuous, and globally bounded subharmonic functions due to the forthcoming Theorem 3.2.18.

The example of a discontinuous subharmonic function in dimension $n \geqslant 2$ does not have a counterpart in dimension 1 since one-dimensional subharmonic functions are necessarily convex (hence continuous), according to Lemma 3.2.9.

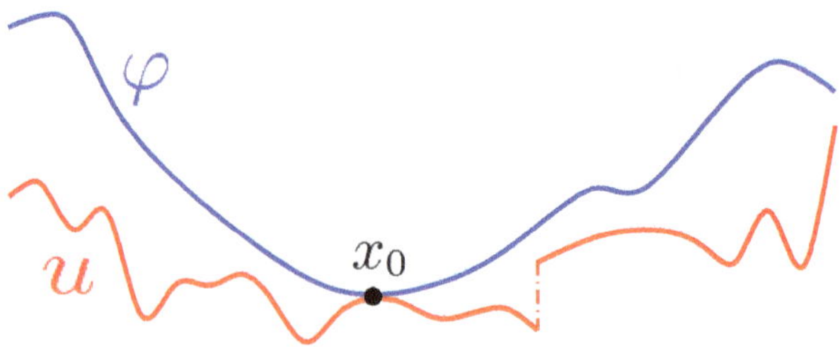

Figure 3.2.2 A function $u - \varphi$ having a local maximum at x_0 (up to a vertical translation, φ is touching u from above at x_0).

and observe that u_k is subharmonic, thanks to (3.2.24), and accordingly,

$$u(x) := -\sum_{j=1}^{+\infty} \frac{1}{2^j} \Gamma(x - q_j) \tag{3.2.38}$$

is subharmonic due to Theorem 3.2.14. We observe that

$$u \in L^1(B_1), \tag{3.2.39}$$

since

$$\int_{B_1} u(x)\,dx \leqslant \sum_{j=1}^{+\infty} \frac{1}{2^j} \int_{B_1} |\Gamma(x - q_j)|\,dx = \sum_{j=1}^{+\infty} \frac{1}{2^j} \int_{B_1(q_j)} |\Gamma(y)|\,dy$$
$$\leqslant \int_{B_2} |\Gamma(y)|\,dy \sum_{j=1}^{+\infty} \frac{1}{2^j} < +\infty.$$

The special feature of u is that it is finite almost everywhere in B_1, thanks to (3.2.39), but, when $n \geqslant 2$, infinite on $\mathbb{Q} \cap B_1$ (hence in a dense set).

We now point out a further characterization of subharmonic functions motivated by the theory of viscosity solutions (see e.g. [CC95] and Figure 3.2.2).

Theorem 3.2.15. *Let $u : \Omega \to [-\infty, +\infty)$ be upper semicontinuous. Then, u is subharmonic in Ω if and only if, for every $x_0 \in \Omega$, every $\rho > 0$ such that $B_\rho(x_0) \subseteq \Omega$ and every $\varphi \in C^2(B_\rho(x_0))$ such that the function $u - \varphi$ has a local maximum at x_0, it holds that $\Delta\varphi(x_0) \geqslant 0$.*

Proof. Assume first that u is subharmonic, and let x_0, ρ and φ be as in the statement of Theorem 3.2.15. Let $r \in (0, \rho)$ and h be the harmonic function coinciding with φ along $\partial B_r(x_0)$ (the existence and continuity of h being warranted by the Poisson kernel representation theory discussed in Theorems 2.11.1 and 2.11.2). Let also $\widetilde{h} := h + (u - \varphi)(x_0)$. Note that, if $y \in \partial B_r(x_0)$,

$$u(y) = (u - \varphi)(y) + \varphi(y) \leqslant (u - \varphi)(x_0) + \varphi(y) = (u - \varphi)(x_0) + h(y) = \widetilde{h}(y).$$

Consequently, by Theorem 3.2.1(vii), we know that $u \leqslant \widetilde{h}$ in $B_r(x_0)$. Therefore, using the mean value property in Theorem 2.1.2(ii),

$$\begin{aligned}\varphi(x_0) &= u(x_0) + (\varphi - u)(x_0) \leqslant \widetilde{h}(x_0) + (\varphi - u)(x_0) = h(x_0)\\ &= \fint_{\partial B_r(x_0)} h(y)\, d\mathcal{H}_y^{n-1} = \fint_{\partial B_r(x_0)} \varphi(y)\, d\mathcal{H}_y^{n-1}.\end{aligned}$$

From this and Theorem 1.1.2, we deduce that

$$0 \leqslant \lim_{r \searrow 0} \frac{1}{r^2} \left(\fint_{\partial B_r(x_0)} \varphi(x)\, d\mathcal{H}_x^{n-1} - \varphi(x_0) \right) = \frac{1}{2(n+2)} \Delta\varphi(x_0).$$

This proves that the first claim in Theorem 3.2.15 implies the second one. We now show that, conversely, the second claim in Theorem 3.2.15 implies the first one. To this end, we take $x \in \Omega$ and $r_0 > 0$ such that $B_{r_0}(x) \Subset \Omega$. Let also $h \in C^2(B_{r_0}(x)) \cap C(\overline{B_{r_0}(x)})$ be a harmonic function such that $u \leqslant h$ on $\partial B_{r_0}(x)$. In light of Theorem 3.2.1(viii), to prove that u is subharmonic and thus complete the proof of Theorem 3.2.15, it suffices to check that $u \leqslant h$ in $B_{r_0}(x)$. For this, we argue by contradiction, and suppose that there exist $x_\star \in B_{r_0}(x)$ and $a > 0$ such that $h(x_\star) + a \leqslant u(x_\star)$. Given $\varepsilon > 0$, to be taken sufficiently small, we define

$$\varphi_\varepsilon(y) := h(y) - \frac{\varepsilon\, |y - x_\star|^2}{n}.$$

We stress that the function $u - \varphi_\varepsilon$ is upper semicontinuous and therefore, in view of Lemma 3.1.4, attains a maximum in $\overline{B_{r_0}(x)}$. Namely, there exists $X \in \overline{B_{r_0}(x)}$ such that $(u - \varphi_\varepsilon)(X) \geqslant (u - \varphi_\varepsilon)(y)$ for all $y \in \overline{B_{r_0}(x)}$ and, in particular,

$$(u - \varphi_\varepsilon)(X) \geqslant (u - \varphi_\varepsilon)(x_\star) = (u - h)(x_\star) \geqslant a,$$

while, if $y \in \partial B_{r_0}(x)$,

$$(u - \varphi_\varepsilon)(y) = (u - h)(y) + \frac{\varepsilon\, |y - x_\star|^2}{n} \leqslant \frac{\varepsilon\, |y - x_\star|^2}{n} \leqslant \frac{\varepsilon\, r_0^2}{n} < a,$$

as long as $\varepsilon > 0$ is sufficiently small. These observations yield that $X \notin \partial B_{r_0}(x)$, and hence X is a local maximum for $u - \varphi_\varepsilon$. Since we are assuming that the second statement in Theorem 3.2.15 is satisfied, we infer that $\Delta\varphi_\varepsilon(X) \geqslant 0$. That is,

$$0 \leqslant \Delta\varphi_\varepsilon(X) = \Delta h(X) - 2\varepsilon = -2\varepsilon.$$

This is a contradiction, and the proof of Theorem 3.2.15 is thereby complete. □

Since subharmonic functions can exhibit quite complicated behaviors (recall the examples in (3.2.36) and (3.2.38)), it is often technically convenient to reduce the analysis to smooth subharmonic functions that approach "as well as possible" a given subharmonic function. For instance, we have the following result.

Theorem 3.2.16. *Every subharmonic function can be locally approximated by a decreasing sequence of C^∞ subharmonic functions.*

Proof. Let u be subharmonic in Ω. By possibly reducing to connecting components, we can suppose that Ω is connected. We can also assume that u is not identically equal to $-\infty$; otherwise, constant functions will be sufficient for the desired approximation. As a result, recalling Theorem 3.2.1(v), we have that $u \in L^1_{\text{loc}}(\Omega)$, and we consider the mollification u_η as in (3.2.15). Given $\Omega' \Subset \Omega$, if $\eta > 0$ is sufficiently small, we recall that

$$u_\eta \geqslant u \text{ in } \Omega', \tag{3.2.40}$$

in light of (3.2.16). Furthermore, by (3.2.18), we know that u_η is subharmonic in Ω'.

Additionally, recalling (3.2.17),

$$u_\eta(x) = \int_0^\eta \tau_\eta(\rho e_1) \left(\int_{\partial B_\rho} u(x-\zeta)\, d\mathcal{H}^{n-1}_\zeta \right) d\rho, \tag{3.2.41}$$

and therefore, making use of (3.2.25),

$$\begin{aligned} u_\eta(x) &\leqslant \int_0^\eta \tau_\eta(\rho e_1)\mathcal{H}^{n-1}(\partial B_r)\, d\rho \fint_{\partial B_\eta(x)} u(\zeta)\, d\mathcal{H}^{n-1}_\zeta \\ &= \int_0^\eta \left(\int_{\partial B_\rho} \tau_\eta(y)\, d\mathcal{H}^{n-1}_y \right) d\rho \fint_{\partial B_\eta(x)} u(\zeta)\, d\mathcal{H}^{n-1}_\zeta \\ &= \int_{B_\eta} \tau_\eta(y)\, dy \fint_{\partial B_\eta(x)} u(\zeta)\, d\mathcal{H}^{n-1}_\zeta \\ &= \fint_{\partial B_\eta(x)} u(\zeta)\, d\mathcal{H}^{n-1}_\zeta. \end{aligned}$$

Consequently, recalling Lemma 3.2.7,

$$\limsup_{\eta\searrow 0} u_\eta(x) \leqslant \limsup_{\eta\searrow 0} \fint_{\partial B_\eta(x)} u(\zeta)\, d\mathcal{H}^{n-1}_\zeta = u(x).$$

This and (3.2.40) give that u_η converges pointwise to u everywhere in Ω'.

Therefore, to complete the proof of the desired result, it only remains to prove that u_η is monotone, namely that $u_\eta \leqslant u_{\eta'}$ whenever $\eta \leqslant \eta'$. For this, we let $\kappa := \frac{\eta'}{\eta} \geqslant 1$, and we remark that

$$\frac{1}{\kappa^n}\tau_\eta\left(\frac{x}{\kappa}\right) = \frac{1}{\kappa^n\eta^n}\tau\left(\frac{x}{\kappa\eta}\right) = \tau_{\kappa\eta}(x);$$

consequently, using again (3.2.41) and changing variable $r := \kappa\rho$,

$$\begin{aligned} u_\eta(x) &= \frac{1}{\kappa}\int_0^{\kappa\eta} \tau_\eta\left(\frac{re_1}{\kappa}\right)\left(\int_{\partial B_{r/\kappa}} u(x-\zeta)\,d\mathcal{H}^{n-1}_\zeta\right) dr \\ &= \frac{1}{\kappa}\int_0^{\kappa\eta} \tau_\eta\left(\frac{re_1}{\kappa}\right)\mathcal{H}^{n-1}(\partial B_{r/\kappa})\left(\fint_{\partial B_{r/\kappa}} u(x-\zeta)\,d\mathcal{H}^{n-1}_\zeta\right) dr \\ &= \int_0^{\kappa\eta} \tau_{\kappa\eta}(re_1)\,\mathcal{H}^{n-1}(\partial B_r)\left(\fint_{\partial B_{r/\kappa}} u(x-\zeta)\,d\mathcal{H}^{n-1}_\zeta\right) dr. \end{aligned}$$

This and (3.2.25), with a further use of (3.2.41), lead to

$$u_\eta(x) \leqslant \int_0^{\kappa\eta} \tau_{\kappa\eta}(re_1)\,\mathcal{H}^{n-1}(\partial B_r)\left(\fint_{\partial B_r} u(x-\zeta)\,d\mathcal{H}^{n-1}_\zeta\right) dr = u_{\kappa\eta}(x),$$

which establishes the desired monotonicity property. □

Subharmonic functions in two-dimensional domains enjoy some special properties. For instance, the classical Hadamard three-circle theorem states that the maximum on a circle of radius ρ of a function that is subharmonic in a planar annulus is a convex function of $\ln\rho$. More explicitly, we have the following.

Theorem 3.2.17. *Let $R > r > 0$. Let Ω be an open subset of $\mathbb{R}^2$ such that*

$$\left\{x \in \mathbb{R}^2 \text{s.t. } |x| \in (r, R)\right\} \Subset \Omega.$$

Let u be subharmonic in Ω, and for every $t \in [\ln r, \ln R]$, let

$$\mathcal{M}(t) := \sup_{\partial B_{e^t}} u.$$

Then, $\mathcal{M}$ is a convex function.

Furthermore, if, for $\rho \in [r, R]$,

$$M(\rho) := \sup_{\partial B_\rho} u, \tag{3.2.42}$$

we have that, for every $r_2 > \rho > r_1$, with $R > r_2 > r_1 > r$,

$$M(\rho) \leqslant \frac{M(r_1)(\ln r_2 - \ln\rho) + M(r_2)(\ln\rho - \ln r_1)}{\ln r_2 - \ln r_1}. \tag{3.2.43}$$

Proof. For every $x \in \mathbb{R}^2 \setminus \{0\}$, we define

$$v(x) := u(x) - \frac{M(r_1)(\ln r_2 - \ln|x|) + M(r_2)(\ln|x| - \ln r_1)}{\ln r_2 - \ln r_1}.$$

Since $\Delta(\ln|x|) = 0$, if $x \in \mathbb{R}^2 \setminus \{0\}$, we see that v is subharmonic in $\Omega \setminus \{0\}$. Accordingly, by the maximum principle in Corollary 3.2.6, for every $x \in A := \{x \in \mathbb{R}^2 \text{ s.t. } |x| \in (r_1, r_2)\}$,

$$\begin{aligned} v(x) \leqslant \sup_{y \in \partial A} v(y) &= \max\left\{ \sup_{y \in \partial B_{r_1}} v(y), \ \sup_{y \in \partial B_{r_2}} v(y) \right\} \\ &= \max\left\{ \sup_{y \in \partial B_{r_1}} u(y) - M(r_1), \ \sup_{y \in \partial B_{r_2}} u(y) - M(r_2) \right\} = 0. \end{aligned}$$

This leads to

$$\begin{aligned} M(\rho) &- \frac{M(r_1)\,(\ln r_2 - \ln \rho) + M(r_2)\,(\ln \rho - \ln r_1)}{\ln r_2 - \ln r_1} \\ &= \sup_{x \in \partial B_\rho} u(x) - \frac{M(r_1)\,(\ln r_2 - \ln|x|) + M(r_2)\,(\ln|x| - \ln r_1)}{\ln r_2 - \ln r_1} \\ &= \sup_{x \in \partial B_\rho} v(x) \leqslant 0, \end{aligned}$$

which is (3.2.43).

Now, if $t_1, t_2 \in [\ln r, \ln R]$, with $t_2 > t_1$, we let $r_1 := e^{t_1}$ and $r_2 := e^{t_2}$. Let also $\vartheta \in [0,1]$ and $\rho := e^{(1-\vartheta)t_1 + \vartheta t_2}$. We also note that $\mathcal{M}(t) = M(e^t)$. Then, from (3.2.43),

$$\begin{aligned} 0 &\leqslant \frac{M(r_1)\,(\ln r_2 - \ln \rho) + M(r_2)\,(\ln \rho - \ln r_1)}{\ln r_2 - \ln r_1} - M(\rho) \\ &= (1-\vartheta)\,M(e^{t_1}) + \vartheta\,M(e^{t_2}) - M(e^{(1-\vartheta)t_1 + \vartheta t_2}) \\ &= (1-\vartheta)\,\mathcal{M}(t_1) + \vartheta\,\mathcal{M}(t_2) - \mathcal{M}((1-\vartheta)t_1 + \vartheta t_2), \end{aligned}$$

thus establishing the convexity of $\mathcal{M}$. □

The name of the three-circle theorem is likely due to inequality (3.2.43), which involves the circles of radii r_2, ρ and r_1.

An interesting consequence of the Hadamard three-circle theorem is a reinforcement of Liouville's theorem (see Theorem 2.20.1) in dimension 2, as follows.

Theorem 3.2.18. *A function which is subharmonic in the whole of $\mathbb{R}^2$ and bounded from above is necessarily constant.*

Proof. Let u be subharmonic in the whole of $\mathbb{R}^2$ with $u \leqslant C$ for some $C > 0$, and let M be as in (3.2.42). On the one hand, by the maximum principle in Corollary 3.2.6, if $\rho_2 > \rho_1 > 0$,

$$M(\rho_2) = \sup_{\partial B_{\rho_2}} u = \sup_{\overline{B_{\rho_2}}} u \geqslant \sup_{\overline{B_{\rho_1}}} u = \sup_{\partial B_{\rho_1}} u = M(\rho_1),$$

hence

$$M \text{ is nondecreasing.} \tag{3.2.44}$$

On the other hand, in light of (3.2.43) (applied here by sending $r_2 \longrightarrow +\infty$), we see that, if $\rho > r_1$,

$$\begin{aligned} M(\rho) &\leqslant \lim_{r_2 \to +\infty} \frac{M(r_1)(\ln r_2 - \ln \rho) + M(r_2)(\ln \rho - \ln r_1)}{\ln r_2 - \ln r_1} \\ &\leqslant \lim_{r_2 \to +\infty} \frac{M(r_1)(\ln r_2 - \ln \rho) + C(\ln \rho - \ln r_1)}{\ln r_2 - \ln r_1} = M(r_1). \end{aligned}$$

This and (3.2.44) yield that, for all $\rho > r_1$,

$$\sup_{\overline{B_\rho}} u = M(\rho) = M(r_1) = \sup_{\overline{B_{r_1}}} u;$$

therefore, by the strong maximum principle in Lemma 3.2.5, we have that u is constant. □

We observe that the assumption on the boundedness from above in Theorem 3.2.18 cannot be replaced by a boundedness from below (for example, the function $u(x) = |x|^2$ is subharmonic in $\mathbb{R}^2$ and bounded below, without being constant). Note that this is an interesting structural difference with respect to Liouville's theorem (Theorem 2.20.1).

Furthermore, Theorem 3.2.18 also holds true in dimension 1, thanks to the convexity equivalence in Lemma 3.2.9, but it does not carry over to dimension $n \geqslant 3$ (not even when assuming bounds from both above and below). For instance, for $n \geqslant 3$, the function $u(x) := -\frac{1}{(1+|x|^2)^{\frac{n-2}{2}}}$ satisfies

$$\Delta u(x) = \frac{n(n-2)}{\left(1+|x|^2\right)^{\frac{n-2}{2}}} \geqslant 0;$$

hence, u is subharmonic everywhere and bounded, but not constant.

For further readings on subharmonic functions, see also [Rad71, HK76, Hör94, Far07, Pon16] and the references therein.

For interesting connections between the three-circle theorem and the Riemann conjecture see [Edw01, p. 188], [Tit86, p. 332] and [Pat88, pp. 70–72].

In connection with maximum principles, we point out some further properties of the Green function in the following.

Lemma 3.2.19. *Let $n \geqslant 2$, and consider an open set $\Omega \subset \mathbb{R}^n$ with C^1 boundary.*

Let G be the Green function of Ω, as introduced in Section 2.10. Let also Γ be the fundamental solution, as introduced in Section 2.7.

Then, for all x, $x_0 \in \Omega$,

$$G(x, x_0) \geqslant 0 \tag{3.2.45}$$

and

$$G(x, x_0) \leqslant \Gamma(|x - x_0|). \tag{3.2.46}$$

Proof. Using the notation in (2.7.3) and (2.7.4) together with (2.10.2) and Lemma 2.10.1, we have that

$$G(x_0, x) = G(x, x_0) = \lim_{\rho \searrow 0} G_\rho(x, x_0),$$

where

$$G_\rho(x, x_0) := \Gamma_\rho(x - x_0) - \Psi^{(x_0)}(x). \tag{3.2.47}$$

Also, when ρ is smaller than the distance of x_0 from $\partial\Omega$, for all $x \in \partial\Omega$, we have that $\Gamma_\rho(x - x_0) = \Gamma(x - x_0)$ and consequently $G_\rho(\cdot, x_0) = 0$ along $\partial\Omega$.

Since G_ρ is superharmonic due to (2.7.5) and (2.10.1), we infer from the maximum principle in Corollary 3.2.6 that, for every $x \in \Omega$,

$$G_\rho(x, x_0) \geqslant \inf_{y \in \partial\Omega} G_\rho(y, x_0) = 0;$$

therefore, (3.2.45) plainly follows.

Also, from 2.10.1 and the maximum principle, we infer that $\Psi^{(x_0)}(x) \geqslant 0$ and accordingly, by (3.2.47), we find that $G_\rho(x, x_0) \leqslant \Gamma_\rho(x - x_0)$. Passing to the limit in ρ, this gives (3.2.46), as desired. □

In view of the above observations, we now point out a bound in Lebesgue spaces. For more general results, see [GT01, Theorems 8.15, 8.16, 8.17, 8.18, 8.25 and 8.26] (here, the situation is much simpler than the general case, since we are focusing on functions vanishing along the boundary of a given domain).

Proposition 3.2.20. *Let $n \geqslant 2$ and Ω be a bounded and open subset of $\mathbb{R}^n$ with boundary of class C^1. Let $p > \frac{n}{2}$ and $f : \Omega \to \mathbb{R}$ with $f^- \in L^p(\Omega)$.*

Let $u \in C^2(\Omega) \cap C^1(\overline{\Omega})$ be a solution of

$$\begin{cases} \Delta u \geqslant f & \text{in } \Omega, \\ u = 0 & \text{on } \partial\Omega. \end{cases}$$

Then,

$$\sup_\Omega u \leqslant C\|f^-\|_{L^p(\Omega)},$$

for some $C > 0$ depending only on n, p and Ω.

Proof. Up to considering connected components, we may suppose that Ω is connected. Let $x_0 \in \Omega'$. By Theorem 2.10.2,

$$-u(x_0) = \int_\Omega G(x_0, y)\, \Delta u(y)\, dy, \tag{3.2.48}$$

where G is the Green function of Ω.

On that account, we infer from (3.2.45) and (3.2.48) that

$$-u(x_0) \geqslant \int_\Omega G(x_0, y)\, f(y)\, dy \geqslant -\int_\Omega G(x_0, y)\, |f^-(y)|\, dy.$$

Additionally, we observe that Γ is locally in L^q, where $q \in [1, +\infty)$ is the dual exponent of p, thanks to the assumption on the range of p, and consequently, by (3.2.46) and the Hölder inequality,

$$u(x_0) \leqslant \int_\Omega G(x_0, y)\, |f^-(y)|\, dy \leqslant \int_\Omega \Gamma(x_0 - y)\, |f^-(y)|\, dy \leqslant C\, \|f^-\|_{L^p(\Omega)},$$

for some $C > 0$ depending only on n, p and Ω. □

One can compare Proposition 3.2.20 with Theorem 2.18.3, which dealt with harmonic functions, but without any assumption on the boundary data.

It is interesting to observe that the exponent p in Proposition 3.2.20 is[5] essentially optimal. As an example, one can consider $\rho \in \left(0, \frac{1}{4}\right)$ and the approximation of the fundamental solution Γ_ρ as in (2.7.4) and thus define $u_\rho(x) := \Gamma_\rho(x) - \Gamma_\rho(e_1)$. In this way, we have that $u_\rho \in C^{1,1}(\overline{B_1})$ with $u_\rho = 0$ on ∂B_1 and, recalling (2.7.5),

$$\Delta u_\rho = \begin{cases} -\dfrac{1}{|B_\rho|} & \text{in } B_\rho, \\ 0 & \text{in } \mathbb{R}^n \setminus \overline{B_\rho}. \end{cases}$$

In this situation, recalling the notation in (2.7.1), we see that

$$\frac{\sup_{B_{1/2}} u_\rho}{\|(\Delta u_\rho)^-\|_{L^{n/2}(B_1)}} = \frac{u_\rho(0)}{|B_\rho|^{\frac{2-n}{n}}} = \frac{1}{|B_\rho|^{\frac{2-n}{n}}} \left(\frac{\rho^2}{2n|B_\rho|} + c_n \big(\overline{v}(\rho) - \overline{v}(1)\big) \right).$$

In particular, when $n = 2$,

$$\frac{\sup_{B_{1/2}} u_\rho}{\|(\Delta u_\rho)^-\|_{L^{n/2}(B_1)}} = \frac{1}{4\pi} - c_2 \ln \rho,$$

[5]For completeness, we mention that in the case of operators in nondivergence form, the threshold for the exponent p somehow switches from $\frac{n}{2}$ to n, see [GT01, Theorems 9.20 and 9.22] and [CC95, Theorem 3.2]. The search for the optimal exponent for general elliptic equations is an active field of research related to a conjecture by Carlo Pucci [Puc66a], see e.g. [AIM09, Tru20] for further details.

which diverges as $\rho \searrow 0$, and this implies the optimality of the exponent p in Proposition 3.2.20.

Also when $n \geqslant 3$, this exponent cannot be improved, as showcased by the following example. For $\varepsilon \in (0,1)$, we consider

$$f_\varepsilon(x) := \begin{cases} -\dfrac{1}{|x|^2|\ln(|x|/2)|} & \text{if } x \in B_1 \setminus B_\varepsilon, \\ 0 & \text{if } x \in B_\varepsilon, \end{cases}$$

and we define

$$B_1 \ni x \longmapsto u_\varepsilon(x) = -\int_{B_1} f_\varepsilon(y) \left[\Gamma(x-y) - \Gamma\left(|y|\left(x - \frac{y}{|y|^2}\right)\right)\right] dy.$$

By Corollary 2.10.3 and Theorem 2.10.4, we have that

$$\begin{cases} \Delta u_\varepsilon = f_\varepsilon & \text{in } B_1, \\ u_\varepsilon = 0 & \text{on } \partial B_1. \end{cases}$$

Note that, on the one hand, if $n \geqslant 3$,

$$\begin{aligned} \|f_\varepsilon^-\|_{L^{n/2}(B_1)} &= \left(\int_{B_1 \setminus B_\varepsilon} \frac{1}{|x|^n|\ln(|x|/2)|^{\frac{2}{n}}}\, dx\right)^{\frac{2}{n}} \\ &\leqslant C\left(\int_\varepsilon^1 \frac{1}{r|\ln(r/2)|^{\frac{n}{2}}}\, dr\right)^{\frac{2}{n}} \\ &= C\left(\int_{\ln 2}^{-\ln(\varepsilon/2)} \frac{1}{t^{\frac{n}{2}}}\, dt\right)^{\frac{2}{n}} \\ &\leqslant C\left(\int_{\ln 2}^{+\infty} \frac{1}{t^{\frac{n}{2}}}\, dt\right)^{\frac{2}{n}} \\ &\leqslant C, \end{aligned}$$

up to renaming C at each line.

On the other hand, if $n \geqslant 3$ and $\rho > 0$ is small enough (and ε is appropriately small possibly depending on ρ),

$$\begin{aligned} u_\varepsilon(0) &= -\int_{B_1} f_\varepsilon(y)\left[\Gamma(y) - \Gamma(e_1)\right] dy \\ &\geqslant \int_{B_\rho \setminus B_\varepsilon} \frac{\Gamma(y) - \Gamma(e_1)}{|y|^2|\ln(|y|/2)|}\, dy \\ &\geqslant \frac{1}{C}\int_{B_\rho \setminus B_\varepsilon} \frac{dy}{|y|^n|\ln(|y|/2)|} \\ &= \frac{1}{C}\int_\varepsilon^\rho \frac{dr}{r|\ln(r/2)|} \end{aligned}$$

$$= \frac{1}{C} \int_{-\ln(\rho/2)}^{-\ln(\varepsilon/2)} \frac{dt}{t}$$
$$\geqslant \frac{\ln|\ln(\varepsilon/2)|}{C}.$$

From these observations, we arrive at

$$\frac{\sup_{B_{1/2}} u_\varepsilon}{\|(\Delta u_\varepsilon)^-\|_{L^{n/2}(B_1)}} \geqslant \frac{\ln|\ln(\varepsilon/2)|}{C},$$

which diverges as $\varepsilon \searrow 0$, and this implies the optimality of the exponent p in Proposition 3.2.20 also when $n \geqslant 3$.

An extension of Proposition 3.2.20 that does not assume vanishing data along the boundary is considered in Proposition 5.3.6, exploiting (differently from the classical literature) the forthcoming Calderón–Zygmund estimates presented in Section 5.3 (actually, only the conceptually simpler portion of these estimates in class $W^{1,p}$).

3.3 The Perron Method

We address here the problem of the existence of harmonic functions with prescribed boundary conditions. Several approaches can be taken for this problem, including, for more general ones, leveraging variational methods and functional analysis (see e.g. Sections 2.2.5(b) and 6.2 in [Eva98]). Here, we follow instead a classical intuition by[6] Henri Poincaré [Poi90], as developed by Oskar Perron [Per23] and Robert Remak [Rem24], formalized in what is nowadays called the "Perron method." In

[6]Poincaré is often considered the last universalist, i.e. the last scientist who was capable of deeply understanding and revolutionizing the set of contemporary knowledge seen as a whole. He was a mathematician, theoretical physicist, engineer and philosopher of science, often relying on his outstanding intuition and capability of visual representation. He laid the foundations of chaos theory and topology. His work contributed to the birth of special relativity, having detected relativistic velocity transformations and a perfect invariance of all of Maxwell's equations and having proposed the existence of gravitational waves.

One of his conjectures became a Millennium Prize Problem (see the footnote in [DV23, Chapter 5]), and actually the first (and, up to today, only) question to receive a complete answer (by Grigori Yakovlevich, a.k.a. Grisha, Perelman [Per03], who rejected the prize of about USD 1 million).

We think that a nice example of enthusiasm and scientific fervor is depicted in Figures 3.2.3 and 3.3.1: rather than posing for the official photograph, Marie Skłodowska (discoverer of the elements polonium and radium, 1903 Nobel Prize in Physics, 1911 Nobel Prize in Chemistry) and Henri Poincaré remain intensively discussing science.

By the way, Poincaré's multifaceted scientific talents were for his peers allegedly as proverbial as his clumsiness and his inability of drawing; it seems that his lack of drawing skill might even have jeopardized Poincaré's admittance to the École Polytechnique, due to a perfect 0 in the exam of wash drawing (fortunately, his examiners were sufficiently magnanimous or impressed by the marks obtained in the mathematical tests to change the 0 into a 1, which was sufficient for an overall pass).

Figure 3.2.3 Photograph of the 1911 Solvay Conference on Physics. Seated: W. Nernst, M. Brillouin, E. Solvay, H. Lorentz, E. Warburg, J. Perrin, W. Wien, M. Skłodowska Curie and H. Poincaré. Standing: R. Goldschmidt, M. Planck, H. Rubens, A. Sommerfeld, F. Lindemann, M. de Broglie, M. Knudsen, F. Hasenöhrl, G. Hostelet, E. Herzen, J. H. Jeans, E. Rutherford, H. Kamerlingh Onnes, A. Einstein and P. Langevin.
Source: Public Domain image from Wikipedia.

a nutshell, the idea is to obtain a harmonic function by an increasing sequence of subharmonic ones in a way which is similar to that of creating a stable electrostatic potential by sweeping out charges from inside the domain. The bottom line of the method is therefore to consider the "largest subharmonic function," check that it is indeed harmonic and that, under suitable conditions, it meets the boundary datum.

From a technical point of view, an important step in this construction is the fact that we can already construct harmonic functions given a boundary datum in balls,

---

Despite some poor drawings by Poincaré, which the reader can find, for example, in [Bar13], Poincaré heavily relied on figures to favor intuition and claimed that "figures first of all make up for the infirmity of our intellect by calling on the aid of our senses; but not only this. It is worth repeating that geometry is the art of reasoning well from badly drawn figures," see [Poi95].

Poincaré was certainly conscious of his limits as an illustrator: when, during his studies of celestial mechanics in [Poi57], he discovered the complex web arising from the intersections of stable and unstable manifolds (ultimately leading to chaotic patterns in dynamical systems), Poincaré wrote, "The complexity of this figure, which I will not even attempt to draw, is striking."

Figure 3.3.1 A detail from Figure 3.2.3.
Source: Public Domain image from Wikipedia.

thanks to the Poisson kernel method developed in Theorems 2.11.1 and 2.11.2 (this will allow us to "lift" a subharmonic function in a ball by replacing it with the harmonic function with the same boundary datum). The construction also relies on the maximum principle, which, among the other useful information, allows one to conclude that the above-mentioned harmonic lifting, when glued together with a subharmonic function, produces a subharmonic function as well.

Here are the technical details related to the above-mentioned "harmonic lift" technique.

Lemma 3.3.1. *Let Ω be an open and bounded subset of $\mathbb{R}^n$. Let $R > 0$, and assume that $B_R \Subset \Omega$. Let $w \in C(\Omega)$ be subharmonic in Ω.*

Then, there exists a unique function $W \in C(\Omega)$ which is harmonic in B_R, subharmonic in Ω and such that

$$\begin{cases} W = w \text{ in } \Omega \setminus B_R, \\ W \geqslant w \text{ in } \Omega. \end{cases} \tag{3.3.1}$$

Proof. The uniqueness claim follows by applying Corollary 2.9.3 in B_R.

We now focus on the existence claim. To this end, we utilize the Poisson kernel of B_R, in light of Theorems 2.11.1 and 2.11.2. In this way, we can find a harmonic function h in B_R with datum w along ∂B_R. We observe that

$$h \text{ is continuous in } \overline{B_R} \tag{3.3.2}$$

since if $\zeta \in \partial B_R$ and $\eta \in B_R$ (with $\rho := |\eta - \zeta|$ to be taken conveniently small), then, given $\delta \in \left[2\rho, \frac{R}{8}\right]$,

$$
\begin{aligned}
&|h(\eta) - w(\zeta)| \\
&= \left| \int_{\partial B_R} w(y) \frac{R^2 - |\eta|^2}{n\,|B_1|\,R\,|\eta - y|^n}\, d\mathcal{H}^{n-1}_y - w(\zeta) \right| \\
&\leqslant \int_{\partial B_R} |w(y) - w(\zeta)| \frac{R^2 - |\eta|^2}{n\,|B_1|\,R\,|\eta - y|^n}\, d\mathcal{H}^{n-1}_y \\
&\leqslant C\,(R^2 - |\eta|^2) \left[\int_{(\partial B_R)\cap B_\delta(\eta)} |w(y) - w(\zeta)| \frac{d\mathcal{H}^{n-1}_y}{|\eta - y|^n} \right. \\
&\quad \left. + \frac{1}{\delta^n} \int_{\partial B_R \setminus B_\delta(\eta)} |w(y) - w(\zeta)|\, d\mathcal{H}^{n-1}_y \right] \\
&\leqslant C\,(R - |\eta|) \left[\sup_{x \in B_{2\delta}(\zeta)} |w(x) - w(\zeta)| \int_{(\partial B_R)\cap B_\delta(\eta)} \frac{d\mathcal{H}^{n-1}_y}{|\eta - y|^n} + \frac{1}{\delta^n} \right] \\
&\leqslant C\sigma \left[\sup_{x \in B_{2\delta}(\zeta)} |w(x) - w(\zeta)| \sum_{j=0}^{\ln(\delta/\sigma)} \int_{(\partial B_R)\cap(B_{\delta/e^j}(\eta)\setminus B_{\delta/e^{j+1}}(\eta))} \frac{d\mathcal{H}^{n-1}_y}{|\eta - y|^n} + \frac{1}{\delta^n} \right] \\
&\leqslant C\sigma \left[\sup_{x \in B_{2\delta}(\zeta)} |w(x) - w(\zeta)| \sum_{j=0}^{\ln(\delta/\sigma)} \frac{(\delta/e^j)^{n-1}}{(\delta/e^{j+1})^n} + \frac{1}{\delta^n} \right] \\
&\leqslant C\sigma \left[\sup_{x \in B_{2\delta}(\zeta)} |w(x) - w(\zeta)| \sum_{j=0}^{\ln(\delta/\sigma)} \frac{e^j}{\delta} + \frac{1}{\delta^n} \right] \\
&\leqslant C\sigma \left[\frac{1}{\sigma} \sup_{x \in B_{2\delta}(\zeta)} |w(x) - w(\zeta)| + \frac{1}{\delta^n} \right]
\end{aligned}
$$

for some $C > 0$ depending only on n, R and $\|w\|_{L^\infty(\partial B_R)}$ and possibly varying from line to line, where $\sigma := R - |\eta|$ denotes the distance between η and ∂B_R.

That is, taking a sequence $\eta_m \in B_R$ such that $\eta_m \to \zeta \in \partial B_R$ as $m \to +\infty$, letting $\sigma_m := R - |\eta_m|$ and noting that $\sigma_m \to 0$,

$$
\begin{aligned}
\lim_{m\to+\infty} |h(\eta_m) - w(\zeta)| &\leqslant \lim_{m\to+\infty} C \left[\sup_{x \in B_{2\delta}(\zeta)} |w(x) - w(\zeta)| + \frac{\sigma_m}{\delta^n} \right] \\
&= C \sup_{x \in B_{2\delta}(\zeta)} |w(x) - w(\zeta)|,
\end{aligned}
$$

for all $\delta \in (0, \frac{R}{8}]$. We can thus send $\delta \searrow 0$ and conclude that

$$\lim_{m\to+\infty} |h(\eta_m) - w(\zeta)| = 0,$$

which proves (3.3.2).

Now, we define

$$W(x) := \begin{cases} h(x) & \text{if } x \in B_R, \\ w(x) & \text{if } x \in \Omega \setminus B_R, \end{cases} \tag{3.3.3}$$

and we remark that

$$W \in C(\overline{\Omega}), \tag{3.3.4}$$

owing to (3.3.2).

Also, the function $w - h$ is subharmonic in B_R. Hence, by Corollary 3.2.6,

$$\sup_{B_R}(w-h) = \lim_{r\nearrow R} \sup_{B_r}(w-h) = \lim_{r\nearrow R} \sup_{\partial B_r}(w-h) = \sup_{\partial B_R}(w-h) = 0.$$

As a consequence, we have that $w \leqslant h$ in B_R, whence

$$W \geqslant w \text{ in } \Omega. \tag{3.3.5}$$

We also remark that

$$W \text{ is subharmonic in } \Omega. \tag{3.3.6}$$

To check this, let $\widetilde{x} \in \Omega$ and $\widetilde{r} > 0$ such that $\widetilde{B} := B_{\widetilde{r}}(\widetilde{x}) \Subset \Omega$. Consider a harmonic function $\widetilde{h}$ which is continuous in the closure of $\widetilde{B}$ such that $W \leqslant \widetilde{h}$ on $\partial\widetilde{B}$. By (3.3.5), we know that $w \leqslant W \leqslant \widetilde{h}$ on $\partial\widetilde{B}$. Hence, recalling Theorem 3.2.1(viii) (as well as Definition 3.2.2), we deduce that $w \leqslant \widetilde{h}$ in the whole of $\widetilde{B}$. As a result,

$$\text{in } \widetilde{B} \setminus B_R, \text{ we have that } W = w \leqslant \widetilde{h}. \tag{3.3.7}$$

Additionally, since $\partial(\widetilde{B} \cap B_R)$ is contained in the closure of $\widetilde{B}$, we know that $W \leqslant \widetilde{h}$ on $\partial(\widetilde{B} \cap B_R)$; therefore, using again the maximum principle of Corollary 3.2.6 (up to the boundary of $\widetilde{B} \cap B_R$), we find that

$$\sup_{\widetilde{B}\cap B_R} (W - \widetilde{h}) = \sup_{\partial(\widetilde{B}\cap B_R)} (W - \widetilde{h}) \leqslant 0.$$

Combining this and (3.3.7), we find that $W \leqslant \widetilde{h}$ in $\widetilde{B}$. This and Theorem 3.2.1(viii) lead to (3.3.6), as desired.

The existence claim is thereby a consequence of (3.3.3), (3.3.4), (3.3.5) and (3.3.6). □

With this, we can address the Perron method for constructing harmonic functions, as follows.

Theorem 3.3.2. *Let $\Omega \subset \mathbb{R}^n$ be open and bounded. Let g be a bounded function on $\partial\Omega$, and define*

$$u(x) := \sup v(x), \tag{3.3.8}$$

where the above supremum is taken over all functions $v \in C(\Omega)$ which are subharmonic in Ω and such that

$$\limsup_{\Omega \ni x \to p} v(x) \leqslant g(p) \qquad \text{for every } p \in \partial\Omega. \tag{3.3.9}$$

Then, u is harmonic in Ω.

Proof. Without loss of generality, up to restricting ourselves to connected components, we may suppose that Ω is connected. We show that if $v \in C(\Omega)$ is subharmonic in Ω and satisfies (3.3.9), then

$$\sup_{\Omega} v \leqslant \sup_{\partial\Omega} g. \tag{3.3.10}$$

For this, we consider a sequence of points $p_j \in \Omega$ such that

$$\lim_{j \to +\infty} v(p_j) = \sup_{\Omega} v.$$

Up to a subsequence, we can suppose that $p_j \to p_\star$ as $j \to +\infty$, for some $p_\star \in \overline{\Omega}$.

Now, if $p_\star \in \partial\Omega$, we deduce from (3.3.9) that $\sup_\Omega v \leqslant g(p_\star)$, from which (3.3.10) follows at once; therefore, we can focus on the case in which $p_\star \in \Omega$.

In this case,

$$v(p_\star) = \lim_{j \to +\infty} v(p_j) = \sup_{\Omega} v. \tag{3.3.11}$$

Furthermore, for every $j \in \mathbb{N}$, we define

$$\Omega_j := \big\{x \in \Omega \text{ s.t. } B_{1/j}(x) \Subset \Omega\big\},$$

and we note that there exists $j_0 \in \mathbb{N}$ such that $p_\star \in \Omega_j$ for all $j \geqslant j_0$. Consequently, using the maximum principle in Corollary 3.2.6, we find that

$$v(p_\star) \leqslant \sup_{\partial\Omega_j} v. \tag{3.3.12}$$

Furthermore, for every $\varepsilon > 0$, we have that there exists $p_\star^{(j)} \in \partial\Omega_j$ such that

$$\sup_{\partial\Omega_j} v \leqslant v\big(p_\star^{(j)}\big) + \varepsilon. \tag{3.3.13}$$

Note that $p_\star^{(j)}$ converges to some $\overline{p}_\star \in \partial\Omega$ as $j \to +\infty$. From this, (3.3.9), (3.3.12) and (3.3.13), we deduce that

$$v(p_\star) \leqslant \limsup_{j \to +\infty} v\big(p_\star^{(j)}\big) + \varepsilon \leqslant \sup_{\partial\Omega} g + \varepsilon.$$

Hence, sending ε to zero and recalling (3.3.11), we obtain (3.3.10), as desired.

As a consequence of (3.3.10), we also have that

$$\begin{aligned}&\text{the supremum in (3.3.8) is finite,}\\ &\text{hence } u \text{ is well defined as a function in the reals.}\end{aligned} \tag{3.3.14}$$

Now, we pick $x_0 \in \Omega$ and $R > 0$ such that $B_R(x_0) \Subset \Omega$, and we show that

$$\begin{aligned}&\text{there exist a sequence of functions } W_k \in C(\Omega)\\ &\text{which are subharmonic in } \Omega, \text{ harmonic in } B_R(x_0),\\ &\text{with } \limsup_{\Omega \ni x \to p} W_k(x) \leqslant g(p) \text{ for every } p \in \partial\Omega\\ &\text{and such that } W_k \leqslant u \text{ in } \Omega,\\ &\text{and a function } H \text{ which is harmonic in } B_R(x_0)\\ &\text{such that } W_k \text{ converges to } H \text{ locally uniformly in } B_R(x_0)\\ &\text{and } u(x_0) = H(x_0).\end{aligned} \tag{3.3.15}$$

To prove this, up to a translation, we can suppose that $x_0 = 0$. Then, by (3.3.8), we can take a sequence of functions $v_k \in C(\Omega)$ which are subharmonic in Ω, such that (3.3.9) is satisfied and

$$v_k(0) \leqslant u(0) \leqslant v_k(0) + \frac{1}{k}. \tag{3.3.16}$$

Let $w_k(x) := \max\{v_k(x), \inf_{\partial\Omega} g\}$. We observe that w_k is also continuous in Ω, that

$$\limsup_{\Omega \ni x \to p} w_k(x) \leqslant g(p)$$

for every $p \in \partial\Omega$, and also that w_k is subharmonic in Ω, thanks to Lemma 3.2.11. In particular, we have that $u \geqslant w_k$. Since in addition $w_k \geqslant v_k$, we deduce from (3.3.16) that

$$w_k(0) \leqslant u(0) \leqslant w_k(0) + \frac{1}{k}. \tag{3.3.17}$$

Now, we take $R > 0$ sufficiently small that $B_R \Subset \Omega$. We take W_k to be the harmonic lift of w_k in B_R, according to Lemma 3.3.1. Using (3.3.1), we see that W_k satisfies (3.3.9), and so we have that

$$u \geqslant W_k \quad \text{in } \Omega. \tag{3.3.18}$$

and in particular $u(0) \geqslant W_k(0)$. This, (3.3.1) and (3.3.17) lead to

$$W_k(0) \leqslant u(0) \leqslant W_k(0) + \frac{1}{k}. \tag{3.3.19}$$

Now, we exploit Cauchy's estimates in Theorem 2.17.1 to see that, for every α, $\beta \in (0, 1)$ with $\alpha < \beta$,

$$\sup_{B_{\alpha R}} |\nabla W_k| \leqslant C \|W_k\|_{L^\infty(B_{\beta R}(0))}, \tag{3.3.20}$$

for some $C > 0$ depending only on n, R, α and β.

By construction, we know that

$$\inf_{\partial\Omega} g \leqslant \max\left\{ v_k(x), \inf_{\partial\Omega} g \right\} = w_k(x).$$

Moreover, recalling (3.3.10),

$$w_k(x) = \max\left\{ v_k(x), \inf_{\partial\Omega} g \right\} \leqslant \sup_{\partial\Omega} g.$$

These observations and the maximum principle in Corollary 2.9.2 yield that

$$\sup_{B_R} |W_k| = \sup_{\partial B_R} |W_k| = \sup_{\partial B_R} |w_k| \leqslant \|g\|_{L^\infty(\partial\Omega)}. \tag{3.3.21}$$

Hence, utilizing (3.3.20), we deduce that W_k is uniformly equicontinuous and equibounded in $B_{\alpha R}$. As a consequence, according to the Arzelà–Ascoli theorem, up to a subsequence, we can assume that W_k converges locally uniformly in $B_{\alpha R}$ to some function H. We also know that $u(0) = H(0)$, due to (3.3.19), and that H is harmonic in $B_{\alpha R}$, thanks to Corollary 2.2.2. The proof of (3.3.15) is thereby completed by collecting all the previous information.

Now, given $x_0 \in \Omega$, we claim that

$$u \text{ is harmonic in a neighborhood of } x_0. \tag{3.3.22}$$

Note that this would complete the proof of Theorem 3.3.2 since harmonicity is a local property, thanks to Theorem 2.1.2(iii).

Hence, to complete the desired result in (3.3.22), it suffices to take R and H as in (3.3.15) and show that

$$u = H \quad \text{in } B_R(x_0). \tag{3.3.23}$$

By (3.3.15), we already know that

$$u \geqslant \lim_{k \to +\infty} W_k(x) = H(x)$$

for every $x \in B_R(x_0)$, and hence, to prove (3.3.23), we can argue by contradiction, and suppose that there exists a point $x_0^\sharp \in B_R(x_0)$ such that $u(x_0^\sharp) \geqslant H(x_0^\sharp) + 2a$, for some $a > 0$.

In particular, by (3.3.8), we can find a function $v^\sharp \in C(\Omega)$ which is subharmonic in Ω such that

$$\limsup_{\Omega \ni x \to p} v^\sharp(x) \leqslant g(p) \quad \text{for every } p \in \partial\Omega$$

and satisfying

$$v^\sharp(x_0^\sharp) \geqslant H(x_0^\sharp) + a. \tag{3.3.24}$$

Let $w_k^\sharp(x) := \max\{v^\sharp(x), W_k(x)\}$, with W_k as in (3.3.15). We observe that $w_k^\sharp \in C(\Omega)$, that

$$\limsup_{\Omega \ni x \to p} w_k^\sharp(x) \leqslant g(p) \quad \text{for every } p \in \partial\Omega$$

and also that $w_k^\sharp$ is subharmonic in Ω, thanks to Lemma 3.2.11.

Now, we take $W_k^\sharp$ as the harmonic lift of $w_k^\sharp$ in the ball $B_R(x_0)$, according to Lemma 3.3.1. As in (3.3.20) and (3.3.21), we see that $W_k^\sharp$ is locally uniformly equicontinuous and equibounded in $B_R(x_0)$; therefore, up to a subsequence, it converges locally uniformly in $B_R(x_0)$ to some function $H^\sharp$ which is harmonic in $B_R(x_0)$.

We stress that, for every $x \in B_R(x_0)$,

$$W_k(x) \leqslant w_k^\sharp(x) \leqslant W_k^\sharp(x)$$

and therefore

$$H(x) \leqslant H^\sharp(x) \quad \text{for every } \; x \in B_R(x_0). \tag{3.3.25}$$

Moreover, by (3.3.24),

$$H(x_0^\sharp) + a \leqslant v^\sharp(x_0^\sharp) \leqslant w_k^\sharp(x_0^\sharp) \leqslant W_k^\sharp(x_0^\sharp),$$

from which we arrive at

$$H(x_0^\sharp) < H(x_0^\sharp) + a \leqslant H^\sharp(x_0^\sharp).$$

Combining this information with (3.3.25) and the strong maximum principle in Theorem 2.9.1(iii), we deduce that

$$H(x) < H^\sharp(x) \quad \text{for every } x \in B_R(x_0). \tag{3.3.26}$$

Furthermore, by the properties of the harmonic lift in Lemma 3.3.1 and (3.3.8), we know that $u \geqslant W_k^\sharp$ in Ω. Consequently, we have that $u \geqslant H^\sharp$ in $B_R(x_0)$. From this, (3.3.15) and (3.3.26), we deduce that

$$u(x_0) = H(x_0) < H^\sharp(x_0) \leqslant u(x_0),$$

which provides the desired contradiction. □

We want to emphasize that Theorem 3.3.2 in itself does not ensure that the harmonic function constructed there attains continuously the boundary datum g along $\partial\Omega$. This however is guaranteed if Ω is sufficiently regular. We introduce some terminology to clarify this point.

We say that an open set $\Omega \subset \mathbb{R}^n$ satisfies the exterior cone condition if for every point $x \in \partial\Omega$, there exists a finite right circular[7] cone K with vertex at x such that

$$\overline{K} \cap \overline{\Omega} = \{x\}. \tag{3.3.27}$$

See Figure 3.3.2 for examples in $\mathbb{R}^2$ and $\mathbb{R}^3$ of domains satisfying the exterior cone condition.

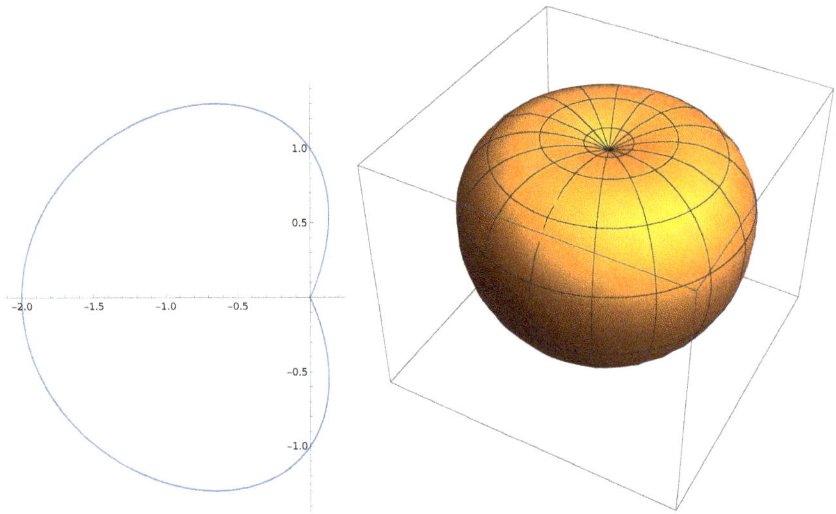

Figure 3.3.2 Domains in $\mathbb{R}^2$ and $\mathbb{R}^3$ satisfying the exterior cone condition. These are parametric plots of $(x(t), y(t)) = ((1 - \cos t)\sin|t/2|\cos t, (1 - \cos t)\sin t)$ and $(x(s,t), y(s,t), z(s,t)) = ((1 - \cos t)\sin t \sin s, (1 - \cos t)\sin t\cos s, (1 - \cos t)\sin|t/2|\cos t)$, respectively, with $s \in (-\pi/2, \pi/2]$, $t \in (-\pi, \pi]$.

[7] As usual, given $b \in (0, \pi)$ and $e \in \partial B_1$, a right circular cone of opening b in direction e is an object of the form

$$\left\{x \in \mathbb{R}^n \text{ s.t. } \frac{x \cdot e}{|x|} > \cos b\right\}.$$

A finite right circular cone then takes the form

$$\left\{x \in B_\rho \text{ s.t. } \frac{x \cdot e}{|x|} > \cos b\right\},$$

for some $\rho > 0$.

We also say that an open set $\Omega \subset \mathbb{R}^n$ is accessible by simple arcs from the exterior if for every $x \in \partial\Omega$, there exists an injective continuous map $\gamma : [0, 1] \to \mathbb{R}^n$ such that $\gamma(t) \in \mathbb{R}^n \setminus \overline{\Omega}$ for all $t \in [0, 1)$ and $\gamma(1) = x$.

As a side remark, we point out that if Ω satisfies the exterior ball condition, then it is also accessible by simple arcs from the exterior.

With this notation, we have the following.

Theorem 3.3.3. *Let $\Omega \subset \mathbb{R}^n$ be open, bounded and connected. Assume that either $n = 2$ and Ω is accessible by simple arcs from the exterior or $n \geqslant 3$ and Ω satisfies the exterior cone*[8] *condition.*

Let g be a continuous function on $\partial\Omega$, and let u be as in (3.3.8). *Then, $u \in C(\overline{\Omega})$ and*

$$\lim_{\Omega \ni x \to p} u(x) = g(p) \tag{3.3.28}$$

for all $p \in \partial\Omega$.

Proof. We begin with a general observation related to superharmonic functions. For this, suppose that $\Omega \subset \mathbb{R}^n$ is open and bounded, and let $p \in \partial\Omega$. Let $\rho > 0$ and $\beta : \overline{\Omega \cap B_\rho(p)} \to \mathbb{R}$ such that $\beta \in C(\Omega \cap B_\rho(p))$. Assume that β is superharmonic in $\Omega \cap B_\rho(p)$, and let

$$m := \inf_{\Omega \cap (B_\rho(p) \setminus B_{\rho/2}(p))} \beta$$

$$\text{and} \quad \omega(x) := \begin{cases} \min\{m, \beta(x)\} & \text{if } x \in \overline{\Omega} \cap B_{\rho/2}(p), \\ m & \text{if } x \in \overline{\Omega} \setminus B_{\rho/2}(p). \end{cases} \tag{3.3.29}$$

We point out that

$$\omega \in C(\Omega). \tag{3.3.30}$$

We claim that

$$\omega \text{ is superharmonic in } \Omega. \tag{3.3.31}$$

---

[8] For completeness, we observe that the proof of Theorem 3.3.3 becomes technically easier for domains with additional regularity, see e.g. [GT01, equation (2.34)] for a barrier in the case of domains satisfying the exterior ball condition. In this respect, one says that an open set $\Omega \subset \mathbb{R}^n$ satisfies the exterior ball condition if for every $x \in \partial\Omega$, there exist $R > 0$ and $y \in \mathbb{R}^n$ such that $\overline{B_R(y)} \cap \overline{\Omega} = \{x\}$. We observe that domains with boundary of class $C^{1,1}$ satisfy the exterior ball condition, see e.g. [ROV16, Lemma A.1]. Note also that this condition is somewhat the counterpart of the interior ball condition that was assumed in the Hopf lemma (namely, Lemma 2.16.1). We also stress that domains satisfying the exterior ball condition obviously satisfy the exterior cone condition as well.

For additional results on the Perron method for domains satisfying the exterior cone condition, see [Puc64, Mil67, Mil71, Mic77, Mic81, CKLŠ99].

For this, we first employ Lemma 3.2.11 to see that

$$\omega \text{ is superharmonic in } \Omega \cap B_{\rho/2}(p). \tag{3.3.32}$$

Furthermore, we observe that ω is constant and hence superharmonic in $\Omega \setminus \overline{B_{\rho/2}(p)}$. This, (3.3.32) and the fact that superharmonicity is a local property (recall Theorem 3.2.1(viii)) gives that, to prove (3.3.31), it suffices to consider an arbitrary point $q \in \Omega \cap \partial B_{\rho/2}(p)$ and an arbitrary small radius r_0 (in particular, $B_{r_0}(q) \Subset \Omega$) and check that, for each $r \in (0, r_0]$,

$$\omega(q) \geqslant \fint_{B_r(q)} \omega(x)\, dx. \tag{3.3.33}$$

To establish this, we observe that $\omega \leqslant m$ and thus

$$\fint_{B_r(q)} \omega(x)\, dx \leqslant m = \omega(q),$$

which gives (3.3.33) and thus (3.3.31), as desired.

Now, we point out some general remarks about the existence of a barrier (for this, we specify the functions β and ω introduced above). Namely, suppose again that $\Omega \subset \mathbb{R}^n$ is open and bounded, and let g be a bounded and continuous function on $\partial\Omega$ and u be as in (3.3.8). Let $p \in \partial\Omega$. We have that

$$\begin{aligned}
&\text{if there exist } \rho > 0 \text{ and a function } \beta : \overline{\Omega \cap B_\rho(p)} \to \mathbb{R},\\
&\text{which is continuous in } \Omega \cap B_\rho(p),\\
&\text{superharmonic in } \Omega \cap B_\rho(p), \text{ strictly positive in } \overline{\Omega \cap B_\rho(p)} \setminus \{p\}\\
&\text{and such that } \lim_{\Omega \ni x \to p} \beta(x) = 0,\\
&\text{then } \lim_{\Omega \ni x \to p} u(x) = g(p).
\end{aligned} \tag{3.3.34}$$

To prove this, we adopt the setting in (3.3.29). We have that

$$m > 0 = \lim_{\Omega \ni x \to p} \beta(x);$$

therefore,

$$\text{if } q \in \overline{\Omega} \setminus \{p\}, \text{ then } \liminf_{\Omega \ni x \to q} \omega(x) > \lim_{\Omega \ni x \to p} \beta(x) = \lim_{\Omega \ni x \to p} \omega(x) = 0. \tag{3.3.35}$$

Now, given $\varepsilon > 0$, we exploit the continuity of g to pick a $\delta > 0$ such that

$$\text{if } x \in (\partial\Omega) \cap B_\delta(p), \text{ we have that } |g(x) - g(p)| \leqslant \varepsilon. \tag{3.3.36}$$

Also, since $\iota := \inf_{\Omega \setminus B_\delta(p)} \omega > 0$, due to (3.3.35), by taking $K := \frac{2\|g\|_{L^\infty(\partial\Omega)}}{\iota}$, we have that, for all $x \in \Omega \setminus B_\delta(p)$,

$$K\omega(x) \geqslant K\iota = 2\|g\|_{L^\infty(\partial\Omega)}. \tag{3.3.37}$$

We now define

$$\omega_\pm(x) := g(p) \pm \varepsilon \pm K\omega(x).$$

We claim that

$$\limsup_{\Omega\ni x\to q} \omega_-(x) \leqslant g \leqslant \liminf_{\Omega\ni x\to q} \omega_+(x) \qquad \text{for every } q \in \partial\Omega. \tag{3.3.38}$$

For this, we take $q \in \partial\Omega$, and we distinguish two cases. If $q \in B_\delta(p)$, we have that

$$\begin{aligned}\limsup_{\Omega\ni x\to q} \omega_-(x) - g(q) &\leqslant \limsup_{\Omega\ni x\to q} \big(g(p) - \varepsilon - K\omega(x)\big) - g(p) + \varepsilon \\ &= -K \liminf_{\Omega\ni x\to q} \omega(x) \leqslant 0,\end{aligned}$$

thanks to (3.3.35) and (3.3.36), and similarly,

$$\begin{aligned}\liminf_{\Omega\ni x\to q} \omega_+(x) - g(q) &\geqslant \liminf_{\Omega\ni x\to q} \big(g(p) + \varepsilon + K\omega(x)\big) - g(p) - \varepsilon \\ &= K \liminf_{\Omega\ni x\to q} \omega(x) \geqslant 0.\end{aligned}$$

These considerations give that (3.3.38) holds true in $(\partial\Omega) \cap B_\delta(p)$; hence, we consider now the case in which $q \in (\partial\Omega) \setminus B_\delta(p)$. In this situation, we utilize (3.3.37) to see that

$$\begin{aligned}\limsup_{\Omega\ni x\to q} \omega_-(x) - g(q) &= \limsup_{\Omega\ni x\to q} \big(g(p) - \varepsilon - K\omega(x)\big) - g(q) \\ &\leqslant g(p) - \varepsilon - 2\|g\|_{L^\infty(\partial\Omega)} - g(q) \leqslant -\varepsilon < 0,\end{aligned}$$

and, in the same way,

$$\begin{aligned}\liminf_{\Omega\ni x\to q} \omega_+(x) - g(q) &= \limsup_{\Omega\ni x\to q} \big(g(p) + \varepsilon + K\omega(x)\big) - g(q) \\ &\geqslant g(p) + \varepsilon + 2\|g\|_{L^\infty(\partial\Omega)} - g(q) \geqslant \varepsilon > 0.\end{aligned}$$

These observations complete the proof of (3.3.38).

Furthermore, ω_+ is superharmonic in Ω and ω_- is subharmonic in Ω, thanks to (3.3.31). Also, these functions are continuous in $\overline{\Omega}$ due to (3.3.30). By (3.3.8) and (3.3.38), we infer that $u \geqslant \omega_-$, and as a result,

$$\lim_{\Omega\ni x\to p} u(x) \geqslant \lim_{\Omega\ni x\to p} \omega_-(x) = g(p) - \varepsilon.$$

By taking ε arbitrarily small, we arrive at

$$\lim_{\Omega\ni x\to p} u(x) \geqslant g(p). \tag{3.3.39}$$

Additionally, for every function $v \in C(\Omega)$ which is subharmonic in Ω and satisfies (3.3.9), we have that $v - \omega_+$ is subharmonic in Ω.

Also, we claim that

$$\sup_{\Omega}(v - \omega_+) \leqslant 0. \tag{3.3.40}$$

To prove it, we argue by contradiction and suppose that there exists a sequence of points $p_k \in \Omega$ such that

$$\lim_{k\to+\infty}(v - \omega_+)(p_k) = \sup_{\Omega}(v - \omega_+) \geqslant a > 0.$$

Since Ω is bounded, we have that p_k converges, up to a subsequence, to some $\overline{p} \in \overline{\Omega}$. We observe that $\overline{p} \in \Omega$; otherwise, if $\overline{p} \in \partial\Omega$, in light of (3.3.9) and (3.3.38), we would have

$$0 < a \leqslant \lim_{k\to+\infty}(v - \omega_+)(p_k) \leqslant \limsup_{\Omega\ni x\to\overline{p}}(v - \omega_+)(x) \leqslant g(\overline{p}) - g(\overline{p}) = 0,$$

which is a contradiction. Hence, exploiting Lemma 3.2.5, we conclude that $v - \omega_+$ is constant in Ω and, in particular, $(v - \omega_+)(x) \geqslant a$ for every $x \in \Omega$.

Accordingly, for every $q \in \partial\Omega$,

$$a \leqslant \limsup_{\Omega\ni x\to q}(v - \omega_+)(x) \leqslant g(q) - g(q) = 0,$$

which is a contradiction; therefore, the proof of (3.3.40) is complete.

Therefore, thanks to (3.3.40), we have that $v \leqslant \omega_+$ in Ω. Taking the supremum over such functions v and recalling the definition of u in (3.3.8), we thereby find that $u \leqslant \omega_+$ in Ω.

Accordingly,

$$\lim_{\Omega\ni x\to p} u(x) \leqslant \lim_{\Omega\ni x\to p} \omega_+(x) = g(p) + \varepsilon.$$

By taking ε as small as we like, we conclude that

$$\lim_{\Omega\ni x\to p} u(x) \leqslant g(p).$$

Combining this with (3.3.39), we obtain (3.3.34), as desired.

Now, in light of (3.3.34), in order to complete the proof of Theorem 3.3.3, it suffices to check that the domains in the hypothesis of Theorem 3.3.3 allow the construction of a barrier β, as required in (3.3.34). To this end, we treat separately the cases $n = 2$ and $n \geqslant 3$.

When $n = 2$, let $p \in \Omega$, and suppose that, up to a translation, p is the origin. Since Ω is accessible by simple arcs from the exterior, there exists an injective continuous map $\gamma : [0,1] \to \mathbb{R}^n$ such that $\gamma(t) \in \mathbb{R}^n \setminus \overline{\Omega}$ for all $t \in [0,1)$ and $\gamma(1) = 0$. In particular, we have that $\gamma(0) \neq \gamma(1) = 0$, and hence we can take $\rho \in (0, \min\{1, |\gamma(0)|\})$.

We use the polar representation in complex coordinates $z = re^{i\vartheta}$ for points $z \in \Omega \cap B_\rho$, with $r > 0$ and $\vartheta \in \mathbb{R}$. We note that the curve γ provides a branch cut for the definition of the polar angle ϑ and thus for one of the complex logarithms in $\Omega \cap B_\rho$. Note also that if $z = re^{i\vartheta} \in \Omega \cap B_\rho$, then $|z| < \rho < 1$, whence $\ln z \neq 0$. As a result, we can define the holomorphic function $-\frac{1}{\ln z}$. This gives rise (see e.g. [Rud66, Chapter 11]) to the harmonic (hence superharmonic) function

$$\Omega \cap B_\rho \ni z = re^{i\vartheta} \mapsto \beta(z) := \Re\left(-\frac{1}{\ln z}\right) = -\frac{\ln r}{\ln^2 r + \vartheta^2}.$$

Note that

$$\lim_{z\to 0} \beta(z) = 0 < \frac{-\ln \widetilde{r}}{\ln^2 \widetilde{r} + \widetilde{\vartheta}^2} = \lim_{z\to\widetilde{z}} \beta(z),$$

for every $\Omega \cap B_\rho \ni \widetilde{z} = \widetilde{r}e^{i\widetilde{\vartheta}} \neq 0$. These considerations show that β is an appropriate barrier in the sense of (3.3.34), whence (3.3.28) follows in this case.

Let us now deal with the case of $n \geqslant 3$. For this, given $b \in (0, \pi)$, we consider[9] the cone

$$C_b := \left\{x = (x_1, \dots, x_n) \in \mathbb{R}^n \text{ s.t. } \frac{x_n}{|x|} > \cos b\right\}, \tag{3.3.41}$$

see Figures 3.3.3 and 3.3.4, and we deal with a preliminary computation about the Laplacian of a given function w of the form

$$C_b \ni x \mapsto w(x) = w_0(|x|, \Theta(x)),$$

where $\Theta(x) \in [0, b)$ is such that

$$x_n = |x| \cos(\Theta(x)) \quad \text{and} \quad |x'| = |x| \sin(\Theta(x)). \tag{3.3.42}$$

We pick a point $\overline{x} \neq 0$. Up to a rotation (recall the rotational invariance of the Laplacian according to Corollary 1.1.4), we suppose that $\overline{x}$ lies on the plane spanned by the first and last elements of the Euclidean basis, namely $\overline{x} = (\overline{x}_1, 0, \dots, 0, \overline{x}_n)$. We define $\overline{x}^\perp := (\overline{x}_n, 0, \dots, 0, -\overline{x}_1)$ and take an orthonormal frame $\{\eta_1, \dots, \eta_n\}$ with

$$\eta_1 := \frac{\overline{x}}{|\overline{x}|} \quad \text{and} \quad \eta_2 := \frac{\overline{x}^\perp}{|\overline{x}|}.$$

[9]For the sake of completeness, we point out that another approach to the proof of Theorem 3.3.3 when $n \geqslant 3$ consists of constructing barriers of the form $|x|^\lambda f\left(\frac{x}{|x|}\right)$ for a suitable $\lambda > 0$, with f being the first eigenfunction for the Laplace–Beltrami operator on the sphere with vanishing datum outside a given cone. See also [Hel69, Theorem 8.27 on pp. 173–174] for a different barrier in the exterior cone condition setting.

We also mention that dealing with the exterior cone condition in dimension $n = 2$ is technically easier since one can rely on complex analysis and seek barriers of the form $\Im z^\lambda = r^\lambda \sin(\lambda\vartheta)$ with $\lambda > 0$ large enough.

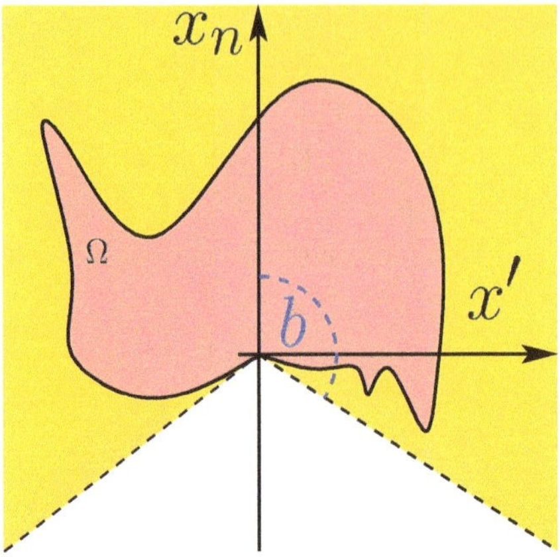

Figure 3.3.3 A domain satisfying the exterior cone condition at the origin.

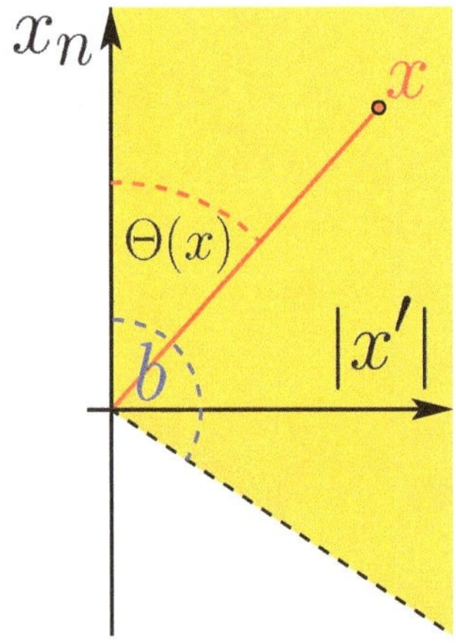

Figure 3.3.4 The cone in (3.3.41).

We remark that if $j \in \{3, \dots, n\}$, then $\overline{x} \cdot \eta_j = |\overline{x}|\eta_1 \cdot \eta_j = 0$, whence, for ε small,

$$|\overline{x} + \varepsilon\eta_j| = \sqrt{|\overline{x}|^2 + 2\varepsilon\overline{x} \cdot \eta_j + \varepsilon^2} = |\overline{x}|\sqrt{1 + \frac{\varepsilon^2}{|\overline{x}|^2}} = |\overline{x}| + \frac{\varepsilon^2}{2|\overline{x}|} + o(\varepsilon^2).$$

Also,

$$\begin{aligned}\overline{x}_n\overline{x} - \overline{x}_1\overline{x}^\perp &= (\overline{x}_1\overline{x}_n, 0, \dots, 0, \overline{x}_n^2) - (\overline{x}_1\overline{x}_n, 0, \dots, 0, -\overline{x}_1^2)\\ &= (0, 0, \dots, 0, |\overline{x}|^2) = |\overline{x}|^2 e_n,\end{aligned}$$

and accordingly

$$e_n = \frac{\overline{x}_n\eta_1 - \overline{x}_1\eta_2}{|\overline{x}|}.$$

This gives that $\eta_j \cdot e_n = 0$ for all $j \in \{3, \dots, n\}$, leading to

$$(\overline{x} + \varepsilon\eta_j) \cdot e_n = \overline{x}_n.$$

As a result,

$$\begin{aligned}
\Theta(\overline{x} + \varepsilon\eta_j) &= \arccos\left(\frac{\overline{x}_n}{|\overline{x}| + \frac{\varepsilon^2}{2|\overline{x}|} + o(\varepsilon^2)}\right) = \arccos\left(\frac{\overline{x}_n}{|\overline{x}|} - \frac{\varepsilon^2\overline{x}_n}{2|\overline{x}|^3} + o(\varepsilon^2)\right) \\
&= \arccos\left(\frac{\overline{x}_n}{|\overline{x}|}\right) + \frac{\varepsilon^2\overline{x}_n}{2|\overline{x}|^2\sqrt{|\overline{x}|^2 - \overline{x}_n^2}} + o(\varepsilon^2) \\
&= \Theta(\overline{x}) + \frac{\varepsilon^2\overline{x}_n}{2|\overline{x}|^2|\overline{x}_1|} + o(\varepsilon^2).
\end{aligned}$$

From these observations we deduce that, for each $j \in \{3, \dots, n\}$,

$$\begin{aligned}
&\partial^2_{\eta_j\eta_j} w(\overline{x}) \\
&= \lim_{\varepsilon\to 0} \frac{w_0(|\overline{x} + \varepsilon\eta_j|, \Theta(\overline{x} + \varepsilon\eta_j)) + w_0(|\overline{x} - \varepsilon\eta_j|, \Theta(\overline{x} - \varepsilon\eta_j)) - 2w_0(|\overline{x}|, \Theta(\overline{x}))}{\varepsilon^2} \\
&= 2\lim_{\varepsilon\to 0} \frac{w_0\left(|\overline{x}| + \frac{\varepsilon^2}{2|\overline{x}|} + o(\varepsilon^2), \Theta(\overline{x}) + \frac{\varepsilon^2\overline{x}_n}{2|\overline{x}|^2|\overline{x}_1|} + o(\varepsilon^2)\right) - w_0(|\overline{x}|, \Theta(\overline{x}))}{\varepsilon^2} \\
&= \frac{\partial_1 w_0(|\overline{x}|, \Theta(\overline{x}))}{|\overline{x}|} + \frac{\overline{x}_n\,\partial_2 w_0(|\overline{x}|, \Theta(\overline{x}))}{|\overline{x}|^2|\overline{x}_1|}.
\end{aligned}$$

Moreover,

$$|\overline{x} + \varepsilon\eta_1| = \left|\frac{|\overline{x}|\overline{x} + \varepsilon\overline{x}}{|\overline{x}|}\right| = |\overline{x}| + \varepsilon.$$

Also, $(\overline{x} + \varepsilon\eta_1) \cdot e_n = \overline{x}_n + \frac{\varepsilon\overline{x}_n}{|\overline{x}|}$, whence

$$\begin{aligned}
\Theta(\overline{x} + \varepsilon\eta_1) &= \arccos\left(\frac{(\overline{x} + \varepsilon\eta_1) \cdot e_n}{|\overline{x} + \varepsilon\eta_1|}\right) \\
&= \arccos\left(\frac{\overline{x}_n + \frac{\varepsilon\overline{x}_n}{|\overline{x}|}}{|\overline{x}| + \varepsilon}\right) = \arccos\left(\frac{\overline{x}_n}{|\overline{x}|}\right) = \Theta(\overline{x}).
\end{aligned}$$

Consequently,

$$\begin{aligned}
&\partial^2_{\eta_1\eta_1} w(\overline{x})\\
&= \lim_{\varepsilon\to 0} \frac{w_0(|\overline{x}+\varepsilon\eta_1|, \Theta(\overline{x}+\varepsilon\eta_1)) + w_0(|\overline{x}-\varepsilon\eta_1|, \Theta(\overline{x}-\varepsilon\eta_1)) - 2w_0(|\overline{x}|, \Theta(\overline{x}))}{\varepsilon^2}\\
&= \lim_{\varepsilon\to 0} \frac{w_0(|\overline{x}|+\varepsilon, \Theta(\overline{x})) + w_0(|\overline{x}|-\varepsilon, \Theta(\overline{x})) - 2w_0(|\overline{x}|, \Theta(\overline{x}))}{\varepsilon^2}\\
&= \partial^2_{11} w_0(|\overline{x}|, \Theta(\overline{x})).
\end{aligned}$$

We also note that

$$\begin{aligned}
|\overline{x}+\varepsilon\eta_2| &= \left|\overline{x}+\frac{\varepsilon\overline{x}^\perp}{|\overline{x}|}\right| = \sqrt{|\overline{x}|^2 + \frac{2\varepsilon\overline{x}\cdot\overline{x}^\perp}{|\overline{x}|} + \frac{\varepsilon^2|\overline{x}^\perp|^2}{|\overline{x}|^2}}\\
&= \sqrt{|\overline{x}|^2+\varepsilon^2} = |\overline{x}| + \frac{\varepsilon^2}{2|\overline{x}|} + o(\varepsilon^2).
\end{aligned}$$

Furthermore, $(\overline{x}+\varepsilon\eta_2)\cdot e_n = \overline{x}_n - \frac{\varepsilon\overline{x}_1}{|\overline{x}|}$, whence

$$\begin{aligned}
\Theta(\overline{x}+\varepsilon\eta_2) &= \arccos\left(\frac{\overline{x}_n - \frac{\varepsilon\overline{x}_1}{|\overline{x}|}}{|\overline{x}| + \frac{\varepsilon^2}{2|\overline{x}|} + o(\varepsilon^2)}\right)\\
&= \arccos\left(\left(\frac{\overline{x}_n}{|\overline{x}|} - \frac{\varepsilon\overline{x}_1}{|\overline{x}|^2}\right)\left(1 - \frac{\varepsilon^2}{2|\overline{x}|^2} + o(\varepsilon^2)\right)\right)\\
&= \arccos\left(\frac{\overline{x}_n}{|\overline{x}|} - \frac{\varepsilon\overline{x}_1}{|\overline{x}|^2} - \frac{\varepsilon^2\overline{x}_n}{2|\overline{x}|^3} + o(\varepsilon^2)\right)\\
&= \Theta(\overline{x}) + \frac{\varepsilon\overline{x}_1}{|\overline{x}||\overline{x}_1|} + \frac{\varepsilon^2\overline{x}_n}{2|\overline{x}|^2|\overline{x}_1|^2} + o(\varepsilon^2).
\end{aligned}$$

This yields that

$$\begin{aligned}
&\partial^2_{\eta_2\eta_2} w(\overline{x})\\
&= \lim_{\varepsilon\to 0} \frac{w_0(|\overline{x}+\varepsilon\eta_2|, \Theta(\overline{x}+\varepsilon\eta_2)) + w_0(|\overline{x}-\varepsilon\eta_2|, \Theta(\overline{x}-\varepsilon\eta_2)) - 2w_0(|\overline{x}|, \Theta(\overline{x}))}{\varepsilon^2}\\
&= \lim_{\varepsilon\to 0} \frac{1}{\varepsilon^2}\left(w_0\left(|\overline{x}| + \frac{\varepsilon^2}{2|\overline{x}|} + o(\varepsilon^2), \Theta(\overline{x}) + \frac{\varepsilon\overline{x}_1}{|\overline{x}||\overline{x}_1|} + \frac{\varepsilon^2\overline{x}_n}{2|\overline{x}|^2|\overline{x}_1|^2} + o(\varepsilon^2)\right)\right.\\
&\quad + w_0\left(|\overline{x}| + \frac{\varepsilon^2}{2|\overline{x}|} + o(\varepsilon^2), \Theta(\overline{x}) - \frac{\varepsilon\overline{x}_1}{|\overline{x}||\overline{x}_1|} + \frac{\varepsilon^2\overline{x}_n}{2|\overline{x}|^2|\overline{x}_1|^2} + o(\varepsilon^2)\right)\\
&\quad \left. - 2w_0(|\overline{x}|, \Theta(\overline{x}))\right)\\
&= \frac{\partial_1 w_0(\overline{x}, \Theta(\overline{x}))}{|\overline{x}|} + \frac{\partial_2 w_0(\overline{x}, \Theta(\overline{x}))\,\overline{x}_n}{|\overline{x}|^2|\overline{x}_1|} + \frac{\partial^2_{22} w_0(\overline{x}, \Theta(\overline{x}))}{|\overline{x}|^2}.
\end{aligned}$$

Gathering these observations, we conclude that

$$\begin{aligned}\Delta w(\overline{x}) &= \sum_{j=1}^{n} \partial^2_{\eta_j\eta_j} w(\overline{x})\\ &= \partial^2_{11} w_0(\overline{x}, \Theta(\overline{x})) + \frac{\partial^2_{22} w_0(\overline{x}, \Theta(\overline{x}))}{|\overline{x}|^2}\\ &\quad + \sum_{j=2}^{n} \left[\frac{\partial_1 w_0(|\overline{x}|, \Theta(\overline{x}))}{|\overline{x}|} + \frac{\overline{x}_n\, \partial_2 w_0(|\overline{x}|, \Theta(\overline{x}))}{|\overline{x}|^2 |\overline{x}_1|} \right]\\ &= \partial^2_{11} w_0(\overline{x}, \Theta(\overline{x})) + \frac{\partial^2_{22} w_0(\overline{x}, \Theta(\overline{x}))}{|\overline{x}|^2}\\ &\quad + (n-1) \left[\frac{\partial_1 w_0(|\overline{x}|, \Theta(\overline{x}))}{|\overline{x}|} + \frac{\overline{x}_n\, \partial_2 w_0(|\overline{x}|, \Theta(\overline{x}))}{|\overline{x}|^2 |\overline{x}_1|} \right].\end{aligned}$$

In particular, if $w(x) = |x|^\lambda f(\Theta(x))$ for some $\lambda \in \mathbb{R}$ (i.e. $w_0(r, \theta) = r^\lambda f(\theta)$), we get that

$$\begin{aligned}\Delta w(\overline{x}) &= \lambda(\lambda-1)|\overline{x}|^{\lambda-2} f(\Theta(\overline{x})) + |\overline{x}|^{\lambda-2} f''(\Theta(\overline{x}))\\ &\quad + (n-1) \left[\lambda |\overline{x}|^{\lambda-2} f(\Theta(\overline{x})) + \frac{\overline{x}_n\, |\overline{x}|^{\lambda-2} f'(\Theta(\overline{x}))}{|\overline{x}_1|} \right].\end{aligned}$$

From this and (3.3.42), we obtain that

$$\begin{aligned}&|\overline{x}|^{2-\lambda} \Delta w(\overline{x})\\ &\quad = \lambda(\lambda+n-2) f(\Theta(\overline{x})) + f''(\Theta(\overline{x})) + (n-1)\cot(\Theta(\overline{x})) f'(\Theta(\overline{x})).\end{aligned} \tag{3.3.43}$$

We remark that $\cot(\Theta(\overline{x})) > \cot b \geqslant -|\cot b|$ for all $\overline{x} \in C_b$.

Now, we set $K := 1 + 2(n-1)|\cot b|$, and we consider a small parameter $\mu \in (0, 1)$. We choose

$$f(\theta) := 1 - \frac{2\mu}{K^2}\left(e^{K\theta} + 1\right) + \frac{2\mu\theta}{K}$$

and observe that, for all $\theta \in [0, \pi)$,

$$|f(\theta)| \leqslant 2, \qquad f'(\theta) = \frac{2\mu}{K}\left(1 - e^{K\theta}\right) \quad \text{and} \quad f''(\theta) = -2\mu e^{K\theta},$$

provided that μ is sufficiently small.

Plugging this information into (3.3.43) and assuming $\lambda \in (0, \mu^2]$ and μ small enough, we obtain that

$$\begin{aligned}|\overline{x}|^{2-\lambda} \Delta w(\overline{x}) &\leqslant 2\lambda n - 2\mu e^{K\Theta(\overline{x})} + \frac{2\mu(n-1)\cot(\Theta(\overline{x}))}{K}\left(1 - e^{K\Theta(\overline{x})}\right)\\ &\leqslant 2\mu^2 n - 2\mu e^{K\Theta(\overline{x})} + \frac{2\mu(n-1)\,|\cot b|}{K}\left(e^{K\Theta(\overline{x})} - 1\right)\end{aligned}$$

$$\begin{aligned}&\leqslant 2\mu^2 n - 2\mu e^{K\Theta(\overline{x})} + \mu\big(e^{K\Theta(\overline{x})} - 1\big)\\&= 2\mu^2 n - \mu e^{K\Theta(\overline{x})} - \mu\\&\leqslant -\mu e^{K\Theta(\overline{x})},\end{aligned}$$

which is negative.

Also, $w > 0$ in $\overline{C_b} \setminus \{0\}$ since

$$f(\theta) \geqslant 1 - 2\mu\left(e^{K\pi} + 1\right) - 2\mu\pi \geqslant \frac{1}{2}$$

as long as μ is small enough. This gives that w satisfies the requirements in (3.3.34), as desired. □

From Theorems 3.3.2 and 3.3.3, we deduce the following[10] existence result for the Dirichlet problem in sufficiently regular domains.

Corollary 3.3.4. *Let $\Omega \subset \mathbb{R}^n$ be open, bounded and connected. Assume either that $n = 2$ and Ω is accessible by arcs from the exterior or that $n \geqslant 3$ and Ω satisfies the exterior cone condition.*

Let $f \in C^1(\Omega')$ for some open set $\Omega' \Supset \Omega$ and $g \in C(\partial\Omega)$.

Then, there exists a unique solution $u \in C^2(\Omega)\cap C(\overline{\Omega})$ to the Dirichlet problem

$$\begin{cases}\Delta u = f & \text{in } \Omega,\\ u = g & \text{on } \partial\Omega.\end{cases}$$

Proof. The uniqueness claim is a consequence of Corollary 2.9.3.

As for the existence claim, without loss of generality, we can suppose that Ω' is bounded, and we consider another bounded open set Ω'' such that $\Omega \Subset \Omega'' \Subset \Omega'$. We take $\tau \in C_0^\infty(\Omega')$ with $\tau = 1$ in Ω'' and extend f to the whole of $\mathbb{R}^n$ by setting

$$\widetilde{f}(x) := \begin{cases}\tau(x)\, f(x) & \text{if } x \in \Omega',\\ 0 & \text{if } x \in \mathbb{R}^n \setminus \Omega'.\end{cases}$$

We apply Proposition 2.7.4 (with f there replaced by $\widetilde{f}$ here), and we construct a function $v \in C^2(\mathbb{R}^n)$ such that $\Delta v = \widetilde{f}$ in $\mathbb{R}^n$. In particular, we have that $\Delta v = f$ in Ω.

Now, for each $x \in \partial\Omega$, we set $\widetilde{g}(x) := g(x) - v(x)$. We remark that $\widetilde{g} \in C(\partial\Omega)$, and hence we can employ Theorems 3.3.2 and 3.3.3 and find a function $\widetilde{u} \in C(\overline{\Omega})$ which is harmonic in Ω and equal to $\widetilde{g}$ on $\partial\Omega$. Thus, we set $u := v + \widetilde{u}$. We see that $u \in C^2(\Omega) \cap C(\overline{\Omega})$ since both v and $\widetilde{u}$ belong to these spaces, that $\Delta u = \Delta v + \Delta\widetilde{u} = f$ in Ω, and that $u = v + \widetilde{g} = g$ along $\partial\Omega$. □

[10] A more precise version of Corollary 3.3.4 follows from the Schauder estimates in Proposition 4.2.3.

Concerning the result in Corollary 3.3.4, we stress that some geometric conditions on the domain are needed to ensure the continuity of the solution at the boundary. Indeed, as an example, we remark that

$$\begin{aligned}&\text{there exists no harmonic function } u \text{ in } B_1 \setminus \{0\}\\&\text{that is continuous in } \overline{B_1 \setminus \{0\}} = \overline{B_1}\\&\text{and such that } u = 0 \text{ on } \partial B_1 \text{ and } u(0) = 1;\end{aligned} \tag{3.3.44}$$

otherwise, by Theorem 2.12.1, such a function could be extended to a harmonic function in B_1, which should necessarily vanish identically in view of the maximum principle in Corollary 2.9.2.

For completeness, we point out that another proof of (3.3.44) can be obtained by exploiting spherical averages and the classification result in Lemma 2.7.1. Indeed, given $r \in (0, 1)$ and $\vartheta \in \partial B_1$, let $u_0(r, \vartheta) := u(r\vartheta)$. For all $x \in \overline{B_1}$, let also

$$v(x) := \fint_{\partial B_1} u(|x|\vartheta)\, d\mathcal{H}^{n-1}_\vartheta = \fint_{\partial B_1} u_0(|x|, \vartheta)\, d\mathcal{H}^{n-1}_\vartheta.$$

By the spherical representation of the Laplacian in Theorem 2.5.2, we know that, for all $x \in B_1 \setminus \{0\}$,

$$\begin{aligned}\Delta v(x) &= \fint_{\partial B_1} \left(\partial_{rr} u_0(|x|, \vartheta) + \frac{n-1}{|x|}\, \partial_r u_0(|x|, \vartheta) + \frac{1}{|x|^2}\, \Delta_{\partial B_1} u_0(|x|, \vartheta) \right) d\mathcal{H}^{n-1}_\vartheta\\&= \fint_{\partial B_1} \Delta u(|x|\vartheta)\, d\mathcal{H}^{n-1}_\vartheta\end{aligned} \tag{3.3.45}$$

and accordingly $\Delta v(x) = 0$. Thus, since v is rotationally symmetric, we can apply Lemma 2.7.1 and gather that, for every $x \in B_1 \setminus \{0\}$,

$$v(x) = \begin{cases} \dfrac{a}{|x|^{n-2}} + b & \text{if } n \neq 2,\\ a \ln |x| + b & \text{if } n = 2 \end{cases} \tag{3.3.46}$$

for some $a, b \in \mathbb{R}$. Now, if u were continuously attaining the datum 1 at the origin, for every $\varepsilon > 0$, there would exist $\delta > 0$ such that if $x \in B_\delta$, then $u(x) \in [1 - \varepsilon, 1 + \varepsilon]$, and hence $v(x) \in [1 - \varepsilon, 1 + \varepsilon]$ too, namely

$$\lim_{x \to 0} v(x) = 1.$$

This and (3.3.46) yield that necessarily $a = 0$ and also $b = 1$. As a result, we have that v is identically equal to 1 in B_1. On the other hand, if u were continuous in $\overline{B}_1$ and vanishing along ∂B_1, there would exist $\rho \in (0,1)$ such that $u(x) \leqslant \frac{1}{2}$ for all $|x| \in [\rho, 1]$, leading to

$$1 = v(\rho e_n) = \fint_{\partial B_1} u(\rho\vartheta)\, d\mathcal{H}^{n-1}_{\vartheta} \leqslant \frac{1}{2}.$$

This contradiction provides an alternative proof of (3.3.44).

Note that (3.3.44) highlights that the assumption of being accessible by arcs from the exterior for a planar domain in Corollary 3.3.4 cannot be removed. It is interesting to ponder over the physical counterpart of this example: one might translate this construction into the phenomenon of a perfectly elastic planar membrane that is constrained along a circle and with a sharp needle placed at the center of the circle. The mathematical example suggests that the needle would break the membrane and pass through it.

In the same spirit, we stress that when $n \geqslant 3$, classical examples due to Henri Lebesgue (and sometimes referred to by the name "Lebesgue spine", see e.g. [CH89, p. 304, figure 19], [DiB95, Section 7.2] and [Hel69, pp. 175–176]) exhibit cuspidal domains in $\mathbb{R}^3$ with a prescribed continuous datum at the boundary in which the solution cannot meet the boundary datum in a continuous way. Let us briefly recall here one of these examples. The main idea underpinning this example is to look at the electrostatic potential of a charged segment with a suitable choice of charge distribution. Then, a cuspidal domain with this segment in its complement will do the job. To make this idea concrete, we observe that, given $a, b \in \mathbb{R}$ with $a^2 + (b+s)^2 > 0$, if

$$\begin{aligned}\Phi(s,a,b) &:= \sqrt{a^2+(b+s)^2} + \frac{b}{2}\ln\left(1 - \frac{b+s}{\sqrt{a^2+(b+s)^2}}\right)\\ &\quad - \frac{b}{2}\ln\left(1 + \frac{b+s}{\sqrt{a^2+(b+s)^2}}\right),\end{aligned}$$

then

$$\frac{\partial\Phi}{\partial s}(s,a,b) = \frac{s}{\sqrt{a^2+(s+b)^2}}.$$

Therefore, if we define $r := \sqrt{x_1^2 + x_2^2}$ and

$$\begin{aligned}v(x) = v(x_1,\dots,x_n) &:= \int_0^1 \frac{s}{|(x_1,x_2,x_3)-(0,0,s)|}\,ds\\ &= \int_0^1 \frac{s}{\sqrt{r^2+(s-x_3)^2}}\,ds,\end{aligned} \tag{3.3.47}$$

we have that

$$\begin{aligned}
v(x) &= \int_0^1 \frac{\partial \Phi}{\partial s}(s, r, -x_3)\, ds \\
&= \Phi(1, r, -x_3) - \Phi(0, r, -x_3) \\
&= \sqrt{r^2 + (1 - x_3)^2} - \frac{x_3}{2} \ln\left(1 - \frac{1 - x_3}{\sqrt{r^2 + (1 - x_3)^2}}\right) \\
&\quad + \frac{x_3}{2} \ln\left(1 + \frac{1 - x_3}{\sqrt{r^2 + (1 - x_3)^2}}\right) \\
&\quad - \sqrt{r^2 + x_3^2} + \frac{x_3}{2} \ln\left(1 + \frac{x_3}{\sqrt{r^2 + x_3^2}}\right) - \frac{x_3}{2} \ln\left(1 + - \frac{x_3}{\sqrt{r^2 + x_3^2}}\right) \\
&= \sqrt{r^2 + (1 - x_3)^2} - \sqrt{r^2 + x_3^2} \\
&\quad + \frac{x_3}{2} \ln \frac{\left(\sqrt{r^2 + (1 - x_3)^2} + 1 - x_3\right)\left(\sqrt{r^2 + x_3^2} + x_3\right)}{\left(\sqrt{r^2 + (1 - x_3)^2} - 1 + x_3\right)\left(\sqrt{r^2 + x_3^2} - x_3\right)}.
\end{aligned} \tag{3.3.48}$$

See Figure 3.3.5 for the graph of v and of its level sets as a function of (r, x_3).

Now, we consider a bounded domain Ω whose boundary is C^∞ outside the origin and such that

$$\Omega \cap B_{1/2} = \left\{ x = (x_1, \ldots, x_n) \in \mathbb{R}^n \text{ s.t. } x_3 < -\frac{1}{2 \ln |(x_1, x_2)|} \right\}. \tag{3.3.49}$$

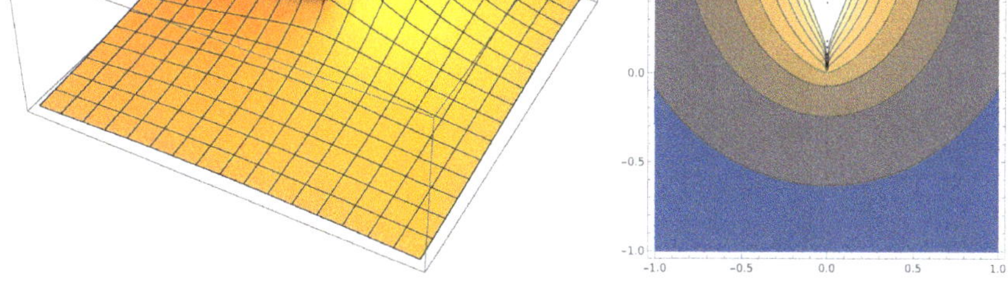

Figure 3.3.5 The graphs of v in (3.3.48) and its level sets as a function of (r, x_3).

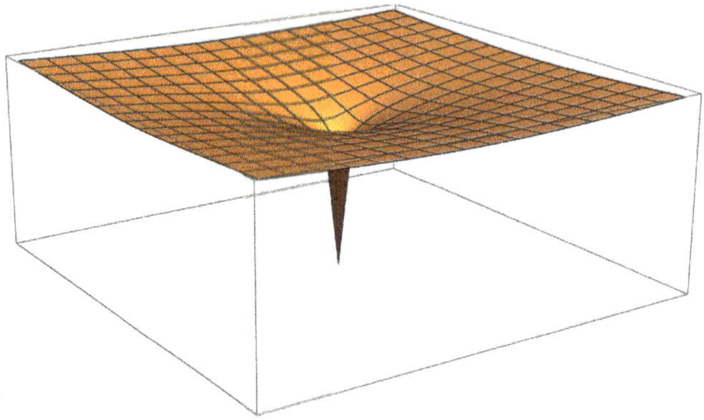

Figure 3.3.6 The boundary of the domain in (3.3.49) near the origin (when $n = 3$).

See Figure 3.3.6 for a sketch of the cuspidal boundary of this domain in the proximity of the origin.

We point out that v is continuous on $\partial\Omega$ because, by (3.3.48),

$$
\begin{aligned}
&\lim_{\substack{x\to 0\\ x_3=-1/(2\ln|(x_1,x_2)|)}} v(x)\\
&= 1-\lim_{r\to 0}\frac{1}{4\ln r}\ln\frac{\left(\sqrt{r^2+\left(1+\frac{1}{2\ln r}\right)^2}+1+\frac{1}{2\ln r}\right)\left(\sqrt{r^2+\frac{1}{4\ln^2 r}}-\frac{1}{2\ln r}\right)}{\left(\sqrt{r^2+\left(1+\frac{1}{2\ln r}\right)^2}-1-\frac{1}{2\ln r}\right)\left(\sqrt{r^2+\frac{1}{4\ln^2 r}}+\frac{1}{2\ln r}\right)}\\
&= 1-\lim_{r\to 0}\frac{1}{4\ln r}\ln\frac{\left(\sqrt{r^2+\left(1+\frac{1}{2\ln r}\right)^2}+1+\frac{1}{2\ln r}\right)^2\left(\sqrt{r^2+\frac{1}{4\ln^2 r}}-\frac{1}{2\ln r}\right)^2}{r^4}\\
&= 1-\lim_{r\to 0}\frac{1}{4\ln r}\ln\frac{\left(2+O\left(\frac{1}{\ln r}\right)\right)^2\frac{1}{4\ln^2 r}\left(\sqrt{4r^2\ln^2 r+1}+1\right)^2}{r^4}\\
&= 1-\lim_{r\to 0}\frac{1}{4\ln r}\ln\frac{\left(2+O\left(\frac{1}{\ln r}\right)\right)^2\frac{1}{4\ln^2 r}\left(2+O(r^2\ln^2 r)\right)^2}{r^4}\\
&= 1-\lim_{r\to 0}\frac{1}{4\ln r}\left\{\ln\left[\left(2+O\left(\frac{1}{\ln r}\right)\right)^2\frac{1}{4\ln^2 r}\left(2+O(r^2\ln^2 r)\right)^2\right]-\ln r^4\right\}\\
&= 2-\lim_{r\to 0}\frac{1}{4\ln r}\ln\left[\left(2+O\left(\frac{1}{\ln r}\right)\right)^2\frac{1}{4\ln^2 r}\left(2+O(r^2\ln^2 r)\right)^2\right]
\end{aligned}
$$

$$= 2 - \lim_{r\to 0} \frac{1}{4\ln r} \ln \frac{4+o(1)}{4\ln^2 r}$$
$$= 2 + \lim_{r\to 0} \frac{\ln(4\ln^2 r)}{4\ln r}$$
$$= 2. \tag{3.3.50}$$

Now, let $I := \{te_n, t \in [0,1]\}$. By (3.3.47), noting that v depends only on $X := (x_1, x_2, x_3)$ and recognizing the fundamental solution in (2.7.6) in dimension $n = 3$, we see that $v \in C^\infty(\mathbb{R}^n \setminus I) \subset C^\infty(\Omega)$ and, for every $x \in \mathbb{R}^n \setminus I$,

$$\Delta v(x) = \Delta_X v(x) = \int_0^1 \Delta_X \frac{s}{|X - (0,0,s)|}\, ds = 0.$$

These observations and the uniqueness result in Corollary 2.9.3 give that v is the unique harmonic function in Ω coinciding with its own boundary values along $\partial\Omega$.

However, recalling (3.3.47), if $x_3 < 0$, then

$$v(x) = \int_0^1 \frac{s}{\sqrt{r^2 + (s + |x_3|)^2}}\, ds \leqslant \int_0^1 \frac{s}{\sqrt{s^2}}\, ds = 1;$$

hence, comparing with (3.3.50), we see that v is not continuous at the origin. This example shows that the exterior cone condition in Corollary 3.3.4 cannot be removed.

A natural question arising from Corollary 3.3.4 is whether one can characterize precisely the domains for which a solution of the Dirichlet problem that is continuous up to the boundary always exists. We address this question using a criterion proposed by Norbert Wiener in the forthcoming Theorem 6.2.1.

As a technical remark, it is interesting to observe that the existence results obtained by the Perron method are conceptually different from the ones obtained by energy or variational methods (compare e.g. Sections 2.2.5(b) and 6.2 in [Eva98]). In particular, solutions constructed via the Perron method do not necessarily possess finite energy. To highlight this phenomenon, we recall a classical example due to Jacques Hadamard [Had06] (see also [MS98] for a comprehensive historical perspective). For this, we use polar coordinates (r, ϑ) in the plane and consider the function on $B_1 \subset \mathbb{R}^2$ defined by

$$u(r,\vartheta) := \sum_{k=1}^{+\infty} \frac{r^{k!}}{k^2} \sin(k!\vartheta).$$

Recalling the representation of the Laplacian in polar coordinates discussed in Theorem 2.5.2, we know that, for every $k \in \mathbb{N}$,

$$\Delta\Big(r^{k!}\sin(k!\vartheta)\Big) = \Big(\partial_{rr} + \frac{1}{r}\partial_r + \frac{1}{r^2}\partial_{\vartheta\vartheta}\Big)\Big(r^{k!}\sin(k!\vartheta)\Big)$$
$$= \Big(k!(k!-1) + k! - (k!)^2\Big) r^{k!-2} \sin(k!\vartheta) = 0,$$

and consequently the function

$$u_N(r,\vartheta) := \sum_{k=1}^{N} \frac{r^{k!}}{k^2} \sin(k!\vartheta)$$

is harmonic in B_1 for all $N \in \mathbb{N} \cap [1,+\infty)$.

Moreover,

$$|u(r,\vartheta) - u_N(r,\vartheta)| \leqslant \sum_{k=N+1}^{+\infty} \frac{r^{k!}}{k^2} \leqslant \sum_{k=N+1}^{+\infty} \frac{1}{k^2},$$

which is infinitesimal as $N \to +\infty$. This shows that u_N converges uniformly to u in $\overline{B_1}$; therefore, u is continuous in $\overline{B_1}$ and harmonic in B_1 (recall Corollary 2.2.2, and in fact it is the only function with these properties due to the uniqueness result in Corollary 2.9.3).

But u has infinite energy, that is,

$$\int_{B_1} |\nabla u|^2 = +\infty. \tag{3.3.51}$$

To check this, we argue by contradiction, assuming the converse, say $\int_{B_1} |\nabla u|^2 \leqslant M$ for some $M \in [0,+\infty)$. Then, since $|\nabla u| \geqslant \frac{1}{r}|\partial_\vartheta u|$,

$$M \geqslant \iint_{(0,1)\times(0,2\pi)} \frac{1}{r}|\partial_\vartheta u(r,\vartheta)|^2 \, dr \, d\vartheta$$

and thus, a.e. $r \in (0,1)$,

$$\int_{(0,2\pi)} |\partial_\vartheta u(r,\vartheta)|^2 \, d\vartheta < +\infty.$$

This states that, a.e. $r \in (0,1)$, the map $(0,2\pi) \ni \vartheta \mapsto f(\vartheta) := \partial_\vartheta u(r,\vartheta)$ belongs to $L^2((0,2\pi))$, and we can therefore consider the corresponding Fourier coefficients (up to normalization)

$$f_\ell := \int_{(0,2\pi)} f(\vartheta)\, e^{-i\vartheta\ell} \, d\vartheta$$

and according to the Plancherel theorem (see e.g. [WZ15]), we find that

$$\sum_{\ell\in\mathbb{Z}} |f_\ell|^2 < +\infty. \tag{3.3.52}$$

Note that, via an integration by parts,

$$\begin{aligned} f_\ell &= \int_{(0,2\pi)} \partial_\vartheta u(r,\vartheta)\, e^{-i\vartheta\ell} \, d\vartheta \\ &= i\ell \int_{(0,2\pi)} u(r,\vartheta)\, e^{-i\vartheta\ell} \, d\vartheta \end{aligned}$$

$$
\begin{aligned}
&= i\ell \lim_{N\to+\infty} \int_{(0,2\pi)} u_N(r,\vartheta)\, e^{-i\vartheta\ell}\, d\vartheta \\
&= i\ell \lim_{N\to+\infty} \sum_{k=1}^{N} \int_{(0,2\pi)} \frac{r^{k!}}{k^2} \sin(k!\vartheta)\, e^{-i\vartheta\ell}\, d\vartheta \\
&= \pi \lim_{N\to+\infty} \sum_{k=1}^{N} \frac{\ell\, \delta_{\ell,k!}\, r^{k!}}{k^2}.
\end{aligned}
$$

In particular, for every $m \in \mathbb{N}$,

$$
|f_{m!}| = \left| \pi \lim_{N\to+\infty} \sum_{k=1}^{N} \frac{m!\, \delta_{m!,k!}\, r^{k!}}{k^2} \right| = \pi \frac{m!\, r^{m!}}{m^2},
$$

which entails that

$$
\sum_{\ell\in\mathbb{Z}} |f_\ell|^2 \geqslant \sum_{m\in\mathbb{N}} |f_{m!}|^2 = +\infty.
$$

This is in contradiction with (3.3.52), and the proof of (3.3.51) is thus complete.

For a thorough comparison between energy techniques and the Perron method, see [AD08].

For additional observations on the Perron method and its relation with the "obstacle problem," see for example [Kel67, Hel69, GT01, Pon16]. A different method for solving the Dirichlet problem was also proposed by Poincaré, which relied on subsequent harmonic replacements in chains of balls covering the domain; see e.g. [GM12, Section 1.4.3] for further details on this technique.

Chapter 4

Equations in Nondivergence Form: $C^{2,\alpha}$-Regularity Theory

4.1 Hints and Limitations for a Smooth Regularity Theory

We discuss here some bits of the regularity theory for elliptic equations (for a thorough presentation, see e.g. [Eva98, GT01, HL11] and the references therein).

As a first observation, we remark that the situation for partial differential equations is structurally very different from that of ordinary differential equations. Indeed, for (say, linear) ordinary differential equations, the regularity theory is mostly straightforward: for instance, if I is an interval in $\mathbb{R}$ and $u : I \to \mathbb{R}$ solves $\ddot{u}(t) = f(t)$ for all $t \in I$, then "u is two derivatives better than f," as can be seen by integrating twice the equation.

For partial differential equations, the situation can be significantly more complicated. For general equations, one cannot expect any kind of regularity at all: for example, given any function $u_0 : \mathbb{R} \to \mathbb{R}$, the function $u(x_1, x_2) := u_0(x_1 - x_2)$ is a solution of

$$\partial_{11}u - \partial_{22}u = 0. \tag{4.1.1}$$

While in principle to write such an equation, one needs to assume that u_0 is twice differentiable, no quantitative estimate on the second derivative of u_0 comes into play; therefore, smooth solutions of (4.1.1) can exhibit arbitrarily large derivatives. This suggests that, in suitable jargon, equation (4.1.1) can be interpreted in a "weak form" which does not require the solution u to possess basically any regularity at all, and no regularity can be inferred for this type of weak solution of (4.1.1).

The situation for the Laplace operator is, on the other hand, radically different. For instance, the weak solutions of $\Delta u = 0$, as discussed in Section 2.2, happen to be as smooth as we wish. This suggests that the Laplace operator possesses some kind of "regularizing effect." Indeed, even if the "algebraic difference" between the Laplacian and the operator in (4.1.1) seems negligible (just a minus sign!), the "geometric structure" underpinning the Laplace operator tends to "average out" differences and oscillations (recall for example Theorem 1.1.1 to appreciate how the Laplacian tends to revert the pointwise values of a function to the average nearby).

In dimension 2, another strong indication that solutions of equations driven by the Laplace operator enjoy additional rigidity and regularity properties comes from the fact that harmonic functions in planar domains can be seen as real (or imaginary) parts of holomorphic functions in complex domains (with the identification of $(x, y) \in \mathbb{R}^2$ with $x + iy \in \mathbb{C}$): with respect to this, we recall that the existence of a complex derivative in a neighborhood entails that the complex function is actually infinitely differentiable and, in fact, real analytic.

For global solutions (i.e. solutions in the whole of $\mathbb{R}^n$), another quantitative hint of the regularity enjoyed by solutions of $\Delta u(x) = f(x)$, for all $x \in \mathbb{R}^n$ and for f in the Schwartz space of smooth functions, whose derivatives are rapidly decreasing at infinity comes from Fourier analysis: indeed, after a Fourier transform, the equation becomes $-4\pi^2|\xi|^2\widehat{u}(\xi) = \widehat{f}(\xi)$. Thus, recalling that, for each $j, m \in \{1, \dots, n\}, \mathcal{F}\left(\partial^2_{jm}u\right)(\xi) = -4\pi^2\xi_j\xi_m\widehat{u}(\xi)$, it follows from the Plancherel theorem that

$$
\begin{aligned}
\|D^2u\|_{L^2(\mathbb{R}^n)} &= \sqrt{\sum_{j,m=1}^{n}\int_{\mathbb{R}^n}(\partial^2_{jm}u(x))^2\,dx} = \sqrt{\sum_{j,m=1}^{n}\int_{\mathbb{R}^n}\left|\mathcal{F}\left(\partial^2_{jm}u\right)(\xi)\right|^2 d\xi} \\
&= 4\pi^2\sqrt{\sum_{j,m=1}^{n}\int_{\mathbb{R}^n}|\xi_j|^2|\xi_m|^2|\widehat{u}(\xi)|^2\,d\xi} \\
&\leqslant 2\sqrt{2}\pi^2\sqrt{\sum_{j,m=1}^{n}\int_{\mathbb{R}^n}(|\xi_j|^4+|\xi_m|^4)|\widehat{u}(\xi)|^2\,d\xi} \\
&= 4\pi^2\sqrt{\sum_{j,m=1}^{n}\int_{\mathbb{R}^n}|\xi_j|^4|\widehat{u}(\xi)|^2\,d\xi} \leqslant 4\pi^2 n\sqrt{\int_{\mathbb{R}^n}|\xi|^4|\widehat{u}(\xi)|^2\,d\xi} \\
&= n\sqrt{\int_{\mathbb{R}^n}|\widehat{f}(\xi)|^2\,d\xi} = n\sqrt{\int_{\mathbb{R}^n}|f(x)|^2\,dx} = n\,\|f\|_{L^2(\mathbb{R}^n)},
\end{aligned}
$$

which can be seen as a global bound in $L^2(\mathbb{R}^n)$ for the second derivative of the solution u with respect to the bound $L^2(\mathbb{R}^n)$ of the source term f.

While these heuristic (or very specific) considerations suggest that solutions of Laplace-like (in jargon, elliptic) equations should possess additional regularity, detecting this regularity is typically an extremely hard task, it relies on a number of technical (albeit beautiful) arguments and, in general, it is very difficult to "guess the right results" only based[1] on a "physical intuition of what's going on." As an example (see e.g. [GT01, Problem 4.9] and [HL11, Remark 3.14]), we remark that the solutions of $\Delta u(x) = f(x)$ for a continuous function f do not need[2] to be continuously twice differentiable (and hence u is not "always two derivatives better than f"!).

Theorem 4.1.1. *There exists $f \in C(\overline{B_1})$ for which the Dirichlet problem*

$$\begin{cases} \Delta u = f & \text{in } B_1, \\ u = 0 & \text{on } \partial B_1 \end{cases} \tag{4.1.2}$$

does not admit any solution $u \in C^2(B_1) \cap C(\overline{B_1})$.

[1]For instance, the example pointed out in Theorem 4.1.1 reveals that the classical notion of continuity alone is not the appropriate one to deal effectively with partial differential equations (even in the more "amenable" case of the Laplace operator): in simple terms, differently from what happens for ordinary differential equations, derivatives in different directions may end up being singular, but their singularities cancel out after summation. However, this in itself does not fully explain the complexity of the situation since cancellations of singularities are not allowed when the source term is zero (since harmonic functions, no matter how weak their definition is, end up being real analytic; recall Theorems 2.1.2 and 2.14.1, as well as Lemma 2.2.1).

A fact that will be apparent from the forthcoming pages is that the appropriate functional spaces to account for the continuous regularity theory of equations driven by the Laplacian, and by more general elliptic operators, are the Hölder spaces $C^{k,\alpha}$ with $\alpha \in (0,1)$. In this sense, one needs to not only quantify precisely the notion of continuity but also take into account a sort of "fractional" regularity encoded in the power α (for harmonic functions, or for solutions of equations with a source term of C^∞, one does not really face this issue since the equation can be differentiated infinitely many times, making any fractional-type regularity incorporated into the subsequent continuous space with integer regularity: say, when the source term can be differentiated infinitely many times, the $C^{k,\alpha}$ structure ends up being merged into the stronger C^{k+1} framework, making the fractional exponent scenario much less visible).

The appearance and significance of this intrinsically fractional structure within the framework of classical partial differential equations are perhaps surprising and even mysterious. To be honest, we do not have a compelling and exhaustive explanation on why "intuitively" the appearance of fractional-type regularity should have been expected just from our "basic perception of the universe," i.e. without relying on the technical mathematical aspects of the equations. However, it may be suggestive to think that the universe is perhaps much more fractal and fractional than what it appears to the naked eye. Many fundamental patterns of the world arise from fractional equations (see e.g. the introduction of [CDV19]), and the Laplace operator is, after all, one operator of a broader family of objects which share a number of geometric properties. In this sense, the classical Laplacian can be seen as a limit of fractional operators for which regularity in fractional spaces appears naturally. By accepting this kind of reasoning, one can suggestively think that the fractional regularity in the Hölder spaces for the Laplacian is simply the "vestigial feature" of this limit process.

[2]For further reference, we point out that the examples provided in the two proofs of Theorem 4.1.1 that we give here also show that there exist $f \in L^\infty(B_1 \setminus \{0\})$ and $u \in C^2(B_1 \setminus \{0\})$ for which $\Delta u = f$ in $B_1 \setminus \{0\}$ but $D^2 u \notin L^\infty(B_1 \setminus \{0\})$.

The analogue of this observation in $L^1(B_1 \setminus \{0\})$ is discussed in the forthcoming Theorem 5.1.1.

Proof. Up to a scaling argument, we prove the desired result for $B_{1/2}$ instead of B_1. We let

$$f(x) := \frac{(x_1^2 - x_2^2)(2(n+2)\ln|x| - 1)}{4\,|x|^2(-\ln|x|)^{3/2}}, \tag{4.1.3}$$

for every $(x_1, x_2, \dots, x_n) \in B_{1/2}$, and we observe that $f \in C(\overline{B_{1/2}})$ since

$$\lim_{x\to 0} |f(x)| \leqslant \lim_{x\to 0} \frac{2(n+2)|\ln|x|| + 1}{4(-\ln|x|)^{3/2}} = 0.$$

We claim that the function f in (4.1.3) does not admit any solution $u \in C^2(B_{1/2}) \cap C(\overline{B_{1/2}})$ of the Dirichlet problem in (4.1.2). To prove this, we argue by contradiction, and we suppose that a solution exists. We consider the function

$$B_{1/2} \ni x = (x_1, x_2, \dots, x_n) \longmapsto v(x) := \sqrt{-\ln|x|}\,(x_1^2 - x_2^2), \tag{4.1.4}$$

and we point out that $v \in C^\infty(B_{1/2} \setminus \{0\}) \cap C(\overline{B_{1/2}})$. Also, by a direct computation, we see that $\Delta v = f$ in $B_{1/2} \setminus \{0\}$. Consequently, the function $w := v - u$ belongs to $C^2(B_{1/2} \setminus \{0\}) \cap C(\overline{B_{1/2}})$ and satisfies

$$\begin{cases} \Delta w = 0 & \text{in } B_{1/2} \setminus \{0\}, \\ w = v & \text{on } \partial B_{1/2}. \end{cases}$$

By Theorem 2.12.1, we can extend w to a harmonic function in the whole of $B_{1/2}$: in particular, there exists $w_\star \in C^2(B_{1/2})$ such that $w_\star = w$ in $B_{1/2} \setminus \{0\}$. As a result,

$$\text{the function } v_\star := u + w_\star \text{ belongs to } C^2(B_{1/2}), \tag{4.1.5}$$

and it is such that $v_\star = v$ in $B_{1/2} \setminus \{0\}$.

Therefore,

$$\begin{aligned} \lim_{x_2 \searrow 0} \partial_{11} v_\star(0, x_2, 0, \dots, 0) &= \lim_{x_2 \searrow 0} \partial_{11} v(0, x_2, 0, \dots, 0) \\ &= \lim_{x_2 \searrow 0} \frac{1 - 4|\ln x_2|}{2\sqrt{-\ln x_2}} = +\infty, \end{aligned}$$

in contradiction with (4.1.5). □

Another Proof of Theorem 4.1.1. There are other possibilities for the functions f and v in (4.1.3) and (4.1.4). An instructive alternative is provided by a superposition method, which goes as follows. One can replace (4.1.4) with

$$v(x) := \sum_{k=1}^{+\infty} \frac{Q(e^k x)}{e^{2k}\,k},$$

where $Q(x) := \eta(x)h(x)$, with $\eta \in C_0^\infty(B_1, [0,1])$, $\eta = 1$ in $B_{1/2}$ and $h(x) := x_1^2 - x_2^2$. Also, one replaces (4.1.3) with

$$f(x) := \Delta v(x).$$

To establish Theorem 4.1.1, according to the previous proof, one has to check that

$$\text{the definitions of } v \text{ and } f \text{ are well-posed,} \tag{4.1.6}$$

$$f \in C(\overline{B_1}), \tag{4.1.7}$$

$$v \notin C^2(B_1). \tag{4.1.8}$$

For this, we observe that the definition of v is well-posed since

$$\sum_{k=1}^{+\infty} \frac{1}{e^{2k}\,k} < +\infty.$$

Furthermore, we have that $\Delta Q = h\Delta\eta + \eta\Delta h + 2\nabla\eta \cdot \nabla h = h\Delta\eta + 2\nabla\eta \cdot \nabla h$. As a result, if

$$v_m(x) := \sum_{k=1}^{m} \frac{Q(e^k x)}{e^{2k}\,k},$$

we have that

$$\Delta v_m(x) = \sum_{k=1}^{m} \frac{\Delta Q(e^k x)}{k} = \sum_{k=1}^{m} \frac{h(e^k x)\Delta\eta(e^k x) + 2\nabla\eta(e^k x) \cdot \nabla h(e^k x)}{k}.$$

Noting that $\eta(e^k x) = 0$ for all $x \in \mathbb{R}^n \setminus B_{e^{-k}}$, we have that, for all $x \in B_2$,

$$\begin{aligned}&\left|h(e^k x)\Delta\eta(e^k x) + 2\nabla\eta(e^k x) \cdot \nabla h(e^k x)\right| \\ &\quad\leqslant C\big(\|h\|_{L^\infty(B_{e^{-k}})} + \|\nabla h\|_{L^\infty(B_{e^{-k}})}\big) \leqslant Ce^{-k},\end{aligned}$$

for some $C > 0$, entailing that Δv_m converges uniformly in B_2 to a continuous function $\widetilde{f}$.

As a result, for every $\varphi \in C_0^\infty(B_1)$,

$$\begin{aligned}\int_{B_1} v(x)\,\Delta\varphi(x)\,dx &= \lim_{m\to+\infty} \int_{B_1} v_m(x)\,\Delta\varphi(x)\,dx \\ &= \lim_{m\to+\infty} \int_{B_1} \Delta v_m(x)\,\varphi(x)\,dx = \int_{B_1} \widetilde{f}(x)\,\varphi(x)\,dx.\end{aligned}$$

This gives that f is indeed well defined and coincides with $\widetilde{f}$, thus completing the proof of (4.1.6) and (4.1.7).

In addition, since $\sum_{k=1}^{+\infty} \frac{1}{e^k\,k} < +\infty$, we have that

$$\partial_1 v(x) = \sum_{k=1}^{+\infty} \frac{\partial_1 Q(e^k x)}{e^k\,k}.$$

In particular, since $h(0) = 0$ and $\partial_1 h(0) = 0$, whence $\partial_1 Q(0) = 0$, we have that $\partial_1 v(0) = 0$. As a consequence, if $\varepsilon \in (0, 1)$, to be taken as small as we wish in what follows, we see that

$$\begin{aligned}
\left|\partial_1 v(\varepsilon e_1) - \partial_1 v(0)\right| = \left|\partial_1 v(\varepsilon e_1)\right| &= \left|\sum_{k=1}^{+\infty} \frac{\partial_1 Q(e^k \varepsilon e_1)}{e^k\, k}\right| \\
&= \left|\sum_{k=1}^{+\infty} \frac{\partial_1 \eta(e^k \varepsilon e_1)\, h(e^k \varepsilon e_1) + \partial_1 h(e^k \varepsilon e_1)\, \eta(e^k \varepsilon e_1)}{e^k\, k}\right| \\
&= \left|\sum_{1 \leqslant k \leqslant |\ln \varepsilon|} \frac{e^{2k}\varepsilon^2\, \partial_1 \eta(e^k \varepsilon e_1) + 2e^k \varepsilon\, \eta(e^k \varepsilon e_1)}{e^k\, k}\right| \\
&= \varepsilon \left|\sum_{1 \leqslant k \leqslant |\ln \varepsilon|} \frac{e^k \varepsilon\, \partial_1 \eta(e^k \varepsilon e_1) + 2\eta(e^k \varepsilon e_1)}{k}\right|.
\end{aligned}$$

Now, let $C_0 := 2 + \|\eta\|_{C^1(\mathbb{R}^n)}$ and note that

$$\begin{aligned}
&\sum_{|\ln \varepsilon| - \ln C_0 \leqslant k \leqslant |\ln \varepsilon|} \frac{|e^k \varepsilon\, \partial_1 \eta(e^k \varepsilon e_1) + 2\eta(e^k \varepsilon e_1)|}{k} \\
&\qquad \leqslant \sum_{|\ln \varepsilon| - \ln C_0 \leqslant k \leqslant |\ln \varepsilon|} \frac{C_0}{k} \leqslant \frac{C}{|\ln \varepsilon|},
\end{aligned}$$

for some $C > 0$.

Also, if $k \in [1, |\ln \varepsilon| - \ln C_0)$, then $e^k \varepsilon \leqslant \frac{1}{C_0} \leqslant \frac{1}{2}$ and therefore $\eta(e^k \varepsilon e_1) = 1$, leading to

$$e^k \varepsilon\, \partial_1 \eta(e^k \varepsilon e_1) + 2\eta(e^k \varepsilon e_1) \geqslant 2 - \frac{|\partial_1 \eta(e^k \varepsilon e_1)|}{C_0} \geqslant 1.$$

Thanks to these observations, we find that

$$\begin{aligned}
\frac{\left|\partial_1 v(\varepsilon e_1) - \partial_1 v(0)\right|}{\varepsilon} &\geqslant \left|\sum_{1 \leqslant k < |\ln \varepsilon| - \ln C_0} \frac{e^k \varepsilon\, \partial_1 \eta(e^k \varepsilon e_1) + 2\eta(e^k \varepsilon e_1)}{k}\right| - \frac{C}{|\ln \varepsilon|} \\
&\geqslant \sum_{1 \leqslant k < |\ln \varepsilon| - \ln C_0} \frac{1}{k} - \frac{C}{|\ln \varepsilon|}.
\end{aligned}$$

This proves (4.1.8); otherwise, if v belonged to $C^2(B_1)$, we would have that

$$\begin{aligned}
\partial_{11} v(0) &= \lim_{\varepsilon \searrow 0} \frac{\left|\partial_1 v(\varepsilon e_1) - \partial_1 v(0)\right|}{\varepsilon} \geqslant \lim_{\varepsilon \searrow 0} \sum_{1 \leqslant k < |\ln \varepsilon| - \ln C_0} \frac{1}{k} - \frac{C}{|\ln \varepsilon|} \\
&= \sum_{k=1}^{+\infty} \frac{1}{k} = +\infty,
\end{aligned}$$

which is a contradiction. □

In view of Theorem 4.1.1, the regularity theory for elliptic equations (and even for the "simplest possible" case given by solutions of the Poisson equation $\Delta u = f$) relies first on the detection of the "appropriate functional spaces" to take into account. The regularity theory that we present here is the so-called Schauder theory, named after its inventor[3] Juliusz Schauder [Sch34]). In a nutshell, the bottom line of this theory is that, while Theorem 4.1.1 says that if $\Delta u = f \in C$ does not yield $u \in C^2$, this statement becomes essentially correct if one works instead in the Hölder spaces, namely $\Delta u = f \in C^\alpha$ yields $u \in C^{2,\alpha}$, provided that $\alpha \in (0, 1)$. The following are the technical details of this fascinating method.

4.2 Potential Theory and Schauder Estimates for the Laplace Operator

To develop the regularity theory in Hölder spaces for elliptic equations, we begin by considering the case of equations driven by the Laplace operator. The main results[4] that we want to address are the following ones, dealing with[5]

---

[3]Being a Polish mathematician of Jewish origin, after the invasion of German troops in 1941, Juliusz Schauder wrote to the German mathematician Ludwig Bieberbach, pleading for his support. Instead, Bieberbach passed his letter to the Gestapo, which arrested and executed Schauder. The regularity theory of elliptic partial differential equations happens to present a number of tragic events related to World War II, when scientific creativity, sense of responsibility, heroism and nobility of spirit had to confront pusillanimity, treacheries, greediness and pettiness. See also footnote 3 on p. 431.

[4]As customary, for every $k \in \mathbb{N}$ and $\alpha \in (0, 1]$, we use the norm notation

$$[u]_{C^\alpha(\Omega)} := \sup_{\substack{x,y\in\Omega\\x\neq y}} \frac{|u(x) - u(y)|}{|x-y|^\alpha}$$

$$\text{and} \quad \|u\|_{C^{k,\alpha}(\Omega)} := \sum_{\substack{\beta\in\mathbb{N}^n\\|\beta|\leqslant k}} \|D^\beta u\|_{L^\infty(\Omega)} + \sum_{\substack{\beta\in\mathbb{N}^n\\|\beta|=k}} [D^\beta u]_{C^\alpha(\Omega)}.$$

We also use the notation

$$[u]_{C^{k,\alpha}(\Omega)} := \sum_{\substack{\beta\in\mathbb{N}^n\\|\beta|=k}} [D^\beta u]_{C^\alpha(\Omega)}.$$

A function is said to belong to $C^{k,\alpha}(\Omega)$ when $\|u\|_{C^{k,\alpha}(\Omega)} < +\infty$. Also, the space $C^{0,\alpha}(\Omega)$ is often denoted as $C^\alpha(\Omega)$ for short. We stress that, in our notation, these Hölder spaces are meant in a "global" sense, and instead we say that $u \in C^{k,\alpha}_{\rm loc}(\Omega)$ when $u \in C^{k,\alpha}(\Omega')$ for every open and bounded set Ω' such that $\overline{\Omega'} \subset \Omega$.

See for example [Kry96, Chapter 3], [Fio16], [FRRO22] and the references therein for a thorough introduction to Hölder spaces (keeping in mind that the notation used is not necessarily uniform across the existing literature).

[5]For completeness, we mention that the choices of the balls, say B_1 and $B_{1/2}$, in the regularity results are somewhat arbitrary and done for the sake of simplicity; one can state regularity results for instance in balls B_R and B_r with $R > r > 0$, see for example the forthcoming Corollary 4.2.12 for future details.

interior estimates (Theorem 4.2.1) and local estimates[6] at the boundary (Theorem 4.2.2).

Theorem 4.2.1. *Let $f \in C^\alpha(B_1)$ for some $\alpha \in (0,1)$. Let $u \in C^2(B_1)$ be a solution of*

$$\Delta u = f \quad in\ B_1. \tag{4.2.1}$$

Then, there exists $C > 0$, depending only on n and α, such that

$$\|u\|_{C^{2,\alpha}(B_{1/2})} \leqslant C\left(\|u\|_{L^\infty(B_1)} + \|f\|_{C^\alpha(B_1)}\right).$$

Theorem 4.2.2. *Let*

$$\begin{aligned} B_r^+ &:= \{x = (x', x_n) \in B_r \ s.t.\ x_n > 0\} \\ and \quad B_r^0 &:= \{x = (x', x_n) \in B_r \ s.t.\ x_n = 0\}. \end{aligned} \tag{4.2.2}$$

Let $f \in C^\alpha(B_1^+)$ and $g \in C^{2,\alpha}(B_1^0)$ for some $\alpha \in (0,1)$. Let $u \in C^2(B_1^+) \cap C(B_1^+ \cup B_1^0)$ be a solution of

$$\begin{cases} \Delta u = f & in\ B_1^+, \\ u = g & on\ B_1^0. \end{cases} \tag{4.2.3}$$

Then, there exists $C > 0$, depending only on n and α, such that

$$\|u\|_{C^{2,\alpha}(B_{1/2}^+)} \leqslant C\left(\|u\|_{L^\infty(B_1^+)} + \|g\|_{C^{2,\alpha}(B_1^0)} + \|f\|_{C^\alpha(B_1^+)}\right).$$

Results such as the ones in Theorems 4.2.1 and 4.2.2 are important, especially considering that, in general, it is difficult to express the solutions of partial differential equations in a simple and explicit form; nonetheless, concrete estimates such as the ones above are, say, "almost as good" as explicit formulas for solutions in the sense that they are suitable for continuity methods (as developed in the forthcoming Theorem 4.5.4) and for stable numerical schemes.

The proofs of Theorems 4.2.1 and 4.2.2 rely on several deep observations of independent interest. To address the proof of Theorem 4.2.1, the strategy is to try to reduce the problem to the case of harmonic functions (and this is advantageous because we already know a lot about the regularity theory of harmonic functions by exploiting for instance Cauchy's estimates in Theorem 2.17.1).

A similar concept for reducing to harmonic functions was exploited in Corollary 3.3.4, which used the idea of subtracting from the given solution a "special" one for which some explicit computations come in handy. Namely, the special solution that one can consider is the Newtonian potential $-\Gamma * f$ in Proposition 2.7.4.

[6]For boundary regularity results under various boundary conditions, see e.g. [Mik78].

On the one hand, we have already encountered in the proof of Proposition 2.7.4 one of the main advantages of working with the Newtonian potential: namely, one can utilize its convolution structure to take derivatives. On the other hand, the proof of Proposition 2.7.4 also highlighted a conceptual limitation of this approach due to the singular behavior of the fundamental solution Γ: indeed, derivatives of the order k of Γ behave at the origin like $|x|^{2-n-k}$, which is integrable only when $k < 2$. This hurdle suggests that a similar procedure becomes problematic if one is interested (as in Theorem 4.2.1) in derivatives of order 2: as a matter of fact, a new ingredient is needed to deal with this significant hindrance, which will be bypassed thanks to a careful analysis of the singular integral induced by the second derivatives of the fundamental solution (as we will see, this new analysis will leverage the symmetry properties of the singular integral kernel associated with second derivatives in order to detect suitable cancellations).

Here are the technical details to make this strategy work and prove Theorem 4.2.1. First, we explicitly express second derivatives of the Newtonian potential in terms of a singular integral kernel (this is not obvious since, due to the discussion above, we cannot simply place two derivatives on the fundamental solution, owing to the lack of local integrability, and indeed the final result also makes it appear as an additional Dirac delta function).

Proposition 4.2.3. *Let Γ be the fundamental solution in* (2.7.6) *and $f \in C_0^\alpha(\mathbb{R}^n)$, for some $\alpha \in (0, 1]$. Set $v := -\Gamma * f$. Then, for all $i, j \in \{1, \dots, n\}$,*

$$\partial_{ij} v(x) = \int_{\mathbb{R}^n} \partial_{ij}\Gamma(x - y)\big(f(x) - f(y)\big)\, dy + \frac{\delta_{ij}}{n} f(x). \tag{4.2.4}$$

Here and in what follows, improper and singular integrals are taken in the principal value sense by averaging outside the singularity: for instance,

$$\begin{aligned}&\int_{\mathbb{R}^n} \partial_{ij}\Gamma(x - y)\big(f(x) - f(y)\big)\, dy \\ &\qquad := \lim_{R\to+\infty} \int_{B_R(x)} \partial_{ij}\Gamma(x - y)\big(f(x) - f(y)\big)\, dy.\end{aligned} \tag{4.2.5}$$

Note that $\partial_{ij}\Gamma$ is indeed not integrable, and hence the integral on the left-hand side of the previous equation is not a standard Lebesgue integral and instead requires the limit definition introduced on the right-hand side. Also, of course, no confusion should arise in equation (4.2.4) between the second-order partial derivative ∂_{ij} and the Kronecker notation

$$\delta_{ij} = \begin{cases} 1 & \text{if } i = j, \\ 0 & \text{if } i \neq j. \end{cases}$$

Interestingly, it follows from the harmonicity of Γ outside its singular point and the assumption that $f \in C_0^\alpha(\mathbb{R}^n)$,

$$\left|\int_{\mathbb{R}^n} \Delta\Gamma(x-y)\big(f(x)-f(y)\big)\,dy\right| = \lim_{r\searrow 0}\left|\int_{B_r(x)} \Delta\Gamma(x-y)\big(f(x)-f(y)\big)\,dy\right|$$
$$\leqslant \lim_{r\searrow 0} C\left|\int_{B_r(x)} |x-y|^{\alpha-n}\,dy\right| = \lim_{r\searrow 0} Cr^\alpha = 0;$$

therefore, (4.2.4) also entails that

$$\text{if } f \in C_0^\alpha(\mathbb{R}^n) \text{ then } -\Delta(\Gamma * f) = f. \tag{4.2.6}$$

This observation shows that the result in Proposition 2.7.4 can be sharpened by replacing the assumption $f \in C_0^1(\mathbb{R}^n)$ with $f \in C_0^\alpha(\mathbb{R}^n)$ for some $\alpha \in (0,1)$. Correspondingly,

$$\begin{gathered}\text{the assumption } f \in C^1(\Omega) \text{ in Corollary 3.3.4}\\ \text{can be relaxed to } f \in C^\alpha(\Omega) \text{ for some } \alpha \in (0,1).\end{gathered} \tag{4.2.7}$$

When needed, the observation in (4.2.6) can also be sharpened since it actually holds that

$$\text{if } x_0 \in \mathbb{R}^n,\ r>0 \text{ and } f \in C^\alpha(B_r(x_0)) \text{ then } -\Delta(\Gamma * f)(x_0) = f(x_0). \tag{4.2.8}$$

To prove this, it is enough to consider a cutoff function $\varphi \in C_0^\infty(B_{3r/4}(x_0))$ with $\varphi = 1$ in $B_{r/2}(x_0)$, define $f_1 := f\varphi$ and $f_2 := f(1-\varphi)$ and note that $f_1 \in C_0^\alpha(\mathbb{R}^n)$. We can also extend f_2 as a bounded function on the whole of $\mathbb{R}^n$, say, by setting $f_2 := 0$ outside $B_r(x_0)$. Accordingly, by (4.2.6) and the fact that $f_2 = 0$ in $B_{r/2}(x_0)$, for all $x \in B_{r/4}(x_0)$, we have that

$$\begin{aligned}-\Delta(\Gamma * f)(x) &= -\Delta(\Gamma * f_1)(x) - \Delta(\Gamma * f_2)(x)\\ &= f_1(x) - \Delta\int_{\mathbb{R}^n} \Gamma(x-y)\,f_2(y)\,dy\\ &= f(x) - \Delta\int_{\mathbb{R}^n\setminus B_{r/2}(x_0)} \Gamma(x-y)\,f_2(y)\,dy.\end{aligned} \tag{4.2.9}$$

Now, if $x \in B_{r/4}(x_0)$ and $y \in \mathbb{R}^n \setminus B_{r/2}(x_0)$, then $|x-y| \geqslant |y| - |x| \geqslant \frac{r}{4}$. Consequently, by (4.2.9) and the harmonicity of Γ away from its singularity, for all $x \in B_{r/4}(x_0)$, we have that

$$-\Delta(\Gamma * f)(x) = f(x) - \int_{\mathbb{R}^n\setminus B_{r/2}(x_0)} \Delta\Gamma(x-y)\,f_2(y)\,dy = f(x),$$

from which one obtains (4.2.8).

Proof of Proposition 4.2.3. This argument is a refinement of that produced in the proof of (2.7.13): namely, given $\rho > 0$, we use the regularization Γ_ρ of the fundamental solution introduced in (2.7.3) and (2.7.4) and use a limit argument. The technical details are as follows. Let $V := \partial_i v = -\partial_i(\Gamma * f)$ and $V_\rho := -\partial_i(\Gamma_\rho * f)$. We know from (2.7.13) that $V = -(\partial_i \Gamma) * f$.

Moreover, for all $x \in \mathbb{R}^n$,

$$
\begin{aligned}
&|V_\rho(x) - V(x)| \\
&= |(\partial_i \Gamma) * f(x) - (\partial_i \Gamma_\rho) * f(x)| \\
&\leqslant \frac{C}{\rho^n} \int_{B_\rho} |y|\,|f(x-y)|\,dy + C \int_{B_\rho} |y|^{1-n}\,|f(x-y)|\,dy \\
&\leqslant C\,\|f\|_{L^\infty(\mathbb{R}^n)}\,\rho,
\end{aligned} \tag{4.2.10}
$$

for some constant $C > 0$ depending only on n.

Additionally, for all $j \in \{1, \dots, n\}$,

$$
\begin{aligned}
&\left| \partial_j V_\rho(x) - \int_{\mathbb{R}^n} \partial_{ij}\Gamma(x-y)\big(f(x) - f(y)\big)\,dy - \frac{\delta_{ij}}{n} f(x) \right| \\
&= \left| (\partial_{ij}\Gamma_\rho) * f(x) + \int_{\mathbb{R}^n} \partial_{ij}\Gamma(x-y)\big(f(x) - f(y)\big)\,dy + \frac{\delta_{ij}}{n} f(x) \right| \\
&= \left| \int_{\mathbb{R}^n} \partial_{ij}\Gamma_\rho(x-y)\, f(y)\,dy + \int_{\mathbb{R}^n} \partial_{ij}\Gamma(x-y)\big(f(x) - f(y)\big)\,dy + \frac{\delta_{ij}}{n} f(x) \right| \\
&= \left| \int_{\mathbb{R}^n} \partial_{ij}\Gamma_\rho(x-y)\, f(x)\,dy - \int_{\mathbb{R}^n} \partial_{ij}\Gamma_\rho(x-y)\,\big(f(x) - f(y)\big)\,dy \right. \\
&\qquad \left. + \int_{\mathbb{R}^n} \partial_{ij}\Gamma(x-y)\big(f(x) - f(y)\big)\,dy + \frac{\delta_{ij}}{n} f(x) \right| \\
&= \left| \int_{\mathbb{R}^n} \partial_{ij}\Gamma_\rho(x-y)\, f(x)\,dy \right. \\
&\qquad \left. + \int_{B_\rho(x)} \big(\partial_{ij}\Gamma(x-y) - \partial_{ij}\Gamma_\rho(x-y)\big)\big(f(x) - f(y)\big)\,dy + \frac{\delta_{ij}}{n} f(x) \right| \\
&\leqslant \left| \int_{\mathbb{R}^n} \partial_{ij}\Gamma_\rho(x-y)\, f(x)\,dy + \frac{\delta_{ij}}{n} f(x) \right| + C\,\|f\|_{C^\alpha(\mathbb{R}^n)} \int_{B_\rho(x)} |x-y|^{\alpha-n}\,dy \\
&\leqslant \left| -\int_{B_\rho(x)} \frac{\delta_{ij} f(x)}{n|B_\rho|}\,dy + \int_{\mathbb{R}^n \setminus B_\rho(x)} \partial_{ij}\Gamma(x-y)\, f(x)\,dy + \frac{\delta_{ij}}{n} f(x) \right| \\
&\qquad + C\,\|f\|_{C^\alpha(\mathbb{R}^n)}\,\rho^\alpha \\
&= \left| \int_{\mathbb{R}^n \setminus B_\rho(x)} \partial_{ij}\Gamma(x-y)\, f(x)\,dy \right| + C\,\|f\|_{C^\alpha(\mathbb{R}^n)}\,\rho^\alpha,
\end{aligned} \tag{4.2.11}
$$

up to renaming $C > 0$ from line to line.

We also let

$$c'_n := \begin{cases} 2c_n & \text{if } n = 2, \\ c_n\, n(n-2) & \text{if } n \neq 2, \end{cases}$$

where c_n is the constant in (2.7.2), and we point out that if $i \neq j$ and $R > \rho$, then

$$\int_{B_R(x)\setminus B_\rho(x)} \partial_{ij}\Gamma(x-y)\,dy = c'_n \int_{B_R(x)\setminus B_\rho(x)} \frac{(x_i-y_i)(x_j-y_j)}{|x-y|^{n+2}}\,dy = 0,$$

due to odd symmetry. Moreover,

$$\int_{B_R(x)\setminus B_\rho(x)} \partial_{ii}\Gamma(x-y)\,dy = \frac{1}{n}\int_{B_R(x)\setminus B_\rho(x)} \Delta\Gamma(x-y)\,dy = 0,$$

thanks to the harmonicity of Γ outside its singularity.

It follows from these observations that, for all $i, j \in \{1,\dots,n\}$,

$$\int_{B_R(x)\setminus B_\rho(x)} \partial_{ij}\Gamma(x-y)\,dy = 0,$$

and thus, in the notation of (4.2.5),

$$\int_{\mathbb{R}^n\setminus B_\rho(x)} \partial_{ij}\Gamma(x-y)\,dy = 0.$$

From this and (4.2.11), we obtain that

$$\left|\partial_j V_\rho(x) - \int_{\mathbb{R}^n} \partial_{ij}\Gamma(x-y)\big(f(x)-f(y)\big)\,dy - \frac{\delta_{ij}}{n}\,f(x)\right| \leqslant C\,\|f\|_{C^\alpha(\mathbb{R}^n)}\,\rho^\alpha.$$

It follows from this estimate and (4.2.10) that V_ρ converges uniformly to V and $\partial_j V_\rho$ converges uniformly to the right-hand side of (4.2.4). That is, recalling (2.7.14), the function $-\Gamma_\rho * f$ converges uniformly together with its derivatives up to the second order (the function converging to $-\Gamma * f$ and the second derivatives converging to the right-hand side of (4.2.4)), and this establishes (4.2.4), as desired. □

In view of Proposition 4.2.3, it is now convenient to study the property of a singular integral kernel $K : \mathbb{R}^n \setminus \{0\} \to \mathbb{R}$ satisfying the following properties:

$$\begin{aligned}&\text{there exists } g \in L^\infty(\partial B_1) \text{ such that for every } x \in \mathbb{R}^n\setminus\{0\}\\ &\text{we have that } K(x) = \frac{1}{|x|^n}\,g\left(\frac{x}{|x|}\right),\end{aligned} \tag{4.2.12}$$

and

$$\begin{aligned}&\text{there exists } C > 0 \text{ such that for every } x \in \mathbb{R}^n\setminus\{0\}\\ &\text{we have that } |\nabla K(x)| \leqslant \frac{C}{|x|^{n+1}}.\end{aligned} \tag{4.2.13}$$

In short, given an open set $\mathcal{V} \subseteq \mathbb{R}^n$ and a function $\Phi \in L^1_{\text{loc}}(\mathbb{R}^n \times \mathbb{R}^n)$, we also use the principal value notation

$$\int_{\mathcal{V}} K(x-y)\,\Phi(x,y)\,dy := \lim_{\substack{R\to+\infty\\ \varepsilon\searrow 0}} \int_{(\mathcal{V}\setminus B_\varepsilon(x))\cap B_R(x)} K(x-y)\,\Phi(x,y)\,dy, \tag{4.2.14}$$

whenever the above limits exist.

We also set

$$Tf(x) := \int_{\mathcal{V}} K(x-y)\,\big(f(x)-f(y)\big)\,dy. \tag{4.2.15}$$

The main step in proving the interior regularity result in Theorem 4.2.1 and the local result at the boundary in Theorem 4.2.2 is now the following singular integral regularity estimate.

Lemma 4.2.4. *Assume that* K *satisfies* (4.2.12) *and* (4.2.13). *Suppose also that, for each* $z_0 \in \mathbb{R}^n$ *and* $\rho > 0$,

$$\left|\int_{\mathcal{V}\setminus B_\rho(z_0)} K(z_0-y)\,dy\right| \leqslant C_0, \tag{4.2.16}$$

for a suitable $C_0 > 0$.

If $f \in C^\alpha(\mathcal{V})$ *for some* $\alpha \in (0,1)$, *then*

$$[Tf]_{C^\alpha(\mathcal{V})} \leqslant C\,[f]_{C^\alpha(\mathcal{V})},$$

for a suitable constant $C > 0$ *depending only on* n, α, K *and* C_0.

Proof. Let $z_0, z_1 \in \mathcal{V}$. Let $\delta := |z_0 - z_1|$ and $z_2 := \frac{z_0+z_1}{2}$. The gist of the following estimate is that in regions of order δ around z_2, all contributions are of order δ^α, and for the complement of these regions, the computation becomes more delicate, and further cancellations must be taken into account. More precisely, we have

$$\begin{aligned}
&|Tf(z_0) - Tf(z_1)|\\
&\quad= \left|\int_{\mathcal{V}} K(z_0-y)\big(f(z_0)-f(y)\big)\,dy - \int_{\mathcal{V}} K(z_1-y)\big(f(z_1)-f(y)\big)\,dy\right|\\
&\quad\leqslant \left|\int_{\mathcal{V}\setminus B_{2\delta}(z_2)} K(z_0-y)\big(f(z_0)-f(y)\big)\,dy\right.\\
&\qquad\left. - \int_{\mathcal{V}\setminus B_{2\delta}(z_2)} K(z_1-y)\big(f(z_1)-f(y)\big)\,dy\right|\\
&\qquad+ \left|\int_{\mathcal{V}\cap B_{2\delta}(z_2)} K(z_0-y)\big(f(z_0)-f(y)\big)\,dy\right|\\
&\qquad+ \left|\int_{\mathcal{V}\cap B_{2\delta}(z_2)} K(z_1-y)\big(f(z_1)-f(y)\big)\,dy\right|.
\end{aligned} \tag{4.2.17}$$

Now, for $j \in \{0, 1\}$, if $y \in \mathcal{V} \cap B_{2\delta}(z_2)$, then $|z_j - y| \leqslant |z_j - z_2| + |z_2 - y| < 3\delta$, whence

$$\begin{aligned}
&\left|\int_{\mathcal{V}\cap B_{2\delta}(z_2)} K(z_j - y)\big(f(z_j) - f(y)\big)\,dy\right| \\
&\qquad \leqslant [f]_{C^\alpha(\mathcal{V})} \int_{B_{2\delta}(z_2)} |K(z_j - y)|\,|z_j - y|^\alpha\,dy \\
&\qquad \leqslant [f]_{C^\alpha(\mathcal{V})} \|g\|_{L^\infty(\partial B_1)} \int_{B_{2\delta}(z_2)} |z_j - y|^{\alpha-n}\,dy \\
&\qquad \leqslant [f]_{C^\alpha(\mathcal{V})} \|g\|_{L^\infty(\partial B_1)} \int_{B_{3\delta}} |x|^{\alpha-n}\,dx \\
&\qquad \leqslant C\,[f]_{C^\alpha(\mathcal{V})} \|g\|_{L^\infty(\partial B_1)}\,\delta^\alpha,
\end{aligned} \tag{4.2.18}$$

for some $C > 0$, thanks to (4.2.12).

Furthermore, making use of (4.2.13), if $y \in \mathcal{V} \setminus B_{2\delta}(z_2)$ and $t \in [0, 1]$, then

$$|z_1 + t(z_0 - z_1) - y| \geqslant |y - z_2| - |z_2 - z_1| - |z_0 - z_1| \geqslant |y - z_2| - \frac{3\delta}{2} \geqslant \frac{3|y - z_2|}{4},$$

from which, up to renaming C, we deduce that

$$\begin{aligned}
|K(z_0 - y) - K(z_1 - y)| &\leqslant \int_0^1 |\nabla K(z_1 + t(z_0 - z_1) - y)|\,dt\,|z_0 - z_1| \\
&\leqslant \sup_{t\in[0,1]} \frac{C\,|z_0 - z_1|}{|z_1 + t(z_0 - z_1) - y|^{n+1}} \leqslant \frac{C\,\delta}{|y - z_2|^{n+1}}.
\end{aligned}$$

For this reason,

$$\begin{aligned}
\Xi := &\left|\int_{\mathcal{V}\setminus B_{2\delta}(z_2)} K(z_0 - y)\big(f(z_0) - f(y)\big)\,dy\right. \\
&\quad \left. - \int_{\mathcal{V}\setminus B_{2\delta}(z_2)} K(z_1 - y)\big(f(z_1) - f(y)\big)\,dy\right| \\
\leqslant &\left|\int_{\mathcal{V}\setminus B_{2\delta}(z_2)} K(z_0 - y)\big(f(z_0) - f(y)\big)\,dy\right. \\
&\quad \left. - \int_{\mathcal{V}\setminus B_{2\delta}(z_2)} K(z_0 - y)\big(f(z_1) - f(y)\big)\,dy\right| \\
&\quad + \left|\int_{\mathcal{V}\setminus B_{2\delta}(z_2)} K(z_0 - y)\big(f(z_1) - f(y)\big)\,dy\right. \\
&\quad \left. - \int_{\mathcal{V}\setminus B_{2\delta}(z_2)} K(z_1 - y)\big(f(z_1) - f(y)\big)\,dy\right| \\
\leqslant &\left|\int_{\mathcal{V}\setminus B_{2\delta}(z_2)} K(z_0 - y)\big(f(z_0) - f(z_1)\big)\,dy\right| + C\,\delta \int_{\mathcal{V}\setminus B_{2\delta}(z_2)} \frac{|f(z_1) - f(y)|}{|y - z_2|^{n+1}}\,dy
\end{aligned}$$

$$\leqslant \left|f(z_0)-f(z_1)\right| \left(\left| \int_{B_{8\delta}(z_0)\setminus B_{2\delta}(z_2)} K(z_0-y)\,dy \right| + \left| \int_{\mathcal{V}\setminus B_{8\delta}(z_0)} K(z_0-y)\,dy \right| \right)$$
$$+C\,[f]_{C^\alpha(\mathcal{V})}\,\delta \int_{\mathbb{R}^n\setminus B_{2\delta}(z_2)} \frac{|z_1-y|^\alpha}{|y-z_2|^{n+1}}\,dy.$$

We also point out that, in view of (4.2.12),

$$\int_{B_{8\delta}(z_0)\setminus B_{2\delta}(z_2)} |K(z_0-y)|\,dy \leqslant \|g\|_{L^\infty(\partial B_1)} \int_{B_{8\delta}(z_0)\setminus B_{\delta/8}(z_0)} \frac{dy}{|z_0-y|^n}$$
$$\leqslant C\,\|g\|_{L^\infty(\partial B_1)}.$$

Recalling (4.2.16), from the observations above, we arrive at

$$\begin{aligned}
\Xi &\leqslant \big(C\,\|g\|_{L^\infty(\partial B_1)}+C_0\big)\big|f(z_0)-f(z_1)\big| \\
&\quad +C\,[f]_{C^\alpha(\mathcal{V})}\,\delta \int_{\mathbb{R}^n\setminus B_{2\delta}(z_2)} \frac{|z_1-y|^\alpha}{|y-z_2|^{n+1}}\,dy \\
&\leqslant \big(C\|g\|_{L^\infty(\partial B_1)}+C_0\big)\,[f]_{C^\alpha(\mathcal{V})}\delta^\alpha \\
&\quad +C\,[f]_{C^\alpha(\mathcal{V})}\,\delta \int_{\mathbb{R}^n\setminus B_{2\delta}(z_2)} \frac{\big(|z_1-z_2|+|y-z_2|\big)^\alpha}{|y-z_2|^{n+1}}\,dy \\
&\leqslant \big(C\|g\|_{L^\infty(\partial B_1)}+C_0\big)\,[f]_{C^\alpha(\mathcal{V})}\,\delta^\alpha \\
&\quad +C\,[f]_{C^\alpha(\mathcal{V})}\,\delta \int_{\mathbb{R}^n\setminus B_{2\delta}(z_2)} |y-z_2|^{\alpha-n-1}\,dy \\
&\leqslant \big(C\|g\|_{L^\infty(\partial B_1)}+C_0\big)\,[f]_{C^\alpha(\mathcal{V})}\,\delta^\alpha + C\,[f]_{C^\alpha(\mathcal{V})}\,\delta^\alpha.
\end{aligned}$$

Combined with (4.2.17) and (4.2.18), this yields that

$$\begin{aligned}
|Tf(z_0)-Tf(z_1)| &\leqslant \Xi + C\,[f]_{C^\alpha(\mathcal{V})}\|g\|_{L^\infty(\partial B_1)}\,\delta^\alpha \\
&\leqslant C\,[f]_{C^\alpha(\mathcal{V})}\big(\|g\|_{L^\infty(\partial B_1)}+1\big)\,\delta^\alpha,
\end{aligned}$$

providing the desired result. □

With this, we can prove the following estimate.

Corollary 4.2.5. *Let $f \in C^\alpha_0(\mathbb{R}^n)$ for some $\alpha \in (0,1)$ and $v := -\Gamma * f$. Then, for all $i, j \in \{1,\dots,n\}$,*

$$[\partial_{ij}v]_{C^\alpha(\mathbb{R}^n)} \leqslant C\,[f]_{C^\alpha(\mathbb{R}^n)},$$

for some positive constant C depending only on n and α.

Proof. By Proposition 4.2.3,

$$\partial_{ij}v(x) = \int_{\mathbb{R}^n} K(x-y)\big(f(x)-f(y)\big)\,dy + \frac{\delta_{ij}}{n}\,f(x),$$

with

$$K(x) := \partial_{ij}\Gamma(x). \tag{4.2.19}$$

That is, in the notation of (4.2.15),

$$\partial_{ij}v(x) = Tf(x) + \frac{\delta_{ij}}{n} f(x). \tag{4.2.20}$$

In light of (2.7.6), we know that $K(x)$ agrees, up to a dimensional constant, with $\frac{x_i x_j}{|x|^{n+2}}$ if $i \neq j$, and with $\frac{|x|^2 - nx_i^2}{|x|^{n+2}}$ if $i = j$. Consequently, assumption (4.2.12) is satisfied with

$$\partial B_1 \ni \omega \mapsto g(\omega) := \begin{cases} \omega_i \omega_j & \text{if } i \neq j, \\ 1 - n\omega_i^2 & \text{if } i = j. \end{cases} \tag{4.2.21}$$

Also, assumption (4.2.13) holds true.

As for (4.2.16), we obtain it as a byproduct of a general strategy based on cancellations. Namely, in the notation of (4.2.12) and (4.2.19), we have that

$$\int_{\partial B_1} g(\omega)\, d\mathcal{H}^{n-1}_\omega = 0. \tag{4.2.22}$$

As a matter of fact, we see from (4.2.21) that (4.2.22) holds true when $i \neq j$ due to an odd symmetry argument as well as when $i = j$ since in this case,

$$\begin{aligned} \fint_{\partial B_1} (1 - n\omega_i^2)\, d\mathcal{H}^{n-1}_\omega &= 1 - n \fint_{\partial B_1} \omega_i^2\, d\mathcal{H}^{n-1}_\omega = 1 - \sum_{k=1}^{n} \fint_{\partial B_1} \omega_k^2\, d\mathcal{H}^{n-1}_\omega \\ &= 1 - \fint_{\partial B_1} |\omega|^2\, d\mathcal{H}^{n-1}_\omega = 0. \end{aligned}$$

Having established (4.2.22), we exploit it to verify that, for each $z_0 \in \mathbb{R}^n$ and $\rho > 0$,

$$\int_{\mathbb{R}^n \setminus B_\rho(z_0)} K(z_0 - y)\, dy = 0. \tag{4.2.23}$$

Indeed,

$$\begin{aligned} \int_{\mathbb{R}^n \setminus B_\rho(z_0)} K(z_0 - y)\, dy &= \int_{\mathbb{R}^n \setminus B_\rho(z_0)} g\left(\frac{z_0 - y}{|z_0 - y|}\right) \frac{dy}{|z_0 - y|^n} \\ &= \iint_{[\rho, +\infty) \times (\partial B_1)} g(\omega) \frac{dr\, d\mathcal{H}^{n-1}_\omega}{r} = 0, \end{aligned}$$

which gives (4.2.23).

In turn, (4.2.23) entails (4.2.16). We can thereby exploit Lemma 4.2.4 and find that

$$[Tf]_{C^\alpha(\mathbb{R}^n)} \leqslant C\, [f]_{C^\alpha(\mathbb{R}^n)}.$$

From this estimate and (4.2.20), the desired result follows. □

In order to prove Theorem 4.2.1, we also need the following L^∞-estimate.

Lemma 4.2.6. *Let $f \in C_0^\alpha(\mathbb{R}^n)$ for some $\alpha \in (0,1)$ and $v := -\Gamma * f$.*

Then, for all $i, j \in \{1, \dots, n\}$,

$$\|\partial_{ij} v\|_{L^\infty(\mathbb{R}^n)} \leqslant C\, \|f\|_{C^\alpha(\mathbb{R}^n)},$$

for some positive constant C depending only on n, α and the support of f.

Proof. By Proposition 4.2.3,

$$\partial_{ij} v(x) = \int_{\mathbb{R}^n} \partial_{ij}\Gamma(x-y)\big(f(x) - f(y)\big)\, dy + \frac{\delta_{ij}}{n} f(x). \tag{4.2.24}$$

We suppose that the support of f is contained in B_R for some $R > 0$; therefore, we have that there exists $x_0 \in B_{R+1}$ such that $f(x_0) = 0$.

We also set

$$\phi(x) := \int_{\mathbb{R}^n} \partial_{ij}\Gamma(x-y)\big(f(x) - f(y)\big)\, dy,$$

and we see that

$$\begin{aligned} |\phi(x_0)| &\leqslant C \int_{\mathbb{R}^n} \frac{|f(y)|}{|x_0 - y|^n}\, dy = C \int_{B_R} \frac{|f(y)|}{|x_0 - y|^n}\, dy = C \int_{B_R} \frac{|f(x_0) - f(y)|}{|x_0 - y|^n}\, dy \\ &\leqslant C[f]_{C^\alpha(\mathbb{R}^n)} \int_{B_R} |x_0 - y|^{\alpha - n}\, dy \leqslant C[f]_{C^\alpha(\mathbb{R}^n)}, \end{aligned}$$

up to renaming $C > 0$. As a consequence, for every $x \in B_{R+2}$,

$$\begin{aligned} |\phi(x)| &\leqslant |\phi(x_0)| + |\phi(x) - \phi(x_0)| \\ &\leqslant C[f]_{C^\alpha(\mathbb{R}^n)} + [\varphi]_{C^\alpha(B_{R+2})} |x - x_0|^\alpha \leqslant C[f]_{C^\alpha(\mathbb{R}^n)}, \end{aligned}$$

where we have used Lemma 4.2.4 (recall (4.2.19) and (4.2.23) to conclude that (4.2.16) is satisfied).

Furthermore, we note that if $x \in \mathbb{R}^n \setminus B_{R+2}$ and $y \in B_R$, then $|x-y| \geqslant |x| - |y| \geqslant R + 2 - R = 2$, and thus

$$|\phi(x)| \leqslant C \int_{B_R} \frac{|f(y)|}{|x-y|^n}\, dy \leqslant C\|f\|_{L^\infty(\mathbb{R}^n)} \int_{B_R} \frac{dy}{2^n} \leqslant C\|f\|_{L^\infty(\mathbb{R}^n)}.$$

These considerations show that

$$\|\phi\|_{L^\infty(\mathbb{R}^n)} \leqslant C\|f\|_{C^\alpha(\mathbb{R}^n)}.$$

From this and (4.2.24), we obtain the desired estimate. □

With this preliminary work, we can now address[7] the proof of the desired interior regularity result.

Proof of Theorem 4.2.1. Let $\varphi \in C_0^\infty(B_{8/9}, [0,1])$, with $\varphi = 1$ in $B_{3/4}$ and $|\nabla\varphi| \leqslant 10$. Let also $\widetilde{f} := f\varphi$. In this way, we have that $\widetilde{f} \in C_0^\alpha(B_1)$. We define

$$v := -\Gamma * \widetilde{f}, \tag{4.2.25}$$

[7] A conceptually different proof of Theorem 4.2.1, not directly relying on singular integral estimates for the Newtonian potential, is presented on p. 355.

and we deduce from (4.2.6) that

$$\Delta v = \widetilde{f} \text{ in } \mathbb{R}^n. \tag{4.2.26}$$

Also, by Corollary 4.2.5, for all $i, j \in \{1, \dots, n\}$,

$$[\partial_{ij} v]_{C^\alpha(\mathbb{R}^n)} \leqslant C\,[\widetilde{f}]_{C^\alpha(\mathbb{R}^n)} \leqslant C\,\|f\|_{C^\alpha(B_1)}, \tag{4.2.27}$$

up to renaming C.

Moreover, for every $x \in B_1$,

$$|v(x)| \leqslant \|\widetilde{f}\|_{L^\infty(\mathbb{R}^n)} \int_{B_1} \Gamma(x-y)\,dy \leqslant \|f\|_{L^\infty(B_1)} \int_{B_2} \Gamma(z)\,dz \leqslant C\,\|f\|_{L^\infty(B_1)}$$

and

$$|\nabla v(x)| \leqslant \int_{B_1} |\nabla\Gamma(x-y)|\,|\widetilde{f}(y)|\,dy \leqslant \|f\|_{L^\infty(B_1)} \int_{B_2} |\nabla\Gamma(z)|\,dz \leqslant C\,\|f\|_{L^\infty(B_1)},$$

thanks to (2.7.13) and (4.2.25).

Also, by Lemma 4.2.6, for all $i, j \in \{1, \dots, n\}$,

$$|\partial_{ij} v(x)| \leqslant C\|\widetilde{f}\|_{C^\alpha(\mathbb{R}^n)} \leqslant C\|f\|_{C^\alpha(B_1)}.$$

Accordingly, we deduce from (4.2.27) that

$$\|v\|_{C^{2,\alpha}(B_1)} \leqslant C\,\|f\|_{C^\alpha(B_1)}, \tag{4.2.28}$$

up to renaming C once again.

We now define $w := u - v$ and exploit (4.2.1) and (4.2.26) to conclude that $\Delta w = 0$ in $B_{3/4}$. Thus, it follows from Cauchy's estimates (see Theorem 2.17.1) that

$$\|w\|_{C^3(B_{1/2})} \leqslant C\,\|w\|_{L^\infty(B_{3/4})} \leqslant C\left(\|u\|_{L^\infty(B_{3/4})} + \|v\|_{L^\infty(B_{3/4})}\right).$$

Since, for every $x, y \in B_{1/2}$,

$$|\partial_{ij} w(x) - \partial_{ij} w(y)| \leqslant C\|w\|_{C^3(B_{1/2})}|x-y| \leqslant C\|w\|_{C^3(B_{1/2})}|x-y|^\alpha,$$

up to renaming C, we thus conclude that

$$\|w\|_{C^{2,\alpha}(B_{1/2})} \leqslant C\left(\|u\|_{L^\infty(B_{3/4})} + \|v\|_{L^\infty(B_{3/4})}\right).$$

In light of this estimate and (4.2.28), we have that

$$\begin{aligned} \|u\|_{C^{2,\alpha}(B_{1/2})} &\leqslant \|v\|_{C^{2,\alpha}(B_{1/2})} + \|w\|_{C^{2,\alpha}(B_{1/2})} \leqslant C\left(\|u\|_{L^\infty(B_{3/4})} + \|v\|_{C^{2,\alpha}(B_{1/2})}\right) \\ &\leqslant C\left(\|u\|_{L^\infty(B_{3/4})} + \|f\|_{C^\alpha(B_1)}\right), \end{aligned}$$

as desired. □

Now, we turn our attention to the local estimates at the boundary to prove Theorem 4.2.2. To this end, we point out a variation of the result in Proposition 4.2.3 (full details of this modification are provided for the facility of the reader).

Proposition 4.2.7. *Let Γ be the fundamental solution in (2.7.6). Let $\Omega_0 \subset \mathbb{R}^n$ be an open and bounded set with Lipschitz boundary. Let $\Omega \subseteq \Omega_0$ be open and bounded, and $f \in L^\infty(\mathbb{R}^n)$. Suppose that $f \in C^\alpha(\Omega)$ and $f = 0$ in $\mathbb{R}^n \setminus \Omega$.*

Set

$$v(x) := -\int_\Omega \Gamma(x-y)\, f(y)\, dy.$$

Then, for all $i, j \in \{1, \dots, n\}$ and all $x \in \Omega$,

$$\begin{aligned}\partial_{ij} v(x) = &\int_{\Omega_0} \partial_{ij}\Gamma(x-y)\big(f(x)-f(y)\big)\, dy \\ &+ f(x)\int_{\partial\Omega_0} \partial_i\Gamma(x-y)\, \nu(y)\cdot e_j\, d\mathcal{H}^{n-1}_y. \end{aligned} \tag{4.2.29}$$

Proof. As in the proof of Proposition 4.2.3, given $\rho > 0$ to be taken as small as we wish, we use the regularization Γ_ρ of the fundamental solution introduced in (2.7.3) and (2.7.4). We define $V(x) := \partial_i v(x)$ and

$$V_\rho(x) := -\partial_i \left(\int_\Omega \Gamma_\rho(x-y)\, f(y)\, dy \right).$$

We know from (4.2.10) (applied here with $f\chi_\Omega$ instead of f) that

$$V_\rho \text{ converges uniformly to } V. \tag{4.2.30}$$

Additionally, if $x \in \Omega' \Subset \Omega$,

$$\begin{aligned} &\bigg| \partial_j V_\rho(x) - \int_{\Omega_0} \partial_{ij}\Gamma(x-y)\big(f(x)-f(y)\big)\, dy \\ &\qquad - f(x)\int_{\partial\Omega_0} \partial_i\Gamma(x-y)\, \nu(y)\cdot e_j\, d\mathcal{H}^{n-1}_y \bigg| \\ &= \bigg| \int_\Omega \partial_{ij}\Gamma_\rho(x-y)\, f(y)\, dy + \int_{\Omega_0} \partial_{ij}\Gamma(x-y)\big(f(x)-f(y)\big)\, dy \\ &\qquad + f(x)\int_{\partial\Omega_0} \partial_i\Gamma(x-y)\, \nu(y)\cdot e_j\, d\mathcal{H}^{n-1}_y \bigg| \\ &= \bigg| \int_{B_\rho(x)} \partial_{ij}\Gamma_\rho(x-y)\, f(y)\, dy + \int_{\Omega\setminus B_\rho(x)} \partial_{ij}\Gamma(x-y)\, f(y)\, dy \\ &\qquad + \int_{B_\rho(x)} \partial_{ij}\Gamma(x-y)\big(f(x)-f(y)\big)\, dy \\ &\qquad + \int_{\Omega_0\setminus B_\rho(x)} \partial_{ij}\Gamma(x-y)\big(f(x)-f(y)\big)\, dy \\ &\qquad + f(x)\int_{\partial\Omega_0} \partial_i\Gamma(x-y)\, \nu(y)\cdot e_j\, d\mathcal{H}^{n-1}_y \bigg| \end{aligned}$$

$$
\begin{aligned}
&= \left| -\delta_{ij} \int_{B_\rho(x)} \frac{f(y)}{n\,|B_\rho|}\,dy + f(x) \int_{\Omega_0 \setminus B_\rho(x)} \partial_{ij}\Gamma(x-y)\,dy \right.\\
&\qquad \left. + f(x) \int_{\partial\Omega_0} \partial_i\Gamma(x-y)\,\nu(y)\cdot e_j\,d\mathcal{H}^{n-1}_y \right|\\
&\qquad + C[f]_{C^\alpha(B_\rho(x))} \int_{B_\rho(x)} |x-y|^{\alpha-n}\,dy.
\end{aligned}
$$

We also note that

$$
\begin{aligned}
&\int_{\Omega_0 \setminus B_\rho(x)} \partial_{ij}\Gamma(x-y)\,dy\\
&\qquad = -\int_{\Omega_0 \setminus B_\rho(x)} \operatorname{div}_y\big(\partial_i\Gamma(x-y)\,e_j\big)\,dy\\
&\qquad = -\int_{\partial(\Omega_0 \setminus B_\rho(x))} \partial_i\Gamma(x-y)\,\nu(y)\cdot e_j\,d\mathcal{H}^{n-1}_y\\
&\qquad = -\int_{\partial\Omega_0} \partial_i\Gamma(x-y)\,\nu(y)\cdot e_j\,d\mathcal{H}^{n-1}_y\\
&\qquad\qquad + \bar{c}_n \int_{\partial B_\rho(x)} \frac{(x_i-x_j)(x_j-y_j)}{|x-y|^{n+1}}\,d\mathcal{H}^{n-1}_y,
\end{aligned}
$$

where, recalling (2.7.2),

$$
\begin{aligned}
\bar{c}_n &:= \begin{cases} c_n & \text{if } n=2,\\ c_n(n-2) & \text{if } n\neq 2 \end{cases}\\
&= \frac{1}{n\,|B_1|}.
\end{aligned}
\tag{4.2.31}
$$

These observations lead to

$$
\begin{aligned}
&\left| \partial_j V_\rho(x) - \int_{\Omega_0} \partial_{ij}\Gamma(x-y)\big(f(x)-f(y)\big)\,dy \right.\\
&\qquad \left. - f(x) \int_{\partial\Omega_0} \partial_i\Gamma(x-y)\,\nu(y)\cdot e_j\,d\mathcal{H}^{n-1}_y \right|\\
&\leqslant \left| -\delta_{ij} \int_{B_\rho(x)} \frac{f(y)}{n\,|B_\rho|}\,dy + \bar{c}_n\,f(x) \int_{\partial B_\rho(x)} \frac{(x_i-y_i)(x_j-y_j)}{|x-y|^{n+1}}\,d\mathcal{H}^{n-1}_y \right|\\
&\qquad + C\,[f]_{C^\alpha(B_\rho(x))}\,\rho^\alpha.
\end{aligned}
$$

Now, we note that

$$
\begin{aligned}
\bar{c}_n \int_{\partial B_\rho(x)} \frac{(x_i-y_i)(x_j-y_j)}{|x-y|^{n+1}}\,d\mathcal{H}^{n-1}_y &= \bar{c}_n\delta_{ij} \int_{\partial B_\rho(x)} \frac{(x_i-y_i)^2}{|x-y|^{n+1}}\,d\mathcal{H}^{n-1}_y\\
= \frac{\bar{c}_n\delta_{ij}}{n} \int_{\partial B_\rho(x)} \frac{|x-y|^2}{|x-y|^{n+1}}\,d\mathcal{H}^{n-1}_y &= \frac{\bar{c}_n\delta_{ij}\,\mathcal{H}^{n-1}(\partial B_\rho)}{n\,\rho^{n+1}} = \frac{\delta_{ij}}{n},
\end{aligned}
$$

thanks to (1.1.6).

From this, we arrive at

$$\begin{aligned}
&\left|\partial_j V_\rho(x) - \int_{\Omega_0} \partial_{ij}\Gamma(x-y)\big(f(x)-f(y)\big)\,dy\right.\\
&\qquad\left. - f(x)\int_{\partial\Omega_0} \partial_i\Gamma(x-y)\,\nu(y)\cdot e_j\,d\mathcal{H}^{n-1}_y\right|\\
&\leqslant \left|-\frac{\delta_{ij}}{n}\fint_{B_\rho(x)} f(y)\,dy + \frac{\delta_{ij}\,f(x)}{n}\right| + C\,[f]_{C^\alpha(B_\rho(x))}\,\rho^\alpha\\
&\leqslant C\,[f]_{C^\alpha(B_\rho(x))}\,\rho^\alpha
\end{aligned}$$

up to renaming $C > 0$.

This entails that $\partial_j V_\rho$ converges locally uniformly to the right-hand side of (4.2.29). The desired result thus follows by recalling (4.2.30). □

Now, we point out two useful regularity results for the Newtonian potential set in a halfspace. For this, we use the notation

$$\mathbb{R}^n_+ := \{x = (x', x_n) \in \mathbb{R}^{n-1}\times\mathbb{R} \text{ s.t. } x_n > 0\}.$$

Proposition 4.2.8. *Let Γ be the fundamental solution in* (2.7.6) *and $f \in C^\alpha(\mathbb{R}^n_+)$. Suppose that $f(x) = 0$ for all $x \in \mathbb{R}^n_+$ with $|x| > \frac{99}{100}$.*

Set

$$v(x) := -\int_{\mathbb{R}^n_+} \Gamma(x-y)\,f(y)\,dy.$$

Then, for all $i, j \in \{1,\dots,n\}$,

$$[\partial_{ij}v]_{C^\alpha(\mathbb{R}^n_+)} \leqslant C\,[f]_{C^\alpha(\mathbb{R}^n_+)}, \tag{4.2.32}$$

for a suitable constant $C > 0$ depending only on n and α.

Proof. We point out that

$$\Delta v(x) = f(x) \quad \text{for all } x \in \mathbb{R}^n_+, \tag{4.2.33}$$

owing to (4.2.8) (applied here to the function $f\chi_{\mathbb{R}^n_+}$).

We also note that

$$\begin{aligned}&\text{it suffices to establish (4.2.32) under the additional assumption}\\ &\text{that either } i \neq n \text{ or } j \neq n.\end{aligned} \tag{4.2.34}$$

Indeed, if (4.2.32) holds true with $(i, j) \neq (n, n)$, then one can exploit (4.2.33) to write that, in $\mathbb{R}^n_+$,

$$\partial_{nn}v = f - \sum_{i=1}^{n-1}\partial_{ii}v$$

and thus

$$[\partial_{nn}v]_{C^\alpha(\mathbb{R}^n_+)} \leqslant [f]_{C^\alpha(\mathbb{R}^n_+)} + \sum_{i=1}^{n-1}[\partial_{ii}v]_{C^\alpha(\mathbb{R}^n_+)} \leqslant C\,[f]_{C^\alpha(\mathbb{R}^n_+)},$$

up to renaming C. This observation establishes (4.2.34).

Thanks to (4.2.34), up to exchanging i and j, we can now focus on the case of

$$i \in \{1,\dots,n\} \quad \text{and} \quad j \in \{1,\dots,n-1\}. \tag{4.2.35}$$

In this setting, given $r > 0$, we exploit Proposition 4.2.7 with $\Omega := B_r^+$ and $\Omega_0 := B_R^+$ with $R > r$. Noting that along $\{x_n = 0\}$, we have that $\nu \cdot e_j = 0$, due to (4.2.35), for every $x \in B_r^+$, we have that

$$\begin{aligned}
\partial_{ij}v(x) &= \int_{B_R^+} \partial_{ij}\Gamma(x-y)\big(f(x)-f(y)\big)\,dy \\
&\quad + f(x)\int_{\partial B_R^+} \partial_i\Gamma(x-y)\,\nu(y)\cdot e_j\,d\mathcal{H}^{n-1}_y \\
&= \int_{B_R^+} \partial_{ij}\Gamma(x-y)\big(f(x)-f(y)\big)\,dy \\
&\quad + f(x)\int_{\partial B_R\cap\{y_n>0\}} \partial_i\Gamma(x-y)\,\nu(y)\cdot e_j\,d\mathcal{H}^{n-1}_y. \qquad (4.2.36)
\end{aligned}$$

Now, we observe that, if $x \in B_r^+$ and $y \in \partial B_R$, as $R \to +\infty$,

$$\frac{1}{|x-y|^n} = \frac{1}{R^n\left|\frac{x}{R}-\frac{y}{R}\right|^n} = \frac{1+o(1)}{R^n};$$

therefore, using also that $j \neq n$,

$$\begin{aligned}
&\int_{\partial B_R\cap\{y_n>0\}} \partial_i\Gamma(x-y)\,\nu(y)\cdot e_j\,d\mathcal{H}^{n-1}_y \\
&\qquad = \frac{c_n(n-2)}{R}\int_{\partial B_R\cap\{y_n>0\}} \frac{(x_i-y_i)y_j}{|x-y|^n}\,d\mathcal{H}^{n-1}_y \\
&\qquad = \frac{c_n(n-2)(1+o(1))}{R^{n+1}}\int_{\partial B_R\cap\{y_n>0\}} (x_i-y_i)y_j\,d\mathcal{H}^{n-1}_y \\
&\qquad = -\frac{c_n(n-2)(1+o(1))}{R^{n+1}}\int_{\partial B_R\cap\{y_n>0\}} y_iy_j\,d\mathcal{H}^{n-1}_y.
\end{aligned}$$

Thus, if $i \neq j$, by odd symmetry, we have that

$$\int_{\partial B_R\cap\{y_n>0\}} \partial_i\Gamma(x-y)\,\nu(y)\cdot e_j\,d\mathcal{H}^{n-1}_y = 0$$

while if $i = j \in \{1,\dots,n-1\}$,

$$\int_{\partial B_R\cap\{y_n>0\}} \partial_i\Gamma(x-y)\,\nu(y)\cdot e_j\,d\mathcal{H}^{n-1}_y$$

$$
\begin{aligned}
&= -\frac{c_n(n-2)(1+o(1))}{R^{n+1}} \int_{\partial B_R \cap \{y_n>0\}} y_i^2 \, d\mathcal{H}_y^{n-1} \\
&= -\frac{c_n(n-2)(1+o(1))}{2R^{n+1}} \int_{\partial B_R} y_i^2 \, d\mathcal{H}_y^{n-1} \\
&= -\frac{c_n(n-2)(1+o(1))}{2nR^{n+1}} \int_{\partial B_R} |y|^2 \, d\mathcal{H}_y^{n-1} \\
&= -\frac{c_n(n-2)\mathcal{H}^{n-1}(\partial B_1)(1+o(1))}{2n},
\end{aligned}
$$

as $R \longrightarrow +\infty$.

Hence, by taking the limit as $R \longrightarrow +\infty$ in (4.2.36) and using these observations, we have that, for every $x \in B_r^+$,

$$
\partial_{ij} v(x) = \int_{\mathbb{R}^n_+} \partial_{ij}\Gamma(x-y)\big(f(x)-f(y)\big)\,dy - f(x)\delta_{ij}\frac{c_n(n-2)\mathcal{H}^{n-1}(\partial B_1)}{2n}. \tag{4.2.37}
$$

The aim is now to exploit Lemma 4.2.4, with $K := \partial_{ij}\Gamma$. Since K satisfies (4.2.12) and (4.2.13), in order to use Lemma 4.2.4, we have to check that the condition in (4.2.16) is satisfied. To this end, recalling the principal value notation in (4.2.14), for every $z_0 \in \mathbb{R}^n$ and every $\rho > 0$,

$$
\begin{aligned}
-\int_{\mathbb{R}^n_+ \setminus B_\rho(z_0)} \partial_{ij}\Gamma(z_0-y)\,dy &= -\lim_{R \to +\infty} \int_{\mathbb{R}^n_+ \cap (B_R(z_0)\setminus B_\rho(z_0))} \partial_{ij}\Gamma(z_0-y)\,dy \\
&= -\lim_{R \to +\infty} \int_{\mathbb{R}^n_+ \cap (B_R(z_0)\setminus B_\rho(z_0))} \operatorname{div}_y \big(\partial_i\Gamma(z_0-y)\,e_j\big)\,dy.
\end{aligned}
$$

As a result, according to the divergence theorem,

$$
\begin{aligned}
&\left| \int_{\mathbb{R}^n_+ \setminus B_\rho(z_0)} \partial_{ij}\Gamma(z_0-y)\,dy \right| \\
&\quad \leqslant \int_{\partial B_R(z_0)} |\nabla\Gamma(z_0-y)|\,d\mathcal{H}_y^{n-1} + \int_{\partial B_\rho(z_0)} |\nabla\Gamma(z_0-y)|\,d\mathcal{H}_y^{n-1}.
\end{aligned} \tag{4.2.38}
$$

We also observe that, for every $r > 0$,

$$
\int_{\partial B_r(z_0)} |\nabla\Gamma(z_0-y)|\,d\mathcal{H}_y^{n-1} \leqslant C \int_{\partial B_\rho(z_0)} \frac{d\mathcal{H}_y^{n-1}}{|z_0-y|^{n-1}} \leqslant C,
$$

up to renaming $C > 0$.

Combining this with (4.2.38), we see that

$$
\text{(4.2.16) holds true with } K = \partial_{ij}\Gamma. \tag{4.2.39}
$$

Thus, recalling (4.2.37) and exploiting Lemma 4.2.4 with $\mathcal{V} := \mathbb{R}^n_+$, we obtain that

$$
\begin{aligned}
[\partial_{ij} v]_{C^\alpha(B_r^+)} &\leqslant [Tf]_{C^\alpha(B_r^+)} + C[f]_{C^\alpha(B_r^+)} \\
&\leqslant [Tf]_{C^\alpha(\mathbb{R}^n_+)} + C[f]_{C^\alpha(\mathbb{R}^n_+)} \leqslant C[f]_{C^\alpha(\mathbb{R}^n_+)},
\end{aligned}
$$

up to renaming C. This and the arbitrariness of r give the desired result. □

Proposition 4.2.9. *Let Γ be the fundamental solution in* (2.7.6) *and $f \in C^\alpha(\mathbb{R}^n_+)$. Suppose that $f(x) = 0$ for all $x \in \mathbb{R}^n_+$ with $|x| > \frac{99}{100}$.*

Set

$$v(x) := -\int_{\mathbb{R}^n_+} \Gamma(x-y)\, f(y)\, dy.$$

Then, for all $i, j \in \{1, \dots, n\}$,

$$\|\partial_{ij} v\|_{L^\infty(\mathbb{R}^n_+)} \leqslant C\, \|f\|_{C^\alpha(\mathbb{R}^n_+)}, \tag{4.2.40}$$

for a suitable constant $C > 0$ depending only on n and α.

Proof. We argue as in the proof of Proposition 4.2.8, exploiting (4.2.33) to conclude that it suffices to establish (4.2.40) under the additional assumption that either $i \neq n$ or $j \neq n$. In this setting, one can employ (4.2.37) to find that, for every $r > 0$, in B^+_r,

$$\partial_{ij} v(x) = \int_{\mathbb{R}^n_+} \partial_{ij}\Gamma(x-y)\big(f(x) - f(y)\big)\, dy - f(x)\delta_{ij} \frac{c_n(n-2)\mathcal{H}^{n-1}(\partial B_1)}{2n}. \tag{4.2.41}$$

Now, we set

$$\phi(x) := \int_{\mathbb{R}^n_+} \partial_{ij}\Gamma(x-y)\big(f(x) - f(y)\big)\, dy.$$

We observe that there exists $x_0 \in B^+_1$ such that $f(x_0) = 0$, and therefore

$$\begin{aligned} |\phi(x_0)| &\leqslant C \int_{\mathbb{R}^n_+} \frac{|f(y)|}{|x_0 - y|^n}\, dy = C \int_{B^+_1} \frac{|f(y)|}{|x_0 - y|^n}\, dy = C \int_{B^+_1} \frac{|f(x_0) - f(y)|}{|x_0 - y|^n}\, dy \\ &\leqslant C[f]_{C^\alpha(\mathbb{R}^n_+)} \int_{B^+_1} |x_0 - y|^{\alpha - n}\, dy \leqslant C[f]_{C^\alpha(\mathbb{R}^n_+)}, \end{aligned}$$

up to renaming $C > 0$. Consequently, thanks to Lemma 4.2.4, applied here with $\mathcal{V} := \mathbb{R}^n_+$ and $K := \partial_{ij}\Gamma$ (recall (4.2.39)), we have that for every $x \in B^+_4$,

$$\begin{aligned} |\phi(x)| &\leqslant |\phi(x_0)| + |\phi(x) - \phi(x_0)| \\ &\leqslant C[f]_{C^\alpha(\mathbb{R}^n_+)} + [\phi]_{C^\alpha(B^+_4)} |x - x_0|^\alpha \leqslant C[f]_{C^\alpha(\mathbb{R}^n_+)}, \end{aligned}$$

up to relabeling $C > 0$.

Furthermore, we note that if $x \in \mathbb{R}^n_+ \setminus B_4$ and $y \in B^+_1$, then $|x - y| \geqslant |x| - |y| \geqslant 3$, and thus

$$|\phi(x)| \leqslant C \int_{B^+_1} \frac{|f(y)|}{|x - y|^n}\, dy \leqslant C\|f\|_{L^\infty(B^+_1)} \int_{B^+_1} \frac{dy}{3^n} \leqslant C\|f\|_{L^\infty(B^+_1)}.$$

These considerations show that

$$\|\phi\|_{L^\infty(\mathbb{R}^n_+)} \leqslant C\|f\|_{C^\alpha(\mathbb{R}^n_+)}.$$

From this and (4.2.41), we obtain that

$$\|\partial_{ij} v\|_{L^\infty(B^+_r)} \leqslant C\|f\|_{C^\alpha(\mathbb{R}^n_+)}.$$

Since r is any arbitrary positive real number, this gives the desired estimate. □

We now complete the proof of the desired local estimates[8] at the boundary.

Proof of Theorem 4.2.2. We first observe that

$$\text{it suffices to prove Theorem 4.2.2 when } g \text{ vanishes identically.} \tag{4.2.42}$$

Indeed, let $\widetilde{u}(x) = \widetilde{u}(x', x_n) := u(x) - g(x')$ and $\widetilde{f}(x) = \widetilde{f}(x', x_n) := f(x) - \Delta g(x')$. By (4.2.3), we have that

$$\begin{cases} \Delta \widetilde{u} = \widetilde{f} & \text{in } B_1^+, \\ \widetilde{u} = 0 & \text{on } B_1^0. \end{cases}$$

Consequently, if Theorem 4.2.2 holds true when g vanishes identically, we have

$$\begin{aligned} \|u\|_{C^{2,\alpha}(B_{1/2}^+)} &\leqslant \|\widetilde{u}\|_{C^{2,\alpha}(B_{1/2}^+)} + \|g\|_{C^{2,\alpha}(B_1^0)} \\ &\leqslant C \left(\|\widetilde{u}\|_{L^\infty(B_1^+)} + \|\widetilde{f}\|_{C^\alpha(B_1^+)} \right) + \|g\|_{C^{2,\alpha}(B_1^0)} \\ &\leqslant C \left(\|u\|_{L^\infty(B_1^+)} + \|g\|_{C^{2,\alpha}(B_1^0)} + \|f\|_{C^\alpha(B_1^+)} \right), \end{aligned}$$

up to renaming C at each step of the calculation, and this observation establishes (4.2.42).

Thus, in light of (4.2.42), we focus now on the proof of Theorem 4.2.2 by taking the additional assumption that

$$g \text{ vanishes identically.} \tag{4.2.43}$$

Furthermore, by multiplying by a cutoff function $\varphi \in C_0^\infty(B_{8/9}, [0,1])$, with $\varphi = 1$ in $B_{3/4}$ and $|\nabla \varphi| \leqslant 10$, as done at the beginning of the proof of Theorem 4.2.1, we can suppose that

$$f = 0 \quad \text{in} \quad \mathbb{R}^n_+ \setminus B_{8/9}^+. \tag{4.2.44}$$

We use the following handy notation to deal with reflections across the horizontal hyperplane: given $x = (x', x_n) \in \mathbb{R}^{n-1} \times \mathbb{R}$, we let $x^\star := (x', -x_n)$. Also, since f is uniformly continuous in B_1^+, it can be extended continuously to B_1^0. With this notation in mind, for all $x = (x', x_n) \in \mathbb{R}^n$, we define

$$f^\star(x) := \begin{cases} f(x) & \text{if } x_n \geqslant 0, \\ f(x^\star) & \text{if } x_n < 0, \end{cases} \tag{4.2.45}$$

and we claim that

$$[f^\star]_{C^\alpha(\mathbb{R}^n)} \leqslant [f]_{C^\alpha(\mathbb{R}^n_+)}. \tag{4.2.46}$$

[8]A conceptually different proof of Theorem 4.2.2 is presented on p. 358.

To check this, let $x, y \in \mathbb{R}^n$. If both x_n and y_n are nonnegative, then

$$|f^\star(x) - f^\star(y)| = |f(x) - f(y)| \leqslant [f]_{C^\alpha(\mathbb{R}^n_+)}|x-y|^\alpha,$$

and (4.2.46) plainly follows. Similarly, if both x_n and y_n are negative, then

$$|f^\star(x) - f^\star(y)| = |f(x^\star) - f(y^\star)| \leqslant [f]_{C^\alpha(\mathbb{R}^n_+)}|x^\star - y^\star|^\alpha = [f]_{C^\alpha(\mathbb{R}^n_+)}|x-y|^\alpha,$$

from which we obtain (4.2.46). Therefore, up to exchanging x and y, we may assume that $x_n \geqslant 0 > y_n$. In this case, we have that $|x_n + y_n| \leqslant |x_n| + |y_n| = x_n - y_n \leqslant |x_n - y_n|$ and consequently

$$\begin{aligned}|f^\star(x) - f^\star(y)| &= |f(x) - f(y^\star)| \leqslant [f]_{C^\alpha(\mathbb{R}^n_+)}|x-y^\star|^\alpha \\ &= [f]_{C^\alpha(\mathbb{R}^n_+)}\left(|x'-y'|^2 + |x_n+y_n|^2\right)^{\frac{\alpha}{2}} \\ &\leqslant [f]_{C^\alpha(\mathbb{R}^n_+)}\left(|x'-y'|^2 + |x_n-y_n|^2\right)^{\frac{\alpha}{2}} = [f]_{C^\alpha(\mathbb{R}^n_+)}|x-y|^\alpha,\end{aligned}$$

which completes the proof of (4.2.46).

Let now

$$w(x) := \int_{\mathbb{R}^n_+} \left(\Gamma(x-y) - \Gamma(x-y^\star)\right) f(y)\,dy.$$

We claim that

$$\lim_{x_n \searrow 0} w(x', x_n) = 0. \tag{4.2.47}$$

Indeed,

$$|x-y^\star| = \sqrt{|x'-y'|^2 + |x_n+y_n|^2} = |x^\star - y|;$$

therefore, for every $y \in B_1$,

$$\left|\Gamma(x-y) - \Gamma(x-y^\star)\right| = \left|\Gamma(x-y) - \Gamma(x^\star - y)\right| \leqslant C\,|x_n| \int_{-1}^{1} \frac{dt}{|(x'-y', tx_n - y_n)|^{n-1}},$$

for some $C > 0$. As a result, by (4.2.44) and exploiting Fubini's theorem and the change of variable $z := (x'-y', tx_n - y_n)$,

$$\begin{aligned}|w(x', x_n)| &\leqslant \|f\|_{L^\infty(\mathbb{R}^n_+)} \int_{B_1} \left|\Gamma(x-y) - \Gamma(x-y^\star)\right| dy \\ &\leqslant C\,|x_n| \int_{-1}^{1} \left(\int_{B_1} \frac{dy}{|(x'-y', tx_n - y_n)|^{n-1}} \right) dt \\ &\leqslant C\,|x_n| \int_{-1}^{1} \left(\int_{B_3} \frac{dz}{|z|^{n-1}} \right) dt \\ &\leqslant C\,|x_n|,\end{aligned}$$

up to renaming $C > 0$, from which (4.2.47) follows.

Let also

$$v(x) := \int_{\mathbb{R}^n_+} \Gamma(x-y)\, f(y)\, dy \quad \text{and} \quad W(x) := \int_{\mathbb{R}^n} \Gamma(x-y)\, f^\star(y)\, dy.$$

Using the notation $\mathbb{R}^n_- := \{x = (x', x_n) \in \mathbb{R}^n \text{ s.t. } x_n < 0\}$ and the substitution $z := y^\star$, we remark that

$$\begin{aligned} w(x) &= \int_{\mathbb{R}^n_+} \Gamma(x-y)\, f(y)\, dy - \int_{\mathbb{R}^n_+} \Gamma(x-y^\star)\, f(y)\, dy \\ &= \int_{\mathbb{R}^n_+} \Gamma(x-y)\, f(y)\, dy - \int_{\mathbb{R}^n_-} \Gamma(x-z)\, f(z^\star)\, dz \\ &= \int_{\mathbb{R}^n_+} \Gamma(x-y)\, f(y)\, dy - \int_{\mathbb{R}^n_-} \Gamma(x-z)\, f^\star(z)\, dz \\ &= 2\int_{\mathbb{R}^n_+} \Gamma(x-y)\, f(y)\, dy - \int_{\mathbb{R}^n} \Gamma(x-z)\, f^\star(z)\, dz \\ &= 2v(x) - W(x). \end{aligned} \tag{4.2.48}$$

Thus, since, by Corollary 4.2.5 and (4.2.46),

$$[\partial_{ij} W]_{C^\alpha(\mathbb{R}^n)} \leqslant C\,[f^\star]_{C^\alpha(\mathbb{R}^n)} \leqslant C\,[f]_{C^\alpha(\mathbb{R}^n_+)},$$

and, by Proposition 4.2.8,

$$[\partial_{ij} v]_{C^\alpha(\mathbb{R}^n_+)} \leqslant C\,[f]_{C^\alpha(\mathbb{R}^n_+)},$$

we conclude that, for all $i, j \in \{1, \dots, n\}$,

$$[\partial_{ij} w]_{C^\alpha(\mathbb{R}^n_+)} \leqslant 2[\partial_{ij} v]_{C^\alpha(\mathbb{R}^n_+)} + [\partial_{ij} W]_{C^\alpha(\mathbb{R}^n)} \leqslant C\,[f]_{C^\alpha(\mathbb{R}^n_+)}. \tag{4.2.49}$$

Furthermore, owing to (4.2.8), for all $x \in \mathbb{R}^n_+$,

$$-\Delta w(x) = -2\Delta v(x) + \Delta W(x) = 2f(x) - f^\star(x) = f(x). \tag{4.2.50}$$

We also point out that, for all $x \in \mathbb{R}^n_+$,

$$|w(x)| + |\nabla w(x)| \leqslant C\|f\|_{L^\infty(\mathbb{R}^n_+)}, \tag{4.2.51}$$

thanks to (2.7.13) and (4.2.25).

Moreover, by (4.2.46), (4.2.48), Lemma 4.2.6 and Proposition 4.2.9, we have that, for all $x \in \mathbb{R}^n_+$,

$$\begin{aligned} |\partial_{ij} w(x)| &\leqslant 2|v(x)| + |W(x)| \leqslant C\left(\|f\|_{C^\alpha(\mathbb{R}^n_+)} + \|f^\star\|_{C^\alpha(\mathbb{R}^n)}\right) \\ &\leqslant C\|f\|_{C^\alpha(\mathbb{R}^n_+)}. \end{aligned} \tag{4.2.52}$$

Now, we define $V := u + w$. In light of (4.2.47) and (4.2.50), we see that $V \in C^2(B_1^+) \cap C(B_1^+ \cap B_1^0)$, and, recalling (4.2.3) and (4.2.43),

$$\begin{cases} \Delta V = 0 & \text{in } B_1^+, \\ V = 0 & \text{on } B_1^0. \end{cases}$$

For this reason and Lemma 2.1.7, we have that the odd reflection of V across B_1^0, which we denote by $V^\sharp$, is harmonic in B_1. As a result, exploiting Cauchy's estimates in Theorem 2.17.1 and employing (4.2.51),

$$\|V\|_{C^3(B^+_{1/2})} \leqslant \|V^\sharp\|_{C^3(B_{1/2})} \leqslant C\|V^\sharp\|_{L^1(B_{7/8})} \leqslant C\left(\|u\|_{L^\infty(B^+_{7/8})} + \|w\|_{L^\infty(B^+_{7/8})}\right)$$
$$\leqslant C\left(\|u\|_{L^\infty(B^+_{7/8})} + \|f\|_{L^\infty(\mathbb{R}^n_+)}\right).$$

The desired result now follows from this estimate and (4.2.49) (recall also (4.2.51) and (4.2.52)). □

By combining the interior estimates of Theorem 4.2.1 and the local estimates at the boundary of Theorem 4.2.2, one can promptly obtain global estimates up to the boundary when[9] the domain is a ball, as pointed out in the following result.

Theorem 4.2.10. *Let $f \in C^\alpha(B_1)$ and $\varphi \in C^{2,\alpha}(\partial B_1)$ for some $\alpha \in (0,1)$. Let $u \in C^2(B_1)$ be a solution of*

$$\begin{cases} \Delta u = f & \text{in } B_1, \\ u = \varphi & \text{on } \partial B_1. \end{cases}$$

Then, there exists $C > 0$, depending only on n and α, such that

$$\|u\|_{C^{2,\alpha}(B_1)} \leqslant C\left(\|f\|_{C^\alpha(B_1)} + \|\varphi\|_{C^{2,\alpha}(\partial B_1)}\right).$$

Proof. Consider a point $\overline{x} \in \partial B_1$. By a (translation of a) Kelvin transform $\mathcal{K}$, a neighborhood $\mathcal{B}(\overline{x})$ of $\overline{x}$ in B_1 is sent to a halfball B^+ centered at $\mathcal{K}(\overline{x})$, see Figure 2.6.3 and Proposition 2.6.3. Let us denote $u_\star(x) := u(\mathcal{K}(x))$ and exploit (2.6.10) to see that $\Delta u_\star = f_\star$ in B^+ for a suitable $f_\star$ with $\|f_\star\|_{C^\alpha(B^+)} \leqslant C\,\|f\|_{C^\alpha(B_1)}$. Also, if B^0 denotes the flat part of ∂B^+, we have that $u_\star(x) = \varphi(\mathcal{K}(x)) =: \varphi_\star(x)$ for all $x \in B^0$. Note also that $\|\varphi_\star\|_{C^{2,\alpha}(B^0)} \leqslant C\,\|\varphi\|_{C^{2,\alpha}(\partial B_1)}$. As a result, making use of the local estimates at the boundary of Theorem 4.2.2,

$$\|u\|_{C^{2,\alpha}(\widetilde{\mathcal{B}}(\overline{x}))} \leqslant C\,\|u_\star\|_{C^{2,\alpha}(\widetilde{B}^+)}$$

[9]We point out that global regularity results on balls are technically easier than those in general domains (compare for example with the forthcoming Theorem 4.5.1). Indeed, for general domains, a natural strategy is to "straighten the boundary" via a diffeomorphism (as sketched in Figure 4.5.1 on p. 385) and then exploit the local estimates at the boundary obtained in halfballs; however, this procedure changes the operator (in particular, the corresponding equation obtained after straightening the boundary is no longer driven by the Laplace operator). This is for us one of the main motivations to study operators that are "slightly more general than the Laplacian": the complications arising from this further generality will be well compensated by the fact that this class of operator is stable for diffeomorphism, thus allowing to straighten the boundary (compare with the forthcoming Lemma 4.4.1).

The reason for which balls are special domains in this setting is that one can use, in this case, the Kelvin transform, which locally straightens the boundary of the ball and also preserves the Laplace operator: this is in fact the core of the proof of Theorem 4.2.10.

$$\leqslant C\left(\|u_\star\|_{L^\infty(B^+)} + \|\varphi_\star\|_{C^{2,\alpha}(B^0)} + \|f_\star\|_{C^\alpha(B^+)}\right)$$
$$\leqslant C\left(\|u\|_{L^\infty(B_1)} + \|\varphi\|_{C^{2,\alpha}(\partial B_1)} + \|f\|_{C^\alpha(B_1)}\right),$$

where $\widetilde{\mathcal{B}}(\overline{x}) \Subset \mathcal{B}(\overline{x})$ and $\widetilde{\mathcal{B}}(\overline{x})$ is sent to $\widetilde{B}^+ \Subset B^+$.

By combining this with the interior estimates of Theorem 4.2.1, we arrive at

$$\|u\|_{C^{2,\alpha}(B_1)} \leqslant C\left(\|u\|_{L^\infty(B_1)} + \|f\|_{C^\alpha(B_1)} + \|\varphi\|_{C^{2,\alpha}(\partial B_1)}\right). \tag{4.2.53}$$

Now, we consider the function

$$w(x) := u(x) - \|f\|_{L^\infty(B_1)}\, |x|^2.$$

Noting that $\Delta w = f - 2n\|f\|_{L^\infty(B_1)} \leqslant 0$, we deduce from the weak maximum principle in Corollary 2.9.2(ii) that, for all $x \in B_1$,

$$w(x) \geqslant \inf_{\partial B_1} w \geqslant -\|\varphi\|_{L^\infty(\partial B_1)} - \|f\|_{L^\infty(B_1)}$$

and therefore

$$u(x) \geqslant -\|\varphi\|_{L^\infty(\partial B_1)} - 2\|f\|_{L^\infty(B_1)}.$$

Similarly, by considering the function

$$\widetilde{w}(x) := u(x) + \|f\|_{L^\infty(B_1)}\, |x|^2$$

and using the weak maximum principle in Corollary 2.9.2(i), one sees that, for all $x \in B_1$,

$$u(x) \leqslant \|\varphi\|_{L^\infty(\partial B_1)} + 2\|f\|_{L^\infty(B_1)}.$$

These observations give that

$$\|u\|_{L^\infty(B_1)} \leqslant \|\varphi\|_{L^\infty(\partial B_1)} + 2\|f\|_{L^\infty(B_1)} \leqslant \|\varphi\|_{C^{2,\alpha}(\partial B_1)} + 2\|f\|_{C^\alpha(B_1)}.$$

The desired result follows from this inequality and (4.2.53). □

We now mention that the interior estimates in Theorem 4.2.1 can also be recast to obtain global estimates in terms of "scaled norms." For this, given an open set $\Omega \subset \mathbb{R}^n$, it is convenient to consider the distance functions

$$d_x := \operatorname{dist}(x, \partial\Omega) = \inf_{p\in\partial\Omega} |x - p| \qquad \text{and} \qquad d_{xy} := \min\{d_x, d_y\},$$

and, for $k \in \mathbb{N}$ and $\alpha \in (0, 1]$, take into account the weighted quantities

$$\|u\|_{C^k_\star(\Omega)} := \sum_{\substack{\beta\in\mathbb{N}^n\\|\beta|\leqslant k}} \sup_{x\in\Omega} d_x^{|\beta|}\, |D^\beta u(x)|,$$

$$[u]_{C^{k,\alpha}_\star(\Omega)} := \sum_{\substack{\beta\in\mathbb{N}^n\\|\beta|=k}} \sup_{\substack{x,y\in\Omega\\x\neq y}} d_{xy}^{k+\alpha}\, \frac{|D^\beta u(x) - D^\beta u(y)|}{|x-y|^\alpha}$$

$$\text{and} \quad \|u\|_{C^{k,\alpha}_\star(\Omega)} := \|u\|_{C^k_\star(\Omega)} + [u]_{C^{k,\alpha}_\star(\Omega)}.$$

Given $m \in \mathbb{N}$, it is also convenient[10] to define

$$\|u\|_{C^{\alpha}_{m,\star}(\Omega)} := \sup_{x\in\Omega} d_x^m\,|u(x)| + \sup_{\substack{x,y\in\Omega\\ x\neq y}} d_{xy}^{m+\alpha}\,\frac{|u(x)-u(y)|}{|x-y|^{\alpha}}.$$

These weighted norms should be compared with the classical ones in footnote 4 on p. 301. With this notation, we have the following weighted and global regularity result.

Theorem 4.2.11. *Let $\Omega \subset \mathbb{R}^n$ be an open set. Let $f \in C^{\alpha}(\Omega)$ for some $\alpha \in (0,1)$. Let $u \in C^2(\Omega)$ be a solution of*

$$\Delta u = f \quad \textit{in } \Omega.$$

Then, there exists $C > 0$, depending only on n, α and Ω, such that

$$\|u\|_{C^{2,\alpha}_{\star}(\Omega)} \leqslant C\left(\|u\|_{L^{\infty}(\Omega)} + \|f\|_{C^{\alpha}_{2,\star}(\Omega)}\right).$$

We remark that, despite their "global" flavor, regularity results as the one in Theorem 4.2.11 are not quite "regularity results up to the boundary" since the weighted norm $\|\cdot\|_{C^{2,\alpha}_{\star}(\Omega)}$ allows a possible degeneracy in the vicinity of the boundary of Ω; nonetheless, Theorem 4.2.11 does provide an interesting global information by giving a uniform bound on the "most singular behavior that can possibly happen at the boundary," regardless of the regularity of the boundary datum itself (note indeed that no regularity for the datum of the solution along $\partial\Omega$ is assumed in Theorem 4.2.11).

Proof of Theorem 4.2.11. Let $x_0 \in \Omega$ and $R_0 := \frac{d_{x_0}}{8}$. Let also $u_0(x) := u(x_0 + R_0 x)$. Then, $\partial_i u_0(x) = R_0 \partial_i u(x_0 + R_0 x)$ and $\partial_{ij} u_0(x) = R_0^2 \partial_{ij} u(x_0 + R_0 x)$.

As a result,

$$\begin{aligned} d_{x_0}\,|\nabla u(x_0)| + d_{x_0}^2\,|D^2 u(x_0)| &= \frac{d_{x_0}}{R_0}\,|\nabla u_0(0)| + \frac{d_{x_0}^2}{R_0^2}\,|D^2 u_0(0)| \\ &\leqslant C\left(|\nabla u_0(0)| + |D^2 u_0(0)|\right). \end{aligned} \tag{4.2.54}$$

We also remark that $\Delta u_0(x) = R_0^2 \Delta u(x_0 + R_0 x) = R_0^2 f(x_0 + R_0 x) =: f_0(x)$ for every $x \in B_1$. Therefore, in view of Theorem 4.2.1,

$$\|u_0\|_{C^{2,\alpha}(B_{1/2})} \leqslant C\left(\|u_0\|_{L^{\infty}(B_1)} + \|f_0\|_{C^{\alpha}(B_1)}\right). \tag{4.2.55}$$

[10]One sees that $\|u\|_{C^{\alpha}_{0,\star}(\Omega)} = \|u\|_{C^{0,\alpha}_{\star}(\Omega)}$. However, the norms $\|\cdot\|_{C^{k,\alpha}_{\star}(\Omega)}$ and $\|\cdot\|_{C^{\alpha}_{m,\star}(\Omega)}$ are conceptually different in general, since $\|\cdot\|_{C^{k,\alpha}_{\star}(\Omega)}$ involves derivatives up to order k, scaled with their own order of differentiation, while $\|\cdot\|_{C^{\alpha}_{m,\star}(\Omega)}$ only involves the function, not its derivatives; however, an additional scaling factor of order m takes place in this case.

We also observe that if $x \neq y \in B_1$, then

$$\begin{aligned}
&|f_0(x) - f_0(y)| \\
&= R_0^2 |f(x_0 + R_0 x) - f(x_0 + R_0 y)| \\
&\leqslant [f]_{C^\alpha(B_{R_0}(x_0))} R_0^2 \,|(x_0 + R_0 x) - (x_0 + R_0 y)|^\alpha \\
&= [f]_{C^\alpha(B_{R_0}(x_0))} R_0^{2+\alpha} \,|x - y|^\alpha \\
&= \sup_{\substack{X,Y \in B_{R_0}(x_0) \\ X \neq Y}} \frac{|f(X) - f(Y)| \, R_0^{2+\alpha} \,|x - y|^\alpha}{|X - Y|^\alpha}.
\end{aligned} \tag{4.2.56}$$

Also, if $X \in B_{R_0}(x_0)$ and $p \in \partial\Omega$, we have that

$$|X - p| \geqslant |x_0 - p| - R_0 \geqslant d_{x_0} - R_0 \geqslant \frac{7 d_{x_0}}{8}$$

and therefore $d_X \geqslant \frac{7d_{x_0}}{8}$. Accordingly, if $X, Y \in B_{R_0}(x_0)$, we have that $d_{XY} \geqslant \frac{7d_{x_0}}{8} = 7R_0$; therefore, we infer from (4.2.56) that

$$\begin{aligned}
|f_0(x) - f_0(y)| &\leqslant C \sup_{\substack{X,Y \in B_{R_0}(x_0) \\ X \neq Y}} \frac{|f(X) - f(Y)| \, d_{XY}^{2+\alpha} \,|x - y|^\alpha}{|X - Y|^\alpha} \\
&\leqslant C \,\|f\|_{C^\alpha_{2,\star}(\Omega)} \,|x - y|^\alpha.
\end{aligned}$$

This gives that $\|f_0\|_{C^\alpha(B_1)} \leqslant C \,\|f\|_{C^\alpha_{2,\star}(\Omega)}$, and thus, recalling (4.2.55), we arrive at

$$\|u_0\|_{C^{2,\alpha}(B_{1/2})} \leqslant C \left(\|u\|_{L^\infty(\Omega)} + \|f\|_{C^\alpha_{2,\star}(\Omega)} \right), \tag{4.2.57}$$

up to renaming C repeatedly.

From (4.2.54) and (4.2.57), we deduce that

$$d_{x_0} \,|\nabla u(x_0)| + d_{x_0}^2 \,|D^2 u(x_0)| \leqslant C \left(\|u\|_{L^\infty(\Omega)} + \|f\|_{C^\alpha_{2,\star}(\Omega)} \right)$$

and therefore, taking the supremum over $x_0 \in \Omega$,

$$\|u\|_{C^2_\star(\Omega)} \leqslant C \left(\|u\|_{L^\infty(\Omega)} + \|f\|_{C^\alpha_{2,\star}(\Omega)} \right). \tag{4.2.58}$$

Let now $x_0 \neq y_0 \in \Omega$. We claim that

$$d_{x_0 y_0}^{2+\alpha} \frac{|D^2 u(x_0) - D^2 u(y_0)|}{|x_0 - y_0|^\alpha} \leqslant C \left(\|u\|_{L^\infty(\Omega)} + \|f\|_{C^\alpha_{2,\star}(\Omega)} \right). \tag{4.2.59}$$

For this, up to exchanging these points, we may suppose that $d_{y_0} \geqslant d_{x_0} = d_{x_0 y_0}$. Using the above notation about R_0 and u_0, we distinguish two cases. If $|x_0 - y_0| \geqslant \frac{R_0}{4}$, we have that

$$d_{x_0 y_0}^{2+\alpha} \frac{|D^2 u(x_0) - D^2 u(y_0)|}{|x_0 - y_0|^\alpha} \leqslant d_{x_0}^{2+\alpha} \frac{|D^2 u(x_0)| + |D^2 u(y_0)|}{(R_0/4)^\alpha}$$

$$\leqslant Cd_{x_0}^2\Big(|D^2u(x_0)| + |D^2u(y_0)|\Big)$$
$$\leqslant C\Big(d_{x_0}^2|D^2u(x_0)| + d_{y_0}^2|D^2u(y_0)|\Big) \leqslant C\,\|u\|_{C^2_\star(\Omega)}.$$

Combining this with (4.2.58), we obtain (4.2.59) in this case.

Thus, to complete the proof of (4.2.59), we can now focus on the case in which $|x_0 - y_0| < \frac{R_0}{4}$. In this situation, we set $z_0 := \frac{y_0 - x_0}{R_0}$ and note that

$$|z_0| \leqslant \frac{|y_0 - x_0|}{R_0} < \frac{1}{4}.$$

Hence, we find that

$$d_{x_0y_0}^{2+\alpha}\frac{|D^2u(x_0) - D^2u(y_0)|}{|x_0 - y_0|^\alpha} = \frac{d_{x_0}^{2+\alpha}}{R_0^2}\frac{|D^2u_0(0) - D^2u_0(z_0)|}{|x_0 - y_0|^\alpha}$$
$$\leqslant C\,d_{x_0}^\alpha\frac{\|D^2u_0\|_{C^\alpha(B_{1/4})}|z_0|^\alpha}{|x_0 - y_0|^\alpha}$$
$$\leqslant C\,d_{x_0}^\alpha\frac{\|D^2u_0\|_{C^\alpha(B_{1/4})}}{R_0^\alpha} \leqslant C\,\|D^2u_0\|_{C^\alpha(B_{1/4})}.$$

From this estimate and (4.2.57), we obtain that (4.2.59) holds true in this case as well.

Hence, taking the supremum in (4.2.59), we conclude that

$$[u]_{C^{2,\alpha}_\star(\Omega)} \leqslant C\Big(\|u\|_{L^\infty(\Omega)} + \|f\|_{C^\alpha_{2,\star}(\Omega)}\Big).$$

The desired result then follows by combining this inequality with (4.2.58). □

For the sake of completeness, as already remarked in footnote 5 on p. 301, we point out that typical regularity results can also be stated with balls of different radii (but the different statements turn out to be typically equivalent, possibly up to an appropriate covering argument). Though we do not linger too much on this technical detail, to highlight this situation, we formulate a variant of the interior estimates in Theorem 4.2.1 in balls of general radii.

Corollary 4.2.12. *Let $R > r > 0$. Let $f \in C^\alpha(B_R)$ for some $\alpha \in (0,1)$. Let $u \in C^2(B_R)$ be a solution of*

$$\Delta u = f \quad \text{in } B_R.$$

Then, there exists $C > 0$, depending only on n, α, R and r, such that

$$\|u\|_{C^{2,\alpha}(B_r)} \leqslant C\Big(\|u\|_{L^\infty(B_R)} + \|f\|_{C^\alpha(B_R)}\Big).$$

The proof of Corollary 4.2.12 leverages a covering argument of general use which is presented as follows.

Lemma 4.2.13. *Let $k \in \mathbb{N}$ and $\alpha \in (0, 1]$. Let R_1, $R_2 > 0$, P_1, $P_2 \in \mathbb{R}^n$, with*

$$\rho := |P_1 - P_2| < R_1 + R_2. \tag{4.2.60}$$

Assume that $u \in C^{k,\alpha}(B_{R_1}(P_1)) \cap C^{k,\alpha}(B_{R_2}(P_2))$. Then, $u \in C^{k,\alpha}(B_{R_1}(P_1) \cup B_{R_2}(P_2))$ and

$$\|u\|_{C^{k,\alpha}(B_{R_1}(P_1)\cup B_{R_2}(P_2))} \leqslant C\left(\|u\|_{C^{k,\alpha}(B_{R_1}(P_1))} + \|u\|_{C^{k,\alpha}(B_{R_2}(P_2))}\right),$$

for some $C > 0$ depending only on n, R_1, R_2 and ρ.

Proof. Certainly,

$$\|u\|_{C^k(B_{R_1}(P_1)\cup B_{R_2}(P_2))} = \max\left\{\|u\|_{C^k(B_{R_1}(P_1))}, \|u\|_{C^k(B_{R_2}(P_2))}\right\}.$$

Also, if $j \in \{1, 2\}$ and both x and $y \in B_{R_j}(P_j)$, then

$$|D^k u(x) - D^k u(y)| \leqslant C\|u\|_{C^{k,\alpha}(B_{R_j}(P_j))}|x - y|^\alpha.$$

In view of these observations, to complete the proof of the desired result, we can focus on the case of

$$x \in B_{R_1}(P_1) \setminus B_{R_2}(P_2) \quad \text{and} \quad y \in B_{R_2}(P_2) \setminus B_{R_1}(P_1). \tag{4.2.61}$$

In this situation, we claim that there exist $C > 0$, depending only on n, R_1, R_2 and ρ, and $z \in B_{R_1}(P_1) \cap B_{R_2}(P_2)$ such that

$$|x - z| + |z - y| \leqslant C|x - y|. \tag{4.2.62}$$

To check this, up to swapping indices, we can suppose that $R_1 \geqslant R_2$. Besides, up to a translation, we can suppose that P_1 is the origin. Also, up to a rotation, we can assume that $P_2 = |P_2|e_n$. Therefore, $P_2 = |P_1 - P_2|e_n = \rho e_n$, see Figure 4.2.1 (in particular, the dependence on P_1 and P_2 can be dropped, and the constant C will depend naturally only on n, R_1, R_2 and ρ). We also take $h > 0$ such that $\partial B_{R_1} \cap \partial B_{R_2}(P_2)$ is contained in the hyperplane $\{x_n = h\}$, and we denote by

$$\mathcal{B}_h := \overline{B_{R_1} \cap B_{R_2}(P_2)} \cap \{x_n = h\}.$$

In this setting, given x and y as in (4.2.61), we consider the geodesics connecting x and y contained in $\overline{B_{R_1} \cup B_{R_2}(P_2)}$. We remark that, in light of (4.2.61), the geodesics passes through $\mathcal{B}_h$, and we let $\tilde{z}$ be a point of the geodesics lying on $\mathcal{B}_h$.

We point out that

$$\begin{aligned}&\text{the geodesics is either a segment passing through } x,\ y \text{ and } \tilde{z},\\ &\text{or the union of two segments joining } x \text{ to } \tilde{z} \text{ and } \tilde{z} \text{ to } y,\end{aligned} \tag{4.2.63}$$

see Figure 4.2.2. To prove this, suppose that the geodesics is not a segment joining x and y. Then, we consider the geodesics joining x to $\tilde{z}$ and observe that it

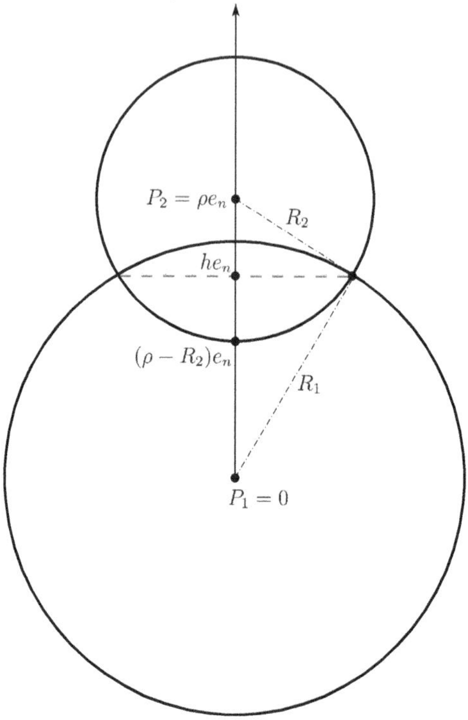

Figure 4.2.1 The geometry appearing in the proof of Lemma 4.2.13 (1/3).

has to be a segment due to the convexity of the ball (recall that $\widetilde{z} \in \overline{B_{R_1}}$). Similarly, the geodesics joining $\widetilde{z}$ to y has to be a segment, since $\widetilde{z}$, $y \in \overline{B_{R_2}(P_2)}$. These considerations establish (4.2.63).

In light of (4.2.63), we consider two situations: the geodesic is either a segment or the union of two segments. In the first case, we have that $|x - \widetilde{z}| + |y - \widetilde{z}| = |x - y|$. Hence, if $\widetilde{z} \in \big(B_{R_1} \cap B_{R_2}(P_2)\big) \cap \{x_n = h\}$, then (4.2.62) follows by taking $z := \widetilde{z}$. If instead $\widetilde{z} \in \partial B_{R_1} \cap \partial B_{R_2}(P_2)$, then we observe that $|x - \widetilde{z}| + |y - \widetilde{z}| < 2|x - y|$, and then, by continuity, we can find $z \in \big(B_{R_1} \cap B_{R_2}(P_2)\big) \cap \{x_n = h\}$ such that $|x - z| + |z - y| \leqslant \frac{3}{2}|x - y|$, which gives (4.2.62) in this case as well.

Hence, we are left with the case in which the geodesic is the union of two segments. In this case, we have that $\widetilde{z} \in \partial B_{R_1} \cap \partial B_{R_2}(P_2)$ (indeed, if not, we could find a shorter path connecting x and y, thus contradicting the minimizing property of geodesics, see Figure 4.2.3). We consider the angle γ between the vectors $x - \widetilde{z}$ and $y - \widetilde{z}$, and we observe that $\gamma \in (\gamma_0, \pi]$, for some $\gamma_0 \in (0, \pi)$.

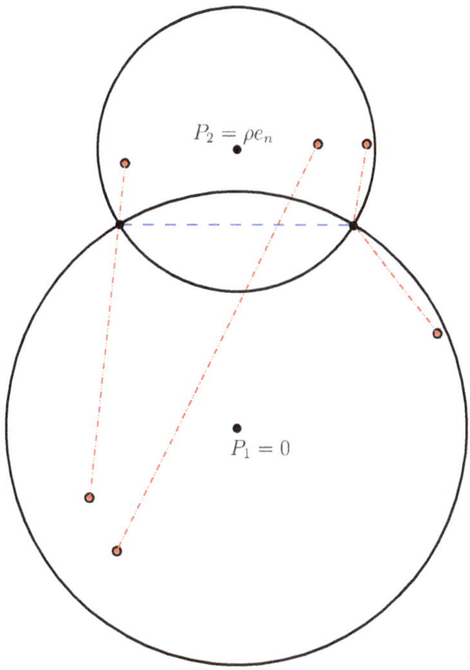

Figure 4.2.2 The geometry appearing in the proof of Lemma 4.2.13 (2/3).

Then,

$$\begin{aligned}|x-y|^2 &= |x-\overline{z}|^2 + |\overline{z}-y|^2 - 2|x-\overline{z}|\,|\overline{z}-y|\ \cos\gamma\\ &\geqslant |x-\overline{z}|^2 + |\overline{z}-y|^2 - 2|x-\overline{z}|\,|\overline{z}-y|\ \cos\gamma_0\\ &\geqslant |x-\overline{z}|^2 + |\overline{z}-y|^2 - \left(|x-\overline{z}|^2 + |\overline{z}-y|^2\right)\cos\gamma_0\\ &= \left(1-\cos\gamma_0\right)\left(|x-\overline{z}|^2 + |\overline{z}-y|^2\right).\end{aligned}$$

As a consequence,

$$\frac{|x-y|}{\sqrt{1-\cos\gamma_0}} \geqslant \sqrt{|x-\overline{z}|^2 + |\overline{z}-y|^2} \geqslant \frac{1}{\sqrt{2}}\left(|x-\overline{z}| + |\overline{z}-y|\right).$$

Thus, by continuity, we can find $z \in \left(B_{R_1} \cap B_{R_2}(P_2)\right) \cap \{x_n = h\}$ such that (4.2.62) holds true for a suitable positive constant C. This completes the proof of (4.2.62).

From (4.2.62), we deduce that

$$\begin{aligned}|D^k u(x) - D^k u(y)| &\leqslant |D^k u(x) - D^k u(z)| + |D^k u(z) - D^k u(y)|\\ &\leqslant C\|u\|_{C^{k,\alpha}(B_{R_1}(P_1))}|x-z|^\alpha + C\|u\|_{C^{k,\alpha}(B_{R_2}(P_2))}|z-y|^\alpha\\ &\leqslant C\left(\|u\|_{C^{k,\alpha}(B_{R_1}(P_1))} + C\|u\|_{C^{k,\alpha}(B_{R_2}(P_2))}\right)|x-y|^\alpha,\end{aligned}$$

thus completing the proof of Lemma 4.2.13. □

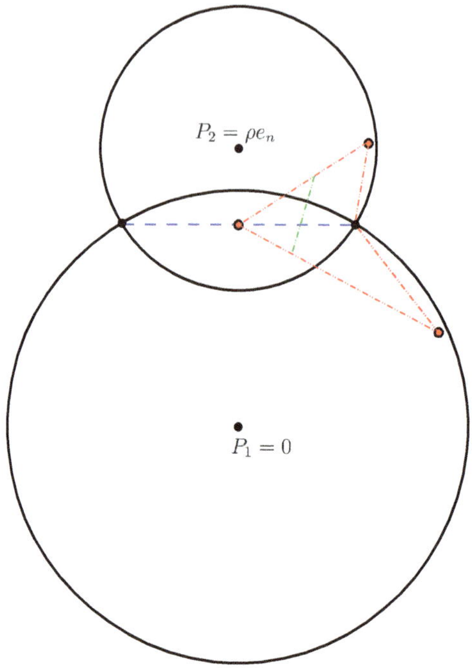

Figure 4.2.3 The geometry appearing in the proof of Lemma 4.2.13 (3/3).

With this, we can now proceed with the proof of Corollary 4.2.12.

Proof of Corollary 4.2.12. We first observe that

$$\text{it suffices to establish Corollary 4.2.12 with } R := 1. \tag{4.2.64}$$

To check this, for all $x \in B_1$ let $\widetilde{u}(x) := u(Rx)$ and $\widetilde{f}(x) := R^2 f(Rx)$. Let also $\widetilde{r} := \frac{r}{R}$. Noting that $\Delta\widetilde{u}(x) = R^2 \Delta u(Rx) = R^2 f(Rx) = \widetilde{f}(x)$ for all $x \in B_1$, if we know that Corollary 4.2.12 holds true with $R = 1$, we find that

$$\begin{aligned}\|u\|_{C^{2,\alpha}(B_r)} &\leqslant C\|\widetilde{u}\|_{C^{2,\alpha}(B_{\widetilde{r}})} \leqslant C\left(\|\widetilde{u}\|_{L^\infty(B_1)} + \|\widetilde{f}\|_{C^\alpha(B_1)}\right)\\ &\leqslant C\left(\|u\|_{L^\infty(B_R)} + \|f\|_{C^\alpha(B_R)}\right),\end{aligned}$$

where C is a positive constant, possibly varying from step to step and depending only on n, α, R and r. This proves (4.2.64).

Hence, by (4.2.64), we can suppose that $R := 1$. Also, if $r \leqslant \frac{1}{2}$, the desired result follows directly from Theorem 4.2.1; therefore, we can suppose that $r \in \left(\frac{1}{2}, 1\right)$. In this setting, we observe that Corollary 4.2.12 with $R = 1$ is proved once we show that for all $j \in \mathbb{N}$, there exists $C > 0$, depending only on n, α and j, such that

$$\|u\|_{C^{2,\alpha}(B_{r_j})} \leqslant C\Big(\|u\|_{L^\infty(B_1)} + \|f\|_{C^\alpha(B_1)}\Big),$$
$$\text{where } r_j := 1 - \frac{1}{2}\left(\frac{3}{4}\right)^j. \tag{4.2.65}$$

Indeed, once (4.2.65) is established, it is enough to note that $r_0 = \frac{1}{2} < r$ and $r_j \to 1 > r$ as $r \to +\infty$, and then pick $j \in \mathbb{N} \cap [1, +\infty)$ such that $r_{j-1} < r \leqslant r_j$: in this way, Corollary 4.2.12 with $R = 1$ would follow directly from (4.2.65).

Thanks to this observation, we focus on the proof of (4.2.65). To this end, we argue by induction over j. When $j = 0$, the claim in (4.2.65) follows from Theorem 4.2.1. To perform the inductive step, we now suppose that the statement in (4.2.65) holds true for an index j, and we aim at showing its validity for the index $j + 1$. For this, we pick a point $p \in \partial B_{r_j}$, and for every $x \in B_1$, we define

$$\rho_j := \frac{1 - r_j}{2}, \quad u_j(x) := u(p + \rho_j x) \quad \text{and} \quad f_j(x) := \rho_j^2 f(p + \rho_j x).$$

We remark that if $x \in B_1$, then $|p+\rho_j x| \leqslant |p|+\rho_j \leqslant r_j+\rho_j = \frac{1+r_j}{2} < 1$; therefore, for every $x \in B_1$, we have that $\Delta u_j(x) = \rho_j^2 f(p + \rho_j x) = f_j(x)$. We can thereby employ Theorem 4.2.1 and infer that

$$\begin{aligned}\|u\|_{C^{2,\alpha}(B_{\rho_j/2}(p))} &\leqslant C\|u_j\|_{C^{2,\alpha}(B_{1/2})} \leqslant C\Big(\|u_j\|_{L^\infty(B_1)} + \|f_j\|_{C^\alpha(B_1)}\Big)\\ &\leqslant C\Big(\|u\|_{L^\infty(B_1)} + \|f\|_{C^\alpha(B_1)}\Big),\end{aligned} \tag{4.2.66}$$

where C is a positive constant, possibly varying from step to step and depending only on n, α and j.

Now, we note that

$$\bigcup_{p\in\partial B_{r_j}} B_{\rho_j/2}(p) = B_{r_j+\rho_j/2} \setminus B_{r_j-\rho_j/2},$$

and thus we can find $N \in \mathbb{N}$ depending only on n and j and $p_1, \ldots, p_N \in \partial B_{r_j}$ such that

$$B_{r_j+\frac{\rho_j}{2}} \subseteq \left(\bigcup_{\ell=1}^{N} B_{\rho_j/2}(p_\ell)\right) \cup B_{r_j}.$$

Thus, since

$$r_j + \frac{\rho_j}{2} = r_j + \frac{1 - r_j}{4} = \frac{1}{4} + \frac{3}{4}r_j = \frac{1}{4} + \frac{3}{4} - \frac{1}{2}\left(\frac{3}{4}\right)^{j+1} = r_{j+1},$$

we have that

$$B_{r_{j+1}} \subseteq \left(\bigcup_{\ell=1}^{N} B_{\rho_j/2}(p_\ell) \right) \cup B_{r_j}.$$

We can therefore exploit (4.2.66) at the points $p_1, \ldots, p_N \in \partial B_{r_j}$ and combine it with the estimate in B_{r_j}: in this way, utilizing Lemma 4.2.13, we conclude that

$$\|u\|_{C^{2,\alpha}(B_{r_{j+1}})} \leqslant C \left(\|u\|_{L^\infty(B_1)} + \|f\|_{C^\alpha(B_1)} \right),$$

up to renaming C, which concludes the inductive step. This completes the proof of (4.2.65). □

For further reference, we recall that a useful technical variation of Theorem 4.2.2 considers the case in which the halfball B_1 is replaced by the intersection of a ball and a halfspace (whose boundary does not necessarily pass through the center of the ball). For this, we use the notation

$$\begin{aligned} & B_r^+(p) := \{x = (x', x_n) \in B_r(p) \text{ s.t. } x_n > 0\} \\ \text{and} \quad & B_r^0(p) := \{x = (x', x_n) \in B_r(p) \text{ s.t. } x_n = 0\}, \end{aligned} \tag{4.2.67}$$

and we have the following theorem.

Theorem 4.2.14. *Let $p = (p', p_n) \in \mathbb{R}^{n-1} \times \mathbb{R}$ with $p' = 0$. Let $f \in C^\alpha(B_1^+(p))$ and $g \in C^{2,\alpha}(B_1^0(p))$ for some $\alpha \in (0,1)$. Let $u \in C^2(B_1^+(p)) \cap C(B_1^+(p) \cup B_1^0(p))$ be a solution of*

$$\begin{cases} \Delta u = f & \text{in } B_1^+(p), \\ u = g & \text{on } B_1^0(p). \end{cases}$$

Then, there exists $C > 0$, depending only on n and α, such that

$$\|u\|_{C^{2,\alpha}(B_{1/2}^+(p))} \leqslant C \left(\|u\|_{L^\infty(B_1^+(p))} + \|g\|_{C^{2,\alpha}(B_1^0(p))} + \|f\|_{C^\alpha(B_1^+(p))} \right).$$

Proof. If $p_n < -\frac{1}{2}$ then $B_{1/2}^+(p) = \varnothing$ and there is nothing to prove, hence we can assume that

$$p_n \geqslant -\frac{1}{2}. \tag{4.2.68}$$

Also, if $p_n \geqslant \frac{3}{4}$, then for all $x = (x', x_n) \in B_{3/4}(p)$, we have that

$$x_n \geqslant p_n - |x_n - p_n| \geqslant \frac{3}{4} - |x - p| > 0$$

and therefore $B_{3/4}(p) \subseteq B_1^+(p)$. Thus, in this case, we can utilize Corollary 4.2.12 (centered at p instead of the origin) and deduce that

$$\|u\|_{C^{2,\alpha}(B_{1/2}^+(p))} = \|u\|_{C^{2,\alpha}(B_{1/2}(p))} \leqslant C \left(\|u\|_{L^\infty(B_{3/4}(p))} + \|f\|_{C^\alpha(B_{3/4}(p))} \right).$$

In view of this observation and (4.2.68), we can focus on the case of

$$|p_n| \leqslant \frac{3}{4}.$$

In this situation, we observe that

$$\lim_{\varrho \searrow 0} \left(\frac{3}{4}\right)^2 + \left(\varrho + \frac{1}{4}\right)^2 + \frac{3}{2}\varrho = \frac{10}{16} < 1;$$

hence we can fix a constant $\varrho \in \left(0, \frac{1}{10}\right)$ such that

$$\left(\frac{3}{4}\right)^2 + \left(\varrho + \frac{1}{4}\right)^2 + \frac{3}{2}\varrho < 1.$$

We have that, for every $q' \in \mathbb{R}^{n-1}$ with $|q'| \leqslant \frac{1}{4}$,

$$B^+_\varrho(q', 0) \subseteq B^+_1(p). \tag{4.2.69}$$

Indeed, if $x \in B^+_\varrho(q', 0)$, then $x_n > 0$ and

$$|x| \leqslant |x - (q', 0)| + |(q', 0)| < \varrho + |q'| \leqslant \varrho + \frac{1}{4}.$$

This gives that

$$p_n^2 + |x|^2 < \left(\frac{3}{4}\right)^2 + \left(\varrho + \frac{1}{4}\right)^2$$

and therefore

$$\begin{aligned} |x - p|^2 &= |x'|^2 + (x_n - p_n)^2 = |x'|^2 + x_n^2 + p_n^2 - 2x_n p_n \leqslant |x|^2 + p_n^2 + \frac{3}{2}|x_n| \\ &\leqslant |x|^2 + p_n^2 + \frac{3}{2}|x - (q', 0)| < \left(\frac{3}{4}\right)^2 + \left(\varrho + \frac{1}{4}\right)^2 + \frac{3}{2}\varrho < 1. \end{aligned}$$

This proves (4.2.69).

As a consequence of (4.2.69), we consider a finite family of halfballs $B^+_\varrho(q'_1, 0), \dots, B^+_\varrho(q'_N, 0)$, with $|q'_j| \leqslant \frac{1}{4}$ for every $j \in \{1, \dots, N\}$, with N universal, such that

$$\mathcal{S} := \left(-\frac{1}{4}, \frac{1}{4}\right)^{n-1} \times \left(0, \frac{\varrho}{4}\right) \subseteq \bigcup_{i=1}^{N} B^+_{\varrho/2}(q'_i, 0),$$

and we utilize Theorem 4.2.2 in each of these halfballs, finding that

$$\|u\|_{C^{2,\alpha}(B^+_{\varrho/2}(q'_i,0))} \leqslant C\left(\|u\|_{L^\infty(B^+_1(p))} + \|g\|_{C^{2,\alpha}(B^0_1(p))} + \|f\|_{C^\alpha(B^+_1(p))}\right),$$

and thus, up to renaming constants,

$$\|u\|_{C^{2,\alpha}(\mathcal{S})} \leqslant C\left(\|u\|_{L^\infty(B^+_1(p))} + \|g\|_{C^{2,\alpha}(B^0_1(p))} + \|f\|_{C^\alpha(B^+_1(p))}\right). \tag{4.2.70}$$

Similarly, one considers a finite family of balls $B_{\varrho/9}(Q_1), \dots, B_{\varrho/8}(Q_M)$, with M universal, such that

$$\mathcal{T} := \left\{ x = (x', x_n) \in B^+_{1-2\varrho}(p) \text{ s.t. } x_n > \frac{\varrho}{8} \right\} \subseteq \bigcup_{i=1}^{M} B_{\varrho/9}(Q_i).$$

One can employ Corollary 4.2.12 (centered at Q_i instead of the origin) in each of these balls and conclude that

$$\|u\|_{C^{2,\alpha}(B_{\varrho/9}(Q_i))} \leqslant C \left(\|u\|_{L^\infty(B_{\varrho/8}(Q_i))} + \|f\|_{C^\alpha(B_{\varrho/8}(Q_i))} \right)$$

and therefore, up to renaming C,

$$\|u\|_{C^{2,\alpha}(\mathcal{T})} \leqslant C \left(\|u\|_{L^\infty(B^+_1(p))} + \|g\|_{C^{2,\alpha}(B^0_1(p))} + \|f\|_{C^\alpha(B^+_1(p))} \right).$$

From this and (4.2.70), up to renaming C, we arrive at

$$\|u\|_{C^{2,\alpha}(\mathcal{S}\cup\mathcal{T})} \leqslant C \left(\|u\|_{L^\infty(B^+_1(p))} + \|g\|_{C^{2,\alpha}(B^0_1(p))} + \|f\|_{C^\alpha(B^+_1(p))} \right). \quad (4.2.71)$$

We also have that

$$\mathcal{S} \cup \mathcal{T} \supseteq B^+_{1/4}(p). \quad (4.2.72)$$

Indeed, take $x = (x', x_n) \in B^+_{1/4}(p)$. If $x_n > \frac{\varrho}{8}$, we use that $|x - p| < \frac{1}{4} < 1 - 2\varrho$ and thus $x \in \mathcal{T}$. If instead $x_n \in \left(0, \frac{\varrho}{8}\right]$, we have that, for every $m \in \{1, \dots, n-1\}$, $|x_m| \leqslant |x'| = |x' - p'| < \frac{1}{4}$, whence $x \in \mathcal{S}$, which completes the proof of (4.2.72).

By (4.2.71) and (4.2.72) we deduce that

$$\|u\|_{C^{2,\alpha}(B^+_{1/4}(p)} \leqslant C \left(\|u\|_{L^\infty(B^+_1(p))} + \|g\|_{C^{2,\alpha}(B^0_1(p))} + \|f\|_{C^\alpha(B^+_1(p))} \right),$$

leading to the desired result up to scaling and covering. □

4.3 Pointwise Hölder Spaces

It is often convenient to investigate the Hölder regularity theory in light of the so-called "pointwise Hölder spaces" (see e.g. [And97], [AGLV09, Section 9.2] and the references therein), that is, in simple terms, to detect the Hölder regularity of a function via the pointwise approximation of its Taylor polynomials.

Definition 4.3.1. Let $\Omega \subseteq \mathbb{R}^n$ be open. Let $k \in \mathbb{N}$ and $\alpha \in (0, 1]$. Let $u : \Omega \to \mathbb{R}$ and $x_0 \in \Omega$. We say that u is $C^{k,\alpha}$ at the point x_0 in Ω (and we write $u \in C^{k,\alpha}_\Omega(x_0)$) if there exist $M > 0$ and a polynomial P of degree at most k such that

$$|u(x) - P(x)| \leqslant M|x - x_0|^{k+\alpha} \quad \text{for all } x \in \Omega.$$

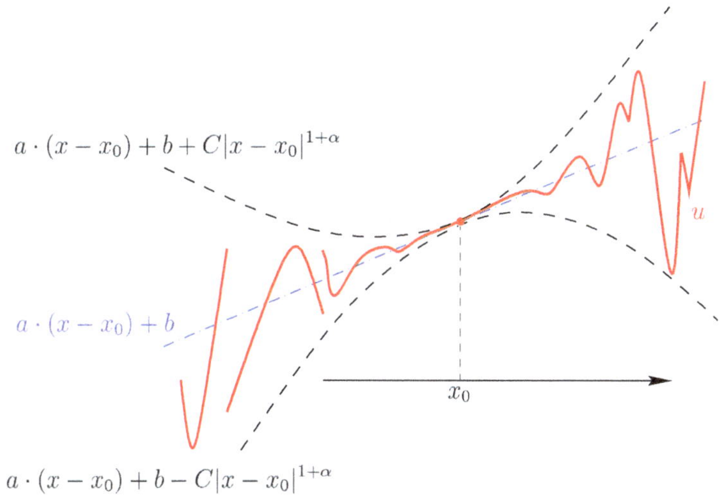

Figure 4.3.1 A function in $C^{1,\alpha}_{\mathbb{R}}(x_0)$.

See Figure 4.3.1 for a pictorial representation of Definition 4.3.1 when $k = 1$. When the constant M becomes uniform within a Lipschitz domain, one recovers from Definition 4.3.1 the usual notion of Hölder spaces, according to the following result.

Proposition 4.3.2. *Let $k \in \mathbb{N}$ and $\alpha \in (0, 1]$. Let $\Omega \subseteq \mathbb{R}^n$ be open, bounded and convex. Let $u : \Omega \to \mathbb{R}$. Then, the following conditions are equivalent:*

(i) $u \in C^{k,\alpha}(\Omega)$,
(ii) *There exists $M \geqslant 0$ such that for every $x_0 \in \Omega$, there exists a polynomial P_{x_0} of degree at most k such that*

$$|D^\gamma P_{x_0}(x_0)| \leqslant M \quad for\ all\ \gamma \in \mathbb{N}^n \text{ with } |\gamma| = m \quad and \quad m \in \{0, \dots, k\}, \tag{4.3.1}$$

and

$$|u(x) - P_{x_0}(x)| \leqslant M|x - x_0|^{k+\alpha} \quad for\ all\, x \in \Omega. \tag{4.3.2}$$

Additionally, if (i) holds true, then M in (ii) can be taken of the form $M := C\,\|u\|_{C^{k,\alpha}(\Omega)}$ for some $C > 0$ depending only on n, k, α and Ω.

Similarly, if (ii) holds true, then $\|u\|_{C^{k,\alpha}(\Omega)} \leqslant CM$, for some $C > 0$ depending only on n, k, α and Ω.

Proof. Assume that (i) holds true. Given $x_0 \in \Omega$, we let P_{x_0} be the Taylor polynomial of degree k of u, namely

$$P_{x_0}(x) := \sum_{\substack{\beta \in \mathbb{N}^n \\ |\beta| \leqslant k}} \frac{D^\beta u(x_0)}{\beta!} (x - x_0)^\beta.$$

In particular, for all $m \in \mathbb{N} \cap [0, k]$ and $\beta \in \mathbb{N}^n$ with $|\beta| = m$,

$$|D^\beta P_{x_0}(x_0)| = |D^\beta u(x_0)| \leqslant \|u\|_{C^{k,\alpha}(\Omega)}. \tag{4.3.3}$$

Also, since Ω is convex, for every $x \in \Omega$, there exists $\xi(x) \in \Omega$ lying on the segment joining x and x_0 such that

$$\begin{aligned} \left|u(x) - P_{x_0}(x)\right| &= \left| \sum_{\substack{\beta \in \mathbb{N}^n \\ |\beta| = k}} \frac{D^\beta u(\xi(x)) - D^\beta u(x_0)}{\beta!} (x - x_0)^\beta \right| \\ &\leqslant \sum_{\substack{\beta \in \mathbb{N}^n \\ |\beta| = k}} \frac{|D^\beta u(\xi(x)) - D^\beta u(x_0)|}{\beta!} |x - x_0|^k \\ &\leqslant \|u\|_{C^{k,\alpha}(\Omega)} \sum_{\substack{\beta \in \mathbb{N}^n \\ |\beta| = k}} \frac{|\xi(x) - x_0|^\alpha}{\beta!} |x - x_0|^k \\ &\leqslant C \|u\|_{C^{k,\alpha}(\Omega)} |x - x_0|^{k+\alpha}, \end{aligned}$$

which, together with (4.3.3), gives (ii).

Now, conversely, suppose that (ii) is satisfied. We observe that

$$\begin{aligned} &\text{for every polynomial } P \text{ in } \mathbb{R}^n \text{ of degree at most } k, \\ &\text{we have that } \|P\|_{C^k(\Omega)} \leqslant C \|P\|_{L^\infty(\Omega)}, \end{aligned} \tag{4.3.4}$$

for some $C > 0$ depending only on n, k and Ω. To check this, we let N be the number of multi-indices $\beta \in \mathbb{N}^n$ such that $|\beta| \leqslant k$. We consider the family $\mathcal{P}_k$ of polynomials of n variables with degree at most k and the map

$$\mathbb{R}^N \ni a = \{a_\beta\}_{\substack{\beta \in \mathbb{N}^n \\ |\beta| \leqslant k}} \longmapsto P_a(x) := \sum_{\substack{\beta \in \mathbb{N}^n \\ |\beta| \leqslant k}} a_\beta x^\beta \in \mathcal{P}_k.$$

We let $\|a\|_1 := \|P_a\|_{C^k(\Omega)}$ and $\|a\|_2 := \|P_a\|_{L^\infty(\Omega)}$. We stress that both $\|\cdot\|_1$ and $\|\cdot\|_2$ are norms in $\mathbb{R}^N$. Since all norms in $\mathbb{R}^N$ are equivalent (see e.g. [Por81, Theorem 15.26]), we find that there exists C as above such that, for every $a \in \mathbb{R}^N$, it holds that $\|P_a\|_{C^k(\Omega)} = \|a\|_1 \leqslant C\|a\|_2 = C\|P_a\|_{L^\infty(\Omega)}$, from which we obtain (4.3.4).

We also claim that, for all $p \in \mathbb{R}^n$ and $r > 0$, if P is a polynomial of degree at most k and $j \in \mathbb{N} \cap [0, k]$, then, for all $\beta \in \mathbb{N}^n$ with $|\beta| = j$,

$$\|D^\beta P\|_{L^\infty(B_r(p))} \leqslant \frac{C}{r^j} \|P\|_{L^\infty(B_r(p))}, \tag{4.3.5}$$

for some $C > 0$ depending only on n and k. To check this, we let $\widetilde{P}(x) := P(p+rx)$, and we make use of (4.3.4) to find that

$$\begin{aligned} r^j \sup_{y \in B_r(p)} |D^\beta P(y)| &= r^j \sup_{x \in B_1} |D^\beta P(p + rx)| = \sup_{x \in B_1} |D^\beta \widetilde{P}(x)| \\ &\leqslant \|\widetilde{P}\|_{C^k(B_1)} \leqslant C\|\widetilde{P}\|_{L^\infty(B_1)} = C\|P\|_{L^\infty(B_r(p))}, \end{aligned}$$

proving (4.3.5).

Now, we observe that for all $x_0, \widetilde{x}_0 \in \Omega$, setting $p := \frac{x_0+\widetilde{x}_0}{2}$ and $r := |x_0 - \widetilde{x}_0| \in (0, 1)$, if $x \in B_r(p)$ we have that $|x - x_0| \leqslant |x - p| + |p - x_0| \leqslant 2r$, and similarly $|x - \widetilde{x}_0| \leqslant 2r$, giving that

$$\begin{aligned} |P_{x_0}(x) - P_{\widetilde{x}_0}(x)| &\leqslant |P_{x_0}(x) - u(x)| + |u(x) - P_{\widetilde{x}_0}(x)| \\ &\leqslant M|x - x_0|^{k+\alpha} + M|x - \widetilde{x}_0|^{k+\alpha} \leqslant CMr^{k+\alpha}. \end{aligned}$$

From this and (4.3.5) (used here with $P := P_{x_0} - P_{\widetilde{x}_0}$) it follows that for all x_0, $\widetilde{x}_0 \in \Omega$, if $p := \frac{x_0+\widetilde{x}_0}{2}$ and $r := |x_0 - \widetilde{x}_0| \in (0, 1)$, then, for all $\beta \in \mathbb{N}^n$ with $|\beta| = j$,

$$\|D^\beta P_{x_0} - D^\beta P_{\widetilde{x}_0}\|_{L^\infty(B_r(p))} \leqslant CMr^{k-j+\alpha}. \tag{4.3.6}$$

Now, we prove that

$$\begin{aligned} &u \in C^{k,\alpha}(\Omega) \text{ and, for every } x_0 \in \Omega \text{ and } \beta \in \mathbb{N}^n \text{ with } |\beta| \leqslant k, \\ &\text{we have that } D^\beta u(x_0) = D^\beta P_{x_0}(x_0). \end{aligned} \tag{4.3.7}$$

This is accomplished by checking that, for all $j \in \mathbb{N} \cap [0, k]$,

$$\begin{aligned} &u \in C^{j,\alpha}(\Omega) \text{ and, for every } x_0 \in \Omega \text{ and } \beta \in \mathbb{N}^n \text{ with } |\beta| \leqslant j, \\ &\text{we have that } D^\beta u(x_0) = D^\beta P_{x_0}(x_0). \end{aligned} \tag{4.3.8}$$

We prove this claim by induction over j. For this, we first observe that, for every $x_0 \in \Omega$, condition (ii) entails that

$$\begin{aligned} \lim_{x \to x_0} |u(x) - P_{x_0}(x_0)| &\leqslant \lim_{x \to x_0} \Big(|u(x) - P_{x_0}(x)| + |P_{x_0}(x) - P_{x_0}(x_0)|\Big) \\ &= \lim_{x \to x_0} |u(x) - P_{x_0}(x)| \leqslant \lim_{x \to x_0} M|x - x_0|^{k+\alpha} = 0. \end{aligned}$$

This says that u is continuous at x_0 and $u(x_0) = P_{x_0}(x_0)$. Furthermore, writing

$$P_{x_0}(x) = \sum_{\substack{\beta \in \mathbb{N}^n \\ |\beta| \leqslant k}} \frac{D^\beta P_{x_0}(x_0)}{\beta!} (x - x_0)^\beta, \tag{4.3.9}$$

if $\delta \in (0,1)$ and $e \in \partial B_1$ we have that

$$|P_{x_0}(x_0+\delta e) - P_{x_0}(x_0)| = \left| \sum_{\substack{\beta\in\mathbb{N}^n\\ |\beta|\leqslant k}} \frac{D^\beta P_{x_0}(x_0)}{\beta!} (\delta e)^\beta - P_{x_0}(x_0)\right|$$
$$= \left| \sum_{\substack{\beta\in\mathbb{N}^n\\ 1\leqslant|\beta|\leqslant k}} \frac{\delta^{|\beta|} D^\beta P_{x_0}(x_0)}{\beta!} e^\beta \right| \leqslant CM\delta,$$

thanks to (4.3.1).

For this reason,

$$\begin{aligned}|u(x_0+\delta e) - u(x_0)| &= |u(x_0+\delta e) - P_{x_0}(x_0)|\\ &\leqslant |u(x_0+\delta e) - P_{x_0}(x_0+\delta e)| + |P_{x_0}(x_0+\delta e) - P_{x_0}(x_0)|\\ &\leqslant M\delta^{k+\alpha} + CM\delta \leqslant CM\delta^\alpha\end{aligned}$$

and therefore $u \in C^\alpha(\Omega)$. This is the desired claim in (4.3.8) when $j = 0$.

Now, if $k \geqslant 1$ and $i \in \{1,\dots,n\}$

$$\begin{aligned}&\lim_{\varepsilon\to 0}\left|\frac{u(x_0+\varepsilon e_i)-u(x_0)}{\varepsilon} - \partial_i P_{x_0}(x_0)\right|\\
&\leqslant \lim_{\varepsilon\to 0}\left(\left|\frac{u(x_0+\varepsilon e_i)-u(x_0)}{\varepsilon} - \frac{P_{x_0}(x_0+\varepsilon e_i)-P_{x_0}(x_0)}{\varepsilon}\right|\right.\\
&\qquad \left.+\left|\frac{P_{x_0}(x_0+\varepsilon e_i)-P_{x_0}(x_0)}{\varepsilon} - \partial_i P_{x_0}(x_0)\right|\right)\\
&= \lim_{\varepsilon\to 0}\left|\frac{u(x_0+\varepsilon e_i)-u(x_0)}{\varepsilon} - \frac{P_{x_0}(x_0+\varepsilon e_i)-P_{x_0}(x_0)}{\varepsilon}\right|\\
&= \lim_{\varepsilon\to 0}\left|\frac{u(x_0+\varepsilon e_i)-P_{x_0}(x_0+\varepsilon e_i)}{\varepsilon}\right|\\
&\leqslant \lim_{\varepsilon\to 0}\frac{M|\varepsilon|^{k+\alpha}}{|\varepsilon|}\\
&= 0.\end{aligned}$$

This gives that u can be differentiated at least once at x_0 and $\nabla u(x_0) = \nabla P_{x_0}(x_0)$. Additionally, differentiating (4.3.9), for all $i \in \{1,\dots,n\}$,

$$\partial_i P_{x_0}(x) = \sum_{\substack{\beta\in\mathbb{N}^n\\ 1\leqslant|\beta|\leqslant k}} \frac{\beta_i\, D^\beta P_{x_0}(x_0)}{\beta!}(x-x_0)^{\beta-e_i}.$$

Hence, recalling (4.3.1), if $\delta \in (0,1)$ and $e \in \partial B_1$,

$$\begin{aligned}
&|\partial_i P_{x_0}(x_0+\delta e) - \partial_i P_{x_0}(x_0)| \\
&= \left| \sum_{\substack{\beta\in\mathbb{N}^n \\ 1\leqslant|\beta|\leqslant k}} \frac{\beta_i\, D^\beta P_{x_0}(x_0)}{\beta!} (\delta e)^{\beta-e_i} - \partial_i P_{x_0}(x_0) \right| \\
&= \left| \sum_{\substack{\beta\in\mathbb{N}^n \\ 2\leqslant|\beta|\leqslant k}} \frac{\beta_i\, D^\beta P_{x_0}(x_0)}{\beta!} (\delta e)^{\beta-e_i} + \sum_{j=1}^{n} \delta_{ij} \partial_j P_{x_0}(x_0)\,(\delta e)^{e_j-e_i} - \partial_i P_{x_0}(x_0) \right| \\
&= \left| \sum_{2\leqslant|\beta|\leqslant k} \frac{\delta^{|\beta|-1}\beta_i\, D^\beta P_{x_0}(x_0)}{\beta!}\, e^{\beta-e_i} \right| \leqslant CM\delta. \qquad (4.3.10)
\end{aligned}$$

Moreover,

$$\begin{aligned}
&\left| \frac{u(x_0+\delta e+\varepsilon e_i) - u(x_0+\delta e)}{\varepsilon} - \frac{u(x_0+\varepsilon e_i) - u(x_0)}{\varepsilon} \right| \\
&\quad \leqslant \left| \frac{P_{x_0+\delta e}(x_0+\delta e+\varepsilon e_i) - P_{x_0+\delta e}(x_0+\delta e)}{\varepsilon} - \frac{P_{x_0}(x_0+\varepsilon e_i) - P_{x_0}(x_0)}{\varepsilon} \right| \\
&\quad\quad + \frac{1}{|\varepsilon|}\Big(|u(x_0+\delta e+\varepsilon e_i) - P_{x_0+\delta e}(x_0+\delta e+\varepsilon e_i)| \\
&\quad\quad + |u(x_0+\delta e) - P_{x_0+\delta e}(x_0+\delta e)| \\
&\quad\quad + |u(x_0+\varepsilon e_i) - P_{x_0}(x_0+\varepsilon e_i)| + |u(x_0) - P_{x_0}(x_0)| \Big) \\
&\quad \leqslant \left| \frac{P_{x_0+\delta e}(x_0+\delta e+\varepsilon e_i) - P_{x_0+\delta e}(x_0+\delta e)}{\varepsilon} - \frac{P_{x_0}(x_0+\varepsilon e_i) - P_{x_0}(x_0)}{\varepsilon} \right| \\
&\quad\quad + \frac{1}{|\varepsilon|}\Big(M|\varepsilon|^{k+\alpha} + 0 + M|\varepsilon|^{k+\alpha} + 0 \Big),
\end{aligned}$$

leading to, as $\varepsilon \to 0$,

$$|\partial_i u(x_0+\delta e) - \partial_i u(x_0)| \leqslant |\partial_i P_{x_0+\delta}(x_0+\delta e) - \partial_i P_{x_0}(x_0)|.$$

This, together with (4.3.6) and (4.3.10), gives that, if $x_0 \in \Omega$, $e \in \partial B_1$ and $\delta \in (0,1)$ is sufficiently small such that $x_0 + \delta e \in \Omega$, then

$$\begin{aligned}
&|\partial_i u(x_0+\delta e) - \partial_i u(x_0)| \\
&\quad \leqslant |\partial_i P_{x_0+\delta}(x_0+\delta e) - \partial_i P_{x_0}(x_0+\delta e)| + |\partial_i P_{x_0}(x_0+\delta e) - \partial_i P_{x_0}(x_0)| \\
&\quad \leqslant CM\delta^{k-1+\alpha} + CM\delta \leqslant CM\delta^\alpha.
\end{aligned}$$

This gives that $\partial_i u \in C^\alpha(\Omega)$, which completes the proof of (4.3.8) when $j = 1$.

Suppose now recursively that $j \in \mathbb{N} \cap [0, k-1]$ and $u \in C^{j,\alpha}(\Omega)$ is such that at each point $x_0 \in \Omega$, we have $D^\beta u(x_0) = D^\beta P_{x_0}(x_0)$ as long as $|\beta| \leqslant j$. Then, recalling (4.3.6), for all $i \in \{1, \dots, n\}$,

$$\begin{aligned}
&\lim_{\varepsilon\to 0}\left|\frac{D^\beta u(x_0+\varepsilon e_i) - D^\beta u(x_0)}{\varepsilon} - D^{\beta+e_i}P_{x_0}(x_0)\right| \\
&\leqslant \lim_{\varepsilon\to 0}\left(\left|\frac{D^\beta u(x_0+\varepsilon e_i) - D^\beta u(x_0)}{\varepsilon} - \frac{D^\beta P_{x_0}(x_0+\varepsilon e_i) - D^\beta P_{x_0}(x_0)}{\varepsilon}\right|\right. \\
&\qquad \left.+\left|\frac{D^\beta P_{x_0}(x_0+\varepsilon e_i) - D^\beta P_{x_0}(x_0)}{\varepsilon} - D^{\beta+e_i}P_{x_0}(x_0)\right|\right) \\
&= \lim_{\varepsilon\to 0}\left|\frac{D^\beta u(x_0+\varepsilon e_i) - D^\beta u(x_0)}{\varepsilon} - \frac{D^\beta P_{x_0}(x_0+\varepsilon e_i) - D^\beta P_{x_0}(x_0)}{\varepsilon}\right| \\
&= \lim_{\varepsilon\to 0}\left|\frac{D^\beta P_{x_0+\varepsilon e_i}(x_0+\varepsilon e_i) - D^\beta P_{x_0}(x_0)}{\varepsilon} - \frac{D^\beta P_{x_0}(x_0+\varepsilon e_i) - D^\beta P_{x_0}(x_0)}{\varepsilon}\right| \\
&= \lim_{\varepsilon\to 0}\left|\frac{D^\beta P_{x_0+\varepsilon e_i}(x_0+\varepsilon e_i) - D^\beta P_{x_0}(x_0+\varepsilon e_i)}{\varepsilon}\right| \\
&\leqslant \lim_{\varepsilon\to 0} CM|\varepsilon|^{k-j-1+\alpha} \\
&\leqslant \lim_{\varepsilon\to 0} CM|\varepsilon|^{\alpha} \\
&= 0.
\end{aligned}$$

This gives that $D^{\beta+e_i}u(x_0) = D^{\beta+e_i}P_{x_0}(x_0)$.

Furthermore, from (4.3.9), we see that, for every $\gamma \in \mathbb{N}^n$,

$$D^\gamma P_{x_0}(x) = \sum_{\substack{\beta\in\mathbb{N}^n \\ |\beta|\leqslant k \\ \gamma\leqslant\beta}} \prod_{\substack{1\leqslant j\leqslant n \\ \gamma_j\geqslant 1}} \beta_j(\beta_j-1)\dots(\beta_j-\gamma_j+1)\frac{D^\beta P_{x_0}(x_0)}{\beta!}(x-x_0)^{\beta-\gamma}.$$

In the above, we use the notation that $\gamma = (\gamma_i)_{1\leqslant i\leqslant n}$, $\beta = (\beta_i)_{1\leqslant i\leqslant n}$, and we say that $\gamma \leqslant \beta$ if $\gamma_i \leqslant \beta_i$ for every $i \in \{1, \dots, n\}$.

Accordingly, if $x_0 \in \Omega$, $e \in \partial B_1$ and $\delta \in (0, 1)$,

$$\begin{aligned}
&\left|D^\gamma P_{x_0}(x_0+\delta e) - D^\gamma P_{x_0}(x_0)\right| \\
&= \left|\sum_{\substack{\beta\in\mathbb{N}^n \\ |\beta|\leqslant k \\ \gamma\leqslant\beta}} \prod_{\substack{1\leqslant j\leqslant n \\ \gamma_j\geqslant 1}} \beta_j(\beta_j-1)\dots(\beta_j-\gamma_j+1)\frac{D^\beta P_{x_0}(x_0)}{\beta!}(\delta e)^{\beta-\gamma} - D^\gamma P_{x_0}(x_0)\right|
\end{aligned}$$

$$= \left| \sum_{\substack{\beta\in\mathbb{N}^n \\ |\gamma|+1\leqslant|\beta|\leqslant k \\ \gamma\leqslant\beta}} \prod_{\substack{1\leqslant j\leqslant n \\ \gamma_j\geqslant 1}} \beta_j(\beta_j-1)\ldots(\beta_j-\gamma_j+1)\frac{D^\beta P_{x_0}(x_0)}{\beta!}(\delta e)^{\beta-\gamma} \right.$$

$$+ \sum_{\substack{\beta\in\mathbb{N}^n \\ |\gamma|=|\beta|\leqslant k \\ \gamma\leqslant\beta}} \prod_{\substack{1\leqslant j\leqslant n \\ \gamma_j\geqslant 1}} \beta_j(\beta_j-1)\ldots(\beta_j-\gamma_j+1)\frac{D^\beta P_{x_0}(x_0)}{\beta!}(\delta e)^{\beta-\gamma}$$

$$\left. - D^\gamma P_{x_0}(x_0) \right|. \tag{4.3.11}$$

Now, we note that if $|\gamma| = |\beta|$ and $\gamma \leqslant \beta$, then $\beta_i = \gamma_i$ for all $i \in \{1, \ldots, n\}$, and therefore $\beta = \gamma$ and

$$\prod_{\substack{1\leqslant j\leqslant n \\ \gamma_j\geqslant 1}} \beta_j(\beta_j-1)\ldots(\beta_j-\gamma_j+1) = \beta!.$$

This gives that

$$\sum_{\substack{\beta\in\mathbb{N}^n \\ |\gamma|=|\beta|\leqslant k \\ \gamma\leqslant\beta}} \prod_{\substack{1\leqslant j\leqslant n \\ \gamma_j\geqslant 1}} \beta_j(\beta_j-1)\ldots(\beta_j-\gamma_j+1)\frac{D^\beta P_{x_0}(x_0)}{\beta!}(\delta e)^{\beta-\gamma} = D^\gamma P_{x_0}(x_0).$$

Plugging this information into (4.3.11) and recalling (4.3.1), we conclude that

$$\begin{aligned} &\left|D^\gamma P_{x_0}(x_0+\delta e) - D^\gamma P_{x_0}(x_0)\right| \\ &\quad = \left| \sum_{\substack{\beta\in\mathbb{N}^n \\ |\gamma|+1\leqslant|\beta|\leqslant k \\ \gamma\leqslant\beta}} \prod_{\substack{1\leqslant j\leqslant n \\ \gamma_j\geqslant 1}} \beta_j(\beta_j-1)\ldots(\beta_j-\gamma_j+1)\frac{|\delta|^{|\beta|-|\gamma|}\, D^\beta P_{x_0}(x_0)}{\beta!} e^{\beta-\gamma} \right| \\ &\quad \leqslant CM\delta. \end{aligned} \tag{4.3.12}$$

Consequently, using (4.3.12) with $\gamma := \beta + e_i$ and (4.3.6), if $x_0 \in \Omega$, $e \in \partial B_1$ and $\delta \in (0, 1)$ is sufficiently small such that $x_0 + \delta e \in \Omega$, then

$$\begin{aligned} &|D^{\beta+e_i}u(x_0) - D^{\beta+e_i}u(x_0+\delta e)| \\ &\quad = |D^{\beta+e_i}P_{x_0}(x_0) - D^{\beta+e_i}P_{x_0+\delta e}(x_0+\delta e)| \\ &\quad \leqslant |D^{\beta+e_i}P_{x_0}(x_0) - D^{\beta+e_i}P_{x_0}(x_0+\delta e)| \\ &\qquad + |D^{\beta+e_i}P_{x_0}(x_0+\delta e) - D^{\beta+e_i}P_{x_0+\delta e}(x_0+\delta e)| \end{aligned}$$

$$\begin{aligned}&\leqslant CM\delta + CM\delta^{k-j-1+\alpha}\\&\leqslant CM\delta^{\alpha}.\end{aligned} \tag{4.3.13}$$

Accordingly, we have that $D^{\beta+e_i}u \in C^{\alpha}(\Omega)$, which completes the inductive step. The claim in (4.3.8) is thereby established, and from this, we also obtain (4.3.7).

From (4.3.1) and (4.3.7), it also follows that

$$\sup_{\substack{x_0\in\Omega\\|\gamma|\leqslant k}} |D^{\gamma}u(x_0)| = \sup_{\substack{x_0\in\Omega\\|\gamma|\leqslant k}} |D^{\gamma}P_{x_0}(x_0)| \leqslant M.$$

Additionally, by (4.3.13), we see that $|D^{\gamma}u(x_0) - D^{\gamma}u(x_0+\delta e)| \leqslant CM\delta^{\alpha}$ for all $\gamma \in \mathbb{N}^n$ such that $|\gamma| \leqslant k$, and therefore $\|u\|_{C^{k,\alpha}(\Omega)} \leqslant CM$. This gives that (i) holds true, as desired. □

It is worth pointing out that, in view of Definition 4.3.1, condition (4.3.2) gives[11] that $u \in C^{k,\alpha}_{\Omega}(x_0)$ for every $x_0 \in \Omega$, with a uniform M.

Without the uniformity in M, one cannot precisely reconstruct $C^{k,\alpha}(\Omega)$ out of $C^{k,\alpha}_{\Omega}(x_0)$ for $x_0 \in \Omega$, as detailed in the following result.

Lemma 4.3.3. *Let $k \in \mathbb{N}$ and $\alpha \in (0,1]$. Let $\Omega \subseteq \mathbb{R}^n$ be open, bounded and convex. We have that*

$$C^{k,\alpha}(\Omega) \subseteq \bigcup_{x_0\in\Omega} C^{k,\alpha}_{\Omega}(x_0), \tag{4.3.14}$$

and the above inclusion is strict.

Proof. If $u \in C^{k,\alpha}(\Omega)$, then condition (i) in Proposition 4.3.2 holds true, leading to (4.3.2), whence $u \in C^{k,\alpha}_{\Omega}(x_0)$ for all $x_0 \in \Omega$, in light of Definition 4.3.1.

---

[11] Condition (4.3.1) is instead mostly technical and can also be obtained from (4.3.2) (if valid at every point) if one knows $P_{\widetilde{x}}$ for a point $\widetilde{x} \in \Omega$: indeed, it follows from (4.3.2) and the boundedness of Ω that

$$\|u\|_{L^\infty(\Omega)} = \sup_{x\in\Omega} |u(x)| \leqslant \sup_{x\in\Omega}\Big(|P_{\widetilde{x}}(x)| + M|x-\widetilde{x}|^{k+\alpha}\Big) \leqslant \|P_{\widetilde{x}}\|_{L^\infty(\Omega)} + CM,$$

whence

$$\begin{aligned}\|P_{x_0}\|_{L^\infty(\Omega)} &= \sup_{x\in\Omega}|P_{x_0}(x)| \leqslant \sup_{x\in\Omega}\Big(|u(x)| + M|x-x_0|^{k+\alpha}\Big)\\&\leqslant \|u\|_{L^\infty(\Omega)} + CM \leqslant C\Big(\|P_{\widetilde{x}}\|_{L^\infty(\Omega)} + M\Big).\end{aligned}$$

From this and the finite-dimensional structure of the polynomials (see e.g. (4.3.4)), one would obtain that $|D^m P_{x_0}(x_0)| \leqslant C\Big(\|P_{\widetilde{x}}\|_{L^\infty(\Omega)} + M\Big)$; therefore, a uniform information on (4.3.2) would entail a uniform information on (4.3.1).

A small catch in the above argument is that M would be replaced, up to constants, by $\|P_{\widetilde{x}}\|_{L^\infty(\Omega)}+M$, which is a quantity not immediately related to the original function u. This small inconvenience can be fixed, for instance, if one is willing to work with bounded functions: this idea will be clarified in the forthcoming Lemma 4.3.4.

This shows the desired inclusion. To check that the inclusion is strict, we consider a sequence of mutually disjoint intervals $I_j \subset (0, +\infty)$ of the form $I_j = [p_j - \ell_j, p_j + \ell_j]$, with $p_j \in (0, 1)$ and $p_j \searrow 0$ as $j \to +\infty$. By possibly shrinking I_j, we can additionally suppose that

$$\ell_j \leqslant \left(\frac{p_j}{2}\right)^{\frac{k+\alpha}{k}}$$

and in particular $\ell_j \leqslant \frac{p_j}{2}$. Therefore, if $x_1 \in I_j$, then

$$|x_1| \geqslant x_1 \geqslant p_j - \ell_j \geqslant \frac{p_j}{2} \geqslant \ell_j^{\frac{k}{k+\alpha}} = \frac{\ell_j}{\ell_j^{\frac{\alpha}{k+\alpha}}} \geqslant \frac{|x_1 - p_j|}{\ell_j^{\frac{\alpha}{k+\alpha}}}. \tag{4.3.15}$$

Let also $s_j := \frac{1}{\ell_j^\alpha}$, and let $\eta_j \in C_0^\infty(I_j, [0, 1])$ be such that $\eta = 1$ in $\left[p_j - \frac{\ell_j}{2}, p_j + \frac{\ell_j}{2}\right]$.

For each $j \in \mathbb{N}$, we define the function

$$\mathbb{R} \ni t \mapsto u_j(t) := s_j |t - p_j|^{k+\alpha} \eta_j(t).$$

Note that the supports of u_j do not intersect, and hence we can define

$$\mathbb{R} \ni t \longmapsto u_\star(t) := \sum_{j \in \mathbb{N}} u_j(t)$$

$$\text{and} \quad \mathbb{R}^n \ni x \longmapsto u(x) = u(x_1, \dots, x_n) := u_\star(x_1). \tag{4.3.16}$$

See Figure 4.3.2 for a sketch of this situation. We stress that $u(x) \geqslant 0$ for all $x \in \mathbb{R}^n$ and $u(x) = 0$ unless x_1 belongs to one of the disjoint intervals I_j. Accordingly, for every x such that $u(x) \neq 0$, we denote by $j_x \in \mathbb{N}$ the unique index for which $x_1 \in I_{j_x}$.

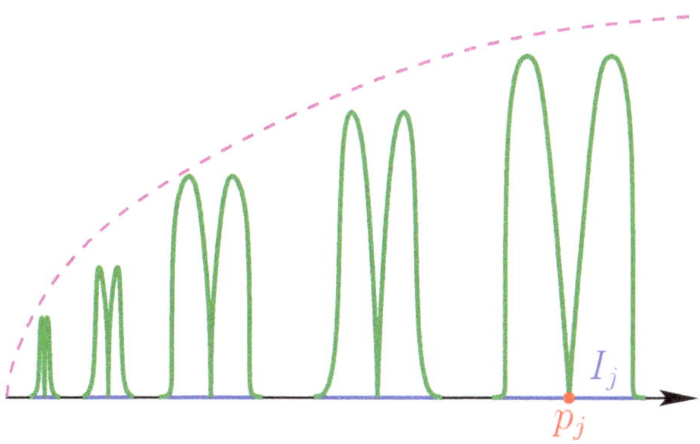

Figure 4.3.2 The function $u_\star$ in (4.3.16) with $k = 0$.

In this way, if $u(x) \neq 0$ then, utilizing (4.3.15),

$$|u(x)| = u(x) = u_{j_x}(x_1) \leqslant s_{j_x}|x_1 - p_{j_x}|^{k+\alpha} = \frac{|x_1 - p_{j_x}|^{k+\alpha}}{\ell_{j_x}^{\alpha}}$$
$$= \left(\frac{|x_1 - p_{j_x}|}{\ell_{j_x}^{\frac{\alpha}{k+\alpha}}}\right)^{k+\alpha} \leqslant |x_1|^{k+\alpha} \leqslant |x|^{k+\alpha}. \tag{4.3.17}$$

In addition, for every $x_0 \in \mathbb{R}^n \setminus \{0\}$, there exists $\rho_{x_0} > 0$ such that $u \in C^{k,\alpha}(B_{\rho_{x_0}}(x_0))$; therefore, in the light of Proposition 4.3.2, there exist $M_{x_0} \geqslant 0$ and polynomial P_{x_0} of degree at most k such that $|u(x) - P_{x_0}(x)| \leqslant M_{x_0}|x - x_0|^{k+\alpha}$ for all $x \in B_{\rho_{x_0}}(x_0)$ (and actually for all $x \in \mathbb{R}^n$, up to changing M_{x_0}, since u is bounded).

It follows from this and (4.3.17) that Definition 4.3.1 is fulfilled for every $x_0 \in \mathbb{R}^n$, taking $P := 0$ and $M := 1$ when $x_0 = 0$ and $P := P_{x_0}$ and $M := M_{x_0}$ when $x_0 \neq 0$. As a result,

$$u \in C^{k,\alpha}_{\mathbb{R}^n}(x_0) \text{ for every } x_0 \in \mathbb{R}^n. \tag{4.3.18}$$

But we claim that, for every $\rho > 0$,

$$u \notin C^{k,\alpha}(B_\rho). \tag{4.3.19}$$

To check this, we argue by contradiction and suppose that $u \in C^{k,\alpha}(B_\rho)$. Using a Taylor expansion for u and (4.3.17), it follows that, for small $|x|$,

$$\sum_{\substack{\beta\in\mathbb{N}^n\\|\beta|\leqslant k}} \frac{D^\beta u(0)}{\beta!} x^\beta + o(|x|^k) = o(|x|^k),$$

and therefore $D^\beta u(0) = 0$ for all $\beta \in \mathbb{N}^n$ with $|\beta| \leqslant k$. Thus, picking $J \in \mathbb{N}$, to be taken as large as we wish,

$$\left|\partial_1^k u\left(p_J + \frac{\ell_J}{8}, 0, \dots, 0\right)\right| = \left|\partial_1^k u\left(p_J + \frac{\ell_J}{8}, 0, \dots, 0\right) - \partial_1^k u(0)\right|$$
$$\leqslant C\left(p_J + \frac{\ell_J}{8}\right)^{\alpha}. \tag{4.3.20}$$

But if $x_1 \in \left(p_J, p_J + \frac{\ell_J}{4}\right)$, we have that

$$\partial_1^k u(x) = c\, s_J\, |x_1 - p_J|^{\alpha}, \quad \text{with } c := \prod_{m=1}^{k}(m+\alpha)$$

and therefore

$$\partial_1^k u\left(p_J + \frac{\ell_J}{8}, 0, \dots, 0\right) = \frac{c\, s_J\, \ell_J^{\alpha}}{8^{\alpha}} = \frac{c}{8^{\alpha}}.$$

This and (4.3.20) entail that

$$\frac{c}{8^{\alpha}} \leqslant \lim_{J\to+\infty} C\left(p_J + \frac{\ell_J}{8}\right)^{\alpha} = 0,$$

which is a contradiction. This proves (4.3.19).

In view of (4.3.18) and (4.3.19), we infer that the inclusion in (4.3.14) is strict. □

One of the technical advantages of the pointwise Hölder spaces introduced in Definition 4.3.1 is that it suffices to check the validity of the polynomial approximation along a geometrically decaying sequence of radii, according to the following result (in which we use the notation of distance function introduced on p. 36).

Lemma 4.3.4. *Let $k \in \mathbb{N}$, $\alpha \in (0,1]$ and $\rho \in (0,1)$. Let $\Omega \subseteq \mathbb{R}^n$ be open and bounded.*

Assume that there exist $\tau_\star$, $\tau_\sharp$, τ_0, τ_1, $\in (0,1)$ such that for all $X \in \Omega$ and all $r > 0$ such that $\frac{d_{\partial\Omega}(X)}{\tau_\sharp} \leqslant r \leqslant \tau_\star$, there exists $q_r \in \partial B_{\tau_1 r}(X)$ such that[12]

$$B_{\tau_0 r}(q_r) \subseteq \Omega \cap B_r(X). \tag{4.3.21}$$

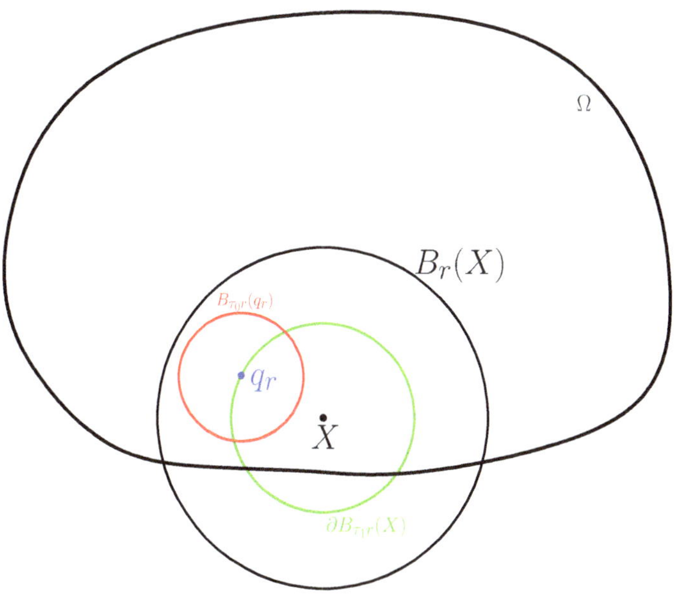

Figure 4.3.3 The geometric condition in (4.3.21).

[12]See Figure 4.3.3 for a sketch of condition (4.3.21).

Let $x_0 \in \Omega$ and $u \in L^\infty(\Omega)$. Then, the following conditions are equivalent:

(i) $u \in C^{k,\alpha}_\Omega(x_0)$, *and there exists* $M \geqslant 0$ *such that*

$$|u(x) - P(x)| \leqslant M|x - x_0|^{k+\alpha} \quad \text{for all } x \in \Omega.$$

(ii) *There exists* $L \geqslant 0$ *such that for every* $j \in \mathbb{N}$, *there exists a polynomial* P_j *of degree at most* k *such that*

$$|u(x) - P_j(x)| \leqslant L\rho^{(k+\alpha)j} \quad \text{for all } x \in \Omega \cap B_{\rho^j}(x_0).$$

Also, if (i) holds true, one can choose $L := M$ *in (ii).*

If instead (ii) holds true, one can choose M *in (i) of the form* $C\Big(\|u\|_{L^\infty(\Omega)} + L\Big)$, *for some* $C > 0$ *depending only on* n, k, α, ρ *and* Ω.

Furthermore, the polynomial P *in (i) satisfies*

$$|D^\beta P(x_0)| \leqslant C\Big(\|u\|_{L^\infty(\Omega)} + M\Big),$$

for all $\beta \in \mathbb{N}^n$ *with* $|\beta| \leqslant k$, *for some* $C > 0$ *depending only on* n, k, α, ρ *and* Ω.

Proof. Up to a translation, we suppose $x_0 = 0$. If (i) holds true, then (ii) also holds true with $P_j := P$ and $L := M$.

Suppose instead that (ii) holds true. The approach to proving (i) is that the polynomials P_j may vary from scale to scale, but they must change "very little" at each iteration due to the geometric decay of the sequence of radii, and this fact entails the existence of a "limit polynomial" with uniform estimates. To make this idea work, let us exploit (ii) to find that

$$\begin{aligned} \|P_{j+1} - P_j\|_{L^\infty(\Omega\cap B_{\rho^{j+1}})} &\leqslant \|P_{j+1} - u\|_{L^\infty(\Omega\cap B_{\rho^{j+1}})} + \|u - P_j\|_{L^\infty(\Omega\cap B_{\rho^j})} \\ &\leqslant 2L\rho^{(k+\alpha)j}. \end{aligned} \tag{4.3.22}$$

We now point out that $\|\cdot\|_{L^\infty(B_1)}$ and $\|\cdot\|_{L^\infty(B_{\tau_0}(\tau_1 e_1))}$ are norms in the finite-dimensional linear space of polynomials of degree at most k: by the equivalence of norms in finite-dimensional spaces (see e.g. [Por81, Theorem 15.26]) it follows that, for every polynomial P of degree at most k,

$$\|P\|_{L^\infty(B_1)} \leqslant C\|P\|_{L^\infty(B_{\tau_0}(\tau_1 e_1))},$$

for some C depending only on n, k, τ_0 and τ_1.

Now, if $p \in \partial B_{\tau_1}$, we take a rotation $\mathcal{R}$ centered at the origin, sending $\tau_1 e_1$ into p. Given a polynomial P of degree at most k, we let $P_{\mathcal{R}}(x) := P(\mathcal{R}x)$, and we observe that $P_{\mathcal{R}}$ is also a polynomial of degree at most k. Consequently, we have that

$$\begin{aligned} \|P\|_{L^\infty(B_1)} &= \sup_{y\in B_1} |P(y)| = \sup_{x\in B_1} |P(\mathcal{R}x)| = \|P_{\mathcal{R}}\|_{L^\infty(B_1)} \\ &\leqslant C\|P_{\mathcal{R}}\|_{L^\infty(B_{\tau_0}(\tau_1 e_1))} = C \sup_{x\in B_{\tau_0}(\tau_1 e_1)} |P(\mathcal{R}x)| \\ &= C \sup_{y\in B_{\tau_0}(\tau_1 e_1)} |P(y)| = C\|P\|_{L^\infty(B_{\tau_0}(p))}. \end{aligned} \tag{4.3.23}$$

Furthermore, using again the equivalence of the norms in finite-dimensional spaces,

$$\|P\|_{L^\infty(B_1)} \leqslant C\|P\|_{L^\infty(B_{\tau_\sharp})}. \tag{4.3.24}$$

Now, we claim that, for every $i \in \mathbb{N}\cap[0,k]$, every $\beta\in\mathbb{N}^n$ with $|\beta| = i$ and every polynomial P of degree at most k,

$$\|D^\beta P\|_{L^\infty(\Omega\cap B_{\rho^{j+1}})} \leqslant \frac{C}{\rho^{i(j+1)}} \|P\|_{L^\infty(\Omega\cap B_{\rho^{j+1}})}. \tag{4.3.25}$$

To check this, we define $\widetilde{P}(x) := P(\rho^{j+1}x)$ and observe that

$$\begin{aligned} \|D^\beta P\|_{L^\infty(\Omega\cap B_{\rho^{j+1}})} &\leqslant \|D^\beta P\|_{L^\infty(B_{\rho^{j+1}})} \\ &\leqslant \frac{C}{\rho^{i(j+1)}} \|P\|_{L^\infty(B_{\rho^{j+1}})} = \frac{C}{\rho^{i(j+1)}} \|\widetilde{P}\|_{L^\infty(B_1)}, \end{aligned} \tag{4.3.26}$$

thanks to (4.3.5).

Now, we distinguish two cases, either $\frac{d_{\partial\Omega}(0)}{\tau_\sharp} \leqslant \rho^{j+1}$ or $\frac{d_{\partial\Omega}(0)}{\tau_\sharp} > \rho^{j+1}$.

Let us first consider the case in which $d_{\partial\Omega}(0) \leqslant \frac{\rho^{j+1}}{\tau_\sharp}$. In this situation, we can exploit (4.3.21) and choose $p\in\partial B_{\tau_1}$ such that $B_{\tau_0\rho^{j+1}}(\rho^{j+1}p) \subseteq \Omega\cap B_{\rho^{j+1}}$. Using this, (4.3.23) and (4.3.26), we obtain that

$$\begin{aligned} \|D^\beta P\|_{L^\infty(\Omega\cap B_{\rho^{j+1}})} &\leqslant \frac{C}{\rho^{i(j+1)}} \|\widetilde{P}\|_{L^\infty(B_1)} \leqslant \frac{C}{\rho^{i(j+1)}} \|\widetilde{P}\|_{L^\infty(B_{\tau_0}(p))} \\ &= \frac{C}{\rho^{i(j+1)}} \|P\|_{L^\infty(B_{\tau_0\rho^{j+1}}(\rho^{j+1}p))} \leqslant \frac{C}{\rho^{i(j+1)}} \|P\|_{L^\infty(\Omega\cap B_{\rho^{j+1}})}, \end{aligned}$$

showing that (4.3.25) holds true in this case.

Now, we focus on the case of $\frac{d_{\partial\Omega}(0)}{\tau_\sharp} > \rho^{j+1}$. In this situation, we have that $B_{\tau_\sharp\rho^{j+1}} \subseteq B_{d_{\partial\Omega}(0)} \subseteq \Omega$. Hence,

$$B_{\tau_\sharp\rho^{j+1}} = B_{\tau_\sharp\rho^{j+1}} \cap B_{\rho^{j+1}} \subseteq \Omega\cap B_{\rho^{j+1}}.$$

As a consequence, by (4.3.24) and (4.3.26),

$$\begin{aligned}\|D^\beta P\|_{L^\infty(\Omega\cap B_{\rho^{j+1}})} &\leqslant \frac{C}{\rho^{i(j+1)}}\|\widetilde{P}\|_{L^\infty(B_1)} \leqslant \frac{C}{\rho^{i(j+1)}}\|\widetilde{P}\|_{L^\infty(B_{\tau_\sharp})}\\ &= \frac{C}{\rho^{i(j+1)}}\|P\|_{L^\infty(B_{\tau_\sharp\rho^{j+1}})} \leqslant \frac{C}{\rho^{i(j+1)}}\|P\|_{L^\infty(\Omega\cap B_{\rho^{j+1}})},\end{aligned}$$

which completes the proof of (4.3.25).

As a result, from (4.3.22) and (4.3.25), for every $i \in \mathbb{N}\cap[0,k]$ and every $\beta \in \mathbb{N}^n$ with $|\beta| = i$,

$$\begin{aligned}\|D^\beta(P_{j+1} - P_j)\|_{L^\infty(\Omega\cap B_{\rho^{j+1}})} &\leqslant \frac{C}{\rho^{i(j+1)}}\|P_{j+1} - P_j\|_{L^\infty(B_{\rho^{j+1}})}\\ &\leqslant CL\rho^{(k+\alpha)j-i(j+1)}\end{aligned}$$

for some $C > 0$ depending only on n, k and Ω.

Consequently, if we write

$$P_j(x) = \sum_{\substack{\beta\in\mathbb{N}^n\\ |\beta|\leqslant k}} a_{\beta,j}x^\beta,$$

we have that

$$\begin{aligned}|a_{\beta,j+1} - a_{\beta,j}| &= \left|\frac{D^\beta P_{j+1}(0)}{\beta!} - \frac{D^\beta P_j(0)}{\beta!}\right|\\ &\leqslant C\|D^\beta(P_{j+1} - P_j)\|_{L^\infty(\Omega\cap B_{\rho^{j+1}})} \leqslant CL\rho^{(k+\alpha)j-|\beta|(j+1)},\end{aligned}$$

which in turn entails that for all $m \in \mathbb{N}\cap[1,+\infty)$,

$$\begin{aligned}|a_{\beta,j+m} - a_{\beta,j}| &\leqslant \sum_{i=0}^{m-1}|a_{\beta,j+i+1} - a_{\beta,j+i}|\\ &\leqslant CL\sum_{i=0}^{m-1}\rho^{(k+\alpha)(j+i)-|\beta|(j+i+1)} \leqslant CL\rho^{(k-|\beta|+\alpha)j-|\beta|}. \qquad (4.3.27)\end{aligned}$$

In particular,

$$|a_{\beta,j+m} - a_{\beta,j}| \leqslant CL\rho^{\alpha j-k},$$

which is infinitesimal as $j \to +\infty$.

This gives that $a_{\beta,j}$ is a converging sequence, say $a_{\beta,j} \to a_\beta$, as $j \to +\infty$. Moreover, passing to the limit as $m \to +\infty$ in (4.3.27),

$$|a_\beta - a_{\beta,j}| \leqslant CL\rho^{(k-|\beta|+\alpha)j-|\beta|}. \qquad (4.3.28)$$

As a result, letting

$$P(x) := \sum_{\substack{\beta\in\mathbb{N}^n\\ |\beta|\leqslant k}} a_\beta x^\beta$$

and using also (ii), we see that, for all $x \in \Omega \cap B_{\rho^j}$,

$$\begin{aligned}
|u(x) - P(x)| &\leqslant |u(x) - P_j(x)| + |P_j(x) - P(x)| \\
&\leqslant L\rho^{(k+\alpha)j} + \left| \sum_{\substack{\beta \in \mathbb{N}^n \\ |\beta| \leqslant k}} (a_{\beta,j} - a_\beta) x^\beta \right| \\
&\leqslant L\rho^{(k+\alpha)j} + CL \sum_{\substack{\beta \in \mathbb{N}^n \\ |\beta| \leqslant k}} \rho^{(k+\alpha)j - |\beta|} \\
&\leqslant L\rho^{(k+\alpha)j} + CL\rho^{(k+\alpha)j-k} \\
&\leqslant CL\rho^{(k+\alpha)j},
\end{aligned} \tag{4.3.29}$$

up to renaming C depending on n, k, α, ρ and Ω.

Now, we take $x \in \Omega$ and distinguish two cases. If $x \in B_1$, we pick $j \in \mathbb{N}$ such that $\rho^{j+1} \leqslant |x| < \rho^j$, and we exploit (4.3.29) to conclude that

$$|u(x) - P(x)| \leqslant CL\rho^{(k+\alpha)j} \leqslant CL|x|^{k+\alpha}, \tag{4.3.30}$$

up to renaming C.

If instead $x \in \Omega \setminus B_1$, then

$$|u(x) - P(x)| \leqslant \|u\|_{L^\infty(\Omega \setminus B_1)} + \sum_{\substack{\beta \in \mathbb{N}^n \\ |\beta| \leqslant k}} |a_\beta|\, |x|^{|\beta|} \leqslant \Big(\|u\|_{L^\infty(\Omega \setminus B_1)} + \sup_{\substack{\beta \in \mathbb{N}^n \\ |\beta| \leqslant k}} |a_\beta| \Big) |x|^{k+\alpha}.$$

Since, by (4.3.4) and (4.3.28),

$$\begin{aligned}
|a_\beta| &\leqslant |a_{\beta,0}| + CL\rho^{-|\beta|} = \left| \frac{D^\beta P_0(0)}{\beta!} \right| + CL\rho^{-|\beta|} \\
&\leqslant \|P_0\|_{C^k(B_1)} + CL \leqslant C\|P_0\|_{L^\infty(B_1)} + CL,
\end{aligned}$$

we thus obtain that, for all $x \in \Omega \setminus B_1$,

$$|u(x) - P(x)| \leqslant C\Big(\|u\|_{L^\infty(\Omega \setminus B_1)} + \|P_0\|_{L^\infty(B_1)} + L \Big) |x|^{k+\alpha}. \tag{4.3.31}$$

Note also that, for all $x \in \Omega \cap B_1$,

$$|P_0(x)| \leqslant |u(x)| + L \leqslant \|u\|_{L^\infty(\Omega)} + L,$$

thanks to (ii), and therefore

$$\|P_0\|_{L^\infty(\Omega \cap B_1)} \leqslant \|u\|_{L^\infty(\Omega)} + L. \tag{4.3.32}$$

Moreover, by the equivalence of the norms on finite-dimensional spaces, we know that $\|P_0\|_{L^\infty(B_1)} \leqslant C\|P_0\|_{L^\infty(\Omega \cap B_1)}$; hence, we infer from (4.3.32) that

$$\|P_0\|_{L^\infty(B_1)} \leqslant C\big(\|u\|_{L^\infty(\Omega)} + L \big).$$

This and (4.3.31) yield that, for all $x \in \Omega \setminus B_1$,

$$|u(x) - P(x)| \leqslant C\Big(\|u\|_{L^\infty(\Omega)} + L\Big)|x|^{k+\alpha}.$$

From this inequality and (4.3.30), we see that (i) holds true with $M := C\Big(\|u\|_{L^\infty(\Omega)} + L\Big)$, as desired.

Additionally, a byproduct of (i) is that, if $x \in \Omega$,

$$|P(x)| \leqslant |u(x)| + CM \leqslant \|u\|_{L^\infty(\Omega)} + CM$$

and thus, by (4.3.4), for all $\beta \in \mathbb{N}^n$ with $|\beta| \leqslant k$,

$$|D^\beta P(0)| \leqslant \|P\|_{C^k(\Omega)} \leqslant C\,\|P\|_{L^\infty(\Omega)} \leqslant C\big(\|u\|_{L^\infty(\Omega)} + M\big),$$

thus concluding the proof of Lemma 4.3.4. □

Corollary 4.3.5. *Let $k \in \mathbb{N}$, $\alpha \in (0, 1]$, $L \geqslant 0$ and $\rho \in (0, 1)$. Let $\Omega \subseteq \mathbb{R}^n$ be open, bounded and convex.*

Let $u \in L^\infty(\mathbb{R}^n)$, and suppose that for every $x_0 \in \Omega$ and every $j \in \mathbb{N}$, there exists a polynomial P_{j,x_0} of degree at most k such that

$$|u(x) - P_{j,x_0}(x)| \leqslant L\rho^{(k+\alpha)j} \quad \textit{for all}\, x \in \Omega \cap B_{\rho^j}(x_0).$$

Then, $u \in C^{k,\alpha}(\Omega)$ and $\|u\|_{C^{k,\alpha}(\Omega)} \leqslant C(L + \|u\|_{L^\infty(\mathbb{R}^n)})$, for some $C > 0$ depending only on n, k, α and Ω.

Proof. We observe that

$$\Omega \text{ satisfies condition (4.3.21).} \tag{4.3.33}$$

For this, to begin with, we show that $\partial\Omega$ can be written locally as the graph of a Lipschitz function, namely

$$\begin{aligned}&\text{for every } p \in \Omega \text{ there exists } \rho_p > 0 \text{ such that}\\&(\partial\Omega) \cap B_{\rho_p}(p) \text{ is the graph, in some direction,}\\&\text{of a convex and Lipschitz function.}\end{aligned} \tag{4.3.34}$$

To establish this, by the boundedness of Ω, we can suppose that $\Omega \Subset B_R$ for some $R > 0$. Hence, given any $\omega \in \partial B_1$, the hyperplane $\Pi_\omega := \{\omega \cdot x = -R\}$ lies outside Ω (otherwise there would be a point $x \in \Omega \Subset B_R$ with $|x| \geqslant |\omega \cdot x| = |-R| = R$, which is a contradiction). We can thus consider the projection P_ω of Ω onto Π_ω, which is open (and convex) in Π_ω since Ω is open (and convex). For all $x \in P_\omega$, we denote by $f_\omega(x)$ the infimal t for which $x + t\omega$ belongs to $\overline{\Omega}$, see Figure 4.3.4.

Note that, by construction,

$$f_\omega(x) \in [0, 2R]. \tag{4.3.35}$$

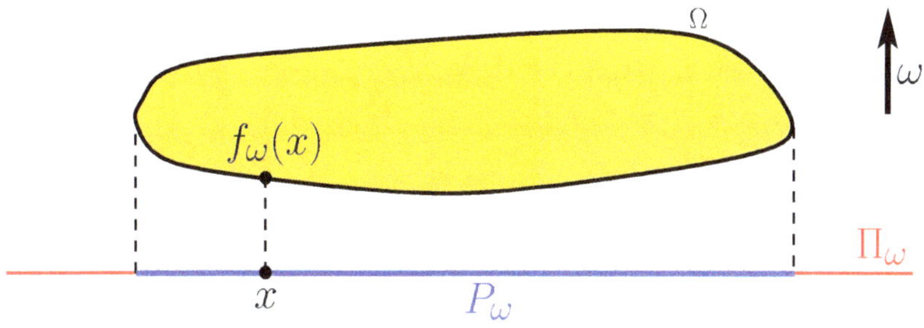

Figure 4.3.4 The geometry involved in (4.3.36) and (4.3.37).

Also,

$$f_\omega \text{ is convex (when we look at it as a function in direction } \omega). \tag{4.3.36}$$

To check this, up to a rotation, we can assume that $\omega = e_n$. We pick $(x', -R)$, $(y', -R) \in P_{e_n}$ and we note that $(x', f_{e_n}(x'))$, $(y', f_{e_n}(y')) \in \partial\Omega$. From the convexity of Ω, we infer that

$$\Big(tx' + (1-t)y',\ tf_{e_n}(x') + (1-t)f_{e_n}(y')\Big) = t(x', f_{e_n}(x')) + (1-t)(y', f_{e_n}(y')) \in \overline{\Omega}$$

for all $t \in [0, 1]$. Hence, by the definition of f_{e_n}, we conclude that

$$f_{e_n}\big(tx' + (1-t)y'\big) \leqslant tf_{e_n}(x') + (1-t)f_{e_n}(y'),$$

which proves (4.3.36).

Additionally, we have that

$$f_\omega \text{ is locally Lipschitz.} \tag{4.3.37}$$

This is a general argument about convexity (see e.g. [EG15, Theorem 6.7]) and goes as follows. Again, we suppose $\omega = e_n$ and take $D \Subset P_{e_n}$. Let $d > 0$ be the distance of D to ∂P_{e_n}, and let also $(x', -R)$ and $(y', -R)$ be two different points in D with $|x' - y'| < d$. We define $e := \frac{y'-x'}{|y'-x'|}$ and, for all $\mu \geqslant 0$,

$$x'_\mu := x' + \mu e.$$

We pick μ such that $(x'_\mu, -R) \in \partial P_{e_n}$, and we stress that $\mu \geqslant d$.

Besides, if $t := \frac{|x'-y'|}{\mu}$, we have that $t \in \left[0, \frac{d}{\mu}\right] \subseteq [0, 1]$. Furthermore,

$$tx'_\mu + (1-t)x' = t(x'_\mu - x') + x' = \frac{|x' - y'|\mu e}{\mu} + x' = (y' - x') + x' = y',$$

and accordingly, by (4.3.36),

$$\begin{aligned} f_{e_n}(y') &= f_{e_n}\big(tx'_\mu + (1-t)x'\big) \leqslant t f_{e_n}(x'_\mu) + (1-t) f_{e_n}(x') \\ &= f_{e_n}(x') + t\big(f_{e_n}(x'_\mu) - f_{e_n}(x')\big). \end{aligned}$$

This and (4.3.35) give that

$$f_{e_n}(y') - f_{e_n}(x') \leqslant t\big(|f_{e_n}(x'_\mu)| + |f_{e_n}(x')|\big) \leqslant \frac{4R|x'-y'|}{\mu} \leqslant \frac{4R|x'-y'|}{d}.$$

By exchanging x' and y', we thereby conclude that, for all $(x', -R)$, $(y', -R) \in D$ with $|x' - y'| < d$,

$$|f_{e_n}(x') - f_{e_n}(y')| \leqslant \frac{4R|x'-y'|}{d}. \tag{4.3.38}$$

Additionally, if $(x', -R)$, $(y', -R) \in D$ with $|x'-y'| \geqslant d$, we deduce from (4.3.35) that

$$|f_{e_n}(x') - f_{e_n}(y')| \leqslant 4R \leqslant \frac{4R|x'-y'|}{d}.$$

Hence, we combine this inequality with (4.3.38), and we see that the proof of (4.3.37) is complete.

From (4.3.36) and (4.3.37) (and reducing P_ω to a slightly smaller subdomain to obtain a global Lipschitz property), we thus obtain (4.3.34), as desired.

Thus, since $\partial\Omega$ is compact, we use (4.3.34) to cover $\partial\Omega$ with a finite number of balls $B_{\rho_1/2}(p_1), \ldots, B_{\rho_N/2}(p_N)$, with $p_1, \ldots, p_N \in \partial\Omega$, such that for all $i \in \{1, \ldots, N\}$, we have that $(\partial\Omega) \cap B_{\rho_i}(p_i)$ can be written as a graph of a convex and Lipschitz function f_i.

With these preliminary observations, we can now address the core of the proof of (4.3.33). For this, we let

$$\tau_\star := \min\left\{1, \frac{\rho_1}{16}, \ldots, \frac{\rho_N}{16}\right\}, \quad \tau_\sharp := \frac{1}{8} \quad \text{and} \quad \tau_1 := \frac{1}{2}, \tag{4.3.39}$$

and we pick $X \in \Omega$. We let $d(X)$ be the distance from X to $\partial\Omega$, and we also pick $r \in \left[\frac{d(X)}{\tau_\sharp}, \tau_\star\right]$.

In this way, we have that $B_{d(X)}(X) \subseteq \Omega$, and there exists $Y \in (\partial B_{d(X)}(X)) \cap (\partial\Omega)$. We observe that if $i \in \{1, \ldots, N\}$ such that $Y \in B_{\rho_i/2}(p_i)$, then

$$B_r(X) \subseteq B_{\rho_i}(p_i) \tag{4.3.40}$$

because if $z \in B_r(X)$, then

$$|z - p_i| \leqslant |z - X| + |X - Y| + |Y - p_i| < r + d(X) + \frac{\rho_i}{2} \leqslant 2r + \frac{\rho_i}{2} \leqslant 2\tau_\star + \frac{\rho_i}{2} < \rho_i,$$

owing to (4.3.39).

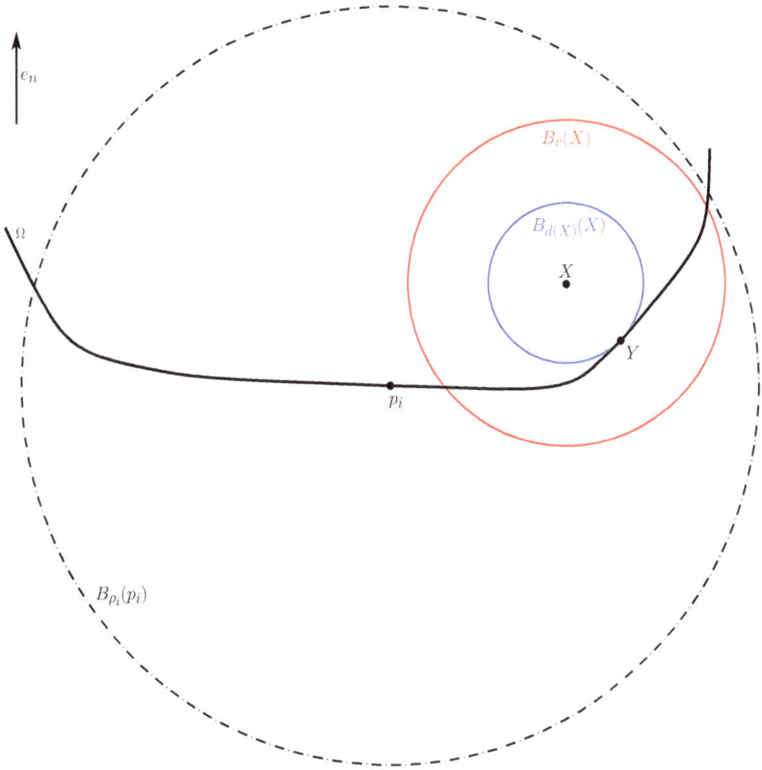

Figure 4.3.5 The geometry involved in (4.3.41).

Up to a rotation, we may assume that f_i is a convex function in the direction of e_n, see Figure 4.3.5. Accordingly, for some $M > 0$ (related to the maximum of the Lipschitz moduli of continuity of the functions $f_1, \ldots, f_N$), we have that

$$\Big\{x_n - Y_n \geqslant M|x' - Y'|\Big\} \cap B_{\rho_i}(p_i) \subseteq \Omega. \tag{4.3.41}$$

Furthermore, since $|Y - X| = d(X) \leqslant \frac{r}{8} < \frac{r}{2}$ and hence $Y \in B_{r/2}(X)$, we can take $\vartheta \geqslant 0$ such that $q_r := Y + \vartheta e_n \in \partial B_{r/2}(X)$. We stress that $q_r \in \partial_{\tau_1 r}(X)$, due to (4.3.39). Also,

$$\vartheta = |\vartheta e_n| = |q_r - Y| \geqslant |q_r - X| - |X - Y| = \frac{r}{2} - d(X) \geqslant \frac{r}{4}. \tag{4.3.42}$$

Similarly,

$$\vartheta = |\vartheta e_n| = |q_r - Y| \leqslant |q_r - X| + |X - Y| = \frac{r}{2} + d(X) \leqslant \frac{5r}{8}. \tag{4.3.43}$$

We also set

$$\tau_0 := \frac{1}{16(M+1)}, \tag{4.3.44}$$

and we claim that

$$B_{\tau_0 r}(q_r) \subseteq \left\{x_n - Y_n \geqslant M|x' - Y'|\right\}. \tag{4.3.45}$$

Indeed, if $\zeta \in B_{\tau_0 r}(q_r)$, then

$$\begin{aligned}\zeta_n - Y_n - M|\zeta' - Y'| &\geqslant q_{r,n} - Y_n - M|q_r' - Y'| - (M+1)\tau_0 r \\ &= \vartheta - 0 - (M+1)\tau_0 r \geqslant \frac{r}{4} - (M+1)\tau_0 r > 0,\end{aligned}$$

thanks to (4.3.42), and this proves (4.3.45).

We also have that

$$B_{\tau_0 r}(q_r) \subseteq B_r(X) \tag{4.3.46}$$

because if $\eta \in B_{\tau_0 r}(q_r)$, then, by (4.3.43),

$$|\eta - X| \leqslant |\eta - q_r| + |q_r - Y| + |Y - X| < \tau_0 r + \vartheta + d(X) \leqslant \tau_0 r + \frac{5r}{8} + \frac{r}{8} < r,$$

which proves (4.3.46).

Now, gathering the results in (4.3.40), (4.3.41), (4.3.45) and (4.3.46), we find that

$$B_{\tau_0 r}(q_r) \subseteq \{x_n - Y_n \geqslant M|x' - Y'|\} \cap B_r(X) \subseteq \Omega \cap B_r(X),$$

which is the desired claim in (4.3.21). The proof of (4.3.33) is thereby complete.

Thus, in light of (4.3.33), we can use Lemma 4.3.4 and find that, for each $x_0 \in \Omega$, we can obtain a polynomial P_{x_0} of degree at most k such that

$$\begin{aligned}&|u(x) - P_{x_0}(x)| \leqslant M|x - x_0|^{k+\alpha} \quad \text{for all } x \in \Omega \\ \text{and} \quad &|D^\beta P_{x_0}(x_0)| \leqslant M \quad \text{for all } \beta \in \mathbb{N}^n \text{ with } |\beta| \leqslant k,\end{aligned}$$

with $M := C(L + \|u\|_{L^\infty(\mathbb{R}^n)})$.

From this and Proposition 4.3.2, the desired result follows. □

We stress that the regularity theory developed through potential analysis, for instance, in the proof of Theorem 4.2.1, relied crucially on a very delicate derivative computation (as performed in Proposition 4.2.3) and in the fine analysis of the singular integral produced by this computation (as carried out in Lemma 4.2.4). Instead, the pointwise Hölder spaces do not require taking derivatives at all, and Corollary 4.3.5 allows one to check the pointwise polynomial approximation only at suitable scales. To highlight the conceptual innovation of the methods leveraging the pointwise Hölder spaces, we provide here a different proof of Theorem 4.2.1.

Another Proof of Theorem 4.2.1. Up to a covering argument (see Lemma 4.2.13), we can prove Theorem 4.2.1 with the ball B_1 replaced by the ball B_4 (changing the initial size of the ball will be convenient in the forthcoming equation (4.3.49), where we will then be allowed to work in the unit ball for the function that will occupy most of the computations needed for this argument).

Also, we observe that it suffices to prove Theorem 4.2.1 under the additional assumptions that

$$\|u\|_{L^\infty(B_4)} \leqslant 1 \quad \text{and} \quad \|f\|_{C^\alpha(B_4)} \leqslant 1. \tag{4.3.47}$$

Indeed, if we know that Theorem 4.2.1 holds true under this additional assumption, we define $\widetilde{u} := \frac{u}{\|u\|_{L^\infty(B_4)}+\|f\|_{C^\alpha(B_4)}}$ and $\widetilde{f} := \frac{f}{\|u\|_{L^\infty(B_4)}+\|f\|_{C^\alpha(B_4)}}$, we observe that $\|\widetilde{u}\|_{L^\infty(B_4)} \leqslant 1$ and $\|\widetilde{f}\|_{C^\alpha(B_4)} \leqslant 1$ and we are thereby in a position to conclude that

$$\frac{\|u\|_{C^{2,\alpha}(B_{1/2})}}{\|u\|_{L^\infty(B_4)} + \|f\|_{C^\alpha(B_4)}} = \|\widetilde{u}\|_{C^{2,\alpha}(B_{1/2})} \leqslant C,$$

thus providing Theorem 4.2.1 in its full generality.

For this reason, we now assume that the additional hypotheses in (4.3.47) are satisfied.

In light of Corollary 4.3.5, we also point out that to establish Theorem 4.2.1, it is enough to check that there exists $\rho \in \left(0, \frac{1}{2}\right)$ such that for every $x_0 \in B_{1/2}$ and every $j \in \mathbb{N}$, there exists a polynomial P_{j,x_0} of degree at most 2 such that

$$|u(x) - P_{j,x_0}(x)| \leqslant M\rho^{(2+\alpha)j} \quad \text{for all } x \in B_{\rho^j}(x_0), \tag{4.3.48}$$

where $M > 0$ depends only on n and α.

For this, we can consider the translation $v(x) := u(x_0 + x)$ and note that, by (4.3.47),

$$\begin{aligned} &\Delta v(x) = f(x_0 + x) =: \phi(x) \text{ for all } x \in B_1, \\ &\|v\|_{L^\infty(B_1)} \leqslant \|u\|_{L^\infty(B_4)} \leqslant 1 \\ \text{and } &\|\phi\|_{C^\alpha(B_1)} \leqslant \|f\|_{C^\alpha(B_4)} \leqslant 1. \end{aligned} \tag{4.3.49}$$

In this setting, to establish (4.3.48), we aim at proving that

$$|v(x) - P_j(x)| \leqslant M\rho^{(2+\alpha)j} \quad \text{for all } x \in B_{\rho^j}, \tag{4.3.50}$$

for a suitable polynomial P_j of degree at most 2 and with $C > 0$ depending only on n and α. As a matter of fact, we prove additionally that we can choose P_j such that

$$\Delta P_j(x) = \phi(0) \quad \text{for every } x \in \mathbb{R}^n. \tag{4.3.51}$$

Quite surprisingly, imposing such an additional constraint will turn out to simplify, rather than complicate, the proof of (4.3.50). In a sense, this additional constraint suggests that an appropriate strategy to detect regularity is "to replace the source by a constant one near the origin," thus reducing to the case of harmonic functions. The technical details are given as follows.

Our goal is to prove (4.3.50) and (4.3.51) by induction over $j \in \mathbb{N}$. For this, we are free to choose $\rho \in \left(0, \frac{1}{2}\right)$ conveniently small and $M \geqslant 1$ conveniently large (in a universal fashion depending only on n and α), and we argue as follows. When $j = 0$ we choose $P_0(x) := \frac{\phi(0)}{2n}|x|^2$; in this way, equation (4.3.51) holds true for $j = 0$ and, for all $x \in B_1$,

$$|v(x) - P_0(x)| \leqslant \|v\|_{L^\infty(B_1)} + \|\phi\|_{L^\infty(B_1)} \leqslant 2,$$

giving that (4.3.50) also holds true for $j = 0$ (as long as we choose $M \geqslant 2$).

Thus, to perform the inductive step, we now suppose that (4.3.50) and (4.3.51) are satisfied for an index j, and we wish to establish them for the index $j + 1$. To this end, it is convenient to consider the rescaled solution

$$B_1 \ni x \longmapsto V(x) := \frac{v(\rho^j x) - P_j(\rho^j x)}{M\rho^{(2+\alpha)j}}.$$

Using the inductive assumption on (4.3.50), we see that

$$\sup_{x\in B_1} |V(x)| = \sup_{y\in B_{\rho^j}} \frac{|v(y) - P_j(y)|}{M\rho^{(2+\alpha)j}} \leqslant 1. \tag{4.3.52}$$

Moreover, using the inductive assumption on (4.3.51), for every $x \in B_1$,

$$\Delta V(x) = \frac{\phi(\rho^j x) - \Delta P_j(\rho^j x)}{M\rho^{\alpha j}} = \frac{\phi(\rho^j x) - \phi(0)}{M\rho^{\alpha j}},$$

and accordingly

$$|\Delta V(x)| \leqslant \frac{\|\phi\|_{C^\alpha(B_1)}\rho^{\alpha j}}{M\rho^{\alpha j}} \leqslant \frac{1}{M}. \tag{4.3.53}$$

We also remark that $B_1(x_0) \subseteq B_2 \Subset B_4$, and hence $V \in C(\overline{B_1})$. We thus consider the solution $w \in C^2(B_1) \cap C(\overline{B_1})$ of

$$\begin{cases} \Delta w = 0 & \text{in } B_1, \\ w = V & \text{on } \partial B_1. \end{cases}$$

We point out that w can be constructed, for instance, by using the Poisson kernel of the ball, as discussed in Theorems 2.11.1 and 2.11.2. Using the weak maximum principle in Corollary 2.9.2(iii), combined with (4.3.52), we know that

$$\|w\|_{L^\infty(B_1)} \leqslant \sup_{\partial B_1} |w| = \sup_{\partial B_1} |V| \leqslant 1.$$

As a consequence, we infer from Cauchy's estimates (recall Theorem 2.17.1) that

$$\|w\|_{C^3(B_{1/2})} \leqslant C\|w\|_{L^\infty(B_1)} \leqslant C. \tag{4.3.54}$$

Let now $W_\pm(x) := w(x) - V(x) \pm \frac{|x|^2}{2nM}$. Recalling (4.3.53), in B_1, we have that $\Delta W_+ = -\Delta V + \frac{1}{M} \geqslant 0$ and, similarly, $\Delta W_- \leqslant 0$. As a result, the weak maximum principle in Corollary 2.9.2 gives that, for every $x \in B_1$,

$$w(x) - V(x) + \frac{|x|^2}{2nM} = W_+(x) \leqslant \sup_{y\in\partial B_1} W_+(y) = \sup_{y\in\partial B_1} \frac{|y|^2}{2nM} \leqslant \frac{1}{M}$$

and

$$w(x) - V(x) - \frac{|x|^2}{2nM} = W_-(x) \geqslant \inf_{y\in\partial B_1} W_-(y) = \inf_{y\in\partial B_1} \left(-\frac{|y|^2}{2nM}\right) \geqslant -\frac{1}{M}.$$

Accordingly,

$$\|w - V\|_{L^\infty(B_1)} \leqslant \frac{2}{M}. \tag{4.3.55}$$

Now, we define the quadratic polynomial

$$\widetilde{P}(x) := w(0) + \nabla w(0) \cdot x + \frac{1}{2} D^2 w(0) x \cdot x,$$

and we observe that $\Delta \widetilde{P}(x) = \Delta w(0) = 0$ for all $x \in \mathbb{R}^n$.

In addition, by (4.3.54) and a Taylor expansion, for all $x \in B_{1/2}$,

$$|w(x) - \widetilde{P}(x)| \leqslant \|D^3 w\|_{L^\infty(B_{1/2})} |x|^3 \leqslant C|x|^3.$$

We stress that the above C depends only on n and α. In particular, since $\alpha \in (0, 1)$, we can choose ρ universally small such that $C\rho^3 \leqslant \frac{\rho^{2+\alpha}}{2}$, with C as above, and conclude that

$$\sup_{x\in B_\rho} |w(x) - \widetilde{P}(x)| \leqslant C\rho^3 \leqslant \frac{\rho^{2+\alpha}}{2}. \tag{4.3.56}$$

We can now choose M sufficiently large such that $\frac{2}{M} \leqslant \frac{\rho^{2+\alpha}}{2}$ and deduce from (4.3.55) and (4.3.56) that

$$\|V - \widetilde{P}\|_{L^\infty(B_\rho)} \leqslant \|V - w\|_{L^\infty(B_\rho)} + \|w - \widetilde{P}\|_{L^\infty(B_\rho)} \leqslant \frac{2}{M} + \frac{\rho^{2+\alpha}}{2} \leqslant \rho^{2+\alpha}. \tag{4.3.57}$$

We now scale back by defining

$$P_{j+1}(x) := P_j(x) + M\rho^{(2+\alpha)j} \widetilde{P}(\rho^{-j} x).$$

In this way we have that

$$\Delta P_{j+1}(x) = \Delta P_j(x) + M\rho^{\alpha j} \Delta \widetilde{P}(\rho^{-j} x) = \phi(0) + 0 = \phi(0). \tag{4.3.58}$$

Furthermore, owing to (4.3.57),

$$\begin{aligned}\sup_{x\in B_{\rho^{j+1}}} |v(x) - P_{j+1}(x)| &= \sup_{x\in B_{\rho^{j+1}}} |M\rho^{(2+\alpha)j} V(\rho^{-j}x) + P_j(x) - P_{j+1}(x)| \\ &= M\rho^{(2+\alpha)j} \sup_{y\in B_\rho} \left| V(y) + \frac{P_j(\rho^j y) - P_{j+1}(\rho^j y)}{M\rho^{(2+\alpha)j}} \right| \\ &= M\rho^{(2+\alpha)j} \sup_{y\in B_\rho} \left| V(y) - \widetilde{P}(y) \right| \\ &\leqslant M\rho^{(2+\alpha)(j+1)}. \end{aligned} \tag{4.3.59}$$

The inductive step is thereby complete: the observations in (4.3.58) and (4.3.59) give (4.3.50) and (4.3.51) for the index $j+1$, as desired. □

The method of polynomial approximation can also be used to obtain local estimates at the boundary: to show this, we give here another proof of Theorem 4.2.2.

Another Proof of Theorem 4.2.2. For computational convenience, up to a covering argument (recall Lemma 4.2.13), we prove Theorem 4.2.2 with B_1^+ replaced by B_4^+. Moreover, as observed in (4.2.42), it suffices to prove Theorem 4.2.2 when g vanishes identically. Also, as done in (4.3.47), it is not restrictive to assume that

$$\|u\|_{L^\infty(B_4^+)} \leqslant 1 \quad \text{and} \quad \|f\|_{C^\alpha(B_4^+)} \leqslant 1. \tag{4.3.60}$$

In addition, by Corollary 4.3.5, we can obtain Theorem 4.2.2 by proving that that there exists $\rho \in \left(0, \frac{1}{2}\right)$ such that for every $x_0 \in B_{1/2}^+$ and every $j \in \mathbb{N}$, there exists a polynomial P_{j,x_0} of degree at most 2 such that

$$\begin{aligned}&|u(x) - P_{j,x_0}(x)| \leqslant M\rho^{(2+\alpha)j} \\ &\qquad \text{for all } x \in B_{\rho^j}^+(x_0) := \{x = (x', x_n) \in \mathbb{R}^n \text{ s.t. } |x - x_0| < \rho^j \text{ and } x_n > 0\},\end{aligned} \tag{4.3.61}$$

where $M > 0$ depends only on n and α.

To prove (4.3.61), we argue by induction over $j \in \mathbb{N}$, and we prove additionally that we can choose P_{j,x_0} such that

$$\begin{aligned}&\Delta P_{j,x_0}(x) = f(x_0) \quad \text{for every } x \in \mathbb{R}^n \\ \text{and, when } x_{0,n} \in (0, \rho^j), \quad &P_{j,x_0}(x', 0) = 0 \quad \text{for every } x' \in \mathbb{R}^{n-1}.\end{aligned} \tag{4.3.62}$$

To perform the inductive argument, we choose $\rho \in \left(0, \frac{1}{2}\right)$ conveniently small and $M \geqslant 1$ conveniently large in what follows. When $j = 0$, we choose $P_{0,x_0}(x) := \frac{f(x_0)\, x_n^2}{2}$; in this way, equation (4.3.62) holds true for $j = 0$ and, for all $x \in B_1^+(x_0)$,

$$|u(x) - P_{0,x_0}(x)| \leqslant \|u\|_{L^\infty(B_1^+(x_0))} + \|f\|_{L^\infty(B_1^+(x_0))} \leqslant 2,$$

thanks to (4.3.60), giving that (4.3.61) also holds true for $j = 0$ provided that $M \geqslant 2$.

We now perform the inductive step toward the proof of (4.3.61) and (4.3.62), assuming that these claims hold true for an index j and aiming at establishing them for the index $j+1$.

To accomplish this goal, we let

$$\overline{B_1} \cap \{x_{0,n} + \rho^j x_n \geqslant 0\} \ni x \longmapsto V(x) := \frac{u(x_0 + \rho^j x) - P_{j,x_0}(x_0 + \rho^j x)}{M\rho^{(2+\alpha)j}}.$$

By the inductive assumption,

$$\sup_{x \in B_1 \cap \{x_{0,n} + \rho^j x_n > 0\}} |V(x)| = \sup_{y \in B^+_{\rho^j}(x_0)} \frac{|u(y) - P_{j,x_0}(y)|}{M\rho^{(2+\alpha)j}} \leqslant 1 \tag{4.3.63}$$

and, for every $x \in B_1 \cap \{x_{0,n} + \rho^j x_n > 0\}$,

$$\Delta V(x) = \frac{f(x_0 + \rho^j x) - \Delta P_j(x_0 + \rho^j x)}{M\rho^{\alpha j}} = \frac{f(x_0 + \rho^j x) - f(x_0)}{M\rho^{\alpha j}},$$

leading to

$$|\Delta V(x)| \leqslant \frac{\|f\|_{C^\alpha(B_4^+)}\, \rho^{\alpha j}}{M\rho^{\alpha j}} \leqslant \frac{1}{M}. \tag{4.3.64}$$

Now, we distinguish two cases, namely Case 1 corresponding to $x_{0,n} \geqslant \rho^j$ and Case 2 corresponding to $x_{0,n} \in (0, \rho^j)$. In Case 2, it is convenient to extend V by an odd reflection across the hyperplane $\{x_{0,n} + \rho^j x_n = 0\}$ by exploiting the second line of (4.3.62) (note that the second line of (4.3.62) becomes void in Case 1). Namely, in Case 2, we observe that, by the inductive assumption, if $x = (x', x_n) \in \overline{B_1} \cap \{x_{0,n} + \rho^j x_n = 0\}$, then

$$V(x) = \frac{u(x_0' + \rho^j x', 0) - P_{j,x_0}(x_0' + \rho^j x', 0)}{M\rho^{(2+\alpha)j}} = 0.$$

As a result, in Case 2, if $x = (x', x_n) \in \{x_{0,n} + \rho^j x_n < 0\}$ with $\left(x', -\frac{2x_{0,n}}{\rho^j} - x_n\right) \in \overline{B_1}$, we define

$$V(x', x_n) := -V\left(x', -\frac{2x_{0,n}}{\rho^j} - x_n\right), \tag{4.3.65}$$

and we have that V is continuous in $\overline{\mathcal{B}}$, where $\mathcal{B}$ is symmetrization of $B_1 \cap \{x_{0,n} + \rho^j x_n > 0\}$ through the hyperplane $\{x_{0,n} + \rho^j x_n = 0\}$, i.e.

$$\begin{aligned}\mathcal{B} :=& \Big(B_1 \cap \{x_{0,n} + \rho^j x_n \geqslant 0\}\Big) \\ & \cup \left\{x = (x', x_n) \in \{x_{0,n} + \rho^j x_n < 0\} \text{ s.t. } \left(x', -\frac{2x_{0,n}}{\rho^j} - x_n\right) \in B_1\right\},\end{aligned} \tag{4.3.66}$$

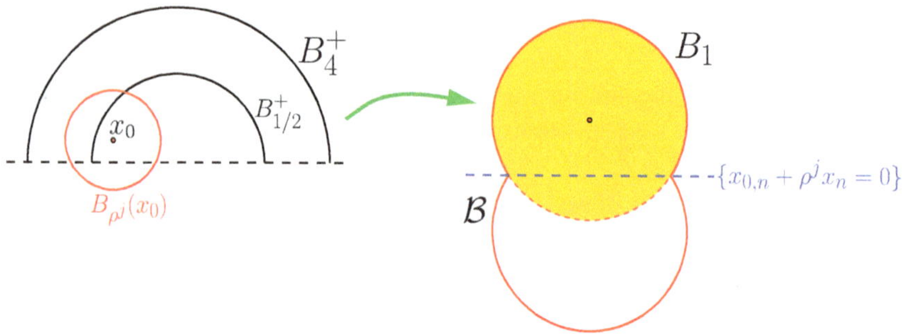

Figure 4.3.6 The geometry involved in the alternative proof of Theorem 4.2.2 when $x_{0,n} \in (0, \rho^j)$ (Case 2).

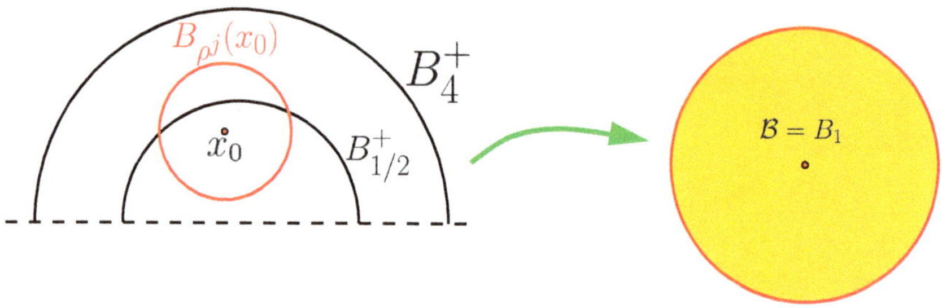

Figure 4.3.7 The geometry involved in the alternative proof of Theorem 4.2.2 when $x_{0,n} \geqslant \rho^j$ (Case 1).

see Figure 4.3.6.

We observe that, by taking $x_n := -\frac{x_{0,n}}{\rho^j}$ in (4.3.65), we find that, in Case 2,

$$V\left(x', -\frac{x_{0,n}}{\rho^j}\right) = -V\left(x', -\frac{x_{0,n}}{\rho^j}\right)$$

and thus

$$V\left(x', -\frac{x_{0,n}}{\rho^j}\right) = 0. \tag{4.3.67}$$

When we are in Case 1, this odd reflection method is not needed: to make the notation uniform between the two cases, we can replace (4.3.66) in Case 1 with $\mathcal{B} := B_1$ (and there is no need to extend V), see Figure 4.3.7.

In both cases, by (4.3.63),

$$\sup_{x\in\mathcal{B}} |V(x)| \leqslant 1. \tag{4.3.68}$$

Now, making use of the Perron method (see Corollary 3.3.4 and observe that $\mathcal{B}$ satisfies the exterior cone condition both in Case 1 and Case 2), we consider the solution $w \in C^2(B_1) \cap C(\overline{B_1})$ of

$$\begin{cases} \Delta w = 0 & \text{in } \mathcal{B}, \\ w = V & \text{on } \partial\mathcal{B}. \end{cases}$$

It follows from the weak maximum principle in Corollary 2.9.2(iii) and (4.3.68) that

$$\|w\|_{L^\infty(\mathcal{B})} \leqslant \sup_{\partial\mathcal{B}} |w| = \sup_{\partial\mathcal{B}} |V| \leqslant 1,$$

and accordingly, by Cauchy's estimates (see Theorem 2.17.1),

$$\|w\|_{C^3(B_{1/2}\cap\{x_{0,n}+\rho^j x_n \geqslant 0\})} \leqslant C\|w\|_{L^\infty(\mathcal{B})} \leqslant C. \tag{4.3.69}$$

We remark that, if we are in Case 2, then w is odd symmetric with respect to the hyperplane $\{x_{0,n} + \rho^j x_n = 0\}$; namely, for all $x \in \mathcal{B}$,

$$w(x', x_n) = -w\left(x', -\frac{2x_{0,n}}{\rho^j} - x_n\right) \tag{4.3.70}$$

Indeed, in Case 2, if

$$W(x', x_n) := w(x', x_n) + w\left(x', -\frac{2x_{0,n}}{\rho^j} - x_n\right),$$

we have that $\Delta W = 0 + 0 = 0$ in $\mathcal{B}$, and in light of (4.3.65), we have that $W(x', x_n) = V(x', x_n) + V\left(x', -\frac{2x_{0,n}}{\rho^j} - x_n\right) = 0$ for all $(x', x_n) \in \partial\mathcal{B}$. From these observations and the uniqueness result in Corollary 2.9.3, we deduce that W vanishes identically in $\mathcal{B}$, from which we obtain (4.3.70), as desired.

As a consequence of (4.3.70), by taking $x_n := -\frac{x_{0,n}}{\rho^j}$ there, we also find that, in Case 2,

$$w\left(x', -\frac{x_{0,n}}{\rho^j}\right) = -w\left(x', -\frac{x_{0,n}}{\rho^j}\right)$$

and therefore

$$w\left(x', -\frac{x_{0,n}}{\rho^j}\right) = 0. \tag{4.3.71}$$

Now, we define $W_\pm(x) := w(x) - V(x) \pm \frac{|x|^2}{2nM}$ and utilize (4.3.64), finding that $\Delta W_+ = -\Delta V + \frac{1}{M} \geqslant 0$, and similarly $\Delta W_- \leqslant 0$, in $B_1 \cap \{x_{0,n} + \rho^j x_n > 0\}$

(the careful reader will note that this set coincides with B_1 in Case 1). These observations and the weak maximum principle in Corollary 2.9.2 give that, for every $x \in B_1 \cap \{x_{0,n} + \rho^j x_n > 0\}$,

$$\begin{aligned} w(x) - V(x) + \frac{|x|^2}{2nM} &= W_+(x) \leqslant \sup_{y\in\partial(B_1\cap\{x_{0,n}+\rho^j x_n>0\})} W_+(y) \\ &= \sup_{y\in\partial(B_1\cap\{x_{0,n}+\rho^j x_n>0\})} \frac{|y|^2}{2nM} \leqslant \frac{1}{M} \end{aligned}$$

and

$$\begin{aligned} w(x) - V(x) - \frac{|x|^2}{2nM} &= W_-(x) \geqslant \inf_{y\in\partial(B_1\cap\{x_{0,n}+\rho^j x_n>0\})} W_-(y) \\ &= \inf_{y\in\partial(B_1\cap\{x_{0,n}+\rho^j x_n>0\})} \left(-\frac{|y|^2}{2nM}\right) \geqslant -\frac{1}{M}. \end{aligned}$$

Note that in the above, we used the fact that $w = V$ on ∂B_1 in Case 1, while in Case 2, we have exploited the facts that $w = V$ on $\partial B_1 \cap \{x_{0,n} + \rho^j x_n \geqslant 0\}$ and $w = V = 0$ on $B_1 \cap \{x_{0,n} + \rho^j x_n = 0\}$, thanks to (4.3.67) and (4.3.71).

As a result,

$$\|w - V\|_{L^\infty(B_1\cap\{x_{0,n}+\rho^j x_n>0\})} \leqslant \frac{2}{M}. \tag{4.3.72}$$

Now, we set

$$\mathbb{R}^n \supset \mathcal{B} \ni \zeta_j := \begin{cases} \left(0, \dots, 0, -\dfrac{x_{0,n}}{\rho^j}\right) & \text{if } x_{0,n} \in (0, \rho^{j+1}), \\ (0, \dots, 0) & \text{otherwise} \end{cases}$$

and define the quadratic polynomial

$$\widetilde{P}(x) := w(\zeta_j) + \nabla w(\zeta_j) \cdot (x - \zeta_j) + \frac{1}{2} D^2 w(\zeta_j)(x - \zeta_j) \cdot (x - \zeta_j).$$

Note that $\Delta \widetilde{P}(x) = \Delta w(\zeta_j) = 0$ for all $x \in \mathbb{R}^n$.

Furthermore, by (4.3.71),

$$w\left(x', -\frac{x_{0,n}}{\rho^j}\right) = \partial_i w\left(x', -\frac{x_{0,n}}{\rho^j}\right) = \partial_{ik} w\left(x', -\frac{x_{0,n}}{\rho^j}\right) = 0$$

for all $i \in \{1, \dots, n-1\}$ and $k \in \{1, \dots, n\}$.

Consequently, if $x_{0,n} \in (0, \rho^{j+1})$, then

$$w(\zeta_j) = \partial_i w(\zeta_j) = \partial_{ik} w(\zeta_j) = 0.$$

For this reason, if $x_{0,n} \in (0, \rho^{j+1})$, then

$$\widetilde{P}\left(x', -\frac{x_{0,n}}{\rho^j}\right) = w(\zeta_j) + \nabla w(\zeta_j) \cdot (x', 0) + \frac{1}{2} D^2 w(\zeta_j)(x', 0) \cdot (x', 0) = 0 \tag{4.3.73}$$

for all $x' \in \mathbb{R}^{n-1}$.

Besides, in view of (4.3.69) and a Taylor expansion around ζ_j, we see that, for all $x \in B_{1/2} \cap \{x_{0,n} + \rho^j x_n > 0\}$,

$$|w(x) - \widetilde{P}(x)| \leqslant \|D^3 w\|_{L^\infty(B_{1/2}\cap\{x_{0,n}+\rho^j x_n>0\})} |x - \zeta_j|^3 \leqslant C|x - \zeta_j|^3. \quad (4.3.74)$$

We stress that the above C depends only on n and α.

We also note that for all $x \in B_\rho$, we have that

$$|x - \zeta_j| \leqslant 2\rho \quad (4.3.75)$$

Indeed, if $x_{0,n} \in (0, \rho^{j+1})$, it holds that

$$\begin{aligned} |x - \zeta_j|^2 &= \left| x + \frac{x_{0,n}}{\rho^j} e_n \right|^2 = |x'|^2 + \left(x_n + \frac{x_{0,n}}{\rho^j} \right)^2 = |x|^2 + \frac{x_{0,n}^2}{\rho^{2j}} + \frac{2x_n x_{0,n}}{\rho^j} \\ &\leqslant \rho^2 + \frac{\rho^{2(j+2)}}{\rho^{2j}} + \frac{2\rho^{j+2}}{\rho^j} = 4\rho^2; \end{aligned}$$

otherwise, $|x - \zeta_j| = |x| \leqslant \rho$, which proves (4.3.75).

Combining (4.3.74) and (4.3.75), we see that, for all $x \in B_\rho \cap \{x_{0,n} + \rho^j x_n > 0\}$,

$$|w(x) - \widetilde{P}(x)| \leqslant C\rho^3,$$

up to renaming C. In particular, since $\alpha \in (0, 1)$, we can choose ρ universally small such that $C\rho^3 \leqslant \frac{\rho^{2+\alpha}}{2}$, with C as above; therefore,

$$\sup_{x \in B_\rho \cap \{x_{0,n} + \rho^j x_n > 0\}} |w(x) - \widetilde{P}(x)| \leqslant C\rho^3 \leqslant \frac{\rho^{2+\alpha}}{2}. \quad (4.3.76)$$

Now, we choose M sufficiently large such that $\frac{2}{M} \leqslant \frac{\rho^{2+\alpha}}{2}$ and deduce from (4.3.72) and (4.3.76) that

$$\begin{aligned} &\|V - \widetilde{P}\|_{L^\infty(B_\rho\cap\{x_{0,n}+\rho^j x_n>0\})} \\ &\leqslant \|V - w\|_{L^\infty(B_\rho\cap\{x_{0,n}+\rho^j x_n>0\})} + \|w - \widetilde{P}\|_{L^\infty(B_\rho\cap\{x_{0,n}+\rho^j x_n>0\})} \\ &\leqslant \frac{2}{M} + \frac{\rho^{2+\alpha}}{2} \leqslant \rho^{2+\alpha}. \end{aligned} \quad (4.3.77)$$

We now scale back by defining

$$P_{j+1,x_0}(x) := P_{j,x_0}(x) + M\rho^{(2+\alpha)j} \widetilde{P}\left(\frac{x - x_0}{\rho^j} \right).$$

Hence, we have that

$$\Delta P_{j+1,x_0}(x) = \Delta P_{j,x_0}(x) + M\rho^{\alpha j} \Delta \widetilde{P}\left(\frac{x - x_0}{\rho^j} \right) = f(x_0) + 0 = f(x_0). \quad (4.3.78)$$

Furthermore, if $x_{0,n} \in (0, \rho^{j+1})$, we can utilize (4.3.73) and conclude that

$$P_{j+1,x_0}(x', 0) = P_{j,x_0}(x', 0) + M\rho^{(2+\alpha)j} \widetilde{P}\left(\frac{x' - x_0'}{\rho^j}, -\frac{x_{0,n}}{\rho^j} \right) = 0. \quad (4.3.79)$$

Furthermore, by (4.3.77),

$$\begin{aligned}
&\sup_{x\in B^+_{\rho^{j+1}}(x_0)} |u(x) - P_{j+1,x_0}(x)| \\
&= \sup_{x\in B^+_{\rho^{j+1}}(x_0)} \left| M\rho^{(2+\alpha)j} V\left(\frac{x-x_0}{\rho^j}\right) + P_{j,x_0}(x) - P_{j+1,x_0}(x)\right| \\
&= M\rho^{(2+\alpha)j} \sup_{y\in B_\rho\cap\{x_{0,n}+\rho^j y_n>0\}} \left| V(y) + \frac{P_{j,x_0}(\rho^j y + x_0) - P_{j+1,x_0}(\rho^j y + x_0)}{M\rho^{(2+\alpha)j}}\right| \\
&= M\rho^{(2+\alpha)j} \sup_{y\in B_\rho\cap\{x_{0,n}+\rho^j y_n>0\}} \left| V(y) - \widetilde{P}(y)\right| \\
&\leqslant M\rho^{(2+\alpha)(j+1)}.
\end{aligned}$$

This and (4.3.78), together with (4.3.79) when $x_{0,n} \in (0, \rho^{j+1})$, establish (4.3.61) and (4.3.62) for the index $j+1$, as desired, thus completing the proof of the inductive step. □

4.4 Schauder Estimates for Equations in Nondivergence Form

Now, we turn our attention to a more general class of elliptic equations. Specifically, rather than looking only at the Laplace operator, we consider equations in the form

$$\sum_{i,j=1}^{n} a_{ij}(x)\partial_{ij}u(x) + \sum_{i=1}^{n} b_i(x)\partial_i u(x) + c(x)u(x) = f(x). \tag{4.4.1}$$

We always suppose that the coefficients a_{ij} give rise to a symmetric matrix, namely

$$a_{ij} = a_{ji}. \tag{4.4.2}$$

This assumption can be taken essentially without loss of generality since if (4.4.2) did not hold, it would suffice to define $\widetilde{a}_{ij} := \frac{a_{ij}+a_{ji}}{2}$ and exploit the symmetry of the second derivatives of u to recast (4.4.1) into a similar equation in which the coefficients a_{ij} are replaced by the symmetric ones $\widetilde{a}_{ij}$.

The structural assumption that we take on equation (4.4.1) is the "ellipticity of the coefficients a_{ij}": more precisely, we always assume that there exist $\lambda \in (0, +\infty)$ and $\Lambda \in [\lambda, +\infty)$ such that for every $\xi = (\xi_1, \dots, \xi_n) \in \partial B_1$,

$$\sum_{i,j=1}^{n} a_{ij}(x)\xi_i\xi_j \in [\lambda, \Lambda]. \tag{4.4.3}$$

From an algebraic viewpoint, this condition states that all the eigenvalues of the symmetric matrix a_{ij} are positive, bounded away from zero and bounded (compare with the notion of ellipticity presented in [DV23, Chapter 1]).

A geometric "visualization" of (4.4.3) can be obtained, at least when $n = 2$, considering the following map E from the space of symmetric matrices into $\mathbb{R}^3$:

$$E : \begin{pmatrix} a_{11} & a_{12} \\ a_{12} & a_{22} \end{pmatrix} \longmapsto \left(\sqrt{2}\, a_{12}, \, \frac{a_{11} - a_{22}}{\sqrt{2}}, \, \frac{a_{11} + a_{22}}{\sqrt{2}} \right).$$

We observe that E is an isometry, in the sense that if A is the matrix corresponding to $\begin{pmatrix} a_{11} & a_{12} \\ a_{12} & a_{22} \end{pmatrix}$, then

$$|E(A)|^2 = 2a_{12}^2 + \frac{(a_{11} - a_{22})^2 + (a_{11} + a_{22})^2}{2} = 2a_{12}^2 + a_{11}^2 + a_{22}^2 = |A|^2.$$

Furthermore, E reverts the trace of the matrix product into the scalar product since

$$\begin{aligned} E(A) \cdot E(\widetilde{A}) &= \left(\sqrt{2}\, a_{12}, \, \frac{a_{11} - a_{22}}{\sqrt{2}}, \, \frac{a_{11} + a_{22}}{\sqrt{2}} \right) \cdot \left(\sqrt{2}\, \widetilde{a}_{12}, \, \frac{\widetilde{a}_{11} - \widetilde{a}_{22}}{\sqrt{2}}, \, \frac{\widetilde{a}_{11} + \widetilde{a}_{22}}{\sqrt{2}} \right) \\ &= 2a_{12}\widetilde{a}_{12} + \frac{(a_{11} - a_{22})(\widetilde{a}_{11} - \widetilde{a}_{22}) + (a_{11} + a_{22})(\widetilde{a}_{11} + \widetilde{a}_{22})}{2} \\ &= 2a_{12}\widetilde{a}_{12} + a_{11}\widetilde{a}_{11} + a_{22}\widetilde{a}_{22} \\ &= \operatorname{tr}(A\widetilde{A}). \end{aligned}$$

In particular, if $\omega := \frac{E(A)}{|E(A)|}$ and $\mathcal{R}$ is the space of the matrices Ξ of the form $\begin{pmatrix} \xi_1^2 & \xi_1\xi_2 \\ \xi_1\xi_2 & \xi_2^2 \end{pmatrix}$, then condition (4.4.3) gives that, for all $\Xi \in \mathcal{R} \setminus \{0\}$,

$$\begin{aligned} \omega \cdot \frac{E(\Xi)}{|E(\Xi)|} &= \frac{E(A) \cdot E(\Xi)}{|E(A)|\,|E(\Xi)|} = \frac{\operatorname{tr}(A\Xi)}{|A|\,|\Xi|} = \frac{a_{11}\,\xi_1^2 + a_{22}\,\xi_2^2 + 2a_{12}\,\xi_1\xi_2}{|A| \sqrt{\xi_1^4 + \xi_2^4 + 2\xi_1^2\xi_2^2}} \\ &= \frac{a_{11}\,\xi_1^2 + a_{22}\,\xi_2^2 + 2a_{12}\,\xi_1\xi_2}{|A|\,(\xi_1^2 + \xi_2^2)} = \frac{1}{|A|\,|\xi|^2} \sum_{i,j=1}^{2} a_{ij}\xi_i\xi_j \in \left[\frac{\lambda}{|A|}, \frac{\Lambda}{|A|} \right]. \end{aligned}$$

Accordingly, the ellipticity condition gives in this situation that E maps $\mathcal{R}$ into the cone

$$K_\omega := \left\{ p \in \mathbb{R}^3 \setminus \{0\} \text{ s.t. } \omega \cdot \frac{p}{|p|} \in \left[\frac{\lambda}{|A|}, \frac{\Lambda}{|A|} \right] \right\} \cup \{0\}, \tag{4.4.4}$$

see Figure 4.4.1.

Equations such as the one in (4.4.1) are often referred to by the name "equations in nondivergence form" (to distinguish them from the ones surfacing from a variational structure, compare for example with equation (1.1.3)). There are sundry reasons for addressing equation (4.4.1) in such a generality. One of the most important, in our personal opinion, is that the Laplace operator is not maintained under "nice transformations of the space" (except for translations and rotations, in

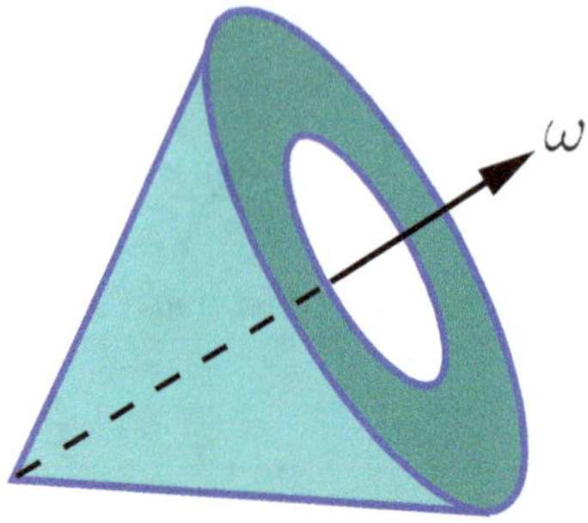

Figure 4.4.1 Graphical intuition for the ellipticity condition in (4.4.3): The cone K_ω in (4.4.4).

view of (1.1.2) and Corollary 1.1.4), and it would instead be desirable to consider a class of equations which is preserved under natural families of transformations.

With this respect, we point out the following result.

Lemma 4.4.1. *Let $\Omega \subseteq \mathbb{R}^n$ be open and $u \in C^2(\Omega)$. Let a_{ij}, b_i, c, $f \in C^\alpha(\Omega)$ for some $\alpha \in [0, 1]$, and assume that, for every $x \in \Omega$,*

$$\sum_{i,j=1}^{n} a_{ij}(x)\partial_{ij}u(x) + \sum_{i=1}^{n} b_i(x)\partial_i u(x) + c(x)u(x) = f(x). \tag{4.4.5}$$

Let $T : \mathbb{R}^n \to \mathbb{R}^n$ be an invertible diffeomorphism of class $C^{2,\alpha}$ with inverse of class $C^{2,\alpha}$. For each $y \in T(\Omega)$, let $\widetilde{u}(y) := u(T^{-1}(y))$.

Then, for every $y \in T(\Omega)$,

$$\sum_{i,j=1}^{n} \widetilde{a}_{ij}(y)\partial_{ij}\widetilde{u}(y) + \sum_{i=1}^{n} \widetilde{b}_i(y)\partial_i\widetilde{u}(y) + \widetilde{c}(y)\widetilde{u}(y) = \widetilde{f}(y) \tag{4.4.6}$$

for suitable $\widetilde{a}_{ij}$, $\widetilde{b}_i$, $\widetilde{c}$, $\widetilde{f} \in C^\alpha(\Omega)$.

Additionally, if $c \geqslant 0$ (respectively, $c \leqslant 0$) in Ω, then $\widetilde{c} \geqslant 0$ (respectively, $\widetilde{c} \leqslant 0$) in $T(\Omega)$.

Furthermore, if there exist $\Lambda \geqslant \lambda > 0$ such that

$$\lambda|\xi|^2 \leqslant \sum_{i,j=1}^{n} a_{ij}\xi_i\xi_j \leqslant \Lambda|\xi|^2 \quad for\ all\ \xi = (\xi_1, \dots, \xi_n) \in \mathbb{R}^n, \tag{4.4.7}$$

then

$$\widetilde{\lambda}|\xi|^2 \leqslant \sum_{i,j=1}^{n} \widetilde{a}_{ij}\xi_i\xi_j \leqslant \widetilde{\Lambda}|\xi|^2 \quad for\ all\ \xi = (\xi_1, \dots, \xi_n) \in \mathbb{R}^n, \tag{4.4.8}$$

for suitable $\widetilde{\Lambda} \geqslant \widetilde{\lambda} > 0$.

Proof. We implicitly assume that u is evaluated at x and $\widetilde{u}$ is evaluated at $y = T(x) = (T_1(x), \dots, T_n(x))$. Also, T is implicitly assumed to be evaluated at x. With this notation,

$$\partial_i u = \sum_{k=1}^{n} \partial_k \widetilde{u}\, \partial_i T_k$$

$$\text{and} \quad \partial_{ij} u = \sum_{k,m=1}^{n} \partial_{km} \widetilde{u}\, \partial_i T_k\, \partial_j T_m + \sum_{k=1}^{n} \partial_k \widetilde{u}\, \partial_{ij} T_k.$$

As a result, setting

$$\begin{aligned} \widetilde{a}_{km}(y) &:= \sum_{i,j=1}^{n} a_{ij}(T^{-1}(y))\, \partial_i T_k\, \partial_j T_m, \\ \widetilde{b}_k(y) &:= \sum_{i,j=1}^{n} a_{ij}(T^{-1}(y))\, \partial_{ij} T_k + \sum_{i=1}^{n} b_i(T^{-1}(y))\, \partial_i T_k, \\ \widetilde{c}(y) &:= c(T^{-1}(y)) \\ \text{and} \quad \widetilde{f}(y) &:= f(T^{-1}(y)), \end{aligned} \tag{4.4.9}$$

in view of (4.4.5) (and omitting the variable by following the previous convention), we have that

$$\begin{aligned} \widetilde{f} - \widetilde{c}\widetilde{u} &= f - cu \\ &= \sum_{i,j=1}^{n} a_{ij} \partial_{ij} u + \sum_{i=1}^{n} b_i \partial_i u \\ &= \sum_{i,j,k,m=1}^{n} a_{ij} \partial_i T_k\, \partial_j T_m\, \partial_{km} \widetilde{u} + \sum_{i,j,k=1}^{n} a_{ij} \partial_{ij} T_k\, \partial_k \widetilde{u} + \sum_{i,k=1}^{n} b_i \partial_i T_k\, \partial_k \widetilde{u} \\ &= \sum_{k,m=1}^{n} \widetilde{a}_{ij} \partial_{km} \widetilde{u} + \sum_{k=1}^{n} \widetilde{b}_k \partial_k \widetilde{u}, \end{aligned}$$

which is (4.4.6).

Also, the sign of $\widetilde{c}$ is clearly the same as the one (if any) of c. Furthermore, if $\xi \in \mathbb{R}^n$, we let $\eta = (\eta_1, \dots, \eta_n) \in \mathbb{R}^n$ be defined as

$$\eta_i := \sum_{k=1}^{n} \partial_i T_k \xi_k.$$

We thus find that

$$\sum_{k,m=1}^{n} \widetilde{a}_{km} \xi_k \xi_m = \sum_{i,j,k,m=1}^{n} a_{ij} \partial_i T_k\, \partial_j T_m\, \xi_k\, \xi_m = \sum_{i,j=1}^{n} a_{ij} \eta_i \eta_j.$$

Hence, if (4.4.7) holds true,

$$\sum_{k,m=1}^{n} \widetilde{a}_{km}\xi_k\xi_m \in [\lambda|\eta|^2, \Lambda|\eta|^2]. \tag{4.4.10}$$

We also observe that, if $(DT)_{ij} = \partial_i T_j$, then $\eta = DT\xi$ and therefore $|\eta| \leqslant (\|T\|_{C^1(\Omega)} + 1)|\xi|$. Moreover, since $\xi = (DT^{-1})\eta$, we have that $|\xi| \leqslant (\|T^{-1}\|_{C^1(\Omega)} + 1)|\eta|$. From these observations and (4.4.10), we deduce that

$$\sum_{k,m=1}^{n} \widetilde{a}_{km}\xi_k\xi_m \in \left[\frac{\lambda}{(\|T^{-1}\|_{C^1(\Omega)} + 1)^2}|\xi|^2, \, (\|T\|_{C^1(\Omega)} + 1)^2\Lambda|\xi|^2\right],$$

which is (4.4.8). □

The fact that the structure of equation (4.4.1) is preserved under suitable transformations (as guaranteed by Lemma 4.4.1) plays a crucial role in the proof of Theorem 4.5.1.

Besides this important property of preservation of elliptic structures, as warranted by Lemma 4.4.1, we recall that equations in the nondivergence form also naturally arise from the linearization of more complicated, and nonlinear, equations, see e.g. the discussion around [DV23, formula (35.7)].

Another natural motivation for equation (4.4.1), which is also helpful to develop some familiarity with this equation, arises in the random movement of a biological population with density u living in some environment Σ (this can be seen as a revisitation of the model presented in [DV23, Chapter 2]). Suppose that Σ presents hills and valleys; for instance, it can be modeled like a graph over $\mathbb{R}^n$ of the form

$$\Sigma = \{X = (x, y) \in \mathbb{R}^n \times \mathbb{R} \text{ s.t. } y = \gamma(x)\},$$

for some (nice) function $\gamma : \mathbb{R}^n \to \mathbb{R}$ (the biological case would correspond to $n = 2$). In this notation, we think that u is a function of time $t \in \mathbb{R}$ and position $X = (x, \gamma(x)) \in \Sigma$ and hence of $t \in \mathbb{R}$ and $x \in \mathbb{R}^n$.

We assume that the biological population reproduces itself; hence, at a given unit of time τ, the population increases proportionally to the existing population: at each time step, the proliferation effects would thus add to the population u an additional term $c\tau u$; the case $c = 0$ would correspond to no reproduction at all, and we may consider that c is also a function of x if the environment presents regions that are either more (higher c) or less (smaller c) favorable for reproduction.

We can also suppose that the population is bred; therefore, at each unit of time τ, some additional population τf is added into the environment. Again, f could depend on x since the breeders may identify strategic zones for inserting the newcomers. Also, in some regions, f could be positive, corresponding to breeders inserting new individuals, or negative, corresponding to breeders killing part of the existing population, for example, for food production (and of course f can change its sign if both of these breeding activities take place). The case of f vanishing identically is also allowed and corresponds to a free biological population with no breeders.

We suppose that the biological population moves randomly across the environment. At each time step, the individuals pick randomly a direction $e \in \partial B_1$ and move by a unit step h along Σ in the direction indicated by e. We stress that, for small h, when an individual moves on the surface Σ by a space length h in the direction indicated by e, its projection on $\mathbb{R}^n$ is moved by a space length $\frac{h}{\sqrt{1+|\partial_e \gamma|^2}}$ in the horizontal direction e, up to $o(h)$, and the spatial displacement of the population in $\mathbb{R}^{n+1}$ is hv, up to $o(h)$, where

$$v(x) := \left(\frac{e}{\sqrt{1+|\partial_e \gamma(x)|^2}}, \frac{\partial_e \gamma(x)}{\sqrt{1+|\partial_e \gamma(x)|^2}} \right)$$

since, using the notation $X(x) := (x, \gamma(x))$,

$$\begin{aligned} & X\left(x + \frac{he}{\sqrt{1+|\partial_e \gamma(x)|^2}}\right) - X(x) \\ &= \left(\frac{he}{\sqrt{1+|\partial_e \gamma(x)|^2}}, \gamma\left(x + \frac{he}{\sqrt{1+|\partial_e \gamma(x)|^2}}\right) - \gamma(x) \right) \\ &= \left(\frac{he}{\sqrt{1+|\partial_e \gamma(x)|^2}}, \frac{h\partial_e \gamma(x)}{\sqrt{1+|\partial_e \gamma(x)|^2}} \right) + o(h) \\ &= hv(x) + o(h), \end{aligned}$$

see Figure 4.4.2.

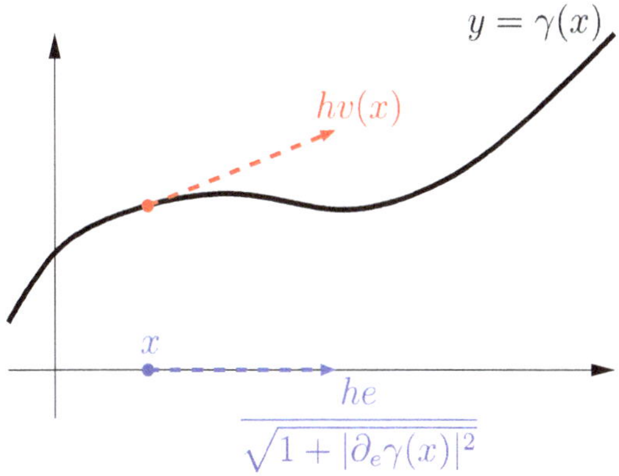

Figure 4.4.2 Random displacement of the population leaving on Σ (with h to be sent to 0 in the limit).

With respect to this observation, up to higher orders that we neglect, the individuals corresponding to the population density $u(x,t)$ are moved at time $t+\tau$ by the random process into $u\left(x+\frac{he}{\sqrt{1+|\partial_e\gamma(x)|^2}},t\right)$, averaged over $e\in\partial B_1$, namely

$$u(x,t+\tau)=\fint_{\partial B_1}u\left(x+\frac{he}{\sqrt{1+|\partial_e\gamma(x)|^2}},t\right)\,d\mathcal{H}_e^{n-1}.$$

For flat surfaces in which $\nabla\gamma=0$, the above would correspond to the classical (isotropic) random walk.

We can also assume that the individuals are mildly subject to gravity, in the sense that the gravity force tends to let them slightly slide down in the direction of the maximal slope of the environment Σ. To model this effect, for instance, we can suppose that at each unit of time, the individuals corresponding to the population density $u(x,t)$ are moved by gravity into $u(x-h^2d(x)\nabla\gamma(x),t+\tau)$. The parameter d models a "grip," which could also vary from region to region; namely, $d=0$ is a perfect adherent situation in which no sliding occurs, whereas a large value of d would provide a slippery surface.

In this complex (but relatively simple, after all) model, the density at time $t+\tau$ is thus the superposition of a number of effects, such as proliferation, breeding, random movement and drifting (one can also remove some of these effects to concentrate on a simpler model), and we can therefore write that

$$\begin{aligned}u(x,t+\tau)=c\tau u(x,t)+\tau f(x,t)+\fint_{\partial B_1}u\left(x+\frac{he}{\sqrt{1+|\partial_e\gamma(x)|^2}},t\right)\,d\mathcal{H}_e^{n-1}\\+u(x+h^2d(x)\nabla\gamma(x),t)-u(x,t).\end{aligned}\tag{4.4.11}$$

By an odd symmetry cancellation, we see that

$$\begin{aligned}&\fint_{\partial B_1}u\left(x+\frac{he}{\sqrt{1+|\partial_e\gamma(x)|^2}},t\right)\,d\mathcal{H}_e^{n-1}\\&=\fint_{\partial B_1}\left[u(x,t)+\nabla u(x,t)\cdot\frac{he}{\sqrt{1+|\partial_e\gamma(x)|^2}}\right.\\&\qquad\left.+\frac{1}{2}D^2u(x,t)\frac{he}{\sqrt{1+|\partial_e\gamma(x)|^2}}\cdot\frac{he}{\sqrt{1+|\partial_e\gamma(x)|^2}}\right]d\mathcal{H}_e^{n-1}+o(h^2)\\&=u(x,t)+\frac{h^2}{2}\sum_{i,j=1}^{n}\fint_{\partial B_1}\frac{\partial_{ij}u(x,t)\,e_i\,e_j}{1+|\partial_e\gamma(x)|^2}\,d\mathcal{H}_e^{n-1}+o(h^2)\\&=u(x,t)+h^2\sum_{i,j=1}^{n}a_{ij}(x)\partial_{ij}u(x,t)+o(h^2),\end{aligned}$$

where

$$
\begin{aligned}
a_{ij}(x) &:= \frac{1}{2} \fint_{\partial B_1} \frac{e_i\, e_j}{1+|\partial_e \gamma(x)|^2}\, d\mathcal{H}^{n-1}_e \\
&= \frac{1}{2}\delta_{ij} \fint_{\partial B_1} \frac{e_i^2}{1+|\partial_e \gamma(x)|^2}\, d\mathcal{H}^{n-1}_e .
\end{aligned} \tag{4.4.12}
$$

We point out that the matrix a_{ij} in this case has a diagonal form[13] (but it is not necessarily constant).

Moreover,

$$
\begin{aligned}
u(x + h^2 d(x)\nabla\gamma(x), t) - u(x,t) &= h^2 d(x)\nabla u(x,t)\cdot\nabla\gamma(x) + o(h^2) \\
&= h^2 \sum_{i=1}^{n} b_i(x)\partial_i u(x,t) + o(h^2),
\end{aligned}
$$

where

$$
b(x) := d(x)\nabla\gamma(x).
$$

By these observations, we can recast (4.4.11) into

$$
\begin{aligned}
u(x,t+\tau) &= c\tau u(x,t) + \tau f(x,t) + u(x,t) \\
&\quad + h^2 \sum_{i,j=1}^{n} a_{ij}(x)\partial_{ij}u(x,t) + h^2 \sum_{i=1}^{n} b_i(x)\partial_i u(x,t) + o(h^2).
\end{aligned}
$$

Hence, by taking the term $u(x,t)$ to the left-hand side, dividing by τ and choosing a quadratic space-time scaling unit, i.e. $\tau := h^2$, sending $h \searrow 0$ (or equivalently $\tau \searrow 0$), we obtain

$$
\partial_t u(x,t) = cu(x,t) + f(x,t) + \sum_{i,j=1}^{n} a_{ij}(x)\partial_{ij}u(x,t) + \sum_{i=1}^{n} b_i(x)\partial_i u(x,t).
$$

With respect to this, we see that equation (4.4.1) describes the stationary states of this model.

We have also encountered a general form of the elliptic equation for biological species possibly subject to drift and chemotaxis: recall [DV23, equation (2.14)].

[13] A variation of the model presented here can also produce matrices a_{ij} that are not diagonal. For instance, one might account for a possible "spatial disorientation" of the individuals by assuming that, instead of moving in the randomly selected direction e, they move in some direction $\eta(e)$, say with $\eta(e) = -\eta(-e)$, for instance $\eta(e) := \frac{(e_1+\varepsilon e_2, e_2, \dots, e_n)}{|(e_1+\varepsilon e_2, e_2, \dots, e_n)|}$ for some $\varepsilon \in \mathbb{R}$. In this way, the matrix in (4.4.12) would be replaced by

$$
\frac{1}{2} \fint_{\partial B_1} \frac{\eta_i(e)\, \eta_j(e)}{1+|\partial_{\eta(e)}\gamma(x)|^2}\, d\mathcal{H}^{n-1}_e ,
$$

which is not necessarily diagonal.

Another example of the general elliptic equation was encountered in the stationary Fokker–Planck setting, recalling equation [DV23, equation (24.2)].

Other motivations for equation (4.4.1) arise in differential geometry, when one studies partial differential equations on manifolds (compare for example with the Laplace–Beltrami operator in local coordinates, as presented in (2.4.9)).

Now, we discuss some interior estimates for equation (4.4.1), which can be seen as a natural extension[14] of Theorem 4.2.1.

Theorem 4.4.2. *Let a_{ij}, b_i, c, $f \in C^\alpha(B_1)$ for some $\alpha \in (0,1)$. Assume the ellipticity condition in* (4.4.3). *Let $u \in C^{2,\alpha}(B_1)$ be a solution of*

$$\sum_{i,j=1}^{n} a_{ij}\partial_{ij}u + \sum_{i=1}^{n} b_i\partial_i u + cu = f \quad \text{in } B_1.$$

Then, there exists $C > 0$, depending only on n, α, a_{ij}, b_i and c, such that

$$\|u\|_{C^{2,\alpha}(B_{1/2})} \leqslant C\left(\|u\|_{L^\infty(B_1)} + \|f\|_{C^\alpha(B_1)}\right). \tag{4.4.13}$$

We also have a counterpart of the local estimates at the boundary given in Theorem 4.2.2, in the halfball notation stated in (4.2.2), according to the following result.

Theorem 4.4.3. *Let a_{ij}, b_i, c, $f \in C^\alpha(B_1^+)$ and $g \in C^{2,\alpha}(B_1^0)$ for some $\alpha \in (0,1)$. Assume the ellipticity condition in* (4.4.3). *Let $u \in C^{2,\alpha}(B_1^+) \cap C(B_1^+ \cup B_1^0)$ be a solution of*

$$\begin{cases} \displaystyle\sum_{i,j=1}^{n} a_{ij}\partial_{ij}u + \sum_{i=1}^{n} b_i\partial_i u + cu = f & \text{in } B_1^+, \\ u = g \quad \text{on } B_1^0. \end{cases}$$

Then, there exists $C > 0$, depending only on n, α, a_{ij}, b_i and c, such that

$$\|u\|_{C^{2,\alpha}(B_{1/2}^+)} \leqslant C\left(\|u\|_{L^\infty(B_1^+)} + \|g\|_{C^{2,\alpha}(B_1^0)} + \|f\|_{C^\alpha(B_1^+)}\right).$$

To deal with the proof of regularity results such as Theorems 4.4.2 and 4.4.3, it is often convenient to deal with intermediate estimates in which weighted norms in a ball are bounded by a small constant times suitable weighted norms in larger balls. While in principle the weighted norms in larger balls do not seem helpful to conclude a meaningful bound, the smallness of the constant in front may allow

[14]The assumption $u \in C^{2,\alpha}(B_1)$ does play a role in the proof of Theorem 4.4.2, but it can be weakened to the more natural hypothesis $u \in C^2(B_1)$. Weakening this assumption at this stage is not completely trivial, and we accomplish this further generality in the forthcoming Corollary 4.5.8.

one to "reabsorb" the inconvenient terms. This methodology relies on some fine analysis of scaling and weighted norms, and it is of general use (see [Sim97, pp. 398–399]): for our scopes, we limit ourselves to the following result.

Proposition 4.4.4. *Let $k \in \mathbb{N}$, $\alpha \in (0,1]$, $\vartheta \in [0,1]$, $\sigma \geqslant 0$ and $R > 0$. There exists $\varepsilon_0 \in (0,1)$ depending only on n and k such that the following holds true.*

Let $u \in C^{k,\alpha}(B_R)$. Assume that for every $x_0 \in B_R$ and every $\rho \in \left(0, \frac{R-|x_0|}{2}\right)$,

$$\mathcal{N}_\vartheta(x_0,\rho) \leqslant \sigma + \varepsilon_0\, \mathcal{N}_\vartheta(x_0,2\rho), \tag{4.4.14}$$

where we used the notation

$$\mathcal{N}_\vartheta(x_0,\rho) := \sum_{j=0}^{k} \rho^j \|D^j u\|_{L^\infty(B_\rho(x_0))} + \vartheta \rho^{k+\alpha}[u]_{C^{k,\alpha}(B_\rho(x_0))}.$$

Then, there exists $C > 0$, depending only on n, k and R, such that

$$\sum_{j=0}^{k} \|D^j u\|_{L^\infty(B_{R/2})} + \vartheta [u]_{C^{k,\alpha}(B_{R/2})} \leqslant C\sigma. \tag{4.4.15}$$

A counterpart of Proposition 4.4.4 holds true in the halfball notation stated in (4.2.2) and generalized in (4.2.67). The technical details go as follows.

Proposition 4.4.5. *Let $k \in \mathbb{N}$, $\alpha \in (0,1]$, $\vartheta \in [0,1]$, $\sigma \geqslant 0$ and $R > 0$. There exists $\varepsilon_0 \in (0,1)$ depending only on n and k such that the following holds true.*

Let $u \in C^{k,\alpha}(B_R^+)$. Assume that for every $x_0 \in B_R^+$ and every $\rho \in \left(0, \frac{R-|x_0|}{2}\right)$,

$$\mathcal{N}_\vartheta(x_0,\rho) \leqslant \sigma + \varepsilon_0\, \mathcal{N}_\vartheta(x_0,2\rho), \tag{4.4.16}$$

where we used the notation

$$\mathcal{N}_\vartheta(x_0,\rho) := \sum_{j=0}^{k} \rho^j \|D^j u\|_{L^\infty(B_\rho^+(x_0))} + \vartheta \rho^{k+\alpha}[u]_{C^{k,\alpha}(B_\rho^+(x_0))}.$$

Then, there exists $C > 0$, depending only on n, k and R, such that

$$\sum_{j=0}^{k} \|D^j u\|_{L^\infty(B_{R/2}^+)} + \vartheta [u]_{C^{k,\alpha}(B_{R/2}^+)} \leqslant C\sigma. \tag{4.4.17}$$

We prove Propositions 4.4.4 and 4.4.5 at the same time by letting $\mathcal{B}_r(x)$ to be equal to $B_r(x)$ in the setting of Proposition 4.4.4 and to be equal to $B_r^+(x)$ in the setting of Proposition 4.4.5.

Proof of Propositions 4.4.4 and 4.4.5. Let

$$Q := \sup_{\substack{x_0 \in \mathcal{B}_R \\ \rho \in \left(0, \frac{R-|x_0|}{2}\right)}} \mathcal{N}_\vartheta(x_0, \rho) \tag{4.4.18}$$

and note that

$$\begin{aligned} Q &\leqslant \sum_{j=0}^{k} \rho^j \|D^j u\|_{L^\infty(\mathcal{B}_R)} + \vartheta \rho^{k+\alpha} [u]_{C^{k,\alpha}(\mathcal{B}_R)} \\ &\leqslant \sum_{j=0}^{k} R^j \|D^j u\|_{L^\infty(\mathcal{B}_R)} + \vartheta R^{k+\alpha} [u]_{C^{k,\alpha}(\mathcal{B}_R)} < +\infty \end{aligned}$$

since $u \in C^{k,\alpha}(\mathcal{B}_R)$.

Now, we take $x_0 \in \mathcal{B}_R$ and $\rho \in \left(0, \frac{R-|x_0|}{2}\right)$. We cover $\mathcal{B}_\rho(x_0)$ by a family of balls $\mathcal{B}_{\frac{\rho}{8}}(p_i)$ with $p_i \in \mathcal{B}_\rho(x_0)$ and $i \in \{1, \ldots, N\}$; we observe that we can bound N from above in dependence only of n.

We note also that

$$|p_i| \leqslant |x_0| + \rho < R - 2\rho + \rho = R - \rho.$$

As a consequence, we have that $p_i \in \mathcal{B}_R$ and $\frac{\rho}{2} \in \left(0, \frac{R-|p_i|}{2}\right)$. Hence, we can use either (4.4.14) or (4.4.16) and find that, for all $i \in \{1, \ldots, N\}$,

$$\mathcal{N}_\vartheta\left(p_i, \frac{\rho}{4}\right) \leqslant \sigma + \varepsilon_0 \mathcal{N}_\vartheta\left(p_i, \frac{\rho}{2}\right) \leqslant \sigma + \varepsilon_0 Q$$

and therefore

$$\sum_{i=1}^{N} \mathcal{N}_\vartheta\left(p_i, \frac{\rho}{4}\right) \leqslant N\sigma + \varepsilon_0 N Q. \tag{4.4.19}$$

Now, we observe that, for all $j \in \{0, \ldots, k\}$,

$$\begin{aligned} \rho^j \|D^j u\|_{L^\infty(\mathcal{B}_\rho(x_0))} &\leqslant \rho^j \sum_{i=1}^{N} \|D^j u\|_{L^\infty(\mathcal{B}_{\rho/8}(p_i))} \\ &= 8^j \sum_{i=1}^{N} \left(\frac{\rho}{8}\right)^j \|D^j u\|_{L^\infty(\mathcal{B}_{\rho/8}(p_i))} \\ &\leqslant 8^j \sum_{i=1}^{N} \mathcal{N}_\vartheta\left(p_i, \frac{\rho}{8}\right) \leqslant 8^k \sum_{i=1}^{N} \mathcal{N}_\vartheta\left(p_i, \frac{\rho}{8}\right). \end{aligned} \tag{4.4.20}$$

We also claim that

$$\vartheta \rho^{k+\alpha} [u]_{C^{k,\alpha}(\mathcal{B}_\rho(x_0))} \leqslant C 8^{k+\alpha} \sum_{i=1}^{N} \mathcal{N}_\vartheta\left(p_i, \frac{\rho}{4}\right), \tag{4.4.21}$$

where C is a positive constant depending only on n and k.

To check (4.4.21), let $x \in \mathcal{B}_\rho(x_0)$ and $i_x \in \{1, \dots, N\}$ be such that $x \in \mathcal{B}_{\frac{\rho}{8}}(p_{i_x})$. We pick $y \in \mathcal{B}_\rho(x_0)$ and distinguish two cases: if $y \in \mathcal{B}_{\frac{\rho}{8}}(x)$, then x, $y \in \mathcal{B}_{\frac{\rho}{4}}(p_{i_x})$ and therefore, if $\beta \in \mathbb{N}^n$ with $|\beta| = k$,

$$\begin{aligned} \vartheta\rho^{k+\alpha}\frac{|D^\beta u(x) - D^\beta u(y)|}{|x-y|^\alpha} &\leqslant \vartheta\rho^{k+\alpha}[u]_{C^{k,\alpha}(\mathcal{B}_{\rho/4}(p_{i_x}))} \\ &= \vartheta 4^{k+\alpha}\left(\frac{\rho}{4}\right)^{k+\alpha}[u]_{C^{k,\alpha}(\mathcal{B}_{\rho/4}(p_{i_x}))} \\ &\leqslant 4^{k+\alpha}\mathcal{N}_\vartheta\left(p_{i_x}, \frac{\rho}{4}\right) \leqslant 4^{k+\alpha}\sum_{i=1}^N \mathcal{N}_\vartheta\left(p_i, \frac{\rho}{4}\right). \end{aligned} \tag{4.4.22}$$

If instead $y \notin \mathcal{B}_{\frac{\rho}{8}}(x)$, we let $i_y \in \{1, \dots, N\}$ be such that $y \in \mathcal{B}_{\frac{\rho}{8}}(p_{i_y})$, and we find that, if $\beta \in \mathbb{N}^n$ with $|\beta| = k$,

$$\begin{aligned} \vartheta\rho^{k+\alpha}\frac{|D^\beta u(x) - D^\beta u(y)|}{|x-y|^\alpha} &\leqslant \rho^{k+\alpha}\frac{|D^\beta u(x)| + |D^\beta u(y)|}{(\rho/8)^\alpha} \\ &\leqslant 8^\alpha\rho^k\Big(\|D^\beta u\|_{L^\infty(\mathcal{B}_{\rho/8}(p_{i_x}))} + \|D^\beta u\|_{L^\infty(\mathcal{B}_{\rho/8}(p_{i_y}))}\Big) \\ &\leqslant 8^\alpha\sum_{i=1}^N \rho^k\|D^\beta u\|_{L^\infty(\mathcal{B}_{\rho/8}(p_i))} \\ &\leqslant 8^{k+\alpha}\sum_{i=1}^N \mathcal{N}_\vartheta\left(p_i, \frac{\rho}{8}\right). \end{aligned}$$

By combining this and (4.4.22), we conclude that, for all $x, y \in \mathcal{B}_\rho(x_0)$,

$$\vartheta\rho^{k+\alpha}\frac{|D^\beta u(x) - D^\beta u(y)|}{|x-y|^\alpha} \leqslant 8^{k+\alpha}\sum_{i=1}^N \mathcal{N}_\vartheta\left(p_i, \frac{\rho}{4}\right).$$

Hence, summing up over all $\beta \in \mathbb{N}^n$ such that $|\beta| = k$ (recall the notation in footnote 4), we obtain the desired result in (4.4.21).

Thus, in light of (4.4.20) and (4.4.21),

$$\begin{aligned} \mathcal{N}_\vartheta(x_0, \rho) &\leqslant C\left((k+1)8^k + 8^{k+\alpha}\right)\sum_{i=1}^N \mathcal{N}_\vartheta\left(p_i, \frac{\rho}{4}\right) \\ &\leqslant C(k+2)8^{k+1}\sum_{i=1}^N \mathcal{N}_\vartheta\left(p_i, \frac{\rho}{4}\right). \end{aligned}$$

Thus, recalling (4.4.19),

$$\mathcal{N}_\vartheta(x_0, \rho) \leqslant C(k+2)8^{k+1}\Big(N\sigma + \varepsilon_0 NQ\Big).$$

Taking the supremum over such x_0 and ρ, in view of (4.4.18), we find that

$$Q \leqslant C(k+2)8^{k+1}\Big(N\sigma + \varepsilon_0 NQ\Big).$$

We now choose $\varepsilon_0 := \frac{1}{2NC(k+2)8^{k+1}}$, and we infer that

$$Q \leqslant C(k+2)8^{k+1}\left(N\sigma + \frac{Q}{2C(k+2)8^{k+1}}\right) = (k+2)8^{k+1}N\sigma + \frac{Q}{2};$$

therefore, reabsorbing one term into the left-hand side,

$$Q \leqslant 2C(k+2)8^{k+1}N\sigma. \tag{4.4.23}$$

Hence, since, for all $\rho \in \left(0, \frac{R}{2}\right)$,

$$Q \geqslant N_\vartheta(0,\rho) = \sum_{j=0}^{k} \rho^j \|D^j u\|_{L^\infty(\mathcal{B}_\rho)} + \vartheta\rho^{k+\alpha}[u]_{C^{k,\alpha}(\mathcal{B}_\rho)},$$

we deduce from (4.4.23) that

$$\begin{aligned} 2C(k+2)8^{k+1}N\sigma &\geqslant \lim_{\rho\nearrow\frac{R}{2}} \sum_{j=0}^{k} \rho^j \|D^j u\|_{L^\infty(\mathcal{B}_\rho)} + \vartheta\rho^{k+\alpha}[u]_{C^{k,\alpha}(\mathcal{B}_\rho)} \\ &= \sum_{j=0}^{k} \left(\frac{R}{2}\right)^j \|D^j u\|_{L^\infty(\mathcal{B}_{R/2})} + \vartheta\left(\frac{R}{2}\right)^{k+\alpha} [u]_{C^{k,\alpha}(\mathcal{B}_{R/2})}. \end{aligned}$$

From this, the claims in (4.4.15) and (4.4.17) follow, as desired. □

With this auxiliary result on scaled estimates, we can proceed with the proof of Theorem 4.4.2 by arguing as follows.

Proof of Theorem 4.4.2. The essence of the proof is to exploit the result for the Laplace operator in Theorem 4.2.1 in a perturbative fashion, since, in simple terms, in tiny balls the coefficients a_{ij} vary very little from constant ones, and the additional lower-order terms induced by b_i and c produce, by scaling, negligible quantities.

To make this argument work, we first observe that Theorem 4.2.1 holds true if the Laplacian is replaced by an operator with constant coefficients of the form

$$\sum_{i,j=1}^{n} \overline{a}_{ij}\partial_{ij}u, \tag{4.4.24}$$

where $\overline{a}_{ij}$ is constant and fulfills the ellipticity condition in (4.4.3), with constants depending also on the elliptic parameters λ and Λ in (4.4.3). Indeed, if u is a solution of $\sum_{i,j=1}^{n} \overline{a}_{ij}\partial_{ij}u = f$ in B_1 and $\overline{a}_{ij}$ is as above, then we can define $\overline{A}$ to be the matrix $\{\overline{a}_{ij}\}_{i,j\in\{1,\dots,n\}}$ and M to be the square root of $\overline{A}$ in the matrix sense, see e.g. [Gen07, p. 125]. We observe that

$$\left|\frac{Mx}{\sqrt{\Lambda}}\right| = \sqrt{\frac{(Mx)\cdot(Mx)}{\Lambda}} = \sqrt{\frac{\overline{A}x\cdot x}{\Lambda}}. \tag{4.4.25}$$

Thus, if we consider the function $\overline{u}(x) := u\left(\frac{Mx}{\sqrt{\Lambda}}\right)$, we have that, for all $x \in B_1$,

$$\left|\frac{Mx}{\sqrt{\Lambda}}\right| < \sqrt{\frac{\Lambda}{\Lambda}} = 1$$

and

$$\begin{aligned}\Delta\overline{u}(x) &= \frac{1}{\Lambda}\sum_{i,j,m=1}^{n} M_{mi}M_{mj}\partial_{ij}u\left(\frac{Mx}{\sqrt{\Lambda}}\right) = \frac{1}{\Lambda}\sum_{i,j=1}^{n}\overline{a}_{ij}\partial_{ij}u\left(\frac{Mx}{\sqrt{\Lambda}}\right)\\ &= \frac{1}{\Lambda}f\left(\frac{Mx}{\sqrt{\Lambda}}\right) =: \overline{f}(x).\end{aligned}$$

We can therefore employ Theorem 4.2.1 and conclude that

$$\|\overline{u}\|_{C^{2,\alpha}(B_{1/2})} \leqslant C\left(\|\overline{u}\|_{L^\infty(B_1)} + \|\overline{f}\|_{C^\alpha(B_1)}\right). \tag{4.4.26}$$

Since, by (4.4.25),

$$\left|\frac{Mx}{\sqrt{\Lambda}}\right| \geqslant \sqrt{\frac{\lambda|x|^2}{\Lambda}} = \sqrt{\frac{\lambda}{\Lambda}}|x|,$$

up to renaming C, we deduce from (4.4.26) that

$$\|u\|_{C^{2,\alpha}\left(B_{\sqrt{\frac{\lambda}{4\Lambda}}}\right)} \leqslant C\left(\|u\|_{L^\infty(B_1)} + \|f\|_{C^\alpha(B_1)}\right).$$

From this and a covering argument (see Lemma 4.2.13), we obtain (4.4.24), as desired.

Now, given $x_0 \in B_{1/2}$ and $\rho \in \left(0, \frac{1}{4}\right)$, to be taken suitably small in the following, we set

$$\widetilde{a}_{ij}(x) := a_{ij}(x_0) - a_{ij}(x)$$
$$\text{and}\quad \widetilde{u}(x) := \frac{u(x_0+\rho x)}{\rho^2}.$$

For all $x \in B_1$, we have that

$$\begin{aligned}\sum_{i,j=1}^{n} a_{ij}(x_0)\partial_{ij}\widetilde{u}(x) &= \sum_{i,j=1}^{n} a_{ij}(x_0)\partial_{ij}u(x_0+\rho x)\\ &= \sum_{i,j=1}^{n} a_{ij}(x_0+\rho x)\partial_{ij}u(x_0+\rho x)\\ &\quad + \sum_{i,j=1}^{n} \widetilde{a}_{ij}(x_0+\rho x)\partial_{ij}u(x_0+\rho x)\end{aligned}$$

$$\begin{aligned}
&= f(x_0+\rho x) - \sum_{i=1}^{n} b_i(x_0+\rho x)\partial_i u(x_0+\rho x) \\
&\qquad - c(x_0+\rho x)u(x_0+\rho x) + \sum_{i,j=1}^{n} \widetilde{a}_{ij}(x_0+\rho x)\partial_{ij}u(x_0+\rho x) \\
&=: \widetilde{f}(x). \qquad (4.4.27)
\end{aligned}$$

As a result, we deduce from (4.4.24) that

$$\|\widetilde{u}\|_{C^{2,\alpha}(B_{1/2})} \leqslant C\left(\|\widetilde{u}\|_{L^\infty(B_1)} + \|\widetilde{f}\|_{C^\alpha(B_1)}\right). \qquad (4.4.28)$$

We remark that if $x \in B_1$, then $|\widetilde{u}(x)| = \frac{|u(x_0+\rho x)|}{\rho^2} \leqslant \frac{\|u\|_{L^\infty(B_\rho(x_0))}}{\rho^2}$. In addition, for every $k \in \mathbb{N}$, we have that $D^k u(y) = \rho^{2-k} D^k \widetilde{u}\left(\frac{y-x_0}{\rho}\right)$; hence, for all y, $z \in B_{\rho/2}(x_0)$,

$$\begin{aligned}
&|u(y)| \leqslant \rho^2 \|\widetilde{u}\|_{L^\infty(B_{1/2})}, \\
&|\nabla u(y)| \leqslant \rho \|\nabla \widetilde{u}\|_{L^\infty(B_{1/2})}, \\
&|D^2 u(y)| \leqslant \|D^2 \widetilde{u}\|_{L^\infty(B_{1/2})}
\end{aligned}$$

and

$$\begin{aligned}
|D^2u(y) - D^2u(z)| &= \left| D^2\widetilde{u}\left(\frac{y-x_0}{\rho}\right) - D^2\widetilde{u}\left(\frac{z-x_0}{\rho}\right)\right| \\
&\leqslant C\rho^{-\alpha}[\widetilde{u}]_{C^{2,\alpha}(B_{1/2})}|x-y|^\alpha.
\end{aligned}$$

It follows from these observations and (4.4.28) that

$$\begin{aligned}
&\sum_{k=0}^{2} \rho^k \|D^k u\|_{L^\infty(B_{\rho/2}(x_0))} + \rho^{2+\alpha}[u]_{C^{2,\alpha}(B_{\rho/2}(x_0))} \\
&\quad \leqslant \rho^2 \|\widetilde{u}\|_{C^{2,\alpha}(B_{1/2})} \\
&\quad \leqslant C\rho^2\left(\|\widetilde{u}\|_{L^\infty(B_1)} + \|\widetilde{f}\|_{C^\alpha(B_1)}\right) \\
&\quad \leqslant C\left(\|u\|_{L^\infty(B_1)} + \rho^2\|\widetilde{f}\|_{C^\alpha(B_1)}\right). \qquad (4.4.29)
\end{aligned}$$

We observe that, for all $x, y \in B_1$ and any function g,

$$|g(x_0+\rho x) - g(x_0+\rho y)| \leqslant \rho^\alpha [g]_{C^\alpha(B_\rho(x_0))}|x-y|^\alpha.$$

As a consequence, letting $b := (b_1, \ldots, b_n)$ and $\widetilde{A} := \{\widetilde{a}_{ij}\}_{i,j\in\{1,\ldots,n\}}$,

$$\begin{aligned}
[\widetilde{f}]_{C^\alpha(B_1)} \leqslant C\rho^\alpha\Big(&[f]_{C^\alpha(B_\rho(x_0))} \\
&+ [b]_{C^\alpha(B_\rho(x_0))}\|\nabla u\|_{L^\infty(B_\rho(x_0))} + \|b\|_{L^\infty(B_\rho(x_0))}[\nabla u]_{C^\alpha(B_\rho(x_0))} \\
&+ [c]_{C^\alpha(B_\rho(x_0))}\|u\|_{L^\infty(B_\rho(x_0))} + \|c\|_{L^\infty(B_\rho(x_0))}[u]_{C^\alpha(B_\rho(x_0))}
\end{aligned}$$

$$
\begin{aligned}
&+ [\widetilde{A}]_{C^\alpha(B_\rho(x_0))}\|D^2u\|_{L^\infty(B_\rho(x_0))}\\
&+ \|\widetilde{A}\|_{L^\infty(B_\rho(x_0))}[D^2u]_{C^\alpha(B_\rho(x_0))}\Big)\\
\leqslant{}& C\rho^\alpha\Big(\|f\|_{C^\alpha(B_\rho(x_0))} + \|u\|_{C^2(B_\rho(x_0))}\\
&+ \|\widetilde{A}\|_{L^\infty(B_\rho(x_0))}[D^2u]_{C^\alpha(B_\rho(x_0))}\Big)\\
\leqslant{}& C\rho^\alpha\Big(\|f\|_{C^\alpha(B_\rho(x_0))} + \|u\|_{C^2(B_\rho(x_0))} + \rho^\alpha[D^2u]_{C^\alpha(B_\rho(x_0))}\Big).
\end{aligned}
\tag{4.4.30}
$$

Additionally,

$$
\begin{aligned}
&\|\widetilde{f}\|_{L^\infty(B_1)}\\
&\leqslant C\Big(\|f\|_{L^\infty(B_\rho(x_0))} + \|u\|_{C^1(B_\rho(x_0))} + \|\widetilde{A}\|_{L^\infty(B_\rho(x_0))}\|D^2u\|_{L^\infty(B_\rho(x_0))}\Big)\\
&\leqslant C\Big(\|f\|_{L^\infty(B_\rho(x_0))} + \|u\|_{C^1(B_\rho(x_0))} + \rho^\alpha\|D^2u\|_{L^\infty(B_\rho(x_0))}\Big).
\end{aligned}
$$

This and (4.4.30) entail that

$$
\begin{aligned}
\|\widetilde{f}\|_{C^\alpha(B_1)} \leqslant C\Big(&\|f\|_{C^\alpha(B_\rho(x_0))} + \|u\|_{C^1(B_\rho(x_0))}\\
&+ \rho^\alpha\|D^2u\|_{L^\infty(B_\rho(x_0))} + \rho^{2\alpha}[D^2u]_{C^\alpha(B_\rho(x_0))}\Big).
\end{aligned}
$$

From this and (4.4.29), we arrive at

$$
\begin{aligned}
&\sum_{k=0}^{2}\rho^k\|D^ku\|_{L^\infty(B_{\rho/2}(x_0))} + \rho^{2+\alpha}[u]_{C^{2,\alpha}(B_{\rho/2}(x_0))}\\
&\leqslant C\Big(\|f\|_{C^\alpha(B_1)} + \|u\|_{L^\infty(B_1)} + \rho^2\|\nabla u\|_{L^\infty(B_\rho(x_0))}\\
&\quad + \rho^{2+\alpha}\|D^2u\|_{L^\infty(B_\rho(x_0))} + \rho^{2+2\alpha}[D^2u]_{C^\alpha(B_\rho(x_0))}\Big)\\
&\leqslant C\Big(\|f\|_{C^\alpha(B_1)} + \|u\|_{L^\infty(B_1)}\Big)\\
&\quad + C\rho^\alpha\left(\sum_{k=0}^{2}\rho^k\|D^ku\|_{L^\infty(B_\rho(x_0))} + \rho^{2+\alpha}[u]_{C^{2,\alpha}(B_\rho(x_0))}\right).
\end{aligned}
\tag{4.4.31}
$$

We are thus in the setting of Proposition 4.4.4 with $k := 2$, $\vartheta := 1$ and $\sigma := C\left(\|f\|_{C^\alpha(B_1)} + \|u\|_{L^\infty(B_1)}\right)$. If ε_0 is as in Proposition 4.4.4, we have that, if ρ is sufficiently small such that $C\rho^\alpha \leqslant \varepsilon_0$, then the framework in (4.4.14) is fulfilled. We thereby deduce from (4.4.15) that

$$
\|u\|_{C^{2,\alpha}(B_{1/4})} \leqslant C\left(\|f\|_{C^\alpha(B_1)} + \|u\|_{L^\infty(B_1)}\right).
\tag{4.4.32}
$$

Accordingly, we can combine (4.4.32) and a covering argument (see Lemma 4.2.13) and obtain (4.4.13), as desired. □

Now we focus on the local estimates at the boundary.

Proof of Theorem 4.4.3. As in (4.2.42), it suffices to prove Theorem 4.4.3 when g vanishes identically. Then, one can repeat verbatim the proof of Theorem 4.4.2 with the following modifications:

- balls B_r are replaced by halfballs B_r^+;
- the use of Theorem 4.2.1 is replaced by that of Theorem 4.2.14;
- the use of Proposition 4.4.4 is replaced by that of Proposition 4.4.5. □

4.5 Global Schauder Estimates and Existence Theory

We now employ the previously developed regularity results to obtain a global, up to the boundary, estimate (in the spirit of the pioneer work by Oliver D. Kellogg on harmonic functions, see [Kel31]), which will be of pivotal use in what follows.

Theorem 4.5.1. *Let $\Omega \subset \mathbb{R}^n$ be a bounded open set with boundary of class $C^{2,\alpha}$ for some $\alpha \in (0, 1)$. Let a_{ij}, b_i, c, $f \in C^\alpha(\Omega)$. Let also $g \in C^{2,\alpha}(\partial\Omega)$. Assume that the ellipticity condition in* (4.4.3) *holds true.*

Let $u \in C^2(\Omega) \cap C(\overline{\Omega})$ be a solution to

$$\begin{cases} \sum_{i,j=1}^{n} a_{ij}\partial_{ij}u + \sum_{i=1}^{n} b_i\partial_i u + cu = f & \text{in } \Omega, \\ u = g & \text{on } \partial\Omega. \end{cases}$$

Then, there exists $C > 0$, depending only on n, α, Ω, a_{ij}, b_i and c, such that[15]

$$\|u\|_{C^{2,\alpha}(\Omega)} \leqslant C\Big(\|g\|_{C^{2,\alpha}(\partial\Omega)} + \|f\|_{C^\alpha(\Omega)} + \|u\|_{L^\infty(\Omega)}\Big). \tag{4.5.1}$$

Additionally, if $c(x) \leqslant 0$ for all $x \in \Omega$, then

$$\|u\|_{C^{2,\alpha}(\Omega)} \leqslant C\Big(\|g\|_{C^{2,\alpha}(\partial\Omega)} + \|f\|_{C^\alpha(\Omega)}\Big). \tag{4.5.2}$$

We stress that, despite their similarity, the estimates in (4.5.1) and (4.5.2) present a fundamental difference since the latter does not involve the solution u on its right-hand side: this "tiny detail" will provide an essential ingredient in the proof of the forthcoming existence result in Theorem 4.5.4.

---

[15] When Ω has boundary of class $C^{2,\alpha}$ and $g : \partial\Omega \to \mathbb{R}$, the notion of $\|g\|_{C^{2,\alpha}(\partial\Omega)}$ can be equivalently defined in various ways. Rather than working with norms for functions defined on curved boundaries, it is usually more handy to consider a boundary function as a restriction of a globally defined function with its appropriate norm. Namely, one can extend g to a function $\widetilde{g} : \overline{\Omega} \to \mathbb{R}$ with $\widetilde{g} = g$ along $\partial\Omega$ and $\|\widetilde{g}\|_{C^{2,\alpha}(\Omega)} < +\infty$, see e.g. [GT01, Lemma 6.38]. Then, one can define $\|g\|_{C^{2,\alpha}(\partial\Omega)}$ as the infimum of $\|\widetilde{g}\|_{C^{2,\alpha}(\Omega)}$ over all possible extensions $\widetilde{g}$, as above. See [GT01, Section 6.2] for further details about this procedure.

The proof of (4.5.2) relies on a general form of the weak maximum principle, which is also of independent interest (compare with Corollary 2.9.2 and see [GT01, Chapters 3, 8 and 9] for more general and sharper forms of this result).

Lemma 4.5.2. *Let $\Omega \subseteq \mathbb{R}^n$ be open and bounded. Let a_{ij}, $b_i \in L^\infty(\Omega)$. Assume that the ellipticity condition in* (4.4.3) *holds true. Let $u \in C^2(\Omega) \cap C(\overline{\Omega})$. Then, we have the following:*

(i) *If*

$$\sum_{i,j=1}^{n} a_{ij}\partial_{ij}u + \sum_{i=1}^{n} b_i\partial_i u \geqslant 0 \quad \text{in } \Omega, \tag{4.5.3}$$

then

$$\sup_{\overline{\Omega}} u = \sup_{\partial\Omega} u.$$

(ii) *If*

$$\sum_{i,j=1}^{n} a_{ij}\partial_{ij}u + \sum_{i=1}^{n} b_i\partial_i u \leqslant 0 \quad \text{in } \Omega,$$

then

$$\inf_{\overline{\Omega}} u = \inf_{\partial\Omega} u.$$

In the above, C is a positive constant depending only on n, λ, $\|b_1\|_{L^\infty(\Omega)}, \dots, \|b_n\|_{L^\infty(\Omega)}$ and Ω.

Proof. We begin by proving (i). To this end, we observe that, as a byproduct of the ellipticity condition in (4.4.3) exploited with $\xi := e_1$, we have that

$$a_{11} \geqslant \lambda. \tag{4.5.4}$$

Now, let

$$\gamma := \frac{1 + \sup_{i\in\{1,\dots,n\}} \|b_i\|_{L^\infty(\Omega)}}{\lambda}$$

and $\eta(x) := e^{\gamma x_1}$. Let also $\varepsilon > 0$ and $u_\varepsilon(x) := u(x) + \varepsilon\eta(x)$. We have that $\partial_i\eta = \delta_{i1}\gamma\eta$ and $\partial_{ij}\eta = \delta_{i1}\delta_{j1}\gamma^2\eta$. Consequently, by (4.5.3) and (4.5.4),

$$\begin{aligned}
&\sum_{i,j=1}^{n} a_{ij}\partial_{ij}u_\varepsilon + \sum_{i=1}^{n} b_i\partial_i u_\varepsilon \\
&\geqslant \varepsilon\left[\sum_{i,j=1}^{n} a_{ij}\partial_{ij}\eta + \sum_{i=1}^{n} b_i\partial_i\eta\right]
\end{aligned}$$

$$
\begin{aligned}
&= \varepsilon \left[\gamma^2 a_{11} + b_1 \gamma\right] \eta \\
&\geqslant \varepsilon \left[\gamma^2 \lambda - \sup_{i\in\{1,\dots,n\}} \|b_i\|_{L^\infty(\Omega)}\gamma\right] \eta \\
&= \varepsilon\gamma\eta.
\end{aligned} \tag{4.5.5}
$$

We claim that

$$\sup_{\overline{\Omega}} u_\varepsilon = \sup_{\partial\Omega} u_\varepsilon. \tag{4.5.6}$$

Indeed, suppose not, then u_ε presents an interior local maximum at some point $\overline{x}_\varepsilon$. Therefore, the Hessian matrix of u_ε at $\overline{x}_\varepsilon$ is nonpositive, and the gradient of u_ε at $\overline{x}_\varepsilon$ vanishes. Thus, we denote by $M = \{M_{ij}\}_{i,j\in\{1,\dots,n\}}$ the square root (in the matrix sense, see e.g. [Gen07, p. 125]) of minus the Hessian matrix of u_ε at $\overline{x}_\varepsilon$. In this way, we have that

$$-\partial_{ij} u_\varepsilon(\overline{x}_\varepsilon) = \sum_{k=1}^n M_{ik} M_{jk}.$$

Now, we use the ellipticity condition in (4.4.3) to see that

$$\sum_{i,j=1}^n a_{ij}(\overline{x}_\varepsilon) M_{ik} M_{jk} \geqslant \lambda \sum_{m=1}^n |M_{mk}|^2.$$

As a result,

$$-\sum_{i,j=1}^n a_{ij}(\overline{x}_\varepsilon)\partial_{ij} u_\varepsilon(\overline{x}_\varepsilon) = \sum_{i,j,k=1}^n a_{ij}(\overline{x}_\varepsilon) M_{ik} M_{jk} \geqslant \lambda \sum_{m,k=1}^n |M_{mk}|^2 \geqslant 0.$$

Recalling (4.5.5), we thereby obtain a contradiction. This completes the proof of (4.5.6): thus, by sending $\varepsilon \searrow 0$ in (4.5.6), we establish the desired claim in (i).

To prove (ii), one can employ (i) on the function $-u$. □

Corollary 4.5.3. *Let $\Omega \subseteq \mathbb{R}^n$ be open and bounded. Let a_{ij}, b_i, c, $f \in L^\infty(\Omega)$. Assume that $c(x) \leqslant 0$ for all $x \in \Omega$ and that the ellipticity condition in* (4.4.3) *holds true. Let $u \in C^2(\Omega) \cap C(\overline{\Omega})$ be a solution of*

$$\sum_{i,j=1}^n a_{ij}\partial_{ij} u + \sum_{i=1}^n b_i \partial_i u + cu \geqslant f \quad \textit{in } \Omega.$$

Then,

$$\sup_{\overline{\Omega}} u \leqslant \sup_{\partial\Omega} u^+ + C\|f\|_{L^\infty(\Omega)}. \tag{4.5.7}$$

In the above, C is a positive constant depending only on n, λ, $\|b_1\|_{L^\infty(\Omega)}, \dots, \|b_n\|_{L^\infty(\Omega)}$ and Ω.

Proof. We observe that

$$\text{it suffices to prove the desired result when } f \text{ vanishes identically.} \tag{4.5.8}$$

Indeed, suppose we want to know the desired result when f vanishes identically. Let $R > 0$ be sufficiently large such that $\Omega \subseteq B_R$, and let

$$v(x) := u(x) + \|f\|_{L^\infty(\Omega)}\eta(x),$$
$$\text{with} \quad \eta(x) := K(e^{\gamma x_1} - e^{\gamma R}), \quad K := \frac{e^{\gamma R}}{\gamma} \quad \text{and} \quad \gamma := \frac{1 + \|b_1\|_{L^\infty(\Omega)}}{\lambda}.$$

We have that $\partial_i \eta(x) = \delta_{i1} K\gamma e^{\gamma x_1}$ and $\partial_{ij}\eta(x) = \delta_{i1}\delta_{j1} K\gamma^2 e^{\gamma x_1}$. As a consequence, noting that $e^{\gamma x_1} - e^{\gamma R} \leqslant 0$ in Ω and recalling (4.5.4), we find that, in Ω,

$$\begin{aligned}
&\sum_{i,j=1}^n a_{ij}\partial_{ij} v + \sum_{i=1}^n b_i \partial_i v + cv \\
&\quad \geqslant f + \|f\|_{L^\infty(\Omega)} \left[K\gamma^2 a_{11} e^{\gamma x_1} + K\gamma b_1 e^{\gamma x_1} + cK(e^{\gamma x_1} - e^{\gamma R})\right] \\
&\quad \geqslant -\|f\|_{L^\infty(\Omega)} + \|f\|_{L^\infty(\Omega)} \left[\lambda K\gamma^2 e^{\gamma x_1} - K\gamma \|b_1\|_{L^\infty(\Omega)} e^{\gamma x_1} + cK(e^{\gamma x_1} - e^{\gamma R})\right] \\
&\quad \geqslant \|f\|_{L^\infty(\Omega)} \left[K\gamma e^{\gamma x_1}(\lambda\gamma - \|b_1\|_{L^\infty(\Omega)}) - 1\right] \\
&\quad = \|f\|_{L^\infty(\Omega)} \left(K\gamma e^{\gamma x_1} - 1\right) \\
&\quad \geqslant \|f\|_{L^\infty(\Omega)} \left(K\gamma e^{-\gamma R} - 1\right) \\
&\quad \geqslant 0.
\end{aligned}$$

Thus, if the desired result holds true when f vanishes identically, we deduce that

$$\sup_{\overline{\Omega}} v \leqslant \sup_{\partial\Omega} v^+.$$

Hence, since $v^+ \leqslant u^+$ and

$$v \geqslant u - \|f\|_{L^\infty(\Omega)} K e^{\gamma R} = u - \frac{\|f\|_{L^\infty(\Omega)}\, e^{2\gamma R}}{\gamma},$$

we obtain that (4.5.7) holds true. This completes the proof of (4.5.8).

Therefore, in light of (4.5.8), we now suppose that f vanishes identically. We let

$$\Omega_\star := \{x \in \Omega \text{ s.t. } u(x) > 0\},$$

and we remark that, if $x \in \Omega_\star$,

$$\sum_{i,j=1}^n a_{ij}\partial_{ij} u + \sum_{i=1}^n b_i \partial_i u \geqslant -cu \geqslant 0.$$

From this and Lemma 4.5.2(i), applied here on $\Omega_\star$ in the place of Ω, we deduce that

$$\sup_{\overline{\Omega_\star}} u = \sup_{\partial\Omega_\star} u. \tag{4.5.9}$$

Now, we prove that (4.5.7) holds true with $f = 0$. Indeed, if not, there exists $X \in \overline{\Omega}$ and $a > 0$ such that $u(X) \geqslant a + \sup_{\partial\Omega} u^+$. In particular, $u(X) > 0$ and thus $X \in \Omega_\star$. This and (4.5.9) yield that

$$a + \sup_{\partial\Omega} u^+ \leqslant u(X) \leqslant \sup_{\overline{\Omega_\star}} u = \sup_{\partial\Omega_\star} u. \tag{4.5.10}$$

This also entails that

$$\sup_{\partial\Omega_\star} u > 0. \tag{4.5.11}$$

Now, since $\partial\Omega_\star$ is compact and u is continuous along $\partial\Omega_\star \subseteq \overline{\Omega}$, the above supremum is attained at some point $Y \in \partial\Omega_\star$. We deduce from (4.5.11) that $u(Y) > 0$, hence $u(Y) = u^+(Y)$.

We claim that

$$Y \in \partial\Omega. \tag{4.5.12}$$

Indeed, if not, Y would be an interior point of Ω and therefore, by continuity, there would exist $\varrho > 0$ such that $B_\varrho(Y) \subset \Omega$ and $u(x) \geqslant \frac{u(Y)}{2} > 0$ for all $x \in B_\varrho(Y)$. But this would give that Y is also an interior point for $\Omega_\star$, in contradiction with the assumption that $Y \in \partial\Omega_\star$. This establishes (4.5.12).

Then, from (4.5.10) and (4.5.12),

$$a + \sup_{\partial\Omega} u^+ \leqslant \sup_{\partial\Omega_\star} u = u(Y) = u^+(Y) \leqslant \sup_{\partial\Omega} u^+.$$

This is a contradiction, which shows that (4.5.7) holds true with $f = 0$, as desired. □

We can now focus on the proof of the global estimates up to the boundary.

Proof of Theorem 4.5.1. The core of the proof consists of straightening the boundary (see Figure 4.5.1) in order to use the local estimates at the boundary given in Theorem 4.4.3 (the remaining part of the domain lies away from the boundary, and one can use there the interior estimates in Theorem 4.4.2). The structural reason why this argument does the business is that domain deformations preserve the class of operators that we are considering, according to Lemma 4.4.1 (as a matter of fact, preserving a suitable class of equations under diffeomorphisms was precisely one of the main motivations for us to study elliptic equations in such a generality, as discussed on p. 365).

The technical details of the proof are as follows. Since Ω is bounded and with boundary of class $C^{2,\alpha}$, we find $\rho > 0$ and a finite family of

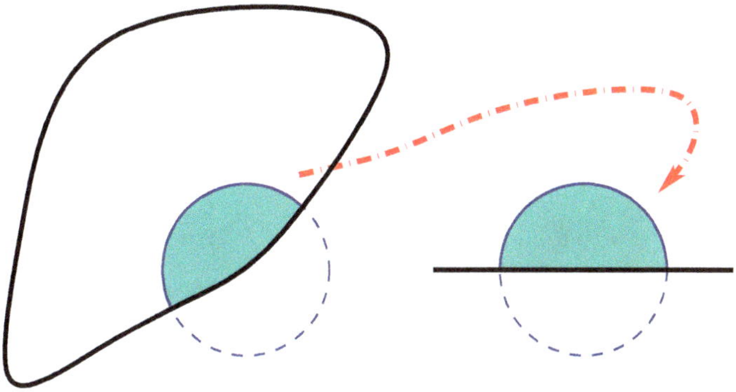

Figure 4.5.1 A domain deformation to locally straighten the boundary.

balls $\{B_\rho(p_i)\}_{i\in\{1,\dots,N\}}$ with $p_i \in \partial\Omega$ and

$$\mathcal{N} := \{x \in \Omega \text{ s.t. } B_{\rho/8}(x) \cap (\partial\Omega) \neq \varnothing\} \subseteq \bigcup_{i=1}^{N} B_{\rho/4}(p_i) \tag{4.5.13}$$

such that $B_\rho(p_i) \cap \Omega$ is equivalent to B_ρ^+ via a diffeomorphism T_i of class $C^{2,\alpha}$ with the inverse of class $C^{2,\alpha}$.

We can thus employ Lemma 4.4.1 in this setting: that is, setting $\widetilde{u}_i(y) := u(T_i^{-1}(y))$, we find from (4.4.6) that $\widetilde{u}_i$ satisfies in B_ρ^+ the same type of equation that u satisfies in $B_\rho(p_i) \cap \Omega$, with the ellipticity constants preserved up to the multiplication by structural constants and with sign of the zero-order coefficient maintained nonpositive.

From these facts, we are in a position to use the local estimates at the boundary in Theorem 4.4.3 on the function $\widetilde{u}_i$, arriving at

$$\|\widetilde{u}_i\|_{C^{2,\alpha}(B^+_{\rho/2})} \leqslant C\left(\|u\|_{L^\infty(\Omega)} + \|g\|_{C^{2,\alpha}(\partial\Omega)} + \|f\|_{C^\alpha(\Omega)}\right).$$

As a consequence,

$$\|u\|_{C^{2,\alpha}(\Omega\cap B_{\rho/2}(p_i))} \leqslant C\left(\|u\|_{L^\infty(\Omega)} + \|g\|_{C^{2,\alpha}(\partial\Omega)} + \|f\|_{C^\alpha(\Omega)}\right)$$

and therefore, by (4.5.13),

$$\|u\|_{C^{2,\alpha}(\mathcal{N})} \leqslant C\left(\|u\|_{L^\infty(\Omega)} + \|g\|_{C^{2,\alpha}(\partial\Omega)} + \|f\|_{C^\alpha(\Omega)}\right), \tag{4.5.14}$$

up to renaming constants from line to line.

Now, we consider a covering of the remaining part of the domain. Namely, we take a finite family of balls such that

$$\mathcal{J} := \{x \in \Omega \text{ s.t. } B_{\rho/9}(x) \cap (\partial\Omega) = \varnothing\} \subseteq \bigcup_{i=1}^{M} B_{\rho/4}(q_i), \tag{4.5.15}$$

for suitable $q_i \in \Omega$. We utilize the interior estimates of Theorem 4.4.2 in each ball $B_{7\rho/20}(q_i) \Supset B_{\rho/4}(q_i)$, and we conclude that

$$\|u\|_{C^{2,\alpha}(B_{\rho/4}(q_i))} \leqslant C\left(\|u\|_{L^\infty(\Omega)} + \|f\|_{C^\alpha(\Omega)}\right).$$

This and (4.5.15) yield that

$$\|u\|_{C^{2,\alpha}(\mathcal{J})} \leqslant C\left(\|u\|_{L^\infty(\Omega)} + \|f\|_{C^\alpha(\Omega)}\right).$$

This, together with (4.5.14) and a covering argument as in Lemma 4.2.13, proves (4.5.1).

Then, the claim in (4.5.2) follows from (4.5.1) and Corollary 4.5.3. □

We point out that the regularity theory developed so far in this section (as well as in Sections 4.2, 4.3 and 4.4) dealt with "*a priori* estimates": namely, we estimated[16] suitable norms (of the derivatives) of a solution of a partial differential equation without knowing that the solution exists at all!

As a matter of fact, there are concrete situations in which the Dirichlet problem

$$\begin{cases} \sum_{i,j=1}^{n} a_{ij}\partial_{ij}u + \sum_{i=1}^{n} b_i\partial_i u + cu = f & \text{in } \Omega, \\ u = g & \text{on } \partial\Omega \end{cases}$$

has no solution $u \in C^2(\Omega) \cap C(\overline{\Omega})$. This possibly unpleasant situation is not an artifact of the generality considered for the equation, and actually explicit examples of nonexistence arise even in low dimensions, even for very smooth domains, and

[16]The Latin wording "*a priori*" means "from the one before" and usually refers to knowledge that is established or developed without being based on previous experience. The use of this wording in the context of partial differential equations has become a standard in the technical jargon after the work of Sergei Natanovich (a.k.a. Serge) Bernstein [Ber06].

As a matter of fact, this business of *a priori* estimates does pose some philosophical puzzles, since we have developed not only a whole regularity theory so far for objects which *could not* exist but actually at a level of generality for which these objects *do not* exist at all, as will be apparent in the forthcoming examples (4.5.16) and (4.5.17), for which, for instance, the application of Theorem 4.5.1 becomes void.

Fortunately, mathematics is not philosophy (see Figure 4.5.2 for a systemic difference), and the hard work done to prove Theorem 4.5.1 will pay off: specifically, we obtain in Theorem 4.5.4 that solutions always exist at least under a sign assumption on the coefficient c. The "magic" of Theorem 4.5.1 is that it does prove uniform estimates in cases in which we know that solutions exist (namely for the Laplace operator), and it thus allows one to use the implicit function theorem (in some form) to find solutions for operators "nearby the Laplacian." But then, one can repeat the argument, precisely because the estimates in Theorem 4.5.1 are uniform, and keep finding solutions for a large class of operators which are "connected" with the Laplacian.

This technique is sometimes called the "continuity method." Making this heuristic idea work will be precisely the content of the proof of Theorem 4.5.4.

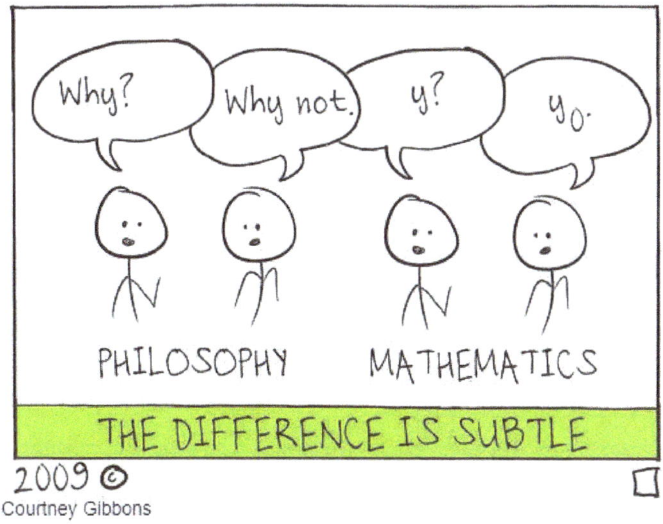

Figure 4.5.2 Courtesy of Courtney Gibbons.

even when the operator $\sum_{i,j=1}^{n} a_{ij}\partial_{ij}$ is just the Laplacian. For example, the Dirichlet problem

$$\begin{cases} u''(x) + u(x) = 1 & \text{for all } x \in (\pi, 2\pi), \\ u(\pi) = 0, \\ u(2\pi) = 0 \end{cases} \tag{4.5.16}$$

possesses no solutions. Indeed, the general solution of the above ordinary differential equation has the form $u(x) = c_1 \sin x + c_2 \cos x + 1$ for constants c_1 and c_2. Imposing $u(\pi) = 0$, we find that $c_2 = 1$. But then, $u(2\pi) = \cos(2\pi) + 1 = 2$, showing that (4.5.16) does not admit any solution.

As for higher-dimensional cases, we observe that for $n \geqslant 2$, the Dirichlet problem

$$\begin{cases} \Delta u(x) - \dfrac{n-1}{|x|^2} \nabla u \cdot x + u(x) = 1 & \text{for all } x \in B_{2\pi} \setminus \overline{B_\pi}, \\ u(x) = 0 & \text{for all } x \in (\partial B_{2\pi}) \cup (\partial B_\pi) \end{cases} \tag{4.5.17}$$

possesses no solutions. To check this, we suppose by contradiction that a solution u exists, and for all $x \in B_{2\pi} \setminus \overline{B_\pi}$, we define

$$v(x) := \fint_{\partial B_1} u(|x|\vartheta)\, d\mathcal{H}^{n-1}_{\vartheta}.$$

We note that, for every $i \in \{1, \dots, n\}$,

$$\partial_i v(x) = \sum_{k=1}^{n} \fint_{\partial B_1} \partial_k u(|x|\vartheta) \frac{\vartheta_k x_i}{|x|} \, d\mathcal{H}^{n-1}_\vartheta.$$

Also, by (3.3.45),

$$\Delta v(x) = \fint_{\partial B_1} \Delta u(|x|\vartheta) \, d\mathcal{H}^{n-1}_\vartheta.$$

Consequently, for every $x \in B_{2\pi} \setminus \overline{B_\pi}$,

$$\begin{aligned}
&\Delta v(x) - \frac{n-1}{|x|^2} \nabla v(x) \cdot x + v(x) \\
&\quad = \fint_{\partial B_1} \left(\Delta u(|x|\vartheta) - \frac{n-1}{|x|^2} \nabla u(|x|\vartheta) \cdot (|x|\vartheta) + u(|x|\vartheta) \right) d\mathcal{H}^{n-1}_\vartheta = 1,
\end{aligned}$$

meaning that v is also a solution of the equation in (4.5.17).

Additionally, v is invariant under rotation; therefore, writing $v(x) = v_0(|x|)$ with $v_0 : \mathbb{R} \to \mathbb{R}$ and recalling the spherical representation of the Laplacian in Theorem 2.5.2, for all $r := |x| \in (\pi, 2\pi)$,

$$\begin{aligned}
\Delta v(x) - \frac{n-1}{|x|^2} \nabla v(x) \cdot x + v(x) &= \left(v_0''(r) + \frac{n-1}{r} v_0'(r) \right) - \frac{n-1}{r^2} v_0'(r)\, r + v_0(r) \\
&= v_0''(r) + v_0(r)
\end{aligned}$$

and therefore $v_0''(r) + v_0(r) = 1$. Since in addition

$$v_0(\pi) = v(\pi e_1) = \fint_{\partial B_1} u(\pi\vartheta) \, d\mathcal{H}^{n-1}_\vartheta = 0$$

and similarly $v_0(2\pi) = 0$, we find that v_0 is a solution of (4.5.16). But we already know that (4.5.16) does not possess any solution; hence, this is a contradiction that shows that (4.5.17) does not admit any solution.

Rather than being discouraged by the examples in (4.5.16) and (4.5.17), we hold onto what we already positively know (such as the existence result obtained via the Perron method in Corollary 3.3.4) and investigate the structure of the previous examples. In particular, in both these examples, the coefficient in front of the zero-order term of the equation is positive. This turns out to be the major obstruction toward the existence of solutions, and in fact when the coefficient in front of the zero-order term is nonnegative, the corresponding Dirichlet problem does admit a (unique) solution. The precise result goes as follows.

Theorem 4.5.4. *Let $\Omega \subset \mathbb{R}^n$ be a bounded open set with boundary of class $C^{2,\alpha}$ for some $\alpha \in (0, 1)$. Let a_{ij}, b_i, c, $f \in C^\alpha(\Omega)$. Let also $g \in C^{2,\alpha}(\partial\Omega)$. Assume that $c(x) \leqslant 0$ for all $x \in \Omega$ and that the ellipticity condition in* (4.4.3) *holds true.*

Then, the Dirichlet problem

$$\begin{cases} \sum_{i,j=1}^{n} a_{ij}\partial_{ij}u + \sum_{i=1}^{n} b_i \partial_i u + cu = f & \text{in } \Omega, \\ u = g & \text{on } \partial\Omega \end{cases} \tag{4.5.18}$$

has a unique solution in $C^2(\Omega) \cap C(\overline{\Omega})$.

Additionally, such a solution u *belongs to* $C^{2,\alpha}(\Omega)$, *and there exists* $C > 0$, *depending only on* n, α, Ω, a_{ij}, b_i *and* c, *such that*

$$\|u\|_{C^{2,\alpha}(\Omega)} \leqslant C\Big(\|g\|_{C^{2,\alpha}(\partial\Omega)} + \|f\|_{C^{\alpha}(\Omega)}\Big). \tag{4.5.19}$$

The proof of Theorem 4.5.4 uses the *a priori* estimate in Theorem 4.5.1 and the following general inversion result for linear operators, which states that small perturbations of invertible operators in Banach spaces remain invertible.

Lemma 4.5.5. *Let* $\mathcal{X}$ *and* $\mathcal{Y}$ *be Banach spaces, and let* $\mathcal{R}$ *and* $\mathcal{S}$ *be bounded linear operators from* $\mathcal{X}$ *to* $\mathcal{Y}$.

Assume that $\mathcal{R}$ *is invertible with a bounded inverse. Assume also that*

$$\|\mathcal{S}\| < \frac{1}{\|\mathcal{R}^{-1}\|}.$$

Then, the operator $\mathcal{R} + \mathcal{S}$ *is invertible and*

$$\left\|(\mathcal{R}+\mathcal{S})^{-1}\right\| \leqslant \frac{\|\mathcal{R}^{-1}\|}{1 - \|\mathcal{R}^{-1}\|\,\|\mathcal{S}\|}.$$

Proof. We denote by $\mathcal{I}$ the identity operator from $\mathcal{X}$ to $\mathcal{X}$. Let $\mathcal{T} := -\mathcal{R}^{-1}\mathcal{S}$, and note that

$$\begin{aligned} \|\mathcal{T}\| = \|\mathcal{R}^{-1}\mathcal{S}\| &= \sup_{x\in\mathcal{X}\setminus\{0\}} \frac{|\mathcal{R}^{-1}\mathcal{S}x|}{|x|} \\ &\leqslant \sup_{x\in\mathcal{X}\setminus\{0\}} \frac{\|\mathcal{R}^{-1}\|\,|\mathcal{S}x|}{|x|} \leqslant \|\mathcal{R}^{-1}\|\,\|\mathcal{S}\| < 1. \end{aligned} \tag{4.5.20}$$

We define

$$\mathcal{L}_j := \mathcal{I} + \mathcal{T} + \mathcal{T}^2 + \cdots + \mathcal{T}^j = \sum_{k=0}^{j} \mathcal{T}^k,$$

and we observe that for all $i \geqslant 1$,

$$\|\mathcal{L}_{j+i} - \mathcal{L}_j\| = \left\|\sum_{k=j+1}^{j+i} \mathcal{T}^k\right\| \leqslant \sum_{k=j+1}^{j+i} \|\mathcal{T}\|^k.$$

This and (4.5.20) yield that $\mathcal{L}_j$ is a Cauchy sequence of bounded linear operators from $\mathcal{X}$ to $\mathcal{X}$, and hence there exists a bounded linear operator $\mathcal{L}$ such that $\|\mathcal{L}_j - \mathcal{L}\| \to 0$ as $j \to +\infty$. In this way, we see that, as $j \to +\infty$,

$$\begin{aligned}\|\mathcal{L}(\mathcal{I} - \mathcal{T}) - \mathcal{I}\| &\leqslant \|\mathcal{L}_j(\mathcal{I} - \mathcal{T}) - \mathcal{I}\| + \|(\mathcal{L} - \mathcal{L}_j)(\mathcal{I} - \mathcal{T})\| \\ &= \left\|\sum_{k=0}^{j} \mathcal{T}^k(\mathcal{I} - \mathcal{T}) - \mathcal{I}\right\| + o(1) = \left\|\sum_{k=0}^{j} \mathcal{T}^k - \sum_{k=0}^{j} \mathcal{T}^{k+1} - \mathcal{I}\right\| + o(1) \\ &= \|\mathcal{T}^{j+1}\| + o(1) = o(1),\end{aligned}$$

thanks again to (4.5.20), and accordingly

$$\mathcal{L}(\mathcal{I} - \mathcal{T}) - \mathcal{I} = 0.$$

This entails that

$$\mathcal{L}\mathcal{R}^{-1}(\mathcal{R} + \mathcal{S}) = \mathcal{L}(\mathcal{I} + \mathcal{R}^{-1}\mathcal{S}) = \mathcal{L}(\mathcal{I} - \mathcal{T}) = \mathcal{I},$$

and as a consequence the operator $\mathcal{R} + \mathcal{S}$ is invertible, with $(\mathcal{R} + \mathcal{S})^{-1} = \mathcal{L}\mathcal{R}^{-1}$.

With this,

$$\begin{aligned}\|(\mathcal{R} + \mathcal{S})^{-1}\| &= \|\mathcal{L}\mathcal{R}^{-1}\| = \lim_{j\to+\infty} \|\mathcal{L}_j\mathcal{R}^{-1}\| \leqslant \lim_{j\to+\infty} \|\mathcal{L}_j\|\,\|\mathcal{R}^{-1}\| \\ &\leqslant \lim_{j\to+\infty} \sum_{k=0}^{j} \|\mathcal{T}\|^k\,\|\mathcal{R}^{-1}\| = \frac{\|\mathcal{R}^{-1}\|}{1 - \|\mathcal{T}\|} = \frac{\|\mathcal{R}^{-1}\|}{1 - \|\mathcal{R}^{-1}\mathcal{S}\|} \leqslant \frac{\|\mathcal{R}^{-1}\|}{1 - \|\mathcal{R}^{-1}\|\,\|\mathcal{S}\|},\end{aligned}$$

as desired. □

Corollary 4.5.6. *Let $\mathcal{X}$ and $\mathcal{Y}$ be Banach spaces, and let $\mathcal{L}_0$ and $\mathcal{L}_1$ be bounded linear operators from $\mathcal{X}$ to $\mathcal{Y}$. For every $t \in [0, 1]$, let*

$$\mathcal{L}_t := t\mathcal{L}_1 + (1 - t)\mathcal{L}_0.$$

Assume that $\mathcal{L}_0$ is surjective and that there exists $K > 0$ such that

$$|x| \leqslant K\,|\mathcal{L}_t x| \quad for\ all\ x \in \mathcal{X}\ and\ t \in [0, 1]. \tag{4.5.21}$$

Then,

$$for\ every\ y \in \mathcal{Y}\ there\ exists\ a\ unique\ x_y \in \mathcal{X}\ such\ that\ \mathcal{L}_1 x_y = y. \tag{4.5.22}$$

In addition,

$$|x_y| \leqslant K\,|y|. \tag{4.5.23}$$

Proof. Up to replacing K in (4.5.21) with $K + \|\mathcal{L}_0\| + \|\mathcal{L}_1\|$, we may assume that

$$\|\mathcal{L}_0\| + \|\mathcal{L}_1\| \leqslant K. \tag{4.5.24}$$

Let $N \in \mathbb{N} \cap \left[2K^2, +\infty\right)$. We claim that

$$\begin{gathered}\text{for all } j \in \{0, \dots, N-1\} \text{ and all } t \in \left[\frac{j}{N}, \frac{j+1}{N}\right]\\ \text{the operator } \mathcal{L}_t : \mathcal{X} \to \mathcal{Y} \text{ is invertible.}\end{gathered} \tag{4.5.25}$$

We prove this by induction over j. If $j = 0$, we let $\mathcal{R} := \mathcal{L}_0$ and $\mathcal{S} := t(\mathcal{L}_1 - \mathcal{L}_0)$. We point out that $\mathcal{R}$ is invertible (we know that it is surjective, and the injectivity follows from (4.5.21) with $t := 0$). Also, the norm of $\mathcal{R}^{-1}$ is bounded from above by K (thanks again to (4.5.21) with $t := 0$). Moreover, for all $t \in \left[0, \frac{1}{N}\right]$, in light of (4.5.24), we have that

$$\begin{aligned}\|\mathcal{S}\| &= \sup_{x \in \mathcal{X}\setminus\{0\}} \frac{\left|t(\mathcal{L}_1 - \mathcal{L}_0)x\right|}{|x|} \leqslant \frac{1}{N} \sup_{x \in \mathcal{X}\setminus\{0\}} \frac{|\mathcal{L}_1 x| + |\mathcal{L}_0 x|}{|x|}\\ &\leqslant \frac{K}{N} \leqslant \frac{K^2}{N\,\|\mathcal{R}^{-1}\|} \leqslant \frac{1}{2\,\|\mathcal{R}^{-1}\|}.\end{aligned}$$

We are therefore in a position to use Lemma 4.5.5, thus finding that $\mathcal{R} + \mathcal{S} = \mathcal{L}_t$ is invertible.

This establishes (4.5.25) when $j = 0$, and we now focus on the inductive step. To this end, we suppose that (4.5.25) holds true for some index $j \in \{0, \dots, N-2\}$, and we aim at proving it for the index $j + 1$. For this, we let $\mathcal{R} := \mathcal{L}_{(j+1)/N}$ and $\mathcal{S} := \left(t - \frac{j+1}{N}\right)(\mathcal{L}_1 - \mathcal{L}_0)$. We know by inductive assumption that $\mathcal{R}$ is invertible and by (4.5.21) with $t := \frac{j+1}{N}$ that $\|\mathcal{R}^{-1}\| \leqslant K$. Additionally, for every $t \in \left[\frac{j+1}{N}, \frac{j+2}{N}\right]$, we have that

$$\begin{aligned}\|\mathcal{S}\| &= \sup_{x \in \mathcal{X}\setminus\{0\}} \frac{\left|\left(t - \frac{j+1}{N}\right)(\mathcal{L}_1 - \mathcal{L}_0)x\right|}{|x|} \leqslant \frac{1}{N} \sup_{x \in \mathcal{X}\setminus\{0\}} \frac{|\mathcal{L}_1 x| + |\mathcal{L}_0 x|}{|x|}\\ &\leqslant \frac{K}{N} \leqslant \frac{K^2}{N\,\|\mathcal{R}^{-1}\|} \leqslant \frac{1}{2\,\|\mathcal{R}^{-1}\|}.\end{aligned}$$

Thus, we are again in a position to use Lemma 4.5.5, and we thereby conclude that $\mathcal{R} + \mathcal{S} = \mathcal{L}_{(j+1)/N} + \left(t - \frac{j+1}{N}\right)(\mathcal{L}_1 - \mathcal{L}_0) = \mathcal{L}_t$ is invertible for every $t \in \left[\frac{j+1}{N}, \frac{j+2}{N}\right]$, which completes the inductive step and proves (4.5.25).

Now, it follows from (4.5.25) that the operator $\mathcal{L}_t$ is invertible for all $t \in [0, 1]$. In particular, the operator $\mathcal{L}_1$ is invertible and thus surjective, which entails (4.5.22).

The claim in (4.5.23) then follows by taking $t := 1$ and $x := x_y$ in (4.5.21). □

With the help of the previously developed setting in functional analysis, we can now establish the existence result for the Dirichlet problem stated in Theorem 4.5.4.

Proof of Theorem 4.5.4. To begin with, we prove that the solution of the Dirichlet problem in (4.5.18) is unique in $C^2(\Omega) \cap C(\overline{\Omega})$. To check this, suppose that there are two solutions in $C^2(\Omega) \cap C(\overline{\Omega})$ and consider their difference v. Then, v is in turn a solution in $C^2(\Omega) \cap C(\overline{\Omega})$ of the Dirichlet problem

$$\begin{cases} \sum_{i,j=1}^{n} a_{ij}\partial_{ij}v + \sum_{i=1}^{n} b_i\partial_i v + cv = 0 & \text{in } \Omega, \\ v = 0 & \text{on } \partial\Omega. \end{cases}$$

We can apply Corollary 4.5.3 to both v and $-v$, thus finding that

$$\sup_{\overline{\Omega}} v \leqslant \sup_{\partial\Omega} v^+ = 0 \qquad \text{and} \qquad \sup_{\overline{\Omega}}(-v) \leqslant \sup_{\partial\Omega}(-v)^+ = 0.$$

This tells us that v vanishes identically, thus establishing the desired uniqueness claim.

Now, we focus on the proof of the solvability of the Dirichlet problem in (4.5.18) and on the regularity estimate in (4.5.19). We first prove these claims under the additional assumption that

$$\begin{gathered} \text{the Dirichlet problem} \\ \begin{cases} \Delta u = f & \text{in } \Omega, \\ u = g & \text{on } \partial\Omega \end{cases} \\ \text{admits a solution in } C^{2,\alpha}(\Omega). \end{gathered} \tag{4.5.26}$$

In this spirit, the idea is that one reduces the solvability of the Dirichlet problem for general elliptic operators in (4.5.18) (as well as the regularity estimate in (4.5.19)) to the case of the Laplace operator, and this strategy somewhat underlies the fact that the general structure of the elliptic operator (with the "right sign of the zero order coefficient") does not provide structural obstructions for the solvability of the Dirichlet problem since, in simple terms, a domain which is "good for the Laplacian" is also "equally good for general operators." Thus, for the moment, we assume the additional hypothesis (4.5.26), and we establish the existence of a solution in $C^{2,\alpha}(\Omega)$ (which is stronger than $C^2(\Omega) \cap C(\overline{\Omega})$ due to the global character of $C^{2,\alpha}(\Omega)$, see the notation in footnote 4 on p. 301) for the Dirichlet problem in (4.5.18), and at the same time we prove that (4.5.19) holds true; after that, we then check that all domains with boundary of $C^{2,\alpha}$ satisfy (4.5.26) (though a bit convoluted, this will turn out to be an effective strategy, since the previously built knowledge obtained in this way will play a pivotal role in the subsequent stages of the proof, as one can appreciate how (4.5.28) will be exploited to check the validity of (4.5.26) by using balls as a further intermediate step).

To complete this plan, as observed in (4.2.42) and keeping in mind the notation of boundary norms discussed in footnote 15 on p. 380, we recall that it suffices to prove the solvability of the Dirichlet problem in (4.5.18) and the regularity estimate in (4.5.19) when

$$g \text{ vanishes identically.} \tag{4.5.27}$$

We let $\mathcal{X}$ be the space of functions v in $C^{2,\alpha}(\Omega)$ such that $v = 0$ on $\partial\Omega$. Let also $\mathcal{Y} := C^{\alpha}(\Omega)$ and

$$Lu := \sum_{i,j=1}^{n} a_{ij}\partial_{ij}u + \sum_{i=1}^{n} b_i\partial_i u + cu.$$

The idea is to "connect L with the Laplace operator"; namely, for each $t \in [0, 1]$, we define

$$L_t u := tLu + (1 - t)\Delta u.$$

We stress that L_t is a bounded linear operator from $\mathcal{X}$ to $\mathcal{Y}$ since $\|L_t u\|_{C^{\alpha}(\Omega)} \leqslant C\,\|u\|_{C^{2,\alpha}(\Omega)}$. Moreover, for all $\xi = (\xi_1, \ldots, \xi_n) \in \partial B_1$,

$$\begin{aligned}\sum_{i,j=1}^{n} \big(ta_{ij} + (1-t)\delta_{ij}\big)\xi_i\xi_j &\in \big[t\lambda + (1-t),\, t\Lambda + (1-t)\big]\\ &\subseteq \big[\min\{1,\lambda\},\, \max\{1,\Lambda\}\big];\end{aligned}$$

therefore, L_t satisfies the ellipticity condition in (4.4.3), up to changing the structural constants. As a result, by the *a priori* estimates in Theorem 4.5.1 (recall especially (4.5.2) and (4.5.27)),

$$\|u\|_{C^{2,\alpha}(\Omega)} \leqslant C\,\|L_t u\|_{C^{\alpha}(\Omega)}.$$

We are therefore in the setting required by (4.5.21): hence, to exploit Corollary 4.5.6, we only need to check that L_0 is surjective. For this, since L_0 is the Laplacian, for every $f \in \mathcal{Y} = C^{\alpha}(\Omega)$ we can apply (4.5.26) to find a function $u_f \in C^{2,\alpha}(\Omega)$ such that $\Delta u_f = f$ in Ω and $u_f = 0$ along $\partial\Omega$.

From this, it follows that $u_f = L_0^{-1} f$; therefore, L_0 is surjective. We can consequently employ Corollary 4.5.6 and deduce that for every $f \in C^{\alpha}(\Omega)$, there exists a unique $u_f \in C^{2,\alpha}(\Omega)$ with $u_f = 0$ on $\partial\Omega$ such that $\mathcal{L}_1 u_f = f$ in Ω and, additionally, $\|u_f\|_{C^{2,\alpha}(\Omega)} \leqslant C\,\|f\|_{C^{\alpha}(\Omega)}$. These observations provide a solution for the Dirichlet problem in (4.5.18) which also satisfies the desired estimate in (4.5.19) (recall (4.5.27)).

Hence,

$$\begin{aligned}&\text{we have completed the proof of Theorem 4.5.4}\\ &\text{under the additional assumption in (4.5.26).}\end{aligned} \tag{4.5.28}$$

To finish, we now need to establish that

$$\text{all domains with boundary of class } C^{2,\alpha} \text{ satisfy (4.5.26).} \tag{4.5.29}$$

To this end, we observe that (4.5.26) holds true when Ω is a ball, thanks to Theorem 4.2.10. From this and (4.5.28), we deduce that

$$\text{Theorem 4.5.4 holds true when } \Omega \text{ is a ball.} \tag{4.5.30}$$

Now, to deal with the proof of (4.5.29), we recall that, in light of Corollary 3.3.4 (recall also (4.2.7)), we can find a solution $u \in C^2(\Omega) \cap C(\overline{\Omega})$ of the Dirichlet problem for the Laplace operator in (4.5.26). Therefore, it only remains to show that

$$\text{this solution belongs to } C^{2,\alpha}(\Omega). \tag{4.5.31}$$

As a matter of fact, we already know from the interior estimates for the Laplace operator given in Theorem 4.2.1 that

$$u \in C^{2,\alpha}_{\rm loc}(\Omega); \tag{4.5.32}$$

therefore, we only need to obtain the uniform estimates of $C^{2,\alpha}$ type in a neighborhood of $\partial\Omega$. To this end, we take $\rho > 0$ sufficiently small, a finite family of points $p_1, \dots, p_N \in \partial\Omega$, and sets with $C^{2,\alpha}$ boundaries $U_1, \dots, U_N$ with $B_{\rho/3}(p_j) \cap \Omega \Subset U_j \Subset B_{\rho/2}(p_j)$, such that

$$\mathcal{N} := \{x \in \Omega \text{ s.t. } B_{\rho/8}(x) \cap (\partial\Omega) \neq \varnothing\} \subseteq \bigcup_{j=1}^{N} B_{\rho/4}(p_j) \tag{4.5.33}$$

and with U_j equivalent via a $C^{2,\alpha}$ diffeomorphism T_j with $C^{2,\alpha}$ inverse to the ball B_1, see Figure 4.5.3.

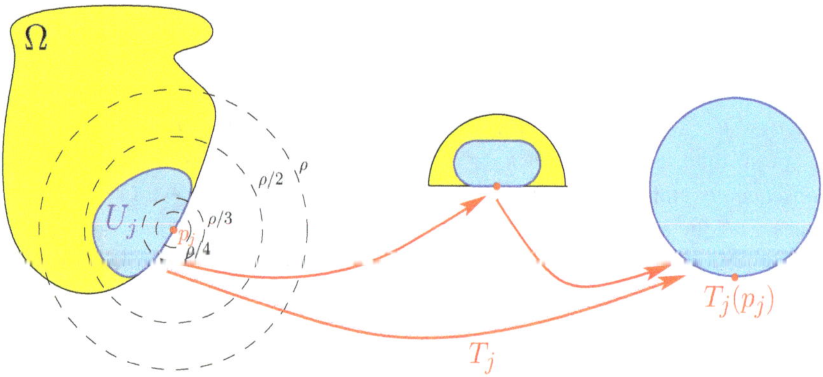

Figure 4.5.3 Reducing to a ball to prove (4.5.29).

Thus, for each $y \in T_j(U_j) = B_1$, we let $\widetilde{u}(y) := u(T_j^{-1}(y))$, and we employ Lemma 4.4.1 to see that $\widetilde{u}$ satisfies an equation of the type

$$\sum_{i,j=1}^{n} \widetilde{a}_{ij}(y)\partial_{ij}\widetilde{u}(y) + \sum_{i=1}^{n} \widetilde{b}_i(y)\partial_i\widetilde{u}(y) = \widetilde{f}(y)$$

for every $y \in B_1$, with $\widetilde{a}_{ij}, \widetilde{b}_i, \widetilde{f} \in C^\alpha(\Omega)$. Additionally, for all $y \in T_j((\partial U_j) \cap (\partial\Omega))$, we have that $\widetilde{u}(y) = g(T_j^{-1}(y)) =: \widetilde{g}(y)$, and we stress that $\widetilde{g}$ is also a $C^{2,\alpha}$ function.

Now, we take a sequence of functions $\widetilde{u}_k \in C^{2,\alpha}(B_1)$ such that $\widetilde{u}_k = \widetilde{u} = \widetilde{g}$ in $T_j(B_{\rho/3}(p_j) \cap \Omega)$ and $\widetilde{u}_k \to \widetilde{u}$ uniformly in $\overline{B_1}$. By (4.5.30), we can solve the Dirichlet problem

$$\begin{cases} \displaystyle\sum_{i,j=1}^{n} \widetilde{a}_{ij}\partial_{ij}\widetilde{v}_k + \sum_{i=1}^{n} \widetilde{b}_i\partial_i\widetilde{v}_k = \widetilde{f} & \text{in } B_1, \\ \widetilde{v}_k = \widetilde{u}_k & \text{on } \partial B_1, \end{cases} \tag{4.5.34}$$

with $\widetilde{v}_k \in C^{2,\alpha}(B_1)$.

Also, the function $\widetilde{w}_{k\ell} := \widetilde{v}_k - \widetilde{v}_\ell$ satisfies

$$\begin{cases} \displaystyle\sum_{i,j=1}^{n} \widetilde{a}_{ij}\partial_{ij}\widetilde{w}_{k\ell} + \sum_{i=1}^{n} \widetilde{b}_i\partial_i\widetilde{w}_{k\ell} = 0 & \text{in } B_1, \\ \widetilde{w}_{k\ell} = \widetilde{u}_k - \widetilde{u}_\ell & \text{on } \partial B_1. \end{cases}$$

Thus, by Corollary 4.5.3 (applied here to both $\widetilde{w}_{k\ell}$ and $-\widetilde{w}_{k\ell}$ with $f := 0$), we see that

$$\|\widetilde{w}_{k\ell}\|_{L^\infty(B_1)} \leqslant \|\widetilde{w}_{k\ell}\|_{L^\infty(\partial B_1)} = \|\widetilde{u}_k - \widetilde{u}_\ell\|_{L^\infty(\partial B_1)}.$$

This gives that $\widetilde{v}_k$ converges uniformly in B_1 to some function $\widetilde{v}_\star$. In particular, we have that $\widetilde{v}_\star \in C(\overline{B_1})$.

Moreover, using Cauchy's estimates (see Theorem 2.17.1), for all $q \in B_1$ and $r \in (0,1)$ such that $B_r(q) \Subset B_1$, we see that

$$\|\widetilde{w}_{k\ell}\|_{C^3(B_{r/2}(q))} \leqslant \frac{C}{r^3}\|\widetilde{w}_{k\ell}\|_{L^\infty(B_1)}.$$

This entails that $\widetilde{v}_k$ is a Cauchy sequence in $C^3(B_{r/2}(q))$ and hence, in $B_{r/2}(q)$,

$$\widetilde{f} = \lim_{k\to+\infty} \sum_{k=1}^{n} \widetilde{a}_{ij}\partial_{ij}\widetilde{v}_k + \sum_{k=1}^{n} \widetilde{b}_i\partial_i\widetilde{v}_k = \sum_{k=1}^{n} \widetilde{a}_{ij}\partial_{ij}\widetilde{v}_\star + \sum_{k=1}^{n} \widetilde{b}_i\partial_i\widetilde{v}_\star.$$

This gives that $\widetilde{v}_\star \in C^2(B_1) \cap C(\overline{B_1})$ solves the Dirichlet problem

$$\begin{cases} \displaystyle\sum_{k=1}^{n} \widetilde{a}_{ij}\partial_{ij}\widetilde{v}_\star + \sum_{k=1}^{n} \widetilde{b}_i\partial_i\widetilde{v}_\star = \widetilde{f} & \text{in } B_1, \\ \widetilde{v}_\star = \widetilde{u} & \text{on } \partial B_1. \end{cases}$$

By the uniqueness claim for the Dirichlet problem (e.g. using again (4.5.30)), we conclude that

$$\widetilde{v}_\star = \widetilde{u}. \tag{4.5.35}$$

In addition, we can apply the local estimates at the boundary given in Theorem 4.4.3 to the function $\widetilde{v}_k$ (applied here to a straightening of ∂B_1 in the vicinity of $T(p_j)$). In this way, we find that

$$\begin{aligned}&\|\widetilde{v}_k\|_{C^{2,\alpha}(T_j(B_{\rho/4}(p_j)\cap\Omega))}\\&\qquad\leqslant C\left(\|\widetilde{v}_k\|_{L^\infty(B_1)}+\|\widetilde{g}\|_{C^{2,\alpha}((T_j((\partial\Omega)\cap B_{\rho/2}(p_j)))}+\|\widetilde{f}\|_{C^\alpha(B_1)}\right).\end{aligned}$$

Consequently, passing to the limit as $k\to+\infty$,

$$\begin{aligned}&\|\widetilde{v}_\star\|_{C^{2,\alpha}(T_j(B_{\rho/4}(p_j)\cap\Omega))}\\&\qquad\leqslant C\left(\|\widetilde{v}_\star\|_{L^\infty(B_1)}+\|\widetilde{g}\|_{C^{2,\alpha}((T_j((\partial\Omega)\cap B_{\rho/2}(p_j)))}+\|\widetilde{f}\|_{C^\alpha(B_1)}\right).\end{aligned}$$

Hence, by (4.5.35),

$$\begin{aligned}&\|\widetilde{u}\|_{C^{2,\alpha}(T_j(B_{\rho/4}(p_j)\cap\Omega))}\\&\qquad\leqslant C\left(\|\widetilde{u}\|_{L^\infty(B_1)}+\|\widetilde{g}\|_{C^{2,\alpha}((T_j((\partial\Omega)\cap B_{\rho/2}(p_j)))}+\|\widetilde{f}\|_{C^\alpha(B_1)}\right).\end{aligned}$$

Transforming back, and possibly renaming constants, we obtain that

$$\|u\|_{C^{2,\alpha}(B_{\rho/4}(p_j)\cap\Omega)}\leqslant C\left(\|u\|_{L^\infty(B_1)}+\|g\|_{C^{2,\alpha}(\partial\Omega)}+\|f\|_{C^\alpha(\Omega)}\right).$$

This and (4.5.33) yield that

$$\|u\|_{C^{2,\alpha}(\mathcal{N})}\leqslant C\left(\|u\|_{L^\infty(B_1)}+\|g\|_{C^{2,\alpha}(\partial\Omega)}+\|f\|_{C^\alpha(\Omega)}\right),$$

up to renaming constants once again.

The latter estimate, together with (4.5.32) (and a covering argument as in Lemma 4.2.13), gives that

$$\|u\|_{C^{2,\alpha}(\Omega)}\leqslant C\left(\|u\|_{L^\infty(B_1)}+\|g\|_{C^{2,\alpha}(\partial\Omega)}+\|f\|_{C^\alpha(\Omega)}\right).$$

Using this and Corollary 4.5.3, we obtain that

$$\|u\|_{C^{2,\alpha}(\Omega)}\leqslant C\left(\|g\|_{C^{2,\alpha}(\partial\Omega)}+\|f\|_{C^\alpha(\Omega)}\right),$$

which completes the proof of (4.5.31). The proof of (4.5.29) is thereby complete as well. □

We observe that the ellipticity assumption in (4.4.3) is an essential ingredient for the solvability of the Dirichlet problem obtained in Theorem 4.5.4. For instance, in this setting, assumption (4.4.3) cannot be weakened by assuming that for every $\xi=(\xi_1,\dots,\xi_n)\in\partial B_1$,

$$\sum_{i,j=1}^n a_{ij}(x)\xi_i\xi_j\in[0,\Lambda] \tag{4.5.36}$$

for some $\Lambda \geqslant 0$ (that is, one cannot take $\lambda = 0$ in (4.4.3)). As an instructive example of this degeneracy, we observe that when $n = 2$ the Dirichlet problem

$$\begin{cases} x_2^2\partial_{11}u + x_1^2\partial_{22}u - 2x_1x_2\partial_{12}u - x_1\partial_1 u - x_2\partial_2 u = 1 & \text{in } B_1 \subset \mathbb{R}^2, \\ u = 0 & \text{on } \partial B_1 \end{cases} \tag{4.5.37}$$

does not possess any solution in $C^{2,\alpha}(B_1)$ for any $\alpha \in (0,1)$. Indeed, if such a solution existed, we deduce from the boundary value prescription that $u(\cos\vartheta, \sin\vartheta) = 0$ for all $\vartheta \in \mathbb{R}$. As a result, taking one derivative in ϑ,

$$0 = -\sin\vartheta\,\partial_1 u(\cos\vartheta, \sin\vartheta) + \cos\vartheta\,\partial_2 u(\cos\vartheta, \sin\vartheta).$$

By taking another derivative, we find that

$$\begin{aligned} 0 &= \sin^2\vartheta\,\partial_{11}u(\cos\vartheta, \sin\vartheta) + \cos^2\vartheta\,\partial_{22}u(\cos\vartheta, \sin\vartheta) \\ &\quad - 2\sin\vartheta\cos\vartheta\,\partial_{12}u(\cos\vartheta, \sin\vartheta) \\ &\quad - \cos\vartheta\,\partial_1 u(\cos\vartheta, \sin\vartheta) - \sin\vartheta\,\partial_2 u(\cos\vartheta, \sin\vartheta). \end{aligned}$$

But the equation in (4.5.37) gives that this quantity is equal to 1, resulting in a contradiction. This shows that no solution in $C^{2,\alpha}(B_1)$ is admissible for the Dirichlet problem in (4.5.37). We remark that this example fulfills the degenerate ellipticity condition in (4.5.36) (but not the strong version in (4.4.3)). Indeed, in this case, $a_{11} = x_2^2$, $a_{22} = x_1^2$ and $a_{12} = a_{21} = -x_1x_2$ and thus, for every $x \in B_1$ and $\xi = (\xi_1, \ldots, \xi_n) \in \partial B_1$,

$$\sum_{i,j=1}^{n} a_{ij}(x)\xi_i\xi_j = x_2^2\xi_1^2 + x_1^2\xi_2^2 - 2x_1x_2\xi_1\xi_2 = (x_2\xi_1 - x_1\xi_2)^2 \in [0,4]$$

with the vanishing threshold attained for $\xi := \frac{x}{|x|}$ when $x \in B_1 \setminus \{0\}$ (and for all $\xi \in \partial B_1$ when $x = 0$).

It is also interesting to observe that the approximation method employed at the end of the proof of Theorem 4.5.4 (namely, the technique adopted for (4.5.34)) can be of general use and leads to a number of extensions of Theorem 4.5.4 in the case of "less regular data." As an example of this feature, we show that for continuous boundary datum, one can still solve the Dirichlet problem continuously up to the boundary, as made precise in the following result.

Theorem 4.5.7. *Let $\Omega \subset \mathbb{R}^n$ be a bounded open set with boundary of class $C^{2,\alpha}$ for some $\alpha \in (0,1)$. Let a_{ij}, b_i, c, $f \in C^\alpha(\Omega)$. Let also $g \in C(\partial\Omega)$. Assume that $c(x) \leqslant 0$ for all $x \in \Omega$ and that the ellipticity condition in* (4.4.3) *holds true.*

Then, the Dirichlet problem

$$\begin{cases} \displaystyle\sum_{i,j=1}^{n} a_{ij}\partial_{ij}u + \sum_{i=1}^{n} b_i\partial_i u + cu = f & \text{in } \Omega, \\ u = g & \text{on } \partial\Omega \end{cases} \tag{4.5.38}$$

has a unique solution in $C^2(\Omega) \cap C(\overline{\Omega})$.

Additionally, such a solution u belongs to $C^{2,\alpha}_{\mathrm{loc}}(\Omega)$, and for every $\Omega' \Subset \Omega$, there exists $C > 0$, depending only on n, α, Ω', Ω, a_{ij}, b_i and c, such that

$$\|u\|_{C^{2,\alpha}(\Omega')} \leqslant C\Big(\|g\|_{L^\infty(\partial\Omega)} + \|f\|_{C^\alpha(\Omega)}\Big). \tag{4.5.39}$$

Proof. The uniqueness claim follows from Corollary 4.5.3, and hence we focus on the existence claim.

Let $g_k \in C^{2,\alpha}(\partial\Omega)$ be such that $g_k \to g$ in $L^\infty(\partial\Omega)$. We can use Theorem 4.5.4 and find $u_k \in C^{2,\alpha}(\Omega)$ that solves the Dirichlet problem

$$\begin{cases} \displaystyle\sum_{i,j=1}^n a_{ij}\partial_{ij}u_k + \sum_{i=1}^n b_i\partial_i u_k + cu_k = f & \text{in } \Omega, \\ u_k = g_k & \text{on } \partial\Omega. \end{cases} \tag{4.5.40}$$

We also let $w_{k\ell} := u_k - u_\ell$, and we observe that

$$\begin{cases} \displaystyle\sum_{i,j=1}^n a_{ij}\partial_{ij}w_{k\ell} + \sum_{i=1}^n b_i\partial_i w_{k\ell} + cw_{k\ell} = 0 & \text{in } \Omega, \\ w_{k\ell} = g_k - g_\ell & \text{on } \partial\Omega. \end{cases}$$

From Corollary 4.5.3, we deduce that

$$\|u_k - u_\ell\|_{L^\infty(\Omega)} = \|w_{k\ell}\|_{L^\infty(\Omega)} \leqslant \|w_{k\ell}\|_{L^\infty(\partial\Omega)} = \|g_k - g_\ell\|_{L^\infty(\partial\Omega)}.$$

As a result, u_k is a Cauchy sequence in $L^\infty(\Omega)$ and hence converges uniformly in Ω to some function $u \in C(\overline{\Omega})$.

Furthermore, for all $x_0 \in \Omega$ and all $r > 0$ such that $B_r(x_0) \Subset \Omega$, we can use the interior estimates in Theorem 4.2.1 and deduce that

$$\begin{aligned} \|u_k\|_{C^{2,\alpha}(B_{r/2}(x_0))} &\leqslant C_r\Big(\|u_k\|_{L^\infty(B_r(x_0))} + \|f\|_{C^\alpha(B_r(x_0))}\Big) \\ &\leqslant C_r\Big(\|u_k\|_{L^\infty(\Omega)} + \|f\|_{C^\alpha(\Omega)}\Big). \end{aligned}$$

with $C_r > 0$, depending only on r, n and α.

Since, using again Corollary 4.5.3, it holds that

$$\|u_k\|_{L^\infty(\Omega)} \leqslant \|u_k\|_{L^\infty(\partial\Omega)} + C\|f\|_{L^\infty(\Omega)} = \|g_k\|_{L^\infty(\partial\Omega)} + C\|f\|_{L^\infty(\Omega)},$$

we find that

$$\|u_k\|_{C^{2,\alpha}(B_{r/2}(x_0))} \leqslant C_r\Big(\|g_k\|_{L^\infty(\partial\Omega)} + \|f\|_{C^\alpha(\Omega)}\Big). \tag{4.5.41}$$

In particular, for large k,

$$\|u_k\|_{C^{2,\alpha}(B_{r/2}(x_0))} \leqslant C_r\Big(1 + \|g\|_{L^\infty(\partial\Omega)} + \|f\|_{C^\alpha(\Omega)}\Big).$$

This gives a uniform bound for $\|u_k\|_{C^{2,\alpha}(B_{r/2}(x_0))}$; therefore, up to a subsequence, we have that $u_k \to u$ in $C^2(\Omega')$ for all $\Omega' \Subset \Omega$. We can therefore pass to the limit as $k \to +\infty$ in (4.5.40) and obtain (4.5.38). Similarly, passing to the limit as $k \to +\infty$ in (4.5.41), we obtain (4.5.39). □

As a side remark, we observe that Theorem 4.5.7 allows one to say that C^2 solutions are automatically $C^{2,\alpha}_{\mathrm{loc}}$ if the data permit, and accordingly the "spurious" assumption $u \in C^{2,\alpha}(B_1)$ in Theorem 4.4.2 can be weakened in favor of $u \in C^2(B_1)$. The details of these observations are contained in the following result.

Corollary 4.5.8. *Let $\Omega \subset \mathbb{R}^n$ be open and bounded. Let a_{ij}, b_i, c, $f \in C^\alpha(\Omega)$ for some $\alpha \in (0,1)$. Assume the ellipticity condition in* (4.4.3). *Let $u \in C^2(\Omega)$ be a solution of*

$$\sum_{i,j=1}^n a_{ij}\partial_{ij}u + \sum_{i=1}^n b_i\partial_i u + cu = f \quad \text{in } \Omega.$$

Then, $u \in C^{2,\alpha}_{\mathrm{loc}}(\Omega)$. and for all $\Omega' \Subset \Omega$. there exists $C > 0$, depending only on n, Ω, Ω', α, a_{ij}, b_i and c, such that

$$\|u\|_{C^{2,\alpha}(\Omega')} \leqslant C\left(\|u\|_{L^\infty(\Omega)} + \|f\|_{C^\alpha(\Omega)}\right). \tag{4.5.42}$$

Proof. We take Ω'' and Ω''' to be open, with $C^{2,\alpha}$ boundary and such that $\Omega' \Subset \Omega'' \Subset \Omega''' \Subset \Omega$. Note that u is continuous along $\partial\Omega'''$ and $\widetilde{f} := f - cu \in C^\alpha(\Omega''')$; hence, we can employ[17] Theorem 4.5.7 and find a solution $v \in C^2(\Omega''') \cap C(\overline{\Omega'''})$ of the Dirichlet problem

$$\begin{cases} \displaystyle\sum_{i,j=1}^n a_{ij}\partial_{ij}v + \sum_{i=1}^n b_i\partial_i v = \widetilde{f} & \text{in } \Omega''', \\ v = u & \text{on } \partial\Omega''' \end{cases} \tag{4.5.43}$$

with $v \in C^{2,\alpha}_{\mathrm{loc}}(\Omega''')$.

Furthermore, we observe that the function $w := u - v$ satisfies

$$\begin{cases} \displaystyle\sum_{i,j=1}^n a_{ij}\partial_{ij}w + \sum_{i=1}^n b_i\partial_i w = 0 & \text{in } \Omega''', \\ w = 0 & \text{on } \partial\Omega''', \end{cases}$$

and accordingly Corollary 4.5.3 entails that

$$\|u - v\|_{L^\infty(\Omega''')} = \|w\|_{L^\infty(\Omega''')} \leqslant \|w\|_{L^\infty(\partial\Omega''')} = 0.$$

Therefore, $u = v$ in Ω''', and as a consequence $u \in C^{2,\alpha}(\Omega'')$.

Hence, we can use Theorem 4.4.2 and a covering argument (recall Lemma 4.2.13) to conclude that

$$\|u\|_{C^{2,\alpha}(\Omega')} \leqslant C\left(\|u\|_{L^\infty(\Omega'')} + \|f\|_{C^\alpha(\Omega'')}\right),$$

which implies the desired estimate in (4.5.42). □

[17] Note that we can apply Theorem 4.5.7 for problem (4.5.43) since $c = 0$ in this setting.

We refer to [GT01, Kic06, Jos13, EE18] for further information about the Schauder estimates and for solvability results of Dirichlet problems. It is worth pointing out that the Schauder estimates can also be used to establish the existence theory of general type of Dirichlet problems by revisiting the Perron method, see e.g. [GT01, Theorem 6.11].

It is also useful to note that the Schauder estimates may be "iterated," or "bootstrapped," in order to achieve higher regularity, whenever the data allow one to do so. In particular, we have the following result.

Theorem 4.5.9. *Let $k \in \mathbb{N}$. Let $\Omega \subset \mathbb{R}^n$ be open and bounded. Let a_{ij}, b_i, c, $f \in C^{k,\alpha}(\Omega)$ for some $\alpha \in (0,1)$. Assume the ellipticity condition in* (4.4.3). *Let $u \in C^2(\Omega)$ be a solution of*

$$\sum_{i,j=1}^{n} a_{ij}\partial_{ij}u + \sum_{i=1}^{n} b_i\partial_i u + cu = f \quad \text{in}\,\Omega. \tag{4.5.44}$$

Then, $u \in C^{k+2,\alpha}_{\text{loc}}(\Omega)$. and for all $\Omega' \Subset \Omega$. there exists $C > 0$, depending only on n, k, Ω, Ω', α, a_{ij}, b_i and c, such that

$$\|u\|_{C^{k+2,\alpha}(\Omega')} \leqslant C\left(\|u\|_{L^\infty(\Omega)} + \|f\|_{C^{k,\alpha}(\Omega)}\right).$$

Proof. When $k = 0$, the result is contained in Corollary 4.5.8. To prove it for every $k \in \mathbb{N}$, we thus proceed by induction, assuming that the desired result holds true for some $\ell \in \mathbb{N} \cap [0, k-1]$ and proving its validity for $\ell + 1$.

The gist of the argument is that one can "differentiate the equation" and observe that the derivative of u satisfies a similar equation, thus reducing the regularity of the derivative of u to the previously known step of the induction. There is only one small catch in this argument: namely, the inductive assumption only gives that $u \in C^{\ell+2,\alpha}_{\text{loc}}$, but to apply the previously known step to the derivative of u, we would need to know that this derivative is in C^2. That is, this strategy would only work when $\ell + 2 \geqslant 3$, that is, $\ell \geqslant 1$. This would create an issue precisely when going from $\ell = 0$ to $\ell = 1$ in the inductive step. To avoid this caveat, it is convenient to work (at least when $\ell = 0$, but also for all $\ell \in \mathbb{N} \cap [0, k-1]$ for the sake of uniformity) with discrete increments rather than derivatives. The technical adjustments needed are as follows.

Let $\beta \in \mathbb{N}^n$ with $|\beta| = \ell \in \mathbb{N}$ and $v := \partial^\beta u$. We stress that this is a good definition since the inductive assumption gives us that $u \in C^{\ell+2,\alpha}_{\text{loc}}(\Omega)$ and in fact $v \in C^{2,\alpha}_{\text{loc}}(\Omega)$. Differentiating (4.5.44) ℓ times, we find that, in Ω,

$$\sum_{i,j=1}^{n} a_{ij}\partial_{ij}v + \sum_{i=1}^{n} \widetilde{b}_i\partial_i v + \widetilde{c}v = \widetilde{f}, \tag{4.5.45}$$

for suitable $\widetilde{b}_i$, $\widetilde{c}$, $\widetilde{f} \in C^{k-\ell,\alpha}(\Omega)$.

Given $h > 0$ and $e \in \partial B_1$, we define

$$w_h(x) := \frac{v(x+he) - v(x)}{h}.$$

Hence, we take Ω'' and Ω''' such that $\Omega' \Subset \Omega'' \Subset \Omega''' \Subset \Omega$, and we deduce that if h is conveniently small, then in Ω''', we have that

$$\sum_{i,j=1}^{n} a_{ij}\partial_{ij}w_h + \sum_{i=1}^{n} \widetilde{b}_{i,h}\partial_i w_h + \widetilde{c}_h w_h = \widetilde{f}_h,$$

for suitable $\widetilde{b}_{i,h}$, $\widetilde{c}_h$, $\widetilde{f}_h \in C^{k-\ell-1,\alpha}(\Omega)$. Since $k - \ell - 1 \geqslant 0$, we can utilize Corollary 4.5.8 and deduce that

$$\|w_h\|_{C^{2,\alpha}(\Omega')} \leqslant C\left(\|w_h\|_{L^\infty(\Omega'')} + \|\widetilde{f}_h\|_{C^\alpha(\Omega'')}\right) \leqslant C\left(\|v\|_{C^1(\Omega''')} + \|f\|_{C^{k,\alpha}(\Omega)}\right).$$

This and the inductive assumption yield that

$$\|w_h\|_{C^{2,\alpha}(\Omega')} \leqslant C\left(\|u\|_{L^\infty(\Omega)} + \|f\|_{C^{k,\alpha}(\Omega)}\right), \tag{4.5.46}$$

up to renaming C.

We can therefore apply the Arzelà–Ascoli theorem and obtain that, up to a subsequence, w_h converges as $h \searrow 0$ in $C^2(\Omega')$. As a result, passing to the limit as $h \searrow 0$ in (4.5.46),

$$\|\partial_e v\|_{C^{2,\alpha}(\Omega')} \leqslant C\left(\|u\|_{L^\infty(\Omega)} + \|f\|_{C^{k,\alpha}(\Omega)}\right).$$

From this, we infer that

$$\|D^{\ell+3}u\|_{L^\infty(\Omega')} + [D^{\ell+3}u]_{C^\alpha(\Omega')} \leqslant C\left(\|u\|_{L^\infty(\Omega)} + \|f\|_{C^{k,\alpha}(\Omega)}\right).$$

This and the inductive assumption lead to

$$\|u\|_{C^{\ell+3,\alpha}(\Omega')} \leqslant C\left(\|u\|_{L^\infty(\Omega)} + \|f\|_{C^{k,\alpha}(\Omega)}\right),$$

which completes the inductive step. □

It follows from Theorem 4.5.9 that solutions are C^∞ when so are a_{ij}, b_i, c and f. A regularity theory in the class of real analytic functions is also possible, see e.g. [Mor08, Section 6.6] and [BJS79, pp. 207–210] (or [Hör03, Theorem 9.5.1] for a more general setting). See also [Nir55, MN57, Mor58a, Mor58b] for additional information on real analytic regularity theory.

A counterpart of Theorem 4.5.9 in terms of boundary regularity holds true as well, see [GT01, Theorem 6.19] for more details.

Chapter 5

Equations in Nondivergence Form: $W^{2,p}$-Regularity Theory

5.1 Hints and Limitations for a Regularity Theory in Lebesgue Spaces

The main motivation for Chapter 4 has been to understand if, and in which sense, solutions of elliptic equations are "two derivatives better than the source term." While in Chapter 4 we focused on Hölder spaces (which are the closest possible replacement of classical spaces of continuous functions, in view of the pathological examples presented in Theorem 4.1.1), our aim is now to consider the Lebesgue spaces L^p: after all, a function is "nice" if either it is continuous, possibly together with its derivatives, or it has some integrability properties, possibly together with its derivatives; therefore, the topic presented in this chapter can certainly be seen as complementary to that of Chapter 4. Of course, in terms of applications, it is absolutely crucial to possess several forms of a regularity theory since different occasions (either related to continuity or integrability properties) naturally occur in several problems of interest.

The first objective of this chapter is thus to understand whether or not solutions of $\Delta u = f$ with $f \in L^p$ happen to be in the Sobolev space $W^{2,p}$ (see e.g. [Eva98, GT01, Leo09, Bre11, DD12] and the references therein for a complete introduction to Sobolev spaces).

We already know (recall footnote 2 on p. 297) that when $p = \infty$, such a regularity theory does not hold. The case $p = 1$ is also out of reach, as pointed out by the following result.

Theorem 5.1.1. *Let $n \geqslant 2$. There exist a set of null measure $\mathcal{Z}$, $f \in L^1(B_1 \setminus \mathcal{Z})$ and $u \in C^2(B_1 \setminus \mathcal{Z})$ such that $\Delta u = f$ in $B_1 \setminus \mathcal{Z}$ but $D^2 u \notin L^1(B_1 \setminus \mathcal{Z})$.*

Proof. Given $x = (x_1, \ldots, x_n) \in \mathbb{R}^n$, we use the notation $\widehat{x} := (x_1, x_2) \in \mathbb{R}^2$ and $\widetilde{x} := (x_3, \ldots, x_n) \in \mathbb{R}^{n-2}$ (of course, when $n = 2$, we have that $\widehat{x} = x$ and the definition of $\widetilde{x}$ can be dismissed). Let

$$\begin{aligned}
&\mathcal{Z} := \left\{x = (\widehat{x}, \widetilde{x}) \in \mathbb{R}^2 \times \mathbb{R}^{n-2} \text{ s.t. } \widehat{x} = 0\right\}, \\
&u_0(r) := \ln\left(\ln\left(\frac{e}{r}\right)\right), \\
&u(x) := u_0(|\widehat{x}|) = \ln\left(\ln\left(\frac{e}{|\widehat{x}|}\right)\right) \\
\text{and}\quad &f(x) := \Delta u(x) = u_0''(|\widehat{x}|) + \frac{1}{|\widehat{x}|}u_0'(|\widehat{x}|) = -\frac{1}{|\widehat{x}|^2 \ln^2\left(\frac{e}{|\widehat{x}|}\right)}.
\end{aligned}$$

Note that $\mathcal{Z}$ has null Lebesgue measure. Thus, using polar coordinates in $\mathbb{R}^2$ and the substitution $t := \ln\left(\frac{e}{r}\right)$, and possibly replacing C from line to line,

$$\begin{aligned}
\|f\|_{L^1(B_1\setminus\mathcal{Z})} &\leqslant \int_{\{|\widehat{x}|\in(0,1)\}\times\{|\widetilde{x}|<1\}} \frac{1}{|\widehat{x}|^2 \ln^2\left(\frac{e}{|\widehat{x}|}\right)}\,dx \\
&= C\int_{\{|\widehat{x}|\in(0,1)\}} \frac{1}{|\widehat{x}|^2 \ln^2\left(\frac{e}{|\widehat{x}|}\right)}\,d\widehat{x} \\
&= C\int_0^1 \frac{dr}{r\ln^2\left(\frac{e}{r}\right)} \\
&= C\int_1^{+\infty} \frac{dt}{t^2},
\end{aligned}$$

which is finite.

We also remark that

$$\begin{aligned}
\mathcal{W} &:= \left\{x = (\widehat{x}, \widetilde{x}) \in \mathbb{R}^2 \times \mathbb{R}^{n-2} \text{ s.t. } |\widehat{x}| \in \left(0, \frac{1}{2}\right) \text{ and } |\widetilde{x}| \in \left[0, \frac{1}{2}\right)\right\} \\
&\subseteq B_1 \setminus \mathcal{Z}.
\end{aligned} \tag{5.1.1}$$

Indeed, if $x = (\widehat{x}, \widetilde{x}) \in \mathcal{W}$, then $\widehat{x} \neq 0$ and

$$|x|^2 = |\widehat{x}|^2 + |\widetilde{x}|^2 < \frac{1}{4} + \frac{1}{4} < 1,$$

proving (5.1.1).

Additionally,

$$\partial_{12}u(x) = \frac{x_1 x_2}{|\widehat{x}|^4 \ln\left(\frac{e}{|\widehat{x}|}\right)}\left(2 - \frac{1}{\ln\left(\frac{e}{|\widehat{x}|}\right)}\right).$$

Since, if $x \in B_1 \setminus \mathcal{Z}$, then $\ln\left(\frac{e}{|\widehat{x}|}\right) \geqslant \ln e = 1$, we infer that

$$|\partial_{12}u(x)| \geqslant \frac{|x_1 x_2|}{|\widehat{x}|^4 \ln\left(\frac{e}{|\widehat{x}|}\right)}.$$

For this reason, making use of (5.1.1), we deduce that

$$\begin{aligned}
\|D^2u\|_{L^1(B_1\setminus\mathcal{Z})} &\geqslant \int_{\mathcal{W}} \frac{|x_1x_2|}{|\widehat{x}|^4 \ln\left(\frac{e}{|\widehat{x}|}\right)}\,dx \\
&= C\int_{\{|\widehat{x}|\in(0,1/2)\}} \frac{|x_1x_2|}{|\widehat{x}|^4 \ln\left(\frac{e}{|\widehat{x}|}\right)}\,d\widehat{x} \\
&= C\iint_{(0,2\pi)\times(0,1/2)} \frac{|\sin\vartheta\,\cos\vartheta|}{r\,\ln\left(\frac{e}{r}\right)}\,d\vartheta\,dr \\
&= C\int_0^{1/2} \frac{dr}{r\,\ln\left(\frac{e}{r}\right)} \\
&= C\int_{\ln(2e)}^{+\infty} \frac{dt}{t},
\end{aligned}$$

which is infinite. □

In the forthcoming pages, we will see that the cases $p = 1$ and $p = \infty$ are the only exceptional cases for a regularity theory in the Lebesgue spaces. That is, in simple terms, solutions of $\Delta u = f$ with $f \in L^p$ do happen to be in the Sobolev space $W^{2,p}$ when $p \in (1, +\infty)$. The techniques used for this regularity theory were chiefly introduced by Alberto Pedro Calderón and Antoni Zygmund [CZ52, CZ56]. The concepts developed in this setting are very deep and fascinating but, as it often happens with truly beautiful ideas, not completely easy to digest at first glance (not by chance Calderón and Zygmund are actually the founders[1] of the

[1]Here is a nice anecdote about the initial meeting between Calderón and Zygmund, see [CKS08]. In 1948, during a scientific visit to Buenos Aires, Zygmund delivered a two-month course based on one of his books. The lectures were attended by many young Argentine mathematicians, including, of course, Alberto Calderón (who had graduated in 1947 with a degree in civil engineering, instead of a degree in mathematics, heeding his father's wishes).

Each of the attendees had to present a topic from the text. Zygmund appeared to be increasingly agitated by Calderón's exposition, until he abruptly interrupted the speaker to ask where he had read the material that he was presenting.

Calderón replied that this material was certainly coming from Zygmund's book, but Zygmund vehemently informed the audience that this material was not sourced from his book.

After the lecture, Zygmund took Calderón aside to further investigate the matter. Calderón finally confessed that he did try to read the material from the book, but after the first couple of lines, instead of turning the page, he had figured out how to develop the arguments by himself in a new and original way.

Obviously, Zygmund immediately recognized Calderón's uncommon mathematical skills and invited him to Chicago to study with him. And this was the birth of one of the most successful collaborations in the recent history of mathematics.

See Figure 5.1.1 for pictures of Calderón and Zygmund. See also [DV23, Chapter 38] for Calderón's prominent contribution to inverse problems.

Figure 5.1.1 Alberto Calderón and Antoni Zygmund.
Source: Photos by Paul Halmos from MacTutor History of Mathematics Archive, licensed under the Creative Commons Attribution-Share Alike 4.0 International license.

so-called "Chicago School of hard Analysis"). We now dive into some details of this construction; for additional readings on the Calderón–Zygmund theory see also [Ste70, Chr90, Ste98, GT01, Hör03, Kry08, Kra09, GM12] and the references therein.

5.2 Potential Theory and Calderón–Zygmund estimates for the Laplace Operator

With respect to the Hölder spaces setting in Chapter 4, the regularity theory in Lebesgue spaces presents an obvious initial difficulty since functions in $W^{2,p}$ are not necessarily twice differentiable; therefore, one cannot give a pointwise meaning to the equation involved and should instead rely on weak formulations. However, to circumvent this technical hurdle, our strategy will be to begin working with smooth objects, obtaining *a priori* estimates (recall footnote 16 on p. 386), with bounds in Sobolev and Lebesgue spaces that are independent of the smoothness assumed on the solution.

As already pointed out on p. 302, a clever strategy in handling regularity theory involves attempting to reduce the problem to the case of harmonic functions by subtracting from the given solution the Newtonian potential $-\Gamma * f$ studied in Proposition 2.7.4. The core of the theory then consists of obtaining suitable bounds on the derivatives of the Newtonian potential (specifically, Section 4.2 focused on such an analysis in Hölder spaces, and we now focus on the case of Lebesgue spaces).

As a preliminary observation, we point out that the regularity theory in L^2 is somewhat special with respect to the general case of L^p with $p \in (1, +\infty)$ since when $p = 2$, one can simply perform an integration by parts, according to the following remarks. First of all, if $R > 0$, $x \in \mathbb{R}^n \setminus B_{2R}$ and $f \in C_0^{0,1}(B_R)$, then

$$\begin{aligned}\left|\int_{\mathbb{R}^n} \partial_{ij}\Gamma(x-y)\big(f(x)-f(y)\big)\,dy\right| &= \left|\int_{\mathbb{R}^n} \partial_{ij}\Gamma(x-y)\,f(y)\,dy\right| \\ &\leqslant C\|f\|_{L^\infty(B_R)}\int_{B_R}\frac{dy}{|x-y|^n} \leqslant \frac{CR^n\|f\|_{L^\infty(B_R)}}{|x|^n},\end{aligned} \tag{5.2.1}$$

up to renaming C at each stage of the computation.

Hence, by (4.2.4), if $v := -\Gamma * f$ is the Newtonian potential of f, then, for all $x \in \mathbb{R}^n \setminus B_{2R}$,

$$|\partial_{ij}v(x)| \leqslant \left|\int_{\mathbb{R}^n} \partial_{ij}\Gamma(x-y)\big(f(x)-f(y)\big)\,dy\right| + \frac{1}{n}\,|f(x)| \leqslant \frac{CR^n\|f\|_{L^\infty(B_R)}}{|x|^n}$$

and therefore

$$\begin{aligned}
\int_{\mathbb{R}^n} |D^2v(x)|^2\,dx &= \lim_{M\to+\infty}\sum_{i,j=1}^n \int_{B_M} |\partial_{ij}v(x)|^2\,dx \\
&= \lim_{M\to+\infty}\sum_{i,j=1}^n \int_{B_M} \Big[\operatorname{div}\big(\partial_{ij}v(x)\partial_j v(x)\,e_i\big) - \partial_{iij}v(x)\partial_j v(x)\Big]\,dx \\
&= \lim_{M\to+\infty}\sum_{i,j=1}^n \left[\int_{\partial B_M} \partial_{ij}v(x)\partial_j v(x)\,e_i\cdot\nu(x)\,d\mathcal{H}_x^{n-1} - \int_{B_M}\partial_{iij}v(x)\partial_j v(x)\,dx\right] \\
&= \lim_{M\to+\infty}\sum_{i,j=1}^n \Bigg[\mathcal{H}^{n-1}(\partial B_M)\,O\left(\frac{1}{M^n}\right) - \int_{B_M}\Big[\operatorname{div}\big(\partial_{ii}v(x)\partial_j v(x)\,e_j\big) \\
&\qquad - \partial_{ii}v(x)\partial_{jj}v(x)\Big]\,dx\Bigg] \\
&= \lim_{M\to+\infty}\sum_{i,j=1}^n \Bigg[O\left(\frac{1}{M}\right) - \int_{\partial B_M}\partial_{ii}v(x)\partial_j v(x)\,e_j\cdot\nu(x)\,d\mathcal{H}_x^{n-1} \\
&\qquad + \int_{B_M}\partial_{ii}v(x)\partial_{jj}v(x)\,dx\Bigg] \\
&= \lim_{M\to+\infty}\sum_{i,j=1}^n \left[O\left(\frac{1}{M}\right) + \int_{B_M}\partial_{ii}v(x)\partial_{jj}v(x)\,dx\right] \\
&= \sum_{i,j=1}^n \int_{\mathbb{R}^n}\partial_{ii}v(x)\partial_{jj}v(x)\,dx \\
&= \int_{\mathbb{R}^n}|\Delta v(x)|^2\,dx.
\end{aligned}$$

This and Proposition 2.7.4 give that, for all $f \in C_0^{0,1}(\mathbb{R}^n)$,

$$\begin{aligned} \|D^2(\Gamma * f)\|^2_{L^2(\mathbb{R}^n)} &= \int_{\mathbb{R}^n} |D^2(\Gamma * f)(x)|^2\, dx \\ &= \int_{\mathbb{R}^n} |f(x)|^2\, dx = \|f\|^2_{L^2(\mathbb{R}^n)}; \end{aligned} \tag{5.2.2}$$

that is, the Newtonian potential is "two derivatives better than f in the sense of L^2."

To develop instead the more complex regularity theory in L^p for all $p \in (1,+\infty)$, in view of Proposition 4.2.3, it is convenient to study the property of a singular integral kernel $K : \mathbb{R}^n \setminus \{0\} \to \mathbb{R}$ satisfying (4.2.12), (4.2.13) and (recalling (4.2.22))

$$\int_{B_\rho(x)\setminus B_\delta(x)} K(x-y)\, dy = 0 \qquad \text{for all } x \in \mathbb{R}^n \text{ and } \rho > \delta > 0, \tag{5.2.3}$$

by defining

$$Tf(x) := \int_{\mathbb{R}^n} K(x-y)\,\big(f(x)-f(y)\big)\, dy, \tag{5.2.4}$$

where the principal value notation in (4.2.14) is implicitly understood.

In this framework, one of the cornerstones of the theory is to establish the following result for singular integrals.

Theorem 5.2.1. *Let $p \in (1,+\infty)$ and $f \in C_0^{0,1}(\mathbb{R}^n)$. Then, there exists a positive constant C, depending only on n, p and K, such that*

$$\|Tf\|_{L^p(\mathbb{R}^n)} \leqslant C\, \|f\|_{L^p(\mathbb{R}^n)}.$$

To address this result, it is opportune to underline the fact that the operator in (5.2.4) is "of order zero." As a matter of fact, in the literature, there are several possible definitions of "order of an operator"; however, what we mean here is simply that the operator in (5.2.4) behaves "neutrally" with respect to scaling. More explicitly, if $f_r(x) := f\left(\frac{x}{r}\right)$, we (informally) say that an operator S has order $\gamma \in \mathbb{R}$ if

$$Sf_r(x) = r^{-\gamma}\, Sf\left(\frac{x}{r}\right).$$

That is, in this setting, operators of order γ pick up a factor $r^{-\gamma}$ under a dilation. Typically, the sign of γ has a significant impact on the possible regularizing effects of a given operator. In rough terms, when $\gamma > 0$, the picture becomes "more irregular at small scales," suggesting a worsening of the regularity by the operator.

For instance, if $Sf(x) := \partial_1 f(x)$, we have that $Sf_r(x) = \frac{1}{r}\partial_1 f\left(\frac{x}{r}\right) = r^{-1} Sf\left(\frac{x}{r}\right)$, and hence in this case, $\gamma = 1$. In general, taking m derivatives of a function produces an operator of order $m \in \mathbb{N}$. Differential operators confirm the idea that operators with order $\gamma > 0$ worsen the regularity of the function to which they apply (by definition, the derivative of a function is less regular than the function itself, having "one derivative less").

Another instructive example is given by integral operators: for instance, if $f : \mathbb{R} \to \mathbb{R}$ and

$$Sf(x) := \int_0^x f(y)\, dy,$$

then

$$Sf_r(x) = \int_0^x f\left(\frac{y}{r}\right) dy = r\int_0^{x/r} f(t)\, dt = rSf\left(\frac{y}{r}\right),$$

showing that in this case, $\gamma = -1$. Thus, integral operators confirm the intuition that their action "improves the regularity" of the function to which the operators apply (by definition, the primitive of a function is more regular than the function itself, having "one derivative more").

As an example of operators of order $\gamma < 0$, one can also consider

$$Sf(x) := \int_{\mathbb{R}^n} |y|^{-\gamma-n} f(x-y)\, dy$$

(the Newtonian potential is of this form when $n \geqslant 3$, with $\gamma = -2$): in this situation,

$$Sf_r(x) = \int_{\mathbb{R}^n} |y|^{-\gamma-n} f\left(\frac{x-y}{r}\right) dy = r^{-\gamma}\int_{\mathbb{R}^n} |z|^{-\gamma-n} f\left(\frac{x}{r}-z\right) dz = r^{-\gamma} Sf\left(\frac{x}{r}\right).$$

With respect to the above terminology, the operator in (5.2.4) is of order zero since it commutes with dilations thanks to (4.2.12): indeed,

$$\begin{aligned}
Tf_r(x) &= \int_{\mathbb{R}^n} K(x-y)\left(f\left(\frac{x}{r}\right) - f\left(\frac{y}{r}\right)\right) dy \\
&= \int_{\mathbb{R}^n} g\left(\frac{x-y}{|x-y|}\right)\left(f\left(\frac{x}{r}\right) - f\left(\frac{y}{r}\right)\right) \frac{dy}{|x-y|^n} \\
&= r^n \int_{\mathbb{R}^n} g\left(\frac{x-rz}{|x-rz|}\right)\left(f\left(\frac{x}{r}\right) - f(z)\right) \frac{dz}{|x-rz|^n} \\
&= \int_{\mathbb{R}^n} g\left(\frac{\frac{x}{r}-z}{\left|\frac{x}{r}-z\right|}\right)\left(f\left(\frac{x}{r}\right) - f(z)\right) \frac{dz}{\left|\frac{x}{r}-z\right|^n} \\
&= \int_{\mathbb{R}^n} K\left(\frac{x}{r}-z\right)\left(f\left(\frac{x}{r}\right) - f(z)\right) dz \\
&= Tf\left(\frac{x}{r}\right).
\end{aligned} \tag{5.2.5}$$

Another classical example of operators of order zero is given by the Hilbert transform of a function $f : \mathbb{R} \to \mathbb{R}$ defined by

$$Hf(x) := \frac{1}{\pi} \int_{\mathbb{R}} \frac{f(x) - f(y)}{x - y}\, dy.$$

The Hilbert transform naturally appears in fractional calculus and complex analysis (since, on the boundary of the complex halfplane, two harmonic conjugated functions of a holomorphic function are related via the Hilbert transform, see e.g. [CDV19, Example 1.10] for further details on this). To confirm that the Hilbert transform has order zero, one computes that

$$\begin{aligned} Hf_r(x) &= \frac{1}{\pi} \int_{\mathbb{R}} \frac{f\left(\frac{x}{r}\right) - f\left(\frac{y}{r}\right)}{x - y}\, dy = \frac{r}{\pi} \int_{\mathbb{R}} \frac{f\left(\frac{x}{r}\right) - f(z)}{x - rz}\, dz \\ &= \frac{1}{\pi} \int_{\mathbb{R}} \frac{f\left(\frac{x}{r}\right) - f(z)}{\frac{x}{r} - z}\, dz = Hf\left(\frac{x}{r}\right). \end{aligned}$$

Interestingly, the Hilbert transform is a "combination of a discrete differential and an integral operator," in the sense that the "incremental quotient" in its integrand acts, in rough terms, with the order of one derivative which is compensated by the integration operation.

Having developed some familiarity with the concept of order of an operator and with its important connection with the regularity issues, we now aim at introducing a notion of Fourier transform for the operator T in (5.2.4). This is not completely straightforward since $K \notin L^1(\mathbb{R}^n)$; therefore, some care is needed in applying Fourier transforms to the convolutions with K. The result that we need in this context is the following.

Lemma 5.2.2. *Let $f \in C_0^{0,1}(\mathbb{R}^n)$. Then,*

$$Tf \in L^2(\mathbb{R}^n). \tag{5.2.6}$$

Moreover, there exists a function $\widetilde{K} \in L^\infty(\mathbb{R}^n)$ such that

$$\widetilde{K}(t\xi) = \widetilde{K}(\xi) \quad \textit{for all } \xi \in \mathbb{R}^n \quad \textit{and} \quad t \in (0, +\infty) \tag{5.2.7}$$

$$\textit{and} \qquad \widehat{Tf}(\xi) = \widetilde{K}(\xi)\, \widehat{f}(\xi). \tag{5.2.8}$$

In the above, we have denoted by $\widehat{f}$ the Fourier transform of f. In this spirit, the claim in (5.2.8) identifies a Fourier multiplier for the operator T (interestingly, Lemma 5.2.2 also says that this multiplier is bounded and positively homogeneous of degree zero).

In our framework, the interest of Lemma 5.2.2 is that it allows us to exploit the Fourier methods to establish a theory in $L^2(\mathbb{R}^n)$ for singular integral operators (this is accomplished in the forthcoming equation (5.2.24), which can be seen as a general counterpart of the estimate on the Newtonian potential obtained in (5.2.2) using the divergence theorem).

Proof of Lemma 5.2.2. To prove (5.2.6), we proceed as in (5.2.1). That is, we take $R > 0$ sufficiently large such that the support of f is contained in B_R. Thus, if $x \in \mathbb{R}^n \setminus B_{2R}$, then

$$\left|\int_{\mathbb{R}^n} K(x-y)\big(f(x)-f(y)\big)\,dy\right| = \left|\int_{\mathbb{R}^n} K(x-y)\,f(y)\,dy\right|$$
$$\leqslant C\|f\|_{L^\infty(B_R)}\int_{B_R}\frac{dy}{|x-y|^n} \leqslant \frac{CR^n\,\|f\|_{L^\infty(B_R)}}{|x|^n},$$

thanks to (4.2.12).

If instead $x \in B_{2R}$, we use (4.2.12) and (5.2.3) to see that

$$\begin{aligned}
&\left|\int_{\mathbb{R}^n} K(x-y)\big(f(x)-f(y)\big)\,dy\right|\\
&\leqslant \left|\int_{B_{4R}(x)} K(x-y)\big(f(x)-f(y)\big)\,dy\right| + \left|\int_{\mathbb{R}^n\setminus B_{4R}(x)} K(x-y)\,f(x)\,dy\right|\\
&\leqslant \|f\|_{C^{0,1}(\mathbb{R}^n)}\int_{B_{4R}(x)} |K(x-y)|\,|x-y|\,dy + 0\\
&\leqslant C\|f\|_{C^{0,1}(\mathbb{R}^n)}\int_{B_{4R}(x)} \frac{dy}{|x-y|^{n-1}}\,dy\\
&\leqslant CR\,\|f\|_{C^{0,1}(\mathbb{R}^n)}.
\end{aligned}$$

From these observations, we infer that

$$\|Tf\|^2_{L^2(\mathbb{R}^n)} \leqslant C\left[R^2\,\|f\|^2_{C^{0,1}(\mathbb{R}^n)}\,|B_{2R}| + R^{2n}\,\|f\|^2_{L^\infty(B_R)}\int_{\mathbb{R}^n\setminus B_{2R}}\frac{dx}{|x|^{2n}}\right] < +\infty,$$

which establishes (5.2.6).

Owing to (5.2.6), we can therefore consider the Fourier transform of Tf in $L^2(\mathbb{R}^n)$. To this end, we let $\varepsilon \in (0,1)$, to be taken arbitrarily small in what follows, and define

$$K_\varepsilon(x) := \chi_{B_{1/\varepsilon}\setminus B_\varepsilon}(x)\,K(x). \tag{5.2.9}$$

Let also

$$T_\varepsilon f(x) := \int_{\mathbb{R}^n} K_\varepsilon(x-y)\big(f(x)-f(y)\big)\,dy. \tag{5.2.10}$$

We claim that

$$\lim_{\varepsilon\searrow 0}\|Tf - T_\varepsilon f\|_{L^2(\mathbb{R}^n)} = 0. \tag{5.2.11}$$

To prove this, we recall (5.2.3) and observe that

$$\begin{aligned}
&|Tf(x) - T_\varepsilon f(x)| \\
&\quad = \left| \int_{\mathbb{R}^n} \big(K(x-y) - K_\varepsilon(x-y)\big)\big(f(x) - f(y)\big)\, dy \right| \\
&\quad = \left| \int_{B_\varepsilon(x)\cup(\mathbb{R}^n\setminus B_{1/\varepsilon}(x))} K(x-y)\big(f(x) - f(y)\big)\, dy \right| \\
&\quad \leqslant \left| \int_{B_\varepsilon(x)} K(x-y)\big(f(x) - f(y)\big)\, dy \right| + \left| \int_{\mathbb{R}^n\setminus B_{1/\varepsilon}(x)} K(x-y)\, f(y)\, dy \right| \\
&\quad \leqslant C\, \|f\|_{C^{0,1}(\mathbb{R}^n)}\, \chi_{B_{R+1}}(x) \int_{B_\varepsilon(x)} \frac{dy}{|x-y|^{n-1}} \\
&\qquad + C\, \|f\|_{L^\infty(\mathbb{R}^n)} \int_{B_R\setminus B_{1/\varepsilon}(x)} \frac{dy}{|x-y|^n} \\
&\quad \leqslant C\varepsilon\, \|f\|_{C^{0,1}(\mathbb{R}^n)}\, \chi_{B_{R+1}}(x) + C\, \|f\|_{L^\infty(\mathbb{R}^n)} \int_{B_R\setminus B_{1/\varepsilon}(x)} \frac{dy}{|x-y|^n}.
\end{aligned}$$

We also observe that if $B_R \setminus B_{1/\varepsilon}(x) \neq \varnothing$, then there exists $z \in B_R$ with $|x-z| \geqslant \frac{1}{\varepsilon}$, and therefore $|x| \geqslant |x-z| - |z| \geqslant \frac{1}{\varepsilon} - R \geqslant \frac{1}{2\varepsilon}$, as long as ε is small enough. As a result,

$$\int_{B_R\setminus B_{1/\varepsilon}(x)} \frac{dy}{|x-y|^n} \leqslant \chi_{\mathbb{R}^n\setminus B_{1/(2\varepsilon)}}(x) \int_{B_R} \frac{dy}{|x-y|^n} \leqslant \frac{CR^n}{|x|^n}\, \chi_{\mathbb{R}^n\setminus B_{1/(2\varepsilon)}}(x).$$

Using these remarks, we arrive at

$$\begin{aligned}
&\|Tf - T_\varepsilon f\|^2_{L^2(\mathbb{R}^n)} \\
&\quad \leqslant C\, \|f\|^2_{C^{0,1}(\mathbb{R}^n)} \left[\int_{\mathbb{R}^n} \left(\varepsilon^2 \chi_{B_{R+1}}(x) + \frac{R^{2n}}{|x|^{2n}}\, \chi_{\mathbb{R}^n\setminus B_{1/(2\varepsilon)}}(x) \right) dx \right] \\
&\quad \leqslant C \|f\|^2_{C^{0,1}(\mathbb{R}^n)} \left[\varepsilon^2 (R+1)^n + \varepsilon^n R^{2n} \right],
\end{aligned}$$

and this proves (5.2.11).

Now, we take into account the Fourier transform of $T_\varepsilon f$ (which we denote by either $\widehat{T_\varepsilon f}$ or $\mathcal{F}(T_\varepsilon f)$. For this, we note that

$$\begin{aligned}
T_\varepsilon f(x) &= \int_{B_{1/\varepsilon}(x)\setminus B_\varepsilon(x)} K(x-y)\big(f(x) - f(y)\big)\, dy \\
&= -\int_{B_{1/\varepsilon}(x)\setminus B_\varepsilon(x)} K(x-y)\, f(y)\, dy \\
&= -\int_{\mathbb{R}^n} K_\varepsilon(x-y)\, f(y)\, dy \\
&= -K_\varepsilon * f(x),
\end{aligned}$$

thanks to (5.2.3), whence

$$\widehat{T_\varepsilon f}(\xi) = -\widehat{K_\varepsilon}(\xi)\,\widehat{f}(\xi). \tag{5.2.12}$$

We let $\xi = |\xi|\omega$ for some $\omega \in \partial B_1$, and we have that

$$\begin{aligned}
\widehat{K_\varepsilon}(\xi) &= \int_{\mathbb{R}^n} K_\varepsilon(x)\, e^{-2\pi i x\cdot\xi}\, dx \\
&= \int_{B_{1/\varepsilon}\setminus B_\varepsilon} K(x)\, e^{-2\pi i|\xi|x\cdot\omega}\, dx \\
&= \frac{1}{|\xi|^n}\int_{B_{|\xi|/\varepsilon}\setminus B_{\varepsilon|\xi|}} K\left(\frac{y}{|\xi|}\right)\, e^{-2\pi i y\cdot\omega}\, dy \\
&= \int_{B_{|\xi|/\varepsilon}\setminus B_{\varepsilon|\xi|}} K(y)\, e^{-2\pi i y\cdot\omega}\, dy.
\end{aligned} \tag{5.2.13}$$

Now, for all $\xi \in \mathbb{R}^n \setminus \{0\}$, we define

$$\widetilde{K}(\xi) := -\int_{\mathbb{R}^n\setminus B_1} K(y)\, e^{-2\pi i y\cdot\omega}\, dy - \int_{B_1} K(y)\,(e^{-2\pi i y\cdot\omega} - 1)\, dy, \tag{5.2.14}$$

where $\omega = \frac{\xi}{|\xi|}$.

We observe that $\widetilde{K}$ is a bounded function. Indeed, using the fact that

$$\omega\cdot\nabla_y(e^{-2\pi i y\cdot\omega}) = -2\pi i e^{-2\pi i y\cdot\omega} \tag{5.2.15}$$

and recalling (4.2.12) and (4.2.13), we see that

$$\begin{aligned}
&\left|\int_{\mathbb{R}^n\setminus B_1} K(y)\, e^{-2\pi i y\cdot\omega}\, dy\right| = C\left|\int_{\mathbb{R}^n\setminus B_1} K(y)\omega\cdot\nabla_y(e^{-2\pi i y\cdot\omega})\, dy\right| \\
&\quad = C\left|\int_{\mathbb{R}^n\setminus B_1}\left[\operatorname{div}\left(K(y)\, e^{-2\pi i y\cdot\omega}\,\omega\right) - \nabla K(y)\cdot\omega\, e^{-2\pi i y\cdot\omega}\right] dy\right| \\
&\quad \leqslant \lim_{\varrho\to+\infty} C\left|\int_{B_\varrho\setminus B_1}\left[\operatorname{div}\left(K(y)\, e^{-2\pi i y\cdot\omega}\,\omega\right) dy\right|\right. + C\int_{\mathbb{R}^n\setminus B_1}\frac{dy}{|y|^{n+1}} \\
&\quad \leqslant \lim_{\varrho\to+\infty} C\left|\int_{\partial(B_\varrho\setminus B_1)} K(y)\, e^{-2\pi i y\cdot\omega}\,\omega\cdot\nu(y)\, d\mathcal{H}^{n-1}_y\right| + C \\
&\quad \leqslant \lim_{\varrho\to+\infty} C\left|\int_{\partial B_\varrho}\frac{d\mathcal{H}^{n-1}_y}{|y|^n} + \int_{\partial B_1}\frac{d\mathcal{H}^{n-1}_y}{|y|^n}\right| + C \\
&\quad \leqslant \lim_{\varrho\to+\infty}\frac{C}{\varrho} + C \\
&\quad \leqslant C,
\end{aligned}$$

up to relabeling $C > 0$ line after line. Consequently, using again (4.2.12),

$$\begin{aligned}|\widetilde{K}(\xi)| &\leqslant \left|\int_{\mathbb{R}^n\setminus B_1} K(y)\,e^{-2\pi i y\cdot\omega}\,dy\right| + C\int_{B_1}\frac{|e^{-2\pi i y\cdot\omega}-1|}{|y|^n}\,dy\\ &\leqslant C + C\int_{B_1}\frac{dy}{|y|^{n-1}} \leqslant C, \qquad (5.2.16)\end{aligned}$$

up to renaming C at each stage of the calculation, as desired.

Also, the homogeneity claimed in (5.2.7) holds true since $\widetilde{K}(\xi)$ only depends on $\omega = \frac{\xi}{|\xi|}$.

We claim that for all $\xi \in \mathbb{R}^n \setminus \{0\}$,

$$\lim_{\varepsilon\searrow 0}\widehat{K_\varepsilon}(\xi) = -\widetilde{K}(\xi). \qquad (5.2.17)$$

To check this, we pick $\xi \in \mathbb{R}^n \setminus \{0\}$, and we take ε small enough such that $1 \in \left(\varepsilon|\xi|, \frac{|\xi|}{\varepsilon}\right)$. Hence, we exploit (5.2.3) and (5.2.13) to see that

$$\begin{aligned}\widehat{K_\varepsilon}(\xi) &= \int_{B_{|\xi|/\varepsilon}\setminus B_{\varepsilon|\xi|}} K(y)\,e^{-2\pi i y\cdot\omega}\,dy\\ &= \int_{B_{|\xi|/\varepsilon}\setminus B_1} K(y)\,e^{-2\pi i y\cdot\omega}\,dy + \int_{B_1\setminus B_{\varepsilon|\xi|}} K(y)\,e^{-2\pi i y\cdot\omega}\,dy\\ &= \int_{B_{|\xi|/\varepsilon}\setminus B_1} K(y)\,e^{-2\pi i y\cdot\omega}\,dy + \int_{B_1\setminus B_{\varepsilon|\xi|}} K(y)\,(e^{-2\pi i y\cdot\omega}-1)\,dy.\end{aligned}$$

From this, (5.2.14) and (5.2.15), we arrive at

$$\begin{aligned}&|\widehat{K_\varepsilon}(\xi) + \widetilde{K}(\xi)|\\ &\leqslant \left|\int_{\mathbb{R}^n\setminus B_{|\xi|/\varepsilon}} K(y)\,e^{-2\pi i y\cdot\omega}\,dy\right| + \left|\int_{B_{\varepsilon|\xi|}} K(y)\,(e^{-2\pi i y\cdot\omega}-1)\,dy\right|\\ &\leqslant C\left(\left|\int_{\mathbb{R}^n\setminus B_{|\xi|/\varepsilon}} K(y)\,\omega\cdot\nabla_y(e^{-2\pi i y\cdot\omega})\,dy\right| + \int_{B_{\varepsilon|\xi|}}\frac{dy}{|y|^{n-1}}\right)\\ &\leqslant C\left(\left|\int_{\mathbb{R}^n\setminus B_{|\xi|/\varepsilon}} \left[\operatorname{div}\left(K(y)\,e^{-2\pi i y\cdot\omega}\,\omega\right) - \nabla K(y)\cdot\omega\,e^{-2\pi i y\cdot\omega}\right]dy\right| + \varepsilon|\xi|\right)\\ &\leqslant \lim_{\varrho\to+\infty} C\left(\left|\int_{B_\varrho\setminus B_{|\xi|/\varepsilon}} \operatorname{div}\left(K(y)\,e^{-2\pi i y\cdot\omega}\,\omega\right)dy\right| + \int_{\mathbb{R}^n\setminus B_{|\xi|/\varepsilon}}\frac{dy}{|y|^{n+1}} + \varepsilon|\xi|\right)\\ &\leqslant \lim_{\varrho\to+\infty} C\left(\left|\int_{\partial(B_\varrho\setminus B_{|\xi|/\varepsilon})} K(y)\,e^{-2\pi i y\cdot\omega}\,\omega\cdot\nu(y)\,d\mathcal{H}^{n-1}_y\right|\right.\\ &\qquad\left. + \int_{\mathbb{R}^n\setminus B_{|\xi|/\varepsilon}}\frac{dy}{|y|^{n+1}} + \varepsilon|\xi|\right)\end{aligned}$$

$$\leqslant \lim_{\varrho \to +\infty} C \left(\int_{\partial B_\varrho} \frac{d\mathcal{H}^{n-1}_y}{|y|^n} + \int_{\partial B_{|\xi|/\varepsilon}} \frac{d\mathcal{H}^{n-1}_y}{|y|^n} + \frac{\varepsilon}{|\xi|} + \varepsilon|\xi| \right)$$
$$\leqslant \lim_{\varrho \to +\infty} C \left(\frac{1}{\varrho} + \frac{\varepsilon}{|\xi|} + \varepsilon|\xi| \right)$$
$$= C \left(\frac{\varepsilon}{|\xi|} + \varepsilon|\xi| \right),$$

whence (5.2.17) plainly follows, as desired.

Now, we claim that

$$\sup_{\varepsilon \in (0,1)} \|\widehat{K_\varepsilon}\|_{L^\infty(\mathbb{R}^n)} < +\infty. \tag{5.2.18}$$

To this end, we repeatedly exploit (5.2.3) and (5.2.13) (together with (4.2.12) and (4.2.13)) by distinguishing three cases, depending on whether $1 > \frac{|\xi|}{\varepsilon}$, or $\frac{|\xi|}{\varepsilon} \geqslant 1 \geqslant \varepsilon|\xi|$, or $\varepsilon|\xi| > 1$.

Let us first assume that $1 > \frac{|\xi|}{\varepsilon}$. Then,

$$|\widehat{K_\varepsilon}(\xi)| = \left| \int_{B_{|\xi|/\varepsilon} \setminus B_{\varepsilon|\xi|}} K(y)\, (e^{-2\pi i y\cdot\omega} - 1)\, dy \right| \leqslant C \int_{B_{|\xi|/\varepsilon}} \frac{dy}{|y|^{n-1}} \leqslant \frac{C\,|\xi|}{\varepsilon} \leqslant C,$$

giving (5.2.18) in this case.

Let us now suppose that $\frac{|\xi|}{\varepsilon} \geqslant 1 \geqslant \varepsilon|\xi|$. In this situation, it is appropriate to use again (5.2.15) to see that

$$\begin{aligned}
&|\widehat{K_\varepsilon}(\xi)| \\
&= \left| \int_{B_{|\xi|/\varepsilon} \setminus B_1} K(y)\, e^{-2\pi i y\cdot\omega}\, dy + \int_{B_1 \setminus B_{\varepsilon|\xi|}} K(y)\, e^{-2\pi i y\cdot\omega}\, dy \right| \\
&\leqslant C \left| \int_{B_{|\xi|/\varepsilon} \setminus B_1} K(y)\, \omega \cdot \nabla_y (e^{-2\pi i y\cdot\omega})\, dy \right| + \left| \int_{B_1 \setminus B_{\varepsilon|\xi|}} K(y)\, (e^{-2\pi i y\cdot\omega} - 1)\, dy \right| \\
&\leqslant C \left| \int_{B_{|\xi|/\varepsilon} \setminus B_1} \left[\operatorname{div}\!\left(K(y)\, e^{-2\pi i y\cdot\omega} \omega \right) - \nabla K(y) \cdot \omega\, e^{-2\pi i y\cdot\omega} \right] dy \right| + C \int_{B_1} \frac{dy}{|y|^{n-1}} \\
&\leqslant C \left| \int_{\partial(B_{|\xi|/\varepsilon} \setminus B_1)} K(y)\, e^{-2\pi i y\cdot\omega} \omega \cdot \nu(y)\, d\mathcal{H}^{n-1}_y \right| + C \int_{\mathbb{R}^n \setminus B_1} \frac{dy}{|y|^{n+1}} + C \\
&\leqslant C \int_{\partial B_{|\xi|/\varepsilon}} \frac{d\mathcal{H}^{n-1}_y}{|y|^n} + C \int_{\partial B_1} \frac{d\mathcal{H}^{n-1}_y}{|y|^n} + C \\
&\leqslant \frac{C\varepsilon}{|\xi|} + C \\
&\leqslant C,
\end{aligned}$$

which establishes (5.2.18) in this case.

It remains to consider the case $\varepsilon|\xi| > 1$: in this situation, we make use again of (5.2.15), and we have that

$$\begin{aligned}
|\widehat{K_\varepsilon}(\xi)| &= \left|\int_{B_{|\xi|/\varepsilon}\setminus B_{\varepsilon|\xi|}} K(y)\, e^{-2\pi i y\cdot\omega}\, dy\right| \\
&\leqslant C\left|\int_{B_{|\xi|/\varepsilon}\setminus B_{\varepsilon|\xi|}} K(y)\, \omega\cdot\nabla_y(e^{-2\pi i y\cdot\omega})\, dy\right| \\
&= C\left|\int_{B_{|\xi|/\varepsilon}\setminus B_{\varepsilon|\xi|}} \left[\operatorname{div}\left(K(y)\, e^{-2\pi i y\cdot\omega}\,\omega\right) - \nabla K(y)\cdot\omega\, e^{-2\pi i y\cdot\omega}\right] dy\right| \\
&\leqslant C\left|\int_{\partial(B_{|\xi|/\varepsilon}\setminus B_{\varepsilon|\xi|})} K(y)\, e^{-2\pi i y\cdot\omega}\,\omega\cdot\nu(y)\, d\mathcal{H}^{n-1}_y\right| + C\int_{\mathbb{R}^n\setminus B_{\varepsilon|\xi|}} \frac{dy}{|y|^{n+1}} \\
&\leqslant C\int_{\partial B_{|\xi|/\varepsilon}} \frac{d\mathcal{H}^{n-1}_y}{|y|^n} + C\int_{\partial B_{\varepsilon|\xi|}} \frac{d\mathcal{H}^{n-1}_y}{|y|^n} + \frac{C}{\varepsilon|\xi|} \\
&\leqslant \frac{C\varepsilon}{|\xi|} + \frac{C}{\varepsilon|\xi|} \\
&\leqslant \frac{C}{\varepsilon|\xi|} \\
&\leqslant C,
\end{aligned}$$

which completes the proof of (5.2.18).

Now, we show that

$$\lim_{\varepsilon\searrow 0} \|\widehat{K_\varepsilon}\,\widehat{f} + \widetilde{K}\,\widehat{f}\|_{L^2(\mathbb{R}^n)} = 0. \tag{5.2.19}$$

For this, we utilize (5.2.16) and (5.2.18) and note that

$$|\widehat{K_\varepsilon}(\xi)\,\widehat{f}(\xi) + \widetilde{K}(\xi)\,\widehat{f}(\xi)|^2 = |\widehat{K_\varepsilon}(\xi) + \widetilde{K}(\xi)|^2\,|\widehat{f}(\xi)|^2 \leqslant C\,|\widehat{f}(\xi)|^2,$$

and the latter function belongs to $L^1(\mathbb{R}^n)$ since $f \in C^{0,1}_0(\mathbb{R}^n)$. From this observation, (5.2.17) and the dominated convergence theorem, we obtain (5.2.19), as desired.

As a consequence, by (5.2.11), (5.2.12) and (5.2.19) and making use of the Plancherel theorem,

$$\begin{aligned}
\|\widehat{Tf} - \widetilde{K}\,\widehat{f}\|_{L^2(\mathbb{R}^n)} &\leqslant \lim_{\varepsilon\searrow 0} \|\widehat{Tf} - \widehat{T_\varepsilon f}\|_{L^2(\mathbb{R}^n)} + \|\widehat{T_\varepsilon f} - \widetilde{K}\,\widehat{f}\|_{L^2(\mathbb{R}^n)} \\
&= \lim_{\varepsilon\searrow 0} \|Tf - T_\varepsilon f\|_{L^2(\mathbb{R}^n)} + \|\widehat{K_\varepsilon}\,\widehat{f} + \widetilde{K}\,\widehat{f}\|_{L^2(\mathbb{R}^n)} = 0,
\end{aligned}$$

which establishes (5.2.8). □

Another crucial ingredient for the proof of Theorem 5.2.1 consists of a deeper understanding of the L^1 case. On the one hand, the counterexample presented in Theorem 5.1.1 prevents us from developing a "completely satisfactory theory in L^1." On the other hand, it still leaves the possibility open for a "weak theory in L^1," in the following sense.

If a function g belongs to $L^1(\mathbb{R}^n)$, then, for all $\lambda > 0$,

$$\|g\|_{L^1(\mathbb{R}^n)} \geqslant \int_{\{|g|\geqslant\lambda\}} |g(x)|\,dx \geqslant \int_{\{|g|\geqslant\lambda\}} \lambda\,dx = \lambda\big|\{|g| \geqslant \lambda\}\big|$$

and therefore

$$\big|\{|g| \geqslant \lambda\}\big| \leqslant \frac{\|g\|_{L^1(\mathbb{R}^n)}}{\lambda}. \tag{5.2.20}$$

This information is often referred to by the name Chebyshev's inequality. When $g := Tf$, we know from the counterexample in Theorem 5.1.1 that (5.2.20) may well become void, simply because Tf may not belong to $L^1(\mathbb{R}^n)$ and the right-hand side of (5.2.20) may accordingly become infinite. However, fortunately, this is not the end of the story since when $g := Tf$, a variant of (5.2.20) holds true, simply by replacing the norm on the right-hand side with $\|f\|_{L^1(\mathbb{R}^n)}$, up to a constant. The precise result in this setting is as follows.

Theorem 5.2.3. *Let $f \in C_0^{0,1}(\mathbb{R}^n)$. Then, for every $\lambda > 0$,*

$$\big|\{|Tf| \geqslant \lambda\}\big| \leqslant \frac{C\,\|f\|_{L^1(\mathbb{R}^n)}}{\lambda},$$

for a suitable constant $C > 0$, depending only on n and K.

This result is very deep, and to prove it in the most efficient way, we can begin with some simple observations which help us "normalize the picture" of Theorem 5.2.3. First of all, by possibly replacing f with $\frac{f}{\lambda}$, we see that

$$\text{it is enough to prove Theorem 5.2.3 when } \lambda = 1. \tag{5.2.21}$$

In addition,

$$\text{it is enough to prove Theorem 5.2.3 when } \|f\|_{L^1(\mathbb{R}^n)} = 1. \tag{5.2.22}$$

To check this, note that if $\|f\|_{L^1(\mathbb{R}^n)} = 0$, then f and Tf vanish identically, and hence the result in Theorem 5.2.3 is obvious. Thus, we can suppose that $\|f\|_{L^1(\mathbb{R}^n)} > 0$, and let $\widetilde{f}(x) := f\big(\|f\|^{1/n}_{L^1(\mathbb{R}^n)}x\big)$. Then, using that T commutes with dilations (recall (5.2.5)),

$$T\widetilde{f}(x) = Tf\big(\|f\|^{1/n}_{L^1(\mathbb{R}^n)}x\big).$$

Also,

$$\|\widetilde{f}\|_{L^1(\mathbb{R}^n)} = \int_{\mathbb{R}^n} \left|f\left(\|f\|_{L^1(\mathbb{R}^n)}^{1/n} x\right)\right| dx = \frac{1}{\|f\|_{L^1(\mathbb{R}^n)}} \int_{\mathbb{R}^n} |f(y)|\, dy = 1.$$

Hence, if we knew that Theorem 5.2.3 held true for functions with unit norm in $L^1(\mathbb{R}^n)$, we would conclude that

$$\frac{\left|\{|Tf| \geqslant \lambda\}|\right\}}{\|f\|_{L^1(\mathbb{R}^n)}} = \left|\{|T\widetilde{f}| \geqslant \lambda\}\right| \leqslant \frac{C\,\|\widetilde{f}\|_{L^1(\mathbb{R}^n)}}{\lambda} = \frac{C}{\lambda};$$

that is, Theorem 5.2.3 would hold for f as well, proving (5.2.22).

Owing to (5.2.21) and (5.2.22), to prove Theorem 5.2.3, it suffices to focus on the renormalized situation in which

$$\begin{aligned}&\text{we assume that } f \in C_0^{0,1}(\mathbb{R}^n) \text{ with } \|f\|_{L^1(\mathbb{R}^n)} = 1\\ &\text{and we aim at showing that } \left|\{|Tf| \geqslant 1\}\right| \leqslant C.\end{aligned} \tag{5.2.23}$$

In this setting, another useful observation is that for "nice functions," one can rely on the L^2 theory. That is, the observation in (5.2.2) can be recast into a general framework of Fourier transforms as follows: first of all, by Lemma 5.2.2 and the Plancherel theorem,

$$\begin{aligned}\|Tf\|_{L^2(\mathbb{R}^n)} &= \|\widehat{Tf}\|_{L^2(\mathbb{R}^n)} = \|\widetilde{K}\,\widehat{f}\|_{L^2(\mathbb{R}^n)}\\ &\leqslant \|\widetilde{K}\|_{L^\infty(\mathbb{R}^n)}\, \|\widehat{f}\|_{L^2(\mathbb{R}^n)} \leqslant C\|f\|_{L^2(\mathbb{R}^n)}.\end{aligned} \tag{5.2.24}$$

As a result,

$$C\|f\|_{L^2(\mathbb{R}^n)}^2 \geqslant \int_{\{|Tf|\geqslant 1\}} |Tf(x)|^2\, dx \geqslant \left|\{|Tf| \geqslant 1\}\right|.$$

In particular, if $\|f\|_{L^\infty(\mathbb{R}^n)} \leqslant 1$ and the setting of (5.2.23) holds true, then

$$\|f\|_{L^2(\mathbb{R}^n)}^2 = \int_{\mathbb{R}^n} |f(x)|^2\, dx \leqslant \int_{\mathbb{R}^n} |f(x)|\, dx = \|f\|_{L^1(\mathbb{R}^n)} = 1$$

and consequently $\left|\{|Tf| \geqslant 1\}\right| \leqslant C$.

Summarizing, in the setting of (5.2.23),

$$\text{if additionally } \|f\|_{L^\infty(\mathbb{R}^n)} \leqslant 1 \text{ then } \left|\{|Tf| \geqslant 1\}\right| \leqslant C, \tag{5.2.25}$$

and we are done.

Thus, in light of (5.2.23) and (5.2.25), to prove Theorem 5.2.3, it only remains to understand the case of smooth and compactly supported functions with unit mass in $L^1(\mathbb{R}^n)$ whose absolute value is not bounded by 1: in simple terms, this is the case in which the function "develops spikes," see Figure 5.2.1. To develop some intuition that comes handy when addressing this case, let us consider a "worse case scenario" of a function developing spikes, that is, one of the sums of

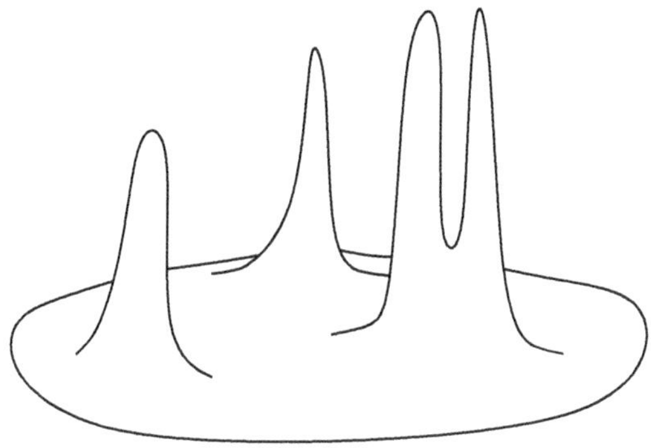

Figure 5.2.1 A function developing spikes.

the Dirac delta functions (well, in strict terms, the Dirac delta function is not a function! but then, so what? At this point, we are just trying to understand what additional difficulties spikes introduce to the setting in (5.2.23) and to unveil a useful device to cope with them). For instance, if f were δ_{x_0}, then, at a formal level, recalling (4.2.12) and (5.2.3), we reduce (5.2.4) to

$$\begin{aligned} Tf(x) &= \int_{\mathbb{R}^n} K(x-y)\left(\delta_{x_0}(x)-\delta_{x_0}(y)\right) dy \\ &= 0 - \int_{\mathbb{R}^n} K(x-y)\,\delta_{x_0}(y)\,dy \\ &= -K(x-x_0) \\ &= -\frac{1}{|x-x_0|^n}\, g\left(\frac{x-x_0}{|x-x_0|}\right). \end{aligned} \tag{5.2.26}$$

As a consequence,

$$\begin{aligned} \{|Tf| \geqslant 1\} &= \left\{x \in \mathbb{R}^n \text{ s.t. } \left|g\left(\frac{x-x_0}{|x-x_0|}\right)\right| \geqslant |x-x_0|^n\right\} \\ &= \left\{y + x_0 \in \mathbb{R}^n \text{ s.t. } \left|g\left(\frac{y}{|y|}\right)\right| \geqslant |y|^n\right\} \\ &= \{te + x_0 \in \mathbb{R}^n \text{ s.t. } e \in \partial B_1,\, t \geqslant 0 \text{ and } |g(e)| \geqslant t^n\} \\ &= \{x + x_0 \in \mathbb{R}^n \text{ s.t. } x \in \mathcal{B}\}, \end{aligned}$$

where

$$\mathcal{B} := \left\{x = te \text{ s.t. } e \in \partial B_1,\, t \geqslant 0 \text{ and } |g(e)| \geqslant t^n\right\}.$$

For instance, when g is identically equal to 1, we have that $\mathcal{B}$ is the closure of the unit ball and $\{|Tf| \geqslant 1\}$ is the closure of the unit ball centered at x_0. So, with an abuse of notation (which we allow for ourselves, as this discussion is already rather heuristic), we consider $\mathcal{B}$ to be a "ball induced by the kernel K," and we realize that when f is the Dirac delta function centered at x_0, then $\{|Tf| \geqslant 1\}$ is the ball induced by the kernel K centered at x_0. Since, in polar coordinates, this cases produces

$$\begin{aligned}|\{|Tf| \geqslant 1\}| = |\mathcal{B}| &= \int_{\partial B_1} \left[\int_0^{|g(e)|^{1/n}} t^{n-1}\,dt \right] d\mathcal{H}_e^{n-1} \\ &\leqslant \int_{\partial B_1} \left[\int_0^{\|g\|_{L^\infty}^{1/n}} t^{n-1}\,dt \right] d\mathcal{H}_e^{n-1} \leqslant C,\end{aligned}$$

we see that the desired result in (5.2.23) would be accomplished in this model case.

We have therefore learned that when f is a single Dirac delta function, the effect on $\{|Tf| \geqslant 1\}$ is that of producing some kind of ball $\mathcal{B}$, and we now consider the subsequent model situation in which f is the sum of two Dirac delta functions (normalized to maintain total unit mass), say $\frac{\delta_{x_0}+\delta_{y_0}}{2}$. In this case, we can imagine that when x_0 and y_0 are extremely close, the situation pretty much reduces to that of a single Dirac delta function. Instead, when x_0 and y_0 are "far away," then one can use (5.2.26) and find that

$$Tf(x) = -\frac{1}{2}\left[\frac{1}{|x-x_0|^n}\, g\left(\frac{x-x_0}{|x-x_0|}\right) + \frac{1}{|x-y_0|^n}\, g\left(\frac{x-y_0}{|x-y_0|}\right)\right], \tag{5.2.27}$$

which is basically the sum of two singularities located considerably far from each other, see e.g. Figure 5.2.2 for the case g identically equal to 1. Hence, for very remote locations of x_0 and y_0, if one neglects, at least[2] for the moment, the "overlapping tails" of the two functions on the right-hand side of (5.2.27), the set $\{|Tf| \geqslant 1\}$ is essentially made by two disjoint balls induced by the kernels (one centered at x_0 and one centered at y_0) of the form

$$\mathcal{B}' := \left\{x = te \text{ s.t. } e \in \partial B_1, t \geqslant 0 \text{ and } |g(e)| \geqslant 2t^n\right\}.$$

Since, substituting $\tau := 2^{\frac{1}{n}} t$,

$$\begin{aligned}|\mathcal{B}'| &= \int_{\partial B_1} \left[\int_0^{(|g(e)|/2)^{1/n}} t^{n-1}\,dt \right] d\mathcal{H}_e^{n-1} \\ &= \frac{1}{2}\int_{\partial B_1} \left[\int_0^{|g(e)|^{1/n}} \tau^{n-1}\,d\tau \right] d\mathcal{H}_e^{n-1} = \frac{|\mathcal{B}|}{2},\end{aligned}$$

[2]We kindly ask the scrupulous reader to hold their fire: we will come back to the issue created by the tails in a second.

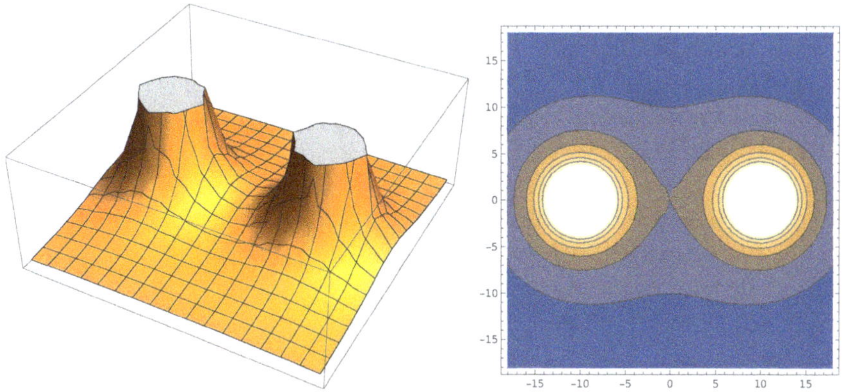

Figure 5.2.2 The function in (5.2.27) and its level sets when $g = 1$, $n = 2$, $x_0 = 10e_1$, $y_0 = -10e_1$ and the factor $-\frac{1}{2}$ is dropped.

we find that for the sum of two Dirac delta functions with singularities located far away, the measure of $\{|Tf| \geqslant 1\}$ is, as a first approximation, essentially twice that of the ball $\mathcal{B}'$, which is precisely that of $\mathcal{B}$. This observation suggests that the case of two Dirac delta functions can also be reduced to that of a single Dirac delta function and is therefore well under control: however, this heuristic also points out that the way to cope with the case of two (or more) Dirac delta functions is to distinguish whether the singularities are "close" (in which case, one has to gather all the singularities together and somewhat replace this cluster with a single Dirac delta function, possibly with an appropriate weight) or "far away" (in which case, each produces a measure comparable, give or take, to that produced by a weighted Dirac delta function).

At this point, however, the scrupulous reader has certainly spotted a noteworthy weakness in the above argument: namely, neglecting the interactions of the tails of remote spikes is too clumsily crude since the decay of the functions in (5.2.27) is too slow and not integrable at infinity. This is a considerable drawback because it entails that, in principle, the more spikes the function develops, the bigger the outcome produced by the tail interactions, which would, again in principle, prevent us from obtaining the desired result (recall that the constant in (5.2.23) is allowed to depend only on n and K, and surely not on the number of spikes of the given function, nor on their precise locations).

But we can nevertheless correct the previous, somewhat sloppy, approximation argument by exploiting again the cancellations of the kernel provided by (5.2.3): to this end, in simple terms, one has to locally subtract from the spike its own average and reabsorb this additional average term into a "universally bounded" part of the function. In this way, on the one hand, the spikes minus their averages provide a bunch of localized functions with zero average, for which the kernel provides an additional cancellation that makes the tail interactions uniformly integrable; on the other hand, for the universally bounded part of the function, one can exploit the observation in (5.2.25) (possibly, with 1 replaced by a universal constant). The combination of these facts will establish the desired result in (5.2.23) after a careful inspection, as we will see in the forthcoming pages.

The *Ultimate Question of Life, the Universe, and Everything* is therefore how to make this heuristic argument work and how to make it quantitative in a sufficiently general setting which includes that in (5.2.23). The answer to this question comes from the so-called Calderón–Zygmund cube decomposition, which is a brilliant method to detect spikes for a given function with unit mass in $L^1(\mathbb{R}^n)$. The idea of this procedure is that one can split the whole space $\mathbb{R}^n$ into cubes, say of unit side. In each of these cubes, the average of the function is less than or equal to 1 (because the total mass of the function is equal to 1). However, in very small cubes located in the vicinity of the spikes, the average of the function can be quite large. Therefore, to locate the position of the possible spikes, one can split dyadically the original cubes and inspect the average in these new cubes. If the average of the function is strictly larger than 1, we place the cube into a new family (in rough terms, in this cube, we have "almost detected a spike," or perhaps more than one spike that are sufficiently close together and that can thereby be clustered in such a cube); if instead the average in the cube is still less than or equal to 1, we keep splitting the cube dyadically, see Figure 5.2.3. This protocol allows us to partition $\mathbb{R}^n$ into two subsets: a subset collecting all the cubes in the new family (in which the average

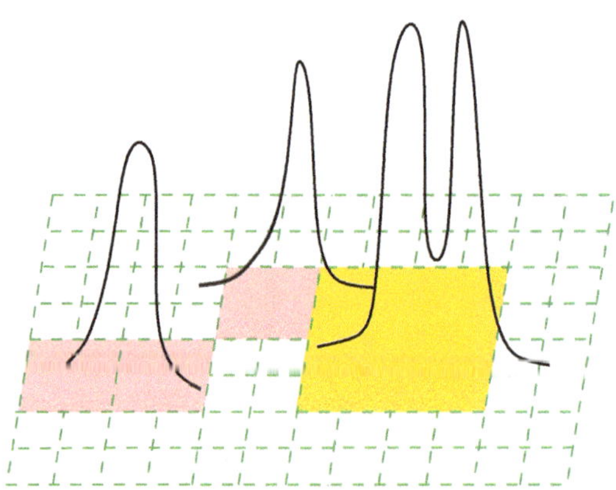

Figure 5.2.3 Cube decomposition to detect and gather together the spikes of Figure 5.2.1.

of the function is strictly larger than 1, though such an average cannot be too large, given the fact that the average in the preceding cube was less than or equal to 1); and a subset containing all the other points (and one can check that the value of the function at these points cannot overcome 1, otherwise the point would fall, sooner or later, after dyadic divisions into one of the cubes of the family).

To make this argument more transparent and precise, we state and prove the following result.

Theorem 5.2.4. *Let $f \in C_0^{0,1}(\mathbb{R}^n)$ with $\|f\|_{L^1(\mathbb{R}^n)} = 1$. Then, there exists a countable family of cubes $\{Q_j\}_{j\in\mathbb{N}}$ with sides parallel to the Cartesian axes such that*

$$\fint_{Q_j} |f(x)|\,dx \in (1, 2^n] \qquad \text{for all } j \in \mathbb{N}, \tag{5.2.28}$$

$$\sum_{j\in\mathbb{N}} |Q_j| \leqslant 1 \tag{5.2.29}$$

$$\text{and} \quad |f(x)| \leqslant 1 \quad \text{for almost all } x \in \mathbb{R}^n \setminus \left(\bigcup_{j\in\mathbb{N}} Q_j\right). \tag{5.2.30}$$

Proof. We begin by considering cubes of side 1 of the form $(m_1, m_1+1) \times \cdots \times (m_n, m_n+1)$, with $m_1, \ldots, m_n \in \mathbb{Z}$. The family of all these cubes will be denoted by $\mathcal{Q}_0$. If Q is any of these cubes, we have that $|Q| = 1$ and

$$\fint_Q |f(x)|\,dx = \int_Q |f(x)|\,dx \leqslant \|f\|_{L^1(\mathbb{R}^n)} = 1.$$

We now denote by $\mathcal{Q}_1$ the cubes of side $\frac{1}{2}$ obtained by dividing into two equal parts each side of the cubes in the family $\mathcal{Q}_0$ (i.e. each cube in $\mathcal{Q}_0$ is split into 2^n cubes to produce the family $\mathcal{Q}_1$; the cubes in $\mathcal{Q}_1$ are also congruent, open and disjoint).

We collect from $\mathcal{Q}_1$ the cubes Q (if any) for which

$$\fint_Q |f(x)|\,dx > 1,$$

and we call $\mathcal{F}_1$ the family containing all the cubes of this type, i.e.

$$\mathcal{F}_1 := \left\{Q \in \mathcal{Q}_1 \text{ s.t. } \fint_Q |f(x)|\,dx > 1\right\}.$$

Then, we keep splitting the remaining cubes in $\mathcal{Q}_1 \setminus \mathcal{F}_1$, collecting them in a family called $\mathcal{Q}_2$, we identify those cubes Q in $\mathcal{Q}_2$ (if any) for which $\fint_Q |f(x)|\,dx > 1$ and we place them into a family $\mathcal{F}_2$; then, we keep splitting the remaining cubes in $\mathcal{Q}_2 \setminus \mathcal{F}_2$, collecting them in a family called $\mathcal{Q}_3$, and so on.

In this way, the cubes in $\mathcal{Q}_j$ have side $\frac{1}{2^j}$ and the cubes Q in $\mathcal{F}_j$ have the property that

$$\fint_Q |f(x)|\,dx > 1. \tag{5.2.31}$$

We define

$$\mathcal{F} := \bigcup_{j=1}^{+\infty} \mathcal{F}_j$$

and write $\mathcal{F} = \{Q_j\}_{j\in\mathbb{N}}$ for suitable cubes Q_j.

We also point out that if Q is in $\mathcal{F}_j$, then its predecessor was in $\mathcal{Q}_{j-1}\setminus\mathcal{F}_{j-1}$ (with the notation that $\mathcal{F}_0$ is void): that is, if Q is in $\mathcal{F}_j$, then there exists $\widetilde{Q} \in \mathcal{Q}_{j-1}\setminus\mathcal{F}_{j-1}$ such that $|\widetilde{Q}| = 2^n|Q|$ and

$$\fint_{\widetilde{Q}} |f(x)|\,dx \leqslant 1.$$

Consequently,

$$\fint_Q |f(x)|\,dx = \frac{2^n}{|\widetilde{Q}|}\int_Q |f(x)|\,dx \leqslant 2^n \fint_{\widetilde{Q}} |f(x)|\,dx \leqslant 2^n.$$

Combining this and (5.2.31), we obtain (5.2.28).

As a byproduct of (5.2.28), we also have that

$$\sum_{j\in\mathbb{N}} |Q_j| \leqslant \sum_{j\in\mathbb{N}} \int_{Q_j} |f(x)|\,dx \leqslant \int_{\mathbb{R}^n} |f(x)|\,dx = \|f\|_{L^1(\mathbb{R}^n)} = 1,$$

which proves (5.2.29), as desired.

Now, we prove (5.2.30). For this, we pick $x \in \mathbb{R}^n \setminus \left(\bigcup_{j\in\mathbb{N}} Q_j\right)$ to be a Lebesgue point for $|f|$: we recall that these points have full measure, thanks to Lebesgue's differentiation theorem, see e.g. [WZ15, Theorem 7.16]. We can also suppose that the coordinates $x_1,\dots,x_n$ of x do not belong to the grid induced by the dyadic cubes (since this is a set of measure zero). In this situation, since x lies outside the cubes of the family $\mathcal{F}$, we have that for every m, there exists a cube K_m of side $\frac{1}{2^m}$ such that $x \in K_m$ and

$$\fint_{K_m} |f(y)|\,dy \leqslant 1.$$

Note that if $K_m^\star := \left(x_1 - \frac{2}{2^m}, x_1 + \frac{2}{2^m}\right) \times \cdots \times \left(x_n - \frac{2}{2^m}, x_n + \frac{2}{2^m}\right)$, we have that $K_m^\star \supseteq K_m$ and $|K_m^\star| = \frac{4^n}{2^{mn}} = 4^n|K_m|$; therefore, K_m shrinks regularly to x (in the terminology of [WZ15, Theorem 7.16]). Accordingly,

$$|f(x)| = \lim_{m\to+\infty} \fint_{K_m} |f(y)|\,dy \leqslant 1,$$

which proves (5.2.30). □

We stress once again that (5.2.28) reveals a special feature of the Calderón–Zygmund cube decomposition since it detects the "correct scale" of the cubes to take into account: if in lieu of this accurate decomposition one considered the simple decomposition obtained by replacing every single spike with a constant function (say, equal to 1) in a cube, with the size of the cubes depending on the height of the given spike, it is very likely that these cubes would end up overlapping between themselves, and the values of the function obtained in this way would run completely out of control. The core of the Calderón–Zygmund cube decomposition, as showcased in (5.2.28), is instead to choose the cubes in such a way that the "measure of all the spikes inside the cube" (as quantified by $\int_{Q_j} |f(x)|\,dx$) is precisely comparable to the measure of the cube itself (that is, $|Q_j|$).

With the aid of Theorem 5.2.4, we can thus retake the heuristic strategy introduced on p. 418 and make it work! The master plan is thus to use a cube decomposition to detect spikes, clustering together spikes that are sufficiently close to lie in the same cube Q_j. One then splits the function f into two functions: on the one side, a "bad part," denoted by $\mathscr{B}$, which collects all the spikes b_j located in all the cubes Q_j, while being careful in subtracting in each of these cubes the average of the corresponding spikes (to make it average zero in its support, which is important to exploit kernel cancellations in the tail interaction and deduce that, away from Q_j, a favorable estimate holds, see the forthcoming equation (5.2.38) and e.g. Figure 2 in [Ste98] for a classical picture about this method); on the other side, a "good part," denoted by $\mathscr{G}$, which accounts for the remaining terms of f coming from (5.2.30), as well as from the average of the spikes (these contributions will be bounded uniformly, allowing us to employ (5.2.25) on $\mathscr{G}$).

The mathematical details of this blueprint are as follows. In the setting of Theorem 5.2.4, we define

$$\begin{aligned} b_j(x) &:= \begin{cases} f(x) - \fint_{Q_j} f(y)\,dy & \text{if } x \in Q_j, \\ 0 & \text{otherwise,} \end{cases} \\ \mathscr{B}(x) &:= \sum_{j\in\mathbb{N}} b_j(x) \\ \text{and} \quad \mathscr{G}(x) &:= f(x) - \mathscr{B}(x). \end{aligned} \tag{5.2.32}$$

Note that

$$\fint_{Q_j} b_j(x)\,dx = 0 \qquad \text{for all } j \in \mathbb{N} \tag{5.2.33}$$

and

$$\mathscr{G}(x) = \begin{cases} f(x) & \text{if } x \in \mathbb{R}^n \setminus \bigcup_{j\in\mathbb{N}} Q_j, \\ \fint_{Q_j} f(y)\,dy & \text{if } x \in Q_j. \end{cases} \tag{5.2.34}$$

With this notation, we have the following lemma.

Lemma 5.2.5. *We have that*

$$|\mathscr{G}(x)| \leqslant 2^n \qquad \text{for all } x \in \mathbb{R}^n, \tag{5.2.35}$$

$$\|\mathscr{G}\|_{L^1(\mathbb{R}^n)} \leqslant 1 \tag{5.2.36}$$

$$\text{and} \quad \fint_{Q_j} |b_j(x)|\,dx \leqslant 2^{n+1} \qquad \text{for all } j \in \mathbb{N}. \tag{5.2.37}$$

Furthermore, if Q'_j denotes the cube with the same center of Q_j and side $(3\sqrt{n}+1)$ times the side of Q_j,

$$\int_{\mathbb{R}^n\setminus Q'_j} |Tb_j(x)|\,dx \leqslant C\,|Q_j|. \tag{5.2.38}$$

Additionally,

$$\int_{\mathbb{R}^n\setminus \bigcup_{j\in\mathbb{N}} Q'_j} |T\mathscr{B}(x)|\,dx \leqslant C. \tag{5.2.39}$$

Here, C is a positive constant depending only on n and K.

Proof. If $x \in Q_j$, we deduce from (5.2.28) and (5.2.34) that

$$|\mathscr{G}(x)| \leqslant \fint_{Q_j} |f(x)|\,dx \leqslant 2^n.$$

From this and (5.2.30), we obtain (5.2.35).

Moreover, using again (5.2.34),

$$\begin{aligned} \int_{\mathbb{R}^n} |\mathscr{G}(x)|\,dx &\leqslant \int_{\mathbb{R}^n\setminus\bigcup_{j\in\mathbb{N}} Q_j} |f(x)|\,dx + \sum_{j\in\mathbb{N}} \int_{Q_j} \left| \fint_{Q_j} f(y)\,dy \right| dx \\ &\leqslant \int_{\mathbb{R}^n\setminus\bigcup_{j\in\mathbb{N}} Q_j} |f(x)|\,dx + \sum_{j\in\mathbb{N}} \int_{Q_j} |f(y)|\,dy \\ &= \|f\|_{L^1(\mathbb{R}^n)}, \end{aligned}$$

which gives (5.2.36).

Furthermore, if $x \in Q_j$,

$$|b_j(x)| \leqslant |f(x)| + \fint_{Q_j} |f(y)| \, dy;$$

therefore, recalling (5.2.28),

$$\begin{aligned} \fint_{Q_j} |b_j(x)| \, dx &\leqslant \fint_{Q_j} |f(x)| \, dx + \fint_{Q_j} \left(\fint_{Q_j} |f(y)| \, dy \right) dx \\ &= 2 \fint_{Q_j} |f(x)| \, dx \leqslant 2^{n+1}, \end{aligned}$$

establishing (5.2.37).

Now, we denote by p_j the center of Q_j and employ (5.2.3), (5.2.4) and (5.2.33) to see that, in the principal value sense,

$$\begin{aligned} Tb_j(x) &= \int_{\mathbb{R}^n} K(x-y) \left(b_j(x) - b_j(y) \right) dy = - \int_{\mathbb{R}^n} K(x-y) \, b_j(y) \, dy \\ &= - \int_{Q_j} K(x-y) \, b_j(y) \, dy = \int_{Q_j} \left(K(x-p_j) - K(x-y) \right) b_j(y) \, dy. \end{aligned}$$

Note that here, we have exploited in a crucial way the fact that b_j averages to zero in Q_j. This allows us to improve the nonintegrable decay of the kernel by using (4.2.13), according to the following calculation. If ℓ_j denotes the length of the side of Q_j, $y \in Q_j$ and $x \in \mathbb{R}^n \setminus Q'_j$, then $|x - p_j| \geqslant \frac{3\sqrt{n}\ell_j}{2}$ and consequently, for all $t \in (0, 1)$,

$$\begin{aligned} |x - (1-t)p_j - ty| &= |x - p_j + t(p_j - y)| \geqslant |x - p_j| - |p_j - y| \\ &\geqslant |x - p_j| - \sqrt{n}\ell_j \geqslant \frac{|x - p_j|}{3}. \end{aligned}$$

As a result, if $y \in Q_j$ and $x \in \mathbb{R}^n \setminus Q'_j$,

$$\begin{aligned} |K(x-p_j) - K(x-y)| &\leqslant \int_0^1 |\nabla K(x - (1-t)p_j - ty)| \, |y - p_j| \, dt \\ &\leqslant C \int_0^1 \frac{|y - p_j|}{|x - (1-t)p_j - ty|^{n+1}} \, dt \\ &\leqslant C\ell_j \int_0^1 \frac{dt}{|x - p_j|^{n+1}} \, dt \\ &\leqslant \frac{C\ell_j}{|x - p_j|^{n+1}}. \end{aligned}$$

We stress that this decay is one power better than the one of the kernel K and, in particular, provides an integrable tail. In this way, we find that, for all $x \in \mathbb{R}^n \setminus Q'_j$,

$$|Tb_j(x)| \leqslant \int_{Q_j} \big|K(x-p_j) - K(x-y)\big|\, |b_j(y)|\, dy \leqslant \frac{C\ell_j}{|x-p_j|^{n+1}} \int_{Q_j} |b_j(y)|\, dy.$$

This and (5.2.37) give that, for all $x \in \mathbb{R}^n \setminus Q'_j$,

$$|Tb_j(x)| \leqslant \frac{C\ell_j\, |Q_j|}{|x-p_j|^{n+1}}.$$

We can now integrate in x and find that

$$\begin{aligned}\int_{\mathbb{R}^n\setminus Q'_j} |Tb_j(x)|\, dx &\leqslant \int_{\mathbb{R}^n\setminus B_{\ell_j/2}(p_j)} \frac{C\ell_j\, |Q_j|\, dx}{|x-p_j|^{n+1}}\\ &\leqslant C\ell_j\, |Q_j| \int_{\ell_j/2}^{+\infty} \frac{1}{\rho^2}\, d\rho \leqslant C\, |Q_j|,\end{aligned}$$

which proves (5.2.38).

From this and (5.2.29), we also arrive at

$$\begin{aligned}\int_{\mathbb{R}^n\setminus\bigcup_{k\in\mathbb{N}} Q'_k} \big|T\mathscr{B}(x)\big|\, dx &\leqslant \sum_{j\in\mathbb{N}} \int_{\mathbb{R}^n\setminus\bigcup_{k\in\mathbb{N}} Q'_k} |Tb_j(x)|\, dx\\ &\leqslant \sum_{j\in\mathbb{N}} \int_{\mathbb{R}^n\setminus Q'_j} |Tb_j(x)|\, dx \leqslant C \sum_{j\in\mathbb{N}} |Q_j| \leqslant C,\end{aligned}$$

thus establishing (5.2.39), as desired. □

With this, we can now complete the proof of Theorem 5.2.3 in the following way.

Proof of Theorem 5.2.3. We work in the renormalized setting provided by (5.2.23), and we use the notation of Lemma 5.2.5. In light of (5.2.35) and (5.2.36), up to renaming constants, we can exploit (5.2.25) for the function $\mathscr{G}$ and obtain that

$$\left|\left\{|T\mathscr{G}| \geqslant \frac{1}{2}\right\}\right| \leqslant C. \tag{5.2.40}$$

Now, we let

$$\mathcal{U} := \bigcup_{j\in\mathbb{N}} Q'_j.$$

By (5.2.29),

$$|\mathcal{U}| \leqslant \sum_{j\in\mathbb{N}} |Q'_j| \leqslant C \sum_{j\in\mathbb{N}} |Q_j| \leqslant C. \tag{5.2.41}$$

Also, by (5.2.39),

$$C \geqslant \int_{\mathbb{R}^n \setminus \mathcal{U}} \left|T\mathscr{B}(x)\right| dx \geqslant \int_{(\mathbb{R}^n \setminus \mathcal{U}) \cap \{|T\mathscr{B}| \geqslant 1/2\}} \left|T\mathscr{B}(x)\right| dx$$
$$\geqslant \frac{1}{2}\left|(\mathbb{R}^n \setminus \mathcal{U}) \cap \left\{|T\mathscr{B}| \geqslant \frac{1}{2}\right\}\right|.$$

This and (5.2.41) yield that

$$\left|\left\{|T\mathscr{B}| \geqslant \frac{1}{2}\right\}\right| \leqslant \left|(\mathbb{R}^n \setminus \mathcal{U}) \cap \left\{|T\mathscr{B}| \geqslant \frac{1}{2}\right\}\right| + |\mathcal{U}| \leqslant C. \tag{5.2.42}$$

We also observe that

$$\{|Tf| \geqslant 1\} \subseteq \left\{|T\mathscr{B}| \geqslant \frac{1}{2}\right\} \cup \left\{|T\mathscr{G}| \geqslant \frac{1}{2}\right\}. \tag{5.2.43}$$

Indeed, suppose not. Then, there exists $x \in \{|Tf| \geqslant 1\}$ with $\left|T\mathscr{B}(x)\right| < \frac{1}{2}$ and $\left|T\mathscr{G}(x)\right| < \frac{1}{2}$. Consequently, by the definition of $\mathscr{G}$ in (5.2.32),

$$1 \leqslant |Tf(x)| = \left|T\mathscr{B}(x) + T\mathscr{G}(x)\right| \leqslant \left|T\mathscr{B}(x)\right| + \left|T\mathscr{G}(x)\right| < \frac{1}{2} + \frac{1}{2} = 1.$$

This is a contradiction; therefore, the claim in (5.2.43) holds true.

Now, by (5.2.40), (5.2.42) and (5.2.43),

$$|\{|Tf| \geqslant 1\}| \leqslant \left|\left\{|T\mathscr{B}| \geqslant \frac{1}{2}\right\}\right| + \left|\left\{|T\mathscr{G}| \geqslant \frac{1}{2}\right\}\right| \leqslant C,$$

up to renaming C as usual. This shows that the claim in (5.2.23) holds true and the proof of Theorem 5.2.3 is thereby complete. □

It is instructive to observe that the estimate in Theorem 5.2.3 is essentially optimal. For example, we can take $\phi \in C_0^\infty(B_1, [0, +\infty))$ with $\int_{B_1} \phi(x)\, dx = 1$ and define $f_j(x) := j^n \phi(jx)$. We also assume that the function g in (4.2.12) satisfies $|g| \geqslant a$ for some $a > 0$ in some region $\Sigma \subseteq \partial B_1$ with $\mathcal{H}^{n-1}(\Sigma) > 0$. We also fix $\lambda_0 > 0$ and take $\lambda \in (0, \lambda_0)$. In this setting, we have that

$$\|f_j\|_{L^1(\mathbb{R}^n)} = j^n \int_{\mathbb{R}^n} \phi(jx)\, dx = \int_{B_1} \phi(y)\, dy = 1.$$

Moreover, by (5.2.3) and (5.2.4),

$$\begin{aligned}
|Tf_j(x)| &= \left|\int_{\mathbb{R}^n} K(x-y)\, f_j(y)\, dy\right| \\
&= \left|j^n \int_{\mathbb{R}^n} K(x-y)\, \phi(jy)\, dy\right| \\
&= \left|\int_{B_1} K\left(x - \frac{z}{j}\right) \phi(z)\, dz\right|
\end{aligned}$$

$$
\begin{aligned}
&\geqslant \left| K(x) \int_{B_1} \phi(z)\, dz \right| - \left| \int_{B_1} \left(K\left(x - \frac{z}{j}\right) - K(x) \right) \phi(z)\, dz \right| \\
&\geqslant |K(x)| - \int_{B_1} \left| K\left(x - \frac{z}{j}\right) - K(x) \right| \phi(z)\, dz \\
&= \frac{1}{|x|^n} \left| g\left(\frac{x}{|x|}\right) \right| - \int_{B_1} \left| K\left(x - \frac{z}{j}\right) - K(x) \right| \phi(z)\, dz.
\end{aligned}
$$

Thus, if

$$
\Sigma_\lambda := \left\{ r\omega, \text{ with } \omega \in \Sigma \text{ and } r \in \left[\left(\frac{a}{4\lambda}\right)^{\frac{1}{n}}, \left(\frac{a}{2\lambda}\right)^{\frac{1}{n}} \right] \right\}
$$

and $x \in \Sigma_\lambda$, we have that $|x|^n \leqslant \frac{a}{2\lambda}$ and $\left| g\left(\frac{x}{|x|}\right) \right| \geqslant a$. From these observations, we arrive at

$$
|Tf_j(x)| \geqslant 2\lambda - \int_{B_1} \left| K\left(x - \frac{z}{j}\right) - K(x) \right| \phi(z)\, dz. \tag{5.2.44}
$$

Also, if $z \in B_1$ and $t \in [0, 1]$,

$$
\left| x - \frac{z}{j} \right| \geqslant |x| - \frac{1}{j} \geqslant \frac{|x|}{2} + \frac{1}{2}\left(\frac{a}{4\lambda}\right)^{\frac{1}{n}} - \frac{1}{j} \geqslant \frac{|x|}{2}
$$

as long as j is large enough. In this setting, using (4.2.13),

$$
\begin{aligned}
&\int_{B_1} \left| K\left(x - \frac{z}{j}\right) - K(x) \right| \phi(z)\, dz \\
&\qquad \leqslant \frac{\|\phi\|_{L^\infty(B_1)}}{j} \iint_{B_1 \times [0,1]} \left| \nabla K\left(x - \frac{z}{j}\right) \right| dz\, dt \\
&\qquad \leqslant \frac{C}{j} \iint_{B_1 \times [0,1]} \frac{dz\, dt}{\left| x - \frac{z}{j} \right|^{n+1}} \\
&\qquad \leqslant \frac{C}{j} \iint_{B_1 \times [0,1]} \frac{dz\, dt}{|x|^{n+1}} \leqslant \frac{C\lambda^{\frac{n+1}{n}}}{a^{\frac{n+1}{n}} j} \leqslant \lambda
\end{aligned}
$$

for sufficiently large j (depending also on λ_0) and consequently, by (5.2.44), $|Tf_j(x)| \geqslant \lambda$.

This gives that $\{|Tf_j| \geqslant \lambda\} \supseteq \Sigma_\lambda$ for j large enough, whence

$$
\begin{aligned}
\frac{\left|\{|Tf_j| \geqslant \lambda\}\right|}{\|f_j\|_{L^1(\mathbb{R}^n)}} &= \left|\{|Tf_j| \geqslant \lambda\}\right| \geqslant |\Sigma_\lambda| \\
&= \iint_{\left[\left(\frac{a}{4\lambda}\right)^{\frac{1}{n}}, \left(\frac{a}{2\lambda}\right)^{\frac{1}{n}}\right] \times \Sigma} r^{n-1}\, dr\, d\mathcal{H}^{n-1}_\omega = \frac{a\,\mathcal{H}^{n-1}(\Sigma)}{4n\lambda},
\end{aligned}
$$

showing the optimality of Theorem 5.2.3.

Another important ingredient in the proof of Theorem 5.2.1 consists of a classical interpolation result[3] due to Józef Marcinkiewicz [Mar39] (see also [Zyg56]). The general idea behind interpolation, in simple terms, is that if one knows something about a functional space X and something about a functional space Z, ideally it is possible to infer some information about a functional space Y which "lies between X and Z." An illustrative example can be taken into account when $X := L^P(\mathbb{R}^n)$ and $Z := L^R(\mathbb{R}^n)$, with $R > P \geqslant 1$, and $Y := L^Q(\mathbb{R}^n)$, with $Q \in (P, R)$. In this case, the Hölder inequality with exponents $\frac{R-P}{R-Q}$ and $\frac{R-P}{Q-P}$ gives that if $v \in L^P(\mathbb{R}^n) \cap L^R(\mathbb{R}^n)$, then $v \in L^Q(\mathbb{R}^n)$, with

$$\begin{aligned} \|v\|_{L^Q(\mathbb{R}^n)} &= \left(\int_{\mathbb{R}^n} |v(x)|^Q\,dx\right)^{\frac{1}{Q}} = \left(\int_{\mathbb{R}^n} |v(x)|^{\frac{(R-Q)P}{R-P}+\frac{(Q-P)R}{R-P}}\,dx\right)^{\frac{1}{Q}} \\ &\leqslant \left(\int_{\mathbb{R}^n} |v(x)|^P\,dx\right)^{\frac{R-Q}{(R-P)Q}} \left(\int_{\mathbb{R}^n} |v(x)|^R\,dx\right)^{\frac{Q-P}{(R-P)Q}} \\ &= \|v\|_{L^P(\mathbb{R}^n)}^{\frac{(R-Q)P}{(R-P)Q}} \|v\|_{L^R(\mathbb{R}^n)}^{\frac{(Q-P)R}{(R-P)Q}}. \end{aligned}$$

The interpolation result that we need in our framework is much more sophisticated than this and, in light of Theorem 5.2.3, addresses a "weak theory in L^p." For this, we say that an operator S acting on the space of functions is *sublinear* if for every functions f and g, it holds that

$$|S(f+g)(x)| \leqslant |Sf(x)| + |Sg(x)| \quad \text{for all } x \in \mathbb{R}^n. \tag{5.2.45}$$

For all $p \in [1, +\infty]$, we say that S is of *weak type* (p, p) if there exists $C > 0$ such that, for all $f \in L^p(\mathbb{R}^n)$ and $\lambda > 0$,

$$\left|\{|Sf| \geqslant \lambda\}\right| \leqslant \frac{C\,\|f\|_{L^p(\mathbb{R}^n)}^p}{\lambda^p}. \tag{5.2.46}$$

Comparing with Chebyshev's inequality in (5.2.20), one may think, heuristically, that when S is of weak type (p, p), one has that "Sf is almost in $L^p(\mathbb{R}^n)$."

One also says that S is of *strong type* (p, p) if there exists $C > 0$ such that, for all $f \in L^p(\mathbb{R}^n)$,

$$\|Sf\|_{L^p(\mathbb{R}^n)} \leqslant C\,\|f\|_{L^p(\mathbb{R}^n)}, \tag{5.2.47}$$

[3] Marcinkiewicz was one of the brilliant students and collaborators of Antoni Zygmund, as well as a collaborator of Juliusz Schauder (this biographical fact also provides a conceptual bridge between the results presented here and those in Chapter 4). In 1939, Marcinkiewicz was taken as a Polish prisoner of war to a Soviet camp. Shortly before dying at the age of 30, he gave his manuscripts to his parents, who unfortunately were also imprisoned and died of hunger in a camp. Unsurprisingly, no trace of these manuscripts was ever found, adding one more tragic story to the uncountable atrocities of World War II. See [Mar64] for a brief description by Zygmund of related events and of the loss of Marcinkiewicz's latest mathematical works. Compare also with footnote 3 on p. 301.

that is, S is a bounded operator on $L^p(\mathbb{R}^n)$.

Using Chebyshev's inequality in (5.2.20), one sees that

$$\text{if } S \text{ is of strong type } (p, p) \text{ then it is also of weak type } (p, p). \qquad (5.2.48)$$

The interpolation result that we use in this framework is the following one.

Theorem 5.2.6. *Let p, $r \in [1, +\infty)$ with $p < r$. Let S be a sublinear operator acting on the space of functions. Assume that S is of weak type (p, p) and of weak type (r, r).*

Then, S is of strong type (q, q) for every $q \in (p, r)$.

Also, the corresponding constant in (5.2.47) for $L^q(\mathbb{R}^n)$ depends only on n, p, q, r and the corresponding constants in (5.2.46) for $L^p(\mathbb{R}^n)$ and $L^r(\mathbb{R}^n)$.

Proof. We repeatedly exploit the layer cake representation formula (valid for all $m \in (0, +\infty)$, see e.g. [WZ15, Theorem 5.51], see Figure 5.2.4 for a schematic representation when $v \geqslant 0$ and $m = 1$)

$$\int_{\mathbb{R}^n} |v(x)|^m \, dx = m \int_0^{+\infty} s^{m-1} \big|\{|v| \geqslant s\}\big| \, ds. \qquad (5.2.49)$$

Let $\lambda > 0$. We split a function f into its "top and bottom parts," which we denote by f_t and f_b, respectively (see Figure 5.2.5); that is, we define

$$f_t := f\chi_{\{|f| \geqslant \lambda\}} \quad \text{and} \quad f_b := f\chi_{\{|f| < \lambda\}}.$$

Note that $f = f_t + f_b$. Since S is sublinear, by (5.2.45), we know that $|Sf| \leqslant |Sf_t| + |Sf_b|$ and consequently

$$\{|Sf| \geqslant \lambda\} \subseteq \left\{|Sf_t| \geqslant \frac{\lambda}{2}\right\} \cup \left\{|Sf_b| \geqslant \frac{\lambda}{2}\right\}.$$

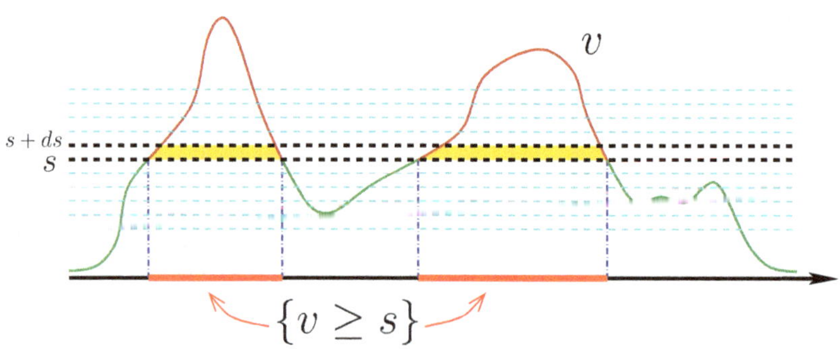

Figure 5.2.4 The layer cake representation formula when $v \geqslant 0$ and $m = 1$.

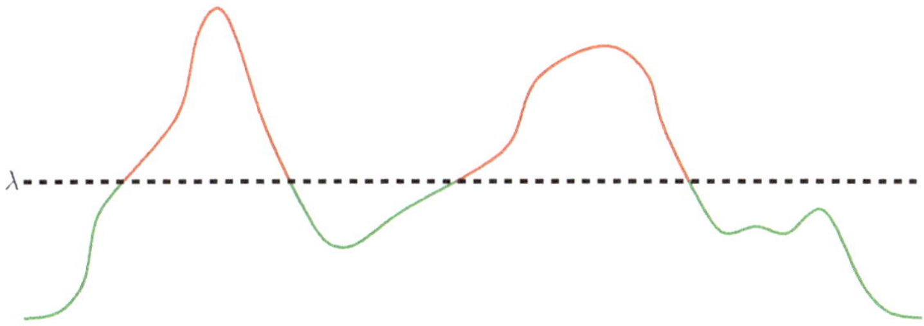

Figure 5.2.5 Splitting a function into its "top" and "bottom" parts.

As a result,

$$\left|\{|Sf| \geqslant \lambda\}\right| \leqslant \left|\left\{|Sf_t| \geqslant \frac{\lambda}{2}\right\}\right| + \left|\left\{|Sf_b| \geqslant \frac{\lambda}{2}\right\}\right|. \tag{5.2.50}$$

Our goal is thus to estimate the two quantities on the right-hand side of (5.2.50) by using the assumptions that S is of weak type (p, p) and of weak type (r, r). Specifically, when dealing with the bottom function f_b, we account for the "tail" of the function f; therefore, to enhance the integrability property as much as possible, it is convenient to take the largest exponent r into consideration (because a small number elevated to a large power becomes even smaller). Instead, when dealing with the top function f_t, we consider the peaks of the function; therefore, it is appropriate to utilize the smallest exponent p (because a large number elevated to a small power becomes less dangerous).

The technical details are given as follows. For the bottom function, using that S is of weak type (r, r) and (5.2.49), we see that

$$\begin{aligned}\left|\left\{|Sf_b| \geqslant \frac{\lambda}{2}\right\}\right| &\leqslant \frac{C\,\|f_b\|^r_{L^r(\mathbb{R}^n)}}{\lambda^r} = \frac{Cr}{\lambda^r}\int_0^{+\infty} s^{r-1}\left|\{|f_b| \geqslant s\}\right| ds\\ &= \frac{Cr}{\lambda^r}\int_0^{\lambda} s^{r-1}\left|\{|f| \geqslant s\}\right| ds.\end{aligned} \tag{5.2.51}$$

Similarly, for the top function, using that S is of weak type (p, p) and (5.2.49), we see that

$$\begin{aligned}\left|\left\{|Sf_t| \geqslant \frac{\lambda}{2}\right\}\right| &\leqslant \frac{C\,\|f_t\|^p_{L^p(\mathbb{R}^n)}}{\lambda^p} = \frac{Cp}{\lambda^p}\int_0^{+\infty} s^{p-1}\left|\{|f_t| \geqslant s\}\right| ds\\ &= \frac{Cp}{\lambda^p}\int_{\lambda}^{+\infty} s^{p-1}\left|\{|f| \geqslant s\}\right| ds.\end{aligned}$$

From this, (5.2.49), (5.2.50), and (5.2.51), we find that

$$\begin{aligned}
&\int_{\mathbb{R}^n} |Sf(x)|^q\,dx \\
&= q\int_0^{+\infty} \lambda^{q-1}\big|\{|Sf|\geqslant\lambda\}\big|\,d\lambda \\
&\leqslant q\int_0^{+\infty} \lambda^{q-1}\left|\left\{|Sf_t|\geqslant\frac{\lambda}{2}\right\}\right|\,d\lambda + q\int_0^{+\infty} \lambda^{q-1}\left|\left\{|Sf_b|\geqslant\frac{\lambda}{2}\right\}\right|\,d\lambda \\
&\leqslant C\left[\int_0^{+\infty}\left(\int_\lambda^{+\infty} \lambda^{q-p-1}s^{p-1}\big|\{|f|\geqslant s\}\big|\,ds\right)d\lambda\right. \\
&\qquad\left.+\int_0^{+\infty}\left(\int_0^{\lambda} \lambda^{q-r-1}s^{r-1}\big|\{|f|\geqslant s\}\big|\,ds\right)d\lambda\right],
\end{aligned}$$

up to renaming constants.

Thus, switching the order of integration and exploiting that $p < q < r$,

$$\begin{aligned}
&\int_{\mathbb{R}^n} |Sf(x)|^q\,dx \\
&\leqslant C\left[\int_0^{+\infty}\left(\int_0^{s} \lambda^{q-p-1}s^{p-1}\big|\{|f|\geqslant s\}\big|\,d\lambda\right)ds.\right. \\
&\qquad\left.+\int_0^{+\infty}\left(\int_s^{+\infty} \lambda^{q-r-1}s^{r-1}\big|\{|f|\geqslant s\}\big|\,d\lambda\right)ds\right] \\
&\leqslant C\left[\int_0^{+\infty} s^{q-p+p-1}\big|\{|f|\geqslant s\}\big|\,ds+\int_0^{+\infty} s^{q-r+r-1}\big|\{|f|\geqslant s\}\big|\,ds\right] \\
&= C\int_0^{+\infty} s^{q-1}\big|\{|f|\geqslant s\}\big|\,ds.
\end{aligned}$$

Hence, making use of (5.2.49) once again,

$$\int_{\mathbb{R}^n} |Sf(x)|^q\,dx \leqslant C\int_{\mathbb{R}^n} |f(x)|^q\,dx,$$

which is the desired result. □

With the previous work and a duality argument, we are now in a position to complete the proof of Theorem 5.2.1.

Proof of Theorem 5.2.1. When $p = 2$, the desired result follows from (5.2.24).

When $p \in (1,2)$, we use the interpolation theory: namely, we know from (5.2.24) that T is of strong type $(2,2)$ and therefore of weak type $(2,2)$, thanks to (5.2.48). Also, we know from Theorem 5.2.3 that T is of weak type $(1,1)$. Hence, it follows from Theorem 5.2.6 that T is of strong type (p,p) for all $p \in (1,2)$, and this proves Theorem 5.2.1 when $p \in (1,2)$.

It remains to prove Theorem 5.2.1 when $p > 2$. For this, we use a duality argument. Let $q := \frac{p}{p-1} \in (1,2)$ be the conjugated exponent of p, and let $g \in C_0^{0,1}(\mathbb{R}^n)$. Let also $\widetilde{K}(x) := K(-x)$ and $\widetilde{T}$ be as in (5.2.4) but with $\widetilde{K}$ replacing K, that is,

$$\widetilde{T}f(x) := \int_{\mathbb{R}^n} \widetilde{K}(x-y)\,\big(f(x)-f(y)\big)\,dy.$$

We stress that $\widetilde{K}$ also satisfies the structural assumptions (4.2.12), (4.2.13) and (5.2.3). Consequently, since we have already established Theorem 5.2.1 when $p \in (1,2)$, we have that

$$\|\widetilde{T}g\|_{L^q(\mathbb{R}^n)} \leqslant C\,\|g\|_{L^q(\mathbb{R}^n)}. \tag{5.2.52}$$

Now, we claim that

$$\int_{\mathbb{R}^n} Tf(x)\,g(x)\,dx = \int_{\mathbb{R}^n} \widetilde{T}g(x)\,f(x)\,dx. \tag{5.2.53}$$

For this, let K_ε be as in (5.2.9) and T_ε be as in (5.2.10). Recalling (5.2.11), we have that

$$\lim_{\varepsilon\searrow 0}\int_{\mathbb{R}^n} |Tf(x)-T_\varepsilon f(x)|\,|g(x)|\,dx \leqslant \lim_{\varepsilon\searrow 0}\|Tf-T_\varepsilon f\|_{L^2(\mathbb{R}^n)}\,\|g\|_{L^2(\mathbb{R}^n)} = 0.$$

Similarly, if $\widetilde{K}_\varepsilon$ and $\widetilde{T}_\varepsilon$ are as in (5.2.9) and (5.2.10) but with $\widetilde{K}$ in place of K, then

$$\lim_{\varepsilon\searrow 0}\int_{\mathbb{R}^n} |\widetilde{T}g(x)-\widetilde{T}_\varepsilon g(x)|\,|f(x)|\,dx = 0.$$

As a result, using (5.2.3),

$$\begin{aligned}
&\int_{\mathbb{R}^n} Tf(x)\,g(x)\,dx - \int_{\mathbb{R}^n} \widetilde{T}g(x)\,f(x)\,dx\\
&= \lim_{\varepsilon\searrow 0}\int_{\mathbb{R}^n} T_\varepsilon f(x)\,g(x)\,dx - \int_{\mathbb{R}^n} \widetilde{T}_\varepsilon g(x)\,f(x)\,dx\\
&= \lim_{\varepsilon\searrow 0}\int_{\mathbb{R}^n}\left(\int_{\mathbb{R}^n} K_\varepsilon(x-y)\big(f(x)-f(y)\big)\,dy\right) g(x)\,dx\\
&\quad - \int_{\mathbb{R}^n}\left(\int_{\mathbb{R}^n} \widetilde{K}_\varepsilon(x-y)\big(g(x)-g(y)\big)\,dy\right) f(x)\,dx\\
&= \lim_{\varepsilon\searrow 0}\left[-\int_{\mathbb{R}^n}\left(\int_{\mathbb{R}^n} K_\varepsilon(x-y)\,f(y)\,dy\right) g(x)\,dx\right.\\
&\qquad\left.+\int_{\mathbb{R}^n}\left(\int_{\mathbb{R}^n} K_\varepsilon(y-x)g(y)\,dy\right) f(x)\,dx\right]\\
&= 0,
\end{aligned}$$

owing to Fubini's theorem. The proof of (5.2.53) is thereby complete.

Thus, by (5.2.52) and (5.2.53),

$$\left|\int_{\mathbb{R}^n} Tf(x)\, g(x)\, dx\right| = \left|\int_{\mathbb{R}^n} \widetilde{T}g(x)\, f(x)\, dx\right| \leqslant \int_{\mathbb{R}^n} |\widetilde{T}g(x)|\, |f(x)|\, dx$$
$$\leqslant \|\widetilde{T}g\|_{L^q(\mathbb{R}^n)}\, \|f\|_{L^p(\mathbb{R}^n)} \leqslant C\|g\|_{L^q(\mathbb{R}^n)}\, \|f\|_{L^p(\mathbb{R}^n)}.$$

By the density of $C_0^{1,0}(\mathbb{R}^n)$ in $L^q(\mathbb{R}^n)$ (see e.g. [DD12, Theorem 1.91]), this estimate remains valid for all $g \in L^q(\mathbb{R}^n)$. Accordingly, the linear functional $\mathcal{L}$: $L^q(\mathbb{R}^n) \to \mathbb{R}$ defined by

$$\mathcal{L}(g) := \int_{\mathbb{R}^n} Tf(x)\, g(x)\, dx$$

is bounded; therefore, by duality (see e.g. [OD18, Theorem 5.12.1]), there exists a unique function $\phi \in L^p(\Omega)$ such that $\mathcal{L}(g) = \int_{\mathbb{R}^n} \phi(x)\, g(x)\, dx$ for all $g \in L^q(\mathbb{R}^n)$ and

$$\|\phi\|_{L^p(\Omega)} = \|\mathcal{L}\| = \sup_{g\in L^q(\Omega)\setminus\{0\}} \frac{|\mathcal{L}(g)|}{\|g\|_{L^q(\Omega)}}.$$

Since necessarily $\phi = Tf$, we obtain that

$$\|Tf\|_{L^p(\Omega)} = \|\phi\|_{L^p(\Omega)}$$
$$= \sup_{g\in L^q(\Omega)\setminus\{0\}} \frac{1}{\|g\|_{L^q(\Omega)}} \left|\int_{\mathbb{R}^n} Tf(x)\, g(x)\, dx\right| \leqslant C\|f\|_{L^p(\mathbb{R}^n)}.$$

This completes the proof of Theorem 5.2.1. □

We observe that Theorem 5.2.1 carries over to all functions $f \in L^p(\mathbb{R}^n)$, by the density of $C_0^{1,0}(\mathbb{R}^n)$ in $L^p(\mathbb{R}^n)$ (see e.g. [DD12, Theorem 1.91]); that is, T extends by continuity to a bounded linear operator from $L^p(\mathbb{R}^n)$ to $L^p(\mathbb{R}^n)$: we stress however that for a general function f in $L^p(\mathbb{R}^n)$, the pointwise definition of Tf in (5.2.4) does not necessarily makes sense, and one has to intend Tf in this setting as the limit in $L^p(\mathbb{R}^n)$ of Tf_j, where $f_j \in C_0^{0,1}(\mathbb{R}^n)$ and $\|f_j - f\|_{L^p(\mathbb{R}^n)} \to 0$ as $j \to +\infty$.

As a consequence of Theorem 5.2.1, we have the following.

Corollary 5.2.7. *Let $p \in (1, +\infty)$. Let Γ be the fundamental solution in* (2.7.6) *and $f \in C_0^{0,1}(\mathbb{R}^n)$. Let $v := -\Gamma * f$. Then, $v \in W^{2,p}(\mathbb{R}^n)$ and*

$$\|D^2 v\|_{L^p(\mathbb{R}^n)} \leqslant C\, \|f\|_{L^p(\mathbb{R}^n)},$$

where $C > 0$ depends only on n and p.

Proof. By (4.2.4), (4.2.21) and (4.2.22), we can write $\partial_{ij}v$ as Tf satisfying conditions (4.2.12), (4.2.13) and (5.2.3). Hence, the desired result is now a direct consequence of Theorem 5.2.1. □

We are now in a position to address the core of the regularity theory in L^p spaces. We begin with the following interior estimates.

Theorem 5.2.8. *Let $p \in (1, +\infty)$ and $f \in C^{0,1}(B_1)$. Assume that $u \in C^2(B_1)$ is a solution of $\Delta u = f$ in B_1.*

Then,

$$\|u\|_{W^{2,p}(B_{1/2})} \leqslant C\left(\|u\|_{L^p(B_1)} + \|f\|_{L^p(B_1)}\right),$$

where $C > 0$ depends only on n and p.

Proof. The attentive reader will find close connections between this proof and that of Theorem 4.2.1. The main additional ingredient here is of course the Newtonian potential estimate in L^p spaces provided by Corollary 5.2.7. We also employ the following useful inequality of general use, valid for all measurable functions $g \in W^{2,p}((0,1)^n)$ and all $\varepsilon \in \left(0, \frac{1}{4}\right)$:

$$\|\nabla g\|_{L^p((0,1)^n)} \leqslant \varepsilon \|D^2 g\|_{L^p((0,1)^n)} + \frac{C}{\varepsilon} \|g\|_{L^p((0,1)^n)}, \tag{5.2.54}$$

with $C > 0$ depending only on n and p. The above is a very particular case of the Gagliardo–Nirenberg interpolation inequality (see [FFRS21, Theorem 1.2], used here with $k := 2$, $j := 1$, $r := q := p$, see also [Gag59, Nir59] for the original articles on this topic and [Leo09, Sections 12.5, 13.3, 16.1 and 16.2] for a thorough presentation of interpolation methods). We give a detailed proof of (5.2.54) for completeness. First of all, we observe that

$$\text{it is enough to establish (5.2.54) when } \frac{1}{3\varepsilon} \in \mathbb{N}. \tag{5.2.55}$$

Indeed, if $\varepsilon \in \left(0, \frac{1}{4}\right)$, we let $N \in \mathbb{N}$ such that $\frac{1}{3\varepsilon} \in [N, N+1)$. Hence, $\frac{1}{3(N+1)} \in (0, \varepsilon] \subseteq \left(0, \frac{1}{4}\right)$, and thus, if (5.2.54) is valid for ε replaced by $\frac{1}{3(N+1)}$, we have that

$$\begin{aligned}
\|\nabla g\|_{L^p((0,1)^n)} &\leqslant \frac{1}{3(N+1)} \|D^2 g\|_{L^p((0,1)^n)} + 3C(N+1)\, \|g\|_{L^p((0,1)^n)} \\
&\leqslant \varepsilon \|D^2 g\|_{L^p((0,1)^n)} + 6CN\, \|g\|_{L^p((0,1)^n)} \\
&\leqslant \varepsilon \|D^2 g\|_{L^p((0,1)^n)} + \frac{2C}{\varepsilon} \|g\|_{L^p((0,1)^n)},
\end{aligned}$$

giving (5.2.54) for ε as well. This establishes (5.2.55).

In view of (5.2.55), hereinafter we suppose that $\frac{1}{3\varepsilon} \in \mathbb{N}$.

Additionally, we point out that we can reduce to dimension $n = 1$, namely

$$\text{it is enough to establish (5.2.54) when } n = 1 \text{ and } g \in W^{2,p}((0,1)). \tag{5.2.56}$$

Indeed, if $g \in W^{2,p}((0,1)^n)$, given $x' = (x_1, \dots, x_{n-1}) \in (0,1)^{n-1}$, we can define $\widetilde{g}(x_n) := g(x', x_n)$. If (5.2.54) holds true when $n = 1$, then, up to renaming constants,

$$\begin{aligned}\int_0^1 |\partial_n g(x', x_n)|^p \, dx_n &= \int_0^1 |\widetilde{g}'(x_n)|^p \, dx_n \\ &\leqslant \varepsilon^p \int_0^1 |\widetilde{g}''(x_n)|^p \, dx_n + \frac{C}{\varepsilon^p} \int_0^1 |\widetilde{g}(x_n)|^p \, dx_n \\ &= \varepsilon^p \int_0^1 |\partial_{nn} g(x', x_n)|^p \, dx_n + \frac{C}{\varepsilon^p} \int_0^1 |g(x', x_n)|^p \, dx_n.\end{aligned}$$

Thus, integrating over $x' \in (0,1)^{n-1}$,

$$\begin{aligned}\int_{(0,1)^n} |\partial_n g(x)|^p \, dx &\leqslant \varepsilon^p \int_{(0,1)^n} |\partial_{nn} g(x)|^p \, dx + \frac{C}{\varepsilon^p} \int_{(0,1)^n} |g(x)|^p \, dx \\ &\leqslant \varepsilon^p \|D^2 g\|^p_{L^p((0,1)^n)} + \frac{C}{\varepsilon^p} \|g\|^p_{L^p((0,1)^n)}.\end{aligned}$$

By exchanging the order of the variables, this entails that, for all $j \in \{1, \dots, n\}$,

$$\int_{(0,1)^n} |\partial_j g(x)|^p \, dx \leqslant \varepsilon^p \|D^2 g\|^p_{L^p((0,1)^n)} + \frac{C}{\varepsilon^p} \|g\|^p_{L^p((0,1)^n)},$$

which yields (5.2.54) in its full generality. This proves (5.2.56).

Hence, in view of (5.2.56), we now focus on the proof of (5.2.54) when $n = 1$. Additionally, by the density of $C^\infty((0,1))$ in $W^{2,p}((0,1))$ (see e.g. [Eva98, Theorem 2, p. 251]), we can assume that $g \in C^\infty((0,1))$. According to the calculus mean value theorem, given $x > y \in (0,1)$, there exists $z_{x,y} \in [y, x]$ such that

$$g(x) - g(y) = g'(z_{x,y})(x - y).$$

Therefore, for all $w \in (0,1)$,

$$g'(w) = g'(w) - g'(z_{x,y}) + \frac{g(x) - g(y)}{x - y} = \int_{z_{x,y}}^{w} g''(t) \, dt + \frac{g(x) - g(y)}{x - y}.$$

In particular, we consider an interval of the form $[a, a+3\varepsilon]$, for some $a \in (0, 1-3\varepsilon)$, we assume that $y \in [a, a+\varepsilon]$ and $x \in [a+2\varepsilon, a+3\varepsilon]$. In this way, $x - y \geqslant \varepsilon$ and thus, for all $w \in [a, a+3\varepsilon]$,

$$|g'(w)| \leqslant \int_a^{a+3\varepsilon} |g''(t)| \, dt + \frac{|g(x)| + |g(y)|}{\varepsilon}.$$

As a consequence, possibly renaming C at each stage of the calculation,

$$\begin{aligned}|g'(w)|^p &\leqslant \left(\int_a^{a+3\varepsilon} |g''(t)|\,dt + \frac{|g(x)|+|g(y)|}{\varepsilon}\right)^p\\ &\leqslant C\left(\left(\int_a^{a+3\varepsilon} |g''(t)|\,dt\right)^p + \frac{|g(x)|^p+|g(y)|^p}{\varepsilon^p}\right)\\ &\leqslant C\left(\varepsilon^{p-1}\int_a^{a+3\varepsilon} |g''(t)|^p\,dt + \frac{|g(x)|^p+|g(y)|^p}{\varepsilon^p}\right).\end{aligned}$$

Integrating over $x \in [a+2\varepsilon, a+3\varepsilon]$, $y \in [a, a+\varepsilon]$ and $w \in [a, a+3\varepsilon]$, and dividing by ε^2, we obtain that

$$\begin{aligned}&\int_a^{a+3\varepsilon} |g'(w)|^p\,dw\\ &\leqslant C\left(\varepsilon^p\int_a^{a+3\varepsilon} |g''(t)|^p\,dt + \frac{1}{\varepsilon^p}\int_{a+2\varepsilon}^{a+3\varepsilon} |g(x)|^p\,dx + \frac{1}{\varepsilon^p}\int_{a+\varepsilon}^{a+\varepsilon} |g(y)|^p\,dy\right)\\ &\leqslant C\left(\varepsilon^p\int_a^{a+3\varepsilon} |g''(t)|^p\,dt + \frac{1}{\varepsilon^p}\int_a^{a+3\varepsilon} |g(s)|^p\,ds\right).\end{aligned}$$

By taking $a := 3\varepsilon j$ with $j \in \left\{0, 1, \dots, \frac{1}{3\varepsilon} - 1\right\}$, we conclude that

$$\begin{aligned}&\int_0^1 |g'(w)|^p\,dw\\ &= \sum_{j\in\left\{0,1,\dots,\frac{1}{3\varepsilon}-1\right\}} \int_{3\varepsilon j}^{3\varepsilon(j+1)} |g'(w)|^p\\ &\leqslant C \sum_{j\in\left\{0,1,\dots,\frac{1}{3\varepsilon}-1\right\}} \left(\varepsilon^p\int_{3\varepsilon j}^{3\varepsilon(j+1)} |g''(t)|^p\,dt + \frac{1}{\varepsilon^p}\int_{3\varepsilon j}^{3\varepsilon(j+1)} |g(s)|^p\,ds\right)\\ &= C\left(\varepsilon^p\int_0^1 |g''(t)|^p\,dt + \frac{1}{\varepsilon^p}\int_0^1 |g(s)|^p\,ds\right),\end{aligned}$$

which completes the proof of (5.2.54).

Now, we deal with the core of the proof of Theorem 5.2.8. First of all, by conveniently extending f outside B_1, see e.g. [GT01, Lemma 6.37],

$$\text{we can suppose that } f \in C_0^{0,1}(B_2), \text{ with } \|f\|_{L^p(\mathbb{R}^n)} \leqslant C\|f\|_{L^p(B_1)}. \tag{5.2.57}$$

Thus, we consider the Newtonian potential $v := -\Gamma * f$. We know from (4.2.6) that $\Delta v = f$ in $\mathbb{R}^n$. As a result, defining $w := u - v$, we have that w is harmonic in B_1. Accordingly, it follows from Cauchy's estimates (see Theorem 2.17.1) that

$$\|D^2 w\|_{L^\infty(B_{1/2})} \leqslant C\,\|w\|_{L^1(B_1)} \leqslant C\Big(\|u\|_{L^1(B_1)} + \|v\|_{L^1(B_1)}\Big).$$

From this inequality and Corollary 5.2.7, recalling (5.2.57), we infer that

$$\begin{aligned}
\|D^2u\|_{L^p(B_{1/2})} &\leqslant \|D^2v\|_{L^p(B_{1/2})} + \|D^2w\|_{L^p(B_{1/2})} \\
&\leqslant C\Big(\|u\|_{L^1(B_1)} + \|v\|_{L^1(B_1)} + \|f\|_{L^p(\mathbb{R}^n)}\Big) \\
&\leqslant C\left(\|u\|_{L^1(B_1)} + \iint_{B_1\times B_2} |\Gamma(x-y)|\,|f(y)|\,dx\,dy + \|f\|_{L^p(\mathbb{R}^n)}\right) \\
&\leqslant C\left(\|u\|_{L^1(B_1)} + \iint_{B_3\times B_2} |\Gamma(z)|\,|f(y)|\,dz\,dy + \|f\|_{L^p(\mathbb{R}^n)}\right) \\
&\leqslant C\Big(\|u\|_{L^1(B_1)} + \|f\|_{L^1(B_2)} + \|f\|_{L^p(\mathbb{R}^n)}\Big) \\
&\leqslant C\Big(\|u\|_{L^1(B_1)} + \|f\|_{L^p(B_1)}\Big).
\end{aligned} \tag{5.2.58}$$

Now, by (5.2.54),

$$\|\nabla u\|_{L^p(B_{1/2})} \leqslant C\left(\|D^2u\|_{L^p(B_{1/2})} + \|u\|_{L^p(B_{1/2})}\right).$$

This and (5.2.58), up to renaming constants, yield that

$$\begin{aligned}
\|u\|_{W^{2,p}(B_{1/2})} &\leqslant C\left(\|D^2u\|_{L^p(B_{1/2})} + \|\nabla u\|_{L^p(B_{1/2})} + \|u\|_{L^p(B_{1/2})}\right) \\
&\leqslant C\left(\|D^2u\|_{L^p(B_{1/2})} + \|u\|_{L^p(B_{1/2})}\right) \\
&\leqslant C\left(\|u\|_{L^1(B_1)} + \|f\|_{L^p(B_1)} + \|u\|_{L^p(B_{1/2})}\right)
\end{aligned}$$

from which the desired result follows. □

As a technical observation, we point out that the assumption that u belongs to $C^2(B_1)$ in the statement of Theorem 5.2.8 has been made mainly to have a pointwise definition of the corresponding equation $\Delta u = f$. This setting can be weakened in several forms; for instance, we point out that in the statement of Theorem 5.2.8, one can relax the conditions $u \in C^2(B_1)$ and $f \in C^{0,1}(B_1)$ to $u \in W^{2,p}(B_1)$ and $f \in L^p(B_1)$, respectively. In this framework, however, the equation is not necessarily satisfied at every point of B_1, but only a.e. in B_1 (which, with $u \in W^{2,p}(B_1)$, also says that the equation provides an identity between functions in $L^p(B_1)$; in jargon, these are called "strong solutions" to distinguish them from the "weak solutions" obtained by integrating by parts as in (2.2.1); the terminology might be confusing anyway since "strong" solutions are "weaker" than classical smooth solutions). The counterpart of Theorem 5.2.8 in this setting is presented as follows.

Corollary 5.2.9. *Let $p \in (1,+\infty)$ and $f \in L^p(B_1)$. Assume that $u \in W^{2,p}(B_1)$ is a solution of $\Delta u = f$ a.e. in B_1.*

Then,

$$\|u\|_{W^{2,p}(B_{1/2})} \leqslant C\Big(\|u\|_{L^p(B_1)} + \|f\|_{L^p(B_1)}\Big),$$

where $C > 0$ depends only on n and p.

Proof. We use a mollification argument. We take $\tau \in C_0^\infty(B_1, [0, +\infty))$ with $\int_{B_1} \tau(x)\,dx = 1$. Given $\eta \in \left(0, \frac{1}{10}\right)$, we let $\tau_\eta(x) := \frac{1}{\eta^n}\tau\left(\frac{x}{\eta}\right)$ and define $u_\eta := u * \tau_\eta$ and $f_\eta := f * \tau_\eta$. We observe that $u_\eta \to u$ in $W^{2,p}(B_{3/4})$ and $f_\eta \to f$ in $L^p(B_{3/4})$ (see e.g. [WZ15, Theorem 9.6]).

Also, for all $\varphi \in C_0^\infty(B_{3/4})$ and $i \in \{1, \dots, n\}$,

$$\begin{aligned}\int_{\mathbb{R}^n} \partial_i u_\eta(x)\,\varphi(x)\,dx &= -\int_{\mathbb{R}^n} u_\eta(x)\,\partial_i\varphi(x)\,dx\\ &= -\iint_{\mathbb{R}^n\times\mathbb{R}^n} u(x-y)\,\tau(y)\,\partial_i\varphi(x)\,dx\,dy\\ &= \iint_{\mathbb{R}^n\times\mathbb{R}^n} \partial_i u(x-y)\,\tau(y)\,\varphi(x)\,dx\,dy\\ &= \int_{\mathbb{R}^n} (\partial_i u * \tau_\eta)(x)\,\varphi(x)\,dx,\end{aligned}$$

that is, $\partial_i u_\eta = \partial_i u * \tau_\eta$ in $B_{3/4}$ (note that this holds pointwise at every point of $B_{3/4}$ since $u_\eta \in C^\infty(B_{3/4})$). Iterating this information, we find that

$$\Delta u_\eta = \Delta u * \tau_\eta = f * \tau_\eta = f_\eta \quad \text{in } B_{3/4}.$$

As a result, we can invoke Theorem 5.2.8 (applied here with $B_{3/4}$ instead of B_1 and renaming constants accordingly), finding that

$$\|u_\eta\|_{W^{2,p}(B_{1/2})} \leqslant C\Big(\|u_\eta\|_{L^p(B_{3/4})} + \|f_\eta\|_{L^p(B_{3/4})}\Big).$$

The desired result thus follows by taking the limit as $\eta \searrow 0$. □

5.3 Calderón–Zygmund Estimates for Equations in Nondivergence Form

As discussed in Section 4.4, once we established a nice regularity estimate for the Poisson equation, it becomes very desirable to extend it to a more general class of elliptic equations belonging to some "natural class" which can be treated as a local perturbation of the Laplace operator. In this setting, we have a natural counterpart of the interior estimates in Corollary 5.2.9 for general elliptic equations, in a form similar to that of Theorem 4.4.2, designed here to address the L^p theory.

Theorem 5.3.1. *Let $a_{ij} \in C(B_1)$ satisfy the ellipticity condition in* (4.4.3). *Let b_i, $c \in L^\infty(B_1)$. Let $f \in L^p(B_1)$ for some $p \in (1, +\infty)$. Let $u \in W^{2,p}(B_1)$ be a* *solution of*

$$\sum_{i,j=1}^{n} a_{ij}\partial_{ij}u + \sum_{i=1}^{n} b_i\partial_i u + cu = f \quad a.e. \text{ n } B_1.$$

Then, there exists $C > 0$*, depending only on* n*,* p*,* a_{ij}*,* b_i *and* c*, such that*

$$\|u\|_{W^{2,p}(B_{1/2})} \leqslant C\left(\|u\|_{L^p(B_1)} + \|f\|_{L^p(B_1)}\right).$$

The proof of Theorem 5.3.1 leverages a result for scaled estimates which can be seen as the counterpart of Proposition 4.4.4 in the L^p framework. This auxiliary result is the following.

Proposition 5.3.2. *Let* $k \in \mathbb{N}$*,* $p \in [1, +\infty)$*,* $\sigma \geqslant 0$ *and* $R > 0$*. There exists* $\varepsilon_0 \in (0, 1)$ *depending only on* n *and* k *such that the following holds true.*

Let $u \in W^{k,p}(B_R)$*. Assume that for every* $x_0 \in B_R$ *and every* $\rho \in \left(0, \frac{R-|x_0|}{2}\right)$,

$$\sum_{j=0}^{k} \rho^j \|D^j u\|_{L^p(B_\rho(x_0))} \leqslant \sigma + \varepsilon_0 \sum_{j=0}^{k} (2\rho)^j \|D^j u\|_{L^p(B_{2\rho}(x_0))}. \tag{5.3.1}$$

Then, there exists $C > 0$*, depending only on* n*,* k *and* R *such that*

$$\|u\|_{W^{k,p}(B_{R/2})} \leqslant C\sigma.$$

Proof. Let

$$Q := \sup_{\substack{x_0 \in B_R \\ \rho \in \left(0, \frac{R-|x_0|}{2}\right)}} \sum_{j=0}^{k} \rho^j \|D^j u\|_{L^p(B_\rho(x_0))},$$

and note that

$$Q \leqslant (k+1)(1+R)^k \|u\|_{W^{k,p}(B_R)} < +\infty.$$

Now, we take $x_0 \in B_R$ and $\rho \in \left(0, \frac{R-|x_0|}{2}\right)$. We cover $B_\rho(x_0)$ by a finite family of balls $B_{\frac{\rho}{4}}(p_i)$ with $p_i \in B_\rho(x_0)$ and $i \in \{1, \ldots, N\}$, where N is bounded from above in dependence only of n.

Since $|p_i| \leqslant |x_0| + \rho < R - 2\rho + \rho = R - \rho$, we have that $p_i \in B_R$ and $\frac{\rho}{2} \in \left(0, \frac{R-|p_i|}{2}\right)$. Thus, for all $i \in \{1, \ldots, N\}$,

$$\sum_{j=0}^{k} \left(\frac{\rho}{4}\right)^j \|D^j u\|_{L^p(B_{\rho/4}(p_i))} \leqslant \sigma + \varepsilon_0 \sum_{j=0}^{k} \left(\frac{\rho}{2}\right)^j \|D^j u\|_{L^p(B_{\rho/2}(p_i))} \leqslant \sigma + \varepsilon_0 Q,$$

thanks to (5.3.1). Therefore,

$$\begin{aligned}\sum_{j=0}^{k} \rho^j \|D^j u\|_{L^p(B_\rho(x_0))} &\leqslant \sum_{j=0}^{k} 4^j \sum_{i=1}^{N} \left(\frac{\rho}{4}\right)^j \|D^j u\|_{L^p(B_{\rho/4}(p_i))} \\ &\leqslant 4^k N\sigma + \varepsilon_0 4^k N Q.\end{aligned}$$

Taking the supremum over such x_0 and ρ, we find that

$$Q \leqslant 4^k N\sigma + \varepsilon_0 4^k NQ.$$

Hence, choosing $\varepsilon_0 := \frac{1}{2^{2k+1}N}$ and reabsorbing one term into the left-hand side,

$$Q \leqslant 2^{2k+1} N\sigma.$$

Since

$$Q \geqslant \lim_{\rho \nearrow R/2} \sum_{j=0}^{k} \rho^j \|D^j u\|_{L^p(B_\rho)} = \sum_{j=0}^{k} \left(\frac{R}{2}\right)^j \|D^j u\|_{L^p(B_{R/2})},$$

the desired result follows. □

With this, we can address the proof of Theorem 5.3.1. The argument is a modification of that presented in the proof of Theorem 4.4.2 (here, we invoke Corollary 5.2.9 instead of Theorem 4.2.1 and Proposition 5.3.2 instead of Proposition 4.4.4), and it utilizes the fact that, zooming in close on a point, the picture reduces to a small perturbation of the Laplace operator.

Proof of Theorem 5.3.1. As observed in (4.4.24), up to an affine transformation, Corollary 5.2.9 holds true if the Laplacian is replaced by an operator with constant coefficients of the form $\sum_{i,j=1}^{n} \overline{a}_{ij}\partial_{ij}u$, where $\overline{a}_{ij}$ is constant and fulfills the ellipticity condition in (4.4.3).

Given $x_0 \in B_{1/2}$ and $\rho \in \left(0, \frac{1}{4}\right)$, to be taken suitably small in the following, we set

$$\begin{aligned} \widetilde{a}_{ij}(x) &:= a_{ij}(x_0) - a_{ij}(x), \\ \widetilde{u}(x) &:= \frac{u(x_0+\rho x)}{\rho^2} \\ \text{and} \quad \widetilde{f}(x) &:= f(x_0+\rho x) - \sum_{i=1}^{n} b_i(x_0+\rho x)\partial_i u(x_0+\rho x) \\ &\quad - c(x_0+\rho x)u(x_0+\rho x) + \sum_{i,j=1}^{n} \widetilde{a}_{ij}(x_0+\rho x)\partial_{ij}u(x_0+\rho x). \end{aligned}$$

By (4.4.27), a.e. $x \in B_1$, we have that

$$\sum_{i,j=1}^{n} a_{ij}(x_0)\partial_{ij}\widetilde{u}(x) = \widetilde{f}(x).$$

Thus, by Corollary 5.2.9,

$$\|\widetilde{u}\|_{W^{2,p}(B_{1/2})} \leqslant C\left(\|\widetilde{u}\|_{L^p(B_1)} + \|\widetilde{f}\|_{L^p(B_1)}\right). \tag{5.3.2}$$

Now, we take $\delta > 0$, to be chosen conveniently small here, and we point out that

$$\sup_{x \in B_1} |\widetilde{a}_{ij}(x_0 + \rho x)| = \sup_{y \in B_\rho(x_0)} |\widetilde{a}_{ij}(y)| = \sup_{y \in B_\rho(x_0)} |a_{ij}(x_0) - a_{ij}(y)| \leqslant \delta,$$

as long as $\rho \in (0, \delta)$ is small enough in dependence of δ, thanks to the continuity assumption on the coefficients a_{ij}.

As a result, for a.e. $x \in B_1$,

$$|\widetilde{f}(x)| \leqslant C\Big(|f(x_0 + \rho x)| + |\nabla u(x_0 + \rho x)| + |u(x_0 + \rho x)| + \delta |D^2 u(x_0 + \rho x)|\Big),$$

leading to

$$\begin{aligned}
&\|\widetilde{f}\|_{L^p(B_1)} \\
&\leqslant C\left(\int_{B_1} \Big(|f(x_0 + \rho x)|^p + |\nabla u(x_0 + \rho x)|^p + |u(x_0 + \rho x)|^p \right. \\
&\quad \left. + \delta^p |D^2 u(x_0 + \rho x)|^p\Big)\, dx\right)^{\frac{1}{p}} \\
&= \frac{C}{\rho^{\frac{n}{p}}}\left(\int_{B_\rho(x_0)} \Big(|f(y)|^p + |\nabla u(y)|^p + |u(y)|^p + \delta^p |D^2 u(y)|^p\Big)\, dy\right)^{\frac{1}{p}} \\
&\leqslant \frac{C}{\rho^{\frac{n}{p}}}\Big(\|f\|_{L^p(B_\rho(x_0))} + \|\nabla u\|_{L^p(B_\rho(x_0))} + \|u\|_{L^p(B_\rho(x_0))} + \delta \|D^2 u\|_{L^p(B_\rho(x_0))}\Big).
\end{aligned}$$

For this reason and (5.3.2),

$$\begin{aligned}
\rho^{\frac{n}{p}} \|\widetilde{u}\|_{W^{2,p}(B_{1/2})} \leqslant C\Big[&\rho^{\frac{n}{p}} \|\widetilde{u}\|_{L^p(B_1)} + \|f\|_{L^p(B_\rho(x_0))} + \|\nabla u\|_{L^p(B_\rho(x_0))} \\
&+ \|u\|_{L^p(B_\rho(x_0))} + \delta \|D^2 u\|_{L^p(B_\rho(x_0))}\Big]. \qquad (5.3.3)
\end{aligned}$$

We also observe that, for every $j \in \mathbb{N}$, we have that $D^j u(y) = \rho^{2-j} D^j \widetilde{u}\left(\frac{y - x_0}{\rho}\right)$, and hence

$$\begin{aligned}
\|D^j u\|_{L^p(B_r(x_0))} &= \rho^{2-j}\left(\int_{B_r(x_0)} \left|D^j \widetilde{u}\left(\frac{y - x_0}{\rho}\right)\right|^p dy\right)^{\frac{1}{p}} \\
&= \rho^{\frac{n}{p}+2-j}\left(\int_{B_{r/\rho}} |D^j \widetilde{u}(x)|^p\, dx\right)^{\frac{1}{p}} = \rho^{\frac{n}{p}+2-j} \|D^j \widetilde{u}\|_{L^p(B_{r/\rho})};
\end{aligned}$$

therefore, (5.3.3) becomes

$$\begin{aligned}
&\sum_{j=0}^{2}\rho^j\|D^ju\|_{L^p(B_{\rho/2}(x_0))}\\
&=\rho^{\frac{n}{p}+2}\sum_{j=0}^{2}\|D^j\widetilde{u}\|_{L^p(B_{1/2})}\\
&\leqslant C\Big[\|u\|_{L^p(B_\rho(x_0))}+\rho^2\|f\|_{L^p(B_\rho(x_0))}+\rho^2\|\nabla u\|_{L^p(B_\rho(x_0))}\\
&\qquad+\rho^2\|u\|_{L^p(B_\rho(x_0))}+\delta\rho^2\|D^2u\|_{L^p(B_\rho(x_0))}\Big]\\
&\leqslant C\Big[\|u\|_{L^p(B_1)}+\|f\|_{L^p(B_1)}+\delta\rho\|\nabla u\|_{L^p(B_\rho(x_0))}+\delta\|u\|_{L^p(B_\rho(x_0))}\\
&\qquad+\delta\rho^2\|D^2u\|_{L^p(B_\rho(x_0))}\Big]\\
&=C\Big[\|u\|_{L^p(B_1)}+\|f\|_{L^p(B_1)}\Big]+C\delta\sum_{j=0}^{2}\rho^j\|D^ju\|_{L^p(B_\rho(x_0))}.
\end{aligned}$$

From this and Proposition 5.3.2 (used here with $\sigma := C\big(\|u\|_{L^p(B_1)}+\|f\|_{L^p(B_1)}\big)$ and $\varepsilon_0 := C\delta$), we obtain the desired result. □

We mention that the L^p theory described here is flexible enough to also address the local estimates at the boundary, thus providing the L^p counterpart of Theorems 4.2.2 and 4.4.3. In this setting, one has the following.

Theorem 5.3.3. *Let the halfball notation in* (4.2.2) *hold true. Let $a_{ij}\in C(\overline{B_1^+})$ satisfy the ellipticity condition in* (4.4.3). *Let b_i, $c\in L^\infty(B_1^+)$. Let $f\in L^p(B_1^+)$ and $g\in W^{2,p}(B_1^+)$ for some $p\in(1,+\infty)$. Let $u\in W^{2,p}(B_1^+)$ be a solution of*

$$\begin{cases}\displaystyle\sum_{i,j=1}^{n}a_{ij}\partial_{ij}u+\sum_{i=1}^{n}b_i\partial_iu+cu=f & \text{a.e. in } B_1^+,\\ u=g & \text{on } B_1^0.\end{cases}\tag{5.3.4}$$

Then, there exists $C>0$, depending only on n, p, a_{ij}, b_i and c, such that

$$\|u\|_{W^{2,p}(B_{1/2}^+)}\leqslant C\Big(\|u\|_{L^p(B_1^+)}+\|g\|_{W^{2,p}(B_1^+)}+\|f\|_{L^p(B_1^+)}\Big).\tag{5.3.5}$$

As usual in the Sobolev space setting, the condition $u=g$ in (5.3.4) is intended in the sense of traces for $W^{1,p}$ functions (see e.g. [Leo09, Chapter 18] for a thorough discussion of the trace operator).

Proof of Theorem 5.3.3. By possibly replacing u by $u-g$,

$$\text{we can reduce ourselves to the case in which } g \text{ vanishes identically.}\tag{5.3.6}$$

Also,

$$\text{it is enough to consider the case in which } a_{ij} = \delta_{ij},\ b_i = 0 \text{ and } c = 0, \text{ that is the case of the Laplace operator.} \tag{5.3.7}$$

Indeed, if (5.3.5) is proved for the Laplace operator (and therefore for operators with constant coefficients, up to an affine transformation), the general case follows by perturbing the coefficients and reabsorbing the lower-order terms, precisely as done in the proof of Theorem 5.3.1 (but using here halfballs instead of balls).

Hence, we work under the simplifying assumptions in (5.3.6) and (5.3.7). In this setting, we can proceed with a reflection and mollification argument. Let $\varphi \in C_0^\infty(B_{7/8}, [0,1])$ with $\varphi = 1$ in $B_{3/4}$. Let also $U := u\varphi \in W^{2,p}(B_1^+)$, and note that $\Delta U = f\varphi + 2\nabla u \cdot \nabla\varphi + u\Delta\varphi =: F$ a.e. in B_1^+. Moreover, we extend u and U to the whole of B_1 by odd reflection, setting $u(x', x_n) := -u(x', -x_n)$ and $U(x', x_n) := -U(x', -x_n)$ for all $x = (x', x_n) \in B_1$ with $x_n < 0$, and also f and F, as done in (4.2.45). Given $\varepsilon > 0$, we note that $U \in W^{2,p}(B_1 \cap \{|x_n| > \varepsilon\})$ and $\Delta U = F$ a.e. in $B_1 \cap \{|x_n| > \varepsilon\}$ (hence also in the L^p sense) and therefore

$$\int_{B_1 \cap \{|x_n| > \varepsilon\}} F(x)\Phi(x)\,dx = -\int_{B_1 \cap \{|x_n| > \varepsilon\}} \nabla U(x) \cdot \nabla\Phi(x)\,dx \tag{5.3.8}$$

for every $\Phi \in C_0^\infty(B_1 \cap \{|x_n| > \varepsilon\})$.

Now, we take $\psi \in C_0^\infty(B_1)$ and $\zeta \in C^\infty(\mathbb{R}, [0,1])$ to be even with $\zeta = 0$ in $[-1,1]$ and $\zeta = 1$ in $\mathbb{R} \setminus (-2,2)$. Let $\zeta_\varepsilon(x) := \zeta\left(\frac{x_n}{\varepsilon}\right)$. Using (5.3.8) with $\Phi := \zeta_\varepsilon\psi$, we note that

$$\begin{aligned}
&\left|\int_{\mathbb{R}^n} F(x)\psi(x)\,dx + \int_{\mathbb{R}^n} \nabla U(x) \cdot \nabla\psi(x)\,dx\right| \\
&\leqslant \left|\int_{\mathbb{R}^n} F(x)\zeta_\varepsilon(x)\psi(x)\,dx + \int_{\mathbb{R}^n} \nabla U(x) \cdot \nabla\big(\zeta_\varepsilon(x)\psi(x)\big)\,dx\right| \\
&\quad + \int_{\mathbb{R}^n} \big|1 - \zeta_\varepsilon(x)\big|\,|F(x)|\,|\psi(x)|\,dx \\
&\quad + \left|\int_{\mathbb{R}^n} \nabla U(x) \cdot \nabla\Big(\big(1 - \zeta_\varepsilon(x)\big)\psi(x)\Big)\,dx\right| \\
&\leqslant 0 + \int_{\mathbb{R}^n \cap \{|x_n| < 2\varepsilon\}} |F(x)|\,|\psi(x)|\,dx \\
&\quad + \left|\int_{\mathbb{R}^n \cap \{|x_n| < 2\varepsilon\}} \nabla U(x) \cdot \nabla\Big(\big(1 - \zeta_\varepsilon(x)\big)\psi(x)\Big)\,dx\right|
\end{aligned}$$

$$\begin{aligned}
&\leqslant C\|F\|_{L^p(\{|x_n|<2\varepsilon\})} + \int_{\mathbb{R}^n\cap\{|x_n|<2\varepsilon\}} |\nabla U(x)|\,|\nabla\psi(x)|\,dx \\
&\quad + \left|\int_{\mathbb{R}^n\cap\{|x_n|<2\varepsilon\}} \partial_n U(x)\partial_n\zeta_\varepsilon(x)\psi(x)\,dx\right| \\
&\leqslant C\|F\|_{L^p(\{|x_n|<2\varepsilon\})} + C\|\nabla U\|_{L^p(\{|x_n|<2\varepsilon\})} \\
&\quad + \left|\int_{\mathbb{R}^n\cap\{x_n\in(0,2\varepsilon)\}} \partial_n U(x)\partial_n\zeta_\varepsilon(x)\psi(x)\,dx\right. \\
&\qquad \left. + \int_{\mathbb{R}^n\cap\{x_n\in(-2\varepsilon,0)\}} \partial_n U(x',-x_n)\partial_n\zeta_\varepsilon(x)\psi(x)\,dx\right|.
\end{aligned}$$

Taking the limit as $\varepsilon \searrow 0$ and noting that

$$\partial_n\zeta_\varepsilon(x) = \frac{1}{\varepsilon}\zeta'\left(\frac{x_n}{\varepsilon}\right),$$

recalling also that ζ is even, whence ζ' is odd, we find that

$$\begin{aligned}
&\left|\int_{\mathbb{R}^n} F(x)\psi(x)\,dx + \int_{\mathbb{R}^n} \nabla U(x)\cdot\nabla\psi(x)\,dx\right| \\
&\leqslant \lim_{\varepsilon\searrow 0}\frac{1}{\varepsilon}\left|\int_{\mathbb{R}^n\cap\{x_n\in(0,2\varepsilon)\}} \partial_n U(x)\zeta'\left(\frac{x_n}{\varepsilon}\right)\psi(x)\,dx\right. \\
&\qquad \left. - \int_{\mathbb{R}^n\cap\{x_n\in(0,2\varepsilon)\}} \partial_n U(x',x_n)\zeta'\left(\frac{x_n}{\varepsilon}\right)\psi(x',-x_n)\,dx\right| \\
&\leqslant \lim_{\varepsilon\searrow 0}\frac{C}{\varepsilon}\int_{\mathbb{R}^n\cap\{x_n\in(0,2\varepsilon)\}} |\partial_n U(x)|\,\big|\psi(x',x_n)-\psi(x',-x_n)\big|\,dx \\
&\leqslant \lim_{\varepsilon\searrow 0}\int_{\mathbb{R}^n\cap\{x_n\in(0,2\varepsilon)\}} |\partial_n U(x)|\,dx \\
&\leqslant \lim_{\varepsilon\searrow 0}\|\nabla U\|_{L^p(\mathbb{R}^n\cap\{x_n\in(0,2\varepsilon)\})} \\
&= 0.
\end{aligned}$$

As a consequence,

$$\int_{\mathbb{R}^n} F(x)\psi(x)\,dx = -\int_{\mathbb{R}^n} \nabla U(x)\cdot\nabla\psi(x)\,dx. \tag{5.3.9}$$

Given $\eta > 0$, we take a mollifier τ_η and the corresponding mollification $U_\eta := U * \tau_\eta \in C_0^\infty(B_{8/9})$ of U. Let also $F_\eta := F * \tau_\eta$. By (5.3.9), for all $\psi \in C_0^\infty(B_{7/8})$, letting $\psi_\eta := \psi * \tau_\eta$ (which is supported in B_1 for η small enough), we have that

$$\begin{aligned}
\int_{\mathbb{R}^n} F_\eta(x)\psi(x)\,dx &= \iint_{\mathbb{R}^n\times\mathbb{R}^n} F(y)\tau_\eta(x-y)\psi(x)\,dx\,dy = \int_{\mathbb{R}^n} F(y)\psi_\eta(y)\,dx \\
&= -\int_{\mathbb{R}^n} \nabla U(y)\cdot\nabla\psi_\eta(y)\,dy = -\int_{\mathbb{R}^n} \nabla U_\eta(x)\cdot\nabla\psi(x)\,dx.
\end{aligned}$$

From this and the fact that U_η is smooth, we arrive at $\Delta U_\eta = F_\eta$ in $B_{7/8}$. We can therefore invoke the interior regularity estimate in Theorem 5.3.1 (or even the simpler formulation given in Corollary 5.2.9) and deduce that

$$\|U_\eta\|_{W^{2,p}(B_{1/2})} \leqslant C\left(\|U_\eta\|_{L^p(B_{7/8})} + \|F_\eta\|_{L^p(B_{7/8})}\right). \tag{5.3.10}$$

Also, since $\Delta(U_\eta - U_{\eta'}) = F_\eta - F_{\eta'}$, we obtain in the same way that

$$\|U_\eta - U_{\eta'}\|_{W^{2,p}(B_{1/2})} \leqslant C\left(\|U_\eta - U_{\eta'}\|_{L^p(B_{7/8})} + \|F_\eta - F_{\eta'}\|_{L^p(B_{7/8})}\right). \tag{5.3.11}$$

Since $U_\eta \to U$ and $F_\eta \to F$ in $L^p(B_{7/8})$ (see e.g. [WZ15, Theorem 9.6]), we deduce from (5.3.11) that $U_\eta \to U$ in $W^{2,p}(B_{1/2})$. This and (5.3.10) yield that

$$\|U\|_{W^{2,p}(B_{1/2})} \leqslant C\left(\|U\|_{L^p(B_{7/8})} + \|F\|_{L^p(B_{7/8})}\right). \tag{5.3.12}$$

Observe in addition that $U = u$ in $B_{1/2}$ and $|U| \leqslant |u|$. Accordingly, we deduce from (5.3.12) that

$$\begin{aligned}\|u\|_{W^{2,p}(B_{1/2})} &\leqslant C\left(\|u\|_{L^p(B_{7/8})} + \|F\|_{L^p(B_{7/8})}\right)\\ &\leqslant C\left(\|u\|_{L^p(B_{7/8})} + \|f\|_{L^p(B_{7/8})} + \|\nabla u\|_{L^p(B_{7/8})}\right).\end{aligned}$$

Consequently, by interpolation (recall (5.2.54)),

$$\|u\|_{W^{2,p}(B_{1/2})} \leqslant C\left(\|u\|_{L^p(B_{7/8})} + \|f\|_{L^p(B_{7/8})}\right) + \varepsilon_0\|D^2u\|_{L^p(B_{7/8})},$$

with ε_0 as in Proposition 5.3.2. By scaling and using Proposition 5.3.2, we conclude that

$$\|u\|_{W^{2,p}(B_{1/2})} \leqslant C\left(\|u\|_{L^p(B_1)} + +\|f\|_{L^p(B_1)}\right),$$

which yields the desired result in (5.3.5) in this case.

This completes the proof of Theorem 5.3.3 when the operator is the Laplacian and g vanishes identically (and therefore in its full generality, thanks to (5.3.6) and (5.3.7)). □

We remark that when g in (5.3.4) is not sufficiently regular, then the local estimate at the boundary in the setting of Theorem 5.3.3 may fail. As an example, one can consider the angular function in the halfplane $\mathbb{R}^2_+ := \mathbb{R} \times (0, +\infty)$ given by

$$u(x) = u(x_1, x_2) := \arccos \frac{x_1}{\sqrt{x_1^2 + x_2^2}}, \tag{5.3.13}$$

see Figure 5.3.1.

We note that $u \in C^2(\mathbb{R}^2_+)$ and $\Delta u = 0$ in $\mathbb{R}^2_+$.

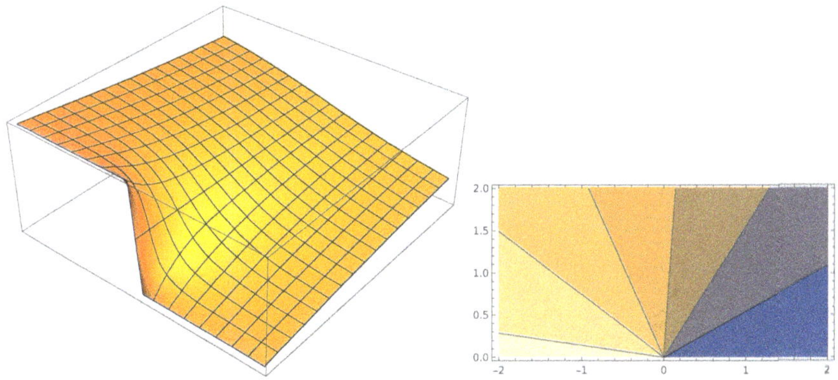

Figure 5.3.1 Graph of the function u in (5.3.13) and of its level sets.

Additionally,

$$\lim_{x_2 \searrow 0} u(x) = \arccos \frac{x_1}{|x_1|} = \begin{cases} 0 & \text{if } x_1 > 0, \\ \pi & \text{if } x_1 < 0. \end{cases}$$

But the estimate in the thesis of Theorem 5.3.3 does not hold since, for some $c > 0$,

$$\begin{aligned} \|u\|^p_{W^{2,p}(B^+_{1/2})} &\geqslant \int_{B^+_{1/2}} |\partial_{11} u(x)|^p \, dx = \int_{B^+_{1/2}} \left| \frac{2x_1 \, |x_2|}{|x|^4} \right|^p dx \\ &= 2^p \int_{B^+_{1/2}} \frac{|x_1|^p \, |x_2|^p}{|x|^{4p}} \, dx = 2^{p-1} \int_{B_{1/2}} \frac{|x_1|^p \, |x_2|^p}{|x|^{4p}} \, dx \\ &= 2^{p-1} \iint_{(0,1/2)\times(\partial B_1)} \rho^{1-2p} \, |\omega_1|^p \, |\omega_2|^p \, d\rho \, d\mathcal{H}^1_\omega \\ &= c \int_0^{1/2} \rho^{1-2p} \, d\rho, \end{aligned}$$

which diverges for all $p \in (1, +\infty)$. See Figure 5.3.2 for a plot of $\partial_{11} u$ and its level sets.

We now recall a global, up to the boundary, version of the regularity estimates in the L^p class that follows from interior and local boundary estimates. This can be seen as the natural counterpart in Lebesgue spaces of Theorems 4.2.10 and 4.5.1.

Theorem 5.3.4. *Let $\Omega \subset \mathbb{R}^n$ be a bounded open set with boundary of class $C^{1,1}$. Let $a_{ij} \in C(\overline{\Omega})$ satisfy the ellipticity condition in (4.4.3). Let b_i, $c \in L^\infty(\Omega)$. Let $f \in L^p(\Omega)$ and $g \in W^{2,p}(\Omega)$ for some $p \in (1, +\infty)$.*

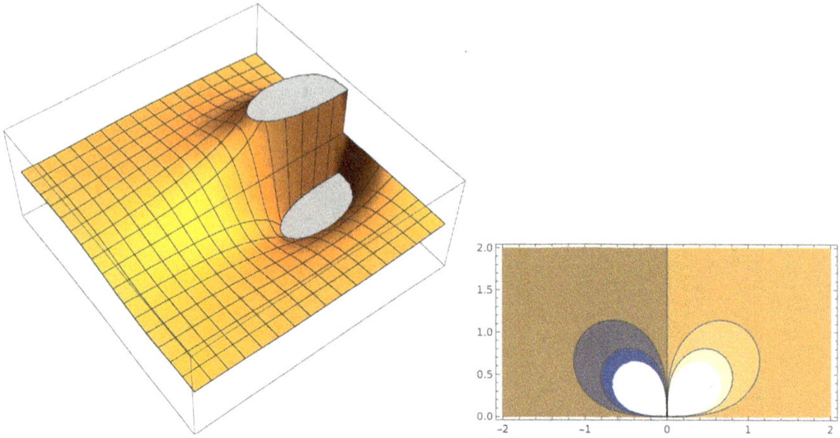

Figure 5.3.2 Graph of the function $\partial_{11}u$ and of its level sets, with u as in (5.3.13).

Let $u \in W^{2,p}(\Omega)$ be a solution of

$$\begin{cases} \displaystyle\sum_{i,j=1}^{n} a_{ij}\partial_{ij}u + \sum_{i=1}^{n} b_i\partial_i u + cu = f & a.e.\ in\ \Omega, \\ u = g & on\ \partial\Omega. \end{cases}$$

Then, there exists $C > 0$, depending only[4] *on n, p, Ω, a_{ij}, b_i and c, such that*

$$\|u\|_{W^{2,p}(\Omega)} \leqslant C\Big(\|g\|_{W^{2,p}(\Omega)} + \|f\|_{L^p(\Omega)} + \|u\|_{L^p(\Omega)}\Big). \tag{5.3.14}$$

Proof. This argument is a variation of that presented in the proof of Theorem 4.5.1, using the regularity results in Lebesgue spaces in place of those in Hölder spaces. Namely, we straighten the boundary of Ω by taking $\rho > 0$ a finite family of balls $\{B_\rho(p_i)\}_{i\in\{1,\dots,N\}}$ with $p_i \in \partial\Omega$ such that

$$\mathcal{N} := \{x \in \Omega \text{ s.t. } B_{\rho/8}(x) \cap (\partial\Omega) \neq \varnothing\} \subseteq \bigcup_{i=1}^{N} B_{\rho/4}(p_i)$$

such that $B_\rho(p_i) \cap \Omega$ is equivalent to B_ρ^+ via a diffeomorphism T_i of class $C^{1,1}$ with the inverse of class $C^{1,1}$.

Recalling (4.4.9), we see that the function $\widetilde{u}_i(y) := u(T_i^{-1}(y))$ satisfies in B_ρ^+ the same type of equation that u satisfies in $B_\rho(p_i) \cap \Omega$, (with structural constants

[4] Actually, browsing through the proof, one can be more precise by saying that the dependence of C here is on n, p, the $C^{1,1}$ regularity parameters of Ω, the modulus of continuity of a_{ij} and its ellipticity constants, $\|b_i\|_{L^\infty(\Omega)}$ and $\|c\|_{L^\infty(\Omega)}$.

of the coefficients possibly multiplied by a different constant as well). Thus, we employ the local estimates at the boundary in (5.3.5), obtaining that

$$\|\widetilde{u}_i\|_{W^{2,p}(B^+_{\rho/2})} \leqslant C\left(\|u\|_{L^p(\Omega)} + \|g\|_{W^{2,p}(\Omega)} + \|f\|_{L^p(\Omega)}\right)$$

and consequently

$$\|u\|_{W^{2,p}(\Omega\cap B_{\rho/2}(p_i))} \leqslant C\left(\|u\|_{L^p(\Omega)} + \|g\|_{W^{2,p}(\Omega)} + \|f\|_{L^p(\Omega)}\right),$$

which in turn leads to

$$\|u\|_{W^{2,p}(\mathcal{N})} \leqslant C\left(\|u\|_{L^p(\Omega)} + \|g\|_{W^{2,p}(\Omega)} + \|f\|_{L^p(\Omega)}\right), \tag{5.3.15}$$

up to renaming constants from line to line.

Now, we consider a finite family of balls such that

$$\mathcal{J} := \{x \in \Omega \text{ s.t. } B_{\rho/9}(x) \cap (\partial\Omega) = \varnothing\} \subseteq \bigcup_{i=1}^{M} B_{\rho/4}(q_i), \tag{5.3.16}$$

for suitable $q_i \in \Omega$. By the interior estimates in Theorem 5.3.1,

$$\|u\|_{W^{2,p}(B_{\rho/4}(q_i))} \leqslant C\left(\|u\|_{L^p(\Omega)} + \|f\|_{L^p(\Omega)}\right).$$

This and (5.3.16) yield that

$$\|u\|_{W^{2,p}(\mathcal{J})} \leqslant C\left(\|u\|_{L^p(\Omega)} + \|f\|_{W^{2,p}(\Omega)}\right).$$

This, together with (5.3.15), yields the desired result. □

It is worth noting that (5.3.14) is the counterpart in Lebesgue spaces of the global estimate in Hölder spaces established in (4.5.1). For the sake of completeness, we mention that a counterpart of (4.5.2) holds true as well; namely, in the setting of Theorem 5.3.4, if $c(x) \leqslant 0$ for a.e. $x \in \Omega$, then

$$\|u\|_{W^{2,p}(\Omega)} \leqslant C\left(\|g\|_{W^{2,p}(\Omega)} + \|f\|_{L^p(\Omega)}\right), \tag{5.3.17}$$

see e.g. [GT01, Lemma 9.17]. This improvement of (5.3.14) is very delicate from a technical point of view[5] since it relies on a maximum principle which is conceptually very different from that in Corollary 4.5.3, which was used to

[5]For the case of the Laplace operator, we observe that (5.3.17) can be obtained by compactness via the standard maximum principle by arguing as follows. By possibly replacing u by $u - g$, we can reduce ourselves to the case in which g vanishes identically. Then, if (5.3.17) were violated, there would exist sequences $u_j \in W^{2,p}(\Omega) \cap W^{1,p}_0(\Omega)$ and $f_j \in L^p(\Omega)$ such that $\Delta u_j = f_j$ in Ω and $\|u_j\|_{L^p(\Omega)} > 2j\|f_j\|_{L^p(\Omega)}$. This and (5.3.14) yield that

$$2j\|f_j\|_{L^p(\Omega)} < \|u_j\|_{W^{2,p}(\Omega)} \leqslant C\left(\|f_j\|_{L^p(\Omega)} + \|u_j\|_{L^p(\Omega)}\right)$$

and therefore $\|u_j\|_{L^p(\Omega)} > j\|f_j\|_{L^p(\Omega)}$ for j large enough.

prove (4.5.2) in the Hölder spaces framework: indeed, the additional difficulty for solutions in $W^{2,p}$ is that they do not need to satisfy the equation at every point of the domain, making the classical maximum principles and the barrier methods unsuitable. To overcome this problem, one needs to develop the so-called Aleksandrov–Bakel′man–Pucci maximum principle [Bak61, Ale61, Puc66b], see e.g. [GT01, Theorems 9.1, 9.5 and 9.15], which is essentially the main technical tool to obtain (5.3.17).

From the global estimate in (5.3.17), one also obtains the following existence (and uniqueness, thanks again to the Aleksandrov–Bakel′man–Pucci maximum principle) result for the Dirichlet problem in Lebesgue spaces.

Theorem 5.3.5. *Let $\Omega \subset \mathbb{R}^n$ be a bounded open set with boundary of class $C^{1,1}$. Let $a_{ij} \in C(\overline{\Omega})$ satisfy the ellipticity condition in* (4.4.3). *Let b_i, $c \in L^\infty(\Omega)$, with $c(x) \leqslant 0$ for a.e. $x \in \Omega$. Let $f \in L^p(\Omega)$ and $g \in W^{2,p}(\Omega)$ for some $p \in (1, +\infty)$.*

Thus, setting $\widetilde{u}_j := \frac{u_j}{\|u_j\|_{L^p(\Omega)}}$ and $\widetilde{f}_j := \frac{f_j}{\|u_j\|_{L^p(\Omega)}}$, noting that

$$\|\widetilde{u}_j\|_{W^{2,p}(\Omega)} = \frac{\|u_j\|_{W^{2,p}(\Omega)}}{\|u_j\|_{L^p(\Omega)}} \leqslant \frac{C\left(\|f_j\|_{L^p(\Omega)} + \|u_j\|_{L^p(\Omega)}\right)}{\|u_j\|_{L^p(\Omega)}} \leqslant C,$$

we find that $\widetilde{u}_j \in W^{2,p}(\Omega) \cap W^{1,p}_0(\Omega)$ solves $\Delta\widetilde{u}_j = \widetilde{f}_j$ in Ω, with $\|\widetilde{u}_j\|_{W^{2,p}(\Omega)} = 1$ and $\|\widetilde{f}_j\|_{L^p(\Omega)} < \frac{1}{j}$. Consequently, up to a subsequence, $\widetilde{u}_j$ converges as $j \to +\infty$ to some $\widetilde{u}$ strongly in $W^{1,p}_0(\Omega)$ and weakly in $W^{2,p}(\Omega)$, and $\widetilde{f}_j \to 0$ in $L^p(\Omega)$.

As a result, for all $\varphi \in C^\infty_0(\Omega)$,

$$\int_\Omega \Delta\widetilde{u}(x)\,\varphi(x)\,dx = \lim_{j\to+\infty}\int_\Omega \Delta\widetilde{u}_j(x)\,\varphi(x)\,dx = \lim_{j\to+\infty}\int_\Omega \widetilde{f}_j(x)\,\varphi(x)\,dx = 0.$$

This gives that for all $\varphi \in C^\infty_0(\Omega)$

$$\int_\Omega \nabla\widetilde{u}(x)\cdot\nabla\varphi(x)\,dx = 0.$$

Taking a sequence of functions $\varphi_k \in C^\infty_0(\Omega)$ such that $\varphi_k \to \widetilde{u}$ in $W^{1,p}_0(\Omega)$, we thereby conclude that

$$\int_\Omega |\nabla\widetilde{u}(x)|^2\,dx = 0$$

and accordingly $\widetilde{u}$ vanishes identically.

But this is a contradiction with the fact that

$$\|\widetilde{u}\|_{L^p(\Omega)} = \lim_{j\to+\infty}\|\widetilde{u}_j\|_{L^p(\Omega)} = 1,$$

and this establishes (5.3.17) for the Laplace operator.

The same argument would mainly carry over if one possessed a convenient maximum principle for the operator under consideration, see the proof of Lemma 9.17 in [GT01].

Then, the Dirichlet problem

$$\begin{cases} \sum_{i,j=1}^{n} a_{ij}\partial_{ij}u + \sum_{i=1}^{n} b_i\partial_i u + cu = f & \text{a.e. in } \Omega, \\ u = g & \text{on } \partial\Omega \end{cases} \tag{5.3.18}$$

admits a unique solution $u \in W^{2,p}(\Omega)$.

Moreover, there exists $C > 0$, depending only on n, p, Ω, a_{ij}, b_i and c, such that

$$\|u\|_{W^{2,p}(\Omega)} \leqslant C\left(\|g\|_{W^{2,p}(\Omega)} + \|f\|_{L^p(\Omega)}\right). \tag{5.3.19}$$

See also [GT01, Section 9.6] for further details on the Dirichlet problem in Lebesgue spaces.

We also mention that a higher regularity result that can be considered the counterpart of Theorem 4.5.9 in the Lebesgue spaces setting holds true: see e.g. [GT01, Theorem 9.19] for a regularity theory in the class $W^{k,p}$.

As an interesting application of the Calderón–Zygmund estimates (in fact, of the technically simpler part dealing just with estimates in class $W^{1,p}$), we present now a variant of Proposition 3.2.20 which does not require vanishing boundary conditions.

Proposition 5.3.6. *Let $n \geqslant 2$ and Ω be a bounded and open subset of $\mathbb{R}^n$.*

Let $p > \frac{n}{2}$ and $f : \Omega \to \mathbb{R}$ with $f \in L^p(\Omega)$.

Let $u \in C^2(\Omega) \cap L^1(\Omega)$ be a solution of

$$\Delta u = f \quad \text{in } \Omega.$$

Then, for every open set $\Omega' \Subset \Omega$,

$$\sup_{\Omega'} |u| \leqslant C\left(\|u\|_{L^1(\Omega)} + \|f\|_{L^p(\Omega)}\right), \tag{5.3.20}$$

for some $C > 0$ depending only on n, p, Ω and Ω'.

Proof. Up to reducing to connected components, we implicitly assume here that the domains under consideration are connected. We consider open sets Ω'' and Ω''' with smooth boundaries and such that $\Omega' \Subset \Omega'' \Subset \Omega''' \Subset \Omega$. We pick $\phi \in C_0^\infty(\Omega''', [0,1])$ with $\phi = 1$ in Ω'' and define $v := \phi u$.

In this way, $v = 0$ on $\partial\Omega'''$ and, in Ω''',

$$\Delta v = \Delta\phi u + 2\nabla\phi \cdot \nabla u + \phi\Delta u =: g.$$

Thus, we can employ Proposition 3.2.20 and deduce that

$$\sup_{\Omega''} v \leqslant C\|g^-\|_{L^p(\Omega''')}. \tag{5.3.21}$$

We stress that, up to renaming C at each level of our calculations,

$$\begin{aligned} g^-(x) &= \max\Big\{0, -\Delta\phi(x)u(x) - 2\nabla\phi(x)\cdot\nabla u(x) - \phi(x)\Delta u(x)\Big\} \\ &\leqslant C\Big(|u(x)| + |\nabla u(x)| + |\Delta u(x)|\Big) \end{aligned}$$

and therefore

$$\|g^-\|_{L^p(\Omega''')} \leqslant C\Big(\|u\|_{L^p(\Omega''')} + \|\nabla u\|_{L^p(\Omega''')} + \|\Delta u\|_{L^p(\Omega''')}\Big).$$

But taking into consideration Theorem 5.3.1 (in its simpler formulation for first derivatives) up to a covering argument, we also know that

$$\|\nabla u\|_{L^p(\Omega''')} \leqslant C\Big(\|u\|_{L^p(\Omega''')} + \|\Delta u\|_{L^p(\Omega''')}\Big)$$

and consequently

$$\|g^-\|_{L^p(\Omega''')} \leqslant C\Big(\|u\|_{L^p(\Omega''')} + \|\Delta u\|_{L^p(\Omega''')}\Big).$$

This inequality, (5.3.21) and the fact that $v = u$ in Ω'' give that

$$\sup_{\Omega''} u \leqslant C\Big(\|u\|_{L^p(\Omega''')} + \|\Delta u\|_{L^p(\Omega''')}\Big).$$

Applying this to $-u$ in lieu of u, we conclude that[6]

$$\sup_{\Omega''} |u| \leqslant C\Big(\|u\|_{L^p(\Omega''')} + \|\Delta u\|_{L^p(\Omega''')}\Big). \tag{5.3.22}$$

This establishes (5.3.20) with the L^1-norm of the solution replaced by its L^p-norm.

To obtain (5.3.20) in its full generality, we proceed as follows. Given $R > r > 0$, we can rescale and apply (5.3.22) to balls of radii $R - r$ centered at points of B_r, finding that

$$\sup_{B_r} |u| \leqslant \frac{C}{(R-r)^{\frac{n}{p}}}\Big(\|u\|_{L^p(B_R)} + (R-r)^2\|\Delta u\|_{L^p(B_R)}\Big). \tag{5.3.23}$$

Now, we claim that for every $\varepsilon > 0$, there exists $C_\varepsilon > 0$ such that

$$\|u\|_{L^p(B_R)} \leqslant \varepsilon(R-r)^{\frac{n}{p}} \sup_{B_R} |u| + \frac{C_\varepsilon \|u\|_{L^1(B_R)}}{(R-r)^{\frac{n(p-1)}{p}}}. \tag{5.3.24}$$

Indeed, we have that

$$\|u\|_{L^p(B_R)} \leqslant \left(\sup_{B_R} |u|^{p-1} \int_{B_R} |u(x)|\,dx\right)^{\frac{1}{p}},$$

and the desired claim follows from Young's inequality with exponents $\frac{p}{p-1}$ and p

---

[6]A quick way to obtain (5.3.22) would also involve using Theorem 5.3.1 in its full standing (rather than its simpler form for first derivatives), obtaining second-derivative estimates such as

$$\|u\|_{W^{2,p}(\Omega'')} \leqslant C\Big(\|u\|_{L^p(\Omega''')} + \|\Delta u\|_{L^p(\Omega''')}\Big).$$

Since $p > \frac{n}{2}$, this and the Sobolev embedding theorem (see e.g. [Eva98, Theorem 6(ii) on p. 270]) lead to (5.3.22).

As a result, if

$$\varphi(\rho) := \sup_{B_\rho} |u|,$$

by choosing $\varepsilon > 0$ conveniently small, we deduce from (5.3.23) and (5.3.24) that

$$\varphi(r) \leqslant \frac{\varphi(R)}{2} + \frac{C}{(R-r)^n}\|u\|_{L^1(B_R)} + \frac{C}{(R-r)^{\frac{n}{p}-2}}\|f\|_{L^p(B_R)}.$$

Since we are dealing with local estimates, we can suppose that $R \leqslant R_0$ for some $R_0 > 0$; therefore, $(R-r)^{\frac{n}{p}-2} = (R-r)^{n-\frac{n(p-1)}{p}-2} \geqslant R_0^{-\frac{n(p-1)}{p}-2}(R-r)^n$, yielding that

$$\varphi(r) \leqslant \frac{\varphi(R)}{2} + \frac{C\big(\|u\|_{L^1(B_R)} + \|f\|_{L^p(B_R)}\big)}{(R-r)^n}.$$

We are thereby in a position to use Lemma 2.18.2 and infer that

$$\sup_{B_r} |u| = \varphi(r) \leqslant \frac{C}{(R-r)^n}\Big(\|u\|_{L^1(B_R)} + \|f\|_{L^p(B_R)}\Big).$$

From this, a covering argument gives (5.3.20), as desired. □

A consequence of the above result is the following[7] observation.

Corollary 5.3.7. *Let $n \geqslant 2$. Let Ω, Ω' be bounded open sets of $\mathbb{R}^n$ with $\Omega' \Subset \Omega$. Let $a \in L^r(\Omega)$ for some $r > \frac{n}{2}$.*

Let $u \in C^2(\Omega) \cap L^1(\Omega)$ be a solution of $\Delta u = au$ in Ω.

Then, there exists a positive constant C, depending only on n, r, Ω, Ω' and $\|a\|_{L^r(\Omega)}$, such that

$$\|u\|_{L^\infty(\Omega')} \leqslant C\|u\|_{L^1(\Omega)}.$$

Proof. Up to a covering argument, we deal with balls, pick $p \in \left(\frac{n}{2}, r\right)$, revisit (5.3.23) in the present setting and use the Hölder inequality with exponents $\frac{r}{p} > 1$ and $\frac{r}{r-p}$ to see that, given $R > r > 0$,

$$\begin{aligned}\sup_{B_r} |u| &\leqslant \frac{C}{(R-r)^{\frac{n}{p}}}\Big(\|u\|_{L^p(B_R)} + (R-r)^2\|au\|_{L^p(B_R)}\Big)\\ &\leqslant \frac{C}{(R-r)^{\frac{n}{p}}}\Big(\|u\|_{L^p(B_R)} + (R-r)^2\|a\|_{L^r(B_R)}\|u\|_{L^{\frac{rp}{r-p}}(B_R)}\Big). \end{aligned} \tag{5.3.25}$$

[7] Interestingly, results such as the one in Corollary 5.3.7 are usually obtained using energy methods, relying on divergence form operators, while the approach presented here is of nondivergence type.

We also utilize (5.3.24) with p replaced by $\frac{rp}{r-p}$, and we find that

$$\|u\|_{L^{\frac{rp}{r-p}}(B_R)} \leqslant \varepsilon(R-r)^{\frac{n(r-p)}{rp}} \sup_{B_R}|u| + \frac{C_\varepsilon \|u\|_{L^1(B_R)}}{(R-r)^{\frac{n(rp-r+p)}{rp}}}. \tag{5.3.26}$$

We also remark that

$$2 + \frac{n(r-p)}{rp} - \frac{n}{p} = \frac{2r-n}{r} > 0.$$

On this account, combining (5.3.23), (5.3.25) and (5.3.26), assuming $R \leqslant R_0$ for some given $R_0 > 0$ and defining

$$\varphi(\rho) := \sup_{B_\rho}|u|,$$

up to redefining constants, we conclude that

$$\begin{aligned}
\varphi(r) &= \sup_{B_r}|u| \\
&\leqslant \frac{C}{(R-r)^{\frac{n}{p}}}\left(\|u\|_{L^p(B_R)} + (R-r)^2\|a\|_{L^r(B_R)}\|u\|_{L^{\frac{rp}{r-p}}(B_R)}\right) \\
&\leqslant \frac{C}{(R-r)^{\frac{n}{p}}}\left[\varepsilon(R-r)^{\frac{n}{p}}\sup_{B_R}|u| + \frac{C_\varepsilon\|u\|_{L^1(B_R)}}{(R-r)^{\frac{n(p-1)}{p}}}\right. \\
&\quad \left. + (R-r)^2\left(\varepsilon(R-r)^{\frac{n(r-p)}{rp}}\sup_{B_R}|u| + \frac{C_\varepsilon\|u\|_{L^1(B_R)}}{(R-r)^{\frac{n(rp-r+p)}{rp}}}\right)\right] \\
&\leqslant C\left(\varepsilon\sup_{B_R}|u| + \frac{\|u\|_{L^1(B_R)}}{(R-r)^n}\right) \\
&\leqslant \frac{\varphi(R)}{2} + \frac{C\|u\|_{L^1(B_R)}}{(R-r)^n},
\end{aligned}$$

as long as we choose ε appropriately small.

We are thereby in a position to use Lemma 2.18.2 and infer that

$$\sup_{B_r}|u| = \varphi(r) \leqslant \frac{C\|u\|_{L^1(B_R)}}{(R-r)^n},$$

from which we obtain the desired result. □

Chapter 6

The Dirichlet Problem in the Light of Capacity Theory

In this chapter, we recall the notion of capacity and exploit it to determine the cases in which the solution to the Dirichlet problem obtained using Perron methods (recall that Corollary 3.3.4) is continuous up to the boundary. This is indeed a delicate issue in light of counterexamples such as the Lebesgue spine discussed on p. 288, see [Kel67, Exercise 10, p. 334].

6.1 Capacitance and Capacity

The mathematical concept of capacity was introduced by Gustave Choquet [Cho53]; see also [Cho86] for a historical account on the creation and development of this theory.

Different versions of the capacity theory are available in the literature. Here, we present a treatment that we find sufficiently close to the physical intuition (though a number of technical details, especially when dealing with general sets, appear to be unavoidable).

Given a bounded and open set $\Omega \subset \mathbb{R}^n$ with boundary of class $C^{2,\alpha}$, for some $\alpha \in (0, 1)$, we define the capacity of Ω as

$$\operatorname{Cap}(\Omega) := \inf_{\substack{v \in C_0^\infty(\mathbb{R}^n) \\ v=1 \text{ in } \Omega}} \int_{\mathbb{R}^n \setminus \overline{\Omega}} |\nabla v(x)|^2 \, dx. \tag{6.1.1}$$

Using the density results in Sobolev spaces (see e.g. [Leo09, Theorem 10.29]), one can also rephrase (6.1.1) in the form

$$\operatorname{Cap}(\Omega) = \inf_{\substack{v \in \mathcal{D}^{1,2}(\mathbb{R}^n) \\ v=1 \text{ in } \Omega}} \int_{\mathbb{R}^n \setminus \overline{\Omega}} |\nabla v(x)|^2 \, dx. \tag{6.1.2}$$

This definition of capacity, which we adopted in (6.1.2), is meaningful only when $n \geqslant 3$, as explained in the following lemma.

Lemma 6.1.1. *Let $n \in \{1, 2\}$. Then,* $\mathrm{Cap}(\Omega) = 0$ *for every bounded and open set $\Omega \subset \mathbb{R}^n$ with boundary of class $C^{2,\alpha}$ for some $\alpha \in (0, 1)$.*

Proof. Let $n = 1$ and $\varepsilon > 0$. Let $\Omega \Subset (-R, R)$ for some $R > 0$ and

$$v_\varepsilon(x) = \begin{cases} 1 & \text{if } x \in [-R, R], \\ \varepsilon\left(R + \dfrac{1}{\varepsilon} - |x|\right) & \text{if } |x| \in \left(R, R + \dfrac{1}{\varepsilon}\right), \\ 0 & \text{if } |x| \in \left[R + \dfrac{1}{\varepsilon}, +\infty\right). \end{cases}$$

Then, by (6.1.2),

$$\mathrm{Cap}(\Omega) \leqslant \int_{\mathbb{R}\setminus\overline{\Omega}} |v'_\varepsilon(x)|^2\, dx = 2\varepsilon.$$

Hence, taking ε as small as we wish, we conclude that $\mathrm{Cap}(\Omega) = 0$.

Let now set $n = 2$ and take $\varepsilon > 0$ so small that $\Omega \Subset B_{1/\varepsilon}$. Let also

$$w_\varepsilon(x) = \begin{cases} 1 & \text{if } |x| \in \left[0, \dfrac{1}{\varepsilon}\right], \\ \dfrac{2\ln\varepsilon - \ln|x|}{\ln\varepsilon} & \text{if } |x| \in \left(\dfrac{1}{\varepsilon}, \dfrac{1}{\varepsilon^2}\right), \\ 0 & \text{if } |x| \in \left[\dfrac{1}{\varepsilon^2}, +\infty\right). \end{cases}$$

Then, by (6.1.2), we have

$$\begin{aligned} \mathrm{Cap}(\Omega) &\leqslant \int_{\mathbb{R}^2\setminus\overline{\Omega}} |\nabla w_\varepsilon(x)|^2\, dx = \frac{1}{\ln^2\varepsilon} \int_{B_{1/\varepsilon^2}\setminus B_{1/\varepsilon}} \frac{dx}{|x|^2} \\ &= \frac{2\pi}{\ln^2\varepsilon} \int_{1/\varepsilon}^{1/\varepsilon^2} \frac{d\rho}{\rho} = \frac{2\pi}{|\ln\varepsilon|}. \end{aligned}$$

Taking ε as small as we wish, we conclude that $\mathrm{Cap}(\Omega) = 0$ in this case too. □

We now clarify the link between the notion of capacity and that of harmonic functions vanishing at infinity and with prescribed value along the boundary of a domain.

Proposition 6.1.2. *Let $n \geqslant 3$ and $\Omega \subset \mathbb{R}^n$ be a bounded and open set with boundary of class $C^{2,\alpha}$ for some $\alpha \in (0, 1)$.*

Then, there exists a unique function $u \in C^2(\mathbb{R}^n \setminus \Omega) \cap \mathcal{D}^{1,2}(\mathbb{R}^n \setminus \Omega)$ such that

$$\begin{cases} \Delta u = 0 & \text{in } \mathbb{R}^n \setminus \overline{\Omega}, \\ u = 1 & \text{on } \partial\Omega. \end{cases}$$

Additionally, $0 \leqslant u(x) \leqslant 1$ *for all* $x \in \mathbb{R}^n \setminus \overline{\Omega}$,

$$\lim_{|x|\to+\infty} u(x) = 0 \tag{6.1.3}$$

and

$$\operatorname{Cap}(\Omega) = \int_{\mathbb{R}^n\setminus\overline{\Omega}} |\nabla u(x)|^2\, dx. \tag{6.1.4}$$

With a slight abuse of notation, we may implicitly assume that the function u in Proposition 6.1.2 is defined in all $\mathbb{R}^n$, simply by setting u to be constantly equal to 1 in Ω.

Proof of Proposition 6.1.2. Let $k_0 \in \mathbb{N}$ be such that $B_{k_0} \Supset \Omega$. Let $v \in C_0^\infty(B_{k_0})$ such that $v = 1$ in Ω. For every $k \in \mathbb{N} \cap [k_0, +\infty)$, we consider the Sobolev space X_k of the functions $w \in \mathcal{D}^{1,2}(\mathbb{R}^n)$ such that $w = v$ in $\Omega \cup (\mathbb{R}^n \setminus B_k)$. We take u_k to be the minimizer in X_k of the functional

$$X_k \ni w \longmapsto \int_{\mathbb{R}^n\setminus\overline{\Omega}} |\nabla w(x)|^2\, dx.$$

We stress that such a minimizer exists, according to the direct method of the calculus of variations, see e.g. [Gia84] and [Leo09, Theorems 11.10, 12.15 and 15.29].

We point out that, if

$$u_k^\star(x) := \begin{cases} 1 & \text{if } u_k(x) > 1,\\ u_k(x) & \text{if } u_k(x) \in [0,1],\\ 0 & \text{if } u_k(x) < 0, \end{cases}$$

then

$$\int_{\mathbb{R}^n\setminus\overline{\Omega}} |\nabla u_k^\star(x)|^2\, dx = \int_{(\mathbb{R}^n\setminus\overline{\Omega})\cap\{0\leqslant u_k\leqslant 1\}} |\nabla u_k(x)|^2\, dx \leqslant \int_{\mathbb{R}^n\setminus\overline{\Omega}} |\nabla u_k(x)|^2\, dx.$$

Thus, up to replacing u_k by $u_k^\star$, we can suppose that

$$0 \leqslant u_k(x) \leqslant 1 \qquad \text{for all } x \in \mathbb{R}^n \setminus \overline{\Omega}. \tag{6.1.5}$$

Moreover, since $X_{k_0} \subseteq X_k$ for all $k \in \mathbb{N} \cap [k_0, +\infty)$, the minimization property in X_k yields that

$$\int_{\mathbb{R}^n\setminus\overline{\Omega}} |\nabla u_k(x)|^2\, dx \leqslant \int_{\mathbb{R}^n\setminus\overline{\Omega}} |\nabla u_{k_0}(x)|^2\, dx =: C_0 \in [0, +\infty); \tag{6.1.6}$$

therefore, up to a subsequence, we can suppose that u_k converges to some function u in $L^2_{\rm loc}(\mathbb{R}^n)$ and ∇u_k converges to ∇u weakly in $L^2(\mathbb{R}^n)$.

Since the minimizer u_k is harmonic in $\mathbb{R}^n \setminus \overline{\Omega}$, so is u, thanks to Corollary 2.2.2. Also, since $u_k = v = 1$ in Ω, $u = 1$ in Ω as well. This gives that u is an admissible function for the infimum procedure on the right-hand side of (6.1.2).

We also point out that, by the above-mentioned weak convergence,

$$\begin{aligned}0 &\leqslant \liminf_{k\to+\infty}\int_{\mathbb{R}^n\setminus\overline{\Omega}}|\nabla u_k(x)-\nabla u(x)|^2\,dx\\&=\liminf_{k\to+\infty}\int_{\mathbb{R}^n\setminus\overline{\Omega}}\left(|\nabla u(x)|^2-2\nabla u_k(x)\cdot\nabla u(x)+|\nabla u_k(x)|^2\right)dx\\&=-\int_{\mathbb{R}^n\setminus\overline{\Omega}}|\nabla u(x)|^2\,dx+\liminf_{k\to+\infty}\int_{\mathbb{R}^n\setminus\overline{\Omega}}|\nabla u_k(x)|^2\,dx.\end{aligned}\tag{6.1.7}$$

In particular, by (6.1.6),

$$\int_{\mathbb{R}^n\setminus\overline{\Omega}}|\nabla u(x)|^2\,dx\leqslant C_0.\tag{6.1.8}$$

Now, we show that

$$\int_{\mathbb{R}^n\setminus\overline{\Omega}}|\nabla u(x)|^2\,dx=\inf_{\substack{\zeta\in\mathcal{D}^{1,2}(\mathbb{R}^n)\\ \zeta=1\text{ in }\Omega}}\int_{\mathbb{R}^n\setminus\overline{\Omega}}|\nabla\zeta(x)|^2\,dx.\tag{6.1.9}$$

Indeed, suppose not, then there exist $b>0$ and $\zeta\in\mathcal{D}^{1,2}(\mathbb{R}^n)$ with $\zeta=1$ in Ω such that

$$\int_{\mathbb{R}^n\setminus\overline{\Omega}}|\nabla\zeta(x)|^2\,dx+b\leqslant\int_{\mathbb{R}^n\setminus\overline{\Omega}}|\nabla u(x)|^2\,dx.$$

Using the density results in Sobolev spaces (see e.g. [Leo09, Theorem 10.29]), we can find $w\in C_0^\infty(\mathbb{R}^n)$ with $w=1$ in Ω such that

$$\int_{\mathbb{R}^n\setminus\overline{\Omega}}|\nabla w(x)|^2\,dx\leqslant\frac{b}{2}+\int_{\mathbb{R}^n\setminus\overline{\Omega}}|\nabla\zeta(x)|^2\,dx.$$

We take $k_\star\in\mathbb{N}\cap[k_0,+\infty)$ sufficiently large such that the support of w is contained in $B_{k_\star}$. As a consequence, we have that $w\in X_k$ for all $k\geqslant k_\star$, and therefore

$$\int_{\mathbb{R}^n\setminus\overline{\Omega}}|\nabla u_k(x)|^2\,dx\leqslant\int_{\mathbb{R}^n\setminus\overline{\Omega}}|\nabla w(x)|^2\,dx.$$

By collecting these items of information and utilizing (6.1.7), we see that

$$\begin{aligned}\frac{b}{2}=b-\frac{b}{2}&\leqslant\left(\int_{\mathbb{R}^n\setminus\overline{\Omega}}|\nabla u(x)|^2\,dx-\int_{\mathbb{R}^n\setminus\overline{\Omega}}|\nabla\zeta(x)|^2\,dx\right)\\&\quad+\left(\int_{\mathbb{R}^n\setminus\Omega}|\nabla\zeta(x)|^2\,dx-\int_{\mathbb{R}^n\setminus\Omega}|\nabla w(x)|^2\,dx\right)\\&\leqslant\liminf_{k\to+\infty}\int_{\mathbb{R}^n\setminus\overline{\Omega}}|\nabla u_k(x)|^2\,dx-\int_{\mathbb{R}^n\setminus\overline{\Omega}}|\nabla w(x)|^2\,dx\leqslant 0.\end{aligned}$$

This contradiction proves (6.1.9).

Thus, the claim in (6.1.4) follows from (6.1.2) and (6.1.9).

We also observe that

$$u \in C(\mathbb{R}^n \setminus \Omega). \tag{6.1.10}$$

To prove this, we let $z \in \partial\Omega$ and use the regularity of Ω to find a ball $B_{\varrho_0}(z_0)$ such that $B_{\varrho_0}(z_0) \subseteq \Omega$ and $z \in (\partial\Omega) \cap (\partial B_{\varrho_0}(z_0))$. We define

$$\Phi(x) := \begin{cases} u(x) - \dfrac{\varrho_0^{n-2}}{|x - z_0|^{n-2}} & \text{if } x \in \mathbb{R}^n \setminus B_{\varrho_0}(z_0), \\ u(x) - 1 & \text{if } x \in B_{\varrho_0}(z_0), \end{cases} \tag{6.1.11}$$

and we observe that $\Phi \in \mathcal{D}^{1,2}(\mathbb{R}^n \setminus \Omega)$; moreover, if $x \in \partial\Omega$, then $|x - z_0| \geqslant \varrho_0$, and hence, in the Sobolev trace sense, we have that $\Phi \geqslant u - 1 = 0$ along $\partial\Omega$. Additionally, Φ is harmonic in $\mathbb{R}^n \setminus \overline{\Omega}$. Therefore, we can exploit Lemma 2.9.5 (used here with $\mathcal{U} := \mathbb{R}^n \setminus \overline{\Omega}$) and gather that $\Phi \geqslant 0$.

This leads to

$$\liminf_{x \to z} u(x) \geqslant \liminf_{x \to z} \frac{\varrho_0^{n-2}}{|x - z_0|^{n-2}} = \frac{\varrho_0^{n-2}}{|z - z_0|^{n-2}} = 1.$$

Since by construction $u \leqslant 1$ (recall (6.1.5)), we thus obtain that

$$\lim_{x \to z} u(x) = 1,$$

thus proving (6.1.10).

Now, we show that

$$\lim_{|x| \to +\infty} u(x) = 0. \tag{6.1.12}$$

To this end, equation (6.1.8) and the Gagliardo–Nirenberg–Sobolev inequality (see e.g. [Leo09, Theorem 11.2]) yield that, since $n \geqslant 3$,

$$\|u\|_{L^{\frac{2n}{n-2}}(\mathbb{R}^n \setminus \overline{\Omega})} \leqslant C_1, \tag{6.1.13}$$

for some $C_1 > 0$.

Furthermore, by (6.1.5), we have that $\|u\|_{L^\infty(\mathbb{R}^n \setminus \overline{\Omega})}$. From this and Cauchy's estimate in Theorem 2.17.1, we infer that, for every $x_0 \in \mathbb{R}^n$ such that $B_2(x_0) \cap \Omega = \varnothing$,

$$|\nabla u(x_0)| \leqslant C, \tag{6.1.14}$$

for some constant $C > 0$ depending only on n.

We now recall that[1]

$$\text{if a uniformly continuous function } \phi \text{ is in } L^q(\mathbb{R}^n) \text{ for some } q \in [1, +\infty), \text{ then } \lim_{|x| \to +\infty} \phi(x) = 0. \tag{6.1.15}$$

From (6.1.13), (6.1.14) and (6.1.15), we infer (6.1.12), as desired.

Furthermore, we claim that the function u that we constructed is unique. For this, suppose that u and v satisfy the theses of Proposition 6.1.2, and let $w := \frac{u+v}{2}$. Since $w = 1$ on $\partial\Omega$, we infer from (6.1.2) that

$$\begin{aligned}
&\int_{\mathbb{R}^n \setminus \overline{\Omega}} |\nabla u(x) - \nabla v(x)|^2 \, dx \\
&= \int_{\mathbb{R}^n \setminus \overline{\Omega}} \Big(|\nabla u(x)|^2 + |\nabla v(x)|^2 - 2\nabla u(x) \cdot \nabla v(x) \Big) \, dx \\
&= 2 \operatorname{Cap}(\Omega) - 2 \int_{\mathbb{R}^n \setminus \overline{\Omega}} \nabla u(x) \cdot \nabla v(x) \, dx \\
&\leqslant 2 \int_{\mathbb{R}^n \setminus \overline{\Omega}} |\nabla w(x)|^2 \, dx - 2 \int_{\mathbb{R}^n \setminus \overline{\Omega}} \nabla u(x) \cdot \nabla v(x) \, dx \\
&= \int_{\mathbb{R}^n \setminus \overline{\Omega}} \frac{|\nabla u(x)|^2 + |\nabla v(x)|^2 + 2\nabla u(x) \cdot \nabla v(x)}{2} \, dx \\
&\quad - 2 \int_{\mathbb{R}^n \setminus \overline{\Omega}} \nabla u(x) \cdot \nabla v(x) \, dx \\
&= \frac{1}{2} \int_{\mathbb{R}^n \setminus \overline{\Omega}} |\nabla u(x) - \nabla v(x)|^2 \, dx.
\end{aligned}$$

This gives that $\int_{\mathbb{R}^n \setminus \overline{\Omega}} |\nabla u(x) - \nabla v(x)|^2 \, dx = 0$; therefore, the function $u - v$ is constant. By looking at the values at infinity, we deduce that $u - v$ is constantly equal to zero, thus establishing the uniqueness claim in Proposition 6.1.2.

[1]To prove (6.1.15), one can argue by contradiction, assuming that there exists a sequence $p_j \in \mathbb{R}^n$ such that $|p_j| \to +\infty$ as $j \to +\infty$ and such that $|\phi(p_j)| \geqslant a$, for some $a > 0$. Then, one takes $\rho > 0$ such that $|\phi(x) - \phi(p_j)| \leqslant \frac{a}{2}$ for every $x \in B_\rho(p_j)$. Since, up to a subsequence, we can assume that the balls $B_\rho(p_j)$ are disjoint, we find that

$$\begin{aligned}
\|\phi\|^q_{L^q(\mathbb{R}^n)} &\geqslant \sum_{j=0}^{+\infty} \int_{B_\rho(p_j)} |\phi(x)|^q \, dx \geqslant \sum_{j=0}^{+\infty} \int_{B_\rho(p_j)} \big(|\phi(p_j)| - |\phi(x) - \phi(p_j)| \big)^q \, dx \\
&\geqslant \sum_{j=0}^{+\infty} \int_{B_\rho(p_j)} \left(\frac{a}{2}\right)^q dx = +\infty,
\end{aligned}$$

providing the desired contradiction.

To complete the proof of Proposition 6.1.2, it remains to show that $u \in C^2(\mathbb{R}^n \setminus \Omega)$. For this, we consider the problem

$$\begin{cases} \Delta v = 0 & \text{in } B_{k_0} \setminus \overline{\Omega}, \\ v = 1 & \text{on } \partial\Omega, \\ v = u & \text{on } \partial B_{k_0}, \end{cases}$$

and exploit Theorem 4.5.4 to say that it admits a unique solution $v \in C^2(B_{k_0} \setminus \overline{\Omega}) \cap C(\overline{B_{k_0}} \setminus \Omega)$. Hence, by (6.1.10) and the uniqueness statement, we deduce that $v = u$. As a consequence of this and (4.5.19) in Theorem 4.5.4, we obtain that $u \in C^2(\mathbb{R}^n \setminus \Omega)$, as desired. □

The statement in formula (6.1.3) of Proposition 6.1.2 can be made more precise, as follows.

Proposition 6.1.3. *Let $n \geqslant 3$ and $\Omega \subset \mathbb{R}^n$ be a bounded and open set with boundary of class $C^{2,\alpha}$ for some $\alpha \in (0, 1)$. Let u be the function given by Proposition 6.1.2. Assume that $\Omega \subset B_{R_0}$, for some $R_0 > 0$. Then, for every $x \in \mathbb{R}^n \setminus B_{R_0}$,*

$$u(x) \leqslant \frac{R_0^{n-2}}{|x|^{n-2}}.$$

Proof. For every $\varepsilon \in (0, 1)$, we set

$$v_\varepsilon(x) := \frac{R_0^{n-2}}{|x|^{n-2}} + \varepsilon.$$

Also, in light of (6.1.3), there exists $R_\varepsilon \geqslant 1/\varepsilon$ such that $|u(x)| \leqslant \frac{\varepsilon}{2}$ for every $x \in \mathbb{R}^n \setminus B_{R_\varepsilon}$. We note that

$$\begin{aligned} & v_\varepsilon(x) = 1 + \varepsilon > 1 \geqslant u(x) && \text{on } \partial B_{R_0} \\ \text{and} \quad & v_\varepsilon(x) \geqslant \varepsilon > \frac{\varepsilon}{2} \geqslant u(x) && \text{on } \partial B_{R_\varepsilon}. \end{aligned}$$

Furthermore, both u and v_ε are harmonic in $B_{R_\varepsilon} \setminus B_{R_0}$; therefore, by the weak maximum principle in Corollary 2.9.2, we have that

$$u(x) \leqslant v_\varepsilon(x) = \frac{R_0^{n-2}}{|x|^{n-2}} + \varepsilon \quad \text{for every } x \in B_{R_\varepsilon} \setminus B_{R_0}.$$

Accordingly, sending $\varepsilon \searrow 0$, we obtain the desired result. □

The result in Proposition 6.1.2 highlights the strong connection (actually, basically, the coincidence, up to physical constants and different terminologies) between the mathematical definition of capacity in (6.1.1) and the physical notion of capacitance (or, more precisely, of self-capacitance, see e.g. [Zan13]) for an isolated conductor, which is the amount of electric charge that must be added to the conductor to raise its electric potential by one unit. To highlight this physical motivation, we make the following observation.

Proposition 6.1.4. *Let $n \geqslant 3$ and $\Omega \subset \mathbb{R}^n$ be a bounded and open set with boundary of class $C^{2,\alpha}$ for some $\alpha \in (0,1)$. Let u be the function given by Proposition 6.1.2. Then,*

$$\mathrm{Cap}(\Omega) = \int_{\mathbb{R}^n \setminus \overline{\Omega}} |\nabla u(x)|^2 \, dx = -\int_{\partial\Omega} \frac{\partial u}{\partial \nu}(x) \, d\mathcal{H}^{n-1}_x. \tag{6.1.16}$$

Proof. Since Ω is bounded, we can consider a radius $R_0 > 0$ such that $\Omega \subset B_{R_0}$. We take $R > 2R_0$ and observe that, by Green's first identity in (2.1.10),

$$\begin{aligned}\int_{B_R \setminus \overline{\Omega}} |\nabla u(x)|^2 \, dx &= \int_{\partial(B_R \setminus \overline{\Omega})} u(x) \frac{\partial u}{\partial \nu}(x) \, d\mathcal{H}^{n-1}_x \\ &= \int_{\partial B_R} u(x) \frac{\partial u}{\partial \nu}(x) \, d\mathcal{H}^{n-1}_x - \int_{\partial\Omega} u(x) \frac{\partial u}{\partial \nu}(x) \, d\mathcal{H}^{n-1}_x \\ &= \int_{\partial B_R} u(x) \frac{\partial u}{\partial \nu}(x) \, d\mathcal{H}^{n-1}_x - \int_{\partial\Omega} \frac{\partial u}{\partial \nu}(x) \, d\mathcal{H}^{n-1}_x.\end{aligned} \tag{6.1.17}$$

We observe that

$$\lim_{R \to +\infty} \int_{B_R \setminus \overline{\Omega}} |\nabla u(x)|^2 \, dx = \int_{\mathbb{R}^n \setminus \overline{\Omega}} |\nabla u(x)|^2 \, dx, \tag{6.1.18}$$

thanks to the dominated convergence theorem.

We claim that

$$\lim_{R \to +\infty} \int_{\partial B_R} u(x) \frac{\partial u}{\partial \nu}(x) \, d\mathcal{H}^{n-1}_x = 0. \tag{6.1.19}$$

To prove this, we consider a point $x_0 \in \mathbb{R}^n \setminus \overline{\Omega}$ such that $|x_0| \geqslant 2R$. In light of Proposition 6.1.3, we have that, for every $x \in \mathbb{R}^n \setminus B_R$,

$$|u(x)| \leqslant \frac{C}{|x|^{n-2}} \leqslant \frac{C}{R^{n-2}} \tag{6.1.20}$$

and accordingly $\|u\|_{L^1(B_{R/4}(x_0))} \leqslant CR^2$, up to renaming $C > 0$. Hence, exploiting Cauchy's estimates in Theorem 2.17.1, we see that

$$|\nabla u(x_0)| \leqslant \frac{C\|u\|_{L^1(B_{R/4}(x_0))}}{R^{n+1}} \leqslant \frac{CR^2}{R^{n+1}} = CR^{1-n}, \tag{6.1.21}$$

up to relabeling $C > 0$.

As a consequence of this and (6.1.20),

$$\left|\int_{\partial B_R} u(x)\frac{\partial u}{\partial \nu}(x)\,d\mathcal{H}_x^{n-1}\right| \leqslant CR^{2-n},$$

from which (6.1.19) follows.

Plugging (6.1.18) and (6.1.19) into (6.1.17), we obtain that

$$\int_{\mathbb{R}^n\setminus\overline{\Omega}} |\nabla u(x)|^2\,dx = -\int_{\partial\Omega}\frac{\partial u}{\partial \nu}(x)\,d\mathcal{H}_x^{n-1}.$$

Then, the desired result follows by recalling (6.1.4). □

Reconsidering the physical motivation discussed before Proposition 6.1.4, we observe that at the equilibrium, the boundary of the conductor Ω is an equipotential energy; thus, denoting by $u : \mathbb{R}^n\setminus\Omega \to \mathbb{R}$ the potential generated by the conductor, up to a normalization, we can suppose that $u = 1$ on $\partial\Omega$ (and u is a harmonic function, recalling the comments after (2.7.6), and we normalize this potential to vanish at infinity). The corresponding electric field is $-\nabla u$. Then, by Gauß' law, the total charge of the conductor can be calculated by taking a large ball $B_R \Supset \Omega$, and omitting physical constants, it is equal to the flow of the electric field through ∂B_R, namely

$$\begin{aligned}
Q &= -\int_{\partial B_R}\frac{\partial u}{\partial \nu}(x)\,d\mathcal{H}^{n-1} = -\int_{\partial(B_R\setminus\Omega)}\frac{\partial u}{\partial \nu}(x)\,d\mathcal{H}^{n-1} - \int_{\partial\Omega}\frac{\partial u}{\partial \nu}(x)\,d\mathcal{H}^{n-1}\\
&= -\int_{B_R\setminus\Omega}\operatorname{div}(\nabla u(x))\,d\mathcal{H}^{n-1} - \int_{\partial\Omega}\frac{\partial u}{\partial \nu}(x)\,d\mathcal{H}^{n-1}\\
&= -\int_{\partial\Omega}\frac{\partial u}{\partial \nu}(x)\,d\mathcal{H}^{n-1}.
\end{aligned}$$

Since the potential V along the surface of the conductor is normalized to 1, this total charge corresponds to the self-capacitance $\frac{Q}{V}$. In turn, this quantity corresponds to the capacity as defined in (6.1.1), thanks to (6.1.16), thus confirming the essential coincidence between the mathematical notion of capacity and the physical concept of self-capacitance. In this spirit, in the physical jargon, the function u in Proposition 6.1.2 is often referred to by the name "conductor potential."

Such a conductor potential, when known, can be efficiently exploited to compute the capacity of a given set, as exemplified by the following result.

Corollary 6.1.5. *Let $n \geqslant 3$, $R > 0$ and c_n be the positive constant introduced in* (2.7.2). *Then,*

$$\operatorname{Cap}(B_R) = \frac{R^{n-2}}{c_n}.$$

Proof. Let $u(x) := \frac{R^{n-2}}{|x|^{n-2}}$. We have that u is harmonic in the complement of B_R and attains the value of 1 along ∂B_R. Thus,

$$\int_{\mathbb{R}^n\setminus\overline{B_R}} |\nabla u(x)|^2\,dx = \mathcal{H}^{n-1}(\partial B_1)\int_R^{+\infty}\left(\frac{(n-2)R^{n-2}}{\rho^{n-1}}\right)^2 \rho^{n-1}\,d\rho$$
$$= n|B_1|(n-2)R^{n-2} = \frac{R^{n-2}}{c_n},$$

thanks to (1.1.6). The desired result thus follows from (6.1.4). □

The pointwise values of the conductor potential can also be conveniently exploited to estimate the value of the capacity of the original set, according to the following observation.

Proposition 6.1.6. *Let $n \geqslant 3$, $\Omega \subset \mathbb{R}^n$ be a bounded and open set with boundary of class $C^{2,\alpha}$ for some $\alpha \in (0,1)$ and u be the conductor potential of Ω, as given in Proposition* 6.1.2.

Let $x \in \mathbb{R}^n \setminus \overline{\Omega}$. Let $R > r > 0$ be such that $B_r(x) \subseteq \mathbb{R}^n \setminus \Omega$ and $B_R(x) \supseteq \Omega$. Then,

$$\frac{c_n\,\mathrm{Cap}(\Omega)}{R^{n-2}} \leqslant u(x) \leqslant \frac{c_n\,\mathrm{Cap}(\Omega)}{r^{n-2}},$$

where c_n is the positive constant in (2.7.2).

Proof. Up to a translation, we suppose that $x = 0$. Take $\Omega' \Supset \Omega$ such that $0 \notin \Omega'$. We let Γ be the fundamental solution in (2.7.6) and exploit Theorem 2.7.3 with $\varphi := u$ in the set $\mathbb{R}^n \setminus \overline{\Omega}$. In this way, we find that[2]

$$\int_{\partial\Omega}\left(\Gamma(x)\frac{\partial u}{\partial\nu}(x) - \frac{\partial\Gamma}{\partial\nu}(x)\right) d\mathcal{H}^{n-1}_x = -u(0).$$

Moreover, according to the divergence theorem,

$$\int_{\partial\Omega}\frac{\partial\Gamma}{\partial\nu}(x)\,d\mathcal{H}^{n-1}_x = \int_\Omega \Delta\Gamma(x)\,dx = 0.$$

As a consequence,

$$u(0) = -\int_{\partial\Omega}\Gamma(x)\frac{\partial u}{\partial\nu}(x)\,d\mathcal{H}^{n-1}_x. \tag{6.1.22}$$

[2]Note that one should argue as in the proof of Proposition 6.1.4 by exploiting Theorem 2.7.3 in $B_R \setminus \overline{\Omega}$ and then sending $R \to +\infty$. The convergence is then guaranteed by the estimates in (6.1.20) and (6.1.21).

Also, since u attains its maximum along $\partial\Omega$, we know that $\frac{\partial u}{\partial \nu} \leqslant 0$ along $\partial\Omega$. Therefore, for every $x \in \partial\Omega$,

$$-\Gamma(x)\frac{\partial u}{\partial \nu}(x) = -\frac{c_n}{|x|^{n-2}}\frac{\partial u}{\partial \nu}(x) \in \left[-\frac{c_n}{R^{n-2}}\frac{\partial u}{\partial \nu}(x),\ -\frac{c_n}{r^{n-2}}\frac{\partial u}{\partial \nu}(x)\right].$$

This, (6.1.16) and (6.1.22) give that

$$\begin{aligned} u(0) &= -\int_{\partial\Omega} \Gamma(x)\frac{\partial u}{\partial \nu}(x)\, d\mathcal{H}_x^{n-1} \\ &\in \left[-\frac{c_n}{R^{n-2}}\int_{\partial\Omega}\frac{\partial u}{\partial \nu}(x)\, d\mathcal{H}_x^{n-1},\ -\frac{c_n}{r^{n-2}}\int_{\partial\Omega}\frac{\partial u}{\partial \nu}(x)\, d\mathcal{H}_x^{n-1}\right] \\ &= \left[\frac{c_n \operatorname{Cap}(\Omega)}{R^{n-2}},\ \frac{c_n \operatorname{Cap}(\Omega)}{r^{n-2}}\right]. \end{aligned}$$

□

The following is a useful property of the capacity as a set function.

Lemma 6.1.7. *Let Ω_1, $\Omega_2 \subset \mathbb{R}^n$ be bounded and open sets with boundary of class $C^{2,\alpha}$ for some $\alpha \in (0,1)$, and suppose that $\Omega_1 \cup \Omega_2$ has boundary of class $C^{2,\alpha}$.*

Then,

$$\operatorname{Cap}(\Omega_1) \leqslant \operatorname{Cap}(\Omega_1 \cup \Omega_2) \leqslant \operatorname{Cap}(\Omega_1) + \operatorname{Cap}(\Omega_2).$$

Proof. Let $\varepsilon > 0$. We use the definition in (6.1.1) to find a function $v_\varepsilon \in C_0^\infty(\mathbb{R}^n)$ with $v_\varepsilon = 1$ in $\Omega_1 \cup \Omega_2$ such that

$$\operatorname{Cap}(\Omega_1 \cup \Omega_2) + \varepsilon \geqslant \int_{\mathbb{R}^n \setminus \overline{\Omega_1 \cup \Omega_2}} |\nabla v_\varepsilon(x)|^2\, dx.$$

Also, since $v_\varepsilon = 1$ in Ω_2, whence $\nabla v_\varepsilon = 0$ a.e. in Ω_2 (see e.g. [LL01, Theorem 6.19]), we have that

$$\int_{\mathbb{R}^n \setminus \overline{\Omega_1 \cup \Omega_2}} |\nabla v_\varepsilon(x)|^2\, dx = \int_{\mathbb{R}^n \setminus \overline{\Omega_1}} |\nabla v_\varepsilon(x)|^2\, dx.$$

From these observations and using again the definition in (6.1.1), we find that

$$\operatorname{Cap}(\Omega_1 \cup \Omega_2) + \varepsilon \geqslant \int_{\mathbb{R}^n \setminus \overline{\Omega_1}} |\nabla v_\varepsilon(x)|^2\, dx \geqslant \operatorname{Cap}(\Omega_1).$$

By taking ε as small as we wish, we obtain

$$\operatorname{Cap}(\Omega_1 \cup \Omega_2) \geqslant \operatorname{Cap}(\Omega_1). \tag{6.1.23}$$

Now, we make use of the definition in (6.1.1) to find $v_{1,\varepsilon}$, $v_{2,\varepsilon} \in C_0^\infty(\mathbb{R}^n)$ with $v_{1,\varepsilon} = 1$ in Ω_1 and $v_{2,\varepsilon} = 1$ in Ω_2 such that

$$\operatorname{Cap}(\Omega_1) + \varepsilon \geqslant \int_{\mathbb{R}^n \setminus \overline{\Omega_1}} |\nabla v_{1,\varepsilon}(x)|^2\, dx$$

and

$$\operatorname{Cap}(\Omega_1) + \varepsilon \geqslant \int_{\mathbb{R}^n \setminus \overline{\Omega_2}} |\nabla v_{2,\varepsilon}(x)|^2 \, dx.$$

As in (6.1.5), without loss of generality, we can suppose that $v_{1,\varepsilon}(x)$, $v_{2,\varepsilon}(x) \in [0,1]$ for all $x \in \mathbb{R}^n$. Let $v_\varepsilon^\star := \max\{v_{1,\varepsilon}, v_{2,\varepsilon}\}$. Then $v_\varepsilon^\star = 1$ in $\Omega_1 \cup \Omega_2$. Furthermore, $v_\varepsilon^\star$ is Lipschitz continuous and compactly supported, which together with (6.1.2) leads to

$$\begin{aligned}
&\int_{\mathbb{R}^n \setminus \overline{\Omega_1}} |\nabla v_{1,\varepsilon}(x)|^2 \, dx + \int_{\mathbb{R}^n \setminus \overline{\Omega_2}} |\nabla v_{2,\varepsilon}(x)|^2 \, dx \\
&\quad = \int_{\mathbb{R}^n} |\nabla v_{1,\varepsilon}(x)|^2 \, dx + \int_{\mathbb{R}^n} |\nabla v_{2,\varepsilon}(x)|^2 \, dx \\
&\quad \geqslant \int_{\{v_{1,\varepsilon} > v_{2,\varepsilon}\}} |\nabla v_{1,\varepsilon}(x)|^2 \, dx + \int_{\{v_{1,\varepsilon} \leqslant v_{2,\varepsilon}\}} |\nabla v_{2,\varepsilon}(x)|^2 \, dx \\
&\quad = \int_{\mathbb{R}^n} |\nabla v_\varepsilon^\star(x)|^2 \, dx \\
&\quad = \int_{\mathbb{R}^n \setminus \overline{\Omega_1 \cup \Omega_2}} |\nabla v_\varepsilon^\star(x)|^2 \, dx \\
&\quad \geqslant \operatorname{Cap}(\Omega_1 \cup \Omega_2).
\end{aligned}$$

Using these observations, we obtain that

$$\operatorname{Cap}(\Omega_1) + \operatorname{Cap}(\Omega_2) + 2\varepsilon \geqslant \operatorname{Cap}(\Omega_1 \cup \Omega_2)$$

and thus, taking ε arbitrarily small,

$$\operatorname{Cap}(\Omega_1) + \operatorname{Cap}(\Omega_2) \geqslant \operatorname{Cap}(\Omega_1 \cup \Omega_2).$$

From this[3] and (6.1.23), the desired result plainly follows. □

A counterpart of the result in Lemma 6.1.7 deals with the pointwise value of the conductor potentials.

Lemma 6.1.8. *Let $n \geqslant 3$, $\Omega_1, \Omega_2 \subset \mathbb{R}^n$ be bounded and open sets with boundary of class $C^{2,\alpha}$ for some $\alpha \in (0,1)$, and suppose that $\Omega_1 \cup \Omega_2$ has boundary of class $C^{2,\alpha}$. For $i \in \{1,2\}$, let u_i be the conductor potential of Ω_i, as given in Proposition* 6.1.2. *Let also u_3 be the conductor potential of $\Omega_1 \cup \Omega_2$.*

Then, for every $x \in \mathbb{R}^n$,

$$u_3(x) \leqslant u_1(x) + u_2(x).$$

[3] Though we do not need this property here, we observe that by considering also the function $\min\{v_{1,\varepsilon}, v_{2,\varepsilon}\}$, this calculation shows that

$$\operatorname{Cap}(\Omega_1) + \operatorname{Cap}(\Omega_2) \geqslant \operatorname{Cap}(\Omega_1 \cup \Omega_2) + \operatorname{Cap}(\Omega_1 \cap \Omega_2).$$

In jargon, this states that the capacity is a "submodular" set function.

Proof. In light of Proposition 6.1.2, the function $u := u_1 + u_2 - u_3$ is harmonic outside $\Omega_1 \cup \Omega_2$, and it belongs to $\mathcal{D}^{1,2}(\mathbb{R}^n)$. Furthermore, on $\partial(\Omega_1 \cup \Omega_2) \subseteq (\partial\Omega_1) \cup (\partial\Omega_2)$, we have that either $u_1 = 1$ or $u_2 = 1$, and therefore $u \geqslant 1 - u_3 \geqslant 0$.

As a result, we can apply Lemma 2.9.5 and deduce that $u \geqslant 0$ in $\mathbb{R}^n$, from which the desired result follows. □

Importantly, one can define the capacity of any bounded (not necessarily open, nor with regular boundary) set Ω by setting

$$\mathrm{Cap}(\Omega) := \inf \mathrm{Cap}(\Omega'), \tag{6.1.24}$$

where the infimum above is taken over all the open and bounded sets Ω' with boundary of class $C^{2,\alpha}$, for some $\alpha \in (0, 1)$, such that $\Omega \subset \Omega'$.

Lemma 6.1.9. *The definition in* (6.1.24) *coincides with that in* (6.1.1) *whenever Ω is open and bounded with boundary of class $C^{2,\alpha}$ for some $\alpha \in (0, 1)$.*

Proof. To avoid confusion, in this proof, we denote by "$\widetilde{\mathrm{Cap}}$" the capacity defined on the left-hand side of (6.1.24) and reserve the notation of "Cap" for the one in (6.1.1).

Suppose that Ω is open and bounded with boundary of class $C^{2,\alpha}$ for some $\alpha \in (0, 1)$. Let also Ω' be open and bounded with boundary of class $C^{2,\alpha}$, with $\Omega \subset \Omega'$. We know that $\mathrm{Cap}(\Omega) \leqslant \mathrm{Cap}(\Omega')$; therefore, by (6.1.24), by taking the infimum over such sets Ω', we have that $\mathrm{Cap}(\Omega) \leqslant \inf \mathrm{Cap}(\Omega') = \widetilde{\mathrm{Cap}}(\Omega)$. On the other hand, by taking $\Omega' := \Omega$ as a candidate in the infimum in (6.1.24), we have that $\widetilde{\mathrm{Cap}}(\Omega) \leqslant \mathrm{Cap}(\Omega)$. These observations show that the definitions in (6.1.1) and (6.1.24) coincide if Ω is open and bounded with boundary of class $C^{2,\alpha}$. □

Additionally, one can show that the results in Lemma 6.1.7 hold true for every bounded sets $\Omega_1, \Omega_2 \subset \mathbb{R}^n$, adopting the definition in (6.1.24). For this, the strategy would be to use the definition in (6.1.24) to find approximating sets $\Omega_{1,j}$ and $\Omega_{2,j}$ which are bounded and open with boundary of class $C^{2,\alpha}$ and contain Ω_1 and Ω_2, respectively. One needs to modify the sets $\Omega_{1,j}$ and $\Omega_{2,j}$ in such a way that $\Omega_{1,j} \cup \Omega_{2,j}$ has boundary of class $C^{2,\alpha}$ as well since in this setting, one can exploit Lemma 6.1.7 and then obtain the desired result by taking the limit. The difficulty is that one would like to perform such modifications to the sets that alter the capacity in a controlled way.

To prove that one can find such sets, we first establish the following observation that relates the capacity of the fattening of a set with the capacity of the set itself.

Lemma 6.1.10. *Let $n \geqslant 3$, Ω be a bounded and open set with boundary of class $C^{2,\alpha}$ for some $\alpha \in (0, 1)$. For every $\rho \in (0, 1)$, let Ω_ρ be a bounded and* *open set with boundary of class* $C^{2,\alpha}$ *such that*

$$\Omega_\rho \subset \bigcup_{p\in\Omega} B_\rho(p).$$

Then, there exists $\rho_\Omega \in (0,1)$ *such that, for every* $\rho \in (0, \rho_\Omega)$,

$$\mathrm{Cap}(\Omega_\rho) \leqslant \left(1+\rho^{1/2}\right)\mathrm{Cap}(\Omega) + C\,\rho^{1/2},$$

for some $C > 0$, *depending on* n *and* Ω.

Proof. We consider the conductor potential u for Ω as given by Proposition 6.1.2, and we take a function $\varphi \in C^\infty(\mathbb{R},[0,1])$ such that $\varphi = 1$ in $(-\infty, 1]$ and $\varphi = 0$ in $(2,+\infty)$. We also recall the definition of the signed distance function $d_{\partial\Omega}$ to $\partial\Omega$ given at the beginning of Section 2.3 (that is, $d_{\partial\Omega} \geqslant 0$ in Ω and $d_{\partial\Omega} \leqslant 0$ in $\mathbb{R}^n \setminus \Omega$), and we set $d_{\mathbb{R}^n\setminus\Omega} := -d_{\partial\Omega}$ and

$$\varphi_\rho(x) := \varphi\left(\frac{d_{\mathbb{R}^n\setminus\Omega}(x)}{\rho}\right).$$

Note that

$$\begin{aligned}
&\text{if } d_{\mathbb{R}^n\setminus\Omega}(x) < \rho, \text{ then } \varphi_\rho(x) = 1,\\
&\text{if } d_{\mathbb{R}^n\setminus\Omega}(x) > 2\rho, \text{ then } \varphi_\rho(x) = 0\\
&\text{and } |\nabla\varphi_\rho| \leqslant \frac{C}{\rho}\chi_{(\rho,2\rho)}(d_{\mathbb{R}^n\setminus\Omega}), \quad \text{for some } C > 0.
\end{aligned}$$

We also set $u_\rho := \varphi_\rho + (1-\varphi_\rho)u$, and we observe that $u_\rho(x) = 1$ if $x \in \Omega_\rho$ and $u_\rho(x) = u(x)$ if $|x|$ is sufficiently large; therefore, $u_\rho \in \mathcal{D}^{1,2}(\mathbb{R}^n)$. As a consequence, exploiting also the Cauchy–Schwarz inequality, for $\varepsilon \in (0,1)$ (to be specified later in the proof),

$$\begin{aligned}
\mathrm{Cap}(\Omega_\rho) &\leqslant \int_{\mathbb{R}^n} |\nabla u_\rho(x)|^2\,dx\\
&= \int_{\mathbb{R}^n} \left|\nabla\varphi_\rho(x)\big(1-u(x)\big) + \big(1-\varphi_\rho(x)\big)\nabla u(x)\right|^2 dx\\
&\leqslant \left(1+\frac{1}{\varepsilon}\right)\int_{\{d_{\mathbb{R}^n\setminus\Omega}(x)\in(\rho,2\rho)\}} |\nabla\varphi_\rho(x)|^2\big(1-u(x)\big)^2\,dx\\
&\quad + (1+\varepsilon)\int_{\mathbb{R}^n}\big(1-\varphi_\rho(x)\big)^2|\nabla u(x)|^2\,dx\\
&\leqslant \frac{C}{\rho^2}\left(1+\frac{1}{\varepsilon}\right)\int_{\{d_{\mathbb{R}^n\setminus\Omega}(x)\in(\rho,2\rho)\}}\big(1-u(x)\big)^2\,dx + (1+\varepsilon)\int_{\mathbb{R}^n}|\nabla u(x)|^2\,dx\\
&= \frac{C}{\rho^2}\left(1+\frac{1}{\varepsilon}\right)\int_{\{d_{\mathbb{R}^n\setminus\Omega}(x)\in(\rho,2\rho)\}}\big(1-u(x)\big)^2\,dx + (1+\varepsilon)\,\mathrm{Cap}(\Omega),
\end{aligned}$$

where we have used (6.1.4) in the last line. Now, we recall that $u \in C^{0,1}(\mathbb{R}^n \setminus \Omega)$; therefore, if $x \in \{d_{\mathbb{R}^n\setminus\Omega}(x) \in (\rho, 2\rho)\}$, then $(1-u(x))^2 \leqslant C\rho^2$, for some $C > 0$. Accordingly,

$$\operatorname{Cap}(\Omega_\rho) \leqslant C\rho\left(1+\frac{1}{\varepsilon}\right) + (1+\varepsilon)\operatorname{Cap}(\Omega),$$

up to renaming constants.

Now, we choose $\varepsilon := \sqrt{\frac{C\rho}{\operatorname{Cap}(\Omega)}}$, and we conclude that

$$\operatorname{Cap}(\Omega_\rho) \leqslant C\rho + \frac{C\rho}{\varepsilon} + (1+\varepsilon)\operatorname{Cap}(\Omega) = C\rho + \operatorname{Cap}(\Omega) + \sqrt{C\rho\operatorname{Cap}(\Omega)},$$

up to relabeling $C > 0$. Thus, by exploiting again the Cauchy–Schwarz inequality,

$$\begin{aligned}\operatorname{Cap}(\Omega_\rho) &\leqslant C\rho + \operatorname{Cap}(\Omega) + \sqrt{(C\rho^{1/2})(\rho^{1/2}\operatorname{Cap}(\Omega))}\\ &\leqslant C\rho + \operatorname{Cap}(\Omega) + C\rho^{1/2} + \rho^{1/2}\operatorname{Cap}(\Omega),\end{aligned}$$

from which the desired result follows. □

We now employ Lemma 6.1.10 to modify the approximating sets while keeping their capacities under control. See Figure 6.1.1 for a sketch of this construction.

Lemma 6.1.11. *Let $n \geqslant 3$, Ω_1, $\Omega_2 \subset \mathbb{R}^n$ be bounded sets. For every $\varepsilon \in (0,1)$, let $\Omega_{1,\varepsilon}$, $\Omega_{2,\varepsilon} \subset \mathbb{R}^n$ be bounded and open sets with boundary of class $C^{2,\alpha}$, for some $\alpha \in (0,1)$, such that $\Omega_i \subset \Omega_{i,\varepsilon}$ and with the property that*

$$\lim_{\varepsilon\searrow 0}\operatorname{Cap}(\Omega_{i,\varepsilon}) = \operatorname{Cap}(\Omega_i), \tag{6.1.25}$$

for every $i \in \{1,2\}$.

Then, for every $\rho \in (0,1)$, there exist bounded and open sets $\Omega_{1,\varepsilon,\rho}$, $\Omega_{2,\varepsilon,\rho} \subset \mathbb{R}^n$ with boundary of class $C^{2,\alpha}$ such that

$$\Omega_{i,\varepsilon} \subset \Omega_{i,\varepsilon,\rho} \subset \bigcup_{p\in\Omega_{i,\varepsilon}} B_\rho(p)$$

and

$$\operatorname{Cap}(\Omega_i) \leqslant \operatorname{Cap}(\Omega_{i,\varepsilon,\rho}) \leqslant \left(1+\rho^{1/2}\right)\operatorname{Cap}(\Omega_i) + C_\varepsilon\,\rho^{1/2} + \phi(\varepsilon), \tag{6.1.26}$$

for some $C_\varepsilon > 0$ and some function ϕ such that $\phi(\varepsilon) \to 0$ as $\varepsilon \searrow 0$, for every $i \in \{1,2\}$, and $\Omega_{1,\varepsilon,\rho} \cup \Omega_{2,\varepsilon,\rho}$ has boundary of class $C^{2,\alpha}$.

Proof. It follows from (6.1.25) that

$$\operatorname{Cap}(\Omega_{i,\varepsilon}) \leqslant \operatorname{Cap}(\Omega_i) + \phi(\varepsilon) \tag{6.1.27}$$

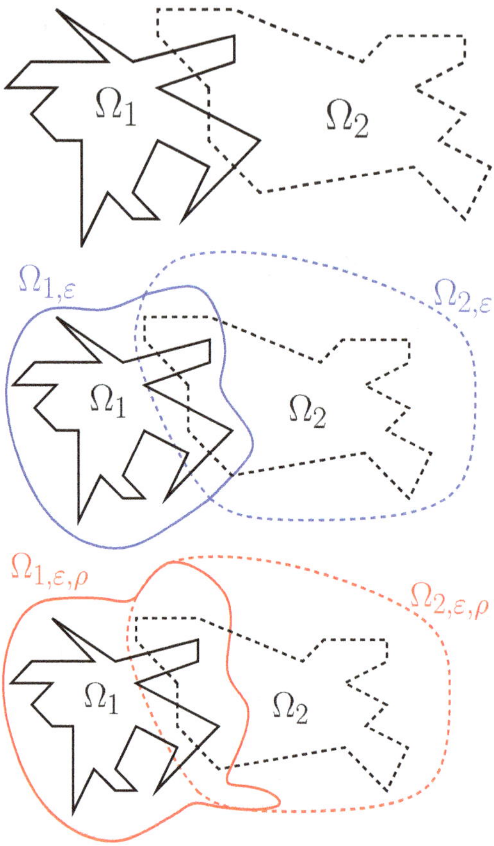

Figure 6.1.1 The sets Ω_1, Ω_2, $\Omega_{1,\varepsilon}$, $\Omega_{2,\varepsilon}$, $\Omega_{1,\varepsilon,\rho}$, $\Omega_{2,\varepsilon,\rho}$ involved in the proof of (6.1.28).

for some function ϕ such that $\phi(\varepsilon) \to 0$ as $\varepsilon \searrow 0$, for every $i \in \{1, 2\}$. Furthermore, for every $\rho \in (0, 1)$, we consider the sets $\Omega_{i,\varepsilon,\rho}$ with boundary of class $C^{2,\alpha}$ such that

$$\Omega_{i,\varepsilon,\rho} \subset \bigcup_{p \in \Omega_{i,\varepsilon}} B_\rho(p),$$

for every $i \in \{1, 2\}$, and $\Omega_{1,\varepsilon,\rho} \cup \Omega_{2,\varepsilon,\rho}$ has boundary of class $C^{2,\alpha}$. See Figure 6.1.1 for a sketch of this construction. We note that $\Omega_{i,\varepsilon} \subset \Omega_{i,\varepsilon,\rho}$ for every $i \in \{1, 2\}$.

Accordingly, in order to complete the proof of Lemma 6.1.11, it remains to prove (6.1.26). For this, we observe that, in light of Lemma 6.1.10, if ρ is sufficiently small,

$$\operatorname{Cap}(\Omega_{i,\varepsilon,\rho}) \leqslant \left(1+\rho^{1/2}\right) \operatorname{Cap}(\Omega_{i,\varepsilon}) + C_\varepsilon \, \rho^{1/2}$$

for some $C_\varepsilon > 0$, for every $i \in \{1,2\}$. This information, together with (6.1.27), gives that

$$\begin{aligned} \operatorname{Cap}(\Omega_{i,\varepsilon,\rho}) &\leqslant \left(1+\rho^{1/2}\right) \left(\operatorname{Cap}(\Omega_i) + \phi(\varepsilon)\right) + C_\varepsilon \, \rho^{1/2} \\ &\leqslant \left(1+\rho^{1/2}\right) \operatorname{Cap}(\Omega_i) + C_\varepsilon \, \rho^{1/2} + \phi(\varepsilon), \end{aligned}$$

up to relabeling C_ε, as desired. □

As a consequence of Lemma 6.1.11, we have the following.

Corollary 6.1.12. *Let $n \geqslant 3$, $\Omega_1, \Omega_2 \subset \mathbb{R}^n$ be bounded sets. Then, for every $j \in \mathbb{N}$, there exist bounded and open sets $\Omega_{1,j}$ and $\Omega_{2,j}$ with boundary of class $C^{2,\alpha}$, for some $\alpha \in (0,1)$, such that $\Omega_i \subset \Omega_{i,j}$ and*

$$\lim_{j\to+\infty} \operatorname{Cap}(\Omega_{i,j}) = \operatorname{Cap}(\Omega_i),$$

for every $i \in \{1,2\}$, and $\Omega_{1,j} \cup \Omega_{2,j}$ has boundary of class $C^{2,\alpha}$.

Proof. By the definition in (6.1.24), for every $\varepsilon \in (0,1)$, there exist bounded and open sets $\Omega_{1,\varepsilon}$, $\Omega_{2,\varepsilon}$ with boundary of class $C^{2,\alpha}$ such that $\Omega_i \subset \Omega_{i,\varepsilon}$ and with the property that

$$\lim_{\varepsilon \searrow 0} \operatorname{Cap}(\Omega_{i,\varepsilon}) = \operatorname{Cap}(\Omega_i),$$

for every $i \in \{1,2\}$.

Hence, we are in a position to apply Lemma 6.1.11 to find, for every $\rho \in (0,1)$, bounded and open sets $\Omega_{1,\varepsilon,\rho}$, $\Omega_{2,\varepsilon,\rho}$ with boundary of class $C^{2,\alpha}$ such that

$$\Omega_{i,\varepsilon} \subset \Omega_{i,\varepsilon,\rho} \subset \bigcup_{p\in\Omega_{i,\varepsilon}} B_\rho(p)$$

and such that (6.1.26) holds true for every $i \in \{1,2\}$. Moreover, $\Omega_{1,\varepsilon,\rho} \cup \Omega_{2,\varepsilon,\rho}$ has boundary of class $C^{2,\alpha}$.

In particular, taking $\varepsilon_j := 1/j$ in (6.1.26), we see that, for every $i \in \{1,2\}$,

$$\operatorname{Cap}(\Omega_i) \leqslant \operatorname{Cap}(\Omega_{i,\varepsilon_j,\rho}) \leqslant \left(1+\rho^{1/2}\right) \operatorname{Cap}(\Omega_i) + C_j \, \rho^{1/2} + \phi\left(\frac{1}{j}\right),$$

for some $C_j > 0$, with $\phi\left(\frac{1}{j}\right) \to 0$ as $j \to +\infty$. Now, choosing

$$\rho_j := \min\left\{\frac{1}{j}, \frac{1}{jC_j^2}\right\},$$

we obtain that

$$\mathrm{Cap}(\Omega_i) \leqslant \mathrm{Cap}(\Omega_{i,\varepsilon_j,\rho_j}) \leqslant \left(1 + \frac{1}{\sqrt{j}}\right)\mathrm{Cap}(\Omega_i) + \frac{1}{\sqrt{j}} + \phi\left(\frac{1}{j}\right),$$

for every $i \in \{1, 2\}$. Consequently,

$$\lim_{j\to+\infty} \mathrm{Cap}(\Omega_{i,\varepsilon_j,\rho_j}) = \mathrm{Cap}(\Omega_i),$$

for every $i \in \{1, 2\}$. Thus, setting $\Omega_{i,j} := \Omega_{i,\varepsilon_j,\rho_j}$ for every $i \in \{1, 2\}$, we obtain the desired result. □

With these observations at hand, we can now prove the following.

Lemma 6.1.13. *The results in Lemma 6.1.7 hold true for every bounded set* Ω_1, $\Omega_2 \subset \mathbb{R}^n$, *adopting the definition in* (6.1.24).

Proof. Let now $\Omega_1, \Omega_2 \subset \mathbb{R}^n$ be bounded. Since $\Omega_1 \subset \Omega_1 \cup \Omega_2$, by the definition in (6.1.24), we have that $\mathrm{Cap}(\Omega_1) \leqslant \mathrm{Cap}(\Omega_1 \cup \Omega_2)$.

Hence, to complete the proof of Lemma 6.1.13, we have to show that

$$\mathrm{Cap}(\Omega_1 \cup \Omega_2) \leqslant \mathrm{Cap}(\Omega_1) + \mathrm{Cap}(\Omega_2). \tag{6.1.28}$$

For this, we let $\varepsilon > 0$, and we see that there exist bounded and open sets $\Omega_{1,\varepsilon}$ and $\Omega_{2,\varepsilon}$ with boundary of class $C^{2,\alpha}$ such that $\Omega_i \subset \Omega_{i,\varepsilon}$ and

$$\lim_{\varepsilon\searrow 0} \mathrm{Cap}(\Omega_{i,\varepsilon}) = \mathrm{Cap}(\Omega_i)$$

for every $i \in \{1, 2\}$. Furthermore, from Lemma 6.1.11, we have that, for every $\rho \in (0, 1)$, there exist bounded and open sets $\Omega_{1,\varepsilon,\rho}$, $\Omega_{2,\varepsilon,\rho} \subset \mathbb{R}^n$ with boundary of class $C^{2,\alpha}$ such that $\Omega_{i,\varepsilon} \subset \Omega_{i,\varepsilon,\rho}$ and

$$\mathrm{Cap}(\Omega_{i,\varepsilon,\rho}) \leqslant \left(1 + \rho^{1/2}\right)\mathrm{Cap}(\Omega_i) + C_\varepsilon\, \rho^{1/2} + \phi(\varepsilon), \tag{6.1.29}$$

for some $C_\varepsilon > 0$, for every $i \in \{1, 2\}$, and $\Omega_{1,\varepsilon,\rho} \cup \Omega_{2,\varepsilon,\rho}$ has boundary of class $C^{2,\alpha}$.

From this and exploiting Lemma 6.1.7, we obtain that

$$\begin{aligned}\mathrm{Cap}(\Omega_1)+\mathrm{Cap}(\Omega_2) &\geqslant \frac{\mathrm{Cap}(\Omega_{1,\varepsilon,\rho})+\mathrm{Cap}(\Omega_{2,\varepsilon,\rho})-2C_\varepsilon\rho^{1/2}-2\phi(\varepsilon)}{1+\rho^{1/2}}\\ &\geqslant \frac{\mathrm{Cap}(\Omega_{1,\varepsilon,\rho}\cup\Omega_{2,\varepsilon,\rho})-2C_\varepsilon\rho^{1/2}-2\phi(\varepsilon)}{1+\rho^{1/2}}\\ &\geqslant \frac{\mathrm{Cap}(\Omega_1\cup\Omega_2)-2C_\varepsilon\rho^{1/2}-2\phi(\varepsilon)}{1+\rho^{1/2}}.\end{aligned}$$

Accordingly, sending $\rho \searrow 0$,

$$\mathrm{Cap}(\Omega_1)+\mathrm{Cap}(\Omega_2)+2\phi(\varepsilon) \geqslant \mathrm{Cap}(\Omega_1\cup\Omega_2)$$

and sending $\varepsilon \searrow 0$, we obtain (6.1.28), as desired. □

It is interesting to observe that the notion of conductor potential introduced in Proposition 6.1.2 can also be extended to arbitrary bounded sets by leveraging the setting in (6.1.24). In this situation, the conductor potential is not necessarily continuous at the boundary of the set, yet it maintains useful harmonicity and energy properties.

Corollary 6.1.14. *Let $n \geqslant 3$ and $\Omega \subset \mathbb{R}^n$ be a bounded set.*

Then, there exists a unique $u \in C^2(\mathbb{R}^n \setminus \overline{\Omega}) \cap \mathcal{D}^{1,2}(\mathbb{R}^n)$ such that

$$\begin{cases}\Delta u = 0 & \text{in } \mathbb{R}^n \setminus \overline{\Omega},\\ u = 1 & \text{a.e. in } \Omega.\end{cases} \tag{6.1.30}$$

Additionally, $0 \leqslant u(x) \leqslant 1$ for a.e. $x \in \mathbb{R}^n$ and

$$\mathrm{Cap}(\Omega) = \int_{\mathbb{R}^n\setminus\overline{\Omega}} |\nabla u(x)|^2\,dx. \tag{6.1.31}$$

Furthermore, if $x \in \mathbb{R}^n \setminus \overline{\Omega}$ and $R > r > 0$ are such that $B_r(x) \subseteq \mathbb{R}^n \setminus \Omega$ and $B_R(x) \supseteq \Omega$, then

$$\frac{c_n\,\mathrm{Cap}(\Omega)}{R^{n-2}} \leqslant u(x) \leqslant \frac{c_n\,\mathrm{Cap}(\Omega)}{r^{n-2}}, \tag{6.1.32}$$

where c_n is the positive constant in (2.7.2).

Proof. For every $j \in \mathbb{N}$, we use (6.1.24) to pick a set Ω_j which is open, bounded and with boundary of class $C^{2,\alpha}$ such that $\Omega \Subset \Omega_j$ and

$$\lim_{j\to+\infty} \mathrm{Cap}(\Omega_j) = \mathrm{Cap}(\Omega). \tag{6.1.33}$$

Without loss of generality, we can suppose that

$$\Omega_j \subset \bigcup_{p\in\Omega} B_{1/j}(p), \tag{6.1.34}$$

in light of Lemma 6.1.7 since reducing the domain also reduces the capacity.

Furthermore, we claim that

$$\text{without loss of generality, we can suppose that } \Omega_{j+1} \subset \Omega_j. \tag{6.1.35}$$

To prove it, we take $\rho_j \in (0, 1)$ (to be specified later in the proof), and we consider bounded and open sets $\Omega_j^\#$ with boundary of class $C^{2,\alpha}$ and such that

$$\bigcup_{p\in\Omega_j} B_{\rho_j/2}(p) \subset \Omega_j^\# \subset \bigcup_{p\in\Omega_j} B_{\rho_j}(p).$$

Now, we define $\widetilde{\Omega}_j := \Omega_1 \cap \cdots \cap \Omega_j$ and note that

$$\widetilde{\Omega}_{j+1} \subset \widetilde{\Omega}_j. \tag{6.1.36}$$

Furthermore, we consider the sets $\widehat{\Omega}_j$ with boundary of class $C^{2,\alpha}$ and such that

$$\bigcup_{p\in\widetilde{\Omega}_j} B_{\rho_j/4}(p) \subset \widehat{\Omega}_j \subset \bigcup_{p\in\widetilde{\Omega}_j} B_{\rho_j/2}(p). \tag{6.1.37}$$

We remark that $\widehat{\Omega}_j \subset \Omega_j^\#$. Using this and Lemma 6.1.10, we see that

$$\text{Cap}(\widehat{\Omega}_j) \leqslant \text{Cap}(\Omega_j^\#) \leqslant \left(1 + \rho_j^{1/2}\right) \text{Cap}(\Omega_j) + C_j\, \rho_j^{1/2},$$

for some $C_j > 0$. Thus, choosing $\rho_0 := 1$ and

$$\rho_j := \min\left\{\frac{1}{j}, \frac{1}{jC_j^2}, \frac{\rho_{j-1}}{10}\right\}, \tag{6.1.38}$$

we obtain that

$$\text{Cap}(\widehat{\Omega}_j) \leqslant \left(1 + \frac{1}{\sqrt{j}}\right) \text{Cap}(\Omega_j) + \frac{1}{\sqrt{j}}.$$

As a consequence of this and (6.1.33),

$$\lim_{j\to+\infty} \text{Cap}(\widehat{\Omega}_j) = \text{Cap}(\Omega).$$

Hence, to complete the proof of (6.1.35), it remains to check that

$$\widehat{\Omega}_{j+1} \subset \widehat{\Omega}_j.$$

Indeed, by (6.1.36), (6.1.37) and (6.1.38),

$$\widehat{\Omega}_{j+1} \subset \bigcup_{p\in\widetilde{\Omega}_{j+1}} B_{\rho_{j+1}/2}(p) \subset \bigcup_{p\in\widetilde{\Omega}_j} B_{\rho_{j+1}/2}(p) \subset \bigcup_{p\in\widetilde{\Omega}_j} B_{\rho_j/20}(p) \subset \widehat{\Omega}_j,$$

as desired.

Accordingly, hereinafter, we assume that $\Omega_{j+1} \subset \Omega_j$. Moreover, we can also suppose that there exists $R > 0$ such that $\Omega_j \subset B_R$ for every $j \in \mathbb{N}$; therefore, if u_j denotes the conductor potential of the set Ω_j, for every $j \in \mathbb{N}$,

$$\int_{\mathbb{R}^n\setminus\overline{\Omega}} |\nabla u_j(x)|^2\, dx = \mathrm{Cap}(\Omega_j) \leqslant \mathrm{Cap}(B_R).$$

Consequently, up to a subsequence, we can assume that

$$\begin{aligned}&\nabla u_j \text{ converges weakly in } L^2(\mathbb{R}^n),\\ &\text{and } u_j \text{ converges strongly in } L^2_{\mathrm{loc}}(\mathbb{R}^n) \text{ and a.e. in } \mathbb{R}^n \text{ to a function } u,\end{aligned} \tag{6.1.39}$$

which is harmonic in $\mathbb{R}^n \setminus \overline{\Omega}$ thanks to Corollary 2.2.2.

Furthermore, we have that $0 \leqslant u \leqslant 1$ since the same holds for u_j, and $u = 1$ a.e. in Ω, thanks to the pointwise convergence in (6.1.39).

Now, to prove (6.1.31), for every $j, k \in \mathbb{N}$, we define $v_{j,k} := \frac{u_{j+k}+u_j}{2}$ and note that $v_{j,k} = 1$ in Ω_{j+k}. Accordingly,

$$\begin{aligned}\mathrm{Cap}(\Omega_{j+k}) &\leqslant \int_{\mathbb{R}^n} |\nabla v_{j,k}(x)|^2\, dx\\ &= \frac{1}{4}\int_{\mathbb{R}^n} \big(|\nabla u_{j+k}(x)|^2 + |\nabla u_j(x)|^2 + 2\nabla u_{j+k}(x)\cdot\nabla u_j(x)\big)\, dx\\ &\leqslant \frac{1}{2}\int_{\mathbb{R}^n} \big(|\nabla u_{j+k}(x)|^2 + |\nabla u_j(x)|^2\big)\, dx\\ &= \frac{1}{2}\big(\mathrm{Cap}(\Omega_{j+k}) + \mathrm{Cap}(\Omega_j)\big).\end{aligned}$$

In light of (6.1.33), this gives that

$$\lim_{j\to+\infty} \int_{\mathbb{R}^n} |\nabla v_{j,k}(x)|^2\, dx = \mathrm{Cap}(\Omega). \tag{6.1.40}$$

Also, we observe that

$$\begin{aligned}
&\frac{1}{4}\int_{\mathbb{R}^n} |\nabla u_{j+k}(x) - \nabla u_j(x)|^2\,dx \\
&= \frac{1}{4}\int_{\mathbb{R}^n} \left(|\nabla u_{j+k}(x)|^2 + |\nabla u_j(x)|^2 - 2\nabla u_{j+k}(x)\cdot\nabla u_j(x)\right)dx \\
&= \frac{1}{4}\int_{\mathbb{R}^n} \left(2|\nabla u_{j+k}(x)|^2 + 2|\nabla u_j(x)|^2 - 4|\nabla v_{j,k}(x)|^2\right)dx \\
&= \frac{1}{2}\left(\mathrm{Cap}(\Omega_{j+k}) + \mathrm{Cap}(\Omega_j)\right) - \int_{\mathbb{R}^n} |\nabla v_{j,k}(x)|^2\,dx.
\end{aligned}$$

From this, (6.1.33) and (6.1.40), we obtain that

$$\begin{aligned}
&\nabla u_j \text{ is a Cauchy sequence in } L^2(\mathbb{R}^n)\\
&\text{and therefore it strongly converges.}
\end{aligned} \tag{6.1.41}$$

Consequently, recalling also (6.1.33),

$$\mathrm{Cap}(\Omega) = \lim_{j\to+\infty} \mathrm{Cap}(\Omega_j) = \lim_{j\to+\infty}\int_{\mathbb{R}^n\setminus\overline{\Omega}} |\nabla u_j(x)|^2\,dx = \int_{\mathbb{R}^n\setminus\overline{\Omega}} |\nabla u(x)|^2\,dx,$$

which establishes (6.1.31), as desired.

We now prove the uniqueness statement. For this, we set

$$X_0 := \Big\{ v \in C_0^\infty(\mathbb{R}^n) \text{ such that there exists an open set } \Omega' \text{ such that } \Omega \Subset \Omega' \text{ and } v = 1 \text{ in } \Omega' \Big\},$$

and we define X to be the closure of X_0 in $\mathcal{D}^{1,2}(\mathbb{R}^n)$. We claim that

$$u \in X. \tag{6.1.42}$$

To prove it, we recall (6.1.39) and note that

$$u_j \in C^{2,\alpha}(\mathbb{R}^n \setminus \Omega_j) \cap \mathcal{D}^{1,2}(\mathbb{R}^n \setminus \Omega_j) \cap L^{\frac{2n}{n-2}}(\mathbb{R}^n \setminus \Omega_j),$$

thanks to the Gagliardo–Nirenberg–Sobolev inequality, see e.g. [Leo09, Theorem 12.4].

We take $R > 0$ such that $\overline{\Omega} \subset \Omega_j \subset B_R$, and we define a function $\tau \in C_0^\infty(\mathbb{R}^n, [0,1])$ such that $\tau = 1$ in B_1 and $\tau = 0$ in $\mathbb{R}^n \setminus B_2$. We set $\tau_R(x) := \tau\left(\frac{x}{R}\right)$ and, for every $j \in \mathbb{N}$, $a_j := \tau_{2R} u_j$ and $b_j := (1 - \tau_{2R}) u_j$.

Note that $a_j \in C_0^{2,\alpha}(\mathbb{R}^n \setminus \overline{\Omega}_j)$. Thus, given $\varepsilon > 0$, taking a mollifier η_ε and the corresponding mollification $a_{j,\varepsilon} := a_j * \eta_\varepsilon$, we have that there exists ε_j sufficiently small such that

$$a_{j,\varepsilon_j} \in X_0. \tag{6.1.43}$$

Indeed, $a_{j,\varepsilon} \in C_0^\infty(\mathbb{R}^n)$. Also,

$$\text{if } y \in \Omega_j \subset B_R, \text{ then } a_j(y) = \tau_{2R}(y)u_j(y) = u_j(y) = 1. \tag{6.1.44}$$

We denote by d_j the distance between Ω and $\partial\Omega_j$, that is,

$$d_j := \inf_{\substack{x\in\Omega\\ y\in\partial\Omega_j}} |x-y| = \min_{\substack{x\in\overline{\Omega}\\ y\in\partial\Omega_j}} |x-y|,$$

and we observe that $d_j > 0$. We take an open set $\widehat{\Omega}_j$ with boundary of class $C^{2,\alpha}$ such that

$$\Omega \Subset \widehat{\Omega}_j \subset \left\{x \in \Omega_j \text{ s.t. } d_{\partial\Omega_j}(x) > \frac{d_j}{2}\right\}.$$

Here, $d_{\partial\Omega_j}$ is the signed distance function to $\partial\Omega_j$ given at the beginning of Section 2.3 (that is, $d_{\partial\Omega_j} \geqslant 0$ in Ω_j and $d_{\partial\Omega_j} \leqslant 0$ in $\mathbb{R}^n\backslash\Omega_j$). In this way, taking $\varepsilon \in (0, d_j/4)$, if $x \in \widehat{\Omega}_j$ and $y \in B_\varepsilon(x)$, then $y \in \Omega_j$; therefore, using (6.1.44),

$$a_{j,\varepsilon}(x) = \int_{B_\varepsilon(x)} a_j(y)\eta_\varepsilon(x-y)\,dy = \int_{B_\varepsilon(x)} \eta_\varepsilon(x-y)\,dy = 1.$$

This establishes (6.1.43).

We also have that

$$a_{j,\varepsilon} \text{ converges to } a_j \text{ in } \mathcal{D}^{1,2}(\mathbb{R}^n) \text{ as } \varepsilon \searrow 0, \tag{6.1.45}$$

see e.g. Theorem 9.6 in [WZ15].

Furthermore, since $b_j \in \mathcal{D}^{1,2}(\mathbb{R}^n)$, there exists $b_{j,k} \in C_0^\infty(\mathbb{R}^n)$ such that

$$b_{j,k} \text{ converges to } b_j \text{ in } \mathcal{D}^{1,2}(\mathbb{R}^n) \text{ as } k \to +\infty. \tag{6.1.46}$$

Without loss of generality, we can suppose that

$$b_{j,k} = 0 \text{ in } B_R \supset \overline{\Omega}. \tag{6.1.47}$$

Indeed, if not, define $\widetilde{b}_{j,k} := (1-\tau_R)b_{j,k}$ and observe that $\widetilde{b}_{j,k} \in C_0^\infty(\mathbb{R}^n)$ and $\widetilde{b}_{j,k} = 0$ in $B_R \supset \overline{\Omega}$. Also, $\widetilde{b}_{j,k} - b_j = b_{j,k} - b_j$ in $\mathbb{R}^n \setminus B_{2R}$ and $\widetilde{b}_{j,k} - b_j = (1-\tau_R)b_{j,k}$ in B_{2R}, and so

$$\begin{aligned}
&\int_{\mathbb{R}^n} |\nabla(\widetilde{b}_{j,k} - b_j)(x)|^2\,dx \\
&= \int_{B_{2R}} |\nabla(\widetilde{b}_{j,k} - b_j)(x)|^2\,dx + \int_{\mathbb{R}^n\backslash B_{2R}} |\nabla(\widetilde{b}_{j,k} - b_j(x))|^2\,dx \\
&= \int_{B_{2R}\backslash B_R} |\nabla\big((1-\tau_R(x))b_{j,k}\big)(x)|^2\,dx + \int_{\mathbb{R}^n\backslash B_{2R}} |\nabla(b_{j,k} - b_j)(x)|^2\,dx
\end{aligned}$$

$$\leqslant 2\left(\int_{B_{2R}\setminus B_R} |\nabla(1-\tau_R(x))|^2|b_{j,k}(x)|^2\,dx + \int_{B_{2R}\setminus B_R} |\nabla b_{j,k}(x)|^2\,dx\right)$$
$$+\int_{\mathbb{R}^n\setminus B_{2R}} |\nabla(b_{j,k}-b_j)(x)|^2\,dx$$
$$\leqslant 2\left(\int_{B_{2R}\setminus B_R} |\nabla(1-\tau_R(x))|^2|b_{j,k}(x)|^2\,dx + \int_{\mathbb{R}^n} |\nabla(b_{j,k}-b_j)(x)|^2\,dx\right)$$
$$\leqslant C\left(\frac{1}{R^2}\int_{B_{2R}\setminus B_R} |b_{j,k}(x)-b_j(x)|^2\,dx + \int_{\mathbb{R}^n} |\nabla(b_{j,k}-b_j)(x)|^2\,dx\right),$$

for some $C > 0$. Hence, thanks to (6.1.46), we deduce from this computation that $\widetilde{b}_{j,k}$ converges to b_j in $\mathcal{D}^{1,2}(\mathbb{R}^n)$ as $k \to +\infty$, and this completes the proof of (6.1.47).

Now, in light of (6.1.45) and (6.1.46), we can take ε_j sufficiently small (possibly smaller than ε_j in (6.1.43)) and k_j sufficiently large such that

$$\int_{\mathbb{R}^n} \left|\nabla(a_{j,\varepsilon_j}-a_j)(x)\right|^2 dx \leqslant \frac{1}{j}$$
$$\text{and} \quad \int_{\mathbb{R}^n} \left|\nabla(b_{j,k_j}-b_j)(x)\right|^2 dx \leqslant \frac{1}{j}.$$

As a consequence, noting that $a_j + b_j = u_j$, if ε_j sufficiently small and k_j is sufficiently large,

$$\int_{\mathbb{R}^n} \left|\nabla(a_{j,\varepsilon_j}+b_{j,k_j}-u_j)(x)\right|^2 dx \leqslant \frac{4}{j}.$$

From this and (6.1.41), we conclude that

$$\begin{aligned}
&\lim_{j\to+\infty}\sqrt{\int_{\mathbb{R}^n} \left|\nabla(a_{j,\varepsilon_j}+b_{j,k_j}-u)(x)\right|^2 dx}\\
&\leqslant \lim_{j\to+\infty}\left(\sqrt{\int_{\mathbb{R}^n} \left|\nabla(a_{j,\varepsilon_j}+b_{j,k_j}-u_j)(x)\right|^2 dx}+\sqrt{\int_{\mathbb{R}^n} \left|\nabla(u_j-u)(x)\right|^2 dx}\right)\\
&= 0.
\end{aligned}\tag{6.1.48}$$

Moreover, from (6.1.43) and (6.1.47), it follows that $a_{j,\varepsilon_j} + b_{j,k_j} \in X_0$. This fact and (6.1.48) establish (6.1.42).

Now, we take $u_1,\ u_2 \in C^2(\mathbb{R}^n \setminus \overline{\Omega}) \cap \mathcal{D}^{1,2}(\mathbb{R}^n)$ to be solutions of (6.1.30). Thanks to (6.1.42), we have that $u_1,\ u_2 \in X$; therefore, there exist functions $\varphi_{1,j}$, $\varphi_{2,j} \in C_0^\infty(\mathbb{R}^n)$ such that $\varphi_{1,j}$ and $\varphi_{2,j}$ converge to u_1 and u_2 in $\mathcal{D}^{1,2}(\mathbb{R}^n)$ as $j \to +\infty$, respectively, and open sets $\Omega'_{1,j}$ and $\Omega'_{2,j}$ such that $\Omega \Subset \Omega'_{1,j} \cap \Omega'_{2,j}$ and $\varphi_{1,j} = 1$ in $\Omega'_{1,j}$ and $\varphi_{j,2} = 1$ in $\Omega'_{2,j}$.

We set $\widetilde{u} := u_1 - u_2$ and $\varphi_j := \varphi_{1,j} - \varphi_{2,j}$, and we note that $\varphi_j \in C_0^\infty(\mathbb{R}^n)$ and φ_j converges to $\widetilde{u}$ in $\mathcal{D}^{1,2}(\mathbb{R}^n)$ as $j \to +\infty$. Also, setting $\Omega'_j := \Omega'_{1,j} \cap \Omega'_{2,j}$, we have that Ω'_j is an open set and $\Omega \Subset \Omega'_j$. Moreover, $\varphi_j = 0$ in Ω'_j. We also remark that $\widetilde{u}$ is harmonic in $\mathbb{R}^n \setminus \overline{\Omega}$.

Now, we take an open set $\widetilde{\Omega}_j$ with boundary of class $C^{2,\alpha}$ such that $\Omega \Subset \widetilde{\Omega}_j \Subset \Omega'_j$, and we exploit Green's first identity in (2.1.10) to see that

$$\int_{\mathbb{R}^n \setminus \overline{\Omega}} \nabla \varphi_j(x) \cdot \nabla \widetilde{u}(x)\, dx = \int_{\partial \widetilde{\Omega}_j} \varphi_j(x) \frac{\partial \widetilde{u}}{\partial \nu}(x)\, d\mathcal{H}_x^{n-1} = 0.$$

Consequently, passing to the limit in j,

$$\int_{\mathbb{R}^n \setminus \overline{\Omega}} |\nabla \widetilde{u}(x)|^2\, dx = 0,$$

from which it follows that $\widetilde{u}$ is constant. Since $\widetilde{u} \in L^{\frac{2n}{n-2}}(\mathbb{R}^n \setminus \overline{\Omega})$ (thanks to the Gagliardo–Nirenberg–Sobolev inequality, see e.g. [Leo09, Theorem 12.4]), this gives that $\widetilde{u}$ vanishes identically, whence u_1 coincides with u_2, and the uniqueness statement is thereby proved.

Now, we focus on the proof of (6.1.32). To this end, we employ Proposition 6.1.6 and see that, if $x \in \mathbb{R}^n \setminus \overline{\Omega_j}$ and $B_{r_j}(x) \subseteq \mathbb{R}^n \setminus \Omega_j$ and $B_{R_j}(x) \supseteq \Omega_j$, then

$$\frac{c_n \operatorname{Cap}(\Omega_j)}{R_j^{n-2}} \leqslant u_j(x) \leqslant \frac{c_n \operatorname{Cap}(\Omega_j)}{r_j^{n-2}}. \tag{6.1.49}$$

Now, we take a set $\mathcal{Z} \subset \mathbb{R}^n$ with null Lebesgue measure and such that $u_j \to u$ in $\mathbb{R}^n \setminus \mathcal{Z}$. We pick $x \in \mathbb{R}^n \setminus (\overline{\Omega} \cup \mathcal{Z})$ and consider $B_r(x) \subseteq \mathbb{R}^n \setminus \Omega$ and $B_R(x) \supseteq \Omega$.

Moreover, in view of (6.1.34), we have that $B_{r_j}(x) \subseteq \mathbb{R}^n \setminus \Omega_j$ and $B_{R_j}(x) \supseteq \Omega_j$, where $r_j := r - \frac{2}{j}$ and $R_j := R + \frac{2}{j}$.

In this setting, we can exploit (6.1.49) and then send $j \to +\infty$. In this way, we conclude that (6.1.32) holds true for all $x \in \mathbb{R}^n \setminus (\overline{\Omega} \cup \mathcal{Z})$, that is a.e. in $\mathbb{R}^n \setminus \overline{\Omega}$. Since u is harmonic and hence continuous in $\mathbb{R}^n \setminus \overline{\Omega}$, this guarantees that (6.1.32) holds true for all $x \in \mathbb{R}^n \setminus \overline{\Omega}$, as desired. □

In the spirit of Lemma 6.1.13, we also have the following.

Lemma 6.1.15. *The result in Lemma 6.1.8 holds true a.e. in $\mathbb{R}^n$ for every bounded set Ω_1, $\Omega_2 \subset \mathbb{R}^n$, adopting the definition in* (6.1.24).

Proof. Let u_1, u_2 and u_3 be the conductor potentials relative to the sets Ω_1, Ω_2 and $\Omega_1 \cup \Omega_2$, respectively, according to Corollary 6.1.14.

Moreover, in light of Corollary 6.1.12, for $i \in \{1, 2\}$, we can find open and bounded sets $\Omega_{i,j}$ with boundaries of class $C^{2,\alpha}$ such that $\Omega_i \subset \Omega_{i,j}$ and

$$\lim_{j\to+\infty} \mathrm{Cap}(\Omega_{i,j}) = \mathrm{Cap}(\Omega_i)$$

and $\Omega_{1,j} \cup \Omega_{2,j}$ has boundary of class $C^{2,\alpha}$. We also let $u_{i,j}$ be the conductor potentials relative to $\Omega_{i,j}$, and we exploit (6.1.39) and (6.1.41) to say that $u_{i,j}$ converges to u_i in $\mathbb{R}^n$ (up to a set of measure zero that we disregard) as $j \to +\infty$.

Let now $u := u_{1,j} + u_{2,j} - u_3$. This function is harmonic outside $\Omega_{1,j} \cup \Omega_{2,j}$, and it belongs to $\mathcal{D}^{1,2}(\mathbb{R}^n)$. Furthermore, on $\partial(\Omega_{1,j} \cup \Omega_{2,j}) \subseteq (\partial\Omega_{1,j}) \cup (\partial\Omega_{2,j})$, we have that either $u_{1,j} = 1$ or $u_{2,j} = 1$;therefore, $u \geqslant 1 - u_3 \geqslant 0$.

As a result, we can apply Lemma 2.9.5 and deduce that $u \geqslant 0$ in $\mathbb{R}^n$. This yields that $u_{1,j} + u_{2,j} \geqslant u_3$, whence, sending $j \to +\infty$, we see that $u_1 + u_2 \geqslant u_3$ a.e. in $\mathbb{R}^n$. □

The conductor potentials for general bounded sets (as constructed in Corollary 6.1.14) is not necessarily continuous along the boundary of Ω (differently from the case of sets with boundaries of class $C^{2,\alpha}$ treated in Proposition 6.1.2). Hence, in the general setting of Corollary 6.1.14, the conductor potential may not need to attain a value of 1 continuously along $\partial\Omega$, but only in the Sobolev trace sense. The possible continuity of the conductor potentials (locally in the complement of a given domain) is closely related to the solvability of the Dirichlet problem, since, in broad terms, the conductor potential can play the role of a useful barrier at boundary points. In this framework, we have the following result.

Proposition 6.1.16. *Let $n \geqslant 3$ and $\Omega \subset \mathbb{R}^n$ be open, bounded and connected. Then, the following statements are equivalent:*

(i) *For every $p \in \partial\Omega$, there exists a sequence $\rho_j > 0$ with $\rho_j \to 0$ as $j \to +\infty$ such that if $u_{p,j}$ is the conductor potential of $B_{\rho_j}(p) \setminus \Omega$, as given by Corollary 6.1.14, we have that, for all $j \in \mathbb{N}$,*

$$\lim_{\Omega\ni x\to p} u_{p,j}(x) = 1. \tag{6.1.50}$$

(ii) *For every $p \in \partial\Omega$ and every sequence $\rho_j > 0$, with $\rho_j \to 0$ as $j \to +\infty$ such that if $u_{p,j}$ is the conductor potential of $B_{\rho_j}(p) \setminus \Omega$, as given by Corollary 6.1.14, we have that, for all $j \in \mathbb{N}$,*

$$\lim_{\Omega\ni x\to p} u_{p,j}(x) = 1.$$

(iii) *For every* $g \in C(\partial\Omega)$, *there exists a unique solution* $u \in C^2(\Omega) \cap C(\overline{\Omega})$ *of the Dirichlet problem*

$$\begin{cases} \Delta u = 0 & \textit{in } \Omega, \\ u = g & \textit{on } \partial\Omega. \end{cases}$$

Proof. We prove that (i) implies (iii), which implies (ii) (this completes the proof, since obviously (ii) implies (i)). Assume (i), pick a point $p \in \partial\Omega$ and let

$$\beta := 1 - \sum_{k=1}^{+\infty} \frac{u_{p,k}}{2^k}.$$

We use (6.1.50) to see that

$$\lim_{\Omega \ni x \to p} \beta(x) = 1 - \sum_{k=1}^{+\infty} \frac{u_{p,k}(p)}{2^k} = 1 - \sum_{k=1}^{+\infty} \frac{1}{2^k} = 0.$$

We also observe that β is harmonic (and therefore superharmonic) in Ω.

We claim that

$$\beta > 0 \text{ in } \overline{\Omega \cap B_{\rho_1}(p)} \setminus \{p\}. \tag{6.1.51}$$

Indeed, by construction, $0 \leqslant \beta \leqslant 1$; hence, if (6.1.51) were not true, there would exist $q \in \overline{\Omega \cap B_{\rho_1}(p)} \setminus \{p\}$ such that $\beta(q) = 0$. That is, there would exist a sequence $q_j \in \Omega \cap B_{\rho_1}(p)$ such that $q_j \to q \neq p$ and $\beta(q_j) \to 0$:

$$\lim_{j \to +\infty} \sum_{k=1}^{+\infty} \frac{u_{p,k}(q_j)}{2^k} = 1. \tag{6.1.52}$$

We let $\rho' := |p - q| > 0$ and $j' \in \mathbb{N}$ be such that $|q - q_j| \leqslant \frac{\rho'}{4}$ for all $j \geqslant j'$. We also take $j'' \in \mathbb{N}$ such that $\rho_j \leqslant \frac{\rho'}{8}$ for all $j \geqslant j''$, and we define $J := \max\{j', j''\}$. We observe that if $j \geqslant J$ and $y \in B_{\rho'/4}(q_j)$, then

$$|p - y| \geqslant |p - q| - |q - q_j| - |q_j - y| \geqslant \rho' - \frac{\rho'}{4} - \frac{\rho'}{4} = \frac{\rho'}{2} > \rho_J,$$

whence

$$B_{\rho'/4}(q_j) \subseteq \mathbb{R}^n \setminus B_{\rho_J}(p) \subseteq \mathbb{R}^n \setminus (B_{\rho_J}(p) \setminus \Omega).$$

Therefore, recalling (6.1.32), if $j \geqslant J$,

$$u_{p,J}(q_j) \leqslant \frac{c_n \operatorname{Cap}(B_{\rho_J}(p) \setminus \Omega)}{(\rho'/4)^{n-2}}. \tag{6.1.53}$$

Also, $B_{\rho_J}(p)\setminus\Omega \subseteq B_{\rho_J}(p) \subseteq B_{\rho'/8}(p)$, and we thereby deduce from Lemma 6.1.7 (recall Lemma 6.1.13) that $\mathrm{Cap}(B_{\rho_J}(p)\setminus\Omega) \leqslant \mathrm{Cap}(B_{\rho'/8}(p))$. This and Corollary 6.1.5 yield that

$$\mathrm{Cap}(B_{\rho_J}(p)\setminus\Omega) \leqslant \frac{(\rho'/8)^{n-2}}{c_n}.$$

As a result, using (6.1.53), if $j \geqslant J$,

$$u_{p,J}(q_j) \leqslant \frac{(\rho'/8)^{n-2}}{(\rho'/4)^{n-2}} = \frac{1}{2^{n-2}}.$$

Thus, in light of (6.1.52),

$$\begin{aligned} 1 &= \lim_{j\to+\infty} \sum_{k=1}^{+\infty} \frac{u_{p,k}(q_j)}{2^k} \\ &\leqslant \lim_{j\to+\infty} \sum_{\substack{k\geqslant 1\\ k\neq J}} \frac{u_{p,k}(q_j)}{2^k} + \frac{1}{2^{n-2+J}} \\ &\leqslant \sum_{\substack{k\geqslant 1\\ k\neq J}} \frac{1}{2^k} + \frac{1}{2^{n-2+J}} \\ &= \sum_{k\geqslant 1} \frac{1}{2^k} + \frac{1}{2^{n-2+J}} - \frac{1}{2^J} \\ &= 1 + \frac{1}{2^{n-2+J}} - \frac{1}{2^J} \\ &< 1. \end{aligned}$$

This is a contradiction; therefore, we have completed the proof of (6.1.51).

These pieces of information, together with (3.3.34), show that if u is the harmonic function constructed by the Perron method in Theorem 3.3.2, then u belongs to $C(\overline{\Omega})$, and hence it is a solution of the Dirichlet problem in (iii) (the uniqueness of such a solution is entailed by Corollary 2.9.3).

Let us now assume that (iii) holds true. We take $p \in \partial\Omega$ and $\rho > 0$. Let $u_{p,\rho}$ be the conductor potential of $B_\rho(p)\setminus\Omega$, as given by Corollary 6.1.14. To establish (ii), it suffices to show that

$$\lim_{\Omega\ni x\to p} u_{p,\rho}(x) = 1. \tag{6.1.54}$$

To this end, let $R > \rho$ and $\mathcal{B} := B_\rho(p)\setminus\Omega$. We consider boundary data equal to 0 on $\partial B_R(p)$ and equal to 1 on $\partial\mathcal{B}$, and we construct the corresponding harmonic function v in $B_R(p)\setminus\overline{\mathcal{B}}$, as given by the Perron method in Theorem 3.3.2.

We also define $g(x) := |x - p|$ and use (iii) to find a solution $\beta \in C^2(\Omega) \cap C(\overline{\Omega})$ of the Dirichlet problem

$$\begin{cases} \Delta\beta = 0 & \text{in } \Omega, \\ \beta = g & \text{on } \partial\Omega. \end{cases}$$

We claim that

$$\beta(x) > 0 \quad \text{for every } x \in \overline{\Omega \cap B_\rho(p)} \setminus \{p\}. \tag{6.1.55}$$

Indeed, by the weak maximum principle in Corollary 2.9.2(iii), we know that, for all $x \in \Omega$,

$$\beta(x) \geqslant \inf_{\partial\Omega} g = g(p) = 0.$$

This and the strong maximum principle in Theorem 2.9.1 give that $\beta > 0$ in Ω. Combining this with the fact that $\beta(x) = g(x) > 0$ for all $x \in (\partial\Omega) \setminus \{p\}$, we obtain (6.1.55), as desired.

Moreover,

$$\lim_{\Omega \ni x \to p} \beta(x) = g(p) = 0.$$

This and (6.1.55) allow us to exploit the barrier technique in (3.3.34) for the function v and conclude that

$$\lim_{B_R(p)\setminus\overline{\mathcal{B}} \ni x \to p} v(x) = 1. \tag{6.1.56}$$

We also remark that for every $y \in \partial B_R(p)$.

$$\lim_{B_R(p)\setminus\overline{\mathcal{B}} \ni x \to y} v(x) = 0 \tag{6.1.57}$$

since along $\partial B_R(p)$, we can use the exterior cone condition, exploit the barrier constructed in the proof of Theorem 3.3.3 (see in particular the computations between formula (3.3.41) and the end of the proof) and rely on (3.3.34).

Additionally,

$$\sup_{B_R(p)\setminus\overline{\mathcal{B}}} v(x) \leqslant 1. \tag{6.1.58}$$

Indeed, let $\mathcal{F}$ be the class of functions $w \in C(B_R(p) \setminus \overline{\mathcal{B}})$ which are subharmonic in $B_R(p) \setminus \overline{\mathcal{B}}$ and such that

$$\limsup_{B_R(p)\setminus\overline{\mathcal{B}} \ni x \to y} w(x) \leqslant 0 \quad \text{for every } y \in \partial B_R(p)$$

$$\text{and} \qquad \limsup_{B_R(p)\setminus\overline{\mathcal{B}} \ni x \to z} w(x) \leqslant 1 \quad \text{for every } z \in \partial\mathcal{B}.$$

We know by the Perron method construction in Theorem 3.3.2 that

$$v(x) := \sup_{w\in\mathcal{F}} w(x). \tag{6.1.59}$$

Also, if $w \in \mathcal{F}$, we take a sequence of points $x_k \in B_R(p) \setminus \overline{\mathcal{B}}$ such that

$$\lim_{k\to+\infty} w(x_k) = \sup_{B_R(p)\setminus\overline{\mathcal{B}}} w.$$

Up to a subsequence, we can suppose that x_k converges to some $\overline{x} \in \overline{B_R(p) \setminus \mathcal{B}}$ as $k \to +\infty$. Now, in light of Lemma 3.2.5, we have that $\overline{x} \in \partial\big(B_R(p) \setminus \overline{\mathcal{B}}\big)$, and therefore

$$\sup_{B_R(p)\setminus\overline{\mathcal{B}}} w = \limsup_{k\to+\infty} w(x_k) \leqslant 1.$$

This and (6.1.59) give (6.1.58).

Now, we recall (6.1.39), namely that the conductor potential $u_{p,\rho}$ is obtained as the a.e. limit of functions u_j taking values in $[0, 1]$, which are harmonic outside a set $\mathcal{U}_j$ that approaches $B_\rho(p)\setminus\Omega$ from outside as $j \to +\infty$, with $u_j = 1$ in $\mathcal{U}_j \supseteq \mathcal{B}$. We can also suppose that $\mathcal{U}_j \Subset B_R(p)$.

In particular, the function $\widetilde{u}_j := u_j - v$ is harmonic in $B_R(p) \setminus \overline{\mathcal{U}_j}$ and non-negative on $\partial(B_R(p) \setminus \overline{\mathcal{U}_j})$, owing to (6.1.57) and (6.1.58). Accordingly, by the maximum principle in Corollary 2.9.2(iii), we have that $\widetilde{u}_j \geqslant 0$, i.e. $u_j \geqslant v$, in $B_R(p) \setminus \overline{\mathcal{B}}$.

Also, we can set v to be equal to 1 in $\mathcal{B}$; therefore, since $u_j = 1$ in $\mathcal{B}$, we infer that $u_j \geqslant v$ in $B_R(p)$. In particular, recalling (6.1.56),

$$\lim_{\mathcal{B}\ni x\to p} u_{p,\rho}(x) = \lim_{\mathcal{B}\ni x\to p}\ \lim_{j\to+\infty} u_j(x) \geqslant \lim_{\mathcal{B}\ni x\to p} v(x) = 1.$$

From this, we obtain (6.1.54), as desired. □

Now, we reconsider Lemma 6.1.7: such a result suggests that the capacity is a way of measuring how large a set is: however, in light of Proposition 6.1.2 (and the comments following its proof), this notion of largeness is not quite about the usual size of an object but rather on its capability of holding charge. Nonetheless, the notion of capacity can be compared to Lebesgue and Hausdorff measures (see e.g. [EG15] for an introduction to the concept of Hausdorff measure as well as for further information about capacities and measures; see also [MZ97, Section 2.1.7] for additional details and a general setting).

Theorem 6.1.17. *Let $n \geqslant 3$ and Ω be a bounded subset of $\mathbb{R}^n$. Then,*

$$c|\Omega|^{\frac{n-2}{n}} \leqslant \operatorname{Cap}(\Omega) \leqslant C\mathcal{H}^{n-2}(\Omega), \tag{6.1.60}$$

for suitable positive constants c and C, depending only on n.

Proof. We begin by proving the first inequality in (6.1.60). To this end, we take u_j as the conductor potential of a sequence of bounded open sets Ω_j approximating Ω from outside in the setting of (6.1.33) and (6.1.39), and we use the Sobolev–Gagliardo–Nirenberg inequality (see e.g. [Leo09, Theorem 11.2]) to find that

$$|\Omega|^{\frac{n-2}{2n}} \leqslant |\Omega_j|^{\frac{n-2}{2n}} = \left(\int_{\Omega_j} |u_j(x)|^{\frac{2n}{n-2}}\,dx\right)^{\frac{n-2}{2n}} \leqslant \left(\int_{\mathbb{R}^n} |u_j(x)|^{\frac{2n}{n-2}}\,dx\right)^{\frac{n-2}{2n}}$$
$$\leqslant C\left(\int_{\mathbb{R}^n} |\nabla u_j(x)|^2\,dx\right)^{\frac{1}{2}} = C\left(\mathrm{Cap}(\Omega_j)\right)^{\frac{1}{2}} \leqslant C\left(\mathrm{Cap}(\Omega)+\frac{1}{j}\right)^{\frac{1}{2}},$$

for some $C > 0$. This establishes the first inequality in (6.1.60) by sending $j \to +\infty$.

To prove the second inequality in (6.1.60), let $\delta > 0$, and suppose that

$$\Omega \subseteq \bigcup_{i\in\mathbb{N}} B_{r_i}(x_i),$$

for suitable $x_i \in \mathbb{R}^n$ and $r_i \in (0, \delta]$. Then, given $\varepsilon > 0$,

$$\overline{\Omega} \subseteq \bigcup_{j\in\mathbb{N}} B_{r_i+\varepsilon}(x_i),$$

and thus, by compactness, there exists a finite set of indices $i_1, \dots, i_N \in \mathbb{N}$ such that

$$\Omega \subseteq \overline{\Omega} \subseteq \bigcup_{j=0}^{N} B_{r_{i_j}+\varepsilon}(x_{i_j}).$$

Therefore, by Lemma 6.1.7 (as refined in Lemma 6.1.13),

$$\mathrm{Cap}(\Omega) \leqslant \sum_{j=0}^{N} \mathrm{Cap}(B_{r_{i_j}+\varepsilon}(x_{i_j})).$$

Hence, sending $\varepsilon \searrow 0$,

$$\mathrm{Cap}(\Omega) \leqslant \sum_{j=0}^{N} \mathrm{Cap}(B_{r_{i_j}}(x_{i_j})) \leqslant \sum_{i\in\mathbb{N}} \mathrm{Cap}(B_{r_i}(x_i)).$$

This and Corollary 6.1.5 give that

$$c_n\,\mathrm{Cap}(\Omega) \leqslant \sum_{i\in\mathbb{N}} r_i^{n-2}.$$

Taking the infimum over all these possible coverings of Ω and then sending $\delta \searrow 0$, we obtain the second inequality in (6.1.60), as desired. □

6.2 Wiener's Criterion

Returning to the existence result for the Dirichlet problem in Corollary 3.3.4, we are now in a position to characterize, using the notion of capacity, the domains which always allow the continuous solvability of the Dirichlet problem up to the boundary. This classical result is due to[4] Norbert Wiener [Wie24], and it is often referred to as "Wiener's criterion." The bottom line of this criterion is that the Dirichlet problem being solvable is equivalent to the domain allowing "enough space" outside each of its boundary points (compare, for example, with the conditions of external cone or external ball discussed in Theorem 3.3.3, and keep in mind the cuspidal counterexample presented on pp. 288–291) and that to precisely quantify this notion, one needs to look at the capacities of small external neighborhoods of every boundary point. The formal result can be stated as follows.

Theorem 6.2.1. *Let $n \geqslant 3$ and Ω be an open and bounded subset of $\mathbb{R}^n$. The following conditions are equivalent:*

(i) *For every $g \in C(\partial\Omega)$, there exists a function $u \in C^2(\Omega) \cap C(\overline{\Omega})$ that solves the Dirichlet problem*

$$\begin{cases} \Delta u = 0 & \text{in } \Omega, \\ u = g & \text{on } \partial\Omega. \end{cases}$$

(ii) *For every $p \in \partial\Omega$ and every $\lambda \in (0, 1)$,*

$$\sum_{j=0}^{+\infty} \frac{\operatorname{Cap}\Big(\big(B_{\lambda^j}(p) \setminus B_{\lambda^{j+1}}(p)\big) \setminus \Omega\Big)}{\lambda^{(n-2)j}} = +\infty. \tag{6.2.1}$$

Proof. The idea of the proof is to consider a point $p \in \partial\Omega$ and relate the convergence or divergence of the series in (6.2.1) with the barrier condition in (3.3.34).

---

[4] Besides his eminent contributions to partial differential equations and stochastic processes, Wiener is often considered the originator of cybernetics and a pioneer of artificial intelligence. This is possibly the reason why the synthesizer-player Luigi Tonet composed an experimental, and slightly dystopic, piece of electronic music titled "Dedicated to Norbert Wiener" (with the lyrics "We shall continue our life and thank you for your gift of eternity, thank you Norbert Wiener, thank you for everything").

For a short time, Wiener was also a journalist for the Boston Herald, an American daily newspaper (allegedly, he was fired quite soon for his reluctance to write favorable articles about an influential politician).

Up to a translation, we suppose that p is the origin. We note that given $\eta \in (0,1)$, if $\mu := \eta^2$,

$$\sum_{j=0}^{+\infty} \frac{\operatorname{Cap}\Big((B_{\eta^j} \setminus B_{\eta^{j+1}}) \setminus \Omega\Big)}{\eta^{(n-2)j}} = +\infty$$

$$\text{if and only if} \quad \sum_{j=0}^{+\infty} \frac{\operatorname{Cap}\Big((B_{\mu^j} \setminus B_{\mu^{j+1}}) \setminus \Omega\Big)}{\mu^{(n-2)j}} = +\infty. \tag{6.2.2}$$

To check this, we observe that

$$B_{\mu^j} \setminus B_{\mu^{j+1}} = B_{\eta^{2j}} \setminus B_{\eta^{2j+2}} = \big(B_{\eta^{2j}} \setminus B_{\eta^{2j+1}}\big) \cup \big(B_{\eta^{2j+1}} \setminus B_{\eta^{2j+2}}\big).$$

This entails that, for each $i \in \{0, 1\}$,

$$B_{\eta^{2j+i}} \setminus B_{\eta^{2j+i+1}} \subseteq B_{\mu^j} \setminus B_{\mu^{j+1}} \subseteq \big(B_{\eta^{2j}} \setminus B_{\eta^{2j+1}}\big) \cup \big(B_{\eta^{2j+1}} \setminus B_{\eta^{2j+2}}\big),$$

whence, by Lemma 6.1.7 (in its general formulation, as warranted by Lemma 6.1.13),

$$\begin{aligned}\operatorname{Cap}\Big((B_{\eta^{2j+i}} \setminus B_{\eta^{2j+i+1}}) \setminus \Omega\Big) &\leqslant \operatorname{Cap}\Big((B_{\mu^j} \setminus B_{\mu^{j+1}}) \setminus \Omega\Big)\\ &\leqslant \operatorname{Cap}\Big((B_{\eta^{2j}} \setminus B_{\eta^{2j+1}}) \setminus \Omega\Big)\\ &\quad + \operatorname{Cap}\Big((B_{\eta^{2j+1}} \setminus B_{\eta^{2j+2}}) \setminus \Omega\Big).\end{aligned}$$

Dividing by $\mu^{(n-2)j} = \eta^{2(n-2)j}$ and summing up, we obtain

$$\begin{aligned}&\eta^{(n-2)i} \sum_{j=0}^{+\infty} \frac{\operatorname{Cap}\Big((B_{\eta^{2j+i}} \setminus B_{\eta^{2j+i+1}}) \setminus \Omega\Big)}{\eta^{(n-2)(2j+i)}}\\ &\leqslant \sum_{j=0}^{+\infty} \frac{\operatorname{Cap}\Big((B_{\mu^j} \setminus B_{\mu^{j+1}}) \setminus \Omega\Big)}{\mu^{(n-2)j}}\\ &\leqslant \sum_{j=0}^{+\infty} \frac{\operatorname{Cap}\Big((B_{\eta^{2j}} \setminus B_{\eta^{2j+1}}) \setminus \Omega\Big)}{\eta^{2(n-2)j}} + \eta^{n-2} \sum_{j=0}^{+\infty} \frac{\operatorname{Cap}\Big((B_{\eta^{2j+1}} \setminus B_{\eta^{2j+2}}) \setminus \Omega\Big)}{\eta^{(n-2)(2j+1)}}.\end{aligned} \tag{6.2.3}$$

Now, if the first series in (6.2.2) converges, we use (6.2.3) and the fact that $\mu \leqslant 1$ to find that

$$\begin{aligned}
&\sum_{j=0}^{+\infty} \frac{\operatorname{Cap}\left((B_{\mu^j} \setminus B_{\mu^{j+1}}) \setminus \Omega\right)}{\mu^{(n-2)j}} \\
&\quad \leqslant \sum_{j=0}^{+\infty} \frac{\operatorname{Cap}\left((B_{\eta^{2j}} \setminus B_{\eta^{2j+1}}) \setminus \Omega\right)}{\eta^{2(n-2)j}} + \sum_{j=0}^{+\infty} \frac{\operatorname{Cap}\left((B_{\eta^{2j+1}} \setminus B_{\eta^{2j+2}}) \setminus \Omega\right)}{\eta^{(n-2)(2j+1)}} \\
&\quad = \sum_{j=0}^{+\infty} \frac{\operatorname{Cap}\left((B_{\eta^j} \setminus B_{\eta^{j+1}}) \setminus \Omega\right)}{\eta^{(n-2)j}} < +\infty,
\end{aligned}$$

showing that the second series in (6.2.2) converges as well.

If instead the first series in (6.2.2) diverges, we exploit (6.2.3) by dividing the first inequality there by $\eta^{(n-2)i}$ and summing over $i \in \{0, 1\}$: in this way, we obtain that

$$\begin{aligned}
+\infty &= \sum_{j=0}^{+\infty} \frac{\operatorname{Cap}\left((B_{\eta^j} \setminus B_{\eta^{j+1}}) \setminus \Omega\right)}{\eta^{(n-2)j}} = \sum_{i=0}^{1} \left[\sum_{j=0}^{+\infty} \frac{\operatorname{Cap}\left((B_{\eta^{2j+i}} \setminus B_{\eta^{2j+i+1}}) \setminus \Omega\right)}{\eta^{(n-2)(2j+i)}} \right] \\
&\leqslant \left(1 + \frac{1}{\eta^{n-2}}\right) \sum_{j=0}^{+\infty} \frac{\operatorname{Cap}\left((B_{\mu^j} \setminus B_{\mu^{j+1}}) \setminus \Omega\right)}{\mu^{(n-2)j}},
\end{aligned}$$

and this proves that the second series in (6.2.2) diverges as well, thus completing the proof of (6.2.2).

We also point out that, given $\lambda \in (0,1)$ and $\varepsilon \in (0, \min\{\lambda, 1-\lambda\})$, there exist $\underline{\lambda} \in (0, \varepsilon)$ and $\overline{\lambda} \in (1-\varepsilon, 1)$ such that

$$\begin{aligned}
&\sum_{j=0}^{+\infty} \frac{\operatorname{Cap}\left((B_{\lambda^j} \setminus B_{\lambda^{j+1}}) \setminus \Omega\right)}{\lambda^{(n-2)j}} = +\infty \\
&\text{if and only if} \quad \sum_{j=0}^{+\infty} \frac{\operatorname{Cap}\left((B_{\underline{\lambda}^j} \setminus B_{\underline{\lambda}^{j+1}}) \setminus \Omega\right)}{\underline{\lambda}^{(n-2)j}} = +\infty \\
&\text{and} \quad \sum_{j=0}^{+\infty} \frac{\operatorname{Cap}\left((B_{\overline{\lambda}^j} \setminus B_{\overline{\lambda}^{j+1}}) \setminus \Omega\right)}{\overline{\lambda}^{(n-2)j}} = +\infty; \qquad (6.2.4)
\end{aligned}$$

that is, the convergence or divergence of the series in (6.2.1) for a given $\lambda \in (0,1)$ can be equivalently shifted to that corresponding to another parameter as close as we wish to 0 as well as to 1. Note that the claim in (6.2.4) is a direct consequence of that in (6.2.2), up to an iteration, by taking $\lambda := \eta$ and $\ell \in \mathbb{N}$ sufficiently large so that $\underline{\lambda} := \eta^{2\ell} \in (0,\varepsilon)$ (or by taking $\lambda := \mu$ and $\ell \in \mathbb{N}$ sufficiently large so that $\overline{\lambda} := \eta^{1/(2\ell)} \in (1-\varepsilon,1)$).

We also point out that the capacity behaves nicely under scaling. Namely, if $\kappa > 0$ and $\kappa E := \{\lambda x,\ x \in E\}$, if a function v is equal to 1 in E, then the function $v_\kappa(x) := v\left(\frac{x}{\kappa}\right)$ is equal to 1 in κE, and hence it follows from (6.1.1) that

$$\operatorname{Cap}(\kappa E) := \kappa^{n-2}\operatorname{Cap}(E). \tag{6.2.5}$$

Now, let us suppose that condition (ii) in the statement of Theorem 6.2.1 holds true, namely that the series in (6.2.1) diverges for some $\lambda \in (0,1)$. Let also $\varepsilon > 0$, to be taken arbitrarily small in what follows. We exploit (6.2.4) that allows us to replace the previous λ with a new one, still called λ for simplicity, with the additional property that $\lambda \in (1-\varepsilon,1)$. We also take $K \in \mathbb{N}$ large enough (possibly in dependence of λ and ε) such that $\lambda^K \in \left(0,\frac{\varepsilon}{4}\right)$. For $\ell \in \{1,\dots,K\}$, we consider the series

$$\mathcal{S}_\ell(\Omega) := \sum_{j=0}^{+\infty} \frac{\operatorname{Cap}\left(\left(B_{\lambda^{Kj+\ell-1}} \setminus B_{\lambda^{Kj+\ell}}\right) \setminus \Omega\right)}{\lambda^{(n-2)(Kj+\ell-1)}}.$$

By (6.2.5), we know that

$$\begin{aligned}\operatorname{Cap}\left(\left(B_{\lambda^{Kj+\ell-1}} \setminus B_{\lambda^{Kj+\ell}}\right) \setminus \Omega\right) &= \operatorname{Cap}\left(\lambda^{\ell-1}\left(\left(B_{\lambda^{Kj}} \setminus B_{\lambda^{Kj+1}}\right) \setminus \frac{\Omega}{\lambda^{\ell-1}}\right)\right)\\ &= \lambda^{(n-2)(\ell-1)}\operatorname{Cap}\left(\left(B_{\lambda^{Kj}} \setminus B_{\lambda^{Kj+1}}\right) \setminus \frac{\Omega}{\lambda^{\ell-1}}\right)\end{aligned}$$

and therefore

$$\mathcal{S}_\ell(\Omega) := \sum_{j=0}^{+\infty} \frac{\lambda^{(n-2)(\ell-1)}\operatorname{Cap}\left(\left(B_{\lambda^{Kj}} \setminus B_{\lambda^{Kj+1}}\right) \setminus \frac{\Omega}{\lambda^{\ell-1}}\right)}{\lambda^{(n-2)(Kj+\ell-1)}} = \mathcal{S}_1\left(\frac{\Omega}{\lambda^{\ell-1}}\right).$$

From this, we arrive at

$$+\infty = \sum_{j=0}^{+\infty} \frac{\operatorname{Cap}\left(\left(B_{\lambda^j} \setminus B_{\lambda^{j+1}}\right) \setminus \Omega\right)}{\lambda^{(n-2)j}} = \sum_{\ell=1}^{K} \mathcal{S}_\ell(\Omega) = \sum_{\ell=1}^{K} \mathcal{S}_1\left(\frac{\Omega}{\lambda^{\ell-1}}\right),$$

and as a consequence, we can find $\ell_\star \in \{1,\dots,K\}$ such that if

$$\Omega_\star := \frac{\Omega}{\lambda^{\ell_\star-1}}, \tag{6.2.6}$$

it holds that

$$\mathcal{S}_1(\Omega_\star) = +\infty. \tag{6.2.7}$$

We now remark that

$$\lambda - \lambda^K \leqslant \lambda^{1-K} - 1. \tag{6.2.8}$$

To check this, for all $t \in \mathbb{R}$, we consider the function $\phi(t) := \lambda^{1+t} - \lambda^{K+t}$, and we note that

$$\phi'(t) = \ln\lambda\,(\lambda^{1+t} - \lambda^{K+t}) = -\lambda^{1+t}|\ln\lambda|\,(1 - \lambda^{K-1}) \leqslant 0.$$

As a result, $\lambda^{1-K} - 1 = \phi(-K) \geqslant \phi(0) = \lambda - \lambda^K$, and this gives (6.2.8).

Now, we consider the conductor potential v_j of the set $E_j := \left(B_{\lambda^j} \setminus B_{\lambda^{j+1}}\right) \setminus \Omega_\star$, as constructed in Corollary 6.1.14. Given $m' \geqslant m \in \mathbb{N}$, and taking K as above, we define

$$V_{m,m'} := \sum_{i=m}^{m'} v_{Ki}.$$

Note that

$$V_{m,m'} \text{ is harmonic outside } E_{m,m'} := \bigcup_{i=m}^{m'} E_{Ki}. \tag{6.2.9}$$

Thus, the strategy is to estimate $V_{m,m'}$ in a small neighborhood of $E_{m,m'}$ and then use the maximum principle in Lemma 2.17.2 to obtain a bound for $V_{m,m'}$ outside $E_{m,m'}$.

To employ this strategy, given $\delta \in (0,1)$, we consider a δ-neighborhood of $E_{m,m'}$, and we observe that if x lies in this δ-neighborhood, then there exists $i_0 \in \{m,\dots,m'\}$ such that x lies in a δ-neighborhood of E_{Ki_0}. We claim that, if δ is sufficiently small, for every $i \neq i_0$,

$$B_{(\lambda-\lambda^K)\lambda^{Ki}-\delta}(x) \cap E_{Ki} = \varnothing. \tag{6.2.10}$$

Indeed, suppose by contradiction that there exists $y \in B_{(\lambda-\lambda^K)\lambda^{Ki}-\delta}(x) \cap E_{Ki}$. Then,

$$|x-y| < (\lambda - \lambda^K)\lambda^{Ki} - \delta \quad\text{and}\quad \lambda^{Ki+1} \leqslant |y| < \lambda^{Ki}.$$

Using these inequalities, we see that

$$\lambda^{Ki_0} + \delta > |x| \geqslant |y| - |x-y| \geqslant \lambda^{Ki+1} - (\lambda - \lambda^K)\lambda^{Ki} + \delta = \lambda^{K(i+1)} + \delta,$$

which gives that $i_0 < i + 1$. On the other hand, exploiting also (6.2.8),

$$\begin{aligned}\lambda^{Ki_0+1} - \delta < |x| \leqslant |x-y| + |y| &< (\lambda - \lambda^K)\lambda^{Ki} - \delta + \lambda^{Ki}\\ &= (\lambda - \lambda^K + 1)\lambda^{Ki} - \delta \leqslant \lambda^{1-K}\lambda^{Ki} - \delta,\end{aligned}$$

which gives instead that $i_0 > i - 1$. Hence, we have that $i = i_0$, and this is a contradiction, which establishes (6.2.10).

In light of (6.2.10), we can employ (6.1.32) (with $\Omega := E_{Ki}$, $u := v_{Ki}$ and $r := (\lambda - \lambda^K)\lambda^{Ki} - \delta$) to find that if x lies in a δ-neighborhood of E_{Ki_0} then, as $\delta \searrow 0$,

$$v_{Ki}(x) \leqslant \frac{c_n(1+o(1))\,\mathrm{Cap}(E_{Ki})}{(\lambda-\lambda^K)^{n-2}\,\lambda^{K(n-2)i}}.$$

As a result, since $v_{Ki_0} \leqslant 1$, we obtain that, if x lies in a δ-neighborhood of E_{Ki_0}, then

$$\begin{aligned} V_{m,m'}(x) &\leqslant 1 + \sum_{\substack{m\leqslant i\leqslant m'\\ i\neq i_0}} v_{Ki}(x) \\ &\leqslant 1 + \sum_{\substack{m\leqslant i\leqslant m'\\ i\neq i_0}} \frac{c_n(1+o(1))\,\mathrm{Cap}(E_{Ki})}{(\lambda-\lambda^K)^{n-2}\,\lambda^{K(n-2)i}}. \end{aligned}$$

That is, if $z \in \partial E_{m,m'}$,

$$\begin{aligned} \varlimsup_{\mathbb{R}^n\setminus E_{m,m'}\ni x\to z} V_{m,m'}(x) &\leqslant 1 + \sum_{\substack{m\leqslant i\leqslant m'\\ i\neq i_0}} \frac{c_n\,\mathrm{Cap}(E_{Ki})}{(\lambda-\lambda^K)^{n-2}\,\lambda^{K(n-2)i}} \\ &\leqslant \frac{1}{(\lambda-\lambda^K)^{n-2}}\left(1+\sum_{i=m}^{m'}\frac{c_n\,\mathrm{Cap}(E_{Ki})}{\lambda^{K(n-2)i}}\right). \end{aligned}$$

As a consequence, by (6.2.9) and Lemma 2.17.2, we infer that, for all $x \in \mathbb{R}^n \setminus E_{m,m'}$,

$$V_{m,m'}(x) \leqslant \frac{1}{(\lambda-\lambda^K)^{n-2}}\left(1+\sum_{i=m}^{m'}\frac{c_n\,\mathrm{Cap}(E_{Ki})}{\lambda^{K(n-2)i}}\right).$$

Therefore,

$$W_{m,m'}(x) := \left(1+\sum_{i=m}^{m'}\frac{c_n\,\mathrm{Cap}(E_{Ki})}{\lambda^{K(n-2)i}}\right)^{-1}(\lambda-\lambda^K)^{n-2}\,V_{m,m'}(x) \leqslant 1. \tag{6.2.11}$$

Now, our objective is to exploit Proposition 6.1.16. To this end, given $\rho > 0$, we consider the conductor potential $u^\star_{p,\rho}$ of $B_{\rho/(\lambda^{\ell_\star-1})} \setminus \Omega_\star$, as given by Corollary 6.1.14. We take m' so large that

$$\lambda^{Km'} < \frac{\rho}{2}, \tag{6.2.12}$$

and we claim that, for every $x \in \mathbb{R}^n$,

$$W_{m,m'}(x) \leqslant u^\star_{p,\rho}(x). \tag{6.2.13}$$

To check this, we use an approximating sequence u_j of conductor potentials as in (6.1.39) which are harmonic outside the sets Z_j with boundaries of class $C^{2,\alpha}$, which approach $B_{\rho/(\lambda^{\ell_\star-1})} \setminus \Omega_\star$ from the outside. We note that the function $\phi := u_j - W_{m,m'}$ is harmonic outside Z_j, thanks to (6.2.12). Moreover, $\phi \geqslant 0$ along ∂Z_j (as well as inside Z_j), owing to (6.2.11). Since $\phi \in \mathcal{D}^{1,2}(\mathbb{R}^n)$, we thus obtain from Lemma 2.9.5 that $\phi \geqslant 0$ in $\mathbb{R}^n \setminus Z_j$. From this, we find that $u_j \geqslant W_{m,m'}$ and thus, sending $j \to +\infty$, we obtain (6.2.13), as desired.

As a consequence of (6.2.13), we have that

$$\begin{aligned}\lim_{\Omega_\star \ni x \to 0} u^\star_{p,\rho}(x) &\geqslant \lim_{\Omega_\star \ni x \to 0} W_{m,m'}(x)\\ &= \left(1 + \sum_{i=m}^{m'} \frac{c_n \operatorname{Cap}(E_{Ki})}{\lambda^{K(n-2)i}}\right)^{-1} (\lambda - \lambda^K)^{n-2} \lim_{\Omega_\star \ni x \to 0} V_{m,m'}(x)\\ &= \left(1 + \sum_{i=m}^{m'} \frac{c_n \operatorname{Cap}(E_{Ki})}{\lambda^{K(n-2)i}}\right)^{-1} (\lambda - \lambda^K)^{n-2} \sum_{i=m}^{m'} \lim_{\Omega_\star \ni x \to 0} v_{Ki}(x). \qquad (6.2.14)\end{aligned}$$

Now, we observe that

$$B_{\lambda^{Ki}+|x|}(x) \supseteq B_{\lambda^{Ki}} \supseteq E_{Ki},$$

and therefore, by (6.1.32) (used here with $\Omega := E_{Ki}$, $u := v_{Ki}$ and $R := \lambda^{Ki} + |x|$),

$$v_{Ki}(x) \geqslant \frac{c_n \operatorname{Cap}(E_{Ki})}{(\lambda^{Ki} + |x|)^{n-2}}.$$

From this and (6.2.14), we arrive at

$$\lim_{\Omega_\star \ni x \to 0} u^\star_{p,\rho}(x) \geqslant \left(1 + \sum_{i=m}^{m'} \frac{c_n \operatorname{Cap}(E_{Ki})}{\lambda^{K(n-2)i}}\right)^{-1} (\lambda - \lambda^K)^{n-2} \sum_{i=m}^{m'} \frac{c_n \operatorname{Cap}(E_{Ki})}{\lambda^{K(n-2)i}}. \qquad (6.2.15)$$

Now, we recall (6.2.7), according to which

$$+\infty = \mathcal{S}_1(\Omega_\star) = \sum_{i=0}^{+\infty} \frac{\operatorname{Cap}\left((B_{\lambda^{Ki}} \setminus B_{\lambda^{Ki+1}}) \setminus \Omega_\star\right)}{\lambda^{K(n-2)i}} = \sum_{i=0}^{+\infty} \frac{\operatorname{Cap}(E_{Ki})}{\lambda^{K(n-2)i}}.$$

This gives that

$$\sum_{i=m}^{+\infty} \frac{\operatorname{Cap}(E_{Ki})}{\lambda^{K(n-2)i}} = +\infty;$$

therefore, by sending $m' \to +\infty$ in (6.2.15), we deduce that

$$\lim_{\Omega_\star \ni x \to 0} u^\star_{p,\rho}(x) \geqslant (\lambda - \lambda^K)^{n-2} \geqslant \left(1 - \varepsilon - \frac{\varepsilon}{4}\right)^{n-2}.$$

Hence, sending $\varepsilon \searrow 0$,

$$\lim_{\Omega_\star \ni x \to 0} u^\star_{p,\rho}(x) \geqslant 1.$$

Since conductor potentials are bounded by 1, we thus conclude that

$$\lim_{\Omega_\star \ni x \to 0} u^\star_{p,\rho}(x) = 1.$$

By scaling back and recalling (6.2.6), if $u_{p,\rho}$ is the conductor potential of $B_\rho \setminus \Omega$, as given in Corollary 6.1.14, we find that

$$\lim_{\Omega \ni x \to 0} u_{p,\rho}(x) = 1.$$

We can accordingly employ Proposition 6.1.16 and obtain the continuous solvability of the Dirichlet problem, thus establishing condition (i) in Theorem 6.2.1.

To complete the proof of Theorem 6.2.1, we now establish the converse statement, namely that if the series in (6.2.1) converges, the statement in (i) does not hold. To this end, we argue for a contradiction. Suppose that the statement in (i) of Theorem 6.2.1 holds true, take $p \in \partial\Omega$ (say, $p = 0$ up to a translation) and recall Proposition 6.1.16 to consider the infinitesimal sequence of radii $\rho_j := \lambda^j$ and have that if u_j is the conductor potential of $B_{\rho_j} \setminus \Omega$, as given by Corollary 6.1.14, we have that, for each $j \in \mathbb{N}$,

$$\lim_{\Omega \ni x \to 0} u_j(x) = 1. \tag{6.2.16}$$

In particular, we can use the convergence of the series in (6.2.1) to pick $m \in \mathbb{N}$ large enough such that

$$\sum_{j=m}^{+\infty} \frac{\operatorname{Cap}\Big(\big(B_{\lambda^j} \setminus B_{\lambda^{j+1}}\big) \setminus \Omega\Big)}{\lambda^{(n-2)j}} \leqslant \frac{\lambda^{n-2}}{10\, c_n} \tag{6.2.17}$$

(where c_n is the positive constant in (2.7.2)) and then utilize (6.2.16) to find $r_0 > 0$ sufficiently small such that if $x \in \Omega \cap \overline{B_{r_0}}$, then

$$u_m(x) \geqslant \frac{9}{10}. \tag{6.2.18}$$

Now, we consider $m' \in \mathbb{N} \cap [m+1, +\infty)$ sufficiently large that

$$\frac{\rho_{m'}^{n-2}}{(r_0 - 2\rho_{m'})^{n-2}} \leqslant \frac{1}{10}.$$

In this way, if $x \in \partial B_{r_0}$, then $B_{r_0-2\rho_{m'}}(x) \subseteq \mathbb{R}^n \setminus B_{\rho_{m'}}$ and therefore, by (6.1.32) (applied here with $B_{\rho_{m'}} \setminus \Omega$ in place of Ω, $u := u_{m'}$ and $r := r_0 - 2\rho_{m'}$),

$$u_{m'}(x) \leqslant \frac{c_n \operatorname{Cap}(B_{\rho_{m'}} \setminus \Omega)}{(r_0 - 2\rho_{m'})^{n-2}}.$$

Thus, recalling Lemma 6.1.7 (in light of Lemma 6.1.13) and Corollary 6.1.5, if $x \in \partial B_{r_0}$,

$$u_{m'}(x) \leqslant \frac{c_n \operatorname{Cap}(B_{\rho_{m'}})}{(r_0 - 2\rho_{m'})^{n-2}} = \frac{\rho_{m'}^{n-2}}{(r_0 - 2\rho_{m'})^{n-2}} \leqslant \frac{1}{10}. \tag{6.2.19}$$

Now, we denote by $U_{m,m'}$ the conductor potential, as given by Corollary 6.1.14, with respect to the set $(B_{\rho_m} \setminus B_{\rho_{m'}}) \setminus \Omega$. We make use of Lemma 6.1.8 (see also Lemma 6.1.15) and find that

$$u_m \leqslant u_{m'} + U_{m,m'}.$$

Hence, using (6.2.18) and (6.2.19), if $x \in \partial B_{r_0}$,

$$U_{m,m'}(x) \geqslant u_m(x) - u_{m'}(x) \geqslant \frac{9}{10} - \frac{1}{10} = \frac{4}{5}. \tag{6.2.20}$$

Now, we claim that

$$U_{m,m'}(x) \geqslant \frac{4}{5} \quad \text{a.e. } x \in B_{r_0}. \tag{6.2.21}$$

To establish this, we use the approximation method in (6.1.39), recalling that $U_{m,m'}$ can be seen as the a.e. limit as $j \to +\infty$ of the sequence of conductor potentials $U_{m,m',j}$, corresponding to open and bounded sets $\mathcal{U}_{m,m',j}$ with boundary of class $C^{2,\alpha}$, which approximate $(B_{\rho_m} \setminus B_{\rho_{m'}}) \setminus \Omega$ from the exterior. In this way, the function $\phi_{m,m',j} := U_{m,m',j} - U_{m,m'}$ is harmonic outside $\mathcal{U}_{m,m',j}$, it belongs to $\mathcal{D}^{1,2}(\mathbb{R}^n)$, and along $\partial\mathcal{U}_{m,m',j}$, we have that $\phi_{m,m',j} := 1 - U_{m,m'} \geqslant 0$. Consequently, recalling Lemma 2.9.5, we have that $\phi_{m,m',j} \geqslant 0$ in $\mathbb{R}^n$. This and (6.2.20) give that for every $x \in \partial B_{r_0}$,

$$U_{m,m',j}(x) \geqslant U_{m,m'}(x) \geqslant \frac{4}{5}. \tag{6.2.22}$$

Let now $\varepsilon \in \left(0, \frac{1}{2}\right)$ and

$$U^{\star}_{m,m',j} := \begin{cases} U_{m,m',j} & \text{in } \mathbb{R}^n \setminus B_{r_0}, \\ \max\left\{U_{m,m',j}, \frac{4}{5} - \varepsilon\right\} & \text{in } B_{r_0}. \end{cases}$$

We stress that $U^{\star}_{m,m',j} \in \mathcal{D}^{1,2}(\mathbb{R}^n)$, thanks to (6.2.22). Furthermore,

$$\int_{\mathbb{R}^n\setminus\mathcal{U}_{m,m',j}} |\nabla U^{\star}_{m,m',j}(x)|^2\,dx \leqslant \int_{\mathbb{R}^n\setminus\mathcal{U}_{m,m',j}} |\nabla U_{m,m',j}(x)|^2\,dx.$$

This and the minimality property of $U_{m,m',j}$ (as in (6.1.2) and (6.1.4)) yield that

$$\begin{aligned} 0 &= \int_{\mathbb{R}^n\setminus\mathcal{U}_{m,m',j}} |\nabla U_{m,m',j}(x)|^2\,dx - \int_{\mathbb{R}^n\setminus\mathcal{U}_{m,m',j}} |\nabla U^{\star}_{m,m',j}(x)|^2\,dx \\ &= \int_{B_{r_0}\cap\{U_{m,m',j}<\frac{4}{5}-\varepsilon\}} |\nabla U_{m,m',j}(x)|^2\,dx. \end{aligned}$$

From this, it follows that $U_{m,m',j} \geqslant \frac{4}{5} - \varepsilon$ in B_{r_0}. Thus, sending $\varepsilon \searrow 0$, we conclude that $U_{m,m',j} \geqslant \frac{4}{5}$ in B_{r_0}. Now, sending $j \to +\infty$, we obtain (6.2.21).

Now, we write

$$(B_{\rho_m} \setminus B_{\rho_{m'}}) \setminus \Omega = \bigcup_{i=m}^{m'-1} (B_{\rho_i} \setminus B_{\rho_{i+1}}) \setminus \Omega,$$

and we denote by w_i the conductor potential of $(B_{\rho_i} \setminus B_{\rho_{i+1}}) \setminus \Omega$, as given in Corollary 6.1.14. We then exploit Lemma 6.1.8 (as refined in Lemma 6.1.15) to see that

$$U_{m,m'} \leqslant \sum_{i=m}^{m'-1} w_i. \tag{6.2.23}$$

We also observe that $B_{\rho_{i+1}}$ lies in the complement of $(B_{\rho_i} \setminus B_{\rho_{i+1}}) \setminus \Omega$; therefore, in view of (6.1.32) (used here with $\left(B_{\rho_i} \setminus B_{\rho_{i+1}}\right) \setminus \Omega$ in place of Ω, $u := w_i$ and $r := \rho_{i+1}$),

$$w_i(0) \leqslant \frac{c_n \operatorname{Cap}\left(\left(B_{\rho_i} \setminus B_{\rho_{i+1}}\right) \setminus \Omega\right)}{\rho_{i+1}^{n-2}} = \frac{c_n \operatorname{Cap}\left(\left(B_{\lambda^i} \setminus B_{\lambda^{i+1}}\right) \setminus \Omega\right)}{\lambda^{n-2}\,\lambda^{(n-2)i}}.$$

This, (6.2.17) and (6.2.23) lead to

$$U_{m,m'}(0) \leqslant \frac{1}{\lambda^{n-2}} \sum_{i=m}^{m'-1} \frac{c_n \operatorname{Cap}\left(\left(B_{\lambda^i} \setminus B_{\lambda^{i+1}}\right) \setminus \Omega\right)}{\lambda^{(n-2)i}} \leqslant \frac{1}{10}.$$

This is in contradiction with (6.2.21), and the proof of Theorem 6.2.1 is thereby complete. □

By inspecting its proof, one also sees that Wiener's criterion in Theorem 6.2.1 is actually a "pointwise" statement; namely, it gives a sharp condition on the solvability of the Dirichlet problem by a function that attains a prescribed boundary datum at a given point, depending on condition (6.2.1) (this type of points are sometimes referred to by the name "regular points" for the Dirichlet problem).

We also observe that Wiener's criterion in Theorem 6.2.1 contains the previous existence result for the Dirichlet problem in Theorem 3.3.3 as a byproduct. Indeed, take a bounded open set satisfying the exterior cone condition. Up to a rigid motion, we can assume that the origin lies on the boundary of Ω, and we use the exterior cone condition to write that

$$\left\{x = (x', x_n) \in B_{r_0} \text{ s.t. } x_n \geqslant b|x'|\right\} \subseteq \mathbb{R}^n \setminus \Omega, \tag{6.2.24}$$

for some $r_0 \in (0, 1)$ and $b > 0$.

We let $i \in \mathbb{N} \cap \left(\frac{|\ln r_0|}{\ln 2}, +\infty\right)$ and take $\lambda := \frac{1}{2}$. We observe that $\lambda^i < r_0$.

We define

$$p_i := \frac{3}{2^{i+2}}\, e_n \quad \text{and} \quad r_i := \frac{1}{(1+b)\, 2^{i+3}},$$

and we claim that

$$\left(B_{\lambda^i} \setminus B_{\lambda^{i+1}}\right) \setminus \Omega \supseteq B_{r_i}(p_i). \tag{6.2.25}$$

To check this, let $x = (x', x_n) \in B_{r_i}(p_i)$. We have that

$$|x| \leqslant |p_i| + r_i \leqslant \frac{3}{2^{i+2}} + \frac{1}{2^{i+3}} = \frac{7}{2^{i+3}} < \frac{8}{2^{i+3}} = \frac{1}{2^i} \tag{6.2.26}$$

and

$$|x| \geqslant |p_i| - r_i \geqslant \frac{3}{2^{i+2}} - \frac{1}{2^{i+3}} = \frac{5}{2^{i+3}} > \frac{4}{2^{i+3}} = \frac{1}{2^{i+1}}. \tag{6.2.27}$$

Furthermore,

$$x \cdot e_n - b|x'| \geqslant p_i \cdot e_n - r_i - br_i = \frac{3}{2^{i+2}} - (1+b)r_i = \frac{3}{2^{i+2}} - \frac{1}{2^{i+3}} > 0.$$

From this, (6.2.24), (6.2.26) and (6.2.27), we obtain (6.2.25), as desired.

Now, we combine Lemma 6.1.7 (recall also Lemma 6.1.13), Corollary 6.1.5 and (6.2.25): in this way, we infer that

$$\begin{aligned} \operatorname{Cap}\left(\left(B_{\lambda^i} \setminus B_{\lambda^{i+1}}\right) \setminus \Omega\right) &\geqslant \operatorname{Cap}\left(B_{r_i}(p_i)\right) = \frac{r_i^{n-2}}{c_n} \\ &= \frac{1}{c_n\, (1+b)^{n-2}\, 2^{(i+3)(n-2)}} = \frac{c}{\lambda^{(n-2)i}}, \end{aligned}$$

for some $c > 0$ depending only on n and b.

For this reason,

$$\sum_{i=0}^{+\infty} \frac{\operatorname{Cap}\left(\left(B_{\lambda^i} \setminus B_{\lambda^{i+1}}\right) \setminus \Omega\right)}{\lambda^{(n-2)i}} = +\infty.$$

Hence, we can apply Theorem 6.2.1 and obtain that the Dirichlet problem can be solved using a harmonic function up to the boundary when Ω is an open bounded set satisfying the exterior cone condition.

Wiener's criterion in Theorem 6.2.1 also contains as a byproduct the Lebesgue spine discussed on p. 288, see [Kel67, Exercise 10, p. 334] (see also [Cai07] for related details on ellipsoidal potentials).

For thorough explanations on the notions of capacity, see e.g. [Kel67, Car67, Del72, AH96, Ada97, Tar97, Tur00, Doo01, HKM06, BB11, EG15, Pon16, HP18].

Chapter 7

Some Interesting Problems Arising from the Poisson Equation

We collect in this chapter some classical and fascinating mathematical problems. The kinship between these problems lies in the use of elliptic partial differential equations and the geometric flavor of the problems and their solutions. More specifically, in Section 7.1, we present the soap bubble theorem (according to which soap bubbles are round). This result is closely related to the isoperimetric problem, discussed in Section 7.2 (giving that balls minimize perimeter for a given volume). Then, in Section 7.3, we focus on a classical overdetermined problem (in rough terms, addressing the type of domains which allow for overabundant boundary prescriptions), and in Section 7.4 we deal with the converse of the mean value formula (determining the type of domains for which a claim such as the one in Theorem 2.1.2 holds true).

The proofs use a number of astute integration-by-parts and integral formulas. A different, and even more geometric, approach to soap bubbles and overdetermined problems is presented in Chapter 8.

7.1 The Soap Bubble Theorem

A classical result of geometric analysis due to Aleksandr Danilovič Aleksandrov (see [Ale62]) is that a bounded domain whose boundary has a constant mean curvature is necessarily a ball. Since soap bubbles are formed through the balance between surface tension and air pressure, they serve as nice concrete examples of constant mean curvature surfaces, and this result is thereby often referred to as the "soap bubble theorem" (further details are given in Theorem 7.1.3 and Corollaries 7.2.4 and 7.2.5). Following [Rei82, Ros87, MP20], we give here a proof of this result based only on the existence result for the Dirichlet problem provided by Theorem 3.3.3 and on the geometric identities presented in Section 2.3

(the original approach based on the reflection principle method is also discussed in Section 8.2).

We begin by introducing some notations. Given a C^2 function u, recalling the setting in (1.1.36), we define

$$\mathcal{D}_u := n\,|D^2u|^2 - (\Delta u)^2. \tag{7.1.1}$$

Exploiting the Cauchy–Schwarz inequality, we see that

$$(\Delta u)^2 = \left(\sum_{i=1}^{n} \partial_{ii}u\right)^2 \leqslant \sum_{i=1}^{n}(\partial_{ii}u)^2 \sum_{i=1}^{n} 1 \leqslant n\,|D^2u|^2;$$

therefore,

$$\mathcal{D}_u \geqslant 0, \tag{7.1.2}$$

and

$$\begin{gathered}\text{equality holds if and only if the Hessian of } u\\ \text{is a scalar function times the identity matrix.}\end{gathered} \tag{7.1.3}$$

It is also convenient to consider an auxiliary equation set on a bounded, connected and open set $\Omega \subseteq \mathbb{R}^n$ with $C^{2,\alpha}$ boundary, with $\alpha \in (0,1)$, whose mean curvature is denoted by H. Namely, in this setting, we consider the solutions of

$$\begin{cases}\Delta u = 1 & \text{in } \Omega,\\ u = 0 & \text{on } \partial\Omega.\end{cases} \tag{7.1.4}$$

In this setting, it is useful to observe that, by (1.1.4),

$$\int_{\partial\Omega} \partial_\nu u(x)\, d\mathcal{H}^{n-1}_x = \int_\Omega \Delta u(x)\, dx = |\Omega| \tag{7.1.5}$$

and that, by Corollary 2.3.11, on $\partial\Omega$,

$$1 = \Delta u = H\,\partial_\nu u + \partial_{\nu\nu} u. \tag{7.1.6}$$

Also, we have the following cornerstone result (see e.g. [Rei82]).

Lemma 7.1.1. *Consider a bounded, connected and open set $\Omega \subseteq \mathbb{R}^n$ with $C^{2,\alpha}$ boundary for some $\alpha \in (0,1)$. Let $u \in C^3(\Omega) \cap C^2(\overline{\Omega})$ be a solution of* (7.1.4).

Then,

$$\int_{\partial\Omega} \partial_{\nu\nu}u(x)\,\partial_\nu u(x)\, d\mathcal{H}^{n-1}_x \geqslant \frac{|\Omega|}{n}, \tag{7.1.7}$$

$$\int_\Omega |D^2u(x)|^2\, dx \geqslant \frac{|\Omega|}{n} \tag{7.1.8}$$

and

$$\int_{\partial\Omega} H(x)\,(\partial_\nu u(x))^2\,d\mathcal{H}^{n-1}_x \leqslant \frac{n-1}{n}\,|\Omega|. \tag{7.1.9}$$

Also, the following conditions are equivalent:

$$\text{equality holds instead of inequality in (7.1.7),} \tag{7.1.10}$$

$$\text{equality holds instead of inequality in (7.1.8),} \tag{7.1.11}$$

$$\text{equality holds instead of inequality in (7.1.9),} \tag{7.1.12}$$

$$\mathcal{D}_u \text{ is identically zero in } \Omega, \tag{7.1.13}$$

and

$$\begin{aligned} &\text{there exist } x_0 \in \mathbb{R}^n \quad \text{and} \quad c \in \mathbb{R} \text{ such that} \\ &u(x) = \frac{|x-x_0|^2}{2n} - c \quad \text{for every } x \in \Omega. \end{aligned} \tag{7.1.14}$$

Moreover, if one of the equivalent conditions (7.1.10), (7.1.11), (7.1.12), (7.1.13) *and* (7.1.14) *is satisfied, then* Ω *is necessarily a ball.*

Proof. Multiplying (7.1.6) by $\partial_\nu u$ and integrating over $\partial\Omega$, we obtain that

$$\int_{\partial\Omega} \partial_\nu u(x)\,d\mathcal{H}^{n-1}_x = \int_{\partial\Omega} H(x)\,(\partial_\nu u(x))^2\,d\mathcal{H}^{n-1}_x + \int_{\partial\Omega} \partial_{\nu\nu} u(x)\,\partial_\nu u(x)\,d\mathcal{H}^{n-1}_x.$$

This and (7.1.5) give

$$\int_{\partial\Omega} H(x)\,(\partial_\nu u(x))^2\,d\mathcal{H}^{n-1}_x + \int_{\partial\Omega} \partial_{\nu\nu} u(x)\,\partial_\nu u(x)\,d\mathcal{H}^{n-1}_x = |\Omega|. \tag{7.1.15}$$

Also, since $u = 0$ along $\partial\Omega$, we have that $\nabla u = |\nabla u|\nu$ on $\partial\Omega$. As a result, on $\partial\Omega$,

$$\frac{1}{2}\partial_\nu|\nabla u|^2 = \nabla u \cdot \partial_\nu \nabla u = |\nabla u|\nu \cdot \nabla\partial_\nu u = |\nabla u|\,\partial_{\nu\nu} u = \nabla u \cdot \nu\,\partial_{\nu\nu} u = \partial_\nu u\,\partial_{\nu\nu} u.$$

Therefore, using Green's first identity (2.1.10) and the Bochner identity in Lemma 1.1.5,

$$\begin{aligned} \int_{\partial\Omega} \partial_{\nu\nu} u(x)\,\partial_\nu u(x)\,d\mathcal{H}^{n-1}_x &= \frac{1}{2}\int_{\partial\Omega} \partial_\nu|\nabla u(x)|^2\,d\mathcal{H}^{n-1}_x \\ &= \frac{1}{2}\int_\Omega \Delta|\nabla u(x)|^2\,dx \\ &= \int_\Omega \Big(|D^2 u(x)|^2 + \nabla u(x)\cdot\nabla\Delta u(x)\Big)\,dx \\ &= \int_\Omega |D^2 u(x)|^2\,dx. \end{aligned} \tag{7.1.16}$$

From (7.1.16) and (7.1.2), it follows that

$$\int_{\partial\Omega} \partial_{\nu\nu}u(x)\,\partial_\nu u(x)\,d\mathcal{H}_x^{n-1} = \frac{1}{n}\int_\Omega \Big(\mathcal{D}_u(x) + (\Delta u(x))^2\Big)\,dx$$
$$= \frac{1}{n}\int_\Omega \Big(\mathcal{D}_u(x) + 1\Big)\,dx \geqslant \frac{|\Omega|}{n},$$

with equality holding if and only if $\mathcal{D}_u$ vanishes identically in Ω, that is, if and only if (7.1.3) holds true at any point of Ω. Namely, we have established (7.1.7) and, recalling (7.1.16), also (7.1.8); moreover, we have proved the fact that the equality in (7.1.7) is equivalent to the equality in (7.1.8) and to (7.1.13). That is, (7.1.10), (7.1.11) and (7.1.13) are equivalent.

Furthermore, in light of (7.1.15) and (7.1.16), we have that (7.1.7), (7.1.8) and (7.1.9) are equivalent. We also point out that (7.1.10) (as well as (7.1.11)) is equivalent to (7.1.12), thanks to (7.1.15) (and (7.1.16)).

It remains to show that (7.1.13) is equivalent to (7.1.14). On the one hand, if (7.1.14) holds true, then one can perform a direct computation and show that (7.1.3) is satisfied; hence, (7.1.13) holds true as well. Conversely, if (7.1.13) is satisfied, we have that $D^2u = \varphi\,\mathrm{Id}$, where Id is the identity matrix and φ is some function. As a result, we find that $1 = \Delta u = n\varphi$ and thus φ must be constantly equal to $\frac{1}{n}$. Consequently,

$$\partial_{ij}u(x) = \frac{\delta_{ij}}{n} \quad \text{for all } x \in \Omega.$$

This identity can be integrated, producing that

$$u(x) = \frac{x_1^2 + \cdots + x_n^2}{2n} + \omega\cdot x \quad \text{for all } x \in \Omega,$$

for some $\omega \in \mathbb{R}^n$. We can therefore "complete the square" and obtain

$$u(x) = \left|\frac{x}{\sqrt{2n}} + \frac{\sqrt{n}\,\omega}{\sqrt{2}}\right|^2 - \frac{n\,|\omega|^2}{2} \quad \text{for all } x \in \Omega.$$

This gives (7.1.14) and completes the proof of the equivalence between (7.1.10), (7.1.11), (7.1.12), (7.1.13) and (7.1.14).

Suppose now that one of the equivalent conditions (7.1.10), (7.1.11), (7.1.12), (7.1.13) and (7.1.14) is satisfied. Since these conditions are equivalent, we may assume that (7.1.14) holds true; therefore, the level set $\{u = 0\}$ is the sphere $\partial B_{\sqrt{2nc}}(x_0)$; hence, $\partial\Omega$ is a sphere, and Ω is a ball. □

The result in Lemma 7.1.1 is pivotal for many applications. A classical one is the so-called Heintze–Karcher inequality, which was established in [HK78] and, as pointed out in [Ros87], can be proved directly by using the tools developed so far, as follows.

Lemma 7.1.2. *Let $\alpha \in (0,1)$. Consider a bounded, connected and open set $\Omega \subseteq \mathbb{R}^n$ with $C^{2,\alpha}$ boundary for some $\alpha \in (0,1)$, with a strictly positive mean curvature H.*

Then,

$$\int_{\partial\Omega} \frac{d\mathcal{H}_x^{n-1}}{H(x)} \geqslant \frac{n}{n-1}\,|\Omega|. \tag{7.1.17}$$

Also, equality holds true if and only if Ω is a ball.

Proof. By Theorem 4.5.4, there exists a solution $u \in C^3(\Omega) \cap C^2(\overline{\Omega})$ to (7.1.4).

By (7.1.5) and the Cauchy–Schwarz inequality,

$$|\Omega|^2 = \left(\int_{\partial\Omega} \partial_\nu u(x)\, d\mathcal{H}_x^{n-1}\right)^2 \leqslant \int_{\partial\Omega} H(x)(\partial_\nu u(x))^2\, d\mathcal{H}_x^{n-1} \int_{\partial\Omega} \frac{d\mathcal{H}_x^{n-1}}{H(x)}.$$

This and (7.1.9) give (7.1.17). Also, if equality holds in (7.1.17), then equality must hold in (7.1.9), that is, condition (7.1.12) is fulfilled. Accordingly, by Lemma 7.1.1, Ω is necessarily a ball.

Vice versa, if Ω is a ball of radius R, then $H = \frac{n-1}{R}$; therefore,

$$\int_{\partial\Omega} \frac{d\mathcal{H}_x^{n-1}}{H(x)} - \frac{n\,|\Omega|}{n-1} = \frac{R\,\mathcal{H}^{n-1}(\partial B_R)}{n-1} - \frac{n\,|B_R|}{n-1} = 0. \qquad \square$$

As pointed out in [Ros87], combining the Heintze–Karcher inequality of Lemma 7.1.2 and the Minkowski identity in Corollary 2.3.5, a straightforward proof of the soap bubble theorem can be obtained. The details are as follows.

Theorem 7.1.3. *Let $\alpha \in (0,1)$. Consider a bounded, connected and open set $\Omega \subseteq \mathbb{R}^n$, and assume that $\Sigma := \partial\Omega$ is of class $C^{2,\alpha}$. Assume that the mean curvature of Σ is constant. Then, Ω is necessarily a ball.*

Proof. By Corollary 2.3.9, we deduce that the mean curvature H of Σ is necessarily a positive constant. Hence, in light of the Minkowski identity in Corollary 2.3.5,

$$\begin{aligned}\int_\Sigma \frac{d\mathcal{H}_x^{n-1}}{H(x)} &= \frac{\mathcal{H}^{n-1}(\Sigma)}{H} = \frac{1}{(n-1)H}\int_\Sigma H(x)\, x\cdot\nu(x)\, d\mathcal{H}_x^{n-1}\\ &= \frac{1}{n-1}\int_\Sigma x\cdot\nu(x)\, d\mathcal{H}_x^{n-1}.\end{aligned}$$

This and the divergence theorem yield that

$$\int_{\Sigma} \frac{d\mathcal{H}_x^{n-1}}{H(x)} = \frac{1}{n-1} \int_{\Omega} \operatorname{div}(x)\,dx = \frac{n}{n-1}\,|\Omega|.$$

That is, equality holds in the Heintze–Karcher inequality (7.1.17), and accordingly, Lemma 7.1.2 guarantees that Ω is a ball. □

7.2 The Isoperimetric Problem

The soap bubble theorem has strong connections with the so-called "isoperimetric problem," that is, the problem of finding the set of the prescribed volume with the smallest possible surface area. We discuss this connection in Corollaries 7.2.4 and 7.2.5. To this end, we first point out the following "first variation formula" for the area functional.

Lemma 7.2.1. *Let $\alpha \in (0,1)$. Let $\Omega \subseteq \mathbb{R}^n$ be a bounded and connected set of class $C^{2,\alpha}$. Let $\rho > 0$ and $p \in \partial\Omega$. Let $T > 0$ and $\Phi \in C^{2,\alpha}(\mathbb{R}^n \times [-T,T], \mathbb{R}^n)$ be such that $\Phi(x,0) = x$ for every $x \in \mathbb{R}^n$. Let $\vartheta(x) := \partial_t \Phi(x,0)$ and $\Omega_t := \Phi(\Omega,t)$. Then,*

$$\frac{d}{dt}\mathcal{H}^{n-1}(\partial\Omega_t)\Big|_{t=0} = \int_{\partial\Omega} H(x)\,\vartheta(x)\cdot\nu(x)\,d\mathcal{H}_x^{n-1}.$$

Proof. In local coordinates, we can assume that $\partial\Omega$ is locally parameterized by some diffeomorphism $f : D \to \Sigma$, for some domain $D \subset \mathbb{R}^{n-1}$, see Figure 2.4.1. We consider the metrics g_{ij} as in (2.4.1), corresponding to the matrix $\mathcal{g}$. We let Σ be the portion of $\partial\Omega$ parameterized by this set of coordinates and $\widetilde{\Sigma} := \Phi(\Sigma,t)$ for some t (with $|t|$ to be taken as small as we wish in what follows). The induced metrics on $\widetilde{\Sigma}$ are denoted by $\widetilde{g}_{ij}$ and the corresponding matrix by $\widetilde{\mathcal{g}}$. Note that $\widetilde{\Sigma}$ is parameterized by

$$\widetilde{f}(\eta) := \Phi(f(\eta),t) = f(\eta) + t\vartheta(f(\eta)) + o(t),$$

with $\eta \in D$.

We also let

$$m_{ij} := \sum_{k=1}^{n} \left[\frac{\partial\vartheta}{\partial x_k} \cdot \frac{\partial f}{\partial \eta_i}\frac{\partial f_k}{\partial \eta_j} + \frac{\partial\vartheta}{\partial x_k} \cdot \frac{\partial f}{\partial \eta_j}\frac{\partial f_k}{\partial \eta_i} \right] \tag{7.2.1}$$

and denote by $\mathcal{m}$ the associated matrix. Since, for every $j \in \{1,\dots,n-1\}$,

$$\frac{\partial \widetilde{f}}{\partial \eta_j}(\eta) = \frac{\partial f}{\partial \eta_j}(\eta) + t\sum_{k=1}^{n} \frac{\partial\vartheta}{\partial x_k}(f(\eta))\,\frac{\partial f}{\partial \eta_j}(\eta)\cdot e_k + o(t),$$

we infer from the definition of metrics in (2.4.1) that

$$\widetilde{g}_{ij} := \frac{\partial \widetilde{f}}{\partial \eta_i} \cdot \frac{\partial \widetilde{f}}{\partial \eta_j} = g_{ij} + t \sum_{k=1}^{n} \left[\frac{\partial \vartheta}{\partial x_k} \cdot \frac{\partial f}{\partial \eta_i} \frac{\partial f_k}{\partial \eta_j} + \frac{\partial \vartheta}{\partial x_k} \cdot \frac{\partial f}{\partial \eta_j} \frac{\partial f_k}{\partial \eta_i} \right] + o(t)$$
$$= g_{ij} + t m_{ij} + o(t),$$

where $f_k := f \cdot e_k$ (also, hereinafter, f and its derivatives are evaluated at $\eta \in D$ and ϑ, and its derivatives are evaluated at $f(\eta)$, but we omit this dependence for the sake of brevity). As a result,

$$\widetilde{\mathscr{g}} = \mathscr{g} + t\, \mathscr{m} + o(t) = \mathscr{g}\Big(\mathrm{Id} + t\,\mathscr{g}^{-1}\, \mathscr{m} + o(t)\Big),$$

whence

$$\det \widetilde{\mathscr{g}} = \det \mathscr{g} \det \Big(\mathrm{Id} + t\,\mathscr{g}^{-1}\, \mathscr{m} + o(t)\Big) = \det \mathscr{g} \left(1 + t \sum_{i,j=1}^{n-1} g^{ji} m_{ij} + o(t)\right).$$

Consequently, the surface element of $\widetilde{\Sigma}$ in local coordinates (see e.g. [EG15, pp. 125–126]), which we denote with a slight abuse of notation as $d\widetilde{\Sigma}$, can be written as

$$d\widetilde{\Sigma} = \sqrt{\det \widetilde{\mathscr{g}}}\, d\eta = \sqrt{\det \mathscr{g}} \left(1 + \frac{t}{2} \sum_{j=1}^{n-1} g^{ji} m_{ij} + o(t)\right) d\eta$$
$$= \left(1 + \frac{t}{2} \sum_{i,j=1}^{n-1} g^{ij} m_{ij} + o(t)\right) d\Sigma. \tag{7.2.2}$$

Using this and the definition of m_{ij} in (7.2.1),

$$\frac{1}{2} \sum_{i,j=1}^{n-1} g^{ij} m_{ij} = \sum_{\substack{1 \leqslant i,j \leqslant n-1 \\ 1 \leqslant k \leqslant n}} g^{ij} \frac{\partial \vartheta}{\partial x_k} \cdot \frac{\partial f}{\partial \eta_i} \frac{\partial f_k}{\partial \eta_j}$$
$$= \sum_{\substack{1 \leqslant i,j \leqslant n-1 \\ 1 \leqslant k \leqslant n}} g^{ij} \frac{\partial \vartheta}{\partial x_k} \cdot \frac{\partial f}{\partial \eta_i} \frac{\partial f}{\partial \eta_j} \cdot e_k = \sum_{i,j=1}^{n-1} g^{ij} \left(D\vartheta \frac{\partial f}{\partial \eta_i}\right) \cdot \frac{\partial f}{\partial \eta_j}.$$

Therefore, in light of Corollary 2.4.6,

$$\frac{1}{2} \sum_{j=1}^{n-1} g^{ij} m_{ij} = \mathrm{div}_\Sigma\, \vartheta.$$

This and (7.2.2) lead to

$$d\widetilde{\Sigma} = \big(1 + t\, \mathrm{div}_\Sigma\, \vartheta + o(t)\big)\, d\Sigma.$$

As a result,

$$\mathcal{H}^{n-1}(\partial\Omega_t) = \mathcal{H}^{n-1}(\partial\Omega) + t\int_{\partial\Omega} \operatorname{div}_\Sigma \vartheta \, d\mathcal{H}^{n-1} + o(t).$$

Hence, recalling the tangential divergence Theorem 2.3.4 (used here with $\varphi := 1$),

$$\mathcal{H}^{n-1}(\partial\Omega_t) = \mathcal{H}^{n-1}(\partial\Omega) + t\int_{\partial\Omega} H(x)\,\vartheta(x)\cdot\nu(x)\,d\mathcal{H}^{n-1}_x + o(t),$$

which produces the desired result. □

Now, we provide some general observations on vector fields and one-parameter families of diffeomorphisms. First, we discuss[1] how to construct a one-parameter family of diffeomorphisms that preserves the volume of a set while having as its initial velocity a zero-flux vector field.

Lemma 7.2.2. *Let Ω be a bounded subset of $\mathbb{R}^n$ with C^2 boundary and exterior unit normal ν. Let $\vartheta \in C^1(\partial\Omega, \mathbb{R}^n)$ such that*

$$\int_{\partial\Omega} \vartheta(x)\cdot\nu(x)\,d\mathcal{H}^{n-1}_x = 0. \tag{7.2.3}$$

Then, there exist $T > 0$, a neighborhood $\mathcal{N}$ of $\partial\Omega$ and $\Phi \in C^1((\Omega\cup\mathcal{N})\times[-T,T],\mathbb{R}^n)$ such that

$$\begin{aligned} &\Phi(\cdot,t) \text{ is a diffeomorphism between } \Omega\cup\mathcal{N} \\ &\text{and} \quad \Phi(\Omega\cup\mathcal{N},t) \quad \text{for all } t\in[-T,T], \end{aligned} \tag{7.2.4}$$

$$\frac{\partial}{\partial t}\Phi(x,t)\Big|_{t=0} = \vartheta(x) \quad \text{for all } x\in\partial\Omega \tag{7.2.5}$$

and

$$\begin{aligned} &\text{the Lebesgue measure of the set } \Phi(\Omega,t) \\ &\text{is equal to the Lebesgue measure of } \Omega \quad \text{for all } t\in[-T,T]. \end{aligned} \tag{7.2.6}$$

Proof. We consider the normal extension of ϑ in a suitably small neighborhood $\mathcal{N}$ of $\partial\Omega$, as defined in (2.3.3). Moreover, we take a smaller neighborhood $\mathcal{N}' \Subset \mathcal{N}$ of $\partial\Omega$ and $\vartheta^\star \in C^1(\Omega\cup\mathcal{N},\mathbb{R}^n)$ such that $\vartheta^\star = \vartheta_{\text{ext}}$ in $\mathcal{N}'$ and $\vartheta^\star = 0$ in $\Omega\setminus\mathcal{N}$.

[1] On the one hand, we believe that Lemma 7.2.2 is handy and can be conveniently used on several occasions. On the other hand, we also point out that the use of Lemma 7.2.2 in these notes can be replaced by the flow of an ordinary differential equation with a prescribed velocity combined with a space dilation to preserve a given volume; see, for example, the alternative proof of Corollary 7.2.4 presented on p. 512 and the alternative argument for the proof of Lemma 7.3.2 given in the footnote on p. 518.

Hence, we let X^t be the flow associated with the vector field $\vartheta^\star$, i.e. for $|t|$ sufficiently small, the solution of the ordinary differential equation

$$\begin{cases} \dfrac{d}{dt}X^t(x) = \vartheta^\star(X^t(x)), \\ X^0(x) = x, \end{cases}$$

for all $x \in \Omega \cup \mathcal{N}'$.

We also take $\nu^\star \in C^1(\Omega \cup \mathcal{N}, \mathbb{R}^n)$ such that $\nu^\star = \nu_{\text{ext}}$ in $\mathcal{N}'$ and $\nu^\star = 0$ in $\Omega \setminus \mathcal{N}$. With this notation, for $x \in \Omega \cup \mathcal{N}$, $s \in \mathbb{R}$ and $|t|$ small, we also define

$$f(x,t,s) := X^t(x) + s\nu^\star(x).$$

Let also $F(t,s)$ be the Lebesgue measure of the set $f(\Omega,t,s)$, that is, $F(t,s) := |f(\Omega,t,s)|$. We stress that

$$f(\Omega,0,s) = \{G(x,s),\ \ x \in \Omega\},$$

where $G(x,s) := x + s\nu^\star(x)$. We also remark that, for small $|s|$, we have $D_xG(x,s) = \mathrm{Id} + sD_x\nu^\star(x)$, where Id is the identity matrix, and consequently,

$$\det D_xG(x,s) = 1 + s\,\mathrm{div}\,\nu^\star(x) + o(s).$$

Using the change of variable $y := G(x,s)$, this entails that, for small $|s|$,

$$\begin{aligned} F(0,s) &= \int_{f(\Omega,0,s)} dy = \int_{G(\Omega,s)} dy = \int_\Omega |\det D_xG(x,s)|\,dx \\ &= \int_\Omega \big(1 + s\,\mathrm{div}\,\nu^\star(x)\big)\,dx + o(s). \end{aligned}$$

Hence, from the divergence theorem,

$$\begin{aligned} F(0,s) &= |\Omega| + s\int_\Omega \mathrm{div}\,\nu^\star(x)\,dx + o(s) \\ &= F(0,0) + s\int_{\partial\Omega} \nu^\star(x)\cdot\nu(x)\,d\mathcal{H}^{n-1}_x + o(s) \\ &= F(0,0) + s\mathcal{H}^{n-1}(\partial\Omega) + o(s). \end{aligned}$$

As a result, we have that

$$\partial_s F(0,0) = \mathcal{H}^{n-1}(\partial\Omega) \neq 0; \tag{7.2.7}$$

thus, according to the implicit function theorem, there exists a C^1 function γ such that $\gamma(0) = 0$ and, when $|t|$ is small enough,

$$F(t,\gamma(t)) = F(0,0), \tag{7.2.8}$$

that is, setting $\Phi(x,t) := f(x,t,\gamma(t))$,

$$|\Phi(\Omega,t)| = |f(\Omega,t,\gamma(t))| = F(t,\gamma(t)) = F(0,0) = |f(\Omega,0,0)| = |\Omega|,$$

thus establishing (7.2.6).

We also point out that $D_x\Phi(\Omega \cup \mathcal{N},0) = \mathrm{Id}$; hence, (7.2.4) is a consequence of the inverse function theorem.

In addition, for small $|t|$,

$$D_x X^t(x) = \mathrm{Id} + tD_x\vartheta^\star(x) + o(t);$$

hence, $\det D_x X^t(x) = 1 + t\,\mathrm{div}\,\vartheta^\star(x) + o(t)$, and for this reason,

$$\begin{aligned} F(t,0) = |f(\Omega,t,0)| &= \int_{f(\Omega,t,0)} dy = \int_{X^t(\Omega)} dy \\ &= \int_\Omega \big(1 + t\,\mathrm{div}\,\vartheta^\star(x)\big)\,dx + o(t) = |\Omega| + t\int_\Omega \mathrm{div}\,\vartheta^\star(x)\,dx + o(t) \\ &= F(0,0) + t\int_{\partial\Omega} \vartheta(x)\cdot\nu(x)\,d\mathcal{H}^{n-1}_x + o(t) = F(0,0) + o(t), \end{aligned}$$

where (7.2.3) has been used in the last step.

Therefore, $\partial_t F(0,0) = 0$, and accordingly, in light of (7.2.7) and (7.2.8),

$$0 = \frac{d}{dt}F(t,\gamma(t))\Big|_{t=0} = \partial_t F(0,0) + \partial_s F(0,0)\,\gamma'(0) = 0 + \mathcal{H}^{n-1}(\Omega)\,\gamma'(0),$$

from which we deduce that $\gamma'(0) = 0$.

Making use of this, we conclude that, for all $x \in \partial\Omega$,

$$\begin{aligned} \frac{\partial}{\partial t}\Phi(x,t)\Big|_{t=0} &= \frac{\partial}{\partial t}f(x,t,\gamma(t))\Big|_{t=0} = \partial_t f(x,t,\gamma(t)) + \partial_s f(x,t,\gamma(t))\,\gamma'(t)\Big|_{t=0} \\ &= \frac{d}{dt}X^t(x)\Big|_{t=0} + \nu_{\mathrm{ext}}(x)\,\gamma'(0) = \vartheta(X^t(x))\Big|_{t=0} = \vartheta(x), \end{aligned}$$

and this proves (7.2.5). □

Now, we point out that a function which averages to zero on the boundary of a domain against all normal fluxes of zero-flux vector fields must necessarily be constant.

Lemma 7.2.3. *Let Ω be a bounded subset of $\mathbb{R}^n$ with C^1 boundary and exterior unit normal ν. Let f be a continuous function on $\partial\Omega$ such that*

$$\int_{\partial\Omega} f(x)\vartheta(x)\cdot\nu(x)\,d\mathcal{H}^{n-1}_x = 0$$

for every $\vartheta \in C^1(\partial\Omega,\mathbb{R}^n)$ such that $\int_{\partial\Omega}\vartheta(x)\cdot\nu(x)\,d\mathcal{H}^{n-1}_x = 0$.

Then, f is constant on $\partial\Omega$.

Proof. Let $f_\varepsilon \in C^\infty(\partial\Omega)$ such that $f_\varepsilon \to f$ uniformly on $\partial\Omega$ as $\varepsilon \to 0$. Let also

$$\vartheta_\varepsilon(x) := \left(f_\varepsilon(x) - \fint_{\partial\Omega} f_\varepsilon(y)\, d\mathcal{H}^{n-1}_y \right) \nu(x).$$

Then, we have that

$$\int_{\partial\Omega} \vartheta_\varepsilon(x) \cdot \nu(x)\, d\mathcal{H}^{n-1}_x = \int_{\partial\Omega} \left(f_\varepsilon(x) - \fint_{\partial\Omega} f_\varepsilon(y)\, d\mathcal{H}^{n-1}_y \right) d\mathcal{H}^{n-1}_x = 0,$$

and accordingly,

$$\begin{aligned} 0 &= \fint_{\partial\Omega} f(x)\vartheta_\varepsilon(x) \cdot \nu(x)\, d\mathcal{H}^{n-1}_x \\ &= \fint_{\partial\Omega} f(x) \left(f_\varepsilon(x) - \fint_{\partial\Omega} f_\varepsilon(y)\, d\mathcal{H}^{n-1}_y \right) d\mathcal{H}^{n-1}_x \\ &= \fint_{\partial\Omega} f(x)\, f_\varepsilon(x)\, dx - \left(\fint_{\partial\Omega} f(y)\, d\mathcal{H}^{n-1}_y \right) \left(\fint_{\partial\Omega} f_\varepsilon(y)\, d\mathcal{H}^{n-1}_y \right). \end{aligned}$$

Therefore, passing to the limit as $\varepsilon \to 0$,

$$\begin{aligned} &\fint_{\partial\Omega} \left(f(x) - \fint_{\partial\Omega} f(y)\, d\mathcal{H}^{n-1}_y \right)^2 d\mathcal{H}^{n-1}_x \\ &\quad = \fint_{\partial\Omega} \left[f^2(x) + \left(\fint_{\partial\Omega} f(y)\, d\mathcal{H}^{n-1}_y \right)^2 - 2f(x) \fint_{\partial\Omega} f(y)\, d\mathcal{H}^{n-1}_y \right] d\mathcal{H}^{n-1}_x \\ &\quad = \fint_{\partial\Omega} f^2(x)\, dx - \left(\fint_{\partial\Omega} f(y)\, d\mathcal{H}^{n-1}_y \right)^2 \\ &\quad = \lim_{\varepsilon \to 0} \left[\fint_{\partial\Omega} f(x)\, f_\varepsilon(x)\, dx - \left(\fint_{\partial\Omega} f(y)\, d\mathcal{H}^{n-1}_y \right) \left(\fint_{\partial\Omega} f_\varepsilon(y)\, d\mathcal{H}^{n-1}_y \right) \right] \\ &\quad = 0, \end{aligned}$$

and this shows that $f(x) = \fint_{\partial\Omega} f(y)\, d\mathcal{H}^{n-1}_y$ for every $x \in \partial\Omega$. $\square$

We can now address the isoperimetric problem of determining the shape-minimizing surface for a prescribed volume. Note that we do not address here the delicate problem of the existence (and smoothness) of surface minimizers for a fixed volume (this is one of the classical topics in the calculus of variations, see e.g. [Giu84, Mag12, Mor16]).

Corollary 7.2.4. *Let $\alpha \in (0, 1)$. Let $\Omega \subseteq \mathbb{R}^n$ be a bounded, open and connected set of class $C^{2,\alpha}$ such that $\mathcal{H}^{n-1}(\partial\Omega) \leqslant \mathcal{H}^{n-1}(\partial\widetilde{\Omega})$ for all $\widetilde{\Omega}$ among the bounded, open and connected sets of class $C^{2,\alpha}$ with a given volume. Then, the mean curvature of $\partial\Omega$ is necessarily constant.*

Proof. We take a zero-flux vector field ϑ and the corresponding one-parameter family of diffeomorphisms $\Phi(\cdot, t)$ constructed in Lemma 7.2.2. In light of (7.2.6) and the minimality of Ω, we have that $\mathcal{H}^{n-1}(\partial\Omega) \leqslant \mathcal{H}^{n-1}(\partial\Omega^{(t)})$, for all $t \in (-T, T)$, for a suitable $T > 0$, where $\Omega_t := \Phi(\Omega, t)$. In particular, recalling Lemma 7.2.1,

$$0 = \frac{d}{dt}\mathcal{H}^{n-1}(\partial\Omega_t)\Big|_{t=0} = \int_{\partial\Omega} H(x)\,\vartheta(x)\cdot\nu(x)\,d\mathcal{H}^{n-1}_x.$$

Since this is valid for all zero-flux vector fields ϑ, we can apply Lemma 7.2.3 (used here with $f := H$) and deduce that H is constant along $\partial\Omega$. □

For completeness, we also give a proof of Corollary 7.2.4 that does not rely on the construction of the family of diffeomorphisms in Lemma 7.2.2, which preserves the volume of the domain but rather exploits the special properties of scaling of the problem under consideration.

Another Proof of Corollary 7.2.4. We consider any vector field $\vartheta \in C^1(\mathbb{R}^n, \mathbb{R}^n)$ and the corresponding solution Φ^t of the associated Cauchy problem:

$$\begin{cases} \dfrac{d}{dt}\Phi^t(x) = \vartheta(\Phi^t(x)), \\ \Phi^0(x) = x. \end{cases}$$

We point out that the solution of the above problem is well defined for $t \in [-T, T]$, for a suitable $T > 0$, due to the existence and uniqueness theorem for ordinary differential equations. We set $\Omega_t := \Phi^t(\Omega)$. In this situation, the volume of Ω_t is not necessarily the same as that of Ω; therefore, it is convenient to define $\Omega_t^\star$ as the dilation of Ω_t with the same volume as Ω. To this end, we point out that

$$|\Omega_t| = \int_\Omega (1 + t\,\mathrm{div}\,\vartheta(x))\,dx + o(t) = |\Omega| + t\int_{\partial\Omega}\vartheta(x)\cdot\nu(x)\,d\mathcal{H}^{n-1}_x + o(t). \tag{7.2.9}$$

In particular, we have that $|\Omega_t| > 0$ if $|t|$ is sufficiently small, whence we can define $\mu(t) := \frac{|\Omega|}{|\Omega_t|}$. Let also

$$\Omega_t^\star := \big(\mu(t)\big)^{\frac{1}{n}}\,\Omega_t.$$

Note that $|\Omega_t^\star| = \mu(t)\,|\Omega_t| = |\Omega|$, thanks to the scaling properties of the Lebesgue measure; hence, using the minimality property of Ω, we infer that

$$\frac{d}{dt}\mathcal{H}^{n-1}(\partial\Omega_t^\star)\Big|_{t=0} = 0. \tag{7.2.10}$$

We also remark that

$$\mathcal{H}^{n-1}(\partial\Omega_t^\star) = \big(\mu(t)\big)^{\frac{n-1}{n}}\,\mathcal{H}^{n-1}(\partial\Omega_t), \tag{7.2.11}$$

thanks to the scaling invariance of the Hausdorff measure. Moreover, by (7.2.9),

$$\begin{aligned}(\mu(t))^{\frac{n-1}{n}} &= \left(\frac{|\Omega_t|}{|\Omega|}\right)^{\frac{1-n}{n}} = \left(1 + \frac{t}{|\Omega|}\int_{\partial\Omega} \vartheta(x)\cdot\nu(x)\,d\mathcal{H}^{n-1}_x + o(t)\right)^{\frac{1-n}{n}} \\ &= 1 + \frac{(1-n)\,t}{n\,|\Omega|}\int_{\partial\Omega} \vartheta(x)\cdot\nu(x)\,d\mathcal{H}^{n-1}_x + o(t).\end{aligned}$$

As a result, we have that

$$\frac{d}{dt}(\mu(t))^{\frac{n-1}{n}}\bigg|_{t=0} = \frac{(1-n)}{n\,|\Omega|}\int_{\partial\Omega} \vartheta(x)\cdot\nu(x)\,d\mathcal{H}^{n-1}_x.$$

Besides, we know that $\mu(0) = 1$ and thus, by (7.2.11),

$$\begin{aligned}&\frac{d}{dt}\mathcal{H}^{n-1}(\partial\Omega_t^\star)\bigg|_{t=0} \\ &\quad= \frac{d}{dt}(\mu(t))^{\frac{n-1}{n}}\bigg|_{t=0}\mathcal{H}^{n-1}(\partial\Omega) + (\mu(0))^{\frac{n-1}{n}}\frac{d}{dt}\mathcal{H}^{n-1}(\partial\Omega_t)\bigg|_{t=0} \\ &\quad= \frac{(1-n)\,\mathcal{H}^{n-1}(\partial\Omega)}{n\,|\Omega|}\int_{\partial\Omega} \vartheta(x)\cdot\nu(x)\,d\mathcal{H}^{n-1}_x + \frac{d}{dt}\mathcal{H}^{n-1}(\partial\Omega_t)\bigg|_{t=0}.\end{aligned}$$

Combining this information with (7.2.10) and Lemma 7.2.1, we gather that

$$\begin{aligned}0 &= \frac{(1-n)\,\mathcal{H}^{n-1}(\partial\Omega)}{n\,|\Omega|}\int_{\partial\Omega} \vartheta(x)\cdot\nu(x)\,d\mathcal{H}^{n-1}_x + \int_{\partial\Omega} H(x)\,\vartheta(x)\cdot\nu(x)\,d\mathcal{H}^{n-1}_x \\ &= \int_{\partial\Omega}\left(\frac{(1-n)\,\mathcal{H}^{n-1}(\partial\Omega)}{n\,|\Omega|} + H(x)\right)\vartheta(x)\cdot\nu(x)\,d\mathcal{H}^{n-1}_x.\end{aligned}$$

This and the fact that ϑ is an arbitrary vector field yield that H is constantly equal to $\frac{(n-1)\,\mathcal{H}^{n-1}(\partial\Omega)}{n\,|\Omega|}$ along $\partial\Omega$. □

By combining Corollary 7.2.4 with the soap bubble theorem in 7.1.3, we immediately obtain the desired classification result for the isoperimetric problem.

Corollary 7.2.5. *Let $\alpha \in (0,1)$ and $\Omega \subseteq \mathbb{R}^n$ be a bounded, open and connected set of class $C^{2,\alpha}$ such that $\mathcal{H}^{n-1}(\partial\Omega) \leqslant \mathcal{H}^{n-1}(\partial\widetilde{\Omega})$ among all the bounded, open and connected sets $\widetilde{\Omega}$ of class $C^{2,\alpha}$ with a given volume. Then, Ω is a ball.*

See Andrejs Treibergs' review at http://www.math.utah.edu/~treiberg/isoperim/isop.pdf

for several proofs[2] of the isoperimetric problem. For further readings about the soap bubble theorem and related topics, see also Nicola Garofalo's notes at https://www.math.purdue.edu/~garofalo/Soap_bubble.pdf and the references therein.

7.3 Serrin's Overdetermined Problem

In the celebrated article [Ser71], James Serrin established an important result about the radial symmetry of solutions of an "overdetermined" Poisson equation. Here, the overdetermination comes from the fact that two boundary conditions are prescribed instead of only one. As clearly discussed in [Ser71], the problem arose from considerations in physics (especially fluid dynamics and mechanics). In fact, the work came in response to a question posed by Roger Fosdick, one of Serrin's colleagues at the University of Minnesota. As remarked in [Ser71], from a physical viewpoint, Serrin's result entails that:

- given "a viscous incompressible fluid moving in straight parallel streamlines through a straight pipe of given cross-sectional form [...] the tangential stress on the pipe wall is the same at all points of the wall if and only if the pipe has a circular cross-section";
- given "a solid straight bar subject to torsion, the magnitude of the resulting traction which occurs at the surface of the bar is independent of position if and only if the bar has a circular cross-section".

See [DV23, Chapter 5] for a physical description of the viscous flow in a straight pipe and [DV23, Chapter 16] for the derivation of the equation for straight bars under torsion.

In Corollary 7.3.3, we also recall another motivation for Serrin's overdetermined equation dealing with the shape of a prismatic bar that maximizes the torsional rigidity.

Here, we provide a proof of a specific case of Serrin's overdetermined problem using a strategy developed by Hans Weinberger in [Wei71]: interestingly, this proof came out shortly after the original argument by Serrin, and it was contained in an article following Serrin's and published in the same journal. These early works resulted in an incredibly fruitful research field, which still plays a distinguished role in contemporary research. See also [PS51, PS89, CH98, NT18, MP20] and the references therein for further details about Serrin's problem and alternative points

[2]For completeness, we mention that the smoothness assumption on the set for the isoperimetric problem can be relaxed, see e.g. [DG58, Gug77, Fus04]. See also [Fus15] for a detailed review.

of view about it; the original approach based on the moving plane method is also presented in Section 8.3. See additionally [DPV21] for detailed physical motivations for Serrin's overdetermined problem and [Wag02] for a broad perspective that combines overdetermined problems, the Pohožaev identity (recall Theorem 2.1.9) and domain variation methods (to be compared e.g. with Lemma 7.3.2 here).

Theorem 7.3.1. *Let $\Omega \subseteq \mathbb{R}^n$ be a bounded, open and connected set of class C^2, and let $u \in C^2(\overline{\Omega})$ be a solution of*

$$\begin{cases} \Delta u = 1 & \text{in } \Omega, \\ u = 0 & \text{on } \partial\Omega, \\ \partial_\nu u = c & \text{on } \partial\Omega, \end{cases} \tag{7.3.1}$$

for some $c \in \mathbb{R}$. Then, Ω is necessarily a ball.

We stress that the equation considered in [Ser71] is actually more general than the one presented here; however, we stick to the model case in (7.3.1) since it allows one to develop different approaches and perspectives to the problem (more general equations instead can only be treated by some of the methods presented).

Proof of Theorem 7.3.1. By (7.1.5),

$$|\Omega| = \int_{\partial\Omega} \partial_\nu u(x)\, d\mathcal{H}^{n-1}_x = c\,\mathcal{H}^{n-1}(\partial\Omega). \tag{7.3.2}$$

Accordingly, the Pohožaev identity in Theorem 2.1.9 and the divergence theorem give that

$$\begin{aligned} \frac{|\Omega|^2\, n}{2\,(\mathcal{H}^{n-1}(\partial\Omega))^2} &= \frac{c^2\, n}{2} = \frac{c^2}{2|\Omega|} \int_\Omega \operatorname{div}(x)\, dx \\ &= \frac{c^2}{2|\Omega|} \int_{\partial\Omega} x \cdot \nu(x)\, d\mathcal{H}^{n-1}_x \\ &= \frac{1}{2|\Omega|} \int_{\partial\Omega} (\partial_\nu u(x))^2\, (x \cdot \nu(x))\, d\mathcal{H}^{n-1}_x \\ &= \frac{n-2}{2|\Omega|} \int_\Omega u(x)\, dx - \frac{n}{|\Omega|} \int_\Omega u(x)\, dx \\ &= -\frac{n+2}{2|\Omega|} \int_\Omega u(x)\, dx \end{aligned} \tag{7.3.3}$$

Now, we define

$$\mathcal{P} := \frac{n\,|\nabla u|^2}{2} - u.$$

Note that

$$\mathcal{P} = \frac{n\,c^2}{2} \quad \text{on } \partial\Omega.$$

Furthermore, by the Bochner identity in Lemma 1.1.5 and the setting in (7.1.1), in Ω,

$$\Delta\mathcal{P} = n\,\nabla(\Delta u)\cdot\nabla u + n\,|D^2u|^2 - \Delta u = n|D^2u|^2 - (\Delta u)^2 = \mathcal{D}_u. \tag{7.3.4}$$

Consequently, by the inequality in (7.1.2), we have that $\Delta\mathcal{P} \geqslant 0$ in Ω. From this and the maximum principle in Corollary 2.9.2(i), we conclude that, for every $x \in \Omega$,

$$\mathcal{P}(x) \leqslant \sup_{\partial\Omega}\mathcal{P} = \frac{n\,c^2}{2},$$

and more precisely, by the strong maximum principle in Theorem 2.9.1(i), either

$$\mathcal{P}(x) < \frac{n\,c^2}{2} \quad \text{for every } x \in \Omega \tag{7.3.5}$$

or

$$\mathcal{P}(x) = \frac{n\,c^2}{2} \quad \text{for every } x \in \Omega. \tag{7.3.6}$$

We claim that

$$\text{condition (7.3.5) cannot hold true.} \tag{7.3.7}$$

To prove this statement, suppose, by contradiction, that (7.3.5) is satisfied. Then, using Green's first identity (2.1.10) (with $\varphi := \psi := u$) and formula (7.3.3),

$$\begin{aligned}\frac{n\,c^2\,|\Omega|}{2} &> \int_\Omega \mathcal{P}(x)\,dx = \frac{n}{2}\int_\Omega |\nabla u(x)|^2\,dx - \int_\Omega u(x)\,dx\\ &= -\frac{n+2}{2}\int_\Omega u(x)\,dx = \frac{n|\Omega|^3}{2\,(\mathcal{H}^{n-1}(\partial\Omega))^2},\end{aligned}$$

which is in contradiction with (7.3.2), and this proves (7.3.7).

Accordingly, by (7.3.7), we deduce that necessarily (7.3.6) holds true; therefore, $\Delta\mathcal{P} = 0$ in Ω.

This and (7.3.4) entail that $\mathcal{D}_u$ vanishes identically in Ω. That is, condition (7.1.13) is fulfilled, whence Ω is necessarily a ball, thanks to Lemma 7.1.1. □

As an additional motivation for Serrin's overdetermined equation, we recall the so-called Saint Venant problem (named after Adhémar Jean Claude Barré de Saint

Venant, one of the pioneers of the modern theory of elasticity and hydraulic engineering). The problem that we take into account is that of a bar of cross-section Ω, which is a rigid beam with shape given by $\Omega\times\mathbb{R}$ (in concrete cases, one takes $\Omega\subseteq\mathbb{R}^2$, but we deal with the case of $\Omega\subseteq\mathbb{R}^n$ for the sake of generality). The Saint Venant problem is focused on the cross-sections Ω that maximize, among competitors with the same volume, the "torsional rigidity":

$$\tau(\Omega) := \max_{\substack{v\in H_0^1(\Omega)\\ v\not\equiv 0}} \frac{\left(\int_\Omega v(x)\,dx\right)^2}{\int_\Omega |\nabla v(x)|^2\,dx}. \tag{7.3.8}$$

The above maximum is attained by the Sobolev embedding, see e.g. [Eva98]. We now exploit the fact that, to maximize τ, the cross-section Ω is necessarily stationary under domain variations: this method is widely used in shape optimization theory and leads to the very interesting concept of "domain derivative," see e.g. [HP18]. Instead, we do not address here the problem of the existence (and smoothness) of the maximizing domain for the torsional rigidity, see Example 2.8 in [BDM93]; see also [Buc07] for a general approach toward the existence of extremals of shape optimization problems (as a matter of fact, the forthcoming Lemma 7.3.2 can be seen as a special case of the Hadamard variational formula presented in [HP18, Chapter 5], to which we refer for full details on a very delicate type of computation).

Lemma 7.3.2. *Let $\alpha\in(0,1)$. Let $\Omega\subseteq\mathbb{R}^n$ be a connected open set of class $C^{2,\alpha}$ that maximizes the torsional rigidity τ among the bounded, open and connected sets of class $C^{2,\alpha}$ with a given volume. Then, there exists a function $u\in C^2(\overline{\Omega})$ such that*

$$\begin{cases}\Delta u = 1 & \text{in } \Omega,\\ u=0 & \text{on } \partial\Omega,\\ \partial_\nu u = c & \text{on } \partial\Omega,\end{cases}$$

for some $c\in\mathbb{R}$.

Proof. We let

$$J(v) := \int_\Omega |\nabla v(x)|^2\,dx$$

and consider the minimizer $u_\star$ of J among every candidate $v\in H_0^1(\Omega)$ with $\int_\Omega v(x)\,dx = 1$. Then, according to the theory of Lagrange multipliers (see e.g. [Eva98]), there exists $\lambda\in\mathbb{R}$ such that, for every $\varphi\in C_0^\infty(\Omega)$,

$$\int_\Omega \nabla u_\star(x)\cdot\nabla\varphi(x)\,dx = \lambda\int_\Omega \varphi(x)\,dx.$$

Hence, setting $\widetilde{u}(x) := u_\star(x) + \frac{\lambda|x|^2}{2n}$, it follows that $\widetilde{u}$ is a weak solution of $\Delta\widetilde{u} = 0$ in Ω since

$$\begin{aligned}\int_\Omega \nabla\widetilde{u}(x)\cdot\nabla\varphi(x)\,dx &= \int_\Omega \nabla u_\star(x)\cdot\nabla\varphi(x)\,dx - \int_\Omega \nabla\left(\frac{\lambda|x|^2}{2n}\right)\cdot\nabla\varphi(x)\,dx\\ &= \lambda\int_\Omega\varphi(x)\,dx - \int_\Omega\Delta\left(\frac{\lambda|x|^2}{2n}\right)\varphi(x)\,dx\\ &= \lambda\int_\Omega\varphi(x)\,dx - \lambda\int_\Omega\varphi(x)\,dx\\ &= 0,\end{aligned}$$

for every $\varphi \in C_0^\infty(\Omega)$.

This and Weyl's lemma (recall Lemma 2.2.1) give that $\widetilde{u}$ is actually $C^2(\Omega)$ and harmonic. As a result, $u_\star \in C^2(\Omega)$ and solves $\Delta u_\star = -\lambda$ in Ω. As a matter of fact, we have that $u_\star \in C^2(\overline{\Omega})$ due to Theorem 4.5.4, and thus $u_\star = 0$ along $\partial\Omega$.

We also remark that, for every $v \in H_0^1(\Omega)$, with $v \not\equiv 0$, setting $w := \frac{v}{\int_\Omega v(x)\,dx}$, we have that $\int_\Omega w(x)\,dx = 1$, and thus

$$\frac{\int_\Omega|\nabla v(x)|^2\,dx}{\left(\int_\Omega v(x)\,dx\right)^2} = \int_\Omega|\nabla w(x)|^2\,dx = J(w) \geqslant J(u_\star) = \frac{\int_\Omega|\nabla u_\star(x)|^2\,dx}{\left(\int_\Omega u_\star(x)\,dx\right)^2}.$$

Consequently, by the definition of torsional rigidity in (7.3.8) and Green's first identity in (2.1.10),

$$\tau(\Omega) = \frac{\left(\int_\Omega u_\star(x)\,dx\right)^2}{\int_\Omega|\nabla u_\star(x)|^2\,dx} = \frac{\int_\Omega u_\star(x)\,dx}{\lambda}. \tag{7.3.9}$$

Then, defining $u := -\frac{u_\star}{\lambda}$, we find that

$$\begin{cases}\Delta u = 1 & \text{in } \Omega,\\ u = 0 & \text{on } \partial\Omega,\end{cases} \tag{7.3.10}$$

and by (7.3.9),

$$\tau(\Omega) = -\int_\Omega u(x)\,dx. \tag{7.3.11}$$

We now consider a family[3] of smooth diffeorphisms Φ and a smooth vector field ϑ as in Lemma 7.2.2. Note that for every $x \in \mathbb{R}^n$, we have that

[3]Corresponding to the alternative proof of Corollary 7.2.4 presented on p. 512, we mention that it is also possible in the proof of Lemma 7.3.2 to bypass the construction of the diffeorphisms of Lemma 7.2.2 by an appropriate scaling argument. Namely, one can instead consider any vector field $\vartheta \in C^1(\mathbb{R}^n, \mathbb{R}^n)$ and the corresponding solution Φ^t of the associated Cauchy problem and then look at the competitor set $\left(\frac{|\Omega|}{|\Phi^t(\Omega)|}\right)^{\frac{1}{n}}\Phi^t(\Omega)$. Note that this set comes from a dilation of the flow of Ω

$\Phi(x,t) = x + t\vartheta(x) + o(t)$ as $t \to 0$, and $|\Phi(\Omega,t)| = |\Omega|$. We remark that $D_x\Phi(x,t) = \mathrm{Id} + tD_x\vartheta(x) + o(t)$, where Id is the identity matrix, and consequently, $\det D_x\Phi(x,t) = 1 + t\operatorname{div}\vartheta(x) + o(t)$.

We also consider $u^{(t)}$ to be the corresponding solution of (7.3.10) when Ω is replaced by $\Phi(\Omega,t)$, namely

$$\begin{cases} \Delta u^{(t)} = 1 & \text{in } \Phi(\Omega,t), \\ u^{(t)} = 0 & \text{on } \partial(\Phi(\Omega,t)). \end{cases}$$

From the above setting and the regularity theory for the Poisson equation (recall Theorem 4.5.4), we have that, for all $t \in [-1,1]$,

$$\left\|u^{(t)}\right\|_{C^{2,\alpha}(\Phi(\Omega,t))} \leqslant C, \tag{7.3.12}$$

for some $C > 0$ and $\alpha \in (0,1)$ independent of t. Also, by (7.3.11), we know that

$$\tau(\Phi(\Omega,t)) = -\int_{\Phi(\Omega,t)} u^{(t)}(x)\,dx,$$

whence if Ω is a maximizing set for a given volume constraint, it follows that

$$\int_{\Phi(\Omega,t)} u^{(t)}(x)\,dx - \int_\Omega u(x)\,dx = \tau(\Omega) - \tau(\Phi(\Omega,t)) = o(t). \tag{7.3.13}$$

Now, we point out that

$$\begin{aligned} &\int_{\Phi(\Omega,t)} u^{(t)}(y)\,dy \\ &= \int_\Omega u^{(t)}(\Phi(x,t))\,(1 + t\operatorname{div}\vartheta(x))\,dx + o(t) \\ &= \int_\Omega u^{(t)}(\Phi(x,t))\,\Delta u(x)\,dx + t\int_\Omega u^{(t)}(\Phi(x,t))\,\operatorname{div}\vartheta(x)\,dx + o(t) \\ &= -\int_\Omega \nabla\Big(u^{(t)}(\Phi(x,t))\Big)\cdot\nabla u(x)\,dx + tI_1 + o(t), \end{aligned} \tag{7.3.14}$$

---

with velocity ϑ; hence, it can be written as $\widetilde{\Phi}^t(\Omega)$, with $\widetilde{\Phi}^t(x) := \left(\frac{|\Omega|}{|\Phi^t(\Omega)|}\right)^{\frac{1}{n}} \Phi^t(x)$. We also point out that

$$\begin{aligned} \widetilde{\Phi}^t(x) &= \left(\frac{|\Omega|}{|\Omega| + t\int_\Omega \operatorname{div}\vartheta(x)\,dx}\right)^{\frac{1}{n}} (x + t\vartheta(x)) + o(t) \\ &= \left(1 - \frac{t}{n\,|\Omega|}\int_\Omega \operatorname{div}\vartheta(x)\,dx\right)(x + t\vartheta(x)) + o(t) \\ &= x + t\widetilde{\vartheta}(x) + o(t), \end{aligned}$$

with

$$\widetilde{\vartheta}(x) := \vartheta(x) - \frac{x}{n\,|\Omega|}\int_\Omega \operatorname{div}\vartheta(x)\,dx,$$

and we observe that $\widetilde{\vartheta}$ has zero flux along $\partial\Omega$. Then, the computations presented in these pages remain almost unchanged simply by replacing ϑ with $\widetilde{\vartheta}$.

where

$$I_1 := \int_\Omega u^{(t)}(\Phi(x,t))\,\operatorname{div}\vartheta(x)\,dx.$$

Furthermore, for every $i \in \{1,\dots,n\}$ and $x \in \Omega$,

$$\begin{aligned}\partial_i\Big(u^{(t)}(\Phi(x,t))\Big) &= \nabla u^{(t)}(\Phi(x,t))\cdot\partial_i\Phi(x,t)\\ &= \partial_i u^{(t)}(\Phi(x,t)) + t\nabla u^{(t)}(\Phi(x,t))\cdot\partial_i\vartheta(x) + o(t)\end{aligned}$$

and

$$\begin{aligned}\Delta\Big(u^{(t)}(\Phi(x,t))\Big) &= \sum_{i=1}^n \partial_i\nabla u^{(t)}(\Phi(x,t))\cdot\partial_i\Phi(x,t)\\ &\quad + t\sum_{i=1}^n \big(D^2u^{(t)}(\Phi(x,t))\partial_i\Phi(x,t)\big)\cdot\partial_i\vartheta(x)\\ &\quad + t\nabla u^{(t)}(\Phi(x,t))\cdot\Delta\vartheta(x) + o(t)\\ &= 1 + 2t\sum_{i=1}^n \partial_i\nabla u^{(t)}(\Phi(x,t))\cdot\partial_i\vartheta(x)\\ &\quad + t\nabla u^{(t)}(\Phi(x,t))\cdot\Delta\vartheta(x) + o(t).\end{aligned} \tag{7.3.15}$$

As a result,

$$\begin{aligned}\int_\Omega u(x)\,dx &= \int_\Omega u(x)\,\Delta\Big(u^{(t)}(\Phi(x,t))\Big)\,dx - 2tI_2 - tI_3 + o(t)\\ &= -\int_\Omega \nabla u(x)\cdot\nabla\Big(u^{(t)}(\Phi(x,t))\Big)\,dx - 2tI_2 - tI_3 + o(t),\end{aligned} \tag{7.3.16}$$

where

$$I_2 := \sum_{i=1}^n \int_\Omega u(x)\,\partial_i\nabla u^{(t)}(\Phi(x,t))\cdot\partial_i\vartheta(x)\,dx$$

$$\text{and}\quad I_3 := \int_\Omega u(x)\,\nabla u^{(t)}(\Phi(x,t))\cdot\Delta\vartheta(x)\,dx.$$

Combining (7.3.14) and (7.3.16), after simplifying one term, we find that

$$\int_{\Phi(\Omega,t)} u^{(t)}(x)\,dx - \int_\Omega u(x)\,dx = t(I_1 + 2I_2 + I_3) + o(t).$$

This and (7.3.13) lead to

$$I_1 + 2I_2 + I_3 = o(1) \qquad \text{as } t \to 0. \tag{7.3.17}$$

Now, we consider the function $w^{(t)}(x) := u^{(t)}(\Phi(x,t)) - u(x)$, and we claim that, as $t \to 0$,

$$w^{(t)} \to 0 \text{ in } W^{2,p}(\Omega), \tag{7.3.18}$$

for every $p > 1$. To prove this, we utilize (7.3.10), (7.3.12) and (7.3.15) to see that $|\Delta w^{(t)}| \leqslant \bar{C}t$, for some $\bar{C} > 0$ independent of t. Hence, using the Calderón–Zygmund regularity theory for the Poisson equation (see (5.3.17) and Footnote 5 on p. 451), we obtain (7.3.18), as desired.

By using (7.3.17) and (7.3.18), we thus conclude that

$$\begin{aligned} 0 &= \lim_{t\to 0} I_1 + 2I_2 + I_3 \\ &= \int_\Omega \left(u(x)\, \operatorname{div} \vartheta(x) + 2u(x) \sum_{i=1}^n \partial_i \nabla u(x) \cdot \partial_i \vartheta(x) + u(x)\, \nabla u(x) \cdot \Delta \vartheta(x) \right) dx. \end{aligned} \tag{7.3.19}$$

We also remark that

$$\begin{aligned} \operatorname{div}\big(u(D\vartheta \nabla u)\big) &= \sum_{i=1}^n \partial_i \big(u\, \partial_i \vartheta \cdot \nabla u\big) \\ &= (D\vartheta \nabla u) \cdot \nabla u + u \nabla u \cdot \Delta \vartheta + u \partial_i \nabla u \cdot \partial_i \vartheta. \end{aligned}$$

Comparing with (7.3.19) and using the divergence theorem, we thereby conclude that

$$\begin{aligned} 0 &= \int_\Omega \left(u(x)\, \operatorname{div} \vartheta(x) + u(x) \sum_{i=1}^n \partial_i \nabla u(x) \cdot \partial_i \vartheta(x) \right. \\ &\qquad \left. + \operatorname{div}\Big(u(x)\big(D\vartheta(x)\nabla u(x)\big)\Big) - \big(D\vartheta(x)\nabla u(x)\big) \cdot \nabla u(x) \right) dx \\ &= \int_\Omega \left(u(x)\, \operatorname{div} \vartheta(x) + u(x) \sum_{i=1}^n \partial_i \nabla u(x) \cdot \partial_i \vartheta(x) - \big(D\vartheta(x)\nabla u(x)\big) \cdot \nabla u(x) \right) dx. \end{aligned} \tag{7.3.20}$$

Additionally,

$$\operatorname{div}(u\vartheta) = \nabla u \cdot \vartheta + u \operatorname{div} \vartheta.$$

Using this information and (7.3.20), as well as exploiting once more the divergence theorem, we see that

$$\begin{aligned} 0 &= \int_\Omega \left(\operatorname{div}\big(u(x)\, \vartheta(x)\big) - \nabla u(x) \cdot \vartheta(x) \right. \\ &\qquad \left. + u(x) \sum_{i=1}^n \partial_i \nabla u(x) \cdot \partial_i \vartheta(x) - \big(D\vartheta(x)\nabla u(x)\big) \cdot \nabla u(x) \right) dx \\ &= \int_\Omega \left(-\nabla u(x) \cdot \vartheta(x) + u(x) \sum_{i=1}^n \partial_i \nabla u(x) \cdot \partial_i \vartheta(x) - \big(D\vartheta(x)\nabla u(x)\big) \cdot \nabla u(x) \right) dx. \end{aligned} \tag{7.3.21}$$

Besides, recalling (7.3.10), we observe that

$$\begin{aligned}\operatorname{div}\big((\nabla u\cdot\vartheta)\nabla u\big) &= \sum_{i=1}^{n}\partial_i(\nabla u\cdot\vartheta)\partial_i u + (\nabla u\cdot\vartheta)\Delta u\\ &= \sum_{i=1}^{n}\partial_i\nabla u\cdot\vartheta\partial_i u + (D\vartheta\nabla u)\cdot\nabla u + \nabla u\cdot\vartheta,\end{aligned}$$

and also,

$$\begin{aligned}\sum_{i=1}^{n}\partial_i\nabla u\cdot\vartheta\partial_i u + u\sum_{i=1}^{n}\partial_i\nabla u\cdot\partial_i\vartheta &= \sum_{i=1}^{n}\partial_i\nabla u\cdot\vartheta\partial_i u + \sum_{i=1}^{n}\partial_i\nabla u\cdot\partial_i\vartheta u - \nabla\Delta u\cdot\vartheta u\\ &= \sum_{i=1}^{n}\partial_i\big(\partial_i\nabla u\cdot\vartheta u\big)\\ &= \operatorname{div} V,\end{aligned}$$

where V is the vector field defined as $V\cdot e_i := \partial_i\nabla u\cdot\vartheta u$, for each $i\in\{1,\dots,n\}$. We also remark that $V=0$ along $\partial\Omega$, thanks to the boundary condition in (7.3.10). Consequently, we can rewrite (7.3.21) in the following form:

$$\begin{aligned}0 &= \int_\Omega\Bigg(\sum_{i=1}^{n}\partial_i\nabla u(x)\cdot\vartheta(x)\partial_i u(x) - \operatorname{div}\Big(\big(\nabla u(x)\cdot\vartheta(x)\big)\nabla u(x)\Big)\\ &\qquad + u(x)\sum_{i=1}^{n}\partial_i\nabla u(x)\cdot\partial_i\vartheta(x)\Bigg)\,dx\\ &= \int_\Omega\Big(\operatorname{div} V(x) - \operatorname{div}\Big(\big(\nabla u(x)\cdot\vartheta(x)\big)\nabla u(x)\Big)\Big)\,dx\\ &= \int_{\partial\Omega}\big(\nabla u(x)\cdot\vartheta(x)\big)\big(\nabla u(x)\cdot\nu(x)\big)\,d\mathcal{H}^{n-1}_x.\end{aligned}$$

Since $u=0$ along $\partial\Omega$ and hence $\nabla u = |\nabla u|\,\nu$ along $\partial\Omega$, we thereby infer that

$$\int_{\partial\Omega}\big(\partial_\nu u(x)\big)^2\,\vartheta(x)\cdot\nu(x)\,d\mathcal{H}^{n-1}_x = 0.$$

This and Lemma 7.2.3 give that[4]

$$\partial_\nu u \text{ is constant along } \partial\Omega. \tag{7.3.22}$$

From this fact and (7.3.10), the desired result follows. □

[4]It is interesting to point out that Lemma 7.3.2 also entails a rather explicit motivation for Serrin's overdetermined problem (7.3.1) in terms of optimal heating. Indeed, suppose that a region Ω is given. The external environment lies at a constant temperature, say zero, and Ω is heated in a uniform and homogeneous way. Up to normalizing constants, the equilibrium configuration is therefore described by the equation

$$\begin{cases}-\Delta U_\Omega = C & \text{in } \Omega,\\ U_\Omega = 0 & \text{on } \partial\Omega,\end{cases}$$

Combining Lemma 7.3.2 with Theorem 7.3.1, we obtain the desired classification result for the Saint Venant problem, as follows.

Corollary 7.3.3. *Let $\alpha \in (0,1)$. Let $\Omega \subseteq \mathbb{R}^n$ be a bounded and connected set of class $C^{2,\alpha}$ that maximizes the torsional rigidity τ among the bounded, open and connected sets of class $C^{2,\alpha}$ with a given volume. Then, Ω is necessarily a ball.*

Other proofs of Corollary 7.3.3 can be obtained from (7.3.11) and rearrangement methods (see [PS51, Tal76] and also [BG15], as well as the references therein).

7.4 The Converse of the Mean Value Formula

We address here a natural question arising from the mean value formula for harmonic functions (recall Theorem 2.1.2). Namely, suppose that we are given a set Ω and a point $x_0 \in \Omega$, and we know that the value at x_0 of every harmonic function in Ω coincides with the volume average of the function in Ω (or, alternatively, with the surface average along $\partial\Omega$). Then, what can we say about the set Ω?

Remarkably, the answer is that Ω is necessarily a ball, and the point x_0 is necessarily the center of such a ball. As a result, balls are the only sets for which the mean value formula in Theorem 2.1.2 holds true.

This beautiful result (obtained in [Kur72, PS89]) follows from the following general statement (see in particular (i), (ii) and (v)).

---

where U_Ω represents the temperature in Ω and $C \in (0,+\infty)$ is the heat source term described by the uniform heating of Ω. In this setting, a natural question is to determine the domain Ω which maximizes (among the smooth, bounded and connected sets of a given volume) the average temperature

$$T(\Omega) := \int_\Omega U_\Omega(x)\,dx.$$

Setting $u_\Omega := -\frac{U_\Omega}{C}$ we have that

$$\begin{cases} \Delta u_\Omega = 1 & \text{in } \Omega, \\ u_\Omega = 0 & \text{on } \partial\Omega. \end{cases}$$

Moreover, in view of (7.3.11),

$$T(\Omega) = -C\int_\Omega u_\Omega(x)\,dx = C\tau(\Omega).$$

Therefore, the determination of the optimal shape for Ω aimed at maximizing the average temperature is equivalent to the Saint Venant problem (in particular, the optimally heated domain is a ball, thanks to Corollary 7.3.3). Furthermore, the optimal temperature U_Ω for such a domain satisfies Serrin's problem (7.3.1) since $\partial_\nu U_\Omega = -C\partial_\nu u_\Omega$ is constant along $\partial\Omega$, owing to (7.3.22).

Theorem 7.4.1. *Let $n \geqslant 2$ and Ω be a bounded and open set with boundary of class C^1. Let also $x_0 \in \Omega$ and G be the Green function of Ω, as defined in* (2.10.2). *Then, the following conditions are equivalent:*

(i) *For every harmonic function $u \in L^1(\Omega)$, we have that*

$$u(x_0) = \fint_\Omega u(x)\,dx.$$

(ii) *For every harmonic function $u \in C(\overline{\Omega})$, we have that*

$$u(x_0) = \fint_{\partial\Omega} u(x)\,d\mathcal{H}_x^{n-1}.$$

(iii) *There exist $c_1 \in \mathbb{R}\setminus\{0\}$, $c_2 \in \mathbb{R}$ and $w \in C^2(\Omega)\cap C^1(\overline{\Omega})$ such that*

$$\begin{cases} \Delta w(x) = c_1 & \text{for all } x \in \Omega,\\ w(x) = 0 & \text{for all } x \in \partial\Omega,\\ \nabla w(y)\cdot\nu(y) = c_2\nabla_y G(x_0,y)\cdot\nu(y) & \text{for all } x \in \partial\Omega. \end{cases}$$

(iv) *The function $\partial\Omega \ni y \mapsto \nabla_y G(x_0,y)\cdot\nu(y)$ is constant.*

(v) *Ω is a ball centered at x_0.*

Proof. We show that (iv) implies (ii), which implies (v), which implies (iv), and that (iii) implies (i), which implies (v), which implies (iii).

For this, let us assume that (iv) holds true, namely $\nabla_y G(x_0,y)\cdot\nu(y) = c$ for every $y \in \partial\Omega$, for some $c \in \mathbb{R}$, and let u be harmonic in Ω and continuous in $\overline{\Omega}$. Let also $\Omega' \Subset \Omega$, with $\Omega' \ni x_0$. By Theorem 2.10.2 (and recalling also Lemma 2.2.1),

$$\begin{aligned} -u(x_0) &= \int_{\Omega'} \Delta u(y) G(x_0,y)\,dy + \int_{\partial\Omega'} u(y)\frac{\partial G}{\partial\nu}(x_0,y)\,d\mathcal{H}_y^{n-1}\\ &= \int_{\partial\Omega'} u(y)\frac{\partial G}{\partial\nu}(x_0,y)\,d\mathcal{H}_y^{n-1}. \end{aligned}$$

Hence, taking Ω' as close as we wish to Ω,

$$u(x_0) = -\int_{\partial\Omega} u(y)\frac{\partial G}{\partial\nu}(x_0,y)\,d\mathcal{H}_y^{n-1} = -c\int_{\partial\Omega} u(y)\,d\mathcal{H}_y^{n-1}. \tag{7.4.1}$$

In particular, choosing $u := 1$, we find that $1 = -c\mathcal{H}^{n-1}(\partial\Omega)$. Plugging this information back into (7.4.1), we obtain (ii), as desired.

Let us now suppose that (ii) holds true, and let u be harmonic in Ω such that $u(y) = \frac{1}{\mathcal{H}^{n-1}(\partial\Omega)} + \nabla_y G(x_0,y)\cdot\nu(y)$ for every $y \in \partial\Omega$ (note that such a function exists thanks to Theorem 3.3.3 since the boundary of class C^1 provides the required external cone). By (ii) and Theorem 2.10.2, we have that, for every $\Omega' \Subset \Omega$, with $\Omega' \ni x_0$,

$$0 = u(x_0) + \int_{\partial\Omega'} u(y)\nabla_y G(x_0,y)\cdot\nu(y)\,d\mathcal{H}_y^{n-1}$$

$$
\begin{aligned}
&= \fint_{\partial\Omega'} u(y)\,dy + \int_{\partial\Omega'} u(y)\nabla_y G(x_0, y)\cdot \nu(y)\,d\mathcal{H}_y^{n-1} \\
&= \int_{\partial\Omega'} u(y)\left(\frac{1}{\mathcal{H}^{n-1}(\partial\Omega')} + \nabla_y G(x_0, y)\cdot \nu(y)\right) d\mathcal{H}_y^{n-1}.
\end{aligned}
$$

Therefore, taking Ω' as close as we wish to Ω,

$$
\begin{aligned}
0 &= \int_{\partial\Omega} u(y)\left(\frac{1}{\mathcal{H}^{n-1}(\partial\Omega')} + \nabla_y G(x_0, y)\cdot \nu(y)\right) d\mathcal{H}_y^{n-1} \\
&= \int_{\partial\Omega} \left(\frac{1}{\mathcal{H}^{n-1}(\partial\Omega')} + \nabla_y G(x_0, y)\cdot \nu(y)\right)^2 d\mathcal{H}_y^{n-1},
\end{aligned}
$$

leading to

$$
\nabla_y G(x_0, y)\cdot \nu(y) = -\frac{1}{\mathcal{H}^{n-1}(\partial\Omega)} \quad \text{for each } y \in \partial\Omega. \tag{7.4.2}
$$

Moreover, since $G(x_0, \cdot) = 0$ along $\partial\Omega$, we know that $\nabla_y G(x_0, y) = \pm|\nabla_y G(x_0, y)|\nu(y)$ for all $y \in \partial\Omega$; therefore, we infer from (7.4.2) that $\frac{1}{\mathcal{H}^{n-1}(\partial\Omega)} = |\nabla_y G(x_0, y)|$ for all $y \in \partial\Omega$. For this reason, we have that

$$
\begin{aligned}
&\left|\int_{\partial\Omega} (y - x_0)\cdot \nabla_y G(x_0, y)\,d\mathcal{H}_y^{n-1}\right| \\
&\quad = \left|\int_{\partial\Omega} |\nabla_y G(x_0, y)|\,(y - x_0)\cdot \nu(y)\,d\mathcal{H}_y^{n-1}\right| \\
&\quad = \frac{1}{\mathcal{H}^{n-1}(\partial\Omega)}\left|\int_{\partial\Omega} (y - x_0)\cdot \nu(y)\,d\mathcal{H}_y^{n-1}\right| \\
&\quad = \frac{1}{\mathcal{H}^{n-1}(\partial\Omega)}\left|\int_{\Omega} \operatorname{div}(y - x_0)\,dy\right| \\
&\quad = \frac{n|\Omega|}{\mathcal{H}^{n-1}(\partial\Omega)}.
\end{aligned} \tag{7.4.3}
$$

We now define

$$
h(y) := (y - x_0)\cdot \nabla_y G(x_0, y) + (n-2)G(x_0, y).
$$

By (2.7.6),

$$(y - x_0) \cdot \nabla_y \Gamma(x_0 - y) = \begin{cases} -\dfrac{c_n(n-2)}{|x_0 - y|^{n-2}} & \text{if } n \neq 2, \\ -c_n & \text{if } n = 2 \end{cases}$$
$$= \begin{cases} -(n-2)\Gamma(x_0 - y) & \text{if } n \neq 2, \\ -c_n & \text{if } n = 2. \end{cases}$$

Thus, recalling the Robin function in (2.10.1) and (2.10.2), as well as the value of c_n when $n = 2$ given in (2.7.2), we see that, for every $y \in \Omega$,

$$\begin{aligned} h(y) &= (y - x_0) \cdot \nabla_y \Gamma(x_0 - y) + (n-2)\Gamma(x_0 - y) \\ &\quad - (y - x_0) \cdot \nabla_y \Psi^{(x_0)}(y) - (n-2)\Psi^{(x_0)}(y) \\ &= c_\star - (y - x_0) \cdot \nabla_y \Psi^{(x_0)}(y) - (n-2)\Psi^{(x_0)}(y), \end{aligned}$$

with

$$c_\star = \begin{cases} -\dfrac{1}{2\pi} & \text{if } n = 2, \\ 0 & \text{otherwise,} \end{cases}$$

which in particular gives that h is harmonic in Ω since so is the Robin function.

Consequently, we can apply (ii) to the function h and infer that

$$\begin{aligned} c_\star - (n-2)\Psi^{(x_0)}(x_0) = h(x_0) &= \fint_{\partial\Omega} h(y)\, d\mathcal{H}_y^{n-1} \\ &= \fint_{\partial\Omega} \Big((y - x_0) \cdot \nabla_y G(x_0, y) + (n-2) G(x_0, y) \Big)\, d\mathcal{H}_y^{n-1} \\ &= \fint_{\partial\Omega} (y - x_0) \cdot \nabla_y G(x_0, y)\, d\mathcal{H}_y^{n-1}. \end{aligned}$$

This and (7.4.3) lead to

$$\left| c_\star - (n-2)\Psi^{(x_0)}(x_0) \right| = \frac{n|\Omega|}{\big(\mathcal{H}^{n-1}(\partial\Omega)\big)^2}. \tag{7.4.4}$$

We now distinguish two cases: when $n = 2$, equation (7.4.4) boils down to

$$\frac{1}{2\pi} = \frac{2|\Omega|}{\big(\mathcal{H}^1(\partial\Omega)\big)^2},$$

that is $\big(\mathcal{H}^1(\partial\Omega)\big)^2 = 4\pi|\Omega|$. Hence, the set Ω is a minimizer for the isoperimetric problem in the plane (see Corollary 7.2.5), and accordingly, it is necessarily a ball.

If instead $n \neq 2$, we exploit the lower bound on the Robin function pointed out in Lemma 2.10.7, we deduce from (7.4.4) that

$$\frac{1}{n|B_1|^{\frac{2}{n}}|\Omega|^{\frac{n-2}{n}}} \leqslant \left|(n-2)\Psi^{(x_0)}(x_0)\right| = \frac{n|\Omega|}{\left(\mathcal{H}^{n-1}(\partial\Omega)\right)^2}$$

or, equivalently, in light of (1.1.6),

$$\frac{|B_1|^{n-1}}{\left(\mathcal{H}^{n-1}(\partial B_1)\right)^n} \leqslant \frac{|\Omega|^{n-1}}{\left(\mathcal{H}^{n-1}(\partial\Omega)\right)^n}.$$

Hence, in this case also, Ω is a minimizer for the isoperimetric problem in the plane (see Corollary 7.2.5), and accordingly, it is necessarily a ball.

In view of these observations, we can write that $\Omega = B_R(p)$ for some $p \in \mathbb{R}^n$ and $R > 0$. Plugging this information back into (ii), given $\varpi \in \mathbb{R}^n$ and choosing $u(x) := \varpi \cdot (x - p)$, by odd symmetry, we find that

$$\varpi \cdot (x_0 - p) = u(x_0) = \fint_{\partial B_R(p)} u(x)\, d\mathcal{H}_x^{n-1} = 0,$$

from which we conclude that $p = x_0$. The proof of (v) is thereby complete.

If instead (v) holds true, then (iv) plainly follows from Theorem 2.10.4.

Let us now assume that (iii) holds true, and let u be harmonic in Ω. Then, according to Green's identity in (2.1.11),

$$\begin{aligned} c_1 \int_\Omega u(y)\, dy &= \int_\Omega \Big(u(y)\Delta w(y) - w(y)\Delta u(y)\Big)\, dy \\ &= \int_{\partial\Omega} \left(u(y)\frac{\partial w}{\partial\nu}(y) - w(y)\frac{\partial u}{\partial\nu}(y)\right) d\mathcal{H}_y^{n-1} \\ &= c_2 \int_{\partial\Omega} u(y)\nabla_y G(x_0, y)\cdot\nu(y)\, d\mathcal{H}_y^{n-1}. \end{aligned} \tag{7.4.5}$$

Also, by Theorem 2.10.2,

$$u(x_0) = -\int_{\partial\Omega} u(y)\nabla_y G(x_0, y)\cdot\nu(y)\, d\mathcal{H}_y^{n-1},$$

which, when combined with (7.4.5), gives

$$-\frac{c_2}{c_1}u(x_0) = \int_\Omega u(y)\, dy.$$

In particular, choosing u to be constant, we find that $-\frac{c_2}{c_1} = |\Omega|$, from which we obtain (i).

Now, let us suppose that (i) holds true. Up to a translation, we assume that $x_0 = 0$. Given $i, j \in \{1, \dots, n\}$, we take $\widetilde{u}$ to be the harmonic function in Ω such that

$$\widetilde{u}(x) = x_i \nu_j(x) - x_j \nu_i(x) \quad \text{for all } x \in \partial\Omega. \tag{7.4.6}$$

We observe that, for every $i, j \in \{1, \dots, n\}$,

$$\delta_{ij}\widetilde{u}(x) = 0 \quad \text{for all } x \in \partial\Omega. \tag{7.4.7}$$

We also define $u(x) := x_i \partial_j \widetilde{u}(x) - x_j \partial_i \widetilde{u}(x)$. We stress that, for each $x \in \Omega$,

$$\Delta u(x) = 2\nabla x_i \cdot \partial_j \nabla \widetilde{u}(x) - 2\nabla x_j \cdot \partial_i \nabla \widetilde{u}(x) = 2\partial_{ij}\widetilde{u}(x) - 2\partial_{ij}\widetilde{u}(x) = 0.$$

In this way, by (i),

$$0 = |\Omega|\, u(0) = \int_\Omega u(x)\, dx = \int_\Omega \Big(x_i \partial_j \widetilde{u}(x) - x_j \partial_i \widetilde{u}(x) \Big)\, dx.$$

Since, by the divergence theorem and (7.4.7),

$$\int_\Omega x_i \partial_j \widetilde{u}(x)\, dx = \int_\Omega \Big(\operatorname{div}\big(x_i \widetilde{u}(x)\, e_j\big) - \delta_{ij}\widetilde{u}(x) \Big)\, dx = \int_{\partial\Omega} x_i\, \widetilde{u}(x)\, \nu_j(x)\, d\mathcal{H}^{n-1}_x,$$

we thereby find that

$$\begin{aligned} 0 &= \int_{\partial\Omega} \Big(x_i\, \widetilde{u}(x)\, \nu_j(x) - x_j\, \widetilde{u}(x)\, \nu_i(x) \Big)\, d\mathcal{H}^{n-1}_x \\ &= \int_{\partial\Omega} \widetilde{u}(x) \Big(x_i\, \nu_j(x) - x_j\, \nu_i(x) \Big)\, d\mathcal{H}^{n-1}_x. \end{aligned}$$

Hence, recalling (7.4.6),

$$0 = \int_{\partial\Omega} \Big(x_i\, \nu_j(x) - x_j\, \nu_i(x) \Big)^2\, d\mathcal{H}^{n-1}_x,$$

which entails that $x_i\, \nu_j(x) - x_j\, \nu_i(x) = 0$ for every $x \in \partial\Omega$.

Thus, for every $x \in \partial\Omega$,

$$x \cdot \nu(x)\, \nu_j(x) = \sum_{i=1}^n x_i\, \nu_i(x) \nu_j(x) = \sum_{i=1}^n x_j\, \nu_i^2(x) = x_j$$

and, as a result, x and $\nu(x)$ are parallel, for every $x \in \partial\Omega$.

Consequently, we write $\nu(x) = g(x)x$, for a suitable scalar function g, and we stress that $1 = |g(x)|\,|x|$, from which we see that g never vanishes. We claim that

$$\text{each connected component of } \partial\Omega \text{ is a sphere centered at the origin.} \tag{7.4.8}$$

To check this, let Σ be a connected component of $\partial\Omega$, pick a point $q \in \Sigma$ and let $R := |q|$. Take also another point $\widetilde{q}$ and a curve $\gamma : [0,1] \to \Sigma$ such that $\gamma(0) = q$ and $\gamma(1) = \widetilde{q}$. Using the fact that $\dot{\gamma}(t)$ is a tangent vector, we have that

$$\begin{aligned}|\widetilde{q}|^2 - R^2 &= |\gamma(1)|^2 - |\gamma(0)|^2 = \int_0^1 \frac{d}{dt}|\gamma(t)|^2\, dt \\ &= 2\int_0^1 \gamma(t)\cdot\dot{\gamma}(t)\, dt = 2\int_0^1 \frac{1}{g(\gamma(t))}\, \nu(\gamma(t))\cdot\dot{\gamma}(t)\, dt = 0,\end{aligned}$$

which proves (7.4.8).

From (7.4.8) and the fact that $0 \in \Omega$, we find that either Ω is a ball centered at the origin or

$$\Omega = B_{R_0} \cup \left(\bigcup_{k=1}^{M} B_{R_k} \setminus \overline{B_{r_k}} \right) \tag{7.4.9}$$

for some $M \in \mathbb{N}$ and suitable radii $R_M > r_M > R_{M-1} > r_{m-1} \cdots > R_1 > r_1 > R_0 > 0$. Thus, to establish (v), we need to rule out the possibility of additional annuli in (7.4.9). To this end, we remark that Ω is connected: otherwise, if $\Omega = \Omega_1 \cup \Omega_2$ for suitable open sets Ω_1 and Ω_2 whose closures have empty intersections, we can suppose that $0 \in \Omega_1$ and let $u := \chi_{\Omega_2}$. In this setting, we have that u is harmonic in Ω and thus, by (i),

$$0 = u(0) = \fint_{\Omega} u(x)\, dx = \frac{|\Omega_2|}{|\Omega|} > 0.$$

This is a contradiction; therefore, we have[5] established (v), as desired.

Suppose now that (v) holds true. We take w to be the solution of

$$\begin{cases} \Delta w = -1 & \text{in } \Omega, \\ w = 0 & \text{on } \partial\Omega \end{cases}$$

[5] An alternative proof that (i) implies (v) exploits the Poisson kernel and goes as follows. Up to a translation, one can suppose that $x_0 = 0$. Take $r > 0$ such that $B_r \subseteq \Omega$ with $p \in (\partial B_r) \cap (\partial\Omega)$. Let

$$u(x) := \frac{|x|^2 - r^2}{|x - p|^n} + r^{2-n}.$$

Note that u is harmonic in $\mathbb{R}^n \setminus \{p\} \supseteq \Omega$. Also, if $x \in \mathbb{R}^n \setminus B_r$ then $u(x) \geqslant r^{2-n}$. Furthermore, we have that $u(0) = -\frac{r^2}{|p|^n} + r^{2-n} = 0$. As a result, using (i) and the mean value formula in Theorem 2.1.2(iii),

$$\begin{aligned} 0 = u(0) &= \fint_{\Omega} u(y)\, dy = \frac{1}{|\Omega|}\left(\int_{B_r} u(y)\, dy + \int_{\Omega\setminus B_r} u(y)\, dy \right) \\ &= \frac{1}{|\Omega|}\left(|B_r|\, u(0) + \int_{\Omega\setminus B_r} u(y)\, dy \right) = \frac{1}{|\Omega|}\int_{\Omega\setminus B_r} u(y)\, dy \geqslant \frac{|\Omega \setminus B_r|}{|\Omega|}\, r^{2-n}. \end{aligned}$$

This gives that $|\Omega \setminus B_r| = 0$ and therefore $\Omega = B_r$.

and u be harmonic in Ω such that $u(y) = \partial_\nu w(y) - |\Omega|\nabla_y G(x_0, y) \cdot \nu(y)$ for all $y \in \partial\Omega$ (we recall that such solutions exist thanks to Theorem 3.3.3). Then, using the mean value formula in Theorem 2.1.2(iii) and Green's identity in (2.1.11),

$$\begin{aligned}
-|\Omega|\, u(x_0) &= -\int_\Omega u(y)\, dy = \int_\Omega u(y)\, \Delta w(y)\, dy \\
&= \int_\Omega \Big(u(y)\Delta w(y) - w(y)\Delta u(y)\Big)\, dy \\
&= \int_{\partial\Omega} \left(u(y)\frac{\partial w}{\partial \nu}(y) - w(y)\frac{\partial u}{\partial \nu}(y)\right)\, d\mathcal{H}_y^{n-1} \\
&= \int_{\partial\Omega} \left(\frac{\partial w}{\partial \nu}(y) - |\Omega|\nabla_y G(x_0, y) \cdot \nu(y)\right) \frac{\partial w}{\partial \nu}(y)\, d\mathcal{H}_y^{n-1}. \qquad (7.4.10)
\end{aligned}$$

Additionally, by Theorem 2.10.2,

$$\begin{aligned}
u(x_0) &= -\int_{\partial\Omega} u(y)\nabla_y G(x_0, y) \cdot \nu(y)\, d\mathcal{H}_y^{n-1} \\
&= -\int_{\partial\Omega} \left(\frac{\partial w}{\partial \nu}(y) - |\Omega|\nabla_y G(x_0, y) \cdot \nu(y)\right) \nabla_y G(x_0, y) \cdot \nu(y)\, d\mathcal{H}_y^{n-1}.
\end{aligned}$$

Comparing this identity with (7.4.10), we conclude that

$$\begin{aligned}
0 &= \int_{\partial\Omega} \left(\frac{\partial w}{\partial \nu}(y) - |\Omega|\nabla_y G(x_0, y) \cdot \nu(y)\right) \frac{\partial w}{\partial \nu}(y)\, d\mathcal{H}_y^{n-1} + |\Omega|\, u(x_0) \\
&= \int_{\partial\Omega} \left(\frac{\partial w}{\partial \nu}(y) - |\Omega|\nabla_y G(x_0, y) \cdot \nu(y)\right) \frac{\partial w}{\partial \nu}(y)\, d\mathcal{H}_y^{n-1} \\
&\quad - \int_{\partial\Omega} \left(\frac{\partial w}{\partial \nu}(y) - |\Omega|\nabla_y G(x_0, y) \cdot \nu(y)\right) |\Omega|\nabla_y G(x_0, y) \cdot \nu(y)\, d\mathcal{H}_y^{n-1} \\
&= \int_{\partial\Omega} \left(\frac{\partial w}{\partial \nu}(y) - |\Omega|\nabla_y G(x_0, y) \cdot \nu(y)\right)^2 d\mathcal{H}_y^{n-1}
\end{aligned}$$

and therefore

$$\frac{\partial w}{\partial \nu}(y) - |\Omega|\nabla_y G(x_0, y) \cdot \nu(y) = 0 \quad \text{for every } y \in \partial\Omega.$$

This states that w solves the equation in (iii) with $c_1 := -1$ and $c_2 := |\Omega|$, as desired. □

Chapter 8

The Moving Plane Method

8.1 A Geometric Approach to Symmetry

In this chapter, we revisit some of the previously showcased results, such as Alexandrov's soap bubble theorem presented in Section 7.1 and Serrin's overdetermined problem presented in Section 7.3: differently from the previous exposition, we exhibit here likely the earliest, and possibly most "visual," approach to these questions, namely the so-called moving plane method. In doing so, we also discuss a classical result by Basilis Gidas, Wei-Ming Ni and Louis Nirenberg [GNN79].

As an outline, the quintessence of the method consists in detecting the symmetry with respect to a given hyperplane by comparing the original picture with its reflected one: to do so, one begins by placing the reflection hyperplane "very far" from the region of interest so that the reflection is void. Then, one slowly moves the hyperplane closer and closer toward the domain under investigation, starts looking at initial reflections, which typically (in the best case scenarios) produce an easily visualizable inclusion (thus allowing the method "to start") and then detects the "critical position" at which the picture becomes symmetric.

For instance, in the soap bubble theorem, the critical position is the "first contact" occurrence between the reflected domain and the original one (or, better put, it stops the method at the latest position at which the reflected domain is included in the original one). See Figure 8.1.1 for a sequence of frames trying to dynamically depict the moving plane method in this situation (of course, for different problems, such as the problem of Gidas, Ni and Nirenberg, the critical position may have a different notion, also related to the equation under consideration). At the contact point(s), some clever use of the maximum principle typically allows one to conclude that the reflected and original domains necessarily coincide, thus providing a symmetry result with respect to this special hyperplane (by repeating the procedure using hyperplanes in different directions, one can also deduce rotational symmetry).

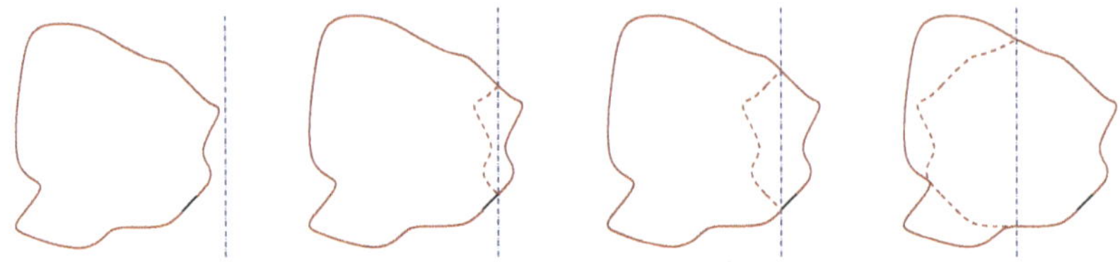

Figure 8.1.1 The moving plane method up and running.

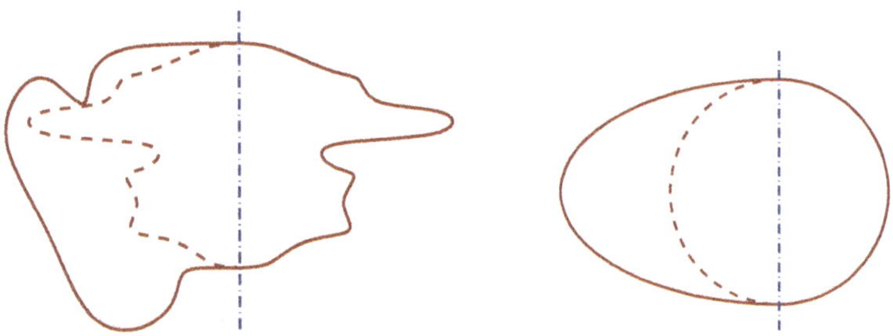

Figure 8.1.2 Interior and boundary touching points.

Of course, a number of conceptual and technical hindrances arise when trying to concretely implement this elegant idea. A common obstacle is that one needs to distinguish between the situation in which an "interior" touching point between the reflected and original domains arises and the case in which a "boundary" obstruction takes place; see the last frames in Figures 8.1.1 and 8.1.2 for possible pictures of these different scenarios.

Throughout this chapter, we employ the notation sketched in Figure 8.1.3. Namely, given a set Ω and $\lambda \in \mathbb{R}$, we let

$$\begin{aligned} T_\lambda &:= \{x = (x_1, \dots, x_n) \in \mathbb{R}^n \text{ s.t. } x_1 = \lambda\}, \\ \Sigma_\lambda &:= \{x = (x_1, \dots, x_n) \in \Omega \text{ s.t. } x_1 > \lambda\}, \\ x_\lambda &:= (2\lambda - x_1, x_2, \dots, x_n) \\ \text{and} \quad \Sigma'_\lambda &:= \{x_\lambda \text{ s.t. } x \in \Sigma_\lambda\}. \end{aligned} \tag{8.1.1}$$

Note that the point x_λ is the reflection of the point $x = (x_1, \dots, x_n)$ with respect to the hyperplane T_λ. Also, Σ'_λ is the reflection of the region Σ_λ of Ω lying to the right of T_λ.

Moreover, given $\omega \in \partial B_1$, we define the reflection of a point p with respect to the plane $\{\omega \cdot x = 0\}$ by

$$\mathcal{R}_\omega p := p - 2(\omega \cdot p)\omega, \tag{8.1.2}$$

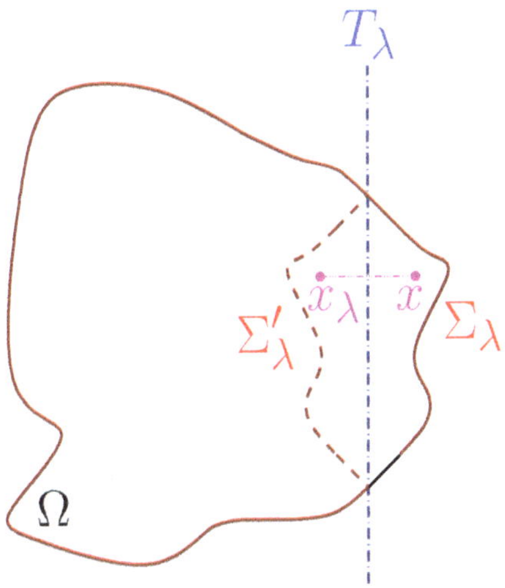

Figure 8.1.3 Notation adopted for the moving plane method.

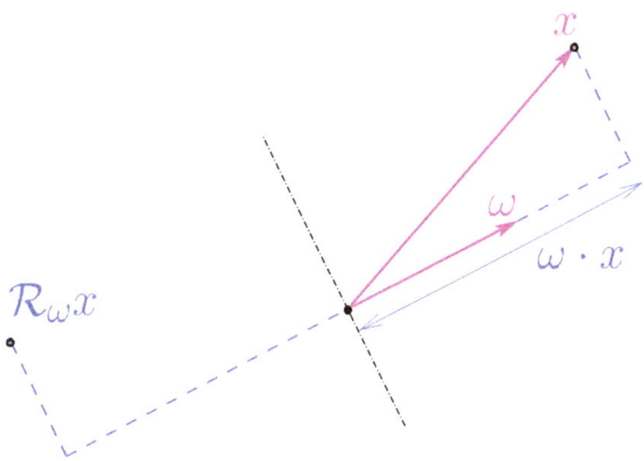

Figure 8.1.4 Reflection of a point x in direction ω.

see Figure 8.1.4 (note also that when $\omega = e_1$, the above definition boils down to the one of x_λ in (8.1.1) with $\lambda = 0$, consistently with the reflection with respect to the hyperplane $\{x_1 = 0\}$).

We refer to [BN91, Dan92b, CR18] and the references therein for further information about the moving plane method and its variants in a number of different situations. Now, let us watch the moving plane method in action in some concrete situations.

8.2 Back to the Soap Bubble Theorem

We now reconsider the discussion on the soap bubble theorem (presented in Theorem 7.1.3), and we outline here a different proof based on the reflection technique depicted in Figure 8.1.1.

For this proof, we first establish the additional (perhaps not completely obvious, see e.g. [Hop89, Lemma 2.2 on p. 147]) result that reflections in every direction are sufficient to prove rotational symmetry, as shown in the following lemma.

Lemma 8.2.1. *Let Ω be as in Theorem 7.1.3, and assume that it admits a hyperplane of symmetry in every direction (that is, for each $\omega \in \partial B_1$, there exists $x_\omega \in \mathbb{R}^n$ such that $\mathcal{R}_\omega(p - x_\omega) + x_\omega \in \Omega$ for all $p \in \Omega$).*

Then, Ω is spherically symmetric (that is, there exists x_0 and a rotation R_0 such that $R_0(p - x_0) + x_0 \in \Omega$ for every $p \in \Omega$).

For this, we point out a general observation[1] relating reflections and rotations.

Lemma 8.2.2. *Let $u : \mathbb{R}^n \to \mathbb{R}$. Suppose that for every $\omega \in \partial B_1$ and $x \in \mathbb{R}^n$, we have that $u(\mathcal{R}_\omega x) = u(x)$.*

Then, u is rotationally invariant, namely there exists $u_0 : \mathbb{R} \to \mathbb{R}$ such that $u(x) = u_0(|x|)$ for all $x \in \mathbb{R}^n$.

Proof. Let $r > 0$ and $p, q \in \partial B_r$. We claim that

$$u(p) = u(q). \tag{8.2.1}$$

To check this, we can suppose that $p \neq q$; otherwise, we are done, and we let $\omega := \frac{p-q}{|p-q|}$. Thus, we have that

$$\begin{aligned} u(p) &= u(\mathcal{R}_\omega p) = u\left(p - 2(\omega \cdot p)\omega\right) = u\left(p - 2\left(\frac{p-q}{|p-q|} \cdot p\right)\frac{p-q}{|p-q|}\right) \\ &= u\left(p - 2\frac{(r^2 - p \cdot q)(p-q)}{|p-q|^2}\right) = u\left(p - 2\frac{(r^2 - p \cdot q)(p-q)}{r^2 + r^2 - 2p \cdot q}\right) \\ &= u\left(p - (p-q)\right) = u(q), \end{aligned}$$

which proves (8.2.1).

Thus, we define $u_0(r) := u(re_1)$, and (8.2.1) comes in handy to see that, for all $x \in \mathbb{R}^n$, $u(x) = u(|x|e_1) = u_0(|x|)$, as desired. □

[1] Lemma 8.2.2 is, in fact, a particular case of the general fact from linear algebra that the reflections generate the orthogonal group, according to the Cartan–Dieudonné theorem. For simplicity, here we present an elementary approach to the problem under consideration.

Proof of Lemma 8.2.1. Up to a translation, we can suppose that the baricenter of Ω is the origin, namely that

$$\int_\Omega x\,dx = 0.$$

Then, using the isometric change of variable induced by the reflection that leaves Ω invariant, that is, looking at the transformation $x = \mathcal{R}_\omega(y - x_\omega) + x_\omega$, we see that

$$\begin{aligned} 0 &= \int_\Omega \big(\mathcal{R}_\omega(y - x_\omega) + x_\omega\big)\,dy = \int_\Omega \big((y - x_\omega) - 2\omega\cdot(y - x_\omega)\omega + x_\omega\big)\,dy \\ &= -|\Omega|\,x_\omega + 2|\Omega|(\omega\cdot x_\omega)\omega + |\Omega|\,x_\omega = 2|\Omega|(\omega\cdot x_\omega)\omega, \end{aligned}$$

and consequently,

$$\omega\cdot x_\omega = 0.$$

As a result,

$$\mathcal{R}_\omega(p - x_\omega) + x_\omega = (p - x_\omega) - 2\omega\cdot(p - x_\omega)\omega + x_\omega = p - 2(\omega\cdot p)\omega = \mathcal{R}_\omega p,$$

and thus the invariance of Ω under hyperplane reflections reads[2]

$$\mathcal{R}_\omega p \in \Omega \qquad \text{for all } p \in \Omega. \tag{8.2.2}$$

This gives that $\chi_\Omega(\mathcal{R}_\omega x) = \chi_\Omega(x)$ for all $x \in \mathbb{R}^n$. We can thereby employ Lemma 8.2.1 with $u := \chi_\Omega$ and deduce that χ_Ω is radial, hence so is Ω. □

Thanks to these preliminary observations, we can now give a proof of the soap bubble theorem by utilizing the moving plane method, looking for symmetry under hyperplane reflections. For this, we follow the procedure described in Figure 8.1.1 (note that since $\partial\Omega$ is of class $C^{2,\alpha}$ at least at the beginning of the reflections, the flipped region Σ'_λ lies inside Ω; hence, we can begin the method). Thus, we can define

$$\lambda_0 := \inf\Big\{\lambda \in \mathbb{R} \text{ s.t. } \Sigma'_t \subset \Omega \text{ for all } t > \lambda\Big\}. \tag{8.2.3}$$

In this critical configuration, we have that Σ'_{λ_0} becomes tangent to $\partial\Omega$ at some point (otherwise the distance between $\partial\Sigma'_{\lambda_0}$ and $\partial\Omega$ would be strictly positive, thus allowing for a small further push toward the left of the hyperplane T_λ, in contradiction with the infimum property in (8.2.3)).

[2]Geometrically, (8.2.2) states the interesting property that the reflection hyperplanes all pass through the baricenter, which was not completely obvious at the beginning.

Hence, as highlighted in the last frame in Figures 8.1.1 and 8.1.2, it is convenient to subdivide the critical case arising in (8.2.3) into two possible[3] subcases:

$$\begin{aligned} &\text{(i) } \Sigma'_{\lambda_0} \text{ is tangent to } \partial\Omega \text{ at a point } p_0 \textit{ not lying} \text{ on the hyperplane } T_{\lambda_0}, \\ &\text{(ii) } \Sigma'_{\lambda_0} \text{ is tangent to } \partial\Omega \text{ at a point } p_0 \textit{ lying} \text{ on the hyperplane } T_{\lambda_0}. \end{aligned} \tag{8.2.4}$$

We could argue that case (i) is somewhat "generic," and one has to be quite unlucky to bump into case (ii); yet, we cannot rule out this perhaps uncommon and less pleasant possibility. For this, we rely on suitable versions of the maximum principle, possibly exploiting different versions for cases (i) and (ii). The strategy is to compare the original set and the reflected one at the tangent point described in (8.2.4). For this, we locally parameterize $\partial\Omega$ over its tangent space by a function that satisfies an elliptic equation (specifically, in this case, the mean curvature equation) and then apply the maximum principle to the function obtained by subtracting the original solution and the reflected one.

To make the argument work, one also needs to investigate a bit more the geometry of Σ'_{λ_0}: on the one hand, in general, Σ'_{λ_0} is not necessarily convex and, even worse, not necessarily connected, see Figure 8.2.1. On the other hand, we have that Σ'_{λ_0} is always "convex in the e_1 direction," as made precise in the following result.

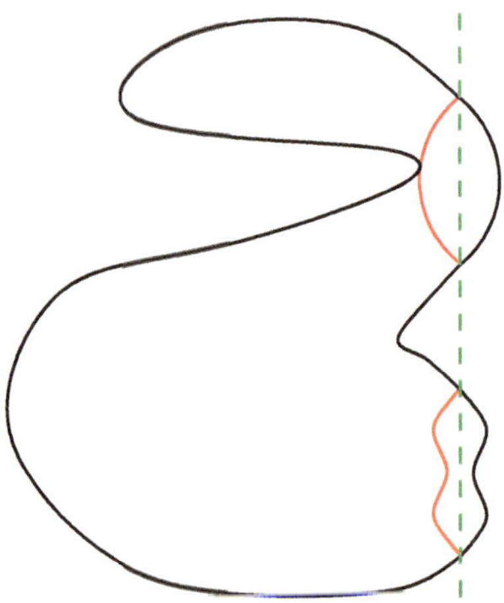

Figure 8.2.1 Σ'_0 is not necessarily convex, nor connected.

[3] As a matter of fact, both cases (i) and (ii) in (8.2.4) can also occur simultaneously; see e.g. the first picture in Figure 8.1.2. This scenario is all the better, as one is free to pick the point $p_0 \in (\partial\Sigma'_{\lambda_0}) \cap (\partial\Omega)$ either on the hyperplane T_{λ_0} or outside it, at their own preference.

Lemma 8.2.3. *Let* $x = (x_1, \dots, x_n) \in \Sigma'_{\lambda_0}$. *Then,* $(\tau, x_2, \dots, x_n) \in \Sigma'_{\lambda_0}$ *for all* $\tau \in [x_1, \lambda_0)$.

Proof. Since Σ'_{λ_0} is open and bounded, we can move horizontally from x toward the left, until we hit the boundary of Σ'_{λ_0}: that is, we take $\zeta_0 < x_1$ such that $(\zeta_0, x_2, \dots, x_n) \in \partial\Sigma'_{\lambda_0}$ and

$$(\zeta, x_2, \dots, x_n) \in \Sigma'_{\lambda_0} \quad \text{for all } \zeta \in (\zeta_0, x_1]. \tag{8.2.5}$$

Now, to prove the desired claim in Lemma 8.2.3, we actually show the following strongest result:

$$(\tau, x_2, \dots, x_n) \in \Sigma'_{\lambda_0} \quad \text{for all } \tau \in (\zeta_0, \lambda_0).$$

For this, it suffices to show that

$$(2\lambda_0 - \tau, x_2, \dots, x_n) \in \Sigma_{\lambda_0} \quad \text{for all } \tau \in (\zeta_0, \lambda_0),$$

or, equivalently, calling $\vartheta := 2\lambda_0 - \tau$,

$$(\vartheta, x_2, \dots, x_n) \in \Omega \quad \text{for all } \vartheta \in (\lambda_0, 2\lambda_0 - \zeta_0). \tag{8.2.6}$$

To check this, we begin by noting that, in light of (8.2.5), $(2\lambda_0 - \zeta, x_2, \dots, x_n) \in \Omega$ for all $\zeta \in (\zeta_0, x_1]$, whence $(\vartheta, x_2, \dots, x_n) \in \Omega$ for all $\zeta \in [2\lambda_0 - x_1, 2\lambda_0 - \zeta_0)$. As a result, we can define

$$\lambda^\star := \inf\Big\{t < 2\lambda_0 - x_1 \text{ s.t. } (\vartheta, x_2, \dots, x_n) \in \Omega \text{ for all } \vartheta \in (t, 2\lambda_0 - \zeta_0)\Big\},$$

we observe that $\lambda^\star \leqslant 2\lambda_0 - x_1$, and to prove (8.2.6), we aim to show that $\lambda^\star \leqslant \lambda_0$.

For this, suppose not. Then,

$$\lambda^\star > \lambda_0. \tag{8.2.7}$$

We know that

$$(\vartheta, x_2, \dots, x_n) \in \Omega \quad \text{for all } \vartheta \in (\lambda^\star, 2\lambda_0 - \zeta_0). \tag{8.2.8}$$

From (8.2.3) and (8.2.8), we see that

$$(2t - \vartheta, x_2, \dots, x_n) \in \Sigma'_t \subseteq \Omega \quad \text{for all } \vartheta \in (\lambda^\star, 2\lambda_0 - \zeta_0) \text{ and } t > \lambda_0. \tag{8.2.9}$$

In particular, we take $\varepsilon > 0$ small enough such that $\vartheta_\varepsilon := \lambda^\star + 2\varepsilon \in (\lambda^\star, 2\lambda_0 - \zeta_0)$ and $t_\varepsilon := \lambda^\star - \varepsilon > \lambda_0$ (we stress that we can fulfill the latter condition owing to (8.2.7)). With these choices, we infer from (8.2.9) that

$$\begin{aligned}\Omega \ni (2t_\varepsilon - \vartheta_\varepsilon, x_2, \dots, x_n) &= \big(2(\lambda^\star - \varepsilon) - (\lambda^\star + 2\varepsilon), x_2, \dots, x_n\big)\\ &= (\lambda^\star - 4\varepsilon, x_2, \dots, x_n),\end{aligned}$$

which is in contradiction with the minimality of $\lambda^\star$. □

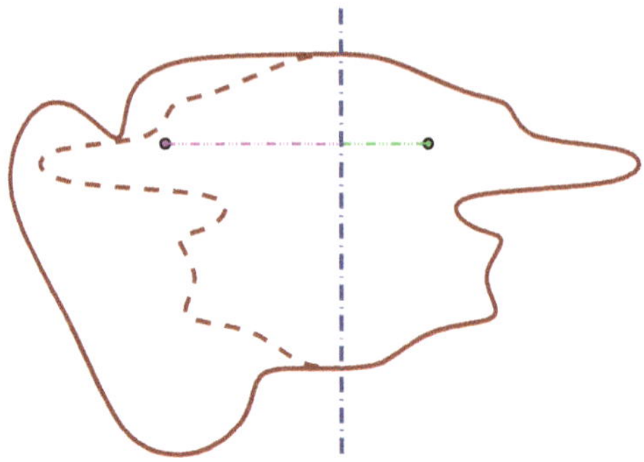

Figure 8.2.2 Σ''_{λ_0} is convex in the e_1 direction.

Corollary 8.2.4. *Let Σ''_{λ_0} be the interior of the closure of $\Sigma_{\lambda_0} \cup \Sigma'_{\lambda_0}$. Assume that $(a, x_2, \dots, x_n)$, $(b, x_2, \dots, x_n) \in \Sigma''_{\lambda_0}$ for some $a < b$. Then, $(\tau, x_2, \dots, x_n) \in \Sigma''_{\lambda_0}$ for all $\tau \in [a, b]$.*

Proof. By Lemma 8.2.3, we have that Σ'_{λ_0} is convex in the e_1 direction, hence so is Σ_{λ_0} by symmetry. This gives the desired result when both $(a, x_2, \dots, x_n)$ and $(b, x_2, \dots, x_n)$ lie either in Σ'_{λ_0} or in Σ_{λ_0}.

It remains to consider the case in which $a \leqslant \lambda_0$ and $b \geqslant \lambda_0$. In this case, one uses Lemma 8.2.3 to connect $(a, x_2, \dots, x_n)$ with $(\lambda_0, x_2, \dots, x_n)$ (possibly connecting it first with $(\lambda_0 - \varepsilon, x_2, \dots, x_n)$ and then sending $\varepsilon \searrow 0$) and similarly to connect $(b, x_2, \dots, x_n)$ with $(\lambda_0, x_2, \dots, x_n)$ (possibly connecting it first with $(\lambda_0 + \varepsilon, x_2, \dots, x_n)$ and then sending $\varepsilon \searrow 0$). See Figure 8.2.2 for a sketch of this configuration. □

With this preliminary work, we can now give the proof of the soap bubble theorem by using the original approach of Aleksandrov [Ale62], via the moving plane method and the reflection principle.

Another Proof of Theorem 7.1.3. Suppose first that we consider case (i) of (8.2.4). Then, in the vicinity of p_0, we can write $\partial\Omega$ and $\partial\Sigma'_{\lambda_0}$ as the graphs of two functions, say u_1 and u_2, respectively, that satisfy the mean curvature equation in (2.3.28). That is, up to a rigid motion, if $H \in \mathbb{R}$ denotes the constant mean curvature of $\partial\Omega$, there exists $\rho > 0$ such that for all $x \in \mathbb{R}^{n-1}$ with $|x| < \rho$, we have that

$$\operatorname{div}\left(\frac{\nabla u_1(x)}{\sqrt{1 + |\nabla u_1(x)|^2}}\right) = \operatorname{div}\left(\frac{\nabla u_2(x)}{\sqrt{1 + |\nabla u_2(x)|^2}}\right) = H, \tag{8.2.10}$$

with

$$u_1(0) = u_2(0), \quad \nabla u_1(0) = \nabla u_2(0) = 0 \quad \text{and} \quad u_1 \geqslant u_2. \tag{8.2.11}$$

By possibly reducing ρ, we can also suppose that

$$|\nabla u_1| + |\nabla u_2| \leqslant 1. \tag{8.2.12}$$

We also remark that, for $m \in \{1, 2\}$,

$$\begin{aligned}
&\operatorname{div}\left(\frac{\nabla u_m(x)}{\sqrt{1 + |\nabla u_m(x)|^2}}\right)\\
&= \frac{\Delta u_m(x)}{\sqrt{1 + |\nabla u_m(x)|^2}} - \frac{D^2 u_m(x) \nabla u_m(x) \cdot \nabla u_m(x)}{(1 + |\nabla u_m(x)|^2)^{\frac{3}{2}}}\\
&= \sum_{i,j=1}^{n} \alpha_{ij}(\nabla u_m(x)) \partial_{ij} u_m(x),
\end{aligned}$$

where

$$\alpha_{ij}(\zeta) := \frac{\delta_{ij}}{\sqrt{1 + |\zeta|^2}} - \frac{\zeta_i \zeta_j}{(1 + |\zeta|^2)^{\frac{3}{2}}}. \tag{8.2.13}$$

Thus, if $w := u_1 - u_2$, we have $w \geqslant 0$ and $w(0) = 0$. Also,

$$\begin{aligned}
0 &= \operatorname{div}\left(\frac{\nabla u_1(x)}{\sqrt{1 + |\nabla u_1(x)|^2}}\right) - \operatorname{div}\left(\frac{\nabla u_2(x)}{\sqrt{1 + |\nabla u_2(x)|^2}}\right)\\
&= \sum_{i,j=1}^{n} \alpha_{ij}(\nabla u_1(x)) \partial_{ij} u_1(x) - \sum_{i,j=1}^{n} \alpha_{ij}(\nabla u_2(x)) \partial_{ij} u_2(x)\\
&= \sum_{i,j=1}^{n} \alpha_{ij}(\nabla u_1(x)) \partial_{ij} w(x) + \sum_{i,j=1}^{n} \Big(\alpha_{ij}(\nabla u_1(x)) - \alpha_{ij}(\nabla u_2(x))\Big) \partial_{ij} u_2(x)\\
&= \sum_{i,j=1}^{n} \alpha_{ij}(\nabla u_1(x)) \partial_{ij} w(x)\\
&\quad + \sum_{i,j=1}^{n} \partial_{ij} u_2(x) \int_0^1 \frac{d}{dt}\Big(\alpha_{ij}\big(t\nabla u_1(x) + (1-t)\nabla u_2(x)\big)\Big)\, dt\\
&= \sum_{i,j=1}^{n} \alpha_{ij}(\nabla u_1(x)) \partial_{ij} w(x)\\
&\quad + \sum_{i,j=1}^{n} \partial_{ij} u_2(x) \int_0^1 \nabla \alpha_{ij}\big(t\nabla u_1(x) + (1-t)\nabla u_2(x)\big)\, dt \cdot \big(\nabla u_1(x) - \nabla u_2(x)\big)
\end{aligned}$$

and therefore

$$\sum_{i,j=1}^{n} a_{ij}(x)\partial_{ij}w(x) + b(x)\cdot\nabla w(x) = 0, \tag{8.2.14}$$

where

$$a_{ij}(x) := \alpha_{ij}(\nabla u_1(x))$$
$$\text{and } b(x) := \sum_{i,j=1}^{n} \partial_{ij}u_2(x)\int_0^1 \nabla\alpha_{ij}\big(t\nabla u_1(x) + (1-t)\nabla u_2(x)\big)\,dt.$$

We point out that

$$\text{the coefficients } a_{ij} \text{ satisfy the ellipticity condition in (2.16.2).} \tag{8.2.15}$$

Indeed, from (8.2.13), we see that, for all $\xi \in \partial B_1$,

$$\begin{aligned}\sum_{i,j=1}^{n} a_{ij}\xi_i\xi_j &= \sum_{i,j=1}^{n} \alpha_{ij}(\nabla u_1)\xi_i\xi_j = \sum_{i,j=1}^{n}\left(\frac{\delta_{ij}}{\sqrt{1+|\nabla u_1|^2}} - \frac{\partial_i u_1\partial_j u_1}{(1+|\nabla u_1|^2)^{\frac{3}{2}}}\right)\xi_i\xi_j\\
&= \sum_{i,j=1}^{n}\left(\frac{\delta_{ij}(1+|\nabla u_1|^2) - \partial_i u_1\partial_j u_1}{(1+|\nabla u_1|^2)^{\frac{3}{2}}}\right)\xi_i\xi_j\\
&= \frac{|\xi|^2(1+|\nabla u_1|^2) - (\nabla u_1\cdot\xi)^2}{(1+|\nabla u_1|^2)^{\frac{3}{2}}}\\
&\leqslant \frac{|\xi|^2}{\sqrt{1+|\nabla u_1|^2}} \leqslant |\xi|^2 = 1,\end{aligned}$$

and similarly, using (8.2.12),

$$\begin{aligned}\sum_{i,j=1}^{n} a_{ij}\xi_i\xi_j &= \frac{|\xi|^2(1+|\nabla u_1|^2) - (\nabla u_1\cdot\xi)^2}{(1+|\nabla u_1|^2)^{\frac{3}{2}}} \geqslant \frac{|\xi|^2(1+|\nabla u_1|^2) - |\nabla u_1|^2|\xi|^2}{(1+|\nabla u_1|^2)^{\frac{3}{2}}}\\
&= \frac{|\xi|^2}{(1+|\nabla u_1|^2)^{\frac{3}{2}}} \geqslant \frac{|\xi|^2}{2^{\frac{3}{2}}} = \frac{1}{2^{\frac{3}{2}}},\end{aligned}$$

thus completing the proof of (8.2.15).

From these considerations and the maximum principle in Theorem 2.16.3, we conclude that w vanishes identically. As a result, we have that $\partial\Omega$ and $\partial\Sigma'_{\lambda_0}$ coincide in a neighborhood of p_0.

Suppose now that we consider case (ii) of (8.2.4). Then, in the vicinity of p_0, we can write $\partial\Omega$ and $\partial\Sigma'_{\lambda_0}$ as the graphs of two functions, say u_1 and u_2, respectively, which touch each other at the boundary of their domain of definition: that is, the setting in (8.2.10), (8.2.11) and (8.2.12) holds true, but the functions u_1 and u_2 are defined in $D := \{x \in \mathbb{R}^{n-1} \text{ s.t. } |x| < \rho \text{ and } x_1 \geqslant 0\}$ and the point of contact between u_1 and u_2, occurring at the origin, lies on the boundary of the domain D. We can therefore define $w := u_1 - u_2$, recall (8.2.14) and invoke Corollary 2.16.5: the latter result, since $\nabla w(0) = 0$, yields that w vanishes identically. Therefore, in this case also, we have proved that $\partial\Omega$ and $\partial\Sigma'_{\lambda_0}$ coincide in a neighborhood of p_0.

These considerations show that, in both cases (i) and (ii) of (8.2.4), the set in which $\partial\Omega$ and $\partial\Sigma'_{\lambda_0}$ coincide is open (as well as closed, by continuity, and nonempty, due to the presence of p_0). This gives that a connected component of $(\partial\Sigma'_{\lambda_0}) \cap \{x_1 < \lambda_0\}$ (namely, the one containing p_0 in its closure) coincides with a connected component of $(\partial\Omega) \cap \{x_1 < \lambda_0\}$. By symmetry and recalling Corollary 8.2.4, we have thereby identified a connected component of Ω which is invariant under the reflection through T_{λ_0}. Since Ω is assumed to be connected, we deduce that Ω is invariant under the reflection through T_{λ_0}.

We can apply this argument in every direction (not only e_1). In this way, we find that Ω is invariant under hyperplane reflections in every direction, and we can therefore invoke Lemma 8.2.1 and conclude that Ω is spherically symmetric. This says that Ω is the union of disjoint rings. But applying again Corollary 8.2.4, we deduce that Ω is convex, and hence it is necessarily a ball. □

8.3 Back to Serrin's Overdetermined Problem

We now reconsider Theorem 7.3.1 and present a different proof of it based on the moving plane method, in line with the original article [Ser71]. The strategy is to slide a hyperplane in a given direction, say the first coordinate direction from right to left, and reach a critical configuration, as stated in (8.2.3). From that, one needs to analyze cases (i) and (ii) in (8.2.4). Case (i) is technically easier to deal with, relying on the Hopf lemma (e.g. Lemma 2.16.1 would do the job). Case (ii) turns out to be more elaborate and hinges on a specific result, known in the literature as Serrin's corner lemma, given as follows:

Lemma 8.3.1. *Let Ω be a bounded, open and connected set of class C^2 with exterior normal ν. Let $D := \Omega \cap \{x_1 > 0\}$. Let $q \in (\partial\Omega) \cap \{x_1 = 0\}$, and suppose that $\nu(q) = e_n$. Let also $\varpi = (\varpi_1, \dots, \varpi_n) \in \partial B_1$, with $\varpi_1 > 0 > \varpi_n$.*

Assume that $w \in C^2(\overline{D})$ satisfies

$$\begin{cases} \Delta w \leqslant 0 & \text{in } D, \\ w \geqslant 0 & \text{in } D, \\ w(q) = 0. \end{cases}$$

Then, either $\partial_\varpi w(q) > 0$ or $\partial_{\varpi\varpi} w(q) > 0$, unless w vanishes identically in D.

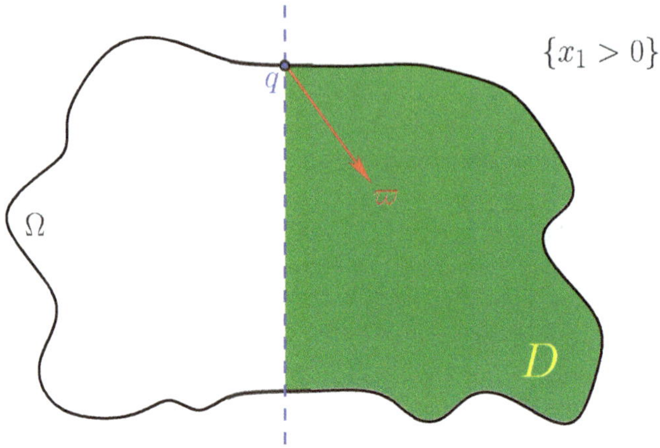

Figure 8.3.1 The geometry involved in Serrin's corner lemma.

See Figure 8.3.1 for a sketch of the geometry involved in Lemma 8.3.1. In a sense, Lemma 8.3.1 addresses one of the cases left out by the theory developed in Section 2.16. Namely, the classical setting for the Hopf lemma requires an interior ball condition (see Figure 2.16.1), which is violated in the assumptions of Lemma 8.3.1, since the domain D presents a right-angled corner at the point q. In this scenario, normal derivatives of harmonic functions reaching their extrema at q may vanish. As an example of this phenomenon, consider the case in which $n = 2$, $\Omega := B_1(-e_2)$, $D := \Omega \cap \{x_1 > 0\}$, and $w(x) := -x_1x_2$. We have that $\Delta w = 0$, $w(0) = 0$ and $w(x) \geqslant 0$ in D, but $\nabla w(0) = 0$, thus violating the normal derivative nondegeneracy of the classical Hopf lemma.

With respect to this occurrence, the power of Lemma 8.3.1 is to detect that this example does provide "the worst possible scenario," since the degeneracy of the first derivative (if it occurs) is compensated by a nondegeneracy of second derivatives: indeed, in the example above, $D^2w = \begin{pmatrix} 0 & -1 \\ -1 & 0 \end{pmatrix}$.

We stress that the vector ϖ in the statement of Lemma 8.3.1 plays the role of a direction entering D nontangentially from q: as a matter of fact, since $\nu(q) = e_n$, the condition $\varpi_n \neq 0$ is a nontangential requirement, and if Ω is parameterized in a neighborhood of q by the sublevel sets of a function Φ, say with $\Phi(q) = 0$, $\nabla\Phi(q) = e_n$ and $\Phi < 0$ in Ω in the vicinity of q, for small $t > 0$, we have that

$$\Phi(q + t\varpi) = t\nabla\Phi(q) \cdot \varpi + o(t) = t\varpi_n + o(t) < 0,$$

hence $q + t\varpi \in \Omega$ and also

$$(q + t\varpi) \cdot e_1 = 0 + t\varpi_1 > 0,$$

giving that $q + t\varpi \in D$. This shows that

$$q + t\varpi \text{ enters } D \quad \text{for small } t > 0. \tag{8.3.1}$$

In a nutshell, the proof of Lemma 8.3.1 consists in revisiting the argument used to establish the classical Hopf lemma by carefully taking into account an additional possible degeneracy of linear type. Namely, we still look at configurations such as the one described in Figure 2.16.3 (here, up to an intersection with a halfspace); however, in the present framework, we "correct" the barrier in (2.16.5) by a linear factor.

The details are as follows.

Proof of Lemma 8.3.1. We use the regularity of Ω to pick a ball contained in Ω and tangent to $\partial\Omega$ at q; that is, we take $\zeta = (\zeta_1, \dots, \zeta_n) \in \Omega$ and $r > 0$ such that $B_r(\zeta) \subseteq \Omega$ and

$$\{q\} = (\partial B_r(\zeta)) \cap (\partial\Omega). \tag{8.3.2}$$

Note that $\zeta_1 = 0$. We let $U := B_r(\zeta) \cap B_{r/2}(q) \cap \{x_1 > 0\}$, see Figure 8.3.2.

We point out that $U \subseteq D$, and we define

$$z(x) = z(x_1, \dots, x_n) := x_1 \left(e^{-\alpha|x-\zeta|^2} - e^{-\alpha r^2}\right),$$

with $\alpha > 1$ to be conveniently chosen in the following calculation. If $x \in U$, then $|x - \zeta| \geqslant |\zeta - q| - |q - x| \geqslant r - \frac{r}{2} = \frac{r}{2}$ and therefore

$$\begin{aligned} \Delta z(x) &= 2\alpha x_1 e^{-\alpha|x-\zeta|^2} \left(2\alpha|x - \zeta|^2 - (n + 2)\right) \\ &\geqslant 2\alpha x_1 e^{-\alpha|x-\zeta|^2} \left(\frac{\alpha}{2} - (n + 2)\right) \geqslant \alpha x_1 e^{-\alpha|x-\zeta|^2} \end{aligned} \tag{8.3.3}$$

so long as α is fixed sufficiently large.

Additionally, we observe that z vanishes along $\{x_1 = 0\}$ as well as along $\partial B_r(\zeta)$. We take $\varepsilon \in (0, 1)$, to be chosen appropriately small here in the following, and we define $v := w - \varepsilon z$. From the previous observations, we arrive at

$$\begin{aligned} &\text{along } \left(\Omega \cap B_{r/2}(q) \cap \{x_1 = 0\}\right) \cup \left((\partial B_r(\zeta)) \cap B_{r/2}(q)\right), \\ &\quad \text{it holds that } v \geqslant -\varepsilon z = 0. \end{aligned} \tag{8.3.4}$$

Moreover, we can suppose that w does not vanish identically; consequently, according to the strong maximum principle in Theorem 2.9.1(ii), it holds that

$$w > 0 \quad \text{in } D. \tag{8.3.5}$$

We claim that there exists $c > 0$ such that

$$w(x) \geqslant c x_1 \quad \text{for all } x \in (\partial B_{r/2}(q)) \cap B_r(\zeta) \cap \{x_1 > 0\}. \tag{8.3.6}$$

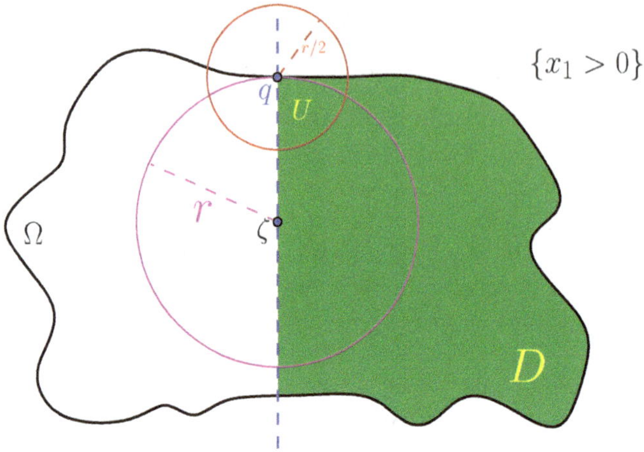

Figure 8.3.2 The geometry involved in the proof of Lemma 8.3.1.

Indeed, suppose not, then there exists a sequence of points

$$x^{(j)} = \left(x_1^{(j)}, \dots, x_n^{(j)}\right) \in (\partial B_{r/2}(q)) \cap B_r(\zeta) \cap \{x_1 > 0\}$$

such that $w(x^{(j)}) \leqslant \frac{x_1^{(j)}}{j}$.

Up to a subsequence, we can suppose that $x^{(j)}$ converges to some point $\widetilde{x} \in (\partial B_{r/2}(q)) \cap \overline{B_r(\zeta)} \cap \{x_1 \geqslant 0\}$ (see Figure 8.3.2) and $w(\widetilde{x}) \leqslant 0$. From this and (8.3.5), we find that $w(\widetilde{x}) = 0$ and $\widetilde{x} \in \{x_1 = 0\} \setminus (\partial\Omega)$.

Hence, on account of (8.3.5) and the Hopf lemma (see Lemma 2.16.1), we infer that $\frac{\partial w}{\partial x_1}(\widetilde{x}) \geqslant c_0$, for some $c_0 > 0$.

As a result, if x lies in a sufficiently small neighborhood of $\widetilde{x}$, we have that $\frac{\partial w}{\partial x_1}(x) \geqslant \frac{c_0}{2}$; consequently, if j is large enough,

$$\begin{aligned}\frac{x_1^{(j)}}{j} &\geqslant w(x^{(j)}) = w(x^{(j)}) - w(0, (x_2^{(j)}, \dots, x_n^{(j)}) \\ &= \int_0^{x_1^{(j)}} \frac{\partial w}{\partial x_1}(\tau, (x_2^{(j)}, \dots, x_n^{(j)}) \, d\tau \geqslant \frac{c_0 \, x_1^{(j)}}{2}.\end{aligned}$$

We have therefore reached a contradiction, which ends the proof of (8.3.6).

From (8.3.6), we deduce that, for all $x \in (\partial B_{r/2}(q)) \cap B_r(\zeta) \cap \{x_1 > 0\}$, we have

$$v(x) \geqslant x_1 \left(c - \varepsilon \sup_{X \in \Omega} \left(X_1 \left(e^{-\alpha|X-\zeta|^2} - e^{-\alpha r^2} \right) \right) \right) \geqslant 0$$

if ε is sufficiently small. Combining this observation and (8.3.4), we conclude that

$$v \geqslant 0 \text{ along } \partial U. \tag{8.3.7}$$

Besides, by (8.3.3), in U, we have that $\Delta v < 0$. Using this, (8.3.7) and again the strong maximum principle in Theorem 2.9.1(ii), we infer that $v > 0$ in U.

As a result, since $q + t\varpi$ lies in D and hence in U, for $t > 0$ sufficiently small (recall (8.3.1)),

$$0 < v(q + t\varpi) = v(q + t\varpi) - v(q) = t\partial_\varpi v(q) + \frac{t^2}{2}\partial_{\varpi\varpi} v(q) + o(t^2). \tag{8.3.8}$$

Now, we distinguish two cases. If $\partial_\varpi v(q) \neq 0$, we write (8.3.8) in the form

$$0 < t\partial_\varpi v(q) + o(t)$$

as $t \searrow 0$, whence we deduce that

$$0 < \partial_\varpi v(q) = \partial_\varpi w(q) - \varepsilon\partial_\varpi z(q) = \partial_\varpi w(q) - \varepsilon\varpi_1\left(e^{-\alpha|q-\zeta|^2} - e^{-\alpha r^2}\right) = \partial_\varpi w(q),$$

which gives the desired result.

If instead $\partial_\varpi v(q) = 0$, then (8.3.8) boils down to

$$0 < \partial_{\varpi\varpi} v(q) + o(1)$$

as $t \searrow 0$, and accordingly,

$$\begin{aligned} 0 \leqslant \partial_{\varpi\varpi} v(q) &= \partial_{\varpi\varpi} w(q) - \varepsilon\partial_{\varpi\varpi} z(q) \\ &= \partial_{\varpi\varpi} w(q) + 2\varepsilon\alpha\varpi_1(q - \zeta) \cdot \varpi e^{-\alpha|q-\zeta|^2} \\ &= \partial_{\varpi\varpi} w(q) + 2\varepsilon\alpha r\varpi_1\varpi_n e^{-\alpha r^2} < \partial_{\varpi\varpi} w(q), \end{aligned}$$

concluding the proof of the desired result. □

With this, we give a proof of Theorem 7.3.1 via the moving plane method, as in the original article [Ser71], by proceeding as follows.

Another Proof of Theorem 7.3.1. Recalling the notation in (8.1.1) and (8.2.3), we claim that

$$\Omega \text{ is symmetric with respect to the reflection across } T_{\lambda_0}. \tag{8.3.9}$$

Once this is proved, we are done: indeed, one would first repeat the argument in any given direction and deduce that Ω is invariant under hyperplane reflections in every direction; from this, one can exploit Lemma 8.2.1 and conclude that Ω is spherically symmetric, hence the union of disjoint rings; thus applying Corollary 8.2.4, one obtains that Ω is a ball.

Therefore, we focus on the proof of (8.3.9). To this end, we note that from (7.3.1) and the weak maximum principle in Corollary 2.9.2(i), it follows that $u \leqslant 0$ in Ω.

Now, let

$$\Sigma_{\lambda_0} \ni x \longmapsto w_{\lambda_0}(x) := u(x) - u(x_{\lambda_0}).$$

We note that $\Delta w_{\lambda_0} = 1 - 1 = 0$ in Σ_{λ_0}. Furthermore, if $x \in \Omega \cap T_{\lambda_0}$, it follows that $x_{\lambda_0} = x$ and thus $w_{\lambda_0}(x) = 0$. If instead $x \in (\partial\Omega) \cap \{x_1 > \lambda_0\}$, then $w_{\lambda_0}(x) = -u(x_{\lambda_0}) \geqslant 0$. These considerations and the weak maximum principle in Corollary 2.9.2(ii) yield that $w_{\lambda_0} \geqslant 0$ in Σ_{λ_0}.

This, in combination with the strong maximum principle in Theorem 2.9.1(ii), gives that

$$\text{either } w_{\lambda_0} > 0 \text{ in } \Sigma_{\lambda_0}, \text{ or } w_{\lambda_0} \text{ vanishes identically} \quad \text{in } \Sigma_{\lambda_0}. \tag{8.3.10}$$

The case in which w_{λ_0} vanishes identically in Σ_{λ_0} is easier to deal with; hence, we briefly discuss this situation now. If w_{λ_0} vanishes identically in Σ_{λ_0}, we have that $u(x) = u(x_{\lambda_0})$ for all $x \in \Sigma_{\lambda_0}$. In this circumstance, the claim in (8.3.9) must hold true; otherwise, there would exist a point $\overline{x} \in \partial\Sigma_{\lambda_0} \cap \{x_1 > \lambda_0\}$ with $\overline{x}_{\lambda_0} \in \Omega$, whence $0 = u(\overline{x}) = u(\overline{x}_{\lambda_0})$. That is, the subharmonic function u would have an interior maximum at $\overline{x}_{\lambda_0}$, and thus it must be constant, thanks to the strong maximum principle in Theorem 2.9.1(i). But this would produce a contradiction with the equation in (7.3.1).

From these observations and (8.3.10), it follows that we can focus on the case in which

$$w_{\lambda_0} > 0 \quad \text{in } \Sigma_{\lambda_0}. \tag{8.3.11}$$

Actually, we show that this case cannot occur at all, reaching a contradiction in what follows. To understand this situation, we distinguish between cases (i) and (ii) in (8.2.4). In all cases, we note that both p_0 and its reflection across T_{λ_0} belong to $\partial\Omega$ and therefore

$$w_{\lambda_0}(p_0) = 0 - 0 = 0. \tag{8.3.12}$$

Now, we show that, in the setting of (8.3.11), case (i) cannot occur. To check this, we argue for a contradiction. If case (i) in (8.2.4) holds true, we let p_0 be the internal point of tangency given in (8.2.4)(i). We point out that, in light of (8.3.11), we have that $w_{\lambda_0} < 0$ in Σ'_{λ_0}. This, equation (8.3.12) and the version of the Hopf lemma given in Lemma 2.16.1 (applied here to a ball $B \subset \Sigma'_{\lambda_0}$ that is tangent to $\partial\Sigma'_{\lambda_0}$ at the point p_0) entail that either $\partial_\nu w_{\lambda_0}(p_0) \neq 0$ or w_{λ_0} is constantly zero in Σ_{λ_0}. But the latter situation cannot occur, thanks to (8.3.11); therefore, necessarily,

$$0 \neq \partial_\nu w_{\lambda_0}(p_0) = c - c = 0,$$

which is a contradiction (note that we have used here the normal prescription along $\partial\Omega$ given in (7.3.1)).

This says that case (i) of (8.2.4), in the setting of (8.3.11), does not occur. We show now that case (ii) of (8.2.4), in the setting of (8.3.11), also does not occur, which will complete this proof of Theorem 7.3.1.

To this end, suppose that case (ii) in (8.2.4) holds true. In this scenario, our goal is to show that w_{λ_0} has a second-order zero at p_0 (this allows us to employ Serrin's corner lemma in Lemma 8.3.1). Namely, we claim that

$$D^k w_{\lambda_0}(p_0) = 0 \quad \text{for all } k \in \{0, 1, 2\}. \tag{8.3.13}$$

As a matter of fact, the case $k = 0$ is given in (8.3.12); hence, we can focus here on the first and second derivatives. To this end, we observe that, for all i, $j \in \{1, \dots, n\}$,

$$\partial_i w_{\lambda_0}(x) = \partial_i u(x) - (-1)^{\delta_{i1}} \partial_i u(x_{\lambda_0})$$

and

$$\partial_{ij} w_{\lambda_0}(x) = \partial_{ij} u(x) - (-1)^{\delta_{i1}+\delta_{j1}} \partial_{ij} u(x_{\lambda_0}).$$

Thus, since p_0 lies on T_{λ_0} (recall that we are dealing here with case (ii) in (8.2.4)) and therefore the reflection of p_0 coincides with p_0 itself,

$$\partial_i w_{\lambda_0}(p_0) = \left(1 - (-1)^{\delta_{i1}}\right)\partial_i u(p_0) = \begin{cases} 0 & \text{if } i \in \{2, \dots, n\}, \\ 2\partial_1 u(p_0) & \text{if } i = 1, \end{cases}$$

and

$$\begin{aligned} \partial_{ij} w_{\lambda_0}(p_0) &= \left(1 - (-1)^{\delta_{i1}+\delta_{j1}}\right)\partial_{ij} u(p_0) \\ &= \begin{cases} 0 & \text{if } i, j \in \{2, \dots, n\}, \\ 0 & \text{if } i = j = 1, \\ 2\partial_{1i} u(p_0) & \text{if } i \in \{2, \dots, n\} \text{ and } j = 1, \\ 2\partial_{1j} u(p_0) & \text{if } j \in \{2, \dots, n\} \text{ and } i = 1. \end{cases} \end{aligned}$$

Therefore, to prove (8.3.13), it remains to show that

$$\partial_1 u(p_0) = 0 \quad \text{and} \quad \partial_{1i} u(p_0) = 0 \quad \text{for all } i \in \{2, \dots, n\}. \tag{8.3.14}$$

To check these items of information, up to a rigid motion, we parameterize $\partial\Omega$ in the vicinity of p_0 by the graph of a function $\phi : \{x' \in \mathbb{R}^{n-1} \text{ s.t. } |x'| < \rho\} \to \mathbb{R}$, for some $\rho > 0$. Given the assumption on Ω in Theorem 7.3.1, we know that ϕ is of class C^2. Since the normal derivative of Ω at p_0 is vertical, we can suppose that the point $(0, \phi(0)) \in \mathbb{R}^{n-1} \times \mathbb{R}$ is p_0 and $\nabla\phi(0) = 0$.

Since $\Sigma'_{\lambda_0} \subseteq \Omega$ (up to reverting the vertical direction), we have that

$$\phi(x_1, x_2, \dots, x_{n-1}) \leqslant \phi(-x_1, x_2, \dots, x_{n-1})$$

provided that $x_1 > 0$ and $|(x_1, x_2, \dots, x_{n-1})| < \rho$, and accordingly

$$\begin{aligned}
0 &\geqslant \phi(x_1, x_2, \dots, x_{n-1}) - \phi(-x_1, x_2, \dots, x_{n-1}) \\
&= \big(\phi(x_1, x_2, \dots, x_{n-1}) - \phi(0)\big) - \big(\phi(-x_1, x_2, \dots, x_{n-1}) - \phi(0)\big) \\
&= \frac{1}{2} D^2\phi(0)(x_1, x_2, \dots, x_{n-1}) \cdot (x_1, x_2, \dots, x_{n-1}) \\
&\quad - \frac{1}{2} D^2\phi(0)(-x_1, x_2, \dots, x_{n-1}) \cdot (-x_1, x_2, \dots, x_{n-1}) + o(|x'|^2).
\end{aligned}$$

In particular, given $\ell \in \{2, \dots, n-1\}$, picking $x_j := 0$ when $j \notin \{1, \ell\}$, we find that

$$\begin{aligned}
0 &\geqslant \frac{1}{2} \left[\partial_{11}\phi(0)x_1^2 + \partial_{\ell\ell}\phi(0)x_\ell^2 + 2\partial_{1\ell}\phi(0)x_1x_\ell\right] \\
&\quad - \frac{1}{2} \left[\partial_{11}\phi(0)x_1^2 + \partial_{\ell\ell}\phi(0)x_\ell^2 - 2\partial_{1\ell}\phi(0)x_1x_\ell\right] \\
&= 2\partial_{1\ell}\phi(0)x_1x_\ell.
\end{aligned}$$

Thus, choosing $x_\ell := \pm x_1$ and dividing by $2x_1^2$,

$$0 \geqslant \pm\partial_{1\ell}\phi(0).$$

Using the arbitrariness of choice in the above sign, we thus obtain that

$$\partial_{1\ell}\phi(0) = 0 \quad \text{for every } \ell \in \{2, \dots, n-1\}. \tag{8.3.15}$$

In this configuration, we again make use of the boundary prescriptions along $\partial\Omega$ given in (7.3.1), thus writing that

$$0 = u\big(x', \phi(x')\big) \quad \text{and} \quad c = \partial_\nu u\big(x', \phi(x')\big) = \nabla u\big(x', \phi(x')\big) \cdot \frac{\big(\nabla\phi(x'), -1\big)}{\sqrt{1 + |\nabla\phi(x')|^2}}.$$

Differentiating these expressions and then setting $x' := 0$, we thus find that $\partial_i u(p_0) = 0$ for all $i \in \{1, \dots, n-1\}$ (which, as an aside, establishes the first claim in (8.3.14) and also gives that $-\partial_n u(p_0) = \partial_\nu u(p_0) = c$) and that

$$\begin{aligned}
0 &= \frac{\partial}{\partial x_i} \left(\nabla u\big(x', \phi(x')\big) \cdot \frac{\big(\nabla\phi(x'), -1\big)}{\sqrt{1 + |\nabla\phi(x')|^2}} \right)\Bigg|_{x'=0} \\
&= -\frac{\partial}{\partial x_i} \Big(\nabla u\big(x', \phi(x')\big) \Big)\Big|_{x'=0} \cdot e_n + \nabla u(p_0) \cdot \frac{\partial}{\partial x_i} \left(\frac{\big(\nabla\phi(x'), -1\big)}{\sqrt{1 + |\nabla\phi(x')|^2}} \right)\Bigg|_{x'=0} \\
&= -\frac{\partial}{\partial x_i} \Big(\partial_n u\big(x', \phi(x')\big) \Big)\Big|_{x'=0} - \partial_n u(p_0) \frac{\partial}{\partial x_i} \left(\frac{1}{\sqrt{1 + |\nabla\phi(x')|^2}} \right)\Bigg|_{x'=0} \\
&= -\partial_{in} u(p_0).
\end{aligned}$$

In particular, we have that $\partial_{1n}u(p_0) = 0$. For this reason, to complete the proof of (8.3.14) and thus of (8.3.13), it only remains to show that

$$\partial_{1i}u(p_0) = 0 \quad \text{for all } i \in \{2, \dots, n-1\}. \tag{8.3.16}$$

To this end, we observe that, for all $i, j \in \{1, \dots, n-1\}$,

$$\begin{aligned} 0 &= \partial_{ij}\Big(u\big(x', \phi(x')\big)\Big)\Big|_{x'=0} \\ &= \partial_i\Big(\partial_j u\big(x', \phi(x')\big) + \partial_n u\big(x', \phi(x')\big)\partial_i \phi(x')\Big)\Big|_{x'=0} \\ &= \partial_{ij}u(p_0) + \partial_n u(p_0)\partial_{ij}\phi(0) \\ &= \partial_{ij}u(p_0) - c\partial_{ij}\phi(0). \end{aligned}$$

In particular, recalling (8.3.15), we have that $\partial_{1i}u(p_0) = c\partial_{1i}\phi(0) = 0$ for all $i \in \{2, \dots, n-1\}$ and the proof of (8.3.16) and hence of (8.3.13) is thus complete.

For our purposes, the result in (8.3.13) is instrumental in allowing us to use Serrin's corner lemma (in the version given here in Lemma 8.3.1). With this, we obtain that w_{λ_0} must vanish identically in Σ_{λ_0}, which provides a contradiction with (8.3.11). □

8.4 The Problem of Gidas, Ni and Nirenberg

The following beautiful result was discovered (or[4] invented?) in [GNN79] (see also [Bre99], [PS07b, Section 8.2], [HL11, Theorem 2.34], [Ni11, Section 3.3] and [Hu11, Sections 6.2 and 6.3] for additional information, as well as [CCPP23] for a recent quantitative approach).

Theorem 8.4.1. *Let $f \in C^1(\mathbb{R})$ and $u \in C^2(\overline{B_1})$ be a solution of*

$$\begin{cases} \Delta u = f(u) & \text{in } B_1, \\ u > 0 & \text{in } B_1, \\ u = 0 & \text{on } \partial B_1. \end{cases} \tag{8.4.1}$$

Then, u is radially symmetric and radially strictly decreasing.

[4]Inventions and discoveries have much in common, but some philosophers make a point in distinguishing between a mathematical realism, in which mathematical entities are understood as existing independently of the human mind (in which case we do not invent mathematics but rather discover it), and an opposed tendency of mathematical anti-realism, holding that mathematical statements do not correspond to a special entity (in which case we really invent mathematics, literally out of the blue). Personally, we do not find such a distinction amazingly thrilling, but if we were forced to take part in this ontological debate, we would opt for "discovered" for the result and "invented" for the proof. Because, as Ennio De Giorgi wisely pointed out [Emm97], "a theorem is something discovered; its proof something invented."

This is a very elegant statement, especially comparing the very mild hypotheses assumed with the very strong result obtained. The result is also of concrete interest, since solutions of the equation under consideration appear in physics (see e.g. [CF77, equation (12)] in the context of the Yang–Mills field theory) and geometry (see e.g. [LN74], in the search for Riemannian metrics that are covariantly connected under the group of Möbius transformations).

The proof of Theorem 8.4.1 relies on the reflection technique, but several specific aspects of the application of the moving plane method arise. For example, while the core of the idea is still to continuously slide a hyperplane to detect symmetry, differently from the sketch in Figure 8.1.1, here the critical situation does not occur when the reflected domain touches the original one but rather when "the reflected function touches the original one" (in a sense that will be made precise in the course of the proof). Moreover, differently from the situation pointed out in Figure 8.1.2, in the course of the forthcoming proof, the reflection of the domain never becomes tangential to the original one, since the domain under consideration is a ball, and therefore the sliding hyperplanes do not cut the domain perpendicularly before reaching the center of the ball.

In rough terms, the use of the moving plane method in the proof of Theorem 8.4.1 consists of the following conceptual steps. First, one considers the difference between the solution u and its reflection: since u vanishes along the boundary and it is positive elsewhere, when the hyperplane is very close to the initial position, one is able to detect a sign for such a difference. Then, one extends the validity of this sign up to a critical position. It thus remains to analyze this critical position in order to detect the symmetry of the solution. Several forms of the maximum principle will come in handy in the above procedure, since the difference between the original solution and its reflection ends up satisfying a nice partial differential equation.

Some further remarks about Theorem 8.4.1: first of all, when f is only Hölder continuous, the result in Theorem 8.4.1 does not hold anymore, since in this case solutions are not necessarily radial, see e.g. [Fra00, Exercise 3.18]. Also, when the domain is an annulus (instead of a ball), Theorem 8.4.1 does not hold, see e.g. [BN83, Dan92a] for the construction of nonradial positive solutions (but see also [PS07b, Theorem 8.2.4] for a radial symmetry result in the annulus under suitable structural assumptions on the nonlinearity f). These observations showcase the rather subtle nature of the radial symmetry results and the several patterns which can lead to symmetry breaking after apparently small modifications of the problem under consideration.

To implement the proof of Theorem 8.4.1, we begin by presenting some auxiliary results.

Lemma 8.4.2. *Let Ω be a bounded open set with boundary of class C^2, with exterior normal ν.*

Let $x_0 \in \partial\Omega$, and suppose that

$$\nu(x_0) \cdot e_1 > 0. \tag{8.4.2}$$

For all $r > 0$, let $\Omega_r := \Omega \cap B_r(x_0)$.

Let $\varepsilon > 0$, $f \in C^1(\mathbb{R})$ and $u \in C^2(\overline{\Omega_\varepsilon})$ be such that $\Delta u = f(u)$ in Ω_ε, $u > 0$ in Ω_ε and $u = 0$ on $(\partial\Omega) \cap B_\varepsilon(x_0)$.

Then, there exists $\delta \in (0, \varepsilon)$ such that $\partial_1 u(x) < 0$ for all $x \in \Omega_\delta$.

Proof. Let $S_r := (\partial\Omega) \cap B_r(x_0)$. In light of (8.4.2), we can find $\varepsilon' \in (0, \varepsilon)$ small enough that $\nu(x) \cdot e_1 > 0$ for all $x \in S_{\varepsilon'}$. As a consequence (see Figure 8.4.1), for all $x \in S_{\varepsilon'}$, there exists $h_x > 0$ such that, for all $h \in (0, h_x)$, we have that $x - he_1 \in \Omega_\varepsilon$ and accordingly,

$$\partial_1 u(x) = \lim_{h \searrow 0} \frac{u(x) - u(x - he_1)}{h} = \lim_{h \searrow 0} \frac{0 - u(x - he_1)}{h} \leqslant 0. \tag{8.4.3}$$

Now, to prove the desired result, we argue by contradiction and suppose that for every $j \in \mathbb{N}$, there exists a point $x^{(j)} \in \Omega_{1/j}$ with

$$\partial_1 u(x^{(j)}) \geqslant 0. \tag{8.4.4}$$

Note that

$$\lim_{j \to +\infty} \text{dist}(x^{(j)}, \partial\Omega) = \text{dist}(x_0, \partial\Omega) = 0;$$

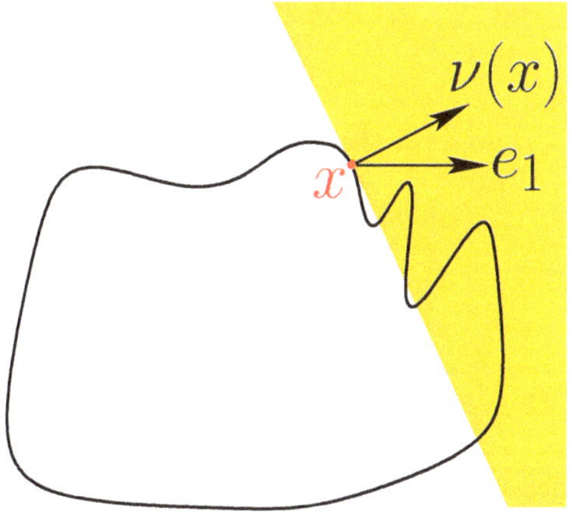

Figure 8.4.1 The geometry involved in the proof of Lemma 8.4.2.

therefore, we can find $t^{(j)} > 0$ such that $x^{(j)} + t^{(j)}e_1 \in \partial\Omega$, with $t^{(j)} \to 0$ as $j \to +\infty$. In particular,

$$\lim_{j\to+\infty} \left|x^{(j)} + t^{(j)}e_1 - x_0\right| = 0,$$

and accordingly, when j is sufficiently large, we have that $x^{(j)} + t^{(j)}e_1 \in S_{\varepsilon'}$.

From this and (8.4.3), we arrive at

$$\partial_1 u\big(x^{(j)} + t^{(j)}e_1\big) \leqslant 0.$$

Comparing with (8.4.4), we thereby find $\tau^{(j)} \in \big[0, t^{(j)}\big]$ such that

$$\partial_1 u\big(x^{(j)} + \tau^{(j)}e_1\big) = 0.$$

Taking the limit as $j \to +\infty$, we conclude that

$$\partial_1 u(x_0) = 0. \tag{8.4.5}$$

Moreover,

$$\begin{aligned} \partial_{11} u(x_0) &= \lim_{j\to+\infty} \frac{\partial_1 u\big(x^{(j)} + \tau^{(j)}e_1\big) - \partial_1 u(x_0)}{x^{(j)} + \tau^{(j)}e_1 - x_0} \\ &= \lim_{j\to+\infty} \frac{0-0}{x^{(j)} + \tau^{(j)}e_1 - x_0} = 0. \end{aligned} \tag{8.4.6}$$

Now, let $\{\zeta_1, \dots, \zeta_{n-1}\}$ be an orthonormal basis for the tangent hyperplane to $\partial\Omega$ at x_0. Since $u = 0$ on $(\partial\Omega) \cap B_\varepsilon(x_0)$, we have that

$$\nabla u(x_0) \cdot \zeta_i = 0 \quad \text{for every } i \in \{1, \dots, n-1\}. \tag{8.4.7}$$

We also note that

$$e_1 \text{ is linearly independent from } \zeta_1, \dots, \zeta_{n-1} \tag{8.4.8}$$

because if, for some $a_1, \dots, a_n \in \mathbb{R}$,

$$a_1\zeta_1 + \dots + a_{n-1}\zeta_{n-1} + a_n e_1 = 0, \tag{8.4.9}$$

then

$$0 = \nu(x_0) \cdot (a_1\zeta_1 + \dots + a_{n-1}\zeta_{n-1} + a_n e_1) = a_n \nu(x_0) \cdot e_1,$$

whence one would deduce, owing to (8.4.2), that $a_n = 0$. Plugging this information back into (8.4.9), we obtain that $a_1\zeta_1 + \dots + a_{n-1}\zeta_{n-1} = 0$ and hence, by the independence of the basis, necessarily, $a_1 = \dots = a_{n-1} = 0$. This proves (8.4.8).

Hence, by (8.4.5), (8.4.7) and (8.4.8), we conclude that

$$\nabla u(x_0) = 0. \tag{8.4.10}$$

Now, for all tangent vectors ζ, we let $\gamma_\zeta \in C^2((-1,1), \partial\Omega \cap B_\varepsilon(x_0))$ with $\gamma_\zeta(0) = x_0$ and $\dot\gamma_\zeta(0) = \zeta$. We have that $u(\gamma_\zeta(t)) = 0$; therefore, by taking derivatives,

$$0 = \sum_{k=1}^{n} \partial_k u(\gamma_\zeta(t))\, \dot\gamma_\zeta(t) \cdot e_k$$

and

$$0 = \sum_{j,k=1}^{n} \partial_{jk} u(\gamma_\zeta(t))\, (\dot\gamma_\zeta(t) \cdot e_j)\, (\dot\gamma_\zeta(t) \cdot e_k) + \sum_{k=1}^{n} \partial_k u(\gamma_\zeta(t))\, \ddot\gamma_\zeta(t) \cdot e_k.$$

Combining this information with (8.4.10), we infer that

$$0 = \sum_{j,k=1}^{n} \partial_{jk} u(x_0)\, (\zeta \cdot e_j)(\zeta \cdot e_k) = \sum_{j,k=1}^{n} \partial_{jk} u(x_0)\, (\zeta \cdot e_j)(\zeta \cdot e_k) = D^2 u(x_0)\zeta \cdot \zeta.$$

This is valid for all tangent vectors ζ. In particular, applying this with $\zeta := \zeta_i - \zeta_j$, $\zeta := \zeta_i$ and $\zeta := \zeta_j$,

$$\begin{aligned} 0 &= D^2 u(x_0)(\zeta_i - \zeta_j) \cdot (\zeta_i - \zeta_j) \\ &= D^2 u(x_0)\zeta_i \cdot \zeta_i + D^2 u(x_0)\zeta_j \cdot \zeta_j - 2D^2 u(x_0)\zeta_i \cdot \zeta_j \\ &= -2D^2 u(x_0)\zeta_i \cdot \zeta_j. \end{aligned} \tag{8.4.11}$$

Hence, by the rotational invariance of the Laplacian (recall Corollary 1.1.4) and (8.4.11),

$$f(0) = f(u(x_0)) = \Delta u(x_0) = \sum_{i=1}^{n-1} D^2 u(x_0)\zeta_i \cdot \zeta_i + \partial_{\nu\nu} u(x_0) = \partial_{\nu\nu} u(x_0). \tag{8.4.12}$$

Now, we show that

$$f(0) = 0. \tag{8.4.13}$$

To this end, we let d be the signed distance function to $\partial\Omega$, with the convention that $d > 0$ in Ω and $d < 0$ in $\mathbb{R}^n \setminus \Omega$ (see the discussion on p. 36 about the signed distance function and, in particular, equation (2.3.2)). We let $t > 0$ to be taken suitably small and $s \in [-t, t]$. Then, for small t,

$$\begin{aligned} d\big(x_0 - t\nu(x_0) + s\zeta_i\big) &= d\big(x_0 - t\nu(x_0)\big) + s\nabla d\big(x_0 - t\nu(x_0)\big) \cdot \zeta_i + O(s^2) \\ &= t - s\nu(x_0) \cdot \zeta_i + O(s^2) = t + O(s^2) = t + O(t^2) > 0. \end{aligned}$$

That is, we have that $x_0 - t\nu(x_0) + s\zeta_i \in \Omega$, and accordingly, by (8.4.10),

$$\begin{aligned}0 &< u\big(x_0 - t\nu(x_0) + s\zeta_i\big)\\ &= u(x_0) + \nabla u(x_0) \cdot \big(- t\nu(x_0) + s\zeta_i\big)\\ &\quad + \frac{1}{2} D^2 u(x_0)\big(- t\nu(x_0) + s\zeta_i\big) \cdot \big(- t\nu(x_0) + s\zeta_i\big) + o(t^2)\\ &= 0 + 0 + \frac{t^2}{2}\partial_{\nu\nu}u(x_0) + \frac{s^2}{2}\partial_{\zeta_i\zeta_i}u(x_0) - stD^2u(x_0)\zeta_i \cdot \nu(x_0) + o(t^2).\end{aligned}$$

Hence, in view of (8.4.12),

$$0 = \frac{t^2 f(0)}{2} - stD^2u(x_0)\zeta_i \cdot \nu(x_0) + o(t^2);$$

therefore, choosing $s := \pm t$ and dividing by t^2,

$$\frac{f(0)}{2} \mp D^2u(x_0)\zeta_i \cdot \nu(x_0) = o(1).$$

By taking the limit as $t \searrow 0$, we conclude that

$$\frac{f(0)}{2} = \pm D^2u(x_0)\zeta_i \cdot \nu(x_0).$$

The possibility of choosing the sign on the right-hand side of the above gives (8.4.13), as desired.

From (8.4.13), letting

$$c(x) := \int_0^{u(x)} f'(t)\, dt,$$

we find that, in Ω_ε,

$$\Delta u(x) = f(u(x)) = f(u(x)) - f(0) = c(x)u(x).$$

We can therefore invoke the version of the Hopf lemma given in Corollary 2.16.4 (applied here to $v := -u$) and deduce that $\partial_\nu u(x_0) \neq 0$. But this is in contradiction with (8.4.10), and the desired result is thus established. □

Lemma 8.4.3. *Let $f \in C^1(\mathbb{R})$ and $u \in C^2(\overline{B_1})$ be a solution of*

$$\begin{cases} \Delta u = f(u) & \text{in } B_1,\\ u > 0 & \text{in } B_1,\\ u = 0 & \text{on } \partial B_1.\end{cases}$$

Then,

$$u(-x_1, x_2, \dots, x_n) = u(x_1, x_2, \dots, x_n) \quad \text{for all } x = (x_1, x_2, \dots, x_n) \in B_1. \tag{8.4.14}$$

Moreover,

$$\partial_1 u(x) < 0 \quad \textit{for all } x = (x_1, x_2, \ldots, x_n) \in B_1 \text{ with } x_1 > 0. \tag{8.4.15}$$

Proof. The moving plane notation in (8.1.1) is adopted here with $\Omega := B_1$. For all $\lambda \in (0, 1)$, we define

$$\Sigma_\lambda \ni x \longmapsto w_\lambda(x) := u(x) - u(x_\lambda). \tag{8.4.16}$$

We note that if $x \in T_\lambda$, then $x_\lambda = x$ and consequently $w_\lambda(x) = 0$. If instead $x \in (\partial\Sigma_\lambda) \cap \{x_1 > \lambda\} \subseteq \partial B_1$, we have that $w_\lambda(x) = 0 - u(x_\lambda) < 0$. From these observations, we arrive at

$$w_\lambda \leqslant 0 \text{ on } \Sigma_\lambda, \text{ but } w_\lambda \text{ does not vanish identically.} \tag{8.4.17}$$

Moreover, we have that $\partial_1 w_\lambda(x) = \partial_1 u(x) + \partial_1 u(2\lambda - x_1, x_2, \ldots, x_n)$, and thus

$$\partial_1 w_\lambda = 2\partial_1 u \quad \text{on } B_1 \cap T_\lambda. \tag{8.4.18}$$

We now let

$$c_\lambda(x) := \int_0^1 f'\big(tu(x) + (1-t)u_\lambda(x)\big)\, dt,$$

and we point out that

$$\|c_\lambda\|_{L^\infty(B_1)} \leqslant \|f'\|_{L^\infty([0, 3\sup_{B_1} u])}. \tag{8.4.19}$$

Also, for all $x \in \Sigma_\lambda$,

$$\Delta w_\lambda(x) = f(u(x)) - f(u(x_\lambda)) = c_\lambda(x)\big(u(x) - u(x_\lambda)\big) = c_\lambda(x)\, w_\lambda(x). \tag{8.4.20}$$

We claim that,

$$\text{for all } \lambda \in (0, 1) \text{ and all } x \in \Sigma_\lambda, \text{ we have that } w_\lambda(x) < 0. \tag{8.4.21}$$

To this end, we observe that when $\varepsilon_0 \in (0, 1)$ is conveniently small and $\lambda \in (1 - \varepsilon_0, 1)$, we can utilize the maximum principle for narrow domains given in Theorem 2.16.6 (or, alternatively, the maximum principle for small volume domains given in Theorem 2.16.7) in combination with (8.4.17) and (8.4.20): this yields $w_\lambda < 0$ in Σ_λ for all $\lambda \in (1 - \varepsilon_0, 1)$.

We can therefore consider[5]

$$\lambda_0 := \inf \Big\{\lambda \in (0, 1) \text{ s.t. } w_\lambda < 0 \text{ in } \Sigma_\lambda\Big\}, \tag{8.4.22}$$

and we know that $\lambda_0 \leqslant 1 - \varepsilon_0$.

---

[5]In a sense, the definition in (8.4.22) is an analytic counterpart of (8.2.3), in which the critical situation instead arose from a geometric situation.

We observe that (8.4.21) is proved once we show that $\lambda_0 = 0$. Hence, we argue by contradiction, and we suppose that $\lambda_0 > 0$. We recall that, by construction,

$$\text{for all } \lambda \in [\lambda_0, 1), \, w_\lambda \leqslant 0 \quad \text{in } \Sigma_\lambda \quad \text{and} \quad w_\lambda = 0 \quad \text{on } T_\lambda. \tag{8.4.23}$$

Moreover,

$$\text{for all } \lambda \in [\lambda_0, 1), \, w_\lambda < 0 \quad \text{in } (\partial B_1) \cap \{x_1 > \lambda\}, \tag{8.4.24}$$

since if $p \in (\partial B_1) \cap \{x_1 > \lambda\}$, then $p_\lambda \in B_1$ (since $\lambda \geqslant \lambda_0 > 0$) and accordingly $w_\lambda(p) = -u(p_\lambda) < 0$, thus proving (8.4.24).

Hence, by (8.4.20), (8.4.23), (8.4.24) and the version of the Hopf lemma given in Corollary 2.16.4, for all $\lambda \in [\lambda_0, 1)$, either

$$\partial_1 w_\lambda < 0 \quad \text{along } B_1 \cap T_\lambda \tag{8.4.25}$$

or w_λ vanishes identically. But the latter possibility cannot hold owing to (8.4.24). These considerations yield that, necessarily, (8.4.25) holds true for all $\lambda \in [\lambda_0, 1)$.

In the same way, using the maximum principle in Theorem 2.16.2, we find that

$$w_{\lambda_0} < 0 \quad \text{in } \Sigma_{\lambda_0}. \tag{8.4.26}$$

From (8.4.18) and (8.4.25), it follows that

$$\text{for all } \lambda \in [\lambda_0, 1), \, \partial_1 u < 0 \quad \text{along } B_1 \cap T_\lambda. \tag{8.4.27}$$

Now, we show that

$$\partial_1 u < 0 \quad \text{along } \overline{B_1} \cap T_{\lambda_0}. \tag{8.4.28}$$

Indeed, by (8.4.27) (applied with $\lambda := \lambda_0$), to check (8.4.28), we only need to take $x_0 \in (\partial B_1) \cap T_{\lambda_0}$ and prove that $\partial_1 u(x_0) < 0$. But this is a consequence of Lemma 8.4.2 (which can be used here since $\nu(x_0) \cdot e_1 = x_0 \cdot e_1 = \lambda_0 > 0$); therefore, (8.4.28) holds true.

Moreover, it follows from the definition of λ_0 in (8.4.22) that there exist a sequence of parameters $\lambda^{(j)} < \lambda_0$ such that $\lambda^{(j)} \nearrow \lambda_0$ as $j \to +\infty$ and a sequence of points $x^{(j)} \in \Sigma_{\lambda^{(j)}}$ such that $w_{\lambda^{(j)}}(x^{(j)}) \geqslant 0$. Up to a subsequence, we suppose that $x^{(j)}$ converges to some point $x^\star$ as $j \to +\infty$, and necessarily, given the convergence of $\lambda^{(j)}$, we have that $x^\star \in \overline{\Sigma_{\lambda_0}}$.

Using the notation $x^{(j)} = \left(x_1^{(j)}, y^{(j)}\right)$ with $y^{(j)} := \left(x_2^{(j)}, \dots, x_n^{(j)}\right)$, we find that

$$\begin{aligned} 0 \leqslant \lim_{j \to +\infty} w_{\lambda^{(j)}}(x^{(j)}) &= \lim_{j \to +\infty} u\left(x_1^{(j)}, y^{(j)}\right) - u\left(2\lambda^{(j)} - x_1^{(j)}, y^{(j)}\right) \\ &= u(x^\star) - u(2\lambda_0 - x_1^\star, x_2^\star, \dots, x_n^\star) = w_{\lambda_0}(x^\star). \end{aligned}$$

Since the opposite inequality holds in $\overline{\Sigma_{\lambda_0}}$, thanks to the definition of λ_0 in (8.4.22), we have that

$$w_{\lambda_0}(x^\star) = 0. \tag{8.4.29}$$

By comparing (8.4.24) and (8.4.29), we deduce that

$$x^\star \notin \partial B_1. \tag{8.4.30}$$

As a consequence of (8.4.30), we have that

$$x^\star \in \left(B_1 \cap T_{\lambda_0}\right) \cup \Sigma_{\lambda_0}. \tag{8.4.31}$$

But $x^\star \notin \Sigma_{\lambda_0}$ due to (8.4.26) and (8.4.29). Therefore, we deduce from (8.4.31) that

$$x^\star \in B_1 \cap T_{\lambda_0}. \tag{8.4.32}$$

We also remark that

$$\begin{aligned} u\left(x_1^{(j)}, y^{(j)}\right) &= u(x^{(j)}) = w_{\lambda^{(j)}}(x^{(j)}) + u(x^{(j)}_{\lambda^{(j)}}) \geqslant 0 + u(x^{(j)}_{\lambda^{(j)}}) \\ &= u\left(2\lambda^{(j)} - x_1^{(j)}, y^{(j)}\right). \end{aligned}$$

Therefore, according to the calculus mean value theorem, there must exist $t^{(j)} \in \left[2\lambda^{(j)} - x_1^{(j)}, x_1^{(j)}\right]$ such that

$$\partial_1 u\left(t^{(j)}, y^{(j)}\right) \geqslant 0. \tag{8.4.33}$$

We stress that, in light of (8.4.32),

$$\lim_{j \to +\infty} x_1^{(j)} = \lambda_0, \quad \text{hence} \lim_{j \to +\infty} t^{(j)} = \lambda_0.$$

We can therefore take the limit as $j \to +\infty$ in (8.4.33) and conclude that $\partial_1 u(x_\star) \geqslant 0$. This and (8.4.32) produce a contradiction with (8.4.28).

Thanks to this contradiction, we have proved that

$$\lambda_0 = 0 \tag{8.4.34}$$

and therefore $w_\lambda \leqslant 0$ in Σ_λ for all $\lambda \in [0, 1)$. Using this with $\lambda := 0$, we obtain that, for all $x \in B_1 \cap \{x_1 > 0\}$,

$$0 \geqslant w_0(x) = u(x_1, x_2, \dots, x_n) - u(-x_1, x_2, \dots, x_n). \tag{8.4.35}$$

On the other hand, setting $v(x_1, x_2, \dots, x_n) := u(-x_1, x_2, \dots, x_n)$, we see that, for all $x \in B_1$,

$$\Delta v(x) = \Delta u(-x_1, x_2, \dots, x_n) = f(u(-x_1, x_2, \dots, x_n)) = f(v(x))$$

and also $v > 0$ in B_1 and $v = 0$ on ∂B_1. Accordingly, from (8.4.35) applied to v, we deduce that, for all $x \in B_1 \cap \{x_1 > 0\}$,

$$0 \geqslant v(x_1, x_2, \dots, x_n) - v(-x_1, x_2, \dots, x_n) = u(-x_1, x_2, \dots, x_n) - u(x_1, x_2, \dots, x_n).$$

Putting together this and (8.4.35), we finally obtain (8.4.14).

Additionally, for all $x = (\lambda, y) \in \mathbb{R} \times \mathbb{R}^{n-1}$ with $x \in B_1$ and $\lambda > 0$,

$$\partial_1 u(\lambda, y) = \lim_{h \searrow 0} \frac{u(\lambda + h, y) - u(\lambda - h, y)}{2h} = \lim_{h \searrow 0} \frac{w_\lambda(\lambda + h, y)}{2h} \leqslant 0.$$

This gives that $\partial_1 u \leqslant 0$ in $B_1 \cap \{x_1 > 0\}$. Hence, since $\Delta(\partial_1 u) = f'(u)\partial_1 u$ in $B_1 \cap \{x_1 > 0\}$, we deduce from the maximum principle in Theorem 2.16.2 that either $\partial_1 u < 0$ or $\partial_1 u$ vanish identically. But the latter cannot be (since $u(0) > 0 = u(e_1)$); therefore, (8.4.15) is proved. □

Now, we utilize Lemma 8.2.2 and 8.4.3 to complete the proof of Theorem 8.4.1.

Proof of Theorem 8.4.1. Since the domain under consideration is a ball, we can make use of Lemma 8.4.3 in every direction, thus finding that for all $\omega \in \partial B_1$ and $x \in \mathbb{R}^n$, we have that $u(\mathcal{R}_\omega x) = u(x)$. Thus, Lemma 8.2.2 gives that $u(x) = u_0(|x|)$ for some $u_0 : \mathbb{R} \to \mathbb{R}$. Hence, reconsidering the last claim in Lemma 8.4.3, for all $i \in \{1, \dots, n\}$, we have that $\partial_i u(x) < 0$ for all $x \in B_1$ with $x_i > 0$. Also, if $x_i < 0$, then

$$\begin{aligned}\partial_i u(x) &= \partial_i \big(u_0(|x|)\big) = \partial_i \Big(u_0\big(|(x_1, \dots, x_{i-1}, -x_i, x_{i+1}, \dots, x_n)|\big)\Big)\\ &= \partial_i \Big(u\big(x_1, \dots, x_{i-1}, -x_i, x_{i+1}, \dots, x_n\big)\Big)\\ &= -\partial_i u(x_1, \dots, x_{i-1}, -x_i, x_{i+1}, \dots, x_n) > 0.\end{aligned}$$

As a result, we have that $x_i \partial_i u(x) < 0$ for all $i \in \{1, \dots, n\}$ and $x \in B_1 \setminus \{0\}$, whence $x \cdot \nabla u(x) < 0$ for all $x \in B_1 \setminus \{0\}$, showing that u is radially strictly decreasing. □

The proof of Theorem 8.4.1 presented above follows the main ideas of the original article in [GNN79]. We showcase also a different proof, following [BN91, Bre99].

Another Proof of Theorem 8.4.1. We provide a different proof of Lemma 8.4.3 (from that, the original proof of Theorem 8.4.1 remains unchanged). We utilize the moving plane method and the notation in (8.4.16) and (8.4.22), aiming to prove (8.4.34), but with a different approach here. The idea is to exploit the maximum principle for small volume domains presented in Theorem 2.16.7 (in the first proof of Theorem 8.4.1, the maximum principle for small volume domains

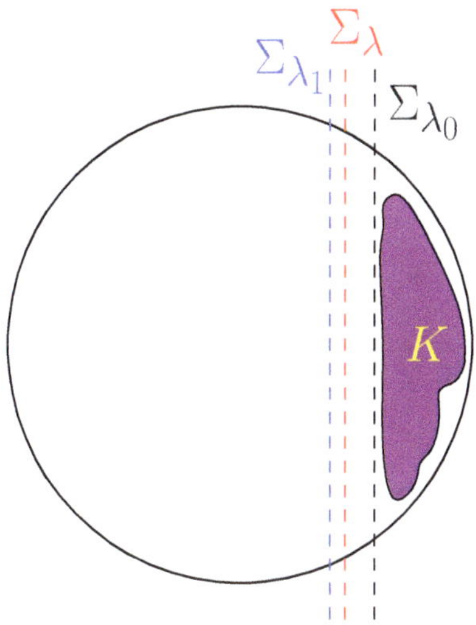

Figure 8.4.2 The geometry involved in (8.4.36).

is not essential since its use can be replaced by the one in narrow domains given in Theorem 2.16.6).

The details of this alternative proof are as follows. Suppose, by contradiction, that (8.4.34) does not hold, hence $\lambda_0 > 0$. Let $\lambda_1 \in (0, \lambda_0)$. Let also $K \Subset \Sigma_{\lambda_0}$ be an open set with C^∞ boundary and such that

$$\|f'\|_{L^\infty(\Omega)} \, |\Sigma_{\lambda_1} \setminus \overline{K}|^{\frac{2}{n}} < c_0, \tag{8.4.36}$$

with c_0 as in (2.16.17), see Figure 8.4.2. In our framework, (8.4.19), (8.4.20) and (8.4.36) allow us to use the maximum principle for small volume domains presented in Theorem 2.16.7, applied here to the function w_λ and the domain $\Sigma_\lambda \setminus \overline{K}$, with $\lambda \in (\lambda_0, \lambda_1)$ conveniently close to λ_0. To this end, recalling (8.4.26), we have that $\max_{\overline{K}} w_{\lambda_0} < 0$, that is, $\max_{\overline{K}} w_{\lambda_0} \leqslant -\delta_0$, for some $\delta_0 > 0$. Hence, for all $\lambda \in (\lambda_0, \lambda_1)$ sufficiently close to λ_0, we have that

$$\max_{\overline{K}} w_\lambda \leqslant 0.$$

Since $w_\lambda = 0$ on $T_\lambda \cap \overline{B_1}$ and $w_\lambda < 0$ on $(\partial B_1) \cap \Sigma_\lambda$, we thus conclude that $w_\lambda \leqslant 0$ on $\partial(\Sigma_\lambda \setminus \overline{K})$. Note also that $\|f'\|_{L^\infty(\Omega)} \, |\Sigma_\lambda \setminus \overline{K}|^{\frac{2}{n}} < c_0$, owing to (8.4.36); hence, by the maximum principle for small volume domains in Theorem 2.16.7, we infer that $w_\lambda < 0$ in $\Sigma_\lambda \setminus \overline{K}$.

Given the arbitrariness in the choice of K (we can freely move it around in Σ_{λ_0}), we thereby conclude that $w_\lambda < 0$ in Σ_λ, which is in contradiction with the infimum property in (8.4.22). □

It is also interesting to recall that a different approach to the proof of Theorem 8.4.1 is possible under the additional assumption that

$$f \text{ is nonpositive and nondecreasing.} \tag{8.4.37}$$

For this, we follow [Ser13], and[6] we argue as follows.

Another Proof of Theorem 8.4.1 under the Additional Assumption (8.4.37). We let

$$M := \max_{\overline{B_1}} u.$$

Given $j \in \{1, \ldots, n\}$, we recall that $\nabla \partial_j u = 0$ a.e. in $\{\partial_j u = 0\}$, see e.g. [LL01, Theorem 6.19]. In particular, we have that $\partial_j^2 u = 0$ a.e. in $\{\partial_j u = 0\}$ and accordingly

$$\text{for a.e. } x \in \{\nabla u = 0\} \text{ we have that } f(u(x)) = \Delta u(x) = 0. \tag{8.4.38}$$

Now, we recall that f is assumed to be nonpositive and nondecreasing by (8.4.37). Hence, either $f < 0$ or there exists $t_0 \in \mathbb{R}$ such that

$$f < 0 \text{ in } (-\infty, t_0) \quad \text{and} \quad f = 0 \text{ in } [t_0, +\infty).$$

This observation and (8.4.38) give that either $\{\nabla u = 0\} = \varnothing$ or

$$\begin{aligned} \varnothing \neq \{\nabla u = 0\} &\subseteq \big\{x \in B_1 \text{ s.t. } f(u(x)) = 0\big\} \cup \mathcal{Z} \\ &\subseteq \big\{x \in B_1 \text{ s.t. } u(x) \geqslant t_0\big\} \cup \mathcal{Z}, \end{aligned} \tag{8.4.39}$$

with $|\mathcal{Z}| = 0$. As a consequence, when $\big|\{\nabla u = 0\}\big| > 0$, we have that $t_0 \leqslant M$.

However, more precisely, in this case, it holds that $t_0 = M$ because otherwise $\{u > t_0\} \neq \varnothing$ and for all $x \in \{u > t_0\}$ one would have that $\Delta u(x) = f(u(x)) = 0$, and thus the weak maximum principle in Corollary 2.9.2 would entail that u is constant in any connected component of $\{u > t_0\}$, leading to

[6] The technique in [Ser13] was also motivated by [Lio81, KP94]; see [DPV22] for a general approach. Here, we actually consider only the simplest possible case, namely that of Theorem 8.4.1 under the additional assumption (8.4.37), but the method introduced in [Ser13] is much more general and also covers the case of bounded but possibly discontinuous nonlinearities f and that of possibly singular or degenerate nonlinear operators. The monotonicity assumption in (8.4.37) can also be weakened (and dropped when $n = 2$, since the parameter β in the forthcoming equation (8.4.52) vanishes in this case, thus making the analysis of the function J in (8.4.42) unnecessary in dimension 2). For the sake of simplicity, here we only confine ourselves to a model case, and we refer to [Ser13] for the method and the results in their full generality.

a contradiction, since at least one of these connected components must contain a point at which the value of u equals $M > t_0$.

We can thus reconsider (8.4.39) and write that

$$\{\nabla u = 0\} \subseteq \big\{x \in B_1 \text{ s.t. } u(x) = M\big\} \cup \mathcal{Z}$$

and consequently, since $\nabla u(x) = 0$ whenever $u(x) = M$,

$$\big|\{\nabla u = 0\}\big| = \Big|\big\{x \in B_1 \text{ s.t. } u(x) = M\big\}\Big|. \tag{8.4.40}$$

It is also convenient to use the coarea formula (see e.g. [EG15]) to note that

$$\begin{aligned} 0 &= \int_{\mathbb{R}^n} |\nabla u(x)|\, \chi_{\{|\nabla u|=0\}}(x)\, dx = \int_{\mathbb{R}} \left[\int_{x\in\{u=t\}} \chi_{\{|\nabla u|=0\}}(x)\, d\mathcal{H}^{n-1}_x \right] dt \\ &= \int_{\mathbb{R}} \Big[\mathcal{H}^{n-1}\Big(\big\{u=t\big\} \cap \big\{|\nabla u| = 0\big\}\Big) \Big]\, dt; \end{aligned}$$

therefore, for a.e. $t \in \mathbb{R}$,

$$\mathcal{H}^{n-1}\Big(\big\{u=t\big\} \cap \big\{|\nabla u| = 0\big\}\Big) = 0. \tag{8.4.41}$$

Now, we set

$$J(t) := \big|\{u > t\}\big| \quad \text{and} \quad J_0(t) := \big|\{u > t\} \cap \{\nabla u \neq 0\}\big|. \tag{8.4.42}$$

Note that J and J_0 are monotone nonincreasing functions of t, and hence their derivative is well-defined for a.e. $t \in (0, M)$. We claim that

$$\begin{aligned} &J_0 \text{ is an absolutely continuous function} \\ &\text{and, for a.e. } t \in (0, M), \quad J_0'(t) = -\int_{\{u=t\}\cap\{\nabla u\neq 0\}} \frac{d\mathcal{H}^{n-1}_x}{|\nabla u(x)|}. \end{aligned} \tag{8.4.43}$$

To check this, let $\varepsilon > 0$. We use the coarea formula (see e.g. [EG15]) to see that

$$\begin{aligned} J_\varepsilon(t) &:= \big|\{u > t\} \cap \{|\nabla u| > \varepsilon\}\big| = \int_{\mathbb{R}^n} \frac{|\nabla u(x)|\, \chi_{\{u>t\}\cap\{|\nabla u|>\varepsilon\}}(x)}{|\nabla u(x)|}\, dx \\ &= \int_{\mathbb{R}} \left[\int_{x\in\{u=\lambda\}} \frac{\chi_{\{u>t\}\cap\{|\nabla u|>\varepsilon\}}(x)}{|\nabla u(x)|}\, d\mathcal{H}^{n-1}_x \right] d\lambda \\ &= \int_t^M \left[\int_{x\in\{u=\lambda\}\cap\{|\nabla u|>\varepsilon\}} \frac{d\mathcal{H}^{n-1}_x}{|\nabla u(x)|} \right] d\lambda. \end{aligned} \tag{8.4.44}$$

Therefore, according to the monotone convergence theorem (see e.g. [WZ15, Theorem 10.27]) we can pass $\varepsilon \searrow 0$ in (8.4.44) and obtain that

$$J_0(t) = \big|\{u > t\} \cap \{|\nabla u| > 0\}\big| = \int_t^M \left[\int_{x\in\{u=\lambda\}\cap\{|\nabla u|>0\}} \frac{d\mathcal{H}^{n-1}_x}{|\nabla u(x)|} \right] d\lambda$$

and this entails (8.4.43), as desired.

As a matter of fact, the result in (8.4.43) can be sharpened in light of (8.4.41), thus producing that, for a.e. $t \in (0, M)$,

$$J_0'(t) = -\int_{\{u=t\}} \frac{d\mathcal{H}_x^{n-1}}{|\nabla u(x)|}. \tag{8.4.45}$$

Also, owing to (8.4.40) and (8.4.42), for all $t \in (0, M)$, we have that

$$J(t) = J_0(t) + \big|\{u = M\}\big|.$$

From this and (8.4.45), one deduces that

$$\begin{gathered} J \text{ is an absolutely continuous function} \\ \text{and, for a.e. } t \in (0, M), \quad J'(t) = -\int_{\{u=t\}} \frac{d\mathcal{H}_x^{n-1}}{|\nabla u(x)|}. \end{gathered} \tag{8.4.46}$$

Now, we define

$$I(t) := -\int_{\{u>t\}} f(u(x))\,dx,$$

and we observe that I is a monotone nonincreasing function of t, thanks to (8.4.37), and hence its derivative is well-defined for a.e. $t \in (0, M)$. Also, as $\tau \nearrow 0$,

$$\begin{aligned} I(t+\tau) - I(t) &= -\int_{\{u>t+\tau\}} f(u(x))\,dx + \int_{\{u>t\}} f(u(x))\,dx \\ &= \int_{\{t+\tau \geqslant u>t\}} f(u(x))\,dx \\ &= -\int_{\{t+\tau \geqslant u>t\}} \big(f(t) + o(1)\big)\,dx \\ &= \big(f(t) + o(1)\big)\big|\{t+\tau \geqslant u > t\}\big| \\ &= \big(f(t) + o(1)\big)\big(J(t) - J(t+\tau)\big). \end{aligned} \tag{8.4.47}$$

This along with the boundedness of f in the range of the solution u and (8.4.46) give that

$$I \text{ is an absolutely continuous function in } (0, M). \tag{8.4.48}$$

Furthermore, dividing by τ in (8.4.47) and taking the limit as $\tau \searrow 0$, we thus conclude that, for a.e. $t \in (0, M)$,

$$I'(t) = -f(t)\,J'(t). \tag{8.4.49}$$

Given α, $\beta \in \mathbb{R}$, to be specified later on, we now take into consideration the function

$$K(t) := \big(I(t)\big)^{\alpha}\big(J(t)\big)^{\beta}.$$

In view of (8.4.46) and (8.4.48), we know that

$$K \text{ is an absolutely continuous function in } (0, M). \tag{8.4.50}$$

Also, we make use of (8.4.49) to find that, a.e. in $(0, M)$,

$$K' = \alpha I^{\alpha-1} J^{\beta} I' + \beta I^{\alpha} J^{\beta-1} J' = \Big(- \alpha I^{\alpha-1} J^{\beta} f + \beta I^{\alpha} J^{\beta-1} \Big) J'. \tag{8.4.51}$$

Now, we choose

$$\alpha := 2 \quad \text{and} \quad \beta := -\frac{n-2}{n}, \tag{8.4.52}$$

and we deduce from (8.4.51) that, a.e. in $(0, M)$,

$$\begin{aligned} K' &= \left(-2IJ^{-\frac{n-2}{n}} f - \frac{n-2}{n} I^2 J^{-\frac{n-2}{n}-1} \right) J' \\ &= \left(-2Jf - \frac{n-2}{n} I \right) IJ^{-\frac{n-2}{n}-1} J'. \end{aligned} \tag{8.4.53}$$

In addition, since f is supposed to be nondecreasing (recall (8.4.37)),

$$I(t) \leqslant - \int_{\{u>t\}} f(t)\, dx = -f(t) \left|\{u > t\}\right| = -f(t)\, J(t).$$

Plugging this information into (8.4.53) and recalling (8.4.46), we obtain that, a.e. in $(0, M)$,

$$K' \leqslant \left(-2Jf + \frac{n-2}{n} Jf \right) IJ^{-\frac{n-2}{n}-1} J' = -\frac{n+2}{n} IJ^{-\frac{n-2}{n}} fJ' \leqslant 0. \tag{8.4.54}$$

Thanks to (8.4.50), we can thus reconstruct the oscillations of K from K' (see e.g. [WZ15, Theorem 7.32(ii)], used here with $f := K$ and $g := 1$), and we have that, for all $a, b \in (0, M)$ with $a < b$,

$$\int_a^b K'(t)\, dt = K(b) - K(a).$$

In particular, by (8.4.54), we deduce that K is nonincreasing in $(0, M)$. Also, recalling (8.4.53), we see that

$$\begin{aligned} K(0^+) - K(M^-) &:= \lim_{\substack{a \searrow 0 \\ b \nearrow M}} K(a) - K(b) = - \int_0^M K'(t)\, dt \\ &= \int_0^M \left(2J(t) f(t) + \frac{n-2}{n} I(t) \right) I(t)\, (J(t))^{-\frac{n-2}{n}-1} J'(t)\, dt. \end{aligned} \tag{8.4.55}$$

Now, we utilize the isoperimetric inequality (see [EG15, Theorem 1 in Section 5.5 and Theorem 2 in Section 5.6.2]) to write that

$$\frac{\mathcal{H}^{n-1}(\{u=t\})}{\big|\{u>t\}\big|^{\frac{n-1}{n}}} \geqslant \frac{\mathcal{H}^{n-1}(\partial B_1)}{|B_1|^{\frac{n-1}{n}}} =: c_\star. \tag{8.4.56}$$

This, in combination with the Cauchy–Schwarz inequality, gives that

$$\begin{aligned}(J(t))^{\frac{n-2}{n}+1} &= \big|\{u>t\}\big|^{\frac{2(n-1)}{n}} \leqslant \left(\frac{\mathcal{H}^{n-1}(\{u=t\})}{c_\star}\right)^2 \\ &= \frac{1}{c_\star^2}\left(\int_{\{u=t\}} \frac{\sqrt{|\nabla u(x)|}}{\sqrt{|\nabla u(x)|}}\, d\mathcal{H}^{n-1}_x\right)^2 \\ &\leqslant \frac{1}{c_\star^2}\int_{\{u=t\}} |\nabla u(x)|\, d\mathcal{H}^{n-1}_x \int_{\{u=t\}} \frac{d\mathcal{H}^{n-1}_x}{|\nabla u(x)|}.\end{aligned} \tag{8.4.57}$$

Furthermore, if ν is the external normal to $\{u>t\}$, we know that $\nu = -\frac{\nabla u}{|\nabla u|}$ outside a set of vanishing $(n-1)$-dimensional Hausdorff measure, thanks to (8.4.41), whence by the divergence theorem (see e.g. [EG15, Theorem 1 in Section 5.8]), we find that

$$I(t) = -\int_{\{u>t\}} f(u(x))\, dx = -\int_{\{u>t\}} \operatorname{div}(\nabla u(x))\, dx = \int_{\{u=t\}} |\nabla u(x)|\, d\mathcal{H}^{n-1}_x. \tag{8.4.58}$$

Plugging this into (8.4.57) and recalling (8.4.46), we conclude that

$$(J(t))^{\frac{n-2}{n}+1} \leqslant \frac{I(t)}{c_\star^2}\int_{\{u=t\}} \frac{d\mathcal{H}^{n-1}_x}{|\nabla u(x)|} = -\frac{I(t)\, J'(t)}{c_\star^2}.$$

We insert this information into (8.4.55), and we obtain

$$K(0^+) - K(M^-) \geqslant -c_\star^2 \int_0^M \left(2J(t)f(t) + \frac{n-2}{n} I(t)\right) dt. \tag{8.4.59}$$

In addition, using the notation

$$F(r) := \int_0^r f(\tau)\, d\tau,$$

we have that

$$\begin{aligned}\int_0^M J(t) f(t)\, dt &= \int_0^M \big|\{u>t\}\big| f(t)\, dt = \int_0^M \left[\int_{B_1} \chi_{\{u>t\}}(x)\, f(t)\, dx\right] dt \\ &= \int_{B_1}\left[\int_0^{u(x)} f(t)\, dt\right] dx = \int_{B_1} F(u(x))\, dx\end{aligned}$$

and

$$
\begin{aligned}
-\int_0^M I(t)\,dt &= \int_0^M \left[\int_{\{u>t\}} f(u(x))\,dx\right] dt \\
&= \int_{B_1}\left[\int_0^{u(x)} f(u(x))\,dt\right] dx = \int_{B_1} u(x)f(u(x))\,dx.
\end{aligned}
$$

These observations and (8.4.59) lead to

$$
\frac{n\left(K(0^+)-K(M^-)\right)}{2c_\star^2} \geqslant -n\int_{B_1} F(u(x))\,dx + \frac{n-2}{2}\int_{B_1} u(x)f(u(x))\,dx.
$$

From this and the Pohozaev identity in Theorem 2.1.9, we arrive at

$$
\begin{aligned}
\frac{n\left(K(0^+)-K(M^-)\right)}{2c_\star^2} &\geqslant \frac{1}{2}\int_{\partial B_1} (\partial_\nu u(x))^2\,(x\cdot\nu(x))\,d\mathcal{H}_x^{n-1} \\
&= \frac{1}{2}\int_{\partial B_1} (\partial_\nu u(x))^2\,d\mathcal{H}_x^{n-1}.
\end{aligned}
$$

This and the Cauchy–Schwarz inequality lead to

$$
\begin{aligned}
&\left(\int_{B_1} f(u(x))\,dx\right)^2 = \left(\int_{B_1} \operatorname{div}(\nabla u(x))\,dx\right)^2 = \left(\int_{\partial B_1} \nabla u(x)\cdot\nu(x)\,d\mathcal{H}_x^{n-1}\right)^2 \\
&= \left(\int_{\partial B_1} |\nabla u(x)|\,d\mathcal{H}_x^{n-1}\right)^2 = \left(\int_{\partial B_1} |\partial_\nu u(x)|\,d\mathcal{H}_x^{n-1}\right)^2 \\
&\leqslant \mathcal{H}^{n-1}(\partial B_1)\int_{\partial B_1} (\partial_\nu u(x))^2\,d\mathcal{H}_x^{n-1} \leqslant \frac{n\mathcal{H}^{n-1}(\partial B_1)\left(K(0^+)-K(M^-)\right)}{c_\star^2}.
\end{aligned}
$$

Thus, since K is nonnegative,

$$
\begin{aligned}
\left(\int_{B_1} f(u(x))\,dx\right)^2 &\leqslant \frac{n\mathcal{H}^{n-1}(\partial B_1)\,K(0^+)}{c_\star^2} \\
&= \frac{n\mathcal{H}^{n-1}(\partial B_1)}{c_\star^2}\lim_{t\searrow 0}\left(I(t)\right)^\alpha\left(J(t)\right)^\beta \\
&= \frac{n\mathcal{H}^{n-1}(\partial B_1)}{c_\star^2}\left(\int_{\{u>0\}} f(u(x))\,dx\right)^\alpha \left|\{u>0\}\right|^\beta \\
&= \frac{n\mathcal{H}^{n-1}(\partial B_1)\,|B_1|^\beta}{c_\star^2}\left(\int_{B_1} f(u(x))\,dx\right)^\alpha \\
&= \frac{n\mathcal{H}^{n-1}(\partial B_1)\,|B_1|^{-\frac{n-2}{n}}}{c_\star^2}\left(\int_{B_1} f(u(x))\,dx\right)^2. \qquad (8.4.60)
\end{aligned}
$$

However, recalling (1.1.6) and (8.4.56),

$$\frac{n\mathcal{H}^{n-1}(\partial B_1)\,|B_1|^{-\frac{n-2}{n}}}{c_\star^2}=\frac{n\mathcal{H}^{n-1}(\partial B_1)\,|B_1|^{-\frac{n-2}{n}}\,|B_1|^{\frac{2(n-1)}{n}}}{\left(\mathcal{H}^{n-1}(\partial B_1)\right)^2}=\frac{n\,|B_1|}{\mathcal{H}^{n-1}(\partial B_1)}=1.$$

With this, we conclude that all the inequalities leading to (8.4.60) must be equalities. In particular, for a.e. $t\in(0,M)$, the set $\{u>t\}$ must optimize the isoperimetric inequality in (8.4.56); hence, it must be a ball, say $\{u>t\}=B_{r(t)}(x(t))$ for some $r(t)>0$ and $x(t)\in B_1$.

Also, the last inequality in (8.4.57) must be an equality, and correspondingly, $|\nabla u(x)|$ must be proportional to $\frac{1}{|\nabla u(x)|}$ on $\{u=t\}$. This gives that $|\nabla u|$ is constant on $\{u=t\}$, for a.e. $t\in(0,M)$, and we thus write in this setting that $|\nabla u|=c(t)$ on $\{u=t\}$.

Hence, for a.e. $t\in(0,M)$, one can use (8.4.58) and infer that

$$\begin{aligned}-\int_{\{u>t\}} f(u(x))\,dx &= \int_{\{u=t\}} |\nabla u(x)|\,d\mathcal{H}^{n-1}_x\\ &= c(t)\,\mathcal{H}^{n-1}\big(\{u=t\}\big)=|\nabla u(p)|\,\mathcal{H}^{n-1}\big(\{u=t\}\big),\end{aligned}\tag{8.4.61}$$

for all $p\in\{u=t\}$.

We now claim that if $u(x)\in(0,M)$, then

$$\nabla u(x)\neq 0.\tag{8.4.62}$$

To check this, suppose by contradiction that $\nabla u(q)=0$ for some $q\in B_1$ with $u(q)\in(0,M)$, and let p_k be a sequence of points such that (8.4.61) holds true and $p_k\to q$ as $k\to+\infty$. In this way, passing to the limit (8.4.62), in this case, we find that

$$-\int_{\{u>u(q)\}} f(u(x))\,dx=|\nabla u(q)|\,\mathcal{H}^{n-1}\big(\{u=t\}\big)=0.$$

Thus, since f is nonpositive (recall (8.4.37)), necessarily $f(u(x))=0$ for all $x\in\{u>u(q)\}$ and accordingly $\Delta u=f=0$ in $\{u>u(q)\}$. This and the weak maximum principle in Corollary 2.9.2 give that u is constant in any connected component of $\{u>u(q)\}$, leading to a contradiction since one of these connected components must contain a point at which the value of u is $M>u(q)$. The proof of (8.4.62) is thereby complete.

From (8.4.62), it follows that for all $t\in(0,M)$,

$$\min_{\{u=t\}}|\nabla u|>0\tag{8.4.63}$$

and accordingly, by (8.4.46), J is locally Lipschitz in $(0,M)$. Hence, since

$$J(t)=\big|\{u>t\}\big|=\big|B_{r(t)}\big|=|B_1|\,(r(t))^n,$$

we also have that

$$(0, M) \ni t \mapsto r(t) \text{ is locally Lipschitz.} \tag{8.4.64}$$

Now, if $T, t \in (0, M)$ with $T > t$, we claim that

$$|x(T) - x(t)| \leqslant r(t) - r(T). \tag{8.4.65}$$

For this, we can suppose that $x(T) \neq x(t)$, otherwise we are done. Also, we have that $B_{r(t)}(x(t)) = \{u > t\} \supseteq \{u > T\} = B_{r(T)}(x(T))$; therefore, for all $\rho \in (0, r(T))$ and all $\varpi \in \partial B_1$, we have that $x(T) + \rho\varpi \in B_{r(T)}(x(T)) \subseteq B_{r(t)}(x(t))$, giving that

$$|x(T) + \rho\varpi - x(t)| < r(t).$$

Picking $\varpi := \frac{x(T)-x(t)}{|x(T)-x(t)|}$, we get that

$$\begin{aligned} |x(T) - x(t)| + \rho &= \left(1 + \frac{\rho}{|x(T) - x(t)|}\right) |x(T) - x(t)| \\ &\leqslant \left|1 + \frac{\rho}{|x(T) - x(t)|}\right| |x(T) - x(t)| \\ &= |x(T) - x(t) + \rho\varpi| < r(t). \end{aligned}$$

Thus, by taking the limit $\rho \nearrow r(T)$, we prove (8.4.65), as desired.

Then, by (8.4.64) and (8.4.65), we obtain that

$$(0, M) \ni t \mapsto x(t) \text{ is locally Lipschitz.} \tag{8.4.66}$$

Now, we show that

$$(0, M) \ni t \mapsto x(t) \text{ is constant.} \tag{8.4.67}$$

Indeed, suppose not: then, there exist t_1 and t_2 in $(0, M)$ with $t_1 < t_2$ such that $x(t_1) \neq x(t_2)$. Hence, by (8.4.66),

$$0 \neq x(t_2) - x(t_1) = \int_{t_1}^{t_2} \frac{dx}{dt}(t)\, dt;$$

consequently, there exists $t_\star \in (t_1, t_2)$ such that $\frac{dx}{dt}(t_\star)$ exists and is different from zero.

We can thereby define $y := \frac{dx}{dt}(t_\star) \neq 0$ and $z := \frac{y}{|y|} \in \partial B_1$. Let also $P_\pm(t) := x(t) \pm r(t)z$. We stress that $P_\pm(t) \in \partial B_{r(t)}(x(t)) = \partial\{u > t\}$. Therefore, we have that $u(P_\pm(t)) = t$ and $\left|\nabla u(P_\pm(t))\right| = c(t)$. Hence, since the exterior normal of $B_{r(t)}(x(t))$ at $P_\pm(t)$ is $\pm z$, we find that

$$\pm\nabla u(P_\pm(t)) \cdot z = -\left|\nabla u(P_\pm(t)\right| = -c(t).$$

As a result, a.e. $t \in (0, M)$, we have that

$$\begin{aligned}
1 &= \frac{d}{dt}t = \frac{d}{dt}u(P_\pm(t)) = \frac{d}{dt}u\big(x(t) \pm r(t)z\big) \\
&= \nabla u\big(x(t) \pm r(t)\, z\big) \cdot \left(\frac{dx}{dt}(t) \pm \frac{dr}{dt}(t)z\right) = \nabla u(P_\pm(t)) \cdot \left(\frac{dx}{dt}(t) \pm \frac{dr}{dt}(t)z\right) \\
&= \nabla u(P_\pm(t)) \cdot \frac{dx}{dt}(t) - c(t)\frac{dr}{dt}(t);
\end{aligned}$$

therefore,

$$\begin{aligned}
0 &= \left[\nabla u(P_+(t)) \cdot \frac{dx}{dt}(t) - c(t)\frac{dr}{dt}(t)\right] - \left[\nabla u(P_-(t)) \cdot \frac{dx}{dt}(t) - c(t)\frac{dr}{dt}(t)\right] \\
&= \Big(\nabla u(P_+(t)) - \nabla u(P_+(t))\Big) \cdot \frac{dx}{dt}(t).
\end{aligned}$$

Consequently, choosing $t := t_\star$,

$$\begin{aligned}
0 &= \Big(\nabla u(P_+(t_\star)) - \nabla u(P_+(t_\star))\Big) \cdot y \\
&= |y|\Big(\nabla u(P_+(t_\star)) - \nabla u(P_+(t_\star))\Big) \cdot z = -2|y|c(t_\star)
\end{aligned}$$

and accordingly $c(t_\star) = 0$. This is in contradiction with (8.4.62), and thus we have completed the proof of (8.4.67).

Then, by (8.4.67), we deduce that $\{u > t\} = B_{r(t)}(x_\star)$ for some $x_\star \in \Omega$; therefore, u is radially symmetric with respect to the point $x_\star$ (which, up to a translation, we can suppose to be the origin).

It remains to show that u is radially strictly decreasing. To this end, it suffices to show that

$$\{u = M\} = \{0\}. \tag{8.4.68}$$

Indeed, if this is true, suppose by contradiction that there exist $a, b \in [0, 1)$ such that $a < b$ and $u(ae_1) \leqslant u(be_1)$. We have that $u(be_1) \neq M$ (otherwise, by (8.4.68), $a < b = 0$, which is a contradiction) and also $u(ae_1) \neq M$ (otherwise $u(be_1) = M$, which we have just excluded). Let $m \in [a, 1]$ be such that

$$u(me_1) = \max_{s\in[a,1]} u(se_1),$$

and note that $m \neq 1$ since, by assumption, $0 = u(e_1) < u(be_1)$. Additionally, we can take $m > a$ since $u(ae_1) \leqslant u(be_1)$. As a result, since u is radial,

$$0 = \left|\frac{d}{ds}u(se_1)\Big|_{s=m}\right| = |\nabla u(se_1) \cdot e_1| = |\nabla u(se_1)|. \tag{8.4.69}$$

Moreover, we have that $u(me_1) \in (0, M)$ due to (8.4.68). This and (8.4.69) are in contradiction with (8.4.63).

This argument shows that u has to be radially strictly decreasing once (8.4.68) is proved. Hence, we now establish (8.4.68). To this end, we observe that it is sufficient to check that

$$\{u = M\} \subseteq \{0\} \tag{8.4.70}$$

since this, together with the fact that $\{u = M\} \neq \varnothing$, would entail (8.4.68). Thus, to prove (8.4.70), we argue by contradiction, supposing that there exist $\omega \in \partial B_1$ and $\ell \in (0, 1)$ such that $u(\ell\omega) = M$. Note that necessarily $\nabla u(\ell\omega) = 0$. Moreover, we have that $u(t\omega) = M$ for all $t \in [0, \ell]$, thanks to (8.4.62). As a consequence,

$$0 = \Delta u(0) = f(u(0)) = f(M). \tag{8.4.71}$$

Now, writing $u(x) = u_0(|x|)$ for some $u_0 : [0, 1] \to \mathbb{R}$, we use the formula for the Laplace operator in spherical coordinates in Theorem 2.5.1, and we find that u_0 is a solution of the following Cauchy problem:

$$\begin{cases} u_0''(r) + \dfrac{n-1}{r} u_0'(r) = f(u(r)), & \text{for all } r \in (0, 1), \\ u_0(\ell) = M, \\ u_0'(\ell) = 0. \end{cases} \tag{8.4.72}$$

We observe that the function constantly equal to M is also a solution of (8.4.72), in light of (8.4.71). Therefore, by the uniqueness[7] of the solutions of initial data

[7] We stress that we are exploiting here the fact that f was taken to be of class $C^1(\mathbb{R})$ in the statement of Theorem 8.4.1 in order to use the classical uniqueness result for the Cauchy problem (see e.g. [Per01, p. 74]). This condition on f can be relaxed to a Lipschitz assumption but cannot be completely dropped without losing the property that u is radially strictly decreasing. As an example, we observe that, given $\varrho \in (0, 1)$ and $\alpha > 2$, the function

$$u(x) := \begin{cases} 1 - \left(\dfrac{|x|^2 - \varrho^2}{1 - \varrho^2}\right)^{\alpha}, & \text{for all } x \in B_1 \setminus B_\varrho, \\ 1 & \text{in } B_\varrho \end{cases}$$

satisfies (8.4.1), with

$$f(t) := \begin{cases} -\dfrac{2\alpha(1-t)^{\frac{\alpha-2}{\alpha}}}{(1-\varrho^2)^2}\Big((n + 2\alpha - 2)(1 - \varrho^2)(1 - t)^{\frac{1}{\alpha}} + 2(\alpha - 1)\varrho^2\Big) & \text{if } t < 1, \\ 0 & \text{if } t \geqslant 1. \end{cases}$$

Note that this u is radial but not radially strictly decreasing since $\{u = 1\} = \overline{B_\varrho}$. Note also that in this case, $f \in C^{\frac{\alpha-2}{\alpha}}(\mathbb{R})$ but $f \notin C^1(\mathbb{R})$. See Figure 8.4.3 for a sketch of such a function f when $\varrho := \frac{1}{2}$, $\alpha := 3$ and $n := 2$. See also Figure 8.4.4 for a sketch of the radial profile of the corresponding solution u.

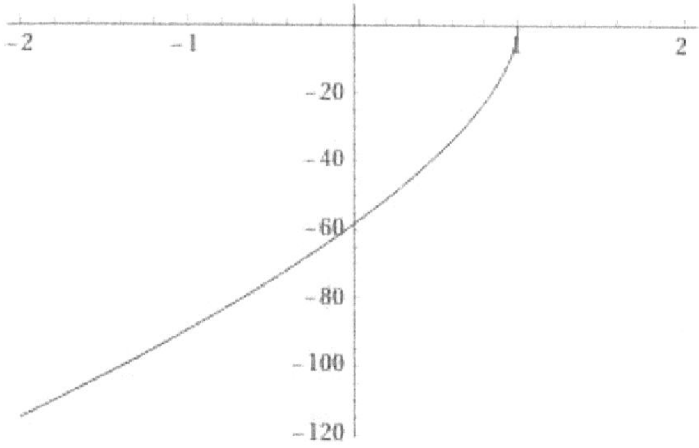

Figure 8.4.3 Sketch of the function f discussed in footnote 7.

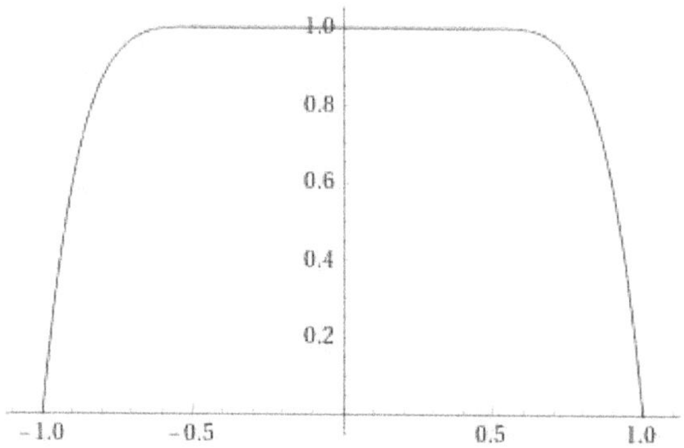

Figure 8.4.4 Sketch of the radial profile of the function u discussed in footnote 7.

problems for ordinary differential equations, we conclude that necessarily u_0 is constant, hence so is u, in contradiction with the setting of Theorem 8.4.1. The proof of (8.4.70) is thereby complete. □

Chapter 9

Local Existence Theory in the Real Analytic Setting

While most of these pages are devoted to the study of partial differential equations, it is now interesting to revisit the classical theory of ordinary differential equations to recognize its influence on a rather delicate aspect of the theory that we presented. Namely, we have developed so far a series of rather specific existence and uniqueness results for some classes of elliptic equations (see e.g. Corollary 3.3.4 and Theorems 4.5.4, 4.5.7, 5.3.5 and 6.2.1); it is therefore natural to wonder about a "general" approach guaranteeing existence results for partial differential equations in rather general scenarios.

One can be hopeful about it since the case of ordinary differential equations is settled in a general form, ensuring the existence and uniqueness of solutions via the Picard–Lindelöf–Cauchy–Lipschitz theorem (see e.g. [MM54, AL93]). For example, in the case of partial differential equations, one may want to consider an equation, or a system of equations, of a given order m (i.e. in which the highest derivative involved is of order m), assign the (say, normal) derivatives up to the order $m-1$ on a hypersurface, and investigate the existence and uniqueness of solutions.

9.1 The Cauchy–Kowalevsky Theorem

To formalize the above-mentioned setting, we can consider a smooth hypersurface Γ described as a level set of a smooth function Φ. That is, we suppose that $\Phi \in C^\infty(\mathbb{R}^n)$ with $\nabla\Phi(x) \neq 0$ for all $x \in \mathbb{R}^n$ and set

$$\Gamma_0 := \{x \in \mathbb{R}^n \text{ s.t. } \Phi(x) = 0\}. \tag{9.1.1}$$

Let Ω be an open subset of $\mathbb{R}^n$, with

$$\Gamma := \Omega \cap \Gamma_0 \neq \varnothing, \tag{9.1.2}$$

see Figure 9.1.1.

Figure 9.1.1 The geometry in (9.1.1) and (9.1.2).

Let also $N \in \mathbb{N}$ with $N \geqslant 1$ and $u : \Omega \to \mathbb{R}^N$. We use the notation $u = (u_1, \dots, u_N)$, with $u_\ell : \Omega \to \mathbb{R}$ for each $\ell \in \{1, \dots, N\}$.

Given $\mu \in \mathbb{N}^N$, we denote by $D^\mu_\star u$ the collection of all the partial derivatives of u_ℓ up to order μ_ℓ, for all $\ell \in \{1, \dots, N\}$, that is,

$$D^\mu_\star u := \left\{\partial^\alpha u_\ell\right\}_{\substack{\alpha\in\mathbb{N}^n\\ |\alpha|\leqslant\mu_\ell\\ \ell\in\{1,\dots,N\}}}.$$

Thus, a system of partial differential equations of order $\mu \in \mathbb{N}^N$ is described by a set of equations of the form

$$F_\ell(x, D^\mu_\star u(x)) = 0, \tag{9.1.3}$$

for $x \in \Omega$ and $\ell \in \{1, \dots, N\}$ and for some smooth functions $F_1, \dots, F_N$. We use the short notation $F := (F_1, \dots, F_N)$.

The prescription along Γ reads $\partial^j_\nu u_\ell = \phi_{j,\ell}$, for all $j \in \mathbb{N}$ with $j \leqslant \mu_\ell - 1$ and $\ell \in \{1, \dots, N\}$ and for suitable assigned functions $\phi_{j,\ell}$.

Here, we are denoting by ν a smooth unit normal vector field along Γ and by $\partial^j_\nu f$ the jth derivative of a function f in the direction of ν, i.e. for each $x \in \Gamma$,

$$\begin{aligned}
\partial^j_\nu f(x) &:= \frac{d^j}{d\tau^j} f(x + \tau\nu(x))\Big|_{\tau=0} \\
&= \frac{d^j}{d\tau^j}\left(\sum_{\substack{0\leqslant m\leqslant j\\ 1\leqslant k_1,\dots,k_m\leqslant n}} \frac{\tau^m}{m!}\,\frac{\partial^m f}{\partial x_{k_1}\dots\partial x_{k_m}}(x)\,\nu_{k_1}(x)\dots\nu_{k_m}(x) + o(\tau^j)\right)\Bigg|_{\tau=0} \\
&= \sum_{1\leqslant k_1,\dots,k_j\leqslant n} \frac{\partial^j f}{\partial x_{k_1}\dots\partial x_{k_j}}(x)\,\nu_{k_1}(x)\dots\nu_{k_j}(x).
\end{aligned} \tag{9.1.4}$$

With this notation, and inspired by the case of ordinary differential equations, one can write a Cauchy problem for partial differential equations in the form

$$\begin{cases} F(x, D^\mu_\star u(x)) = 0 & \text{for all } x \in \Omega, \\ \partial^j_\nu u_\ell(x) = \phi_{j,\ell}(x) & \text{for all } x \in \Gamma,\ \ell \in \{1, \dots, N\} \text{ and } j \in \{0, \dots, \mu_\ell - 1\}. \end{cases} \tag{9.1.5}$$

The investigation of this problem was pioneered by Cauchy in [Cau42a, Cau42b, Cau42c, Cau42d, Cau42e, Cau42f] for nonlinear differential equations of second order and then generalized by Kowalevsky in [Kow75] for general analytic nonlinear systems of differential equations (independently, a rather general case was also treated by Darboux [Dar75] and a simplified strategy was put forth by Goursat in [Gou98]). Nowadays, problems such as the one in (9.1.5), in which one seeks a solution to a partial differential equation satisfying certain conditions that are prescribed on a hypersurface in the domain, are called "Cauchy problems."

The main result related to (9.1.5) is known as[1] the Cauchy–Kowalevsky theorem, which, in a suitable form, is presented in the following[2] in Theorem 9.1.6.

As a special feature, the Cauchy–Kowalevsky theorem deals with a real analytic setting (a complex analytic case can be treated as well, but with minor modifications in the notation). Before going into the details of the theorem, however, it is instructive to look at some negative examples to discover how delicate this matter is. We begin with a one-dimensional[3] heat equation.

---

[1]Sof'ja Vasil'evna Kovalevskaja, born Korvin-Krukovskaja, often used the name of Sophie Kowalevski (or Kowalevsky, or von Kowalevsky) in her academic publications. Her first name is ofttimes replaced by the nickname Sonya, and hence different spellings and transliterations are overabundant. At that time, Russia happened to be a very conservative country, and women were not allowed to attend universities. To study abroad, Kowalevsky needed written permission from her husband. So, in 1868, she contracted a "fictitious marriage" with a paleontology student (and translator of the works of Charles Darwin). She obtained her doctorate (first woman ever) in Germany under the supervision of Weierstraß. The Cauchy–Kowalevsky Theorem is one of the results of her PhD thesis.

The name Kowalevsky is also linked to the study of the precession of a spinning top with special symmetry regarding moments of inertia and the center of gravity, which make this mechanical system integrable (for this study, Kowalevsky was awarded the Prix Bordin from the French Academy of Sciences in 1888).

Besides mathematics, Kowalevsky was very active in the radical movement of Russian nihilism (defined by Russian anarchist Peter Kropotkin as "the symbol of struggle against all forms of tyranny, hypocrisy, and artificiality"), she wrote plays and autobiographical novels, and she served as a nurse in the 1871 Paris Commune.

Her motherland, Russia, never recognized Kowalevsky's academic degrees obtained abroad and never offered her an academic position (but she became a full professor at Stockholm University as well as an editor of the very prestigious Swedish journal *Acta Mathematica*). The Russian Academy of Sciences changed its rules to appoint her a corresponding member. See Figure 9.1.2 for a portrait of Kowalevsky at 18 years and for a commemorative coin and stamp.

[2]We stress that Theorem 9.1.6 is certainly neither the ultimate nor the most general form of the Cauchy–Kowalevsky theorem (yet, it is general enough to consider nonlinear systems of various order).

[3]We observe that the equation in (9.1.6) is precisely the heat equation in [DV23, equation (1.6)] in the absence of sources and in one spatial dimension, with the notation $x_2 := t$.

It is worth remarking that a solution to the heat equation (9.1.6) for small $x_2 > 0$ and small $|x_1|$ does exist, only that it is not real analytic (thus confirming that analyticity is not always the most convenient

Figure 9.1.2 Sophie Kowalevski.
Source: Public Domain images from Wikipedia.

Theorem 9.1.1. *The Cauchy problem*[4]

$$\begin{cases} \partial_{x_1 x_1} u(x) - \partial_{x_2} u(x) = 0 & \text{for all } x = (x_1, x_2) \in (-\varepsilon, \varepsilon) \times (-\varepsilon, \varepsilon), \\ u(x_1, 0) = \dfrac{1}{1 - x_1} & \text{for all } x_1 \in (-\varepsilon, \varepsilon) \end{cases} \tag{9.1.6}$$

does not admit any real analytic solution.

More precisely,

$$\text{there is no solution of (9.1.6) which is real analytic in a neighborhood of the origin.} \tag{9.1.7}$$

Proof. Suppose by contradiction that such a solution exists, and let us write it in a neighborhood of the origin as

$$u(x_1, x_2) = \sum_{i,j=0}^{+\infty} c_{i,j}\, x_1^i\, x_2^j.$$

---

setting for a partial differential equation). Indeed, the function

$$u(x_1, x_2) := \int_{-1/2}^{1/2} \frac{e^{-\frac{(x_1 - y)^2}{4x_2}}}{\sqrt{4\pi x_2}\,(1 - y)}\, dy$$

is a solution as desired, but it is not real analytic in x_2 at $x_2 = 0$.

[4]Note that any prescription of $\partial_{x_2} u$ along $\left(-\frac{1}{2}, \frac{1}{2}\right) \times \{0\}$ would cast (9.1.6) into the general setting of (9.1.5), here with $n := 2$, $N := 1$, $\Omega := (-\varepsilon, \varepsilon) \times (-\varepsilon, \varepsilon)$, $\Gamma_0 := (-1, 1) \times \{0\}$ and $\Gamma := (-\varepsilon, \varepsilon) \times \{0\}$.

We claim that

$$c_{i,j} = \frac{\prod_{k=1}^{2j}(i+k)}{j!}. \tag{9.1.8}$$

A proof of (9.1.8) is obtained by induction over j. Indeed, for small values of x_1, we know that

$$\sum_{i=0}^{+\infty} c_{i,0}\, x_1^i = u(x_1, 0) = \frac{1}{1-x_1} = \sum_{i=0}^{+\infty} x_1^i,$$

showing that $c_{i,0} = 1$, thus confirming (9.1.8) when $j = 0$.

Suppose now that (9.1.8) is true for the index j. Let us prove it for the index $j+1$. To this end, we observe that

$$\begin{aligned}
\sum_{i,j=0}^{+\infty} (i+2)(i+1)c_{i+2,j}\, x_1^i x_2^j &= \sum_{i,j=0}^{+\infty} i(i-1)c_{i,j}\, x_1^{i-2} x_2^j = \partial_{x_1 x_1} u(x) \\
&= \partial_{x_2} u(x) = \sum_{i,j=0}^{+\infty} j c_{i,j}\, x_1^i x_2^{j-1} \\
&= \sum_{i,j=0}^{+\infty} (j+1)c_{i,j+1}\, x_1^i x_2^j,
\end{aligned}$$

showing that $(i+2)(i+1)c_{i+2,j} = (j+1)c_{i,j+1}$.

The inductive hypothesis thereby entails that

$$c_{i,j+1} = \frac{(i+2)(i+1)c_{i+2,j}}{(j+1)} = \frac{(i+2)(i+1)\prod_{k=1}^{2j}(i+2+k)}{(j+1)\, j!} = \frac{\prod_{k=1}^{2(j+1)}(i+k)}{(j+1)!},$$

which completes the proof of (9.1.8).

From the above, we infer that

$$c_{0,j} = \frac{\prod_{k=1}^{2j} k}{j!} = \frac{(2j)!}{j!},$$

and accordingly, for small x_2,

$$u(0, x_2) = \sum_{j=0}^{+\infty} c_{0,j}\, x_2^j = \sum_{j=0}^{+\infty} \frac{(2j)!}{j!}\, x_2^j,$$

which is a divergent series, and the proof of (9.1.7) is complete. □

This is another classical example due to[5] Hans Lewy (see [Lew57]). The example is in the complex setting: in this framework, we denote by $i := \sqrt{-1}$ the imaginary unit (no confusion should arise when, in these pages, we use i as an index). Also, variables in $\mathbb{C}$ correspond to $x + iy$, often identified with $(x, y) \in \mathbb{R}^2$, and an additional real variable is denoted by $t \in \mathbb{R}$.

Theorem 9.1.2. *Let $f \in C^\infty(\mathbb{R}, \mathbb{R})$. Assume that f is not analytic at the origin. Then, there cannot be a smooth, complex-valued solution $u = u(x, y, t)$ to*

$$\partial_x u + i\partial_y u - 2i(x + iy)\partial_t u = f(t) \tag{9.1.9}$$

in any neighborhood of the origin.

Proof. Suppose by contradiction that such a solution u exists in a small neighborhood of the origin. Let C_r be a circle of radius $r > 0$ in the coordinates $(x, y) \in \mathbb{R}^2$, traveled counterclockwise.

Then, for small r and $|t|$, we can integrate (9.1.9) over $(x, y) \in C_r$ (for a given t) and obtain

$$2\pi r f(t) = \oint_{C_r} \Big(\partial_x u + i\partial_y u - 2i(x + iy)\partial_t u\Big). \tag{9.1.10}$$

We introduce polar coordinates (ρ, θ) in the (x, y) variables (in the vicinity of C_r) by considering the function

$$u_0(\rho, \theta, t) := u(\rho\cos\theta, \rho\sin\theta, t).$$

We also look at the quantity $\sigma := \ln\rho$ and the corresponding function

$$v(\sigma, \theta, t) := u_0(e^\sigma, \theta, t) = u(e^\sigma\cos\theta, e^\sigma\sin\theta, t).$$

Note that

$$\lim_{\rho\searrow 0} v(\ln\rho, \theta, t) = \lim_{\rho\searrow 0} u(\rho\cos\theta, \rho\sin\theta, t) = u(0, 0, t). \tag{9.1.11}$$

[5]Lewy also had an interesting life. After having studied mathematics and physics at Göttingen and having worked extensively on partial differential equations, also in collaboration with Richard Courant, he traveled to Rome to study algebraic geometry with Tullio Levi-Civita and Federigo Enriques. He escaped Europe to avoid racial persecutions, and after a short-term position at Brown University, he moved to the University of California, Berkeley, and devoted himself to partial differential equations and complex analysis.

After 15 years, Lewy was fired from Berkeley for refusing to sign a loyalty oath imposed on the faculty by the University of California's Board of Regents (he was later reinstated to his job by the California Supreme Court).

See Figure 9.1.3 for a picture of Lewy.

Figure 9.1.3 Hans Lewy.
Source: Photo by George M. Bergman, image from Wikipedia, licensed under the GNU Free Documentation License, Version 1.2.

We remark that

$$\begin{aligned}\frac{\partial v}{\partial\sigma}+i\frac{\partial v}{\partial\theta}&=e^{\sigma}\cos\theta\,\frac{\partial u}{\partial x}+e^{\sigma}\sin\theta\,\frac{\partial u}{\partial y}-ie^{\sigma}\sin\theta\,\frac{\partial u}{\partial x}+ie^{\sigma}\cos\theta\,\frac{\partial u}{\partial y}\\&=e^{\sigma}(\cos\theta-i\sin\theta)\frac{\partial u}{\partial x}+ie^{\sigma}(\cos\theta-i\sin\theta)\frac{\partial u}{\partial y}\\&=e^{\sigma}(\cos\theta-i\sin\theta)\left(\frac{\partial u}{\partial x}+i\frac{\partial u}{\partial y}\right)\\&=e^{\sigma-i\theta}\left(\frac{\partial u}{\partial x}+i\frac{\partial u}{\partial y}\right),\end{aligned}$$

where the derivatives of v are computed at (σ,θ,t) and those of u at $(e^{\sigma}\cos\theta,e^{\sigma}\sin\theta,t)$.

As a consequence,

$$\begin{aligned}&\frac{\partial u}{\partial x}(e^{\sigma}\cos\theta,e^{\sigma}\sin\theta,t)+i\frac{\partial u}{\partial y}(e^{\sigma}\cos\theta,e^{\sigma}\sin\theta,t)\\&\qquad=e^{-\sigma+i\theta}\left(\frac{\partial v}{\partial\sigma}(\sigma,\theta,t)+i\frac{\partial v}{\partial\theta}(\sigma,\theta,t)\right),\end{aligned}$$

and accordingly,

$$\begin{aligned}\oint_{C_r}\left(\partial_x u+i\partial_y u\right)&=r\int_0^{2\pi}\left(\partial_x u(r\cos\theta,r\sin\theta,t)+i\partial_y u(r\cos\theta,r\sin\theta,t)\right)d\theta\\&=\int_0^{2\pi}e^{i\theta}\left(\partial_\sigma v(\ln r,\theta,t)+i\partial_\theta v(\ln r,\theta,t)\right)d\theta.\end{aligned}$$

Also,

$$\partial_\theta\Big(ie^{i\theta}v(\ln r,\theta,t)\Big)+e^{i\theta}v(\ln r,\theta,t)=ie^{i\theta}\partial_\theta v(\ln r,\theta,t);$$

therefore, by the periodicity of v in the variable θ,

$$\int_0^{2\pi} e^{i\theta}v(\ln r,\theta,t)\,d\theta=i\int_0^{2\pi} e^{i\theta}\partial_\theta v(\ln r,\theta,t)\,d\theta.$$

These observations lead to

$$\oint_{C_r}\Big(\partial_x u+i\partial_y u\Big)=\int_0^{2\pi} e^{i\theta}\Big(\partial_\sigma v(\ln r,\theta,t)+v(\ln r,\theta,t)\Big)\,d\theta.$$

Combining this with (9.1.10), we infer that

$$\begin{aligned}2\pi r f(t)=&\int_0^{2\pi} e^{i\theta}\Big(\partial_\sigma v(\ln r,\theta,t)+v(\ln r,\theta,t)\Big)\,d\theta\\&-2ir^2\int_0^{2\pi} e^{i\theta}\partial_t v(\ln r,\theta,t)\,d\theta.\end{aligned}\tag{9.1.12}$$

Now, we define (for small $|t|$)

$$F(t):=\pi\int_0^t f(\tau)\,d\tau$$

and (for small $Y>0$ and small $|X|$)

$$V(X,Y):=i\int_0^{2\pi} e^{i\theta}\sqrt{Y}\,v(\ln\sqrt{Y},\theta,X)\,d\theta+F(X).$$

We stress that the image of f (and therefore of F) is supposed to be real; therefore, the imaginary part of the complex-valued function V satisfies

$$\Im V(X,Y)=\Im\left(i\int_0^{2\pi} e^{i\theta}\sqrt{Y}\,v(\ln\sqrt{Y},\theta,X)\,d\theta\right).$$

As a result, recalling (9.1.11), we see that

$$\lim_{Y\searrow 0}\Im V(X,Y)=0\tag{9.1.13}$$

Our goal is to deduce from (9.1.12) that

$$V \text{ satisfies the Cauchy–Riemann equations in the variables } (X,Y).\tag{9.1.14}$$

To this end, we calculate that

$$\begin{aligned}
&\partial_X V + i\partial_Y V\\
&= i\int_0^{2\pi} e^{i\theta}\sqrt{Y}\,\partial_t v(\ln\sqrt{Y},\theta,X)\,d\theta + \pi f(X) - \partial_Y\int_0^{2\pi} e^{i\theta}\sqrt{Y}\,v(\ln\sqrt{Y},\theta,X)\,d\theta\\
&= i\int_0^{2\pi} e^{i\theta}\sqrt{Y}\,\partial_t v(\ln\sqrt{Y},\theta,X)\,d\theta + \pi f(X) - \int_0^{2\pi} e^{i\theta}\frac{1}{2\sqrt{Y}}\,v(\ln\sqrt{Y},\theta,X)\,d\theta\\
&\quad - \int_0^{2\pi} e^{i\theta}\frac{1}{2\sqrt{Y}}\,\partial_\sigma v(\ln\sqrt{Y},\theta,X)\,d\theta\\
&= 0,
\end{aligned}$$

which establishes (9.1.14).

Thanks to (9.1.13) and (9.1.14), we can apply the Schwarz reflection principle (see e.g. [Rud66, Theorem 11.14]) and conclude that V can be extended to a holomorphic function (still denoted by V for simplicity) in a neighborhood of the origin. In this way, we can think of $V = V(X,Y)$ as a holomorphic function in $(-r_0,r_0)^2$ for some $r_0 > 0$ small enough.

As a byproduct, the function $V(X,0) = F(X)$ is real analytic for $X \in (-r_0,r_0)^2$. But then, $F' = \pi f$ is also real analytic for $X \in (-r_0,r_0)^2$, against our assumptions. □

From Theorem 9.1.2, one immediately obtains the following interesting example for a system of two real equations.

Corollary 9.1.3. *Let $f \in C^\infty(\mathbb{R},\mathbb{R})$. Assume that f is not analytic at the origin.*

Then, there cannot be a smooth solution $(u_1(x,y,t), u_2(x,y,t))$, with u_1 and u_2 real valued, to

$$\begin{cases} \partial_x u_1 - \partial_y u_2 + 2y\partial_t u_1 + 2x\partial_t u_2 = f(t),\\ \partial_x u_2 + i\partial_y u_1 - 2x\partial_t u_1 + 2y\partial_t u_2 = 0 \end{cases}$$

in any neighborhood of the origin.

Proof. If such a solution existed, the complex-valued function $u := u_1 + iu_2$ would be a solution to (9.1.9), in contradiction with Theorem 9.1.2. □

The following variation of the example in Theorem 9.1.2 is inspired by an example comprising the case of a single real equation, which was put forth by François Trèves (see [Tre62]):

Theorem 9.1.4. *Let $f \in C^\infty(\mathbb{R})$. Assume that f is not analytic at the origin. Then, there cannot be a smooth solution $u = u(x, y, t)$ of*

$$\left(\partial_x^2 + \partial_y^2 + 4(x^2 + y^2)\partial_t^2 + 4y\partial_{xt} - 4x\partial_{yt}\right)^2 u + 16\,\partial_t^2 u = f(t) \tag{9.1.15}$$

in any neighborhood of the origin.

Proof. Suppose by contradiction that such a solution exists. We denote by $\mathcal{L}_0$ the operator on the left-hand side of (9.1.15), thus writing $\mathcal{L}_0 u = f$.

Let also

$$\mathcal{L} := \partial_x + i\partial_y - 2i(x + iy)\partial_t.$$

We observe that $\mathcal{L}$ is the operator dealt with in Theorem 9.1.2. Given a differential operator with complex coefficients, we denote with a superscript "$\star$" the operator obtained by replacing the coefficients by their complex conjugates. For example,

$$\mathcal{L}^\star = \partial_x - i\partial_y + 2i(x - iy)\partial_t.$$

It is worth pointing out that

$$\mathcal{L} = \mathcal{L}_1 + i\mathcal{L}_2 \quad \text{and} \quad \mathcal{L}^\star = \mathcal{L}_1 - i\mathcal{L}_2,$$

where

$$\mathcal{L}_1 = \partial_x + 2y\partial_t \quad \text{and} \quad \mathcal{L}_2 = \partial_y - 2x\partial_t.$$

We also observe that

$$\mathcal{M} := \mathcal{L}\mathcal{L}^\star = \left(\mathcal{L}_1 + i\mathcal{L}_2\right)\left(\mathcal{L}_1 - i\mathcal{L}_2\right) = \mathcal{L}_1^2 + \mathcal{L}_2^2 + i\left(\mathcal{L}_2\mathcal{L}_1 - \mathcal{L}_1\mathcal{L}_2\right);$$

therefore, since the coefficients of $\mathcal{L}_1$ and $\mathcal{L}_2$ are real,

$$\mathcal{M}^\star = \mathcal{L}_1^2 + \mathcal{L}_2^2 - i\left(\mathcal{L}_2\mathcal{L}_1 - \mathcal{L}_1\mathcal{L}_2\right).$$

As a result, setting

$$\mathcal{L}_3 := \mathcal{L}_1^2 + \mathcal{L}_2^2 \quad \text{and} \quad \mathcal{L}_4 := \mathcal{L}_2\mathcal{L}_1 - \mathcal{L}_1\mathcal{L}_2,$$

we find that

$$\mathcal{M} = \mathcal{L}_3 + i\mathcal{L}_4 \quad \text{and} \quad \mathcal{M}^\star = \mathcal{L}_3 - i\mathcal{L}_4.$$

Furthermore,

$$\mathcal{L}_4 = \left(\partial_y - 2x\partial_t\right)\left(\partial_x + 2y\partial_t\right) - \left(\partial_x + 2y\partial_t\right)\left(\partial_y - 2x\partial_t\right) = 4\partial_t.$$

Consequently, since both $\mathcal{L}_1$ and $\mathcal{L}_2$ commute with ∂_t, $\mathcal{L}_3$ also commutes with ∂_t and, for this reason,

$$\mathcal{M}\mathcal{M}^\star = \left(\mathcal{L}_3 + i\mathcal{L}_4\right)\left(\mathcal{L}_3 - i\mathcal{L}_4\right) = \left(\mathcal{L}_3 + 4i\partial_t\right)\left(\mathcal{L}_3 - 4i\partial_t\right) = \mathcal{L}_3^2 + 16\,\partial_t^2.$$

Additionally,

$$\mathcal{L}_3 = \left(\partial_x + 2y\partial_t\right)^2 + \left(\partial_y - 2x\partial_t\right)^2 = \partial_x^2 + \partial_y^2 + 4(x^2+y^2)\partial_t^2 + 4y\partial_{xt} - 4x\partial_{yt},$$

leading to

$$\mathcal{M}\mathcal{M}^\star = \left(\partial_x^2 + \partial_y^2 + 4(x^2+y^2)\partial_t^2 + 4y\partial_{xt} - 4x\partial_{yt}\right)^2 + 16\,\partial_t^2 = \mathcal{L}_0.$$

This gives that, defining $v := \mathcal{L}^\star\mathcal{M}^\star u$,

$$f = \mathcal{L}_0 u = \mathcal{M}\mathcal{M}^\star u = \mathcal{L}\mathcal{L}^\star\mathcal{M}^\star u = \mathcal{L}v,$$

but this is in contradiction with Theorem 9.1.2. □

Now, we introduce one of the fundamental notions of the local existence theory, namely the concept of noncharacteristic hypersurface. In rough terms, we say that a hypersurface is noncharacteristic for the Cauchy problem (9.1.5) if the knowledge of the normal derivatives of order up to $m-1$ along Γ combined with the equation involving the derivatives of order up to m allows one to determine uniquely all the derivatives of order up to m along Γ. A more precise definition is as follows: for all $\ell \in \{1,\dots,N\}$ and $\alpha \in \mathbb{N}^n$ with $|\alpha| \leqslant \mu_\ell$, we denote by $\xi_{\alpha\ell}$ the variable corresponding to $\partial^\alpha u_\ell$ in the function F used in (9.1.3), i.e. we write

$$F_\ell = F_\ell\left(x, \{\xi_{\alpha j}\}_{\substack{\alpha\in\mathbb{N}^n\\ |\alpha|\leqslant\mu_j\\ j\in\{1,\dots,N\}}}\right). \tag{9.1.16}$$

Recalling the setting in (9.1.1), for every $i, j \in \{1,\dots,N\}$, we define

$$\mathcal{M}_{ij} := \sum_{\substack{\alpha\in\mathbb{N}^n\\ |\alpha|=\mu_j}} \frac{\partial F_i}{\partial \xi_{\alpha j}} \left(\frac{\partial\Phi}{\partial x_1}\right)^{\alpha_1} \cdots \left(\frac{\partial\Phi}{\partial x_n}\right)^{\alpha_n}. \tag{9.1.17}$$

In this framework, we can consider the matrix $\mathcal{M} := \{\mathcal{M}_{ij}\}_{i,j\in\{1,\dots,N\}}$. Note also that

$$\mathcal{M} = \mathcal{M}(x,\xi), \quad \text{where } \xi := \{\xi_{\alpha\ell}\}_{\substack{\alpha\in\mathbb{N}^n\\ |\alpha|\leqslant\mu_\ell\\ \ell\in\{1,\dots,N\}}}.$$

Then, we arrive at the following definition.

Definition 9.1.5. We say that Γ is noncharacteristic for the Cauchy problem (9.1.5) if, when $x \in \Gamma$,

$$\det \mathcal{M} \neq 0.$$

A special, but very interesting, situation occurs when the system of equations depends linearly on the derivatives of higher order (the dependence on lower-order derivatives may be nonlinear). This is called a quasilinear system of partial differential equations, and in this case,

$$F_\ell = \sum_{\substack{\alpha\in\mathbb{N}^n\\ |\alpha|=\mu_j\\ j\in\{1,\dots,N\}}} A_{\alpha\ell j}\left(x, \{\xi_{\beta k}\}_{\substack{\beta\in\mathbb{N}^n\\ |\beta|\leqslant\mu_k-1\\ k\in\{1,\dots,N\}}}\right)\xi_{\alpha j} + G_\ell\left(x, \{\xi_{\beta k}\}_{\substack{\beta\in\mathbb{N}^n\\ |\beta|\leqslant\mu_k-1\\ k\in\{1,\dots,N\}}}\right),$$

for some $A_{\alpha\ell j}$ and G_ℓ.

Accordingly, in the quasilinear case,

$$\mathcal{M} = \sum_{\substack{\alpha\in\mathbb{N}^n\\ |\alpha|=\mu_j}} A_{\alpha\ell j}\left(\frac{\partial\Phi}{\partial x_1}\right)^{\alpha_1}\dots\left(\frac{\partial\Phi}{\partial x_n}\right)^{\alpha_n}.$$

The case of systems of second-order linear elliptic equations (see (4.4.1)) can be written in the form

$$\sum_{k,h=1}^{n} a_{kh}(x)\partial_{kh}u(x) + \sum_{k=1}^{n} b_k(x)\partial_k u(x) + c(x)u(x) - f(x) = 0.$$

In this case, $N = 1$, $\mu_1 = 2$ and

$$F = \sum_{\substack{\alpha\in\mathbb{N}^n\\ |\alpha|=2}} a_\alpha\xi_\alpha + \sum_{\substack{\alpha\in\mathbb{N}^n\\ |\alpha|=1}} b_\alpha\xi_\alpha + c\xi_0 - f,$$

where

$$a_\alpha := \begin{cases} a_{kh} & \text{if } |\alpha| = 2 \text{ and } \alpha = e_k + e_h,\\ a_{kk} & \text{if } |\alpha| = 2 \text{ and } \alpha = 2e_k \end{cases}$$

and $b_\alpha := b_k$ when $|\alpha| = 1$ and $\alpha = e_k$.

In this case, $\mathcal{M}$ is simply a scalar function, namely

$$\begin{aligned}\mathcal{M} &= \sum_{\substack{\alpha\in\mathbb{N}^n\\ |\alpha|=2}} a_\alpha\left(\frac{\partial\Phi}{\partial x_1}\right)^{\alpha_1}\dots\left(\frac{\partial\Phi}{\partial x_n}\right)^{\alpha_n}\\ &= \sum_{1\leqslant k\neq h\leqslant n} a_{kh}\frac{\partial\Phi}{\partial x_k}\frac{\partial\Phi}{\partial x_h} + \sum_{k=1}^{n} a_{kk}\left(\frac{\partial\Phi}{\partial x_k}\right)^2\\ &= \sum_{1\leqslant k,h\leqslant n} a_{kh}\frac{\partial\Phi}{\partial x_k}\frac{\partial\Phi}{\partial x_h}.\end{aligned}$$

In particular, in this case, the ellipticity condition in (2.16.2) guarantees that $\mathcal{M} \geqslant \lambda|\nabla\Phi| > 0$; therefore,

$$\text{second-order linear elliptic equations are always noncharacteristic.} \tag{9.1.18}$$

Having got acquainted with the basic notation, we can now state the local existence and uniqueness result in the analytic class for noncharacteristic problems.

Theorem 9.1.6. *Let* Γ *and* F *be real analytic. Let* $\partial_\nu^j u_\ell$ *also be real analytic for all* $\ell \in \{1, \dots, N\}$ *and* $j \in \{0, \dots, m-1\}$.

Assume that Γ *is noncharacteristic for the Cauchy problem* (9.1.5).

Then, the Cauchy problem (9.1.5) *possesses one and only one solution which is real analytic in a neighborhood of* Γ.

That is, under real analyticity and noncharacteristic assumptions, the Cauchy problem (9.1.5) admits one and only one solution which is real analytic in Ω as long as Ω is "sufficiently small."

We stress that Theorem 9.1.6 always applies to second-order linear elliptic equations with analytic coefficients, source and data, thanks to (9.1.18).

The importance of Theorem 9.1.6 is that properly defined, but very general, initial value problems for partial differential equations end up admitting a solution, at least at a local scale, and such a solution is real analytic.

However, the applicability of Theorem 9.1.6 suffers from certain structural limitations. First of all, it necessitates the stringent requirement of working in the class of real analytic functions (and this setting cannot be, in general, relaxed, given the counterexample in Theorem 9.1.4; see also [Lew57, Har59, Hör60, Miz61, Tre62, GS67, Trè70, MU83, Ego93, PS07a] for further details).

Furthermore, while Theorem 9.1.6 guarantees the existence of a unique real analytic function, it does not preclude the existence of other solutions which are not real analytic. For linear systems, the analytic solution is however the unique one (see [Hol01] and [Hör03, Theorem 5.3.1]), whereas for nonlinear equations in real analytic settings, there exist examples of other, nonreal analytic solutions (see [Mét85, Mét93, Hör00]; see also [Zac66, Hör71] for generalizations and further information on this uniqueness problem).

Also, the solution provided by Theorem 9.1.6 is only of local type (i.e. in a small neighborhood of the hypersurface Γ) and cannot be, in general, taken as a global solution: for instance, even for the case of the Laplacian, we know that overdetermined boundary value problems do not admit a global solution (e.g. unless Γ is a sphere, see Theorem 7.3.1). Another interesting example showing that local solutions do not admit a global extension, not even when Γ is a hyperplane and not even when the equation is first order, is[6] the following.

[6]The example in (9.1.19) provides a general paradigm for shock formation due to the intersection of "characteristic curves," i.e. curves along which the solution maintains a constant value (see Figure 9.1.4 for a sketch of this phenomenon), and coincides, up to renaming variables, with the example discussed for the inviscid Bateman–Burgers equation in [DV23, equation (26.9)].

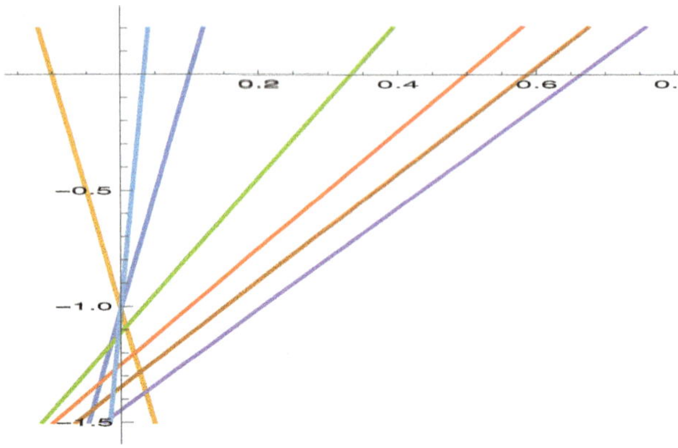

Figure 9.1.4 A shock produced by the intersection of characteristic curves.

Theorem 9.1.7. *The Cauchy problem*

$$\begin{cases} \partial_{x_2} u(x) = u(x)\, \partial_{x_1} u(x) & \text{for all } x = (x_1, x_2) \text{ in a neighborhood of } \mathbb{R} \times \{0\}, \\ u(x_1, 0) = -\dfrac{x_1}{1 + x_1^2} & \text{for all } x_1 \in \mathbb{R} \end{cases} \tag{9.1.19}$$

does not admit a real analytic solution in the whole of $\mathbb{R}^2$.

Proof. Note that the setting of (9.1.19) is real analytic. Also, in this situation, we have that $N := 1$, $n := 2$, $\mu_1 := 1$, $\Gamma = \mathbb{R} \times \{0\}$, $\Phi(x_1, x_2) := x_2$ and $F := \xi_{(0,1)} - \xi_{(0,0)}\xi_{(1,0)}$. Consequently, by (9.1.17), we know that $\mathcal{M}$ is equal to $\frac{\partial F}{\partial \xi_{(0,1)}} = 1 \neq 0$, and hence the assumptions of Theorem 9.1.6 are satisfied.

However, the solution constructed via Theorem 9.1.6 cannot be extended arbitrarily far from $\mathbb{R} \times \{0\}$ because if it were possible to extend it, setting

$$v(t) := u\left(x_1 + \frac{tx_1}{1 + x_1^2}, t\right) + \frac{x_1}{1 + x_1^2} \quad \text{and} \quad a(t) := \partial_{x_1} u\left(x_1 + \frac{tx_1}{1 + x_1^2}, t\right),$$

then $v(0) = 0$ and, in the domain of validity of the equation,

$$\begin{aligned} \dot{v}(t) &= \frac{x_1}{1 + x_1^2}\, \partial_{x_1} u\left(x_1 + \frac{tx_1}{1 + x_1^2}, t\right) + \partial_{x_2} u\left(x_1 + \frac{tx_1}{1 + x_1^2}, t\right) \\ &= \left[\frac{x_1}{1 + x_1^2} + u\left(x_1 + \frac{tx_1}{1 + x_1^2}, t\right)\right] \partial_{x_1} u\left(x_1 + \frac{tx_1}{1 + x_1^2}, t\right) \\ &= v(t)\partial_{x_1} u\left(x_1 + \frac{tx_1}{1 + x_1^2}, t\right) \\ &= a(t)v(t). \end{aligned}$$

The uniqueness result for ordinary differential equations thus entails that v vanishes identically and therefore

$$u\left(x_1+\frac{tx_1}{1+x_1^2},t\right)=-\frac{x_1}{1+x_1^2}. \tag{9.1.20}$$

That is, if the solution could be extended in $(-2\delta,2\delta)\times(-1-2\delta^2,0]$ for some $\delta>0$, then we could pick $x_1:=\delta$ and $t:=-1-\delta^2$ in (9.1.20), finding that

$$u(0,-1-\delta^2)=u\left(\delta+\frac{-(1+\delta^2)\delta}{1+\delta^2},-1-\delta^2\right)=-\frac{\delta}{1+\delta^2}<0.$$

But by picking $x_1:=0$ and $t:=-1-\delta^2$ in (9.1.20), we see that $u(0,-1-\delta^2)=0$, which is a contradiction. □

We also stress that the requirement of the hypersurface Γ to be noncharacteristic cannot in general be dropped from Theorem 9.1.6: indeed, the example in (9.1.6), which does not allow for a local solution, violates the noncharacteristic requirement since in this case, $N:=1$, $n:=2$, $\mu_1:=2$, $\Phi(x_1,x_2):=x_2$ and $F=\xi_{(2,0)}-\xi_{(0,1)}$. In this situation, we deduce from (9.1.17) that $\mathcal{M}$ is equal to $\frac{\partial F}{\partial\xi_{(2,0)}}\left(\frac{\partial\Phi}{\partial x_1}\right)^2=0$, which shows that the noncharacteristic assumption is not satisfied in this case.

Now, we present a proof of Theorem 9.1.6. This proof will leverage Cauchy's original technique, namely the "method of majorants," which Cauchy[7] called *calcul de limites*. The gist of this method is that, on the one hand, one can use formal series expansions to obtain recursive formulas[8] for the coefficients of the solution (if any), and on the other hand, one can estimate these coefficients to

[7]The word "limites" in this terminology used by Cauchy should be understood as "bounds" rather than "limits."

The term "method of majorants" was instead introduced by Poincaré.

[8]The idea of finding a particular solution by recurrence relations dates back to Isaac Newton and is sometimes called the "method of undetermined coefficients." At Newton's time, the common belief was that analytic expressions always make sense, while the precise discussion about convergence came much later, and Cauchy was one of the prominent figures in the introduction of a rigorous and modern approach to calculus and analysis.

See Figure 9.1.5 for a monotype of Newton by the poet and painter William Blake: actually, Blake clearly held a grudge against Newton, who is depicted as sitting naked and crouched at the bottom of the sea, looking down, with his gaze directed only at some blunt geometrical diagrams, so absorbed in his work that he was oblivious to the beautiful rocks around him and unaware of the similarity between his drawing and the arc of his own body as well as the shape of his own eye (thus, Blake seems to suggest that math is incapable of understanding nature and lacks introspection).

About Augustin-Louis Cauchy: he was a very prolific mathematician, maybe second only to Euler, and many theorems and concepts are named after him.

By the way, the scientific publishers at the time may have been flooded by Cauchy's overwhelming productivity: for instance, the journal *Compte Rendus de l'Académie des Sciences* instituted a four-page limit for papers precisely to try to dam Cauchy's exuberance. This page limit was in effect for a long time, though presently the journal seems to have become more flexible, stating in its webpage

Figure 9.1.5 A strapping Isaac Newton as viewed by William Blake.
Source: Public Domain image from Wikipedia.

rigorously establish convergence: to perform this estimate, rather than looking for specific direct bounds, one can construct an explicit "majorant problem" whose relatively simple solution automatically provides the necessary bounds for the formal solution of the original problem.

---

that "there is no page limit for articles submitted to the *Comptes Rendus Mathématique*. However, the maximum recommended length for a manuscript is about 50,000 characters including spaces [...], which is equivalent to about 10 pages."

Cauchy was also a very staunch royalist and Catholic (in fact, a "bigoted Catholic," according to the prominent mathematician Niels Henrik Abel, who however highly appreciated Cauchy's mathematical stature). Cauchy's loyalty to the Bourbons was so firm that he never swore the required oath of allegiance to the new regimes (preferring to go into exile at some point).

Cauchy's cast-iron opinions were broadly known, and they possibly made him unpopular in some academic circles. For instance, a Florentine nobleman called Guglielmo Brutus Icilius Timeleone Libri Carucci dalla Sommaja (known in France simply as Guillaume Libri) was made chair in mathematics at the Collège de France before him. However, the name of this individual is nowadays mostly remembered for the theft of ancient and precious manuscripts: appointed as the Inspector of Libraries in France, Libri fled to England in the company of more than 30,000 precious books. Some of the loot was sold on the English antiques market. Through a couple of auctions, Libri managed to amass over a million francs, in a period when the average daily wage of a worker was about four francs (on top of that, Libri also had the hideous habit of tearing off pages from books that he found of particular interest). Well, *Nomen est Omen*, as "Libri" means "Books" in Italian.

One last curious anecdote about Libri: he is the one credited with popularizing the name of medieval mathematician Leonardo Bonacci, or Leonardo Bigollo Pisano, in the form that is most widespread nowadays, namely "Fibonacci," short for "filius Bonacci," or "son of Bonacci" in Latin (even this is not that original anyway, since the name of Fibonacci was already used to refer to Leonardo Bonacci in documents from the sixteenth century).

See Figure 9.1.6 for the portraits of Cauchy and Libri (unsurprisingly, with Libri in the position of nonchalantly grabbing a book).

Figure 9.1.6 Cauchy and Libri.
Source: Public Domain images from Wikipedia.

To appreciate the method of majorants in action, let us first review Cauchy's approach to ordinary differential equations, by recalling the proof of the local existence and uniqueness results for ordinary differential equations in a real analytic setting.

Theorem 9.1.8. *Let* $(t_0, u_0) \in \mathbb{R} \times \mathbb{R}^N$ *and* $f : \mathbb{R} \times \mathbb{R}^N \to \mathbb{R}^N$. *Assume that* f *is real analytic in a neighborhood of* (t_0, u_0).

Then, the initial value problem

$$\begin{cases} \dot{u}(t) = f(t, u(t)), & \text{for all } t \in (t_0 - \tau, t_0 + \tau), \\ u(t_0) = u_0 \end{cases} \tag{9.1.21}$$

possesses one and only one solution which is real analytic in a neighborhood of t_0.

Proof. Up to replacing $u(t)$ with $u(t + t_0)$, we can suppose that $t_0 = 0$. It is also convenient to define

$$v(t) := (t, u(t)) \quad \text{and} \quad g(v) := (1, f(v)), \tag{9.1.22}$$

thus recasting (9.1.21) into the form

$$\begin{cases} \dot{v}(t) = g(v(t)) & \text{for all } t \in (-\tau, \tau), \\ v(0) = v_0, \end{cases} \tag{9.1.23}$$

where $v_0 := (0, u_0)$.

Actually, up to replacing v with $v - v_0$ and $g(v)$ with $g(v - v_0)$, we can additionally assume that $v_0 = 0$.

Now, we establish the local existence and uniqueness theory of (9.1.23) using the method of majorants. To this end, we let $L := N + 1$, and we formally look for a solution $v = (v_1, \ldots, v_L)$ of (9.1.23), where, for every $k \in \{1, \ldots, L\}$, v_k is in the form

$$v_k(t) := \sum_{j=0}^{+\infty} \vartheta_{k,j}\, t^j, \tag{9.1.24}$$

for suitable $\vartheta_{k,j} \in \mathbb{R}$.

Note that, by the initial condition in (9.1.23),

$$0 = v_k(0) = \vartheta_{k,0}. \tag{9.1.25}$$

We also expand $g = (g_1, \ldots, g_L)$ into its Taylor series near the origin by writing

$$g_k(v) = \sum_{\beta \in \mathbb{N}^L} \mu_{k,\beta}\, v^\beta, \tag{9.1.26}$$

for suitable $\mu_{k,\beta} \in \mathbb{R}$.

Now, we claim that

$$\vartheta_{k,j} = \mathcal{P}_{k,j}\left(\left\{\mu_{k,\beta}\right\}_{\substack{\beta\in\mathbb{N}^L \\ |\beta|\leqslant j-1}}, \left\{\vartheta_{m,i}\right\}_{\substack{1\leqslant m\leqslant L \\ 1\leqslant i\leqslant j-1}} \right), \tag{9.1.27}$$

where $\mathcal{P}_{k,j}$ is a polynomial with *nonnegative coefficients*.

We remark that when $j = 0$, this claim follows from (9.1.25). So, we can suppose that $j \geqslant 1$. Thus, to prove (9.1.27), we plug (9.1.24) and (9.1.26) into the first line of (9.1.23), and we see that

$$\sum_{j=0}^{+\infty} (j+1)\vartheta_{k,j+1}\, t^j = \sum_{j=0}^{+\infty} j\vartheta_{k,j}\, t^{j-1} = \dot{v}_k = g_k(v) = \sum_{\beta\in\mathbb{N}^L} \mu_{k,\beta}\, v^\beta.$$

Hence, since

$$v_k^{\beta_k} = \left(\sum_{j_k=0}^{+\infty} \vartheta_{k,j_k}\, t^{j_k} \right)^{\beta_k} = \sum_{j_{k,1},\ldots,j_{k,\beta_k}=0}^{+\infty} \vartheta_{k,j_{k,1}} \cdots \vartheta_{k,j_{k,\beta_k}}\, t^{j_{k,1}+\cdots+j_{k,\beta_k}},$$

we conclude that

$$\begin{aligned}
&\sum_{j=0}^{+\infty} (j+1)\vartheta_{k,j+1}\, t^j \\
&\quad = \sum_{\substack{\beta\in\mathbb{N}^L \\ j_{1,1},\ldots,j_{1,\beta_1},\ldots,j_{L,1},\ldots,j_{L,\beta_L} \in \mathbb{N}}} \mu_{k,\beta}\, \vartheta_{1,j_{1,1}} \cdots \vartheta_{1,j_{1,\beta_1}} \cdots \vartheta_{L,j_{L,1}} \cdots \vartheta_{L,j_{L,\beta_L}} \\
&\quad\quad \times t^{j_{1,1}+\cdots+j_{1,\beta_1}+\cdots+j_{L,1}+\cdots+j_{L,\beta_L}}.
\end{aligned}$$

From this, we arrive at[9]

$$\begin{aligned}
&(j+1)\vartheta_{k,j+1}\\
&\quad = \sum_{\beta\in\mathbb{N}^L} \mu_{k,\beta}\,\vartheta_{1,j_{1,1}}\dots\vartheta_{1,j_{1,\beta_1}}\dots\vartheta_{L,j_{L,1}}\dots\vartheta_{L,j_{L,\beta_L}},\\
&j_{1,1},\dots,j_{1,\beta_1},\dots,j_{L,1},\dots,j_{L,\beta_L}\in\mathbb{N}\\
&j_{1,1}+\dots+j_{1,\beta_1}+\dots+j_{L,1}+\dots+j_{L,\beta_L}=j
\end{aligned} \tag{9.1.28}$$

for all $j\in\mathbb{N}$.

Hence, to complete the proof of (9.1.27) (with j replaced by $j+1$), we only need to show that, with respect to the indices above,

$$|\beta|\leqslant j \quad\text{and}\quad j_{i,\ell}\leqslant j, \tag{9.1.29}$$

for each $i\in\{1,\dots,L\}$ and $\ell\in\{1,\dots,\beta_i\}$.

The second inequality in (9.1.29) is in fact a consequence of the fact that, in (9.1.28),

$$j_{i,\ell}\leqslant j_{1,1}+\dots+j_{1,\beta_1}+\dots+j_{L,1}+\dots+j_{L,\beta_L}=j.$$

As for the first inequality in (9.1.29), we observe that the indices in (9.1.28) also satisfy $j_{i,\ell}\geqslant 1$ (otherwise, the corresponding term in the sum vanishes due to (9.1.25)) and therefore

$$|\beta|=\beta_1+\dots+\beta_L\leqslant j_{1,1}+\dots+j_{1,\beta_1}+\dots+j_{L,1}+\dots+j_{L,\beta_L}=j.$$

This establishes (9.1.29), and thus (9.1.27), as desired.

We stress that (9.1.27) determines all the coefficients $\vartheta_{k,j}$ uniquely (i.e. by arguing recursively, starting from (9.1.25)). This entails that, if a real analytic solution of (9.1.23) exists, it is uniquely determined by (9.1.27).

Having established this uniqueness result for (9.1.23) and thus for (9.1.21), we now aim at proving the existence result claimed in Theorem 9.1.8. To this end, we show that the formal series in (9.1.24) is indeed convergent. This task is accomplished via the method of majorants. For this, we first use (9.1.25) and (9.1.27) to see that

$$\vartheta_{k,j}=Q_{k,j}\left(\left\{\mu_{k,\beta}\right\}_{\substack{\beta\in\mathbb{N}^L\\|\beta|\leqslant j-1}}\right), \tag{9.1.30}$$

where $Q_{k,j}$ is a polynomial with *nonnegative coefficients*.

[9]Note that, when $j=0$, the only term on the right-hand side of (9.1.28) that is not zero is given by $\beta=(0,\dots,0)$, thanks to (9.1.25). Hence, when $j=0$, we have that $\vartheta_{k,1}=\mu_{k,0}$.

As a consequence,

$$|\vartheta_{k,j}| \leqslant Q_{k,j}\left(\left\{|\mu_{k,\beta}|\right\}_{\substack{\beta\in\mathbb{N}^L\\ |\beta|\leqslant j-1}}\right). \tag{9.1.31}$$

We stress that this inequality is a consequence of the fact that the coefficients of $Q_{k,j}$ are nonnegative. It is interesting to observe how this monotonicity property will come in handy by allowing us to avoid the explosion of combinatorial calculations and reduce ourselves to an explicit solution which will bound all the terms $\vartheta_{k,j}$.

To this end, we employ the analyticity of g_k in (9.1.26), and we deduce (see e.g. [KP02, Corollary 1.1.7]) that

$$|\mu_{k,\beta}| \leqslant \frac{C}{R^{|\beta|}}, \tag{9.1.32}$$

for some $C, R > 0$.

To obtain a more manageable problem (see the forthcoming footnote 10), it is convenient to observe that, for every $\omega = (\omega_1, \dots, \omega_K) \in \mathbb{N}^K$,

$$\frac{(\omega_1 + \cdots + \omega_K)!}{\omega_1! \cdots \omega_K!} \geqslant 1. \tag{9.1.33}$$

This inequality can be proved by induction over K. When $K = 1$, it is actually an identity. When $K = 2$, we can check it by setting $\varpi := \omega_1 + \omega_2$ and noting that $\frac{(\omega_1+\omega_2)!}{\omega_1!\omega_2!} = \frac{\varpi!}{\omega_1!(\varpi-\omega_1)!} = \binom{\varpi}{\omega_1} \geqslant 1$.

To perform the inductive step, we assume (9.1.33) for the index K, and we prove it for $K + 1$ through the following calculation:

$$\begin{aligned}\frac{(\omega_1 + \cdots + \omega_{K+1})!}{\omega_1! \dots \omega_{K+1}!} &= \frac{(\omega_1 + \cdots + \omega_{K-1} + (\omega_K + \omega_{K+1}))!}{\omega_1! \dots \omega_{K-1}!\omega_K!\omega_{K+1}!}\\ &\geqslant \frac{\omega_1! \dots !\omega_{K-1}!(\omega_K + \omega_{K+1})!}{\omega_1! \dots \omega_{K-1}!\omega_K!\omega_{K+1}!} = \frac{(\omega_K + \omega_{K+1})!}{\omega_K!\omega_{K+1}!} \geqslant 1.\end{aligned}$$

The proof of (9.1.33) is thus complete.

Therefore, owing to (9.1.32) and (9.1.33),

$$|\mu_{k,\beta}| \leqslant \frac{C\,|\beta|!}{\beta!\,R^{|\beta|}} =: \mu^\star_{k,\beta}, \tag{9.1.34}$$

and we can thereby majorize once again (9.1.31), finding that

$$|\vartheta_{k,j}| \leqslant Q_{k,j}\left(\left\{\mu^\star_{k,\beta}\right\}_{\substack{\beta\in\mathbb{N}^L\\ |\beta|\leqslant j-1}}\right). \tag{9.1.35}$$

We now cook up a majorization problem for (9.1.23) by defining

$$g_k^\star(v) := \sum_{\beta\in\mathbb{N}^L} \mu_{k,\beta}^\star\, v^\beta \tag{9.1.36}$$

and looking for a solution to the system of ordinary differential equations

$$\begin{cases} \dot v^\star(t) = g^\star(v^\star(t)) & \text{for all } t\in(-\tau,\tau), \\ v^\star(0) = 0. \end{cases} \tag{9.1.37}$$

If we seek a real analytic solution in the form

$$v_k^\star(t) := \sum_{j=0}^{+\infty} \vartheta_{k,j}^\star\, t^j, \tag{9.1.38}$$

we find that all the coefficients $\vartheta_{k,j}^\star$ are determined by the prescription in (9.1.30), whence

$$\vartheta_{k,j}^\star = Q_{k,j}\left(\left\{\mu_{k,\beta}^\star\right\}_{\substack{\beta\in\mathbb{N}^L\\ |\beta|\leqslant j-1}}\right).$$

Combining this and (9.1.35), it follows that

$$|\vartheta_{k,j}| \leqslant \vartheta_{k,j}^\star. \tag{9.1.39}$$

Thus, to complete the proof of Theorem 9.1.8 it suffices to show that a real analytic solution to the majorization problem in (9.1.37) exists. Indeed, by the uniqueness result that we have already proved, this solution necessarily coincides with the formal power series in (9.1.38). Furthermore, the analyticity of this solution entails (see e.g. [KP02, Corollary 1.1.7]) that $|\vartheta_{k,j}^\star| \leqslant \frac{C_\star}{\rho^j}$ for some $C_\star$, $\rho > 0$; therefore, in view of (9.1.39), we have that $|\vartheta_{k,j}| \leqslant \frac{C_\star}{\rho^j}$, which in turn shows that the formal series in (9.1.24) is also convergent, thus providing the desired solution of (9.1.23). (This is the core of the method of majorants in action!)

The advantage of the majorization problem in (9.1.37) with respect to the original problem in (9.1.23) is that the majorization problem is explicitly solvable (which would complete the proof of Theorem 9.1.8, as we have just discussed). To see this, we remark that

$$g_k^\star(v) = \sum_{\beta\in\mathbb{N}^L} \frac{C\,|\beta|!}{\beta!\,R^{|\beta|}}\, v^\beta.$$

We also use the notation

$$\zeta := \frac{v_1+\cdots+v_L}{R}.$$

Since, according to the multinomial theorem, for every $\kappa \in \mathbb{N}$ we have that

$$\zeta^\kappa = \sum_{\substack{\beta\in\mathbb{N}^L \\ \beta_1+\dots+\beta_L=\kappa}} \frac{\kappa!}{\beta!\,R^\kappa} v_1^{\beta_1}\dots v_L^{\beta_L},$$

we find that

$$g_k^\star(v) = C\sum_{\kappa\in\mathbb{N}}\sum_{\substack{\beta\in\mathbb{N}^L \\ |\beta|=\kappa}} \frac{\kappa!}{\beta!\,R^\kappa} v^\beta = C\sum_{\kappa\in\mathbb{N}}\zeta^\kappa = \frac{C}{1-\zeta} = \frac{CR}{R-(v_1+\dots+v_L)}. \tag{9.1.40}$$

As a result, the majorization problem in (9.1.37) becomes

$$\begin{cases} \dot v^\star(t) = \dfrac{CR}{R-(v_1+\dots+v_L)} & \text{for all } t\in(-\tau,\tau), \\ v^\star(0)=0. \end{cases}$$

Since the role played by each $v_k^\star$ is the same in this system, it is conceivable to look for solutions in which all the entries are the same, i.e. we seek solutions in the form

$$v_1^\star(t) = \dots = v_L^\star(t) = V(t),$$

for some scalar function V, thus reducing the problem to

$$\begin{cases} \dot V(t) = \dfrac{CR}{R-LV} & \text{for all } t\in(-\tau,\tau), \\ V(0)=0. \end{cases} \tag{9.1.41}$$

This scalar equation can now be solved explicitly[10] using the method of separation of variables, leading to, for t close to 0,

$$\begin{aligned} RV(t) - \frac{LV^2(t)}{2} &= \int_0^t \frac{d}{ds}\left(RV(s) - \frac{LV^2(s)}{2}\right) ds \\ &= \int_0^s (R-LV(s))\dot V(s)\,ds = CRt. \end{aligned}$$

This is a quadratic equation for $V(t)$, which can be solved (assuming that $V(0)=0$), thus obtaining that

$$V(t) = \frac{R-\sqrt{R(R-2CLt)}}{L}. \tag{9.1.42}$$

In this way, we have found explicitly a solution to (9.1.41), which happens to be real analytic for t in a small neighborhood of the origin. The proof of Theorem 9.1.8 is thereby complete. □

[10]The possibility of solving this problem explicitly was indeed motivating the step of the proof between (9.1.32) and (9.1.34).

Having reviewed our knowledge of real analytic ordinary differential equations, we are now ready to dive into the technical details of the proof of Theorem 9.1.6. In a nutshell, we first perform some reductions of the Cauchy problem in (9.1.5) that exploit the noncharacteristic assumption described in Definition 9.1.5 (here, we utilize the invariance of the problem under real analytic transformations that simplify the geometry under consideration), and then we take advantage of the real analyticity of the problem.

Proof of Theorem 9.1.6. First of all, we utilize the noncharacteristic assumption and the implicit function theorem to write the system of equations in terms of the derivatives of the highest order (this is actually a special case of a quasilinear system). To this end, we focus our local analysis in a small neighborhood of a point $p \in \Gamma$, and up to changing the order of the spatial coordinates, we can assume that the hypersurface Γ is locally described as a graph in the nth direction of a real analytic function $\Psi : \mathbb{R}^{n-1} \to \mathbb{R}$, i.e. $\Phi(x', x_n) = x_n - \Psi(x')$, with $\Psi(p') = p_n$.

Thus, up to a sign change, the normal vector can be written in the form

$$\nu(x') := -\frac{\nabla\Phi(x', x_n)}{|\nabla\Phi(x', x_n)|} = \frac{(\nabla\Psi(x'), -1)}{\sqrt{1 + |\nabla\Psi(x')|^2}}. \tag{9.1.43}$$

With a slight abuse of notation, we also identify normal and tangent vectors on Γ, with their extensions as unit vector fields in a neighborhood of Γ by normal projection (recall (2.3.3)).

Moreover, we adopt an alternative notation for the variables of F corresponding to the derivatives of the highest order. Namely, for every $j \in \{1, \ldots, N\}$ and every $\alpha \in \mathbb{N}^n$ with $|\alpha| = \mu_j$,

$$\begin{aligned} &\text{we identify } \xi_{\alpha j} \text{ with } \zeta_{i_1,\ldots,i_{\mu_j},j}, \text{ with } 1 \leqslant i_1 \leqslant \cdots \leqslant i_{\mu_j} \leqslant n, \\ &\quad \text{where } \alpha_k \text{ corresponds to the number of elements} \\ &\quad \text{of } \{i_1, \ldots, i_{\mu_j}\} \text{ which are equal to } k. \end{aligned} \tag{9.1.44}$$

In terms of derivatives, this corresponds to the identification of $\partial^\alpha = \partial_{x_1}^{\alpha_1} \cdots \partial_{x_n}^{\alpha_n}$ with $\partial_{i_1} \cdots \partial_{i_{\mu_j}}$.

It is also convenient to look at dummy variables $\zeta_{i_1,\ldots,i_{\mu_j},j}$ with $1 \leqslant i_1, \ldots, i_{\mu_j} \leqslant n$: for these variables, we do not require anymore that the indices are ordered; we simply identify $\zeta_{i_1,\ldots,i_{\mu_j},j}$ with $\zeta_{i'_1,\ldots,i'_{\mu_j},j}$ whenever the indices $i'_1, \ldots, i'_{\mu_j}$ constitute a permutation of $i_1, \ldots, i_{\mu_j}$ (for instance, with this notation, if $j := 1$ and $\mu_1 := 3$, we have that $\zeta_{2,2,1,1} = \zeta_{1,2,2,1}$).

Additionally, we consider a real analytic orthonormal vector frame $\{\tau_1, \ldots, \tau_n\}$ such that $\tau_n = \nu$ (i.e. an orthonormal frame having ν as its last vector). As usual, for each $i \in \{1, \dots, n\}$, we write $\tau_i = (\tau_{i,1}, \dots, \tau_{i,n}) \in \mathbb{R}^n$. The orthonormality condition thus gives[11] that, for all $m, k \in \{1, \dots, n\}$,

$$\sum_{i=1}^{n} \tau_{i,m}\tau_{i,k} = \delta_{m,k}. \tag{9.1.45}$$

For every $j \in \{1, \dots, N\}$, we consider the change of variable

$$\overline{\zeta}_{i_1,\dots,i_{\mu_j},j} := \sum_{1 \leqslant k_1,\dots,k_{\mu_j} \leqslant n} \zeta_{k_1,\dots,k_{\mu_j},j}\ \tau_{i_1,k_1} \dots \tau_{i_{\mu_j},k_{\mu_j}}. \tag{9.1.46}$$

The interest in this relation arises from

$$\begin{aligned} \overline{\zeta}_{n,\dots,n,j} &= \sum_{1 \leqslant k_1,\dots,k_{\mu_j} \leqslant n} \zeta_{k_1,\dots,k_{\mu_j},j}\ \tau_{n,k_1} \dots \tau_{n,k_{\mu_j}} \\ &= \sum_{1 \leqslant k_1,\dots,k_{\mu_j} \leqslant n} \zeta_{k_1,\dots,k_{\mu_j},j}\ \nu_{k_1} \dots \nu_{k_{\mu_j}}. \end{aligned}$$

That is, in the notation of (9.1.4),

$$\begin{aligned} &\text{the variable } \overline{\zeta}_{n,\dots,n,j} \text{ corresponds} \\ &\quad \text{to the normal derivative of order } \mu_j, \text{ namely } \partial_\nu^{\mu_j}. \end{aligned} \tag{9.1.47}$$

We also stress that (9.1.46) is indeed an invertible change of variable since, by (9.1.45),

$$\begin{aligned} &\zeta_{m_1,\dots,m_{\mu_j},j} \\ &\quad = \sum_{1 \leqslant k_1,\dots,k_{\mu_j} \leqslant n} \zeta_{k_1,\dots,k_{\mu_j},j}\ \delta_{m_1,k_1} \dots \delta_{m_{\mu_j},k_{\mu_j}} \\ &\quad = \sum_{\substack{1 \leqslant k_1,\dots,k_{\mu_j} \leqslant n \\ 1 \leqslant i_1,\dots,i_{\mu_j} \leqslant n}} \zeta_{k_1,\dots,k_{\mu_j},j}\ \tau_{i_1,k_1} \dots \tau_{i_{\mu_j},k_{\mu_j}}\ \tau_{i_1,m_1} \dots \tau_{i_{\mu_j},m_{\mu_j}} \\ &\quad = \sum_{1 \leqslant i_1,\dots,i_{\mu_j} \leqslant n} \overline{\zeta}_{i_1,\dots,i_{\mu_j},j}\ \tau_{i_1,m_1} \dots \tau_{i_{\mu_j},m_{\mu_j}}. \end{aligned} \tag{9.1.48}$$

---

[11]More explicitly, to obtain (9.1.45), one can consider the matrix P with entries $P_{ij} := \tau_{i,j}$, and we observe that its transpose P^T coincides with its inverse because

$$(PP^T)_{ij} = \sum_{k=1}^{m} \tau_{i,k}\tau_{j,k} = \tau_i \cdot \tau_j = \delta_{i,j}.$$

As a result,

$$\delta_{i,j} = (P^T P)_{ij} = \sum_{k=1}^{n} \tau_{k,i}\tau_{k,j},$$

which is (9.1.45).

An interesting byproduct of (9.1.48) is that

$$\frac{\partial \zeta_{m_1,\dots,m_{\mu_j},j}}{\partial \overline{\zeta}_{n,\dots,n,j}} = \tau_{n,m_1} \dots \tau_{n,m_{\mu_j}} = \nu_{m_1} \dots \nu_{m_{\mu_j}}. \tag{9.1.49}$$

With these observations, we can define $\overline{F}$ as F in the new set of coordinates given by (9.1.46).

Therefore, recalling (9.1.47), in order to make it explicit the equation under consideration in terms of the highest order normal derivatives, we need to use the implicit function theorem on the function $\overline{F}$ with respect to the set of variables $(\overline{\zeta}_{n,\dots,n,1}, \dots, \overline{\zeta}_{n,\dots,n,N})$. To this end, in view of (9.1.49), we observe that, for all $i, j \in \{1, \dots, N\}$,

$$\begin{aligned}\frac{\partial \overline{F}_i}{\partial \overline{\zeta}_{n,\dots,n,j}} &= \sum_{1 \leqslant m_1,\dots,m_{\mu_j} \leqslant n} \frac{\partial F_i}{\partial \zeta_{m_1,\dots,m_{\mu_j},j}} \frac{\partial \zeta_{m_1,\dots,m_{\mu_j},j}}{\partial \overline{\zeta}_{n,\dots,n,j}} \\ &= \sum_{1 \leqslant m_1,\dots,m_{\mu_j} \leqslant n} \frac{\partial F_i}{\partial \zeta_{m_1,\dots,m_{\mu_j},j}} \nu_{m_1} \dots \nu_{m_{\mu_j}}.\end{aligned}$$

Hence, recalling (9.1.44),

$$\frac{\partial \overline{F}_i}{\partial \overline{\zeta}_{n,\dots,n,j}} = \sum_{\substack{\alpha \in \mathbb{N}^n \\ |\alpha| = \mu_j}} \frac{\partial F_i}{\partial \xi_{\alpha j}} \nu_1^{\alpha_1} \dots \nu_n^{\alpha_n}.$$

As a result, in light of (9.1.17), Definition 9.1.5 and (9.1.43), we infer that the matrix $\left\{ \frac{\partial \overline{F}_i}{\partial \overline{\zeta}_{n,\dots,n,j}} \right\}_{i,j \in \{1,\dots,N\}}$ is invertible; consequently, according to the real analytic version of the implicit function theorem (see e.g. [KP02]), the system of equations $F(x, D_\star^\mu u(x)) = 0$ is equivalent to

$$\partial_\nu^{\mu_\ell} u_\ell = G_\ell, \tag{9.1.50}$$

for $\ell \in \{1, \dots, N\}$, where G_ℓ is a real analytic function of x and of the derivatives of $(u_1, \dots, u_N)$ in a tangential/normal frame of reference, being the order of the derivatives of u_m less than or equal to μ_m and the order of the derivatives of u_m in the normal direction less than or equal to $\mu_m - 1$. Recall also that (9.1.50) is complemented with the boundary conditions in (9.1.5).

Now, we want to reduce to the case in which Γ is a hyperplane. For this, we aim at straightening Γ by using a new coordinate system $(y_1, \dots, y_{n-1}, t)$ in which the coordinates $(y_1, \dots, y_{n-1})$ parameterize Γ and correspond to the hyperplane $\{t = 0\}$ and the variable t identifies displacements in the direction normal to Γ. This reference system is precisely that of the "Fermi coordinates," which were mentioned in [DV23, Chapter 28].

Namely, we consider the change of coordinates near the origin described by

$$x = x(y,t) = (y, \Psi(y)) + t\nu(y). \tag{9.1.51}$$

We remark that

$$\det\left(\frac{\partial x}{\partial(y,t)}\bigg|_{(y,t)=(0,0)} \right) = -\sqrt{1+|\nabla\Psi(0)|^2} \neq 0,$$

guaranteeing that (9.1.51) is indeed a real analytic change of variable in a neighborhood of the origin (see e.g. [KP02] for the inverse function theorem in the real analytic setting).

Therefore, we can consider the function $\bar{u}(y,t) := u(x(y,t))$ and set (9.1.50) and the boundary conditions in (9.1.5) into the equivalent problem

$$\begin{cases} \partial_t^{\mu_\ell} u_\ell = H_\ell\left(y, t, \left\{ \partial_y^\alpha \partial_t^j u_m \right\}_{\substack{m\in\{1,\dots,N\}\\ \alpha\in\mathbb{N}^{n-1}, j\in\mathbb{N}\\ |\alpha|+j\leqslant\mu_m\\ j\leqslant\mu_m-1}} \right) \\ \qquad \text{for all } |y|<\rho,\ |t|<\rho \text{ and } \ell\in\{1,\dots,N\}, \\ \\ \partial_t^j u_\ell(y,0) = \psi_{j,\ell}(y) \\ \qquad \text{for all } |y|<\rho,\ \ell\in\{1,\dots,N\} \text{ and } j\in\{0,\dots,\mu_\ell-1\}, \end{cases} \tag{9.1.52}$$

for some $\rho > 0$ (to be chosen conveniently small) and given real analytic functions H_ℓ and $\psi_{j,\ell}$.

Actually, the boundary conditions in (9.1.52) naturally imply the knowledge of all the derivatives

$$\left\{ \partial_y^\alpha \partial_t^j u_m \right\}_{\substack{m\in\{1,\dots,N\}\\ \alpha\in\mathbb{N}^{n-1}, j\in\mathbb{N}\\ |\alpha|+j\leqslant\mu_m\\ j\leqslant\mu_m-1}}$$

at $t = 0$, simply by differentiating in the direction of y_i, for each $i \in \{1,\dots,n-1\}$: more explicitly,

$$\partial_y^\alpha \partial_t^j u_\ell(y,0) = \partial_y^\alpha \psi_{j,\ell}(y) =: \Psi_{\ell,j,\alpha}(y),$$

for all $|y| < \rho$, $\ell \in \{1,\dots,N\}$ and $j \in \{0,\dots,\mu_\ell - 1\}$.

As a consequence, the Cauchy problem in (9.1.52) is equivalent to

$$\begin{cases} \partial_t^{\mu_\ell} u_\ell = H_\ell\left(y, t, \left\{\partial_y^\alpha \partial_t^j u_m\right\}_{\substack{m\in\{1,\dots,N\}\\ \alpha\in\mathbb{N}^{n-1}, j\in\mathbb{N}\\ |\alpha|+j\leqslant\mu_m\\ j\leqslant\mu_m-1}}\right) \\ \qquad \text{for all } |y|<\rho,\ |t|<\rho \text{ and } \ell\in\{1,\dots,N\}, \\ \\ \partial_y^\alpha \partial_t^j u_\ell(y,0) = \Psi_{\ell,j,\alpha}(y) \\ \qquad \text{for all } |y|<\rho,\ \ell\in\{1,\dots,N\} \text{ and } j\in\{0,\dots,\mu_\ell-1\}, \end{cases} \tag{9.1.53}$$

and we can limit ourselves to consider in the latter equation the indices $\alpha \in \mathbb{N}^{n-1}$ with $|\alpha| \leqslant \mu_\ell$.

As a matter of fact, we can also write

$$\partial_t^{\mu_\ell} u_\ell\Big|_{t=0} = H_\ell\left(y, t, \left\{\partial_y^\alpha \partial_t^j u_m\right\}_{\substack{m\in\{1,\dots,N\}\\ \alpha\in\mathbb{N}^{n-1}, j\in\mathbb{N}\\ |\alpha|+j\leqslant\mu_m\\ j\leqslant\mu_m-1}}\right)\Bigg|_{t=0} =: \Psi_{\ell,\mu_\ell,0}, \tag{9.1.54}$$

and we stress that the latter quantity is determined using the boundary data; therefore, (9.1.53) can also be rewritten in the form

$$\begin{cases} \partial_t^{\mu_\ell} u_\ell = H_\ell\left(y, t, \left\{\partial_y^\alpha \partial_t^j u_m\right\}_{\substack{m\in\{1,\dots,N\}\\ \alpha\in\mathbb{N}^{n-1}, j\in\mathbb{N}\\ |\alpha|+j\leqslant\mu_m\\ j\leqslant\mu_m-1}}\right) \\ \qquad \text{for all } |y|<\rho,\ |t|<\rho \text{ and } \ell\in\{1,\dots,N\}, \\ \\ \partial_y^\alpha \partial_t^j u_\ell(y,0) = \Psi_{\ell,j,\alpha}(y) \\ \qquad \text{for all } |y|<\rho,\ \ell\in\{1,\dots,N\} \text{ and } |\alpha|+j\leqslant\mu_\ell. \end{cases} \tag{9.1.55}$$

One can actually reduce (9.1.55) to a quasilinear system of equations. To this end, we let $a_{\ell,m,j,\alpha}$ be the derivative of H_ℓ with respect to the variable corresponding to $\partial_y^\alpha \partial_t^j u_m$, and we consider the system

$$\begin{cases} \partial_t^{\mu_\ell+1} u_\ell = \partial_t H_\ell + \displaystyle\sum_{\substack{m\in\{1,\dots,N\}\\ \beta\in\mathbb{N}^{n-1},j\in\mathbb{N}\\ |\beta|+j\leqslant\mu_m\\ j\leqslant\mu_m-1}} a_{\ell,m,j,\beta}\,\partial_y^\beta \partial_t^{j+1} u_m \\ \qquad \text{for all } |y|<\rho,\ |t|<\rho \text{ and } \ell\in\{1,\dots,N\}, \\ \\ \partial_y^\alpha \partial_t^j u_\ell(y,0) = \Psi_{\ell,j,\alpha}(y) \\ \qquad \text{for all } |y|<\rho,\ \ell\in\{1,\dots,N\} \text{ and } |\alpha|+j\leqslant\mu_\ell. \end{cases} \tag{9.1.56}$$

Indeed, if (9.1.55) holds true, then we obtain (9.1.56) simply by taking a derivative in t of the first line of (9.1.55).

Conversely, if (9.1.56) holds true, then its first line says that $\partial_t\left(\partial_t^{\mu_\ell} u_\ell - H_\ell\right) = 0$, and accordingly, for all $|t|<\rho$ and $|y|<\rho$, we infer that $\partial_t^{\mu_\ell} u_\ell - H_\ell$ is equal to some function of y only. However, this function is necessarily zero due to (9.1.54) and the boundary condition contained in the second line of (9.1.56).

The equivalence between (9.1.55) and (9.1.56) is thereby established.

Additionally, we can reduce (9.1.56) to a Cauchy problem with homogeneous data. Indeed, up to replacing $u_\ell(y,t)$ with

$$u_\ell(y,t) - \sum_{i=0}^{\mu_\ell} \frac{\Psi_{\ell,i,0}(y)}{i!}\, t^i,$$

and eventually modifying H_ℓ, we reduce (9.1.56) to

$$\begin{cases} \partial_t^{\mu_\ell+1} u_\ell = \partial_t H_\ell + \displaystyle\sum_{\substack{m\in\{1,\dots,N\}\\ \beta\in\mathbb{N}^{n-1},j\in\mathbb{N}\\ |\beta|+j\leqslant\mu_m\\ j\leqslant\mu_m-1}} a_{\ell,m,j,\beta}\,\partial_y^\beta \partial_t^{j+1} u_m \\ \qquad \text{for all } |y|<\rho,\ |t|<\rho \text{ and } \ell\in\{1,\dots,N\}, \\ \\ \partial_y^\alpha \partial_t^j u_\ell(y,0) = 0 \\ \qquad \text{for all } |y|<\rho,\ \ell\in\{1,\dots,N\} \text{ and } |\alpha|+j\leqslant\mu_\ell. \end{cases} \tag{9.1.57}$$

It is now convenient to reduce (9.1.57) to a first-order system. For this, if u_ℓ is as in (9.1.57), whenever $|\alpha|+j\leqslant\mu_\ell$, we introduce the functions

$$v_{\ell,j,\alpha} := \partial_y^\alpha \partial_t^j u_\ell \tag{9.1.58}$$

and consider as the new unknown vectorial solution the function

$$v := \left\{ v_{\ell,j,\alpha} \right\}_{\substack{\ell\in\{1,\dots,N\}\\ \alpha\in\mathbb{N}^{n-1},j\in\{1,\dots,\mu_\ell\}\\ |\alpha|\leqslant\mu_\ell}}.$$

For each $\alpha = (\alpha_1, \dots, \alpha_n) \in \mathbb{N}^n \setminus \{0\}$, we also let $i(\alpha) \in \{1, \dots, N\}$ the smallest index for which $\alpha_i \neq 0$.

In this notation, we claim that (9.1.57) is equivalent to the first-order quasilinear system[12]

$$\begin{cases} \partial_t v_{\ell,j,\alpha} = \begin{cases} v_{\ell,j+1,\alpha} & \text{if } j \leqslant \mu_\ell - 1 \\ \partial_t H_\ell + \displaystyle\sum_{\substack{m\in\{1,\dots,N\}\\ \beta\in\mathbb{N}^{n-1},j\in\{1,\dots,\mu_\ell\}\\ |\beta|+j\leqslant\mu_m-1}} a_{\ell,m,j,\beta}\, v_{m,j+1,\beta} \\ \quad + \displaystyle\sum_{\substack{m\in\{1,\dots,N\}\\ \beta\in\mathbb{N}^{n-1},j\in\{1,\dots,\mu_\ell\}\\ |\beta|+j=\mu_m\\ j\leqslant\mu_m-1}} a_{\ell,m,j,\beta}\, \partial_{y_{i(\beta)}} v_{m,j+1,\beta-e_{i(\beta)}} & \text{if } j = \mu_\ell, \end{cases} \\ \qquad \text{for all } |y| < \rho,\ |t| < \rho,\ \ell \in \{1, \dots, N\} \text{ and } |\alpha| + j \leqslant \mu_\ell, \\ v_{\ell,j,\alpha}(y, 0) = 0 \quad \text{for all } |y| < \rho,\ \ell \in \{1, \dots, N\} \text{ and } |\alpha| + j \leqslant \mu_\ell, \end{cases} \tag{9.1.59}$$

[12] Yes, the notation in (9.1.59) is quite cumbersome. However, let us try to understand what's going on through a specific example. Suppose that we are considering the scalar equation

$$\partial_t^3 u = \partial_y^2 \partial_t u + \partial_y u + \partial_t^2 u,$$

with $y \in \mathbb{R}$.

The strategy would be to define

$$v_1 := u, \quad v_2 := \partial_y u, \quad v_3 := \partial_y^2 u, \quad v_4 := \partial_t u, \quad v_5 := \partial_t^2 u \quad \text{and} \quad v_6 := \partial_y \partial_t u.$$

This reduces the above scalar equation to the first-order system

$$\begin{cases} \partial_t v_1 = \partial_t u = v_4, \\ \partial_t v_2 = \partial_y \partial_t u = v_6, \\ \partial_t v_3 = \partial_y^2 \partial_t u = \partial_y v_6, \\ \partial_t v_4 = \partial_t^2 u = v_5, \\ \partial_t v_5 = \partial_t^3 u = \partial_y^2 \partial_t u + \partial_y u + \partial_t^2 u = \partial_y v_6 + v_2 + v_5, \\ \partial_t v_6 = \partial_y \partial_t^2 u = \partial_y v_5. \end{cases}$$

The framework in (9.1.59) is nothing more than applying this strategy to a general system of equations.

where, in light of (9.1.55), the variables of $\partial_t H_\ell$ and $a_{\ell,m,i,\beta}$ are now

$$\left(y,t,\left\{v_{m,i,\gamma}\right\}_{\substack{m\in\{1,\dots,N\}\\ \gamma\in\mathbb{N}^{n-1},i\in\mathbb{N}\\ |\gamma|+i\leqslant\mu_m\\ i\leqslant\mu_m-1}}\right).$$

Indeed, if we have a solution to (9.1.57), then we can obtain a solution to (9.1.59) simply by using the setting in (9.1.58).

Suppose now that (9.1.59) holds true, and choose

$$u_\ell(y,t) := \int_0^t \left(\int_0^{\tau_{\mu_\ell}} \left(\dots\left(\int_0^{\tau_2} v_{\ell,\mu_\ell,0}(y,\tau_1)\,d\tau_1\right)\dots\right)d\tau_{\mu_\ell-1}\right)d\tau_{\mu_\ell}. \tag{9.1.60}$$

We claim that

whenever $\ell\in\{1,\dots,N\}$, $j\in\{0,\dots,\mu_\ell\}$ and $|\alpha|+j\leqslant\mu_\ell$, equation (9.1.58) is satisfied. (9.1.61)

Note that once (9.1.61) is proven, we obtain as a byproduct that (9.1.57) holds true.

We prove (9.1.61) by induction over $p:=\mu_\ell-j\in\{0,\dots,\mu_\ell\}$. For this, let us first suppose that $p=0$. Then, $j=\mu_\ell$, and by taking μ_ℓ derivatives in t in (9.1.60), we have that $\partial_t^{\mu_\ell}u_\ell(y,t)=v_{\ell,\mu_\ell,0}(y,t)$, which is (9.1.58) for $j=\mu_\ell$.

Hence, to perform the inductive step, we now suppose that (9.1.61) holds true for all $p\leqslant p_0$ with $p_0\leqslant\mu_\ell-1$, and we aim at proving it for p_0+1, i.e. for $j=\mu_\ell-p_0-1$. To this end, we take $\alpha\in\mathbb{N}^n$ with $|\alpha|+j\leqslant\mu_\ell$ and we differentiate (9.1.60) to see that

$$\begin{aligned}&\partial_y^\alpha\partial_t^j u_\ell(y,t)\\ &\quad=\int_0^t\left(\int_0^{\tau_{\mu_\ell-j}}\left(\dots\left(\int_0^{\tau_2}\partial_y^\alpha v_{\ell,\mu_\ell,0}(y,\tau_1)\,d\tau_1\right)\dots\right)d\tau_{\mu_\ell-j-1}\right)d\tau_{\mu_\ell-j}.\end{aligned} \tag{9.1.62}$$

We also have that $j=\mu_\ell-p_0-1\leqslant\mu_\ell-1$. Hence, by (9.1.59) and the inductive assumption, we see that

$$\partial_t v_{\ell,j,\alpha}=v_{\ell,j+1,\alpha}=\partial_y^\alpha\partial_t^{j+1}u_\ell=\partial_t(\partial_y^\alpha\partial_t^j u_\ell).$$

Accordingly, the function $v_{\ell,j,\alpha}-\partial_y^\alpha\partial_t^j u_\ell$ is constant in t. Also, this function vanishes at $t=0$ due to the boundary condition in (9.1.59) and the expression in (9.1.62). As a consequence, the function $v_{\ell,j,\alpha}-\partial_y^\alpha\partial_t^j u_\ell$ vanishes identically, giving (9.1.58) in this case. The proof of (9.1.61) is thereby complete. Hence, we have reduced the proof of Theorem 9.1.6 to the analysis of (9.1.59).

So, owing to (9.1.59), to prove Theorem 9.1.6 (using now an easier notation), it suffices to establish, in the real analytic framework, a local existence and uniqueness theory for quasilinear systems of first order with homogeneous boundary data of the form

$$\begin{cases} \partial_t w_k = \sum\limits_{\substack{1\leqslant i\leqslant n-1\\ 1\leqslant m\leqslant M}} b_{i,m,k}(y,t,w)\partial_i w_m + c_k(y,t,w) \\ \qquad \text{for all } |y|<\rho,\ |t|<\rho,\ k\in\{1,\dots,M\}, \\ \\ w_k(y,0)=0 \\ \qquad \text{for all } |y|<\rho \text{ and } k\in\{1,\dots,M\}, \end{cases} \tag{9.1.63}$$

where the unknown vectorial solution is $w(y,t) = (w_1(y,t), \dots, w_M(y,t))$.

One can actually suppose that, up to adding one more equation and unknown, the coefficients $b_{i,m,k}$ and c_k do not depend on t (this trick[13] was already exploited in (9.1.22) for the case of ordinary differential equations). Namely, by defining $L := M+1$ and $w_L(y,t) := t$, we have that $\partial_t w_L = 1$, and the system in (9.1.63) can be written in the form

$$\begin{cases} \partial_t w_k = \sum\limits_{\substack{1\leqslant i\leqslant n-1\\ 1\leqslant m\leqslant L}} B_{i,m,k}(y,w)\partial_i w_m + C_k(y,w) \\ \qquad \text{for all } |y|<\rho,\ |t|<\rho,\ k\in\{1,\dots,L\}, \\ \\ w_k(y,0)=0 \\ \qquad \text{for all } |y|<\rho \text{ and } k\in\{1,\dots,L\}, \end{cases} \tag{9.1.64}$$

where the unknown vectorial solution is $w(y,t) = (w_1(y,t), \dots, w_L(y,t))$.

We now address the analytic part of the proof of Theorem 9.1.6, which relies on the method of majorants and establishes the local unique solvability of (9.1.64) in the real analytic setting. To this end, we use the notation $Y := (y,t)$ and seek $w = (w_1, \dots, w_L)$, with w_k in the form[14]

$$w_k(Y) = \sum_{\alpha\in\mathbb{N}^n} \vartheta_{k,\alpha} Y^\alpha, \tag{9.1.65}$$

for suitable $\vartheta_{k,\alpha} \in \mathbb{R}$. Note also that, under the boundary condition in (9.1.64), for every $\alpha' \in \mathbb{N}^{n-1}$,

$$\vartheta_{k,(\alpha',0)} = 0. \tag{9.1.66}$$

[13]Here, one could also get rid of the dependence on y, again by adding $n-1$ new equations and unknowns, but this would introduce a nonhomogeneous boundary term at $t=0$; therefore, we preferred to maintain the dependence on y in the coefficients.

[14]The counterpart of (9.1.65) in the setting of ordinary differential equations was considered in (9.1.24).

We also expand $B_{i,m,k}$ and C_k as[15]

$$B_{i,m,k}(y,w) = \sum_{\substack{\beta\in\mathbb{N}^L\\ \sigma\in\mathbb{N}^{n-1}}} \lambda_{i,m,k,\beta,\sigma}\, y^\sigma w^\beta$$

$$\text{and}\quad C_k(y,w) = \sum_{\substack{\beta\in\mathbb{N}^L\\ \sigma\in\mathbb{N}^{n-1}}} \mu_{k,\beta,\sigma}\, y^\sigma w^\beta, \tag{9.1.67}$$

for suitable $\lambda_{i,m,k,\beta,\sigma}$, $\mu_{k,\beta,\sigma}\in\mathbb{R}$.

We claim that[16]

$$\vartheta_{k,\alpha} = \mathcal{P}_{k,\alpha}\left(\left\{\lambda_{i,m,k,\beta,\sigma}\right\}_{\substack{1\leqslant i\leqslant n-1\\ 1\leqslant m\leqslant L\\ \beta\in\mathbb{N}^L\\ \sigma\in\mathbb{N}^{n-1}\\ |\beta|+|\sigma|\leqslant|\alpha|-1}}, \left\{\mu_{k,\beta,\sigma}\right\}_{\substack{\beta\in\mathbb{N}^L\\ \sigma\in\mathbb{N}^{n-1}\\ |\beta|+|\sigma|\leqslant|\alpha|-1}}, \left\{\vartheta_{m,\delta}\right\}_{\substack{1\leqslant m\leqslant L\\ \delta\in\mathbb{N}^n\\ |\delta|\leqslant|\alpha|-1}} \right),$$

where $\mathcal{P}_{k,\alpha}$ is a polynomial with *nonnegative coefficients*. (9.1.68)

Note that this statement is true when $\alpha_n = 0$, thanks to (9.1.66); hence, we can focus on the proof of (9.1.68) when $\alpha_n \neq 0$.

For this, we argue as follows: plugging (9.1.65) and (9.1.67) into (9.1.64), we find that

$$\begin{aligned}
&\sum_{\alpha\in\mathbb{N}^n} \alpha_n \vartheta_{k,\alpha} Y^{\alpha-e_n}\\
&= \partial_t w_k\\
&= \sum_{\substack{1\leqslant i\leqslant n-1\\ 1\leqslant m\leqslant L}} B_{i,m,k}(y,w)\partial_i w_m + C_k(y,w)\\
&= \sum_{\substack{1\leqslant i\leqslant n-1\\ 1\leqslant m\leqslant L\\ \beta\in\mathbb{N}^L\\ \sigma\in\mathbb{N}^{n-1}\\ \alpha\in\mathbb{N}^n}} \lambda_{i,m,k,\beta,\sigma}\,\alpha_i \vartheta_{m,\alpha} Y^{\alpha+(\sigma,0)-e_i}\, w^\beta + \sum_{\substack{\beta\in\mathbb{N}^L\\ \sigma\in\mathbb{N}^{n-1}}} \mu_{k,\beta,\sigma}\, y^\sigma w^\beta.
\end{aligned}$$

[15]The counterpart of (9.1.67) in the setting of ordinary differential equations was considered in (9.1.26).

[16]The counterpart of (9.1.68) in the setting of ordinary differential equations was considered in (9.1.27).

Hence, since

$$\begin{aligned} w^{\beta} &= w_1^{\beta_1} \dots w_L^{\beta_L} \\ &= \left(\sum_{\alpha^{(1)} \in \mathbb{N}^n} \vartheta_{1,\alpha^{(1)}} Y^{\alpha^{(1)}} \right)^{\beta_1} \dots \left(\sum_{\alpha^{(L)} \in \mathbb{N}^n} \vartheta_{L,\alpha^{(L)}} Y^{\alpha^{(L)}} \right)^{\beta_L} \\ &= \sum_{\substack{\alpha^{(1,1)},\dots,\alpha^{(1,\beta_1)} \in \mathbb{N}^n \\ \dots \\ \alpha^{(L,1)},\dots,\alpha^{(L,\beta_L)} \in \mathbb{N}^n}} \vartheta_{1,\alpha^{(1,1)}} \dots \vartheta_{1,\alpha^{(1,\beta_1)}} \dots \vartheta_{L,\alpha^{(L,1)}} \dots \vartheta_{L,\alpha^{(1,\beta_L)}} \\ &\quad \times Y^{\alpha^{(1,1)}+\dots+\alpha^{(1,\beta_1)}+\dots+\alpha^{(L,1)}+\dots+\alpha^{(L,\beta_L)}}, \end{aligned}$$

we obtain that

$$\begin{aligned} &\sum_{\alpha \in \mathbb{N}^n} \alpha_n \vartheta_{k,\alpha} Y^{\alpha - e_n} \\ &= \sum_{\substack{1 \leqslant i \leqslant n-1 \\ 1 \leqslant m \leqslant L \\ \beta \in \mathbb{N}^L \\ \sigma \in \mathbb{N}^{n-1} \\ \alpha \in \mathbb{N}^n \\ \alpha^{(1,1)},\dots,\alpha^{(1,\beta_1)} \in \mathbb{N}^n \\ \dots \\ \alpha^{(L,1)},\dots,\alpha^{(L,\beta_L)} \in \mathbb{N}^n}} \alpha_i \lambda_{i,m,k,\beta,\sigma} \vartheta_{1,\alpha^{(1,1)}} \dots \vartheta_{1,\alpha^{(1,\beta_1)}} \dots \vartheta_{L,\alpha^{(L,1)}} \dots \vartheta_{L,\alpha^{(1,\beta_L)}} \vartheta_{m,\alpha} \\ &\quad \times Y^{\alpha^{(1,1)}+\dots+\alpha^{(1,\beta_1)}+\dots+\alpha^{(L,1)}+\dots+\alpha^{(L,\beta_L)}+\alpha+(\sigma,0)-e_i} \\ &\quad + \sum_{\substack{\beta \in \mathbb{N}^L \\ \sigma \in \mathbb{N}^{n-1} \\ \alpha^{(1,1)},\dots,\alpha^{(1,\beta_1)} \in \mathbb{N}^n \\ \dots \\ \alpha^{(L,1)},\dots,\alpha^{(L,\beta_L)} \in \mathbb{N}^n}} \mu_{k,\beta,\sigma} \vartheta_{1,\alpha^{(1,1)}} \dots \vartheta_{1,\alpha^{(1,\beta_1)}} \dots \vartheta_{L,\alpha^{(L,1)}} \dots \vartheta_{L,\alpha^{(1,\beta_L)}} \\ &\quad \times Y^{\alpha^{(1,1)}+\dots+\alpha^{(1,\beta_1)}+\dots+\alpha^{(L,1)}+\dots+\alpha^{(L,\beta_L)}+(\sigma,0)}. \end{aligned}$$

As a result,

$$\begin{aligned} \alpha_n \vartheta_{k,\alpha} = &\sum \gamma_i \lambda_{i,m,k,\beta,\sigma} \vartheta_{1,\alpha^{(1,1)}} \dots \vartheta_{1,\alpha^{(1,\beta_1)}} \dots \vartheta_{L,\alpha^{(L,1)}} \dots \vartheta_{L,\alpha^{(1,\beta_L)}} \vartheta_{m,\gamma} \\ &+ \sum \mu_{k,\beta,\sigma} \vartheta_{1,\alpha^{(1,1)}} \dots \vartheta_{1,\alpha^{(1,\beta_1)}} \dots \vartheta_{L,\alpha^{(L,1)}} \dots \vartheta_{L,\alpha^{(1,\beta_L)}}, \end{aligned} \tag{9.1.69}$$

where the first sum is taken over the indices

$$\begin{gathered} 1 \leqslant i \leqslant n-1, \qquad 1 \leqslant m \leqslant L, \qquad \beta \in \mathbb{N}^L, \qquad \sigma \in \mathbb{N}^{n-1}, \qquad \gamma \in \mathbb{N}^n, \\ \alpha^{(1,1)},\dots,\alpha^{(1,\beta_1)} \in \mathbb{N}^n, \qquad \dots, \qquad \alpha^{(L,1)},\dots,\alpha^{(L,\beta_L)} \in \mathbb{N}^n, \\ \alpha^{(1,1)}+\dots+\alpha^{(1,\beta_1)}+\dots+\alpha^{(L,1)}+\dots+\alpha^{(L,\beta_L)}+\gamma-e_i+(\sigma,0) = \alpha - e_n \end{gathered}$$

and the second sum over the indices

$$\begin{aligned}&\beta \in \mathbb{N}^L, \qquad \sigma \in \mathbb{N}^{n-1},\\ &\alpha^{(1,1)}, \dots, \alpha^{(1,\beta_1)} \in \mathbb{N}^n, \qquad \dots, \qquad \alpha^{(L,1)}, \dots, \alpha^{(L,\beta_L)} \in \mathbb{N}^n\\ &\alpha^{(1,1)} + \dots + \alpha^{(1,\beta_1)} + \dots + \alpha^{(L,1)} + \dots + \alpha^{(L,\beta_L)} + (\sigma, 0) = \alpha - e_n.\end{aligned}$$

This determines $\vartheta_{k,\alpha}$; however, to complete the proof of (9.1.68), we need to check the order of the indices on the right-hand side of (9.1.68).

To this end, we point out that the conditions $\gamma_i \neq 0$ and

$$\alpha^{(1,1)} + \dots + \alpha^{(1,\beta_1)} + \dots + \alpha^{(L,1)} + \dots + \alpha^{(L,\beta_L)} + \gamma - e_i + (\sigma, 0) = \alpha - e_n$$

give that $|\gamma| \geqslant 1$ and

$$|\alpha^{(1,1)}| + \dots + |\alpha^{(1,\beta_1)}| + \dots + |\alpha^{(L,1)}| + \dots + |\alpha^{(L,\beta_L)}| + |\gamma| + |\sigma| = |\alpha|,$$

whence

$$|\alpha^{(1,1)}| + \dots + |\alpha^{(1,\beta_1)}| + \dots + |\alpha^{(L,1)}| + \dots + |\alpha^{(L,\beta_L)}| + |\sigma| \leqslant |\alpha| - 1.$$

Furthermore, in (9.1.69), we have that $|\gamma| \leqslant |\alpha| - 1$, due to (9.1.66).

These observations take care of the order of the indices of $\lambda_{i,m,k,\beta,\sigma}$ on the right-hand side of (9.1.68) as well as the indices of $\vartheta_{m,\delta}$ on the right-hand side of (9.1.68), which come from the first sum on the right-hand side of (9.1.69).

Thus, to complete the proof of (9.1.68), we need to take care, on the right-hand side of (9.1.68), of the order of the indices of $\mu_{k,\beta,\sigma}$ and that of the indices of $\vartheta_{m,\delta}$ which come from the second sum on the right-hand side of (9.1.69).

With that in mind, we observe that the condition

$$\alpha^{(1,1)} + \dots + \alpha^{(1,\beta_1)} + \dots + \alpha^{(L,1)} + \dots + \alpha^{(L,\beta_L)} + (\sigma, 0) = \alpha - e_n \tag{9.1.70}$$

combined with (9.1.66) leads to the fact that, in the second sum on the right-hand side of (9.1.69), one has that

$$\begin{aligned}|\beta| + |\sigma| &= \beta_1 + \dots + \beta_L + |\sigma|\\ &\leqslant |\alpha^{(1,1)}| + \dots + |\alpha^{(1,\beta_1)}| + \dots + |\alpha^{(L,1)}| + \dots + |\alpha^{(L,\beta_L)}| + |\sigma|\\ &= |\alpha| - 1,\end{aligned}$$

which takes care of the indices of $\mu_{k,\beta,\sigma}$.

Moreover, from (9.1.70), we also deduce that

$$|\alpha^{(1,1)}|, \dots, |\alpha^{(1,\beta_1)}|, \dots, |\alpha^{(L,1)}|, \dots, |\alpha^{(L,\beta_L)}| \leqslant |\alpha| - 1,$$

which takes care of the indices of $\vartheta_{m,\delta}$. This completes the proof of (9.1.68).

We stress that (9.1.68) determines all the coefficients $\vartheta_{k,\alpha}$ uniquely (i.e. by arguing recursively, starting from (9.1.66)). This entails that, if a real analytic solution exists, it is uniquely determined by (9.1.65).

Having established the uniqueness result in Theorem 9.1.6, we now aim at proving the existence result. To this end, we show that the formal series in (9.1.65) is indeed convergent. This task is accomplished via the method of majorants. For this, we first use (9.1.66) and (9.1.68) to see that[17]

$$\vartheta_{k,\alpha} = Q_{k,\alpha}\left(\left\{ \lambda_{i,m,k,\beta,\sigma} \right\}_{\substack{1\leqslant i\leqslant n-1 \\ 1\leqslant m\leqslant L \\ \beta\in\mathbb{N}^L \\ \sigma\in\mathbb{N}^{n-1} \\ |\beta|+|\sigma|\leqslant|\alpha|-1}}, \left\{ \mu_{k,\beta,\sigma} \right\}_{\substack{\beta\in\mathbb{N}^L \\ \sigma\in\mathbb{N}^{n-1} \\ |\beta|+|\sigma|\leqslant|\alpha|-1}} \right), \tag{9.1.71}$$

where $Q_{k,\alpha}$ is a polynomial with *nonnegative coefficients*.

As a consequence,

$$|\vartheta_{k,\alpha}| \leqslant Q_{k,\alpha}\left(\left\{ |\lambda_{i,m,k,\beta,\sigma}| \right\}_{\substack{1\leqslant i\leqslant n-1 \\ 1\leqslant m\leqslant L \\ \beta\in\mathbb{N}^L \\ \sigma\in\mathbb{N}^{n-1} \\ |\beta|+|\sigma|\leqslant|\alpha|-1}}, \left\{ |\mu_{k,\beta,\sigma}| \right\}_{\substack{\beta\in\mathbb{N}^L \\ \sigma\in\mathbb{N}^{n-1} \\ |\beta|+|\sigma|\leqslant|\alpha|-1}} \right). \tag{9.1.72}$$

We stress that in this inequality, we have employed the fact that the above coefficients are nonnegative: this is an interesting case in which one can use the monotonicity property encoded in an abstract mathematical structure to deduce useful information hidden behind[18] incendiary combinatorial calculations.

Furthermore, by the analyticity of $B_{i,m,k}$ and C_k, as expressed in (9.1.67), we know (see e.g. [KP02, Corollary 1.1.7]) that

$$|\lambda_{i,m,k,\beta,\sigma}| + |\mu_{k,\beta,\sigma}| \leqslant \frac{C}{R^{|\beta|+|\sigma|}}, \tag{9.1.73}$$

for some $C, R > 0$.

As a byproduct of (9.1.73) and (9.1.33), we obtain that

$$|\lambda_{i,m,k,\beta,\sigma}| + |\mu_{k,\beta,\sigma}| \leqslant \frac{C\,(|\beta|+|\sigma|)!}{\beta!\,\sigma!\,R^{|\beta|+|\sigma|}}. \tag{9.1.74}$$

[17] The counterpart of (9.1.71) in the setting of ordinary differential equations was considered in (9.1.30).

[18] There are however alternative proofs of the Cauchy–Kowalevsky theorem that do not rely on the method of majorants but focus instead on a direct estimation of all the derivatives of the solution, see e.g. [DiB95, pp. 26–35].

We have thus reduced ourselves to the "worst possible scenario." Indeed, on the one hand, if we define

$$\begin{aligned} \lambda^{\star}_{i,m,k,\beta,\sigma} &:= \frac{C\,(|\beta|+|\sigma|)!}{\beta!\,\sigma!\,R^{|\beta|+|\sigma|}}, \\ \mu^{\star}_{k,\beta,\sigma} &:= \frac{C\,(|\beta|+|\sigma|)!}{\beta!\,\sigma!\,R^{|\beta|+|\sigma|}}, \\ B^{\star}_{i,m,k}(y,w) &= \sum_{\substack{\beta\in\mathbb{N}^L\\ \sigma\in\mathbb{N}^{n-1}}} \lambda^{\star}_{i,m,k,\beta,\sigma}\, y^{\sigma} w^{\beta} \\ \text{and}\quad C^{\star}_{k}(y,w) &:= \sum_{\substack{\beta\in\mathbb{N}^L\\ \sigma\in\mathbb{N}^{n-1}}} \mu^{\star}_{k,\beta,\sigma}\, y^{\sigma} w^{\beta}, \end{aligned} \tag{9.1.75}$$

we infer from (9.1.72) and (9.1.74) that

$$|\vartheta_{k,\alpha}| \leqslant Q_{k,\alpha}\left(\left\{\lambda^{\star}_{i,m,k,\beta,\sigma}\right\}_{\substack{1\leqslant i\leqslant n-1\\ 1\leqslant m\leqslant L\\ \beta\in\mathbb{N}^L\\ \sigma\in\mathbb{N}^{n-1}\\ |\beta|+|\sigma|\leqslant|\alpha|-1}}, \left\{\mu^{\star}_{k,\beta,\sigma}\right\}_{\substack{\beta\in\mathbb{N}^L\\ \sigma\in\mathbb{N}^{n-1}\\ |\beta|+|\sigma|\leqslant|\alpha|-1}} \right). \tag{9.1.76}$$

On the other hand, in light of (9.1.71), we also have that if there exists a real analytic solution

$$w^{\star}_k(Y) = \sum_{\alpha\in\mathbb{N}^n} \vartheta^{\star}_{k,\alpha} Y^{\alpha} \tag{9.1.77}$$

to the Cauchy problem

$$\begin{cases} \partial_t w^{\star}_k = \displaystyle\sum_{\substack{1\leqslant i\leqslant n-1\\ 1\leqslant m\leqslant L}} B^{\star}_{i,m,k}(y,w^{\star})\partial_i w^{\star}_m + C^{\star}_k(y,w^{\star}) \\ \qquad \text{for all } |y|<\rho,\ |t|<\rho,\ k\in\{1,\dots,L\}, \\ \\ w^{\star}_k(y,0) = 0 \\ \qquad \text{for all } |y|<\rho \text{ and } k\in\{1,\dots,L\}, \end{cases} \tag{9.1.78}$$

then necessarily

$$\vartheta^{\star}_{k,\alpha} = Q_{k,\alpha}\left(\left\{\lambda^{\star}_{i,m,k,\beta,\sigma}\right\}_{\substack{1\leqslant i\leqslant n-1\\ 1\leqslant m\leqslant L\\ \beta\in\mathbb{N}^L\\ \sigma\in\mathbb{N}^{n-1}\\ |\beta|+|\sigma|\leqslant|\alpha|-1}}, \left\{\mu^{\star}_{k,\beta,\sigma}\right\}_{\substack{\beta\in\mathbb{N}^L\\ \sigma\in\mathbb{N}^{n-1}\\ |\beta|+|\sigma|\leqslant|\alpha|-1}} \right). \tag{9.1.79}$$

Therefore, we have found a "majorization problem" for our original situation in the sense that, by (9.1.76) and (9.1.79),

$$|\vartheta_{k,\alpha}| \leqslant \vartheta^{\star}_{k,\alpha}. \tag{9.1.80}$$

Hence, to show that the formal series in (9.1.65) defines, in fact, an analytic function (and thus complete the proof of Theorem 9.1.6), it suffices to establish suitable bounds on the solution (if any) of our majorization problem (9.1.78). This strategy is advantageous since the majorization problem can be explicitly solved.

Indeed, from (9.1.75), one deduces that

$$B^{\star}_{i,m,k}(y,w) = C^{\star}_{k}(y,w) = \sum_{\substack{\beta\in\mathbb{N}^L\\ \sigma\in\mathbb{N}^{n-1}}} \frac{C\,(|\beta|+|\sigma|)!}{\beta!\,\sigma!\,R^{|\beta|+|\sigma|}}\, y^{\sigma} w^{\beta}.$$

We also set

$$\zeta := \frac{y_1+\dots+y_{n-1}+w_1+\dots+w_L}{R}.$$

Using the multinomial theorem, we observe that, for every $\kappa\in\mathbb{N}$,

$$\begin{aligned}\zeta^{\kappa} &= \sum_{\substack{\omega\in\mathbb{N}^{L+n-1}\\ |\omega|=\kappa}} \frac{\kappa!}{\omega_1!\dots\omega_{L+n-1}!\,R^{|\omega|}}\, y_1^{\omega_1}\dots y_{n-1}^{\omega_{n-1}} w_1^{\omega_n}\dots w_L^{\omega_{L+n-1}}\\ &= \sum_{\substack{\omega=(\beta,\sigma)\in\mathbb{N}^L\times\mathbb{N}^{n-1}\\ |\beta|+|\sigma|=\kappa}} \frac{(|\beta|+|\sigma|)!}{\beta!\sigma!\,R^{|\beta|+|\sigma|}}\, y_1^{\sigma_1}\dots y_{n-1}^{\sigma_{n-1}} w_1^{\beta_1}\dots w_L^{\beta_L}.\end{aligned}$$

Therefore,[19] if $|\zeta|<1$, we find that

$$\begin{aligned}B^{\star}_{i,m,k}(y,w) = C^{\star}_{k}(y,w) &= C\sum_{\kappa\in\mathbb{N}}\zeta^{\kappa} = \frac{C}{1-\zeta}\\ &= \frac{CR}{R-(y_1+\dots+y_{n-1}+w_1+\dots+w_L)}.\end{aligned} \tag{9.1.81}$$

Consequently, the majorization problem (9.1.78) can be written in the form

$$\begin{cases} \dfrac{R-(y_1+\dots+y_{n-1}+w_1+\dots+w_L)}{CR}\,\partial_t w^{\star}_k = \displaystyle\sum_{\substack{1\leqslant i\leqslant n-1\\ 1\leqslant m\leqslant L}} \partial_i w^{\star}_m + 1\\ \qquad \text{for all } |y|<\rho,\ |t|<\rho,\ k\in\{1,\dots,L\},\\ \\ w^{\star}_k(y,0)=0\\ \qquad \text{for all } |y|<\rho \text{ and } k\in\{1,\dots,L\}. \end{cases} \tag{9.1.82}$$

[19] The counterpart of (9.1.81) in the setting of ordinary differential equations was considered in (9.1.40).

We seek a solution for this problem in the form

$$w_1^\star(y,t) = \cdots = w_L^\star(y,t) := W(\eta,t), \quad \text{where } \eta := y_1 + \cdots + y_{n-1}. \tag{9.1.83}$$

With this *ansatz*, up to renaming ρ, we reduce (9.1.82) to

$$\begin{cases} \dfrac{R-(\eta+LW)}{CR}\partial_t W = (n-1)L\partial_\eta W + 1 & \text{for all } |\eta| < \rho,\ |t| < \rho, \\ W(\eta,0) = 0 & \text{for all } |\eta| < \rho. \end{cases} \tag{9.1.84}$$

To find a solution for this, we use the so-called method of characteristics, i.e. we look for a curve $\tau \mapsto (\eta(\tau), t(\tau))$ along which we control the solution. Since

$$\frac{d}{d\tau}\big(W(\eta(\tau),t(\tau))\big) = \partial_\eta W(\eta(\tau),t(\tau))\dot\eta(\tau) + \partial_t W(\eta(\tau),t(\tau))\dot t(\tau),$$

in light of (9.1.84), it is convenient to choose

$$\dot\eta := -(n-1)L \quad \text{and} \quad \dot t = \frac{R-(\eta+LW)}{CR}, \tag{9.1.85}$$

so that

$$\frac{d}{d\tau}\big(W(\eta(\tau),t(\tau))\big) = 1,$$

and accordingly,

$$W(\eta(\tau),t(\tau)) = \tau. \tag{9.1.86}$$

Given $\eta_0 \in \mathbb{R}$, from (9.1.85) and (9.1.86), we arrive at

$$\eta(\tau) = \eta_0 - (n-1)L\tau \tag{9.1.87}$$

and

$$\dot t(\tau) = \frac{R-\eta_0+(n-1)L\tau - L\tau}{CR} = \frac{R-\eta_0+(n-2)L\tau}{CR};$$

therefore,

$$t(\tau) = \frac{R-\eta_0}{CR}\tau + \frac{(n-2)L}{2CR}\tau^2. \tag{9.1.88}$$

Our goal is now to invert these parametric representations, i.e. rather than having η and t as functions of τ (and η_0), we wish to write τ in dependence of η and t. For this, owing to (9.1.87) and (9.1.88), we note that

$$t = \frac{R-(\eta+(n-1)L\tau)}{CR}\tau + \frac{(n-2)L}{2CR}\tau^2.$$

Hence, solving for τ (and picking the solution for which $\tau = 0$ when $t = 0$), we find

$$\tau = \frac{R-\eta-\sqrt{(R-\eta)^2 - 2CnLRt}}{nL}.$$

Overall,

$$W(\eta, t) = \tau = \frac{R - \eta - \sqrt{(R - \eta)^2 - 2CnLRt}}{nL}, \qquad (9.1.89)$$

which provides[20] a real analytic solution for the Cauchy problem in (9.1.84) as long as ρ is small enough.

Recalling (9.1.83), we have thereby constructed explicitly a real analytic solution to (9.1.82). From its analyticity (see e.g. [KP02, Corollary 1.1.7]) and recalling the notation in (9.1.77), we infer that

$$\vartheta^{\star}_{k,\alpha} \leqslant |\vartheta^{\star}_{k,\alpha}| \leqslant \frac{C^{\star}}{\rho^{|\alpha|}},$$

for some $C^{\star}, \rho > 0$, and consequently, by (9.1.80),

$$|\vartheta_{k,\alpha}| \leqslant \frac{C^{\star}}{\rho^{|\alpha|}}.$$

This shows that the formal series in (9.1.65) is convergent near the origin, thus providing locally a real analytic solution to the Cauchy problem in (9.1.64) and completing the proof of Theorem 9.1.6. □

For further discussions on the Cauchy–Kowalevsky theorem, see for example [CP82, Chapter 6], [CH89, Chapter 1, Section 7], [ES92, Chapter 2], [Joh91, Chapter 3, Section 3], [KP02, Sections 2.3, 2.4 and 2.5], [Trè06, Chapter 2, Sections 17, 18, 19, 20 and 21], [Riv16, Sections II, III and IV] and [Tre22, Chapter 5].

For additional information about the method of majorants, see [Hil97] and the references therein.

See also [Nag42, CL55, Fri57, Ler57, Duf59, Lud60, Fri61b, Ovs71, Nir72, Nis77, BG83, Tut05, Spa18] and the references therein for different proofs and generalizations of the Cauchy–Kowalevsky theorem.

As regards the many presentations of the Cauchy–Kowalevsky theorem, the most famous one is probably that by Jacques Hadamard [Had53]. Not only did Hadamard dedicate a whole treatise to the subject, which is, as reported in [Aud11], "the unnatural coupling [...] of the name of the old reactionary Cauchy with that of a young (female) revolutionary (thanks to the hyphen) must be the work of that great progressive Hadamard in lectures he gave at the beginning of the 20th century in New York." More importantly, Hadamard, despite his deep appreciation of the result, pointed out some intrinsic criticalities. First of all, the analytic setting in which the result is posed may be conceptually problematic from a physical point

[20] The counterpart of (9.1.89) in the setting of ordinary differential equations was considered in (9.1.42).

of view. This is not an abstract problem of regularity[21]; the issue for Hadamard was that the real analytic representation of a function involves infinitely many derivatives, while, in practice, one can only measure finitely many parameters.

This is not merely a philosophical problem for Hadamard since he related it to the notion of "well-posedness," which, according to him, should include a good notion of dependence from the initial data. In this, elliptic equations can be "problematic," as shown by the example

$$\mathbb{R}^2 \ni (x, y) \longmapsto u_k(x, y) := e^{ky-\sqrt{k}} \sin(kx),$$

which is harmonic in $\mathbb{R}^2$ and such that, for all $\alpha \in \mathbb{N}^2$,

$$\lim_{k \to +\infty} \sup_{x \in \mathbb{R}} |D^\alpha u_k(x, 0)| = 0.$$

That is, for large k, the harmonic function u_k and all its derivatives are extremely small along the x-axis. However, $u_k(x, y)$ diverges as $k \to +\infty$ for all $x \in \mathbb{R} \setminus \pi\mathbb{Z}$ whenever $y > 0$, thus showing how small variations in the data along the x-axis can dramatically influence the behavior of the solution (examples of this type led Hadamard to consider hyperbolic, instead of elliptic, equations, in which these phenomena are not present).

In a sense, according to Hadamard's observations, the limitation of the Cauchy–Kowalevsky theorem may be found exactly in its broad generality: since the result is true essentially for "every" equation, without any significant restriction on its structure, it cannot separate the equations which have different behaviors in terms of continuity from the data.

---

[21] According to [Gal83, p. 479], "physicists rightly believe that all functions (with, possibly, some exceptions) are analytic."

Appendix A

An Interesting Example

A.1 The Importance of the Interior Ball Condition for the Hopf Lemma

In this appendix, we showcase that the interior ball condition in Lemma 2.16.1 cannot be dropped by recalling the following interesting example in which all the hypotheses of Lemma 2.16.1 are met except the interior ball condition, and the thesis of Lemma 2.16.1 is violated.

Given $(x, y) \in \mathbb{R}^2$, we use the notation $z = x + iy \in \mathbb{C}$. We observe that a branch of the complex logarithmic function Log in the halfplane $[0, +\infty) \times \mathbb{R}$ can be uniquely defined: for instance, using polar coordinates $z = \rho e^{i\vartheta}$ (corresponding to $x = \rho \cos\vartheta$ and $y = \rho \sin\vartheta$), when $\vartheta \in \left[-\frac{\pi}{2}, \frac{\pi}{2}\right]$, we can write that $\operatorname{Log} z := \ln\rho + i\vartheta$. We also set

$$\mathcal{U} := \left\{ z = x + iy \in \mathbb{C} \text{ s.t. } x \in (0, 1) \right\}.$$

In this way, since $\operatorname{Log} z \neq 0$ in $\mathcal{U}$, we have that the function $\mathcal{U} \ni z \mapsto \frac{z}{\operatorname{Log} z}$ is holomorphic, and accordingly, by (2.1.1), the function

$$u(x, y) := \Re\left(\frac{z}{\operatorname{Log} z}\right) \tag{A.1.1}$$

is harmonic when $x + iy \in \mathcal{U}$.

We observe that u can also be written in polar coordinates as

$$\begin{aligned} u(x, y) &= \Re\left(\frac{\rho e^{i\vartheta}}{\ln\rho + i\vartheta}\right) = \Re\left(\frac{\rho(\cos\vartheta + i\sin\vartheta)(\ln\rho - i\vartheta)}{(\ln\rho)^2 + \vartheta^2}\right) \\ &= \frac{\rho\,(\ln\rho\cos\vartheta + \vartheta\sin\vartheta)}{(\ln\rho)^2 + \vartheta^2}. \end{aligned} \tag{A.1.2}$$

See Figure A.1.1 for a sketch of the graph of u.

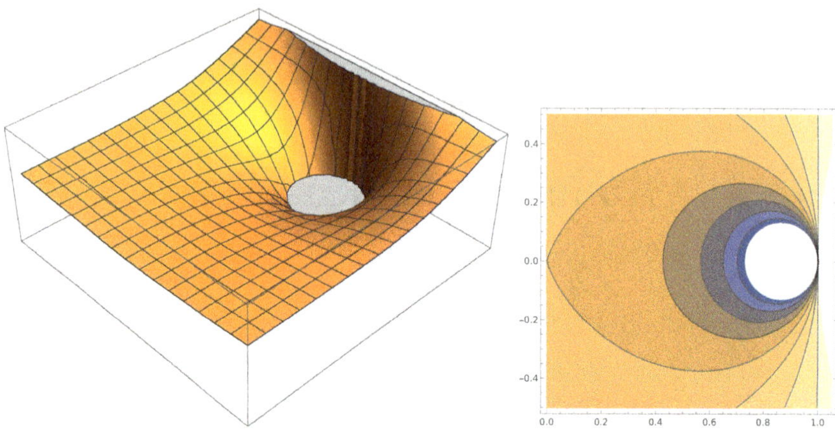

Figure A.1.1 The function u in (A.1.2) and its level sets.

Note also that, for sufficiently small ρ, by (A.1.2),

$$|u(x,y)| \leqslant \frac{2\rho}{\ln\rho} = o(\rho);$$

therefore, $u(0,0) = 0$ and $\nabla u(0,0) = (0,0)$.

Furthermore, we also infer from (A.1.1) that, if $x > 0$,

$$\begin{aligned}\nabla u(x,y) &= \Re\left(\nabla\frac{z}{\operatorname{Log} z}\right) = \Re\left(\frac{(1,i)}{\operatorname{Log} z} + \frac{z}{\operatorname{Log}^2 z}\nabla\operatorname{Log} z\right)\\ &= \Re\left(\frac{(1,i)}{\operatorname{Log} z} + \frac{(1,i)}{\operatorname{Log}^2 z}\right).\end{aligned} \tag{A.1.3}$$

Therefore, if $x > 0$, for small ρ,

$$|\nabla u(x,y)| = O\left(\frac{1}{|\operatorname{Log} z|}\right) = O\left(\frac{1}{|\ln\rho|}\right),$$

yielding that

$$\lim_{\substack{(x,y)\to(0,0)\\ x+iy\in\mathcal{U}}} \nabla u(x,y) = 0 = \nabla u(0,0). \tag{A.1.4}$$

We now define

$$\Omega_o := \left\{(x,y)\in\mathbb{R}^2 \text{ s.t. } x+iy\in\mathcal{U} \text{ and } u(x,y) < 0\right\}. \tag{A.1.5}$$

Then, we have that

$$\begin{aligned}&\Omega_o \text{ is a bounded domain with boundary of class } C^1,\\ &\text{with } (0,0),\,(1,0)\in\partial\Omega.\end{aligned} \tag{A.1.6}$$

Indeed, we can exploit (A.1.2) to write Ω_o in polar coordinates in the form

$$\begin{aligned}&\left\{(\rho,\vartheta)\in(0,+\infty)\times\left(-\frac{\pi}{2},\frac{\pi}{2}\right)\text{ s.t. }\ln\rho\cos\vartheta+\vartheta\sin\vartheta<0\right\}\\&=\left\{(\rho,\vartheta)\in(0,+\infty)\times\left(-\frac{\pi}{2},\frac{\pi}{2}\right)\text{ s.t. }\rho<\frac{1}{\exp(\vartheta\tan\vartheta)}\right\}.\end{aligned}\tag{A.1.7}$$

In particular, if ρ is as above, we have that $\rho<1$, giving that Ω_o is bounded.

Furthermore, since

$$\lim_{\vartheta\to\pm\pi/2}\vartheta\tan\vartheta=+\infty,$$

we deduce from (A.1.7) that

$$\partial\Omega_o=(0,0)\cup\left\{(\rho,\vartheta)\in(0,+\infty)\times\left(-\frac{\pi}{2},\frac{\pi}{2}\right)\text{ s.t. }\rho=\frac{1}{\exp(\vartheta\tan\vartheta)}\right\}.$$

This yields that[1]

$$\partial\Omega_o\setminus B_\varepsilon\text{ is of class }C^1\quad\text{for each }\varepsilon>0.\tag{A.1.8}$$

Hence, to complete the proof of (A.1.6), we only need to take into consideration the regularity of $\partial\Omega_o$ in the vicinity of the origin. To this end, given a small $\delta>0$, we consider $\vartheta\in\left(-\frac{\pi}{2},-\frac{\pi}{2}+\delta\right)\cup\left(\frac{\pi}{2}-\delta,\frac{\pi}{2}\right)$, and we write $\rho=\frac{1}{\exp(\vartheta\tan\vartheta)}$ as

$$x=\frac{\cos\vartheta}{\exp(\vartheta\tan\vartheta)}\quad\text{and}\quad y=\frac{\sin\vartheta}{\exp(\vartheta\tan\vartheta)}.\tag{A.1.9}$$

Let also

$$(-\delta,\delta)\setminus\{0\}\ni\tau\mapsto\varphi(\tau):=\begin{cases}\dfrac{1}{\exp\left(\left(\tau+\frac{\pi}{2}\right)\tan\left(\tau+\frac{\pi}{2}\right)\right)} & \text{if }\tau\in(-\delta,0),\\[2ex] \dfrac{1}{\exp\left(\left(\tau-\frac{\pi}{2}\right)\tan\left(\tau-\frac{\pi}{2}\right)\right)} & \text{if }\tau\in(0,\delta).\end{cases}\tag{A.1.10}$$

[1] A weaker form of (A.1.8) can be proved as follows. From (A.1.3), we have that $\{\nabla u=0\}\cap\{x>0\}=\{\operatorname{Log}z=-1\}$. From this and (A.1.1), we obtain that $\{\nabla u=0\}\cap\{x>0\}\subseteq\{u=-\Re z\}=\{u=-x\}$ and thus $\{\nabla u=0\}\cap\{u=0\}\cap\{x>0\}=\varnothing$. This and the implicit function theorem yield that

$$\partial\Omega_o\setminus(B_\varepsilon\cup B_\varepsilon(1,0))\text{ is of class }C^1\text{ for each }\varepsilon>0.$$

This is weaker than (A.1.8) since the point $(1,0)$ is not captured by this argument (and it cannot be since, as remarked in (A.1.17), the point $(1,0)$ does not belong to $\{u=0\}$, though it lies in $\partial\Omega_o$).

Furthermore, an additional argument is needed to prove (A.1.6) since (A.1.4) prevents us from using the implicit function theorem at the origin.

We note that φ can be continuously extended at $\tau = 0$ since

$$\begin{aligned} &\lim_{\tau\to 0^+} \varphi(\tau) = \lim_{\tau\to 0^+} \frac{1}{\exp\left(\left(\tau - \frac{\pi}{2}\right)\tan\left(\tau - \frac{\pi}{2}\right)\right)} = 0 \\ \text{and} \quad &\lim_{\tau\to 0^-} \varphi(\tau) = \lim_{\tau\to 0^-} \frac{1}{\exp\left(\left(\tau + \frac{\pi}{2}\right)\tan\left(\tau + \frac{\pi}{2}\right)\right)} = 0; \end{aligned}$$

therefore, hereinafter, we consider φ as defined for $\tau \in (-\delta, \delta)$, with $\varphi(0) = 0$.

We also set

$$(-\delta, \delta) \ni \tau \mapsto \zeta(\tau) := \begin{cases} \tau + \dfrac{\pi}{2} & \text{if } \tau \in (-\delta, 0), \\ \\ \tau - \dfrac{\pi}{2} & \text{if } \tau \in (0, \delta) \end{cases}$$

so that (A.1.10) allows for the short notation

$$\varphi(\tau) = \frac{1}{\exp\left(\zeta(\tau)\tan(\zeta(\tau))\right)},$$

with a continuous extension at $\tau = 0$. Thus, by (A.1.9), we see that $\partial\Omega_o$ near the origin can be identified with the parametric curve

$$\begin{cases} x(\tau) = \dfrac{\cos(\zeta(\tau))}{\exp\left((\zeta(\tau))\tan(\zeta(\tau))\right)} = \varphi(\tau)\,\cos(\zeta(\tau)) = \operatorname{sign}\tau\,\sin\tau\,\varphi(\tau), \\ y(\tau) = \dfrac{\sin(\zeta(\tau))}{\exp\left((\zeta(\tau))\tan(\zeta(\tau))\right)} = \varphi(\tau)\,\sin(\zeta(\tau)) = -\operatorname{sign}\tau\,\cos\tau\,\varphi(\tau), \end{cases} \tag{A.1.11}$$

where the sign function has been used.

We also recall that, according to L'Hôpital's rule,

$$\begin{aligned} \lim_{\tau\to 0} \tau \tan\left(\tau - \frac{\operatorname{sign}\tau\,\pi}{2}\right) &= \lim_{\tau\to 0} \frac{\tan\left(\tau - \frac{\operatorname{sign}\tau\,\pi}{2}\right)}{1/\tau} \\ &= \lim_{\tau\to 0} \frac{1/\sin^2\tau}{-1/\tau^2} = -1. \end{aligned} \tag{A.1.12}$$

We now denote by Ξ a function that is smooth outside the origin and such that, for every $k \in \mathbb{N}$,

$$\limsup_{\tau\to 0^+} |D^k \Xi(\tau)| + \limsup_{\tau\to 0^-} |D^k \Xi(\tau)| < +\infty.$$

For typographical convenience, such a function Ξ can vary from line to line in the forthcoming computation. We claim that, for small $\tau \neq 0$, the function $\frac{\tau\tan\left(\tau-\frac{\operatorname{sign}\tau\,\pi}{2}\right)+1}{\tau^2}$ satisfies the above properties; hence, we can write that

$$\frac{\tau\tan\left(\tau-\frac{\operatorname{sign}\tau\,\pi}{2}\right)+1}{\tau^2}=\Xi(\tau). \tag{A.1.13}$$

To check this, we employ (A.1.12) and standard Taylor expansions for trigonometric functions and thus obtain that, for $\tau \neq 0$,

$$\begin{aligned}
\tau\tan\left(\tau-\frac{\operatorname{sign}\tau\,\pi}{2}\right)+1 &= \int_0^\tau \frac{d}{d\ell}\left[\ell\tan\left(\ell-\frac{\operatorname{sign}\ell\,\pi}{2}\right)\right]d\ell \\
&= \int_0^\tau \left[\frac{\ell}{\sin^2\ell}-\frac{\cos\ell}{\sin\ell}\right]d\ell = \int_0^\tau \left[\frac{\ell-\sin\ell\cos\ell}{\sin^2\ell}\right]d\ell \\
&= \int_0^\tau \left[\frac{\ell-(\ell+\ell^3\,\Xi(\ell))(1+\ell^2\,\Xi(\ell))}{\sin^2\ell}\right]d\ell \\
&= \int_0^\tau \frac{\ell^3\,\Xi(\ell)}{\sin^2\ell}\,d\ell = \int_0^\tau \ell\,\Xi(\ell)\,d\ell \\
&= \tau^2\int_0^1 \lambda\,\Xi(\tau\lambda)\,d\lambda = \tau^2\,\Xi(\tau),
\end{aligned}$$

where the possibility of renaming Ξ at each step of the calculation was repeatedly utilized. This proves (A.1.13).

Now, by (A.1.13),

$$\tau\tan\left(\tau-\frac{\operatorname{sign}\tau\,\pi}{2}\right)=-1+\tau^2\,\Xi(\tau),$$

and consequently, by (A.1.10),

$$\begin{aligned}
\varphi(\tau) &= \frac{1}{\exp\left(\left(\tau-\frac{\operatorname{sign}\tau\,\pi}{2}\right)\left(-\frac{1}{\tau}+\tau\,\Xi(\tau)\right)\right)} \\
&= \frac{1}{\exp\left(\frac{\operatorname{sign}\tau\,\pi}{2\tau}-1+\tau\,\Xi(\tau)\right)} \\
&= \frac{1}{\exp\left(\frac{\pi}{2|\tau|}-1+\tau\,\Xi(\tau)\right)}.
\end{aligned}$$

It is thereby convenient to reparameterize this curve by setting $t := \frac{\operatorname{sign}\tau}{\exp\left(\frac{\pi}{2|\tau|}\right)}$. In this way, we have that $\operatorname{sign} t = \operatorname{sign}\tau$ and

$$\tau = -\frac{\operatorname{sign} t\,\pi}{2\ln|t|}.$$

Hence, with a slight abuse of notation, for small t, we can write

$$\begin{aligned}\operatorname{sign}\tau\,\varphi(\tau) &= \frac{\operatorname{sign}\tau}{\exp\left(\frac{\pi}{2|\tau|}\right)\exp\left(-1+\tau\,\Xi(\tau)\right)} = \frac{t}{\exp\left(-1+\frac{1}{\ln|t|}\,\Xi\left(\frac{1}{\ln|t|}\right)\right)} \\ &= \frac{et}{1+\frac{1}{\ln|t|}\,\Xi\left(\frac{1}{\ln|t|}\right)} = et\left(1+\frac{1}{\ln|t|}\,\Xi\left(\frac{1}{\ln|t|}\right)\right).\end{aligned}$$

Moreover,

$$\sin\tau = \tau+\tau^3\,\Xi(\tau) = -\frac{\operatorname{sign}t\,\pi}{2\ln|t|}+\frac{1}{\ln^3|t|}\,\Xi\left(\frac{1}{\ln|t|}\right)$$

and

$$\cos\tau = 1+\tau^2\,\Xi(\tau) = 1+\frac{1}{\ln^2|t|}\,\Xi\left(\frac{1}{\ln|t|}\right).$$

These observations allow us to recast the parametric curve in (A.1.11) into

$$\begin{cases} x(t) &= et\left(1+\frac{1}{\ln|t|}\Xi\left(\frac{1}{\ln|t|}\right)\right)\left(-\frac{\operatorname{sign}t\,\pi}{2\ln|t|}+\frac{1}{\ln^3|t|}\,\Xi\left(\frac{1}{\ln|t|}\right)\right) \\ &= et\left(-\frac{\operatorname{sign}t\,\pi}{2\ln|t|}+\frac{1}{\ln^2|t|}\,\Xi\left(\frac{1}{\ln|t|}\right)\right), \\ y(t) &= et\left(1+\frac{1}{\ln|t|}\,\Xi\left(\frac{1}{\ln|t|}\right)\right)\left(1+\frac{1}{\ln^2|t|}\,\Xi\left(\frac{1}{\ln|t|}\right)\right) \\ &= et\left(1+\frac{1}{\ln|t|}\,\Xi\left(\frac{1}{\ln|t|}\right)\right). \end{cases} \tag{A.1.14}$$

From the latter equation, we deduce that

$$\frac{1}{e}\lim_{t\to0}\dot{y}(t) = \lim_{t\to0}\frac{d}{dt}\left[t\left(1+\frac{1}{\ln|t|}\,\Xi\left(\frac{1}{\ln|t|}\right)\right)\right] = 1,$$

and as a result, according to the inverse function theorem, for small y,

$$t = \frac{y}{e}\left(1+\frac{1}{\ln|y|}\,\Xi\left(\frac{1}{\ln|y|}\right)\right).$$

Plugging this information into the first equation in (A.1.14), we obtain that $\partial\Omega_o$, in the vicinity of the origin, can be written as a graph in the horizontal direction in the form

$$\begin{aligned} x &= y\left(-\frac{\operatorname{sign}y\,\pi}{2\ln|y|}+\frac{1}{\ln^2|y|}\,\Xi\left(\frac{1}{\ln|y|}\right)\right) \\ &= -\frac{\pi|y|}{2\ln|y|}+\frac{1}{\ln^2|y|}\,\Xi\left(\frac{|y|}{\ln|y|}\right) =: F(y). \end{aligned} \tag{A.1.15}$$

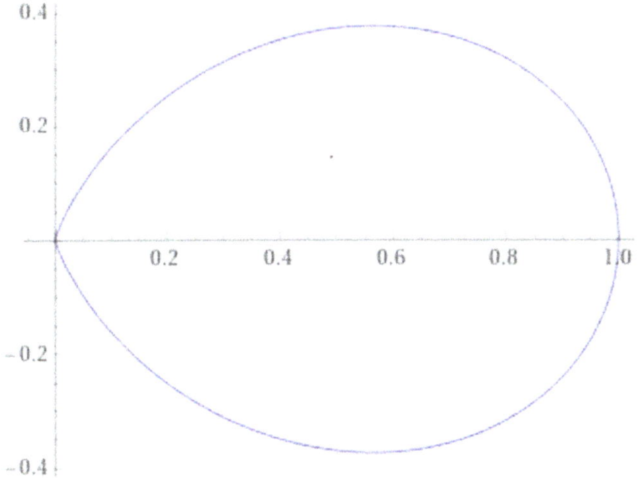

Figure A.1.2 The set Ω_o in (A.1.5).

Since F is of class C^1 near $y = 0$, this proves (A.1.6).

Note also that

$$\text{the interior ball condition for } \Omega_o \text{ is not satisfied at the origin} \tag{A.1.16}$$

because, by (A.1.15),

$$\lim_{y \to 0} |F''(y)| = +\infty.$$

See Figure A.1.2 for a sketch of the domain Ω_o.

Besides, we point out that, by (A.1.2),

$$\begin{aligned} \liminf_{\Omega_o \ni (x,y) \to (1,0)} u(x,y) &\leqslant \lim_{x \nearrow 1} u(x,0) \\ &= \lim_{\rho \nearrow 1} \frac{\rho\,(\ln \rho + 0)}{(\ln \rho)^2 + 0} = \lim_{\rho \nearrow 1} \frac{1}{\ln \rho} = -\infty. \end{aligned} \tag{A.1.17}$$

Therefore, it is convenient to consider a domain $\Omega \subset \Omega_o$ such that Ω is of class C^1 and $\Omega \supset \Omega_o \cap \{x \leqslant 1/2\}$. In this way, we have that $u \in C^2(\Omega) \cap C^1(\overline{\Omega})$ due to (A.1.4), and by construction,

$$\overline{\Omega} \subseteq \overline{\Omega_o} \subseteq \overline{\{u < 0\}}.$$

From this, we infer that if $x \in \partial\Omega$, then $u(x) \leqslant 0 = u(0,0)$, whence

$$u(0,0) = \max_{\partial\Omega} u.$$

These observations yield that all the assumptions of Lemma 2.16.1 are satisfied, with the exception of the inner ball condition. However, the thesis of Lemma 2.16.1 is violated since $\partial_\nu u(0,0) = 0$ owing to (A.1.4), which showcases the fact that the inner ball condition cannot be removed from Lemma 2.16.1.

Appendix B

An Application to Physical Geodesy

B.1 Determination of the Gravitational Potential

In what follows, relying on combined tools from partial differential equations, functional analysis and statistics, we briefly discuss the following two classical problems of physical geodesy (see [Lau73, Lau75, Dia88]):

- Given some experimental data on the gravity potential measured at some specific locations, is it possible to have some *knowledge of the gravity potential at other sites, possibly inaccessible to experiments?*
- is it possible to extrapolate from these data some *knowledge of the mass density of the Earth in unreachable locations, such as the center of the Earth?*

Concrete answers to these two very fascinating, but quite difficult, questions will be provided, respectively, in (B.1.12) and (B.1.17)–(B.1.18).

Let us now dive deeper into details, following [Lau73, Lau75, Dia88], for an application of elliptic partial differential equations, spherical harmonics, reproducing kernels and statistics to the determination of the gravity field. Despite the scholastic approach to the problem, which gets dismissed by stating that the gravity acceleration is everywhere constantly equal to 9.8 m/s^2, the problem is technically more complicated, also in view of the lack of homogeneity of the Earth. Namely, anomalous density distributions within the Earth can produce significant gravity anomalies that need to be taken into account in engineering and technological applications. In rough terms, negative gravity anomalies can be produced by local deficiencies in mass due to rocks which are less dense than the average (such as granite plutons and sedimentary basins), and conversely, positive gravity anomalies can be the outcome of the presence of dense rocks (such as basalt). One of the objectives of physical geodesy is to have a better understanding of the gravity field starting from the millions of experimental measurements taken with an uneven distribution (though, nowadays, measurements of satellite orbits make everything much more precise) to obtain a reliable "educated guess" at points where measurements are not available. For this, mathematics seems indeed to play a pivotal[1] role. Here, we try to provide an exposition accessible to a broad audience, possibly by skating over some technical details, while highlighting some of the brilliant ideas that emerged in the analysis of this fascinating topic (see also [HWM05] for the basics of physical geodesy).

The gravity force at a point $x \in \mathbb{R}^3$ produced by a point mass μ located at a point x_0 is proportional to $-\frac{\mu\,(x-x_0)}{|x-x_0|^2}$, which corresponds to a potential proportional to $\frac{\mu}{|x-x_0|}$ (being the gravitational force obtained as the gradient of the potential). Remarkably, such a gravity potential is harmonic outside the given mass source (recall (2.7.6) and the fact that $n = 3$). The gravity force produced by an extensive mass can be seen as the superposition of the forces produced by all "infinitesimal point masses" that constitute it; hence, this composite force can also be seen as the gradient of a gravitation potential, which is harmonic away from the mass itself. Recalling Section 2.21, we write this potential as a superposition of spherical harmonics, according to the so-called "Laplace expansion of the gravitational potential." For this, we point out that, if N_k is the dimension of the space of harmonic homogeneous polynomials of degree k, as given in (2.21.15), the physical case $n = 3$ produces $N_k = 2k + 1$. Furthermore, if $|y| < |x|$, using the notation $\tau := \frac{|y|}{|x|}$ and $t := \frac{x\cdot y}{|x|\,|y|}$, the generating function expression[2] of Legendre polynomials in (2.22.18) when $n = 3$ gives that

$$\begin{aligned}\frac{1}{|x-y|} &= \frac{1}{\sqrt{|x|^2+|y|^2-2x\cdot y}} = \frac{1}{|x|\sqrt{1-2t\tau+\tau^2}}\\ &= \frac{1}{|x|}\sum_{k=0}^{+\infty} P_k(t)\,\tau^k = \sum_{k=0}^{+\infty} P_k\left(\frac{x\cdot y}{|x|\,|y|}\right)\frac{|y|^k}{|x|^{k+1}}.\end{aligned} \tag{B.1.1}$$

Also, the additional identity for Legendre polynomials in Theorem 2.22.5 says that

$$P_k\left(\frac{x\cdot y}{|x|\,|y|}\right) = \frac{\mathcal{H}^2(\partial B_1)}{N_k}\,F_k\left(\frac{x}{|x|},\frac{y}{|y|}\right) = \frac{4\pi}{2k+1}\sum_{j=1}^{2k+1} Y_{k,j}\left(\frac{x}{|x|}\right)Y_{k,j}\left(\frac{y}{|y|}\right),$$

[1] *"It is a lively world, which can almost certainly benefit from some more mathematics,"* see [Dia88, p. 170]. *"It seems to me dangerous to give too much attention to the non-theoretical approach to physical geodesy [...] An exact theoretical study of the probabilistic background of the method, as I have tried to carry through here, will often tell you, how the problems have been changed and hidden in place, where they are difficult to discover,"* see [Lau73, pp. 92–93].

[2] Concretely, using for instance (2.22.14), (2.22.17) or (2.22.18), one can list the first 10 Legendre

where $\{Y_{k,1}, \ldots, Y_{k,2k+1}\}$ is a basis of spherical harmonics that are orthonormal with respect to the scalar product in $L^2(\partial B_1)$. This and (B.1.1) lead to

$$\frac{\mu}{|x-y|} = \sum_{\substack{k\geqslant 0\\1\leqslant j\leqslant 2k+1}} \frac{4\pi\mu}{2k+1} Y_{k,j}\left(\frac{x}{|x|}\right) Y_{k,j}\left(\frac{y}{|y|}\right) \frac{|y|^k}{|x|^{k+1}}. \tag{B.1.2}$$

If, up to scaling, we assume the Earth to be a (not necessarily homogeneous) ball of radius equal to 1, we can have that the gravity potential at a point $x \in \mathbb{R}^3 \setminus B_1$ is generated by the superposition of infinitesimal masses at y distributed according to some measure $d\mu_y$.

We observe that, up to normalization constants, the homogeneous case corresponding to $d\mu_y = dy$ produces a potential at $x \in \mathbb{R}^3 \setminus B_1$ of the type

$$U_0(x) := \int_{B_1} \frac{dy}{|x-y|} = \frac{4\pi}{3\,|x|}, \tag{B.1.3}$$

see Lemma 2.7.5. In general, without the homogeneity assumption, formula (B.1.2) produces a gravity potential at $x \in \mathbb{R}^3 \setminus B_1$ of the form

$$\begin{aligned} U(x) &:= \int_{B_1} \frac{d\mu_y}{|x-y|} \\ &= \sum_{\substack{k\geqslant 0\\1\leqslant j\leqslant 2k+1}} \frac{4\pi}{(2k+1)\,|x|^{k+1}} Y_{k,j}\left(\frac{x}{|x|}\right) \int_{B_1} Y_{k,j}\left(\frac{y}{|y|}\right) |y|^k\, d\mu_y. \end{aligned} \tag{B.1.4}$$

Since the mass distribution inside the Earth is something we do not know much about, we can formalize our uncertainty about the true value of $d\mu_y$ by treating

polynomials when $n = 3$ as follows:

k	$P_k(t)$
0	1
1	t
2	$\frac{1}{2}\left(3t^2 - 1\right)$
3	$\frac{1}{2}\left(5t^3 - 3t\right)$
4	$\frac{1}{8}\left(35t^4 - 30t^2 + 3\right)$
5	$\frac{1}{8}\left(63t^5 - 70t^3 + 15t\right)$
6	$\frac{1}{16}\left(231t^6 - 315t^4 + 105t^2 - 5\right)$
7	$\frac{1}{16}\left(429t^7 - 693t^5 + 315t^3 - 35t\right)$
8	$\frac{1}{128}\left(6435t^8 - 12012t^6 + 6930t^4 - 1260t^2 + 35\right)$
9	$\frac{1}{128}\left(12155t^9 - 25740t^7 + 18018t^5 - 4620t^3 + 315t\right)$
10	$\frac{1}{256}\left(46189t^{10} - 109395t^8 + 90090t^6 - 30030t^4 + 3465t^2 - 63\right)$

and higher-degree Legendre polynomials can be computed similarly.

this measure as a random quantity. For instance, we could replace $d\mu_y$ with $dy + \mu(y, \varpi)\, dy$, where dy takes care of the first approximation in which the Earth is a uniformly distributed round ball as in (B.1.3), the term $\mu(y, \varpi)\, dy$ addresses the correction needed to this uniform model and the parameter ϖ belongs to a sample space of possible outcomes which somehow encodes our uncertainty on the values of the mass distribution. We take a suitable Gaußian *ansatz* on the random variables, which are normalized to have an expected value of zero and a suitable variance. Namely, we suppose that

$$\mathbb{E}\big(\mu(y, \cdot)\big) = 0;$$

that is, the expected value of the mass distribution of the Earth coincides with that of the uniform density model, and

$$\mathbb{E}\big(\mu(X, \cdot)\, dX \times \mu(Y, \cdot)\, dY\big) = \overline{\mu}(|X|)\, \delta_0\big(X - Y\big)\, dX\, dY, \tag{B.1.5}$$

for some function $\overline{\mu}$, where δ_0 is the Dirac delta function[3] at the origin. In rough terms, the *ansatz* in (B.1.5) states that the mass distribution covariance is expected to be uniform along each sphere ∂B_ρ with $\rho \in (0, 1)$, and the distributions at different layers are expected to be independent. In this way, the gravity potential in (B.1.4) also becomes a random quantity of the form $U(x, \varpi) = U_0(x) + \xi(x, \varpi)$, with U_0 corresponding to the homogeneous case in (B.1.3) and

$$\xi(x, \varpi) = \sum_{\substack{k \geqslant 1 \\ 1 \leqslant j \leqslant 2k+1}} \frac{4\pi}{(2k+1)\, |x|^{k+1}}\, Y_{k,j}\left(\frac{x}{|x|}\right) \int_{B_1} Y_{k,j}\left(\frac{y}{|y|}\right) |y|^k\, \mu(y, \varpi)\, dy. \tag{B.1.6}$$

It is also convenient to introduce the "covariance function"

$$r(x, y) := \mathbb{E}\big(\xi(x, \cdot)\, \xi(y, \cdot)\big). \tag{B.1.7}$$

The idea is now to take a suitable *ansatz* on $r(x, y)$ that can be compared with the available data in order to make "educated guesses" on the gravity potential at places where direct measurements are not available. For this, combining (B.1.5), (B.1.6) and (B.1.7), we see that

$$\begin{aligned} r(x, y) = &\sum_{\substack{m \geqslant 1 \\ 1 \leqslant i \leqslant 2m+1 \\ k \geqslant 1 \\ 1 \leqslant j \leqslant 2k+1}} \frac{16\pi^2}{(2m+1)\,(2k+1)\, |x|^{m+1}\, |y|^{k+1}}\, Y_{m,i}\left(\frac{x}{|x|}\right) Y_{k,j}\left(\frac{y}{|y|}\right) \\ &\times \iint_{B_1 \times B_1} Y_{m,i}\left(\frac{X}{|X|}\right) Y_{k,j}\left(\frac{Y}{|Y|}\right) |X|^m\, |Y|^k\, \mathbb{E}\big(\mu(X, \cdot)\, dX \times \mu(Y, \cdot)\, dY\big) \end{aligned}$$

[3]More precisely, the meaning of $\delta_0(X - Y)\, dX\, dY$ is that, for every $\varphi \in C(B_1 \times B_1)$,

$$\iint_{B_1 \times B_1} \varphi(X, Y)\, \delta_0(X - Y)\, dX\, dY = \int_{B_1} \varphi(X, X)\, dX.$$

$$= \sum_{\substack{m\geqslant 1\\ 1\leqslant i\leqslant 2m+1\\ k\geqslant 1\\ 1\leqslant j\leqslant 2k+1}} \frac{16\pi^2}{(2m+1)(2k+1)|x|^{m+1}|y|^{k+1}} Y_{m,i}\left(\frac{x}{|x|}\right) Y_{k,j}\left(\frac{y}{|y|}\right)$$

$$\times \int_{B_1} Y_{m,i}\left(\frac{X}{|X|}\right) Y_{k,j}\left(\frac{X}{|X|}\right) |X|^{m+k}\,\overline{\mu}(|X|)\,dX. \tag{B.1.8}$$

It is also convenient to recall the orthogonality condition in (2.22.1) and use polar coordinates to see that

$$\begin{aligned}&\int_{B_1} Y_{m,i}\left(\frac{X}{|X|}\right) Y_{k,j}\left(\frac{X}{|X|}\right) |X|^{m+k}\,\overline{\mu}(|X|)\,dX\\ &= \int_0^1 \left(\int_{\partial B_1} Y_{m,i}(\vartheta)\,Y_{k,j}(\vartheta)\,r^{m+k+2}\,\overline{\mu}(r)\,d\mathcal{H}^2_\vartheta\right) dr\\ &= \delta_{mk}\,\delta_{ij}\int_0^1 r^{2k+2}\,\overline{\mu}(r)\,dr.\end{aligned}$$

Therefore, defining

$$\sigma_k := \sqrt{\int_0^1 r^{2k+2}\,\overline{\mu}(r)\,dr}, \tag{B.1.9}$$

we can rewrite (B.1.8) as

$$r(x,y) = \sum_{\substack{k\geqslant 1\\ 1\leqslant j\leqslant 2k+1}} \frac{16\pi^2\,\sigma_k^2}{(2k+1)^2\,|x|^{k+1}|y|^{k+1}} Y_{k,j}\left(\frac{x}{|x|}\right) Y_{k,j}\left(\frac{y}{|y|}\right).$$

Hence, using again the additional identity for Legendre polynomials in Theorem 2.22.5 and recalling the setting in (2.22.2),

$$r(x,y) = \sum_{k\geqslant 1} \frac{4\pi\,\sigma_k^2}{(2k+1)\,|x|^{k+1}|y|^{k+1}} P_k\left(\frac{x}{|x|}\cdot\frac{y}{|y|}\right). \tag{B.1.10}$$

The values of σ_k, which are still to be determined, can be inferred using the data of the measurements of the gravity anomalies[4] (that are the scalar discrepancies between the measured gravity accelerations and the ones predicted by the homogeneous model) at different points $x_1,\dots,x_N$.

[4]In the classical physical geodesy, the gravity anomaly at a point $\overline{x}$ is computed as the difference between the intensity of the actual gravity force field at $\overline{x}$ and the ideal one at the point $\underline{x}$, which is the projection of $\overline{x}$ on the surface $\{U_0 = U(\overline{x})\}$, omitting the random dependence on ϖ for notational simplicity, see Figure B.1.1. Concretely, in view of (B.1.3), let us consider, for instance $\overline{x} \in \left\{U = \frac{4\pi}{3}\right\}$, $\underline{x} \in \partial B_1 = \left\{U_0 = \frac{4\pi}{3}\right\}$ and suppose that $\varepsilon := |\overline{x} - \underline{x}|$ is small. Then, the gravity anomaly $\mathcal{A}$ at $\overline{x}$ is,

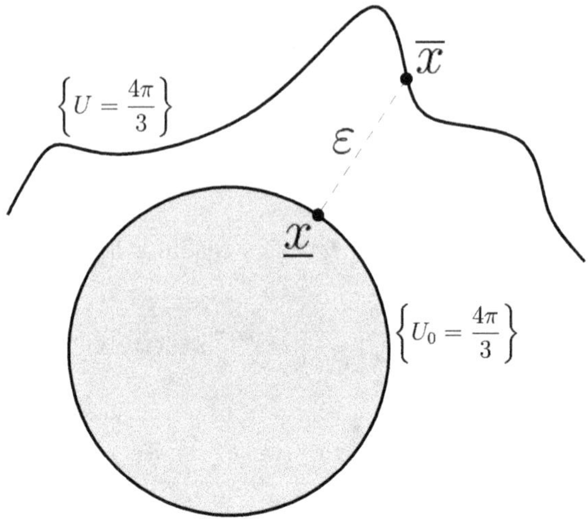

Figure B.1.1 Computing the gravity anomaly.

---

expanding in the perturbation parameters ε and ξ,

$$\begin{aligned}\mathcal{A}(\overline{x}) &:= |\nabla U(\overline{x})| - |\nabla U_0(\underline{x})| = |\nabla U_0(\overline{x}) + \nabla \xi(\overline{x})| - |\nabla U_0(\underline{x})| \\ &= \left| -\frac{4\pi\, \underline{x}}{3(1+\varepsilon)^2} + \nabla \xi(\overline{x}) \right| - \frac{4\pi}{3} = \frac{4\pi}{3}\left(\left| -\frac{\underline{x}}{(1+\varepsilon)^2} + \frac{3\nabla \xi(\overline{x})}{4\pi} \right| - 1 \right) \\ &= \frac{4\pi}{3}\left(\sqrt{\frac{1}{(1+\varepsilon)^4} + \frac{9|\nabla \xi(\overline{x})|^2}{16\pi^2} - \frac{3\nabla \xi(\overline{x}) \cdot \underline{x}}{2(1+\varepsilon)\pi}} - 1 \right) \simeq \frac{4\pi}{3}\left(-2\varepsilon - \frac{3\nabla \xi(\overline{x}) \cdot \underline{x}}{4\pi} \right) \\ &\simeq \frac{4\pi}{3}\left(-\frac{3\nabla \xi(\overline{x}) \cdot \underline{x}}{4\pi} - \frac{2\varepsilon}{|\overline{x}|} \right).\end{aligned}$$

Since

$$\frac{4\pi}{3} = U(\overline{x}) = U_0(\overline{x}) + \xi(\overline{x}) = \frac{4\pi}{3(1+\varepsilon)} + \xi(\overline{x}) \simeq \frac{4\pi(1-\varepsilon)}{3} + \xi(\overline{x}),$$

we find that $\frac{4\pi\varepsilon}{3} \simeq \xi(\overline{x})$, and accordingly, the gravity anomaly in spherical approximation is outlined by

$$\mathcal{A}(\overline{x}) \simeq -\nabla \xi(\overline{x}) \cdot \underline{x} - \frac{2\xi(\overline{x})}{|\overline{x}|}.$$

The identity

$$-\partial_h \xi - \frac{2\xi}{h} = \mathcal{A},$$

where h is the vertical displacement from the center of the Earth is sometimes considered the "fundamental equation of physical geodesy in spherical approximation" and can be concretely used to infer the values of the covariance function in terms of the observed data, see e.g. [Lau75, formulas (3.1) and (3.2)].

Interestingly, since ξ is harmonic outside B_1, the fundamental equation of the physical geodesy in spherical approximation along ∂B_1 can be considered as a (Robin-type) boundary condition for an elliptic problem.

Comparing with the observations, it was proposed in [Lau73] to consider the values

$$\sigma_k^2 := \begin{cases} 0 & \text{if } k \in \{1,2\}, \\ \dfrac{2k+1}{4\pi\,(k-1)(k-2)} & \text{if } k \geqslant 3, \end{cases} \tag{B.1.11}$$

up to normalizing constants, which we omit here for the sake of simplicity. Remarkably, this choice, which appeared to be in good agreement with the measures, also provides a significant mathematical simplification, reducing the series in (B.1.10) to a simpler closed form. Indeed, replacing (B.1.10) with the choice in (B.1.11), one obtains

$$r(x,y) = \sum_{k\geqslant 3} \frac{1}{(k-1)(k-2)\,|x|^{k+1}\,|y|^{k+1}}\, P_k\left(\frac{x}{|x|}\cdot\frac{y}{|y|}\right), \tag{B.1.12}$$

and this has a closed expression since the quantity

$$\sum_{k\geqslant 3} \frac{\tau^k}{(k-1)(k-2)}\, P_k(t) \tag{B.1.13}$$

can be obtained by exploiting the generating function definition[5] of Legendre polynomials in (2.22.18).

It is also interesting to remark that a precise knowledge of the gravitational potential can be inferred by that of the covariance function according to the following reproducing kernel procedure. We use the notation $r^y(x) := r(x,y)$ and define

$$\langle r^y, r^z \rangle := r(y,z).$$

Let us remark that with the use of GPS for gravity measurements, it is expected that the use of the gravity anomaly will be replaced by that of "gravity disturbance" (with the simplification of one term in the fundamental equation, thus reducing to a Neumann-type boundary condition, see for example [HWM05, pp. 93 and 95] for further details on this technical point).

We also stress that in our notes, for the sake of simplicity, we took the liberty of following a purely spherical approximation rather than an ellipsoidal approximation (though of course this approach is too crude for accurate geodesic analysis, see [HWM05, p. 97] for explicit warnings about the careless use of this approximation).

[5]Namely, according to (2.22.19), we note also that $\mu_k = 1$ when $n = 3$; therefore,

$$\frac{1}{\sqrt{1-2t\tau+\tau^2}} = \sum_{k=0}^{+\infty} P_k(t)\,\tau^k.$$

Consequently, recalling that $P_0(t) = 1$, $P_1(t) = t$ and $P_2(t) = \frac{1}{2}(3t^2-1)$ (see for example the table on p. 621), one can obtain a closed-form expression for (B.1.13) through algebraic manipulations and by integrating in τ.

We extend this "scalar product" by linearity, namely, for every $\alpha_1, \dots, \alpha_N, \beta_1, \dots, \beta_N \in \mathbb{R}$,

$$\left\langle \sum_{i=1}^{N} \alpha_i r^{x_i}, \sum_{j=1}^{N} \beta_j r^{x_j} \right\rangle := \sum_{1 \leqslant i,j \leqslant N} \alpha_i \beta_j \langle r^{x_i}, r^{x_j} \rangle = \sum_{1 \leqslant i,j \leqslant N} \alpha_i \beta_j r(x_i, x_j).$$

With this, we can introduce a notion of "norm" by setting

$$\left\| \sum_{i=1}^{N} \alpha_i r^{x_i} \right\| := \sqrt{\left\langle \sum_{i=1}^{N} \alpha_i r^{x_i}, \sum_{j=1}^{N} \alpha_j r^{x_j} \right\rangle} = \sqrt{\sum_{1 \leqslant i,j \leqslant N} \alpha_i \alpha_j r(x_i, x_j)}.$$

The advantage of this setting is that it preserves the (expected value of the) size of the original random variable in the sense that

$$\begin{aligned}
&\mathbb{E}\left(\left| \xi(x, \varpi) - \sum_{i=1}^{N} \alpha_i \, \xi(x_i, \varpi) \right|^2 \right) - \left\| r^x - \sum_{i=1}^{N} \alpha_i r^{x_i} \right\|^2 \\
&= \mathbb{E}\left(\xi(x, \varpi)\xi(x, \varpi) + \sum_{1 \leqslant i,j \leqslant N} \alpha_i \alpha_j \, \xi(x_i, \varpi)\xi(x_j, \varpi) - 2\sum_{i=1}^{N} \alpha_i \, \xi(x, \varpi) \, \xi(x_i, \varpi) \right) \\
&\quad - r(x, x) - \sum_{1 \leqslant i,j \leqslant N} \alpha_i \alpha_j r(x_i, x_j) + 2\sum_{i=1}^{N} \alpha_i r(x, x_i) \\
&= 0.
\end{aligned}$$

Therefore, given $x, x_1, \dots, x_N$,

$$\begin{aligned}
&\text{the coefficients } \alpha_1, \dots, \alpha_N \text{ minimize} \\
&\text{the expected value of } \left| \xi(x, \varpi) - \sum_{i=1}^{N} \alpha_i \, \xi(x_i, \varpi) \right|^2 \\
&\text{if and only if they minimize } \left\| r^x - \sum_{i=1}^{N} \alpha_i r^{x_i} \right\|^2 .
\end{aligned} \tag{B.1.14}$$

In our setting, the points $x_1, \dots, x_N$ correspond to the sites where the measurements have been taken (say, for the sake of simplicity, the values $\xi(x_1, \varpi)$ are known), while the location x is inaccessible to experiments, and the value of $\xi(x, \varpi)$ has to be inferred from the available data.

The minimal α_i in (B.1.14) can be determined by a first variation of the problem; that is, fixing $j \in \{1, \dots, N\}$, the minimal α_i satisfies

$$\begin{aligned}0 &= \lim_{\varepsilon\to 0} \frac{1}{2\varepsilon}\left[\left\| r^x - \sum_{i=1}^{N}(\alpha_i + \varepsilon\delta_{ij}) r^{x_i}\right\|^2 - \left\| r^x - \sum_{i=1}^{N}\alpha_i r^{x_i}\right\|^2\right] \\ &= -r(x, x_j) + \sum_{i=1}^{N}\alpha_i r(x_j, x_i).\end{aligned}$$

Supposing that the symmetric square matrix $\{r(x_j, x_i)\}_{1\leqslant i,j\leqslant N}$ is invertible and denoting by r^{ij} its inverse, we thus obtain that the minimal coefficients satisfy

$$\alpha_k = \sum_{j=1}^{N} r^{jk}\, r(x, x_j). \tag{B.1.15}$$

Accordingly, recalling (B.1.14), one takes as "best approximation" $\xi_\star(x)$, for the value of $\xi(x, \varpi)$, the quantity $\sum_{k=1}^{N}\alpha_k\, \xi(x_k, \varpi)$ with α_k as in (B.1.15). As a result, if the observed values of the potential are denoted by

$$m_i := \xi(x_i, \varpi), \tag{B.1.16}$$

we set

$$b_j := \sum_{k=1}^{N} r^{jk}\, m_k$$

and we find that

$$\xi_\star(x) = \sum_{1\leqslant j,k\leqslant N} r^{jk}\, r(x, x_j)\, \xi(x_k, \varpi) = \sum_{1\leqslant j,k\leqslant N} r^{jk}\, r(x, x_j)\, m_k = \sum_{j=1}^{N} b_j\, r(x, x_j).$$

The problem of an "educated guess" on ξ at the point x is therefore reduced to that of determining with a sufficiently good approximation the covariance function matrix and its inverse, for which explicit expressions can be of great use.

Following [Lau75], it is also interesting to observe that, as a byproduct of the choice of the coefficients of the covariance function in (B.1.11), the measurements of the gravity anomaly seem to indicate a mass distribution within the Earth, which becomes more and more irregular as one approaches the center. To check this intriguing statement, one can consider the case in which

$$\overline{\mu}(r) := \frac{1}{2\pi}\left(\frac{5}{r^7} - \frac{3}{r^5}\right) \tag{B.1.17}$$

and observe that, by (B.1.9), one obtains in this situation,

$$\sigma_k^2 = \frac{1}{2\pi}\int_0^1 \left(5r^{2k-5} - 3r^{2k-3}\right) dr$$
$$= \frac{1}{2\pi}\left(\frac{5}{2(k-2)} - \frac{3}{2(k-1)}\right) = \frac{2k+1}{4\pi(k-1)(k-2)}$$

when $k \geqslant 3$, which coincides precisely with the choice in (B.1.11).

Namely, the covariance choice in (B.1.11) that nicely fits the experimental data is closely related to a covariance measure, as in (B.1.17), which is singular as $r \searrow 0$. This is an indication of the possibility that

the variation of the mass distribution is significantly different between the surface of the Earth and the vicinity of its center. (B.1.18)

A slightly different approach to the problem has been sketched in [Dia88, Section B] using a purely "Bayesian approach." In rough terms, one takes a development of the gravity potential on the surface of the Earth of the form

$$\xi(x,\varpi) = \sum_{k=0}^{+\infty} c_k(\varpi)\, u_k(x),$$

for suitable spherical harmonics u_k and random functions c_k that are assumed to be Gaußian, with zero mean and variance ς_k^2, namely

$$\mathbb{E}(c_k) = 0 \quad \text{and} \quad \mathbb{E}(c_i c_j) = \delta_{ij}\varsigma_i^2. \tag{B.1.19}$$

This set of assumptions can be seen as a "prior," somewhat based on a subjective choice, from which, in light of the N known measurements, one can obtain a "posterior" estimating the potential according to Bayes' law, which in turn computes the probability of an event based on some knowledge of other related events. Since the inference is based on actual data and the predictions can be confronted with newly collected measurements, one has some control on the reliability of this method. The core of this is the idea that the probability that two events, say A and B, have both occurred is the "conditional probability of A given B" (that is, the probability that A has occurred, given that B has occurred) multiplied by the probability that B has occurred. That is, denoting by $P(A)$ and $P(B)$ the probabilities[6] of A and B,

[6]Note that, exchanging the roles of A and B in (B.1.20),

$$P(A \cap B) = P(B|A)\, P(A),$$

leading to

$$P(A|B) = \frac{P(B|A)\, P(A)}{P(B)}.$$

See for example [Bol07, pp. 63–70] for an accessible introduction to Bayesian statistics. In our setting, the left-hand side of this identity represents the posterior probability (that is, the estimated range of values for the gravitational potential at a point x given the evidence of the measurements at $x_1, \ldots, x_N$) and the conditional probability on the right-hand side can be considered a "likelihood function."

respectively, and by $P(A|B)$ the conditional probability of A given B,

$$P(A \cap B) = P(A|B)\, P(B). \tag{B.1.20}$$

Namely, given $\delta > 0$ and $v, v_1, \dots, v_N \in \mathbb{R}$, we set $I := (v - \delta, v + \delta)$, $I_1 := (v_1 - \delta, v_1 + \delta), \dots, I_N := (v_N - \delta, v_N + \delta)$, and we take A as the event "the value of the potential at x lies in I" and B as the event "for all $i \in \{1, \dots, N\}$ the value of the potential at x_i lies in I_i." We also recall that the linear combination of independent Gaußian random variables is also Gaußian (see for example [Ros07, Example 2.46 and p. 73]), and in particular, by (B.1.19), the random variable $\xi(x, \omega)$ is Gaußian with zero mean and variance

$$S(x) := \sum_{k=1}^{N} (u_k(x))^2 \varsigma_i^2.$$

We stress that this quantity is known in dependence on the prior variances ς_i and on the point x (since the spherical harmonics are given, and we suppose that we can compute their values at a given point with sufficient accuracy). That is, for a given i, if we denote by B_i the event "the value of the potential at x_i lies in I_i," we have that $B = B_1 \cap \dots \cap B_N$. We consider the joint probability density function $\gamma_{B_1,\dots,B_N}$ (which, under nondegeneracy assumptions, is explicitly known given the covariances of $\xi(x_1, \omega), \dots, \xi(x_N, \omega)$, see for example [JP03, Theorem 16.1], [KM05, formula (2.27) in Chapter 2] or [Ser09, formula (5.7)]), thus obtaining

$$P(B) = \int_{I_1 \times I_N} \gamma_{B_1,\dots,B_N}(\varpi_1, \dots, \varpi_N)\, d\varpi_1 \dots d\varpi_N. \tag{B.1.21}$$

Similarly, considering the joint probability density function $\gamma_{A,B_1,\dots,B_N}$,

$$P(A \cap B) = \int_{I \times I_1 \times I_N} \gamma_{A,B_1,\dots,B_N}(\varpi_0, \varpi_1, \dots, \varpi_N)\, d\varpi_0 \dots d\varpi_N. \tag{B.1.22}$$

In this way, one can exploit (B.1.20), (B.1.21) and (B.1.22) to determine the probability $P(A|B)$ for the gravity potential at the point x to lie in the given range I of values given the known measures that determine the potential at points $x_1, \dots, x_N$ in the intervals $I_1, \dots, I_N$.

References

[AJS21] N. Abatangelo, S. Jarohs, and A. Saldaña, Fractional Laplacians on ellipsoids. *Math. Eng.* **3**(5) (2021). Paper No. 038, 33. DOI: 10.3934/mine.2021038. MR4181195.

[AV14] N. Abatangelo and E. Valdinoci, A notion of nonlocal curvature. *Numer. Funct. Anal. Optim.* **35**(7–9), 793–815 (2014). DOI: 10.1080/01630563.2014.901837. MR3230079.

[AGLV09] P. Abry, P. Gonçalves, and J. Lévy Véhel (eds.), *Scaling, Fractals and Wavelets*. Digital Signal and Image Processing Series. ISTE, London; John Wiley & Sons, Inc., Hoboken (2009). MR2590673.

[Ada97] D. R. Adams, Potential and capacity before and after Wiener. *Proceedings of the Norbert Wiener Centenary Congress, 1994* (East Lansing, MI, 1994). Proceedings of Symposia in Applied Mathematics, Vol. 52. American Mathematical Society, Providence (1997), pp. 63–83. DOI: 10.1090/psapm/052/1440907. MR1440907.

[AH96] D. R. Adams and L. I. Hedberg, *Function Spaces and Potential Theory*. Grundlehren der Mathematischen Wissenschaften [Fundamental Principles of Mathematical Sciences], Vol. 314. Springer-Verlag, Berlin (1996). MR1411441.

[AL93] R. P. Agarwal and V. Lakshmikantham, *Uniqueness and Nonuniqueness Criteria for Ordinary Differential Equations*. Series in Real Analysis, Vol. 6. World Scientific Publishing Co., Inc., River Edge (1993). MR1336820.

[AKS10] M. Agranovsky, D. Khavinson, and H. S. Shapiro, Malmheden's theorem revisited. *Expo. Math.* **28**(4), 337–350 (2010). DOI: 10.1016/j.exmath.2010.03.002. MR2734448.

[Ahl37] L. V. Ahlfors, On Phragmén-Lindelöf's principle. *Trans. Amer. Math. Soc.* **41**(1), 1–8 (1937). DOI: 10.2307/1989875. MR1501888.

[Aik08] H. Aikawa, Equivalence between the boundary Harnack principle and the Carleson estimate. *Math. Scand.* **103**(1), 61–76 (2008). DOI: 10.7146/math.scand.a-15069. MR2464701.

[Ale61] A. D. Aleksandrov, Certain estimates for the Dirichlet problem. *Soviet Math. Dokl.* **1**, 1151–1154 (1961). MR0147776.

[Ale62] A. D. Aleksandrov, Uniqueness theorems for surfaces in the large. V. *Amer. Math. Soc. Transl.* (2) **21**, 412–416 (1962). MR0150710.

[Alm99] E. Almansi, Sull'integrazione dell'equazione differenziale $\Delta^{2n} = 0$. *Annali di Mat.* (3) **2**, 1–51 (1899) (Italian).

[Anc78] A. Ancona, Principe de Harnack à la frontière et théorème de Fatou pour un opérateur elliptique dans un domaine lipschitzien. *Ann. Inst. Fourier* (Grenoble) **28**(4), 169–213, x (1978) (French, with English summary). MR513885.

[And97] P. Andersson, Characterization of pointwise Hölder regularity. *Appl. Comput. Harmon. Anal.* **4**(4), 429–443 (1997). DOI: 10.1006/acha.1997.0219. MR1474098.

[AN16] D. E. Apushkinskaya and A. I. Nazarov, A counterexample to the Hopf-Oleinik lemma (elliptic case). *Anal. PDE* **9**(2), 439–458 (2016). DOI: 10.2140/apde.2016.9.439. MR3513140.

[AD08] W. Arendt and D. Daners, The Dirichlet problem by variational methods. *Bull. Lond. Math. Soc.* **40**(1), 51–56 (2008). DOI: 10.1112/blms/bdm091. MR2409177.

[AG01] D. H. Armitage and S. J. Gardiner, *Classical Potential Theory*. Springer Monographs in Mathematics. Springer-Verlag London, Ltd., London (2001). MR1801253.

[AIM09] K. Astala, T. Iwaniec, and G. Martin, *Elliptic Partial Differential Equations and Quasiconformal Mappings in the Plane*. Princeton Mathematical Series, Vol. 48. Princeton University Press, Princeton (2009). MR2472875.

[AC85] I. Athanasopoulos and L. A. Caffarelli, A theorem of real analysis and its application to free boundary problems. *Comm. Pure Appl. Math.* **38**(5), 499–502 (1985). DOI: 10.1002/cpa.3160380503. MR803243.

[Aud11] M. Audin, *Remembering Sofya Kovalevskaya*. Springer, London (2011). Translated from the 2008 French original. MR3013678.

[ABR01] S. Axler, P. Bourdon, and W. Ramey, *Harmonic Function Theory*, 2nd ed. Graduate Texts in Mathematics, Vol. 137. Springer-Verlag, New York (2001). MR1805196.

[Bak61] I. J. Bakel′man, On the theory of quasilinear elliptic equations. *Sibirsk. Mat. Ž.* **2**, 179–186 (1961) (Russian). MR0126604.

[BG83] M. S. Baouendi and C. Goulaouic, Sharp estimates for analytic pseudodifferential operators and application to Cauchy problems. *J. Differ. Equ.* **48**(2), 241–268 (1983). DOI: 10.1016/0022-0396(83)90051-7. MR696869.

[Bar13] C. Bartocci, "Reasoning well from badly drawn figures": The birth of algebraic topology. *Lett. Mat. Int.* **1**, 13–22 (2013). DOI: 10.1007/s40329-013-0010-4.

[BCN96] H. Berestycki, L. A. Caffarelli, and L. Nirenberg, Inequalities for second-order elliptic equations with applications to unbounded domains. I. *Duke Math. J.* **81**(2), 467–494 (1996). DOI: 10.1215/S0012-7094-96-08117-X. A celebration of John F. Nash, Jr. MR1395408.

[BN91] H. Berestycki and L. Nirenberg, On the method of moving planes and the sliding method. *Bol. Soc. Brasil. Mat.* (N.S.) **22**(1), 1–37 (1991). DOI: 10.1007/BF01244896. MR1159383.

[BNV94] H. Berestycki, L. Nirenberg, and S. R. S. Varadhan, The principal eigenvalue and maximum principle for second-order elliptic operators in general domains. *Comm. Pure Appl. Math.* **47**(1), 47–92 (1994). DOI: 10.1002/cpa.3160470105. MR1258192.

[Ber06] S. Bernstein, Sur la généralisation du problème de Dirichlet. *Math. Ann.* **62**(2), 253–271 (1906). DOI: 10.1007/BF01449980 (French). MR1511375.

[BJS79] L. Bers, F. John, and M. Schechter, *Partial Differential Equations*. Lectures in Applied Mathematics, Vol. 3. American Mathematical Society, Providence (1979). With supplements by Lars Gårding and A. N. Milgram; With a preface by A. S. Householder; Reprint of the 1964 original. MR598466.

[BB11] A. Björn and J. Björn, *Nonlinear Potential Theory on Metric Spaces*. EMS Tracts in Mathematics, Vol. 17. European Mathematical Society (EMS), Zürich (2011). MR2867756.

[Bôc06] M. Bôcher, Introduction to the theory of Fourier's series. *Ann. Math.* (2) **7**(3), 81–152 (1906). DOI: 10.2307/1967238. MR1502321.

[Bog05] T. Boggio, Sulle funzioni di Green d'ordine m. *Rend. Circ. Mat. Palermo* **20**, 97–135 (1905) (Italian).

[Bol07] W. M. Bolstad, *Introduction to Bayesian Statistics*, 2nd ed. Wiley-Interscience [John Wiley & Sons], Hoboken (2007). MR2352885.

[Boo75] W. M. Boothby, *An Introduction to Differentiable Manifolds and Riemannian Geometry*. Pure and Applied Mathematics, No. 63. Academic Press [A subsidiary of Harcourt Brace Jovanovich, Publishers], New York (1975). MR0426007.

[Bor99] J. L. Borges, *The Library of Babel*. Collected Fictions. Penguin Books, London (1999).

[Bre99] H. Brezis, Symmetry in nonlinear PDE's. *Differential Equations: La Pietra 1996 (Florence)*. Proceedings of Symposia in Pure Mathematics, Vol. 65. American Mathematical Society, Providence (1999), pp. 1–12. DOI: 10.1090/pspum/065/1662746. MR1662746.

[Bre11] H. Brezis, *Functional Analysis, Sobolev Spaces and Partial Differential Equations*. Universitext. Springer, New York (2011). MR2759829.

[BN83] H. Brézis and L. Nirenberg, Positive solutions of nonlinear elliptic equations involving critical Sobolev exponents. *Comm. Pure Appl. Math.* **36**(4), 437–477 (1983). DOI: 10.1002/cpa.3160360405. MR709644.

[Buc07] D. Bucur, Do optimal shapes exist? *Milan J. Math.* **75**, 379–398 (2007). DOI: 10.1007/s00032-007-0074-8. MR2371551.

[BG15] D. Bucur and A. Giacomini, The Saint-Venant inequality for the Laplace operator with Robin boundary conditions. *Milan J. Math.* **83**(2), 327–343 (2015). DOI: 10.1007/s00032-015-0243-0. MR3412286.

[BDM93] G. Buttazzo and G. Dal Maso, An existence result for a class of shape optimization problems. *Arch. Rational Mech. Anal.* **122**(2), 183–195 (1993). DOI: 10.1007/BF00378167. MR1217590.

[Cab95] X. Cabré, On the Alexandroff-Bakel′man-Pucci estimate and the reversed Hölder inequality for solutions of elliptic and parabolic equations. *Comm. Pure Appl. Math.* **48**(5), 539–570 (1995). DOI: 10.1002/cpa.3160480504. MR1329831.

[CV02] V. Cafagna and A. Vitolo, On the maximum principle for second-order elliptic operators in unbounded domains. *C. R. Math. Acad. Sci.* (Paris) **334**(5), 359–363 (2002). DOI: 10.1016/S1631-073X(02)02267-7 (English, with English and French summaries). MR1892934.

[CC95] L. A. Caffarelli and X. Cabré, *Fully Nonlinear Elliptic Equations*. American Mathematical Society Colloquium Publications, Vol. 43. American Mathematical Society, Providence (1995). MR1351007.

[CFMS81] L. Caffarelli, E. Fabes, S. Mortola, and S. Salsa, Boundary behavior of nonnegative solutions of elliptic operators in divergence form. *Indiana Univ. Math. J.* **30**(4), 621–640 (1981). DOI: 10.1512/iumj.1981.30.30049. MR620271.

[CS05] L. Caffarelli and S. Salsa, *A Geometric Approach to Free Boundary Problems*. Graduate Studies in Mathematics, Vol. 68. American Mathematical Society, Providence (2005). MR2145284.

[Cai07] W. Cai, Potential field of a uniformly charged ellipsoid (2007). http://micro.stanford.edu/~caiwei/me340a/A_Ellipsoid_Potential.pdf.

[CZ52] A. P. Calderon and A. Zygmund, On the existence of certain singular integrals. *Acta Math.* **88**, 85–139 (1952). DOI: 10.1007/BF02392130. MR52553.

[CZ56] A. P. Calderón and A. Zygmund, On singular integrals. *Am. J. Math.* **78**, 289–309 (1956). DOI: 10.2307/2372517. MR84633.

[Can01] D. M. Cannell, *George Green: Mathematician & Physicist 1793–1841; The Background to His Life and Work*, 2nd ed. Society for Industrial and Applied Mathematics (SIAM), Philadelphia (2001). With a foreword and an obituary of Cannell by Lawrie Challis. MR1829410.

[CL93] D. M. Cannell and N. J. Lord, George Green, mathematician and physicist 1793–1841. *Math. Gaz.* **77**(478), 26–51 (1993). DOI: 10.2307/3619259.

[CDV19] A. Carbotti, S. Dipierro, and E. Valdinoci, Local density of solutions to fractional equations. *De Gruyter Stud. Math.* **74**, xi + 129 (2019). DOI: 10.1515/9783110664355.

[Car67] L. Carleson, *Selected Problems on Exceptional Sets*. Van Nostrand Mathematical Studies, No. 13. D. Van Nostrand Co., Inc., Princeton (1967). MR0225986.

[Cau42a] A.-L. Cauchy, Mémoire sur un théorème fondamental, dans le calcul intégral. *C. R. Acad. Sci.* (Paris) **39**, 1020–1026 (1842).

[Cau42b] A.-L. Cauchy, Mémoire sur l'emploi du calcul des limites dans l'intégration des équations aux dérivées partielles. *C. R. Acad. Sci.* (Paris) **40**, 44–59 (1842).

[Cau42c] A.-L. Cauchy, Mémoire sur l'application du calcul des limites à l'intégration d'un système d'équations aux dérivées partielles. *C. R. Acad. Sci.* (Paris) **40**, 85–101 (1842).

[Cau42d] A.-L. Cauchy, Mémoire sur les systèmes d'équations aux dérivées partielles d'ordres quelconques, et sur leur réduction à des systèmes d'équations linéares du premier ordre. *C. R. Acad. Sci.* (Paris) **40**, 131–138 (1842).

[Cau42e] A.-L. Cauchy, Note sur divers théorèmes relatifs aux calcul des limites. *C. R. Acad. Sci.* (Paris) **40**, 138–139 (1842).

[Cau42f] A.-L. Cauchy, Mémoire sur les intégrales des systèmes d'équations différentielles et aux dérivées partielles, et sur le devéloppement de ces intégrales en séries ordonnés suivant les puissances ascendantes d'un paramètre que renferment les équations proposées. *C. R. Acad. Sci.* (Paris) **40**, 141–146 (1842).

[CS03] L. Challis and F. Sheard, The Green of Green functions. *Physics Today* **56**(12), 41–46 (2003). DOI: 10.1063/1.1650227.

[CP82] J. Chazarain and A. Piriou, *Introduction to the Theory of Linear Partial Differential Equations*. Studies in Mathematics and Its Applications, Vol. 14. North-Holland Publishing Co., Amsterdam (1982). Translated from the French. MR678605.

[Cho53] G. Choquet, Theory of capacities. *Ann. Inst. Fourier* (Grenoble) **5**, 131–295 (1953/54, 1955). MR00760.

[Cho86] G. Choquet, La naissance de la théorie des capacités: réflexion sur une expérience personnelle. *C. R. Acad. Sci. Sér. Gén. Vie Sci.* **3**(4), 385–397 (1986) (French). MR867115.

[CH98] M. Choulli and A. Henrot, Use of the domain derivative to prove symmetry results in partial differential equations. *Math. Nachr.* **192**, 91–103 (1998). DOI: 10.1002/mana.19981920106. MR1626395.

[Chr90] M. Christ, *Lectures on Singular Integral Operators*. CBMS Regional Conference Series in Mathematics, Vol. 77. Published for the Conference Board of the Mathematical Sciences, Washington, DC; by the American Mathematical Society, Providence (1990). MR1104656.

[CKS08] M. Christ, C. E. Kenig, and C. Sadosky, *Alberto P. Calderón the Mathematician, His Life and Works*. Selected papers of Alberto P. Calderón. American Mathematical Society, Providence (2008), pp. xv–xx. MR2435330.

[CCPP23] G. Ciraolo, M. Cozzi, M. Perugini, and L. Pollastro, A quantitative version of the Gidas-Ni-Nirenberg theorem (2023). *arXiv e-prints*.

[CR18] G. Ciraolo and A. Roncoroni, The method of moving planes: A quantitative approach. *Bruno Pini Mathematical Analysis Seminar 2018*. Bruno Pini Mathematical Analysis Seminar, Vol. 9. University of Bologna, Alma Mater Studiorum, Bologna (2018), pp. 41–77 (English, with English and Italian summaries). MR3932952.

[CGP+15] G. Citti, L. Grafakos, C. Pérez, A. Sarti, and X. Zhong, *Harmonic and Geometric Analysis*. Advanced Courses in Mathematics. CRM Barcelona. Birkhäuser/Springer Basel AG, Basel (2015). Edited by Joan Mateu. MR3380557.

[CR97] B. J. Cole and T. J. Ransford, Subharmonicity without upper semicontinuity. *J. Funct. Anal.* **147**(2), 420–442 (1997). DOI: 10.1006/jfan.1996.3070. MR1454488.

[CF77] E. Corrigan and D. B. Fairlie, Scalar field theory and exact solutions to a classical $\mathrm{SU}(2)$ gauge theory. *Phys. Lett. B* **67**(1), 69–71 (1977). DOI: 10.1016/0370-2693(77)90808-5. MR436814.

[CH89] R. Courant and D. Hilbert, *Methods of Mathematical Physics. Vol. II: Partial Differential Equations*. Wiley Classics Library. John Wiley & Sons, Inc., New York (1989). Reprint of the 1962 original; A Wiley-Interscience Publication. MR1013360.

[CL55] R. Courant and P. Lax, Cauchy's problem for nonlinear hyperbolic differential equations in two independent variables. *Ann. Mat. Pura Appl.* (4) **40**, 161–166 (1955). DOI: 10.1007/BF02416530. MR76161.

[CKLŚ99] M. G. Crandall, M. Kocan, P. L. Lions, and A. Święch, Existence results for boundary problems for uniformly elliptic and parabolic fully nonlinear equations. *Electron. J. Differ. Equ.* **24**, 22 p. (1999). MR1696765.

[Dah77] B. E. J. Dahlberg, Estimates of harmonic measure. *Arch. Rational Mech. Anal.* **65**(3), 275–288 (1977). DOI: 10.1007/BF00280445. MR466593.

[Dan92a] E. N. Dancer, Global breaking of symmetry of positive solutions on two-dimensional annuli. *Differ. Integral Equ.* **5**(4), 903–913 (1992). MR1167503.

[Dan92b] E. N. Dancer, Some notes on the method of moving planes. *Bull. Austral. Math. Soc.* **46**(3), 425–434 (1992). DOI: 10.1017/S0004972700012089. MR1190345.

[Dar75] G. Darboux, Sur l'existence de l'intégrale dans les équations aux dérivées partielles d'ordre Quelconque. *C. R. Acad. Sci.* (Paris) **80**, 317–318 (1875).

[DG57] E. De Giorgi, Sulla differenziabilità e l'analiticità delle estremali degli integrali multipli regolari. *Mem. Accad. Sci. Torino. Cl. Sci. Fis. Mat. Nat.* (3) **3**, 25–43 (1957) (Italian). MR0093649.

[DG58] E. De Giorgi, Sulla proprietà isoperimetrica dell'ipersfera, nella classe degli insiemi aventi frontiera orientata di misura finita. *Atti Accad. Naz. Lincei Mem. Cl. Sci. Fis. Mat. Natur. Sez. Ia* (8) **5**, 33–44 (1958) (Italian). MR98331.

[Del72] C. Dellacherie, *Ensembles analytiques, capacités, mesures de Hausdorff*. Lecture Notes in Mathematics, Vol. 295. Springer-Verlag, Berlin (1972) (French). MR0492152.

[DD12] F. Demengel and G. Demengel, *Functional Spaces for the Theory of Elliptic Partial Differential Equations*. Universitext. Springer, London; EDP Sciences, Les Ulis (2012). Translated from the 2007 French original by Reinie Erné. MR2895178.

[DSS15] D. De Silva and O. Savin, A note on higher regularity boundary Harnack inequality. *Discrete Contin. Dyn. Syst.* **35**(12), 6155–6163 (2015). DOI: 10.3934/dcds.2015.35.6155. MR3393271.

[DSS16] D. De Silva and O. Savin, Boundary Harnack estimates in slit domains and applications to thin free boundary problems. *Rev. Mat. Iberoam.* **32**(3), 891–912 (2016). DOI: 10.4171/RMI/902. MR3556055.

[DSS20] D. De Silva and O. Savin, A short proof of boundary Harnack principle. *J. Differ. Equ.* **269**(3), 2419–2429 (2020). DOI: 10.1016/j.jde.2020.02.004. MR4093736.

[Dia88] P. Diaconis, Bayesian numerical analysis. *Statistical Decision Theory and Related Topics, IV*, Vol. 1 (West Lafayette, Ind., 1986). Springer, New York (1988), pp. 163–175. MR927099.

[DiB95] E. DiBenedetto, *Partial Differential Equations*. Birkhäuser Boston, Inc., Boston (1995). MR1306729.

[DPV21] S. Dipierro, G. Poggesi, and E. Valdinoci, A Serrin-type problem with partial knowledge of the domain. *Nonlinear Anal.* **208**, 112330, 44 (2021). DOI: 10.1016/j.na.2021.112330. MR4230553.

[DPV22] S. Dipierro, G. Poggesi, and E. Valdinoci, Radial symmetry of solutions to anisotropic and weighted diffusion equations with discontinuous nonlinearities. *Calc. Var. Partial Differ. Equ.* **61**(2), 31 (2022). Paper No. 72. DOI: 10.1007/s00526-021-02157-5. MR4380032.

[DSV17] S. Dipierro, O. Savin, and E. Valdinoci, All functions are locally s-harmonic up to a small error. *J. Eur. Math. Soc. (JEMS)* **19**(4), 957–966 (2017). DOI: 10.4171/JEMS/684. MR3626547.

[DV23] S. Dipierro and E. Valdinoci, Elliptic partial differential equations from an elementary viewpoint: Models and motivations (2023). *Preprint.*

[Doo01] J. L. Doob, *Classical Potential Theory and Its Probabilistic Counterpart*. Classics in Mathematics. Springer-Verlag, Berlin (2001). Reprint of the 1984 edition. MR1814344.

[Duf59] G. F. D. Duff, Mixed problems for hyperbolic equations of general order. *Canadian J. Math.* **11**, 195–221 (1959). DOI: 10.4153/CJM-1959-024-1. MR104912.

[Duf57] R. J. Duffin, A note on Poisson's integral. *Quart. Appl. Math.* **15**, 109–111 (1957). DOI: 10.1090/qam/86885. MR86885.

[Duf01] D. G. Duffy, *Green's Functions with Applications*. Studies in Advanced Mathematics. Chapman & Hall/CRC, Boca Raton (2001). MR1888091.

[EE18] D. E. Edmunds and W. D. Evans, *Elliptic Differential Operators and Spectral Analysis*. Springer Monographs in Mathematics. Springer, Cham (2018). MR3931747.

[Edw01] H. M. Edwards, *Riemann's Zeta Function*. Dover Publications, Inc., Mineola (2001). Reprint of the 1974 original [Academic Press, New York; MR0466039 (57 #5922)]. MR1854455.

[EF14] C. Efthimiou and C. Frye, *Spherical Harmonics in p Dimensions*. World Scientific Publishing Co. Pte. Ltd., Hackensack (2014). MR3290046.

[Ego93] Y. V. Egorov, On an example of a linear hyperbolic equation without solutions. *C. R. Acad. Sci. Paris Sér. I Math.* **317**(12), 1149–1153 (1993) (English, with English and French summaries). MR1257229.

[ES92] Y. V. Egorov and M. A. Shubin, *Partial Differential Equations. I. Foundations of the Classical Theory.* Encyclopaedia of Mathematical Sciences, Vol. 30. Springer-Verlag, Berlin (1992). Translation by R. Cooke; Translation edited by Yu. V. Egorov and M. A. Shubin. MR1141630.

[Emm97] M. Emmer, Interview with Ennio De Giorgi. *Notices Am. Math. Soc.* **44**(9), 1097–1101 (1997). Translated from the Italian. MR1470169.

[Eps62] B. Epstein, On the mean-value property of harmonic functions. *Proc. Am. Math. Soc.* **13**, 830 (1962). DOI: 10.2307/2034188. MR140700.

[ES65] B. Epstein and M. M. Schiffer, On the mean-value property of harmonic functions. *J. Anal. Math.* **14**, 109–111 (1965). DOI: 10.1007/BF02806381. MR177124.

[Eva98] L. C. Evans, *Partial Differential Equations.* Graduate Studies in Mathematics, Vol. 19. American Mathematical Society, Providence (1998). MR1625845.

[EG15] L. C. Evans and R. F. Gariepy, *Measure Theory and Fine Properties of Functions*, Revised edn. Textbooks in Mathematics. CRC Press, Boca Raton (2015). MR3409135.

[Far07] A. Farina, Liouville-type theorems for elliptic problems. *Handbook of Differential Equations: Stationary Partial Differential Equations.* Handbook of Differential Equations, Vol. IV. Elsevier/North-Holland, Amsterdam (2007), pp. 61–116. DOI: 10.1016/S1874-5733(07)80005-2. MR2569331.

[FRRO22] X. Fernández-Real and X. Ros-Oton, *Regularity Theory for Elliptic PDE.* Zurich Lectures in Advanced Mathematics, Vol. 28. EMS Press, Berlin (2022). MR4560756.

[Fio16] R. Fiorenza, *Hölder and Locally Hölder Continuous Functions, and Open Sets of Class C^k, $C^{k,\lambda}$.* Frontiers in Mathematics. Birkhäuser/Springer, Cham (2016). MR3588287.

[FFRS21] A. Fiorenza, M. R. Formica, T. G. Roskovec, and F. Soudský, Detailed proof of classical Gagliardo-Nirenberg interpolation inequality with historical remarks. *Z. Anal. Anwend.* **40**(2), 217–236 (2021). DOI: 10.4171/zaa/1681. MR4237368.

[Fra00] L. E. Fraenkel, *An Introduction to Maximum Principles and Symmetry in Elliptic Problems.* Cambridge Tracts in Mathematics, Vol. 128. Cambridge University Press, Cambridge (2000). MR1751289.

[Fri57] V. R. Fridlender, On the problem of Cauchy-Kovalevski for certain partial differential equations. *Uspehi Mat. Nauk* (N.S.) **12**(3)(75), 385–388 (1957) (Russian). MR0091421.

[FP89] S. Friedlander and A. Powell, The mathematical miller of Nottingham. *Math. Intell.* **11**(4), 38–40 (1989). DOI: 10.1007/BF03025884. MR1016105.

[Fri61a] A. Friedman, A strong maximum principle for weakly subparabolic functions. *Pacific J. Math.* **11**, 175–184 (1961). MR123096.

[Fri61b] A. Friedman, A new proof and generalizations of the Cauchy-Kowalewski theorem. *Trans. Am. Math. Soc.* **98**, 1–20 (1961). DOI: 10.2307/1993510. MR166462.

[Fus04] N. Fusco, The classical isoperimetric theorem. *Rend. Accad. Sci. Fis. Mat. Napoli* (4) **71**, 63–107 (2004). MR2147710.

[Fus15] N. Fusco, The quantitative isoperimetric inequality and related topics. *Bull. Math. Sci.* **5**(3), 517–607 (2015). DOI: 10.1007/s13373-015-0074-x. MR3404715.

[Gag59] E. Gagliardo, Ulteriori proprietà di alcune classi di funzioni in più variabili. *Ricerche Mat.* **8**, 24–51 (1959) (Italian). MR109295.

[Gal83] G. Gallavotti, *The Elements of Mechanics*. Texts and Monographs in Physics. Springer-Verlag, New York (1983). Translated from the Italian. MR698947.

[Gar51] P. R. Garabedian, A partial differential equation arising in conformal mapping. *Pacific J. Math.* **1**, 485–524 (1951). MR46440.

[GGS10] F. Gazzola, H.-C. Grunau, and G. Sweers, *Polyharmonic Boundary Value Problems: Positivity Preserving and Nonlinear Higher Order Elliptic Equations in Bounded Domains*. Lecture Notes in Mathematics, Vol. 1991. Springer-Verlag, Berlin (2010). MR2667016.

[Gen07] J. E. Gentle, *Matrix Algebra: Theory, Computations, and Applications in Statistics*. Springer Texts in Statistics. Springer, New York (2007). MR2337395.

[Gia84] M. Giaquinta, Direct methods for regularity in the calculus of variations. *Nonlinear Partial Differential Equations and Their Applications*. Collège de France Seminar, Vol. VI (Paris, 1982/1983). Research Notes in Mathematics, Vol. 109. Pitman, Boston (1984), pp. 258–274. MR772245.

[GM12] M. Giaquinta and L. Martinazzi, *An Introduction to the Regularity Theory for Elliptic Systems, Harmonic Maps and Minimal Graphs*, 2nd edn. Appunti. Scuola Normale Superiore di Pisa (Nuova Serie) [Lecture Notes. Scuola Normale Superiore di Pisa (New Series)], Vol. 11. Edizioni della Normale, Pisa (2012). MR3099262.

[GNN79] B. Gidas, W. M. Ni, and L. Nirenberg, Symmetry and related properties via the maximum principle. *Comm. Math. Phys.* **68**(3), 209–243 (1979). MR544879.

[GT01] D. Gilbarg and N. S. Trudinger, *Elliptic Partial Differential Equations of Second Order*. Classics in Mathematics. Springer-Verlag, Berlin (2001). Reprint of the 1998 edition. MR1814364.

[Giu84] E. Giusti, *Minimal Surfaces and Functions of Bounded Variation*. Monographs in Mathematics, Vol. 80. Birkhäuser Verlag, Basel (1984). MR775682.

[GO71] M. Goldstein and W. H. Ow, On the mean-value property of harmonic functions. *Proc. Am. Math. Soc.* **29**, 341–344 (1971). DOI: 10.2307/2038138. MR279320.

[Gou98] E. Goursat, Sur l'existence des fonctions intégrales d'un système d'équations aux dérivées partielles. *Bull. Soc. Math. France* **26**, 129–134 (1898) (French). MR1504315.

[Gre28] G. Green, *An Essay on the Application of Mathematical Analysis to the Theories of Electricity and Magnetism*. Printed for the author by T. Wheelhouse, Nottingham (1828).

[Gre52] G. Green, An essay on the application of mathematical analysis to the theories of electricity and magnetism. *J. Reine Angew. Math.* **44**, 356–374 (1852). DOI: 10.1515/crll.1852.44.356. MR1578805.

[Gre54] G. Green, An essay on the application of mathematical analysis to the theories of electricity and magnetism. *J. Reine Angew. Math.* **47**, 161–221 (1854). DOI: 10.1515/crll.1854.47.161. MR1578862.

[Gug77] H. W. Guggenheimer, *Differential Geometry*. Dover Books on Advanced Mathematics. Dover Publications, Inc., New York (1977). Corrected reprint of the 1963 edition. MR0493768.

[GS67] V. W. Guillemin and S. Sternberg, The Lewy counterexample and the local equivalence problem for G-structures. *J. Differ. Geom.* **1**, 127–131 (1967). MR222800.

[Had06] J. Hadamard, Sur le principe de Dirichlet. *Bull. Soc. Math. France* **34**, 135–138 (1906) (French). MR1504545.

[Had53] J. Hadamard, *Lectures on Cauchy's Problem in Linear Partial Differential Equations*. Dover Publications, New York (1953). MR51411.

[HL11] Q. Han and F. Lin, *Elliptic Partial Differential Equations*, 2nd edn. Courant Lecture Notes in Mathematics, Vol. 1. Courant Institute of Mathematical Sciences, New York; American Mathematical Society, Providence (2011). MR2777537.

[Har87] A. Harnack, *Die Grundlagen der Theorie des logarithmischen Potentiales und der eindeutigen Potentialfunktion in der Ebene*. V. G. Teubner, Leipzig (1887).

[Har59] P. Hartman, On smooth linear partial differential equations without solutions. *Proc. Am. Math. Soc.* **10**, 252–257 (1959). DOI: 10.2307/2033586. MR125304.

[HK76] W. K. Hayman and P. B. Kennedy, *Subharmonic Functions. Vol. I.* London Mathematical Society Monographs, No. 9. Academic Press [Harcourt Brace Jovanovich, Publishers], London (1976). MR0460672.

[Heb96] E. Hebey, *Sobolev Spaces on Riemannian Manifolds*. Lecture Notes in Mathematics, Vol. 1635. Springer-Verlag, Berlin (1996). MR1481970.

[HJS02] H. Hedenmalm, S. Jakobsson, and S. Shimorin, A biharmonic maximum principle for hyperbolic surfaces. *J. Reine Angew. Math.* **550**, 25–75 (2002). DOI: 10.1515/crll.2002.074. MR1925907.

[HKM06] J. Heinonen, T. Kilpeläinen, and O. Martio, *Nonlinear Potential Theory of Degenerate Elliptic Equations*. Dover Publications, Inc., Mineola (2006). Unabridged republication of the 1993 original. MR2305115.

[HK78] E. Heintze and H. Karcher, A general comparison theorem with applications to volume estimates for submanifolds. *Ann. Sci. École Norm. Sup.* (4) **11**(4), 451–470 (1978). MR533065.

[Hel69] L. L. Helms, *Introduction to Potential Theory*. Pure and Applied Mathematics, Vol. XXII. Wiley-Interscience A Division of John Wiley & Sons, New York (1969). MR0261018.

[HP18] A. Henrot and M. Pierre, *Shape Variation and Optimization: A Geometrical Analysis*. EMS Tracts in Mathematics, Vol. 28. European Mathematical Society (EMS), Zürich (2018). English version of the French publication [MR2512810] with additions and updates. MR3791463.

[Hil97] E. Hille, *Ordinary Differential Equations in the Complex Domain*. Dover Publications, Inc., Mineola (1997). Reprint of the 1976 original. MR1452105.

[HWM05] B. Hofmann-Wellenhof and H. Moritz, *Physical Geodesy*. Springer-Verlag, Vienna (2005).

[Hol01] E. Holmgren, Über Systeme von linearen partiellen Differentialgleichungen. *Stockh. Öfv* **58**, 91–103 (1901) (German).

[Hop89] H. Hopf, *Differential Geometry in the Large*, 2nd edn. Lecture Notes in Mathematics, Vol. 1000. Springer-Verlag, Berlin (1989). Notes taken by Peter Lax and John W. Gray; With a preface by S. S. Chern; With a preface by K. Voss. MR1013786.

[Hör60] L. Hörmander, Differential operators of principal type. *Math. Ann.* **140**, 124–146 (1960). DOI: 10.1007/BF01360085. MR130574.

[Hör71] L. Hörmander, A remark on Holmgren's uniqueness theorem. *J. Differ. Geom.* **6**, 129–134 (1971/72). MR320486.

[Hör94] L. Hörmander, *Notions of Convexity*. Progress in Mathematics, Vol. 127. Birkhäuser Boston, Inc., Boston (1994). MR1301332.

[Hör00] L. Hörmander, A counterexample of Gevrey class to the uniqueness of the Cauchy problem. *Math. Res. Lett.* **7**(5–6), 615–624 (2000). DOI: 10.4310/MRL.2000.v7.n5.a7. MR1809287.

[Hör03] L. Hörmander, *The Analysis of Linear Partial Differential Operators. I: Distribution Theory and Fourier Analysis*. Classics in Mathematics. Springer-Verlag, Berlin (2003). Reprint of the 2nd (1990) edn. [Springer, Berlin; MR1065993 (91m:35001a)]. MR1996773.

[Hu11] B. Hu, *Blow-up Theories for Semilinear Parabolic Equations*. Lecture Notes in Mathematics, Vol. 2018. Springer, Heidelberg (2011). MR2796831.

[JP03] J. Jacod and P. Protter, *Probability Essentials*, 2nd edn. Universitext. Springer-Verlag, Berlin (2003). MR1956867.

[JK82] D. S. Jerison and C. E. Kenig, Boundary behavior of harmonic functions in nontangentially accessible domains. *Adv. in Math.* **46**(1), 80–147 (1982). DOI: 10.1016/0001-8708(82)90055-X. MR676988.

[Joh91] F. John, *Partial Differential Equations*, 4th edn. Applied Mathematical Sciences, Vol. 1. Springer-Verlag, New York (1991). MR1185075.

[Jos13] J. Jost, *Partial Differential Equations*, 3rd edn. Graduate Texts in Mathematics, Vol. 214. Springer, New York (2013). MR3012036.

[Jos17] J. Jost, *Riemannian Geometry and Geometric Analysis*, 7th edn. Universitext. Springer, Cham (2017). MR3726907.

[Kal95] H. Kalf, On the expansion of a function in terms of spherical harmonics in arbitrary dimensions. *Bull. Belg. Math. Soc. Simon Stevin* **2**(4), 361–380 (1995). MR1355826.

[KM05] P. Kall and J. Mayer, *Stochastic Linear Programming: Models, Theory, and Computation*. International Series in Operations Research & Management Science, Vol. 80. Springer-Verlag, New York (2005). MR2118904.

[Kas07] M. Kassmann, Harnack inequalities: An introduction. *Bound. Value Probl.* 21, Art. ID 81415. Posted on 2007. DOI: 10.1155/2007/81415. MR2291922.

[Kat72] T. Kato, Schrödinger operators with singular potentials. *Israel J. Math.* **13**, 135–148 (1972, 1973). DOI: 10.1007/BF02760233. MR333833.

[Kel31] O. D. Kellogg, On the derivatives of harmonic functions on the boundary. *Trans. Am. Math. Soc.* **33**(2), 486–510 (1931). DOI: 10.2307/1989419. MR1501602.

[Kel67] O. D. Kellogg, *Foundations of Potential Theory*. Reprint from the first edition of 1929. Die Grundlehren der Mathematischen Wissenschaften, Band 31. Springer-Verlag, Berlin (1967). MR0222317.

[Kem72] J. T. Kemper, A boundary Harnack principle for Lipschitz domains and the principle of positive singularities. *Comm. Pure Appl. Math.* **25**, 247–255 (1972). DOI: 10.1002/cpa.3160250303. MR293114.

[KP94] S. Kesavan and F. Pacella, Symmetry of positive solutions of a quasilinear elliptic equation via isoperimetric inequalities. *Appl. Anal.* **54**(1–2), 27–37 (1994). DOI: 10.1080/00036819408840266. MR1382205.

[Kic06] S. Kichenassamy, Schauder-type estimates and applications. *Handbook of Differential Equations: Stationary Partial Differential Equations*, Vol. III. Elsevier/North Holland, Amsterdam (2006).

[KV86] S. Kichenassamy and L. Véron, Singular solutions of the p-Laplace equation. *Math. Ann.* **275**(4), 599–615 (1986). DOI: 10.1007/BF01459140. MR859333.

[Kow75] S. V. Kowalevsky, Zur Theorie der partiellen Differentialgleichung. *J. Reine Angew. Math.* **80**, 1–32 (1875) (German). DOI: 10.1515/crll.1875.80.1. MR1579652.

[Kra09] S. G. Krantz, *Explorations in Harmonic Analysis: With Applications to Complex Function Theory and the Heisenberg Group*. Applied and Numerical Harmonic Analysis. Birkhäuser Boston, Ltd., Boston (2009). With the assistance of Lina Lee. MR2508404.

[KP02] S. G. Krantz and H. R. Parks, *A Primer of Real Analytic Functions*, 2nd edn. Birkhäuser Advanced Texts: Basler Lehrbücher. [Birkhäuser Advanced Texts: Basel Textbooks]. Birkhäuser Boston, Inc., Boston (2002). MR1916029.

[Kry96] N. V. Krylov, *Lectures on Elliptic and Parabolic Equations in Hölder Spaces*. Graduate Studies in Mathematics, Vol. 12. American Mathematical Society, Providence (1996). MR1406091.

[Kry08] N. V. Krylov, *Lectures on Elliptic and Parabolic Equations in Sobolev Spaces*. Graduate Studies in Mathematics, Vol. 96. American Mathematical Society, Providence (2008). MR2435520.

[KS79] N. V. Krylov and M. V. Safonov, An estimate for the probability of a diffusion process hitting a set of positive measure. *Dokl. Akad. Nauk SSSR* **245**(1), 18–20 (1979) (Russian). MR525227.

[Kue19] C. Kuehn, *PDE Dynamics: An Introduction*. Mathematical Modeling and Computation, Vol. 23. Society for Industrial and Applied Mathematics (SIAM), Philadelphia (2019). MR3938717.

[Kur72] Ü. Kuran, On the mean-value property of harmonic functions. *Bull. London Math. Soc.* **4**, 311–312 (1972). DOI: 10.1112/blms/4.3.311. MR320348.

[Lau73] S. L. Lauritzen, *The Probabilistic Background of Some Statistical Methods in Physical Geodesy*. Geodaetisk Institut, Copenhagen (1973). Geodaetisk Institut, Meddelelse No. 48. MR0478314.

[Lau75] S. L. Lauritzen, Random orthogonal set functions and stochastic models for the gravity potential of the earth. *Stochastic Process. Appl.* **3**, 65–72 (1975). DOI: 10.1016/0304-4149(75)90007-1. MR414029.

[Leo09] G. Leoni, *A First Course in Sobolev Spaces*. Graduate Studies in Mathematics, Vol. 105. American Mathematical Society, Providence (2009). MR2527916.

[Ler57] J. Leray, Uniformisation de la solution du problème linéaire analytique de Cauchy près de la variété qui porte les données de Cauchy. *C. R. Acad. Sci.* (Paris) **245**, 1483–1488 (1957) (French). MR93634.

[Lew57] H. Lewy, An example of a smooth linear partial differential equation without solution. *Ann. Math.* (2) **66**, 155–158 (1957). DOI: 10.2307/1970121. MR88629.

[LSZH15] A.-M. Li, U. Simon, G. Zhao, and Z. Hu, *Global Affine Differential Geometry of Hypersurfaces*, 2nd revised and extended edn. De Gruyter Expositions in Mathematics, Vol. 11. De Gruyter, Berlin (2015). MR3382197.

[LJRX14] S. Li, H. Jiang, Z. Ren, and C. Xu, Optimal tracking for a divergent-type parabolic PDE system in current profile control. *Abstr. Appl. Anal.* 8, Art. ID 940965. Posted on 2014. DOI: 10.1155/2014/940965. MR3224327.

[LL01] E. H. Lieb and M. Loss, *Analysis*, 2nd edn. Graduate Studies in Mathematics, Vol. 14. American Mathematical Society, Providence (2001). MR1817225.

[Lio81] P.-L. Lions, Two geometrical properties of solutions of semilinear problems. *Appl. Anal.* **12**(4), 267–272 (1981). DOI: 10.1080/00036818108839367. MR653200.

[Lit59] W. Littman, A strong maximum principle for weakly L-subharmonic functions. *J. Math. Mech.* **8**, 761–770 (1959). DOI: 10.1512/iumj.1959.8.58048. MR0107746.

[Lit63] W. Littman, Generalized subharmonic functions: Monotonic approximations and an improved maximum principle. *Ann. Scuola Norm. Sup. Pisa Cl. Sci.* (3) **17**, 207–222 (1963). MR177186.

[LN74] C. Loewner and L. Nirenberg, *Partial Differential Equations Invariant under Conformal or Projective Transformations*. Contributions to Analysis (A Collection of Papers Dedicated to Lipman Bers). Academic Press, New York (1974), pp. 245–272. MR0358078.

[Lud60] D. Ludwig, Exact and asymptotic solutions of the Cauchy problem. *Comm. Pure Appl. Math.* **13**, 473–508 (1960). DOI: 10.1002/cpa.3160130310. MR115010.

[Lüt90] J. Lützen, *Joseph Liouville 1809–1882: Master of Pure and Applied Mathematics*. Studies in the History of Mathematics and Physical Sciences, Vol. 15. Springer-Verlag, New York (1990). MR1066463.

[Mag12] F. Maggi, *Sets of Finite Perimeter and Geometric Variational Problems: An Introduction to Geometric Measure Theory*. Cambridge Studies in Advanced Mathematics, Vol. 135. Cambridge University Press, Cambridge (2012). MR2976521.

[MP20] R. Magnanini and G. Poggesi, Serrin's problem and Alexandrov's soap bubble theorem: Enhanced stability via integral identities. *Indiana Univ. Math. J.* **69**(4), 1181–1205 (2020). DOI: 10.1512/iumj.2020.69.7925. MR4124125.

[Mal34] H. W. Malmheden, Eine neue Lösung des Dirichletschen Problems für sphärische Bereiche. *Fysiogr. Sällsk. Lund Förh.* **4**(17), 1–5 (1934).

[MZ97] J. Malý and W. P. Ziemer, *Fine Regularity of Solutions of Elliptic Partial Differential Equations*. Mathematical Surveys and Monographs, Vol. 51. American Mathematical Society, Providence (1997). MR1461542.

[Mar39] J. Marcinkiewicz, Sur l'interpolation d'opérations. *C. R. Acad. Sci.* (Paris) **208**, 1272–1273 (1939).

[Mar64] J. Marcinkiewicz, *Collected Papers*. Państwowe Wydawnictwo Naukowe, Warsaw (1964). Edited by Antoni Zygmund. With the collaboration of Stanislaw Lojasiewicz, Julian Musielak, Kazimierz Urbanik and Antoni Wiweger. Instytut Matematyczny Polskiej Akademii Nauk. MR0168434.

[Max54] J. C. Maxwell, *A Treatise on Electricity and Magnetism*, 3rd edn. Dover Publications, Inc., New York (1954). Two volumes bound as one. MR0063293.

[MS98] V. Maz'ya and T. Shaposhnikova, *Jacques Hadamard, a Universal Mathematician*. History of Mathematics, Vol. 14. American Mathematical Society, Providence; London Mathematical Society, London (1998). MR1611073.

[MU83] G. A. Mendoza and G. A. Uhlmann, A necessary condition for local solvability for a class of operators with double characteristics. *J. Funct. Anal.* **52**(2), 252–256 (1983). DOI: 10.1016/0022-1236(83)90084-8. MR707206.

[Mét85] G. Métivier, Uniqueness and approximation of solutions of first order nonlinear equations. *Invent. Math.* **82**(2), 263–282 (1985). DOI: 10.1007/BF01388803. MR809715.

[Mét93] G. Métivier, Counterexamples to Hölmgren's uniqueness for analytic nonlinear Cauchy problems. *Invent. Math.* **112**(1), 217–222 (1993). DOI: 10.1007/BF01232431. MR1207483.

[Mic77] J. H. Michael, A general theory for linear elliptic partial differential equations. *J. Differ. Equ.* **23**(1), 1–29 (1977). DOI: 10.1016/0022-0396(77)90134-6. MR425346.

[Mic81] J. H. Michael, Barriers for uniformly elliptic equations and the exterior cone condition. *J. Math. Anal. Appl.* **79**(1), 203–217 (1981). DOI: 10.1016/0022-247X(81)90018-4. MR603385.

[Mik78] V. P. Mikhaĭlov, *Partial Differential Equations*. "Mir", Moscow; Distributed by Imported Publications, Chicago (1978). Translated from the Russian by P. C. Sinha. MR601389.

[Mil67] K. Miller, Barriers on cones for uniformly elliptic operators. *Ann. Mat. Pura Appl.* (4) **76**, 93–105 (1967) (English, with Italian summary). DOI: 10.1007/BF02412230. MR221087.

[Mil71] K. Miller, Extremal barriers on cones with Phragmén-Lindelöf theorems and other applications. *Ann. Mat. Pura Appl.* (4) **90**, 297–329 (1971). DOI: 10.1007/BF02415053. MR316884.

[Miz61] S. Mizohata, Solutions nulles et solutions non analytiques. *J. Math. Kyoto Univ.* **1**, 271–302 (1961/62) (French). DOI: 10.1215/kjm/1250525061. MR142873.

[MR09] S. Montiel and A. Ros, *Curves and Surfaces*, 2nd edn. Graduate Studies in Mathematics, Vol. 69. American Mathematical Society, Providence; Real Sociedad Matemática Española, Madrid (2009). Translated from the 1998 Spanish original by Montiel and edited by Donald Babbitt. MR2522595.

[Mor16] F. Morgan, *Geometric Measure Theory: A Beginner's Guide*, 5th edn. Elsevier/Academic Press, Amsterdam (2016). Illustrated by James F. Bredt. MR3497381.

[Mor08] C. B. Morrey Jr., *Multiple Integrals in the Calculus of Variations*. Classics in Mathematics. Springer-Verlag, Berlin (2008). Reprint of the 1966 edn. [MR0202511]. MR2492985.

[Mor58a] C. B. Morrey Jr., On the analyticity of the solutions of analytic non-linear elliptic systems of partial differential equations. I. Analyticity in the interior. *Am. J. Math.* **80**, 198–218 (1958). DOI: 10.2307/2372830. MR106336.

[Mor58b] C. B. Morrey Jr., On the analyticity of the solutions of analytic non-linear elliptic systems of partial differential equations. II. Analyticity at the boundary. *Am. J. Math.* **80**, 219–237 (1958). DOI: 10.2307/2372831. MR107081.

[MN57] C. B. Morrey Jr. and L. Nirenberg, On the analyticity of the solutions of linear elliptic systems of partial differential equations. *Comm. Pure Appl. Math.* **10**, 271–290 (1957). DOI: 10.1002/cpa.3160100204. MR89334.

[Mor39] A. P. Morse, The behavior of a function on its critical set. *Ann. Math.* (2) **40**(1), 62–70 (1939). DOI: 10.2307/1968544. MR1503449.

[Mos60] J. Moser, A new proof of De Giorgi's theorem concerning the regularity problem for elliptic differential equations. *Comm. Pure Appl. Math.* **13**, 457–468 (1960). DOI: 10.1002/cpa.3160130308. MR170091.

[Mos61] J. Moser, On Harnack's theorem for elliptic differential equations. *Comm. Pure Appl. Math.* **14**, 577–591 (1961). DOI: 10.1002/cpa.3160140329. MR159138.

[MM54] F. J. Murray and K. S. Miller, *Existence Theorems for Ordinary Differential Equations*. New York University Press, New York (1954). MR0064934.

[Mur02] J. D. Murray, *Mathematical Biology. I: An Introduction*, 3rd edn. Interdisciplinary Applied Mathematics, Vol. 17. Springer-Verlag, New York (2002). MR1908418.

[Nag42] M. Nagumo, Über das Anfangswertproblem partieller Differentialgleichungen. *Jpn. J. Math.* **18**, 41–47 (1942) (German). DOI: 10.4099/jjm1924.18.0_41. MR15186.

[Nas58] J. Nash, Continuity of solutions of parabolic and elliptic equations. *Am. J. Math.* **80**, 931–954 (1958). DOI: 10.2307/2372841. MR100158.

[Nee94] T. Needham, The geometry of harmonic functions. *Math. Mag.* **67**(2), 92–108 (1994). DOI: 10.2307/2690683. MR1272823.

[Nee97] T. Needham, *Visual Complex Analysis*. The Clarendon Press, Oxford University Press, New York (1997). MR1446490.

[NV94] I. Netuka and J. Veselý, *Mean Value Property and Harmonic Functions. Classical and Modern Potential Theory and Applications* (Chateau de Bonas, 1993). NATO Advanced Science Institutes Series C: Mathematical and Physical Sciences, Vol. 430. Kluwer Academic Publishers, Dordrecht (1994), pp. 359–398. MR1321628.

[Neu84] C. Neumann, *Vorlesungen über Riemann's Theorie der Abel'schen Integrale. Zweite vollständig umgearbeitete und wesentlich vermehrte Auflage*. Leipzig, Teubner. XIV und 472 S (1884) (German).

[Ni11] W.-M. Ni, *The Mathematics of Diffusion*. CBMS-NSF Regional Conference Series in Applied Mathematics, Vol. 82. Society for Industrial and Applied Mathematics (SIAM), Philadelphia (2011). MR2866937.

[Nir55] L. Nirenberg, Remarks on strongly elliptic partial differential equations. *Comm. Pure Appl. Math.* **8**, 649–675 (1955). DOI: 10.1002/cpa.3160080414. MR75415.

[Nir59] L. Nirenberg, On elliptic partial differential equations. *Ann. Scuola Norm. Sup. Pisa Cl. Sci.* (3) **13**, 115–162 (1959). MR109940.

[Nir72] L. Nirenberg, An abstract form of the nonlinear Cauchy-Kowalewski theorem. *J. Differ. Geom.* **6**, 561–576 (1972). MR322321.

[Nis77] T. Nishida, A note on a theorem of Nirenberg. *J. Differ. Geom.* **12**(4), 629–633 (1977, 1978). MR512931.

[NT18] C. Nitsch and C. Trombetti, The classical overdetermined Serrin problem. *Complex Var. Elliptic Equ.* **63**(7–8), 1107–1122 (2018). DOI: 10.1080/17476933.2017.1410798. MR3802818.

[OD18] J. T. Oden and L. F. Demkowicz, *Applied Functional Analysis*, 3rd edn. Textbooks in Mathematics. CRC Press, Boca Raton (2018). MR3753707.

[Ovs71] L. V. Ovsjannikov, A nonlinear Cauchy problem in a scale of Banach spaces. *Dokl. Akad. Nauk SSSR* **200**, 789–792 (1971) (Russian). MR0285941.

[Pat88] S. J. Patterson, *An Introduction to the Theory of the Riemann Zeta-Function*. Cambridge Studies in Advanced Mathematics, Vol. 14. Cambridge University Press, Cambridge (1988). MR933558.

[PS89] L. E. Payne and P. W. Schaefer, Duality theorems in some overdetermined boundary value problems. *Math. Methods Appl. Sci.* **11**(6), 805–819 (1989). DOI: 10.1002/mma.1670110606. MR1021402.

[Pei83] J. Peiffer, Joseph Liouville (1809–1882): ses contributions à la théorie des fonctions d'une variable complexe. *Revue Hist. Sci.* **36**(3/4), 209–248 (1983).

[Per03] G. Perelman, Finite extinction time for the solutions to the Ricci flow on certain three-manifolds (2003). *arXiv e-prints*.

[Per01] L. Perko, *Differential Equations and Dynamical Systems*, 3rd edn. Texts in Applied Mathematics, Vol. 7. Springer-Verlag, New York (2001). MR1801796.

[Per23] O. Perron, Eine neue Behandlung der ersten Randwertaufgabe für $\Delta u = 0$. *Math. Z.* **18**(1), 42–54 (1923) (German). DOI: 10.1007/BF01192395. MR1544619.

[Poh65] S. I. Pohožaev, On the eigenfunctions of the equation $\Delta u + \lambda f(u) = 0$. *Dokl. Akad. Nauk SSSR* **165**, 36–39 (1965) (Russian). MR0192184.

[Poi90] H. Poincaré, Sur les équations aux dérivées partielles de la physique mathématique. *Am. J. Math.* **12**(3), 211–294 (1890) (French). DOI: 10.2307/2369620. MR1505534.

[Poi95] H. Poincaré, Analysis situs. *J. Éc. Politech.* **2**(3), 1–123 (1895) (French).

[Poi57] H. Poincaré, *Les méthodes nouvelles de la mécanique céleste. Tome III. Invariants intégraux. Solutions périodiques du deuxième genre. Solutions doublement asymptotiques.* Dover Publications, Inc., New York (1957) (French). MR0087814.

[PS51] G. Pólya and G. Szegö, *Isoperimetric Inequalities in Mathematical Physics*. Annals of Mathematics Studies, Vol. 27. Princeton University Press, Princeton (1951). MR0043486.

[Pon16] A. C. Ponce, *Elliptic PDEs, Measures and Capacities: From the Poisson Equations to Nonlinear Thomas-Fermi Problems*. EMS Tracts in Mathematics, Vol. 23. European Mathematical Society (EMS), Zürich (2016). MR3675703.

[Por81] I. R. Porteous, *Topological Geometry*, 2nd edn. Cambridge University Press, Cambridge (1981). MR606198.

[PS07a] K. Pravda-Starov, Explicit examples of nonsolvable weakly hyperbolic operators with real coefficients. *Bull. Braz. Math. Soc.* (N.S.) **38**(3), 397–425 (2007). DOI: 10.1007/s00574-007-0052-3. MR2344206.

[Puc64] C. Pucci, Regolarità alla frontiera di soluzioni di equazioni ellittiche. *Ann. Mat. Pura Appl.* (4) **65**, 311–328 (1964) (Italian, with English summary). DOI: 10.1007/BF02418230. MR171990.

[Puc66a] C. Pucci, Operatori ellittici estremanti. *Ann. Mat. Pura Appl.* (4) **72**, 141–170 (1966) (Italian, with English summary). DOI: 10.1007/BF02414332. MR208150.

[Puc66b] C. Pucci, Limitazioni per soluzioni di equazioni ellittiche. *Ann. Mat. Pura Appl.* (4) **74**, 15–30 (1966) (Italian, with English summary). DOI: 10.1007/BF02416445. MR214905.

[PS07b] P. Pucci and J. Serrin, *The Maximum Principle*. Progress in Nonlinear Differential Equations and their Applications, Vol. 73. Birkhäuser Verlag, Basel (2007). MR2356201.

[Rad71] T. Radó, *On the Problem of Plateau. Subharmonic Functions*. Springer-Verlag, New York (1971) (Reprint). MR0344979.

[Rau14] J. Rauch, Earnshaw's theorem in electrostatics and a conditional converse to Dirichlet's theorem. *Séminaire Laurent Schwartz—Équations aux Dérivées Partielles et Applications.* Année 2013–2014. Ed. École Polytechnique, Palaiseau (2014), Exp. No. XII, 10. MR3445477.

[RS75] M. Reed and B. Simon, *Methods of Modern Mathematical Physics. II. Fourier Analysis, Self-adjointness*. Academic Press [Harcourt Brace Jovanovich, Publishers], New York (1975). MR0493420.

[Rei82] R. C. Reilly, Mean curvature, the Laplacian, and soap bubbles. *Am. Math. Monthly* **89**(3), 180–188, 197–198 (1982). DOI: 10.2307/2320201. MR645791.

[Rel40] F. Rellich, Darstellung der Eigenwerte von $\Delta u + \lambda u = 0$ durch ein Randintegral. *Math. Z.* **46**, 635–636 (1940) (German). DOI: 10.1007/BF01181459. MR2456.

[Rem24] R. Remak, Über potentialkonvexe Funktionen. *Math. Z.* **20**(1), 126–130 (1924) (German). DOI: 10.1007/BF01188075. MR1544666.

[RG12] H. Render and M. Ghergu, Positivity properties for the clamped plate boundary problem on the ellipse and strip. *Math. Nachr.* **285**(8–9), 1052–1062 (2012). DOI: 10.1002/mana.201100045. MR2928398.

[Riv16] T. Rivière, Exploring the unknown: the work of Louis Nirenberg on partial differential equations. *Notices Amer. Math. Soc.* **63**(2), 120–125 (2016). DOI: 10.1090/noti1328. MR3409063.

[Roc13] R. Roccaforte, The volume of non-uniform tubular neighborhoods and an application to the n-dimensional Szegö theorem. *J. Math. Anal. Appl.* **398**(1), 61–67 (2013). DOI: 10.1016/j.jmaa.2012.08.031. MR2984315.

[Ros87] A. Ros, Compact hypersurfaces with constant higher order mean curvatures. *Rev. Mat. Iberoamericana* **3**(3–4), 447–453 (1987). DOI: 10.4171/RMI/58. MR996826.

[Ros97] S. Rosenberg, *The Laplacian on a Riemannian Manifold: An Introduction to Analysis on Manifolds*. London Mathematical Society Student Texts, Vol. 31. Cambridge University Press, Cambridge (1997). MR1462892.

[ROV16] X. Ros-Oton and E. Valdinoci, The Dirichlet problem for nonlocal operators with singular kernels: convex and nonconvex domains. *Adv. Math.* **288**, 732–790 (2016). DOI: 10.1016/j.aim.2015.11.001. MR3436398.

[Ros07] S. M. Ross, *Introduction to Probability Models*, 9th edn. Elsevier/Academic Press, Amsterdam (2007).

[Rud66] W. Rudin, *Real and Complex Analysis*. McGraw-Hill Book Co., New York (1966). MR0210528.

[Sal15] S. Salsa, *Partial Differential Equations in Action: From Modelling to Theory*, 2nd edn. Unitext, Vol. 86. Springer, Cham (2015). La Matematica per il 3+2. MR3362185.

[SV15] S. Salsa and G. Verzini, *Partial Differential Equations in Action: Complements and Exercises*. Unitext, Vol. 87. Springer, Cham (2015). Translated from the 2005 Italian edition by Simon G. Chiossi; La Matematica per il 3+2. MR3380662.

[Sau06] F. Sauvigny, *Partial Differential Equations. 1: Foundations and Integral Representations*. Universitext. Springer-Verlag, Berlin (2006). With consideration of lectures by E. Heinz; Translated and expanded from the 2004 German original. MR2254749.

[Sch34] J. Schauder, Über lineare elliptische Differentialgleichungen zweiter Ordnung. *Math. Z.* **38**(1), 257–282 (1934). DOI: 10.1007/BF01170635 (German). MR1545448.

[Sch91] S. H. Schot, A simple solution method for the first boundary value problem of the polyharmonic equation. *Appl. Anal.* **41**(1–4), 145–153 (1991). DOI: 10.1080/00036819108840020. MR1103852.

[Sch72] H. A. Schwarz, *Gesammelte mathematische Abhandlungen. Band I, II*. Chelsea Publishing Co., Bronx (1972) (German). Nachdruck in einem Band der Auflage von 1890. MR0392470.

[Ser09] R. Serfozo, *Basics of Applied Stochastic Processes*. Probability and Its Applications. Springer-Verlag, Berlin (2009). MR2484222.

[Ser13] J. Serra, Radial symmetry of solutions to diffusion equations with discontinuous nonlinearities. *J. Differ. Equ.* **254**(4), 1893–1902 (2013). DOI: 10.1016/j.jde.2012.11.015. MR3003296.

[Ser71] J. Serrin, A symmetry problem in potential theory. *Arch. Rational Mech. Anal.* **43**, 304–318 (1971). DOI: 10.1007/BF00250468. MR333220.

[Ser11] J. Serrin, Weakly subharmonic function. *Boll. Unione Mat. Ital.* (9) **4**(3), 347–361 (2011). MR2906766.

[ST94] H. S. Shapiro and M. Tegmark, An elementary proof that the biharmonic Green function of an eccentric ellipse changes sign. *SIAM Rev.* **36**(1), 99–101 (1994). DOI: 10.1137/1036005. MR1267051.

[Sim97] L. Simon, Schauder estimates by scaling. *Calc. Var. Partial Differ. Equ.* **5**(5), 391–407 (1997). DOI: 10.1007/s005260050072. MR1459795.

[Spa18] S. Spagnolo, *Il teorema più bello.* In L. Modica and C. Petronio (eds.) Pisa University Press, Pisa (2018).

[Sta66] G. Stampacchia, *Èquations elliptiques du second ordre à coefficients discontinus.* Séminaire de Mathématiques Supérieures, No. 16 (Été), Vol. 1965. Les Presses de l'Université de Montréal, Montreal (1966) (French). MR0251373.

[Ste70] E. M. Stein, *Singular Integrals and Differentiability Properties of Functions.* Princeton Mathematical Series, No. 30. Princeton University Press, Princeton (1970). MR0290095.

[Ste98] E. M. Stein, Singular integrals: The roles of Calderón and Zygmund. *Notices Am. Math. Soc.* **45**(9), 1130–1140 (1998). MR1640159.

[ST18] I. Stewart and D. Tall, *Complex Analysis.* Cambridge University Press, Cambridge (2018). The hitch hiker's guide to the plane; 2nd edn. of [MR0698076]. MR3839273.

[Swe16a] G. Sweers, An elementary proof that the triharmonic Green function of an eccentric ellipse changes sign. *Arch. Math.* (Basel) **107**(1), 59–62 (2016). DOI: 10.1007/s00013-016-0909-z. MR3514727.

[Swe16b] G. Sweers, On sign preservation for clotheslines, curtain rods, elastic membranes and thin plates. *Jahresber. Dtsch. Math.-Ver.* **118**(4), 275–320 (2016). DOI: 10.1365/s13291-016-0147-0. MR3554425.

[Swe19] G. Sweers, Correction to: An elementary proof that the triharmonic Green function of an eccentric ellipse changes sign. *Arch. Math.* (Basel) **112**(2), 223–224 (2019). DOI: 10.1007/s00013-018-1274-x. MR3908840.

[Tal76] G. Talenti, Elliptic equations and rearrangements. *Ann. Scuola Norm. Sup. Pisa Cl. Sci.* (4) **3**(4), 697–718 (1976). MR601601.

[Tar97] N. N. Tarkhanov, *The Analysis of Solutions of Elliptic Equations.* Mathematics and Its Applications, Vol. 406. Kluwer Academic Publishers Group, Dordrecht (1997). Translated from the 1991 Russian original by P. M. Gauthier and revised by the author. MR1447439.

[TG50] W. Thomson and G. Green, An Essay on the Application of mathematical Analysis to the theories of Electricity and Magnetism. *J. Reine Angew. Math.* **39**, 73–89 (1850). DOI: 10.1515/crll.1850.39.73. MR1578654.

[Tit86] E. C. Titchmarsh, *The Theory of the Riemann Zeta-Function*, 2nd edn. The Clarendon Press, Oxford University Press, New York (1986). Edited and with a preface by D. R. Heath-Brown. MR882550.

[Tre62] F. Treves, The equation $(\partial^2/\partial x^2 + \partial^2/\partial y^2 + (x^2 + y^2)(\partial/\partial t))^2 u + \partial^2 u/\partial t^2 = f$, with real coefficients, is "without solutions". *Bull. Am. Math. Soc.* **68**(4), 332 (1962). DOI: 10.1090/S0002-9904-1962-10794-0. MR1566203.

[Trè70] F. Trèves, On local solvability of linear partial differential equations. *Bull. Am. Math. Soc.* **76**, 552–571 (1970). DOI: 10.1090/S0002-9904-1970-12443-0. MR257550.

[Trè06] F. Trèves, *Basic Linear Partial Differential Equations*. Dover Publications, Inc., Mineola (2006). Reprint of the 1975 original. MR2301309.

[Tre22] F. Treves, *Analytic Partial Differential Equations*. Grundlehren der mathematischen Wissenschaften [Fundamental Principles of Mathematical Sciences], Vol. 359. Springer, Cham (2022). MR4436039.

[Tru20] N. S. Trudinger, Remarks on the Pucci conjecture. *Indiana Univ. Math. J.* **69**(1), 109–118 (2020). DOI: 10.1512/iumj.2020.69.8460. MR4077156.

[Tur00] B. O. Turesson, *Nonlinear Potential Theory and Weighted Sobolev Spaces*. Lecture Notes in Mathematics, Vol. 1736. Springer-Verlag, Berlin (2000). MR1774162.

[Tut05] W. Tutschke, The Cauchy-Kovalevskaya theorem—old and new. *Anal. Theory Appl.* **21**(2), 166–175 (2005). DOI: 10.1007/BF02836921. MR2301259.

[Vig92] M. Vignati, A geometric property of functions harmonic in a disk. *Elem. Math.* **47**(1), 33–38 (1992) (English, with German summary). MR1158146.

[Wag02] A. Wagner, Pohozaev's identity from a variational viewpoint. *J. Math. Anal. Appl.* **266**(1), 149–159 (2002). DOI: 10.1006/jmaa.2001.7719. MR1876774.

[Wal04] P. Walker, *Examples and Theorems in Analysis*. Springer-Verlag London, Ltd., London (2004). MR2016192.

[Wei62] H. F. Weinberger, *Symmetrization in Uniformly Elliptic Problems*. Studies in Mathematical Analysis and Related Topics. Stanford University Press, Stanford (1962), pp. 424–428. MR0145191.

[Wei64] H. F. Weinberger, On bounding harmonic functions by linear interpolation. *Bull. Am. Math. Soc.* **70**, 525–529 (1964). DOI: 10.1090/S0002-9904-1964-11183-6. MR162953.

[Wei71] H. F. Weinberger, Remark on the preceding paper of Serrin. *Arch. Rational Mech. Anal.* **43**, 319–320 (1971). DOI: 10.1007/BF00250469. MR333221.

[Wei76] R. Weinstock, On a fallacious proof of Earnshaw's theorem. *Am. J. Phys.* **44**, 392–393 (1976). DOI: 10.1119/1.10449.

[WZ15] R. L. Wheeden and A. Zygmund, *Measure and Integral: An Introduction to Real Analysis*, 2nd edn. Pure and Applied Mathematics. CRC Press, Boca Raton (2015). MR3381284.

[Wie24] N. Wiener, The Dirichlet Problem. *J. Math. Phys.* **3**(3), 127–146 (1924). DOI: 10.1002/sapm192433127.

[Wu78] J. M. G. Wu, Comparisons of kernel functions, boundary Harnack principle and relative Fatou theorem on Lipschitz domains. *Ann. Inst. Fourier* (Grenoble) **28**(4), 147–167 (1978), vi (English, with French summary). MR513884.

[Zac66] E. C. Zachmanoglou, Uniqueness of the Cauchy problem when the initial surface contains characteristic points. *Arch. Rational Mech. Anal.* **23**, 317–326 (1966). DOI: 10.1007/BF00281166. MR217426.

[Zan13] A. Zangwill, *Modern Electrodynamics*. Cambridge University Press, Cambridge (2013). MR3012344.

[Zyg56] A. Zygmund, On a theorem of Marcinkiewicz concerning interpolation of operations. *J. Math. Pures Appl.* **35**(9), 223–248 (1956). MR80887.

Index